Lecture Notes on Data Engineering and Communications Technologies

Volume 159

Series Editor

Fatos Xhafa, Technical University of Catalonia, Barcelona, Spain

The aim of the book series is to present cutting edge engineering approaches to data technologies and communications. It will publish latest advances on the engineering task of building and deploying distributed, scalable and reliable data infrastructures and communication systems.

The series will have a prominent applied focus on data technologies and communications with aim to promote the bridging from fundamental research on data science and networking to data engineering and communications that lead to industry products, business knowledge and standardisation.

Indexed by SCOPUS, INSPEC, EI Compendex.

All books published in the series are submitted for consideration in Web of Science.

More information about this series at https://link.springer.com/bookseries/15362

Zhengbing Hu · Zhiwei Ye ·
Matthew He
Editors

Advances in Artificial Systems for Medicine and Education VI

Springer

Editors
Zhengbing Hu
International Center of Informatics
and Computer Science, Faculty
of Applied Mathematics
National Technical University of Ukraine
"Igor Sikorsky Kyiv Polytechnic Institute"
Kyiv, Ukraine

Zhiwei Ye
School of Computer Science
Hubei University of Technology
Wuhan, China

Matthew He
Halmos College of Arts and Sciences
Nova Southeastern University
Fort Lauderdale, FL, USA

ISSN 2367-4512 ISSN 2367-4520 (electronic)
Lecture Notes on Data Engineering and Communications Technologies
ISBN 978-3-031-24467-4 ISBN 978-3-031-24468-1 (eBook)
https://doi.org/10.1007/978-3-031-24468-1

This Springer imprint is published by the registered company Springer Nature Switzerland AG
The registered company address is: Gewerbestrasse 11, 6330 Cham, Switzerland

Preface

The development of artificial intelligence systems and their applications in a number of industries is one of the most important issues in contemporary science and technology. One of these disciplines is medical engineering, the application of which strives to increase the efficacy of recognizing and treating various diseases. The rapid development of artificial intelligence systems needs a more intensive training program for a growing number of relevant employees. Simultaneously, artificial intelligence systems have significant potential for application within educational technologies to improve the quality of specialized training by incorporating the personal characteristics of specialists and the development of new computer devices.

Scientists are trying to make digital systems with artificial intelligence that have the same intellectual abilities as humans and other living things. A thorough look at genetic systems and biological processes that are passed down from generation to generation could reveal living nature's bio-information patents and open up new ways to make artificial intelligence more powerful. Because of this, there needs to be a lot of biomathematical research on genetic and sensory-motor systems, based on what we already know in math, computer and quantum informatics, biology, physics and other fields. In other words, studying genetic systems and making advances in artificial intelligence should happen at the same time to help each other.

The Sixth International Conference of Artificial Intelligence, Medical Engineering, Education (August 19–21, 2022, Wuhan, China) has its purpose to present new thematic approaches, methods and achievements of mathematicians, biologists, physicians and technologists and also to attract additional interest from different specialists to this prospective theme. Its proceedings additionally includes articles on specific tasks in various fields, where artificial intelligence systems can be applied in the future with great benefit.

The conference is organized jointly by the Wuhan University of Technology and the International Research Association of Modern Education and Computer Science.

The best contributions to the conference were selected by the program committee for inclusion in this book out of all submissions.

Zhengbing Hu
Zhiwei Ye
Matthew He

Organization

Conference Organizers and Supporters

Wuhan University of Technology, China
Hubei University of Technology, China
Polish Operational and Systems Society, Poland
International Center of Informatics and Computer Science, Ukraine
International Research Association of Modern Education and Computer Science, Hong Kong

Contents

Advances in Mathematics and Bio-mathematics

Modeling of Tri-phase PWM Rectifier and Controller Design

Chengquan Liang[1], Zhenpeng Qin[1(✉)], Ying Zhang[1], Deshen Lv[1], and Xiao Lu[2]

[1] School of Intelligent Manufacturing, Nanning University, Nanning 530200, Guangxi, China
759137827@qq.com

[2] School of Information Science and Engineering, Guilin University of Technology, Guilin 541006, Guangxi, China

Abstract. The characteristics of tri-phase voltage source PWM rectifier are similar to those of Boost converter. The transfer function established from active current to output voltage has zero point in the right half plane, which belongs to an unstable system. The existing control methods have the problems of low response speed and weak robustness. In this paper, a control method based on instantaneous power balance is proposed. Firstly, based on the instantaneous power balance principle, the transfer function from the active current to the output voltage of the tri-phase PWM rectifier is obtained. Then the stable PI parameter range of the closed-loop control system is derived by using the Routh stability criterion, and the control signal is applied to the controlled system in the form of feedback. Finally, the PI control simulation model of the tri-phase PWM rectifier is built in Matlab/Simulink, and an experimental prototype of the tri-phase PWM rectifier is made. The simulation and experimental results prove that the control method has fast response speed and good robustness. The results verify the correctness of the theoretical analysis and the effectiveness of the control method.

Keywords: Power balance · PWM rectifier · Routh stability criterion

1 Introduction

With the rapid development of power electronics, high voltage, high current and high efficiency rectifiers are in urgent demand in all walks of life. Various rectifiers have been widely used in power stations, power grids, electroplating, new energy power generation and other fields. Compared with the traditional uncontrollable or phase controlled rectifier, PWM rectifier has obvious advantages, including the sinusoidal current at the grid side and the unit power factor at the grid side, which is friendly to the power grid environment [1]. It also has the advantages of high power density and strong controllability compared with other improved rectifier schemes such as multi pulse rectifier. At the same time, PWM rectifier also has the characteristics of bidirectional energy flow, which is the key power electronic interface for new energy utilization.

To get theoretical guidance for the controller design of PWM rectifier, some literatures have studied the modeling and control of tri-phase voltage source PWM rectifier.

Z. Hu et al. (Eds.): AIMEE 2022, LNDECT 159, pp. 3–14, 2023.
https://doi.org/10.1007/978-3-031-24468-1_1

For example, reference had conducted a detailed study on the modeling, control and performance analysis of the tri-phase voltage source PWM rectifier, built the numerical simulation mathematical model of the tri-phase voltage source PWM rectifier by using the simulation toolbox in MATLAB [2–4]. It obtained the model of the PWM rectifier in the d-q coordinate system by using the method of writing circuit equations. However, the calculation process is complex and the control coefficient is difficult to determine; Literature [5] simplified the single input single output linearization model of tri-phase voltage source PWM rectifier, and some people had proposed new control methods based on this model [6–10], but these control methods had problems such as low control response speed and poor robustness. Reference [11] used the complex vector method to model the tri-phase PWM rectifier, and did Matlab simulation and DSP experiments, which simplified the mathematical model to a certain extent, but increased the difficulty in the implementation of the vector controller. Literature [12, 13] adopted the voltage and current double closed-loop and space vector pulse width modulation methods for control, and the current decoupling control in the d-q coordinate system. Finally, the control method is simulated and verified. While the discrete domain model of the system when the tri-phase PWM rectifier adopts the voltage and current double closed-loop control structure established in literature [14] proposed the engineering design method of the voltage and current controller parameters, and analyzes and verifies the parameters affecting the system stability, However, the accuracy of the control method is low. Literature [15–17] combined with the adaptive control idea and based on the proportional integral control, proposed a control method using fuzzy adaptive PI control in the outer voltage loop and feedforward decoupling in the inner current loop. However, it did not provide the basis for the selection of control parameters. Reference [18] studied the DPC control algorithm for the power fluctuation of three-level PWM rectifier with dual non-zero voltage vector output in one switching cycle. However, the control algorithm is time-consuming and has little practical engineering significance.

Literature 2–14 discusses problems from different perspectives, but they all have different problems, such as difficult control, slow response and low stability. In view of this, this paper first writes the power balance equation based on the principle of instantaneous power balance, obtains the voltage and current flow at the static operating point, and applies small disturbances to the active current and DC voltage, so as to obtain the transfer function from the active current to the output voltage. According to the Routh stability criterion, the value range of PI parameters for the stability of the closed-loop control system and the design basis are obtained. Finally, the correctness and effectiveness of the proposed model and controller design are proved by simulation analysis and experiments. The results show that the system using this method has fast response speed and high operation stability.

2 Instantaneous Power Balance Modeling of PWM Rectifier

At present, the most widely used tri-phase voltage source PWM rectifier [19], which adds a group of bridge arms to the bridge arms of single-phase full bridge voltage source rectifier, can be applied to the circuit of tri-phase power balance. The topology diagram of tri-phase PWM rectifier is shown in the Fig. 1.

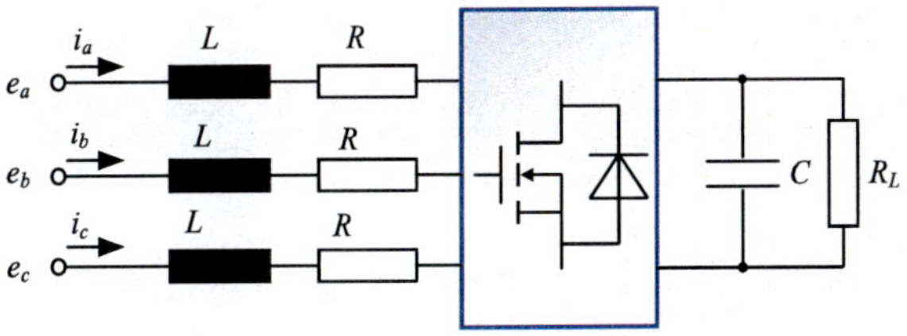

Fig. 1. The structure of tri-phase rectifier

Where *ea, eb* and *ec* are the tri-phase voltage of the power grid, *ia, ib* and *ic* are the tri-phase current from the power grid to the rectifier, R is the equivalent resistance of the inductance L, C is the bus capacitance of the PWM rectifier, and RL is the load resistance.

According to the instantaneous power balance principle, the instantaneous value of the past power viewed from the grid side and the past power viewed from the inverter side remains the same, and the equation is obtained as follows:

$$\begin{aligned} &e_a(t)i_a(t) + e_b(t)i_b(t) + e_c(t)i_c(t) - R\left[i_a^2(t) + i_b^2(t) + i_c^2(t)\right] \\ &-\frac{1}{2}L\frac{d}{dt}\left[i_a^2(t) + i_b^2(t) + i_c^2(t)\right] = Cu_{dc}(t)\frac{du_{dc}(t)}{dt} + \frac{u_{dc}^2(t)}{R_L} \end{aligned} \tag{1}$$

$e_a(t)$, $i_a(t)$, $e_b(t)$, $i_b(t)$, $e_c(t)$ and $i_c(t)$ are the instantaneous values of tri-phase voltage and current of power grid respectively, and $u_{dc}(t)$ is the instantaneous value of DC bus voltage.

Assuming that the grid voltage is pure sine, there is no background harmonic and the circuit structure is tri-phase symmetrical, it is obtained from the instantaneous reactive power balance:

$$e_a(t)i_a(t) + e_b(t)i_b(t) + e_c(t)i_c(t) = 3EI_p \tag{2}$$

where, E is the effective value of grid phase voltage and I_p is the effective value of grid phase current. Therefore, the balance Eq. (1) of instantaneous power can be simplified as:

$$3EI_p - \frac{3}{2}L\frac{d}{dt}I_p^2 - 3RI_p^2 = Cu_{dc}(t)\frac{du_{dc}(t)}{dt} + \frac{u_{dc}^2(t)}{R_L} \tag{3}$$

For the purpose of obtaining the transfer function from active current to DC voltage, a small disturbance is applied at the static operating point. Order:

$$\begin{cases} I_p(t) = I_p + \hat{i}_p \\ u_{dc}(t) = U_{dc} + \hat{u}_{dc} \end{cases} \tag{4}$$

Substituting (4) into (3), ignoring the steady-state operating point and the quadratic term of small disturbance, it is able to get:

$$3E\hat{i}_p - 3LI_p\frac{d\hat{i}_p}{dt} - 6RI_p\hat{i}_p = CU_{dc}\frac{d\hat{u}_{dc}}{dt} + \frac{2U_{dc}\hat{u}_{dc}}{R_L} \tag{5}$$

Laplace transform the above Eq. (5) to obtain:

$$G_{dc}(s) = \frac{\hat{u}_{dc}}{\hat{i}_p} = \frac{3E - 3LI_p s - 6RI_p}{CU_{dc}s + \frac{2U_{dc}}{R_L}} \tag{6}$$

At the steady-state operating point, according to the power conservation, the following can be obtained:

$$3EI_p = \frac{U_{dc}^2}{R_L} \tag{7}$$

Combined with the above formula, it can be got:

$$G_{dc}(s) = \frac{3E - \frac{2RU_{dc}^2}{ER_L} - \frac{LU_{dc}^2}{ER_L}s}{CU_{dc}s + \frac{2U_{dc}}{R_L}} \tag{8}$$

In the above formula (8), $G_{dc}(s)$ is the transfer function from the active current of the PWM rectifier to the DC voltage, and it is also the open-loop transfer function of the PWM rectifier without a controller. The block diagram is shown in the Fig. 2.

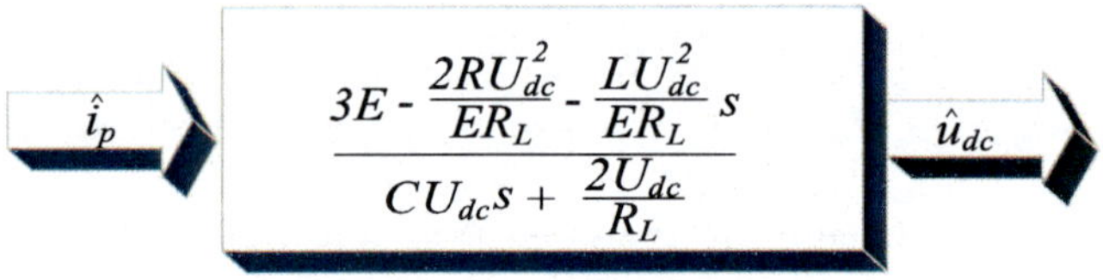

Fig. 2. The open loop structure of tri-phase rectifier

3 Controller Design Based on Instantaneous Power Balance Model

3.1 Closed Loop Model of Proportional Integral Control

In order to realize the stable and fast tracking of the DC bus voltage command value, it is necessary to add a closed-loop controller to correct the original system. The command to be tracked is direct current. According to the internal model principle, PI control can be used to achieve the stability of DC voltage. The transfer function of PI controller is as follows:

$$G_c(s) = k_p + \frac{k_i}{s} \tag{9}$$

The block diagram of DC side voltage closed-loop control of tri-phase PWM rectifier with PI controller is shown in Fig. 3.

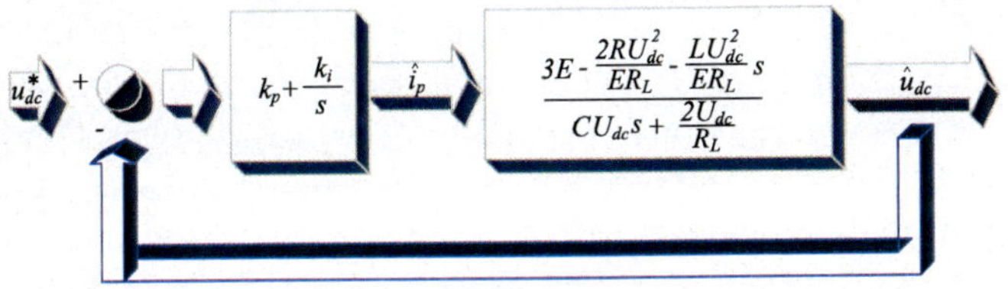

Fig. 3. The close loop control structure of tri-phase rectifier

The voltage closed-loop transfer function with PI control is derived from the block diagram

$$G(s) = \frac{G_c(s)G_{dc}(s)}{1 + G_c(s)G_{dc}(s)} \tag{10}$$

Substituting Eqs. (8) and (9) into Eq. (10) and sorting out to obtain Eq. (11).

$$G(s) = \frac{-LU_{dc}^2k_ps^2 + (3E^2R_Lk_p - 2RU_{dc}^2k_p}{(ECR_LU_{dc} - LU_{dc}^2k_p)s^2 + (2EU_{dc} + 3E^2R_Lk_p}$$
$$\frac{-LU_{dc}^2k_i)s + 3E^2R_Lk_i - 2RU_{dc}^2k_i}{-LU_{dc}^2k_i - 2RU_{dc}^2k_p)s + 3E^2R_Lk_i - 2RU_{dc}^2k_i} \tag{11}$$

The closed-loop characteristic equation of the system is then presented as follows:

$$\begin{aligned}
&\left(ECR_LU_{dc} - LU_{dc}^2k_p\right)s^2 \\
&+\left(2EU_{dc} + 3E^2R_Lk_p - LU_{dc}^2k_i - 2RU_{dc}^2k_p\right)s \\
&+3E^2R_Lk_i - 2RU_{dc}^2k_i = 0
\end{aligned} \tag{12}$$

3.2 Stability Analysis of Proportional Integral Control

The stability of closed-loop control of DC bus voltage of PWM rectifier can be determined by the expression of its characteristic root. According to Routh stability criterion, for a second-order linear system, the necessary and sufficient condition for the stability of the system is that each coefficient is positive. Therefore, the closed-loop control stability condition of the tri-phase PWM rectifier using proportional integral control is obtained, namely:

$$\begin{cases}
ECR_LU_{dc} - LU_{dc}^2k_p > 0 \\
2EU_{dc} + 3E^2R_Lk_p - LU_{dc}^2k_i - 2RU_{dc}^2k_p > 0 \\
\left(3E^2R_L - 2RU_{dc}^2\right)k_i > 0
\end{cases} \tag{13}$$

Since R is the parasitic resistance of the grid connected inductor, its value is small, and the third inequality of Eq. (13) will be automatically satisfied. The value range of PI controller parameters on the premise of ensuring system stability is sorted out, namely:

$$\begin{cases} 0 < k_p < \dfrac{ECR_L}{LU_{dc}} \\ 0 < k_i < \dfrac{2EU_{dc} + \left(3E^2R_L - 2RU_{dc}^2\right)k_p}{LU_{dc}^2} \end{cases} \tag{14}$$

From the above Eq. (14), it can be seen that the proportional parameters are affected by the grid voltage, bus capacitance, load resistance, bus voltage and grid connected inductance. The value range of the integral parameters is not only related to the above electrical parameters, but also restricted by the proportional parameters. Therefore, when designing the parameters of the PI controller, the proportional parameter *kp* is first determined, and then the value of *ki* is obtained according to the constraint relationship between *kp* and *ki*.

4 Analysis of Simulation and Experimental Results

4.1 Simulation Model and Analysis of Influencing Factors

The simulation model of tri-phase PWM rectifier is built in matlab/simulink, and the specific parameters are set as shown in Table 1.

Table 1. Main parameters of tri-phase PWM rectifier simulation circuit

Symbol	Simulation parameters	Explain
E	220 V	Phase voltage of three-phase AC power supply
f	50 Hz	Grid frequency
L	3 mH	Grid connected inductance
C	1000 uF	Bus capacitance
Udc	700 V	DC voltage
R	1 Ω	Parasitic resistance of grid connected inductance
RL	500 Ω	Load resistance

The parameters in Table 1 are substituted into Eq. (14) to obtain the stable boundary of *kp* and *Ki*, as shown in Fig. 4. It can be seen from the results shown in Fig. 4 that as the value of *kp* parameter increases, the stable area of system operation also gradually increases.

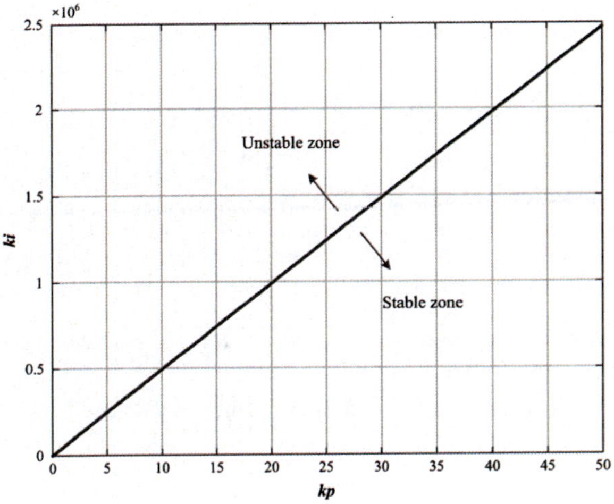

Fig. 4. Stability influenced by k_p and k_i

Using the model based on instantaneous power balance, select different *kp* and *ki* parameters for simulation, and observe the response waveform of DC bus voltage, as shown in Fig. 5. It presens the response results of output voltage when *ki* = *5000* and *kp* take different values. It can be seen that when *kp* = *30*, the voltage response speed is the fastest, the transition process is short, and the voltage is stable at *700* V; When *kp* = *2*, the voltage response speed is lower than that when *kp* = *30*, and the transition time increases. After oscillation, the waveform tends to be stable, and the voltage is stable at *700* V; When *kp* = *0*, the voltage response speed is the lowest, and the waveform always keeps oscillation and cannot keep stable.

It can be seen from the Fig. 4 that after the integral parameter *ki* is determined, the appropriate *kp* parameter can be selected according to Fig. 4 to make the output voltage response index of the system meet the design requirements. The unreasonable design of *kp* parameters will not only increase the number of overshoots and oscillations of the system, but may even lead to the instability of the system. The simulation results show that the PI controller parameters designed by the mathematical model of the tri-phase PWM rectifier based on the instantaneous power balance modeling method can well meet the requirements of the DC voltage loop for steady-state error and response speed, and provide a good reference for the design of the PI parameters of the voltage loop.

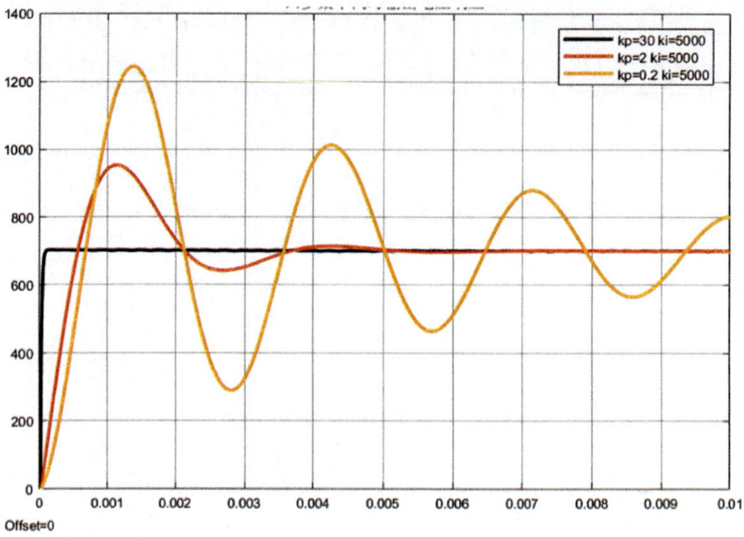

Fig. 5. Voltage response influenced by *kp* and *ki*

4.2 Analysis of Experimental Results

Based on the parameters in Table 1, an experimental prototype of a tri-phase PWM rectifier is built.

The PI parameters of the system are selected according to the principle of Fig. 5, taking into account the stability and rapidity of the system. The steady-state experimental waveforms of grid current and bus voltage are shown in Fig. 6. The time base in the figure is *10* ms/div, the DC bus voltage measured in channel 3 is *200* V/div, and the grid current measured in channel 2 is *2* A/div.

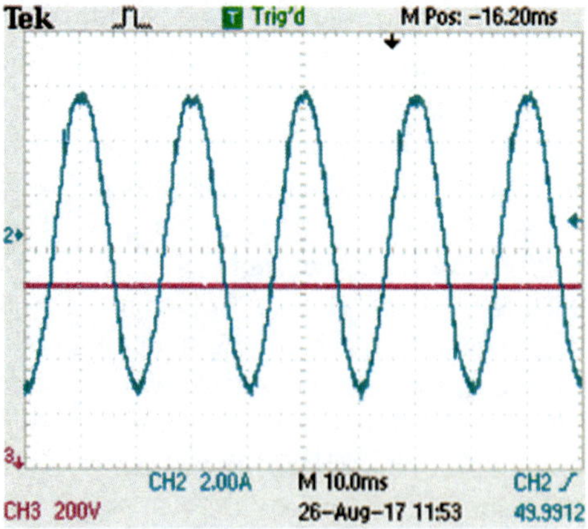

Fig. 6. Waveform of DC bus voltage and grid current

The bus voltage is basically stable at the given DC voltage of *700V*, and the distortion of grid current is very small. The results of THD analysis by importing the grid current data into Matlab are shown in Fig. 7. It can be seen that the current total harmonic distortion rate THD = *2.77%* < *5%,* meeting the needs of the power industry.

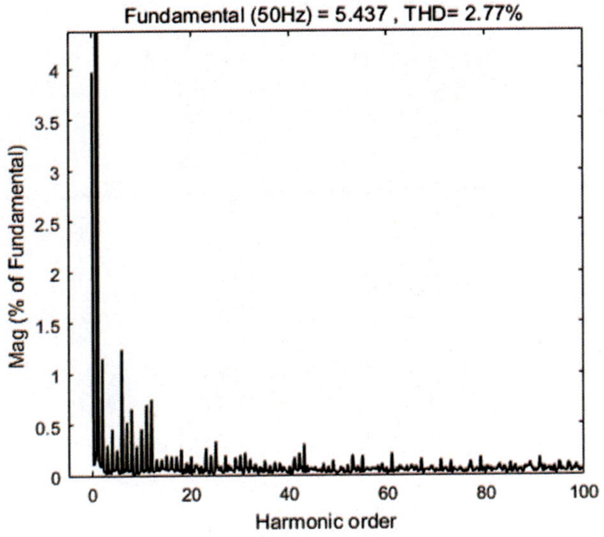

Fig. 7. THD Analysis of grid current

Change the load conditions, carry out sudden load addition and sudden load reduction experiments respectively, and verify the dynamic performance of the PI parameters of the designed voltage loop.

Figure 8 shows the experimental waveform of output voltage and grid current when switching from half load to full load. It can be seen from the figure that at the moment of sudden load application, the DC bus voltage fluctuation increases compared with that at the steady state and decreases to a certain extent, but after about *100* ms, it becomes stable and can be adjusted back to the specified voltage value. The controller has good rapidity and accuracy.

Figure 9 shows the test waveform of output voltage and grid current when switching from full load to half load. It can be seen that the sudden load drop occurs at the starting point of the waveform in Fig. 9. When the load is switched from full load to half load, the DC bus voltage rises, but does not exceed *10* V, and stabilizes to a new steady-state value after about *500* ms. Compared with sudden load increase, the response time of sudden load decrease is relatively long due to the energy storage effect of DC side capacitance.

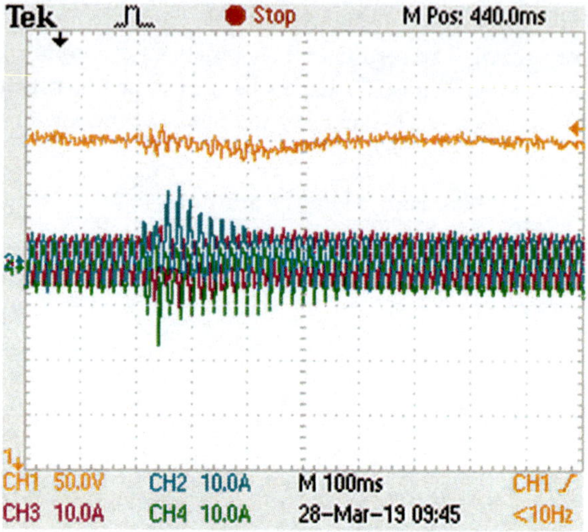

Fig. 8. Load increased experiment

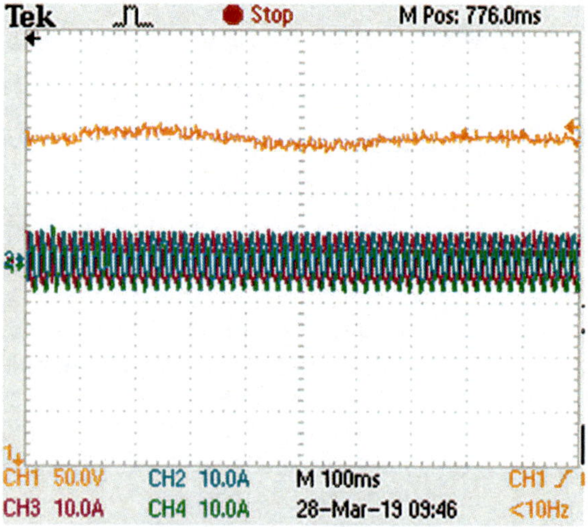

Fig. 9. Load decreased experiment

5 Conclusion

In this paper, a modeling method of tri-phase voltage source PWM rectifier based on instantaneous power balance is proposed. Based on the model obtained by this method, a PI controller is designed. According to the Routh stability criterion, the range of PI parameters for the stability of the closed-loop control system and the design basis are obtained. Finally, the correctness of the proposed model and controller design is proved by simulation analysis and experiments.

The results show that the PI controller parameters designed by the mathematical model of tri-phase PWM rectifier based on the instantaneous power balance modeling method can well meet the requirements of DC voltage loop for steady-state error and response speed, and provide a good reference for the design of voltage loop PI parameters. Change the load conditions, and carry out the load sudden increase and decrease experiments respectively. When the load is suddenly applied, the DC bus voltage tends to be stable after a short transition time, and is stable at the specified voltage value. When the load is suddenly reduced, due to the energy storage effect of the DC side capacitance, the system transition time is increased compared with the sudden load increase, but it can still quickly recover to the stable value, which indicates that the system has fast response speed and high stability. The research results obtained in this paper can provide an effective basis for the design of system control parameters of PWM rectifier.

Acknowledgment. This project is supported by Fund Project: "research and design of three phase PWM rectifier based on adaptive fuzzy PI control" (2021ky1806).

References

1. Singh, B., Singh, B.N., Chandra, A., et al.: A review of three-phase improved power quality AC-DC converters. IEEE Trans. Industr. Electron. **51**(3), 641–660 (2004)
2. Ende, W., Shenghua, H.: A new double closed loop control strategy for three-phase voltage source PWM rectifier. Chinese J. Electr. Eng. 32(15): 24–30+18 (2012) (in Chinese)
3. Marković, N., Bjelić, S., Živanić, J., Bogićević, Z.: Analysis and estimation of values of currents and voltages at the disturbances in induction machine using tested matlab simulation. Int. J. Intell. Syst. Appl. (IJISA) **7**(1), 1–8 (2014)
4. Boora, S., Agarwal, S.K., Sandhu, K.S.: Traditional PI and FPIL based VFCU for an isolated induction generator (IIG). Int. J. Intell. Syst. Appl. (IJISA) **9**(12), 10–19 (2017)
5. Yin, B., Oruganti, R., Panda, S.K., et al.: A simple Single input single output (SISO) model for a three-phase PWM rectifier. IEEE Trans. Power Electron. **24**(3), 620–631 (2009)
6. Xing, Z., Chongwei, Z.G.: Study on a new space voltage vector control method about reversible PWM converter. Proc. CSEE **21**(10), 102–109 (2001)
7. Pinghua, Z., Guijie, Y., Tiecai, L.: Direct voltage control of three-phase PWM rectifier based on feedback linearization. Proc. CSEE **30**(18), 39–46 (2010)
8. Chunjiang, Z., Zhongnan, G., Qin, W., et al.: Analysis of bi-direction operation state of three phase PWM rectifier based on a new phase and amplitude control. Proc. CSEE **26**(11), 167–171 (2006)
9. Xin, Z., Yuan Bo, G., Xiaohua, Z., et al.: Three-phase voltage-type PWM rectifier based on adaptive tracking control. Proc. CSEE **30**(27), 76–82 (2010). **(in Chinese)**
10. Weihua, D., Bo, Z., Dongyuan, Q., et al.: The research of decoupled state variable feedback linearization control method of three-phase voltage source PWM rectifier. Proc. CSEE **25**(7), 97–103 (2005). **(in Chinese)**
11. Dai Peng, F., Xiao, Y.Q., et al.: Research on low switching frequency PWM rectifier based on complex vector regulator. Chinese J. Electr. Eng. **31**(21), 25–31 (2011). **(in Chinese)**
12. Renbiao, T.: Research on tri-phase rectifier based on new double closed loop and SVPWM control strategy. Electr. Measur. Instru. **52**(16), 97–100 (2015). **(in Chinese)**
13. Alawad, N.A., Rahman, N.G.: Design of (FPID) controller for automatic voltage regulator using differential evolution algorithm. Int. J. Modern Educ. Comput. Sci. (IJMECS) **11**(12), 21–28 (2019). **(in Chinese)**

14. Shiming, H., Jian, X., Dayi, D., et al.: Modeling, control and stability analysis of tri-phase voltage source PWM rectifier[J/OL]. Power Grid Technology: 1–9 [2019–04–08] https://doi.org/10.13335/j.1000-3673.pst.2018.1933(in Chinese)
15. Xiao, Y., Jicun, F., Chuanliang, L.: Research on TRI-phase PWM rectifier based on fuzzy adaptive control. Comput. Technol. Autom. 36(04), 29–32+45 (2017) (in Chinese)
16. Yuanjun, P., Lipeng, Y.: Research and application of fuzzy adaptive control in three-phase PWM rectifier. Mining Machinery **09**, 144–146 (2007)
17. Dong, G., Ma, H.: Voltage quality evaluation of distribution network based on probabilistic load flow. Int. J. Modern Educ. Comput. Sci. (IJMECS) **10**(8), 55–62 (2018)
18. Bishuang, F., Guanzheng, T., Shaosheng, F., et al.: A direct power control method for three-level PWM rectifier with dual non-zero voltage vector outputs. Chinese J. Electr. Eng. **35**(22), 5832–5841 (2015)
19. Xin, Q.: Design and experimental study of three-phase PWM rectifier control system. Qinhuangdao: Yanshan University (2015) (in Chinese)

AI as a Catalyst for ESG Investing

Margherita Mori(✉)

Department of Industrial and Information Engineering and of Economics, University of L'Aquila, 67100 L'Aquila, Italy
margherita.mori@univaq.it

Abstract. This paper aims at providing a conceptual framework for analysis of how artificial intelligence (AI) can contribute to impact investing as an area of asset management that is focused on environmental, social and governance (ESG) issues. Actually, companies have been increasingly concerned of discovering what their profile might become by complying with ESG criteria; as they further gain momentum, the need keep surfacing to cope with them more extensively and/or intensively. Meanwhile, the financial arena has been increasingly populated with investors showing highlighted involvement in considering ESG implications before selecting alternative options, as well as in monitoring their financial assets and eventually reallocating their portfolios. A key role can be played by AI: its persistent progress sounds like an invitation to focus on quantitative models to assess ESG ratings, as well as to discuss how technological innovation can reshape sustainable investing, to evoke just a few of the most promising applications and areas of knowledge. Conclusions draw upon lessons learned, best practices and success stories, that are worth sharing, disseminating, and replicating, to benefit not only those who are directly interested in the evolving trends under investigation but society on a global scale.

Keywords: Artificial intelligence · ESG criteria · Impact investing

1 Introduction

The 2030 Agenda, that was adopted by the United Nations (UN) in 2015, has generated heated interest in sustainability issues, such as those linked to ending extreme poverty and reducing inequality: not surprisingly, remarkable efforts have been made in the last few years to comply with this strategic to-do list that provides a shared blueprint for partnerships for peace and prosperity for people and the planet; although some progress has been achieved, much remains to get done in sight of fulfilling the 17 Sustainable Development Goals (SDGs) and the associated 169 targets. Anyway, the debate on these issues has contributed to shed unprecedented light upon ESG factors that by definition deal with environmental, social and governance issues and that are considered essential ingredients for impact and responsible investing.

At the same time, an increased trend has been recorded in the financial sphere of the economy, dealing with technological applications and leading to fintech as an ecosystem that asset allocation and other operating areas can – and should – rely upon. Along this

Z. Hu et al. (Eds.): AIMEE 2022, LNDECT 159, pp. 15–24, 2023.
https://doi.org/10.1007/978-3-031-24468-1_2

pathway, further advances must be accounted for, that stem from AI, due to its ability to improve operational processes, not only for financial firms, and that include machine learning as its prominent subset: it enables data analytics, which implies the use of algorithms and statistical models to imitate the way in which humans learn, gradually improving its accuracy, to draw inferences from patterns in data and to predict new output values; in finance, AI applications allow for new operating solutions, such as those aimed at evaluating risks, and hence have direct impact on performing prediction modelling, quantitative trading and credit rating, to mention just a few areas of interest that are closely linked to asset management.

Based on these thoughts, it sounds appealing to wonder how AI can support ESG investing, which leads to investigate the state of the art and to sketch a tentative picture of what to expect, at least for the foreseeable future. Accordingly, the main goal of this paper is to focus on most relevant topics at the crossroads of technology, innovation and sustainability, in order to provide a conceptual framework of analysis of emerging issues and to hopefully contribute to sustainable development: by common opinion, experiments are reserved to applied sciences and thus excluded from the methodological set available to finance; however, the financial arena looks like a huge laboratory, with both physical and virtual features, and their scrutiny leads to sort out lessons learned that could be usefully shared, to devise success stories that might be fruitfully replicated and best practices that should be conveniently disseminated.

2 Insights into the State-of-the-Art

Due to the underlying interdisciplinary approach, there seems to be no choice but to get access to most recent literature on several topics that fall within the scope of diverse fields of research: on one hand, ESG criteria have been more and more intensively dealt with, by academics and practitioners alike, in an attempt at meeting vital challenges, such as those linked to climate change and growing inequalities, while addressing global issues, including those associated with health and food security; on the other hand, a multifaceted array of subject matters has been debated that refer to the potential for progress based on AI advances, with its applications playing a special role within the financial system, in terms of risk assessment and management, fraud prevention and detection, just to quote some recurrent examples. In between, special attention needs to be paid to asset management as an essential subset of activity, for both individual and institutional investors, with wide-ranging effects that span from the micro- to the macro-economic level and that go beyond the financial side of the coin.

2.1 Spotlight on ESG Criteria

According to a widely shared view, ESG criteria refer to three key factors that have gained momentum since the adoption of the 2030 Agenda for Sustainable Development. To make a long story short, their extraordinary relevance can be perceived by considering a few implications: environmental issues refer to waste and pollution, resource depletion, greenhouse gas emission, deforestation, and climate change, thus leading to examine – among other aspects – how a company performs as a steward of our natural resources;

social factors deal with employee relations and diversity, working conditions (including child labor and slavery), local communities, health and safety, and conflicts, and their impact can be perceived, for instance, while examining how a company manages its relationships with employees, suppliers, customers, and the community it interacts with; finally, governance criteria have to do with tax strategies, executive remuneration, donations and political lobbying, corruption and bribery, board diversity and structure, which support the increasingly felt need to evaluate – broadly speaking, as a typical case – how a company polices itself.

As such, these factors have been increasingly resorted to, in order to appraise environmental, social and governance systems – with companies representing only a part of their potential – in terms of their sustainability impact, thus making an ideal link clearly emerge between that strategic Agenda and the ESG framework. However, differences have been reported as noticeable, that do not allow to identify the criteria at issue with those currently deemed inherent in the concept of sustainability: actually, its distinguishing elements share similarities with ESG criteria, particularly as far as the environmental dimension is concerned, but their pillars imply broader social and corporate governance considerations; in other words, ESG-minded business practices draw upon company stakeholders, identity, and decision-making, while a sustainable footprint would tend to be associated with the narrower-scoped relationship between a company and the environment.

From a practical point of view, it can be argued that sustainability is a vague umbrella term, whereas the acronym ESG is interpreted as referred to much more specific, data-driven and measurable factors [1], though in both cases relevance is attributed to the goal of improving a company's business practices in order to boost profits and win favor from investors, customers, and regulators: no wonder that sustainability can mean different things to different entities, with most common definitions being forged to evoke "going green"; by contrast, ESG criteria are based upon specific standards that have been set by lawmakers, investors and specialized institutions and that companies can measure and report against. Therefore, these criteria are commonly presented as making up an investment framework that helps to assess a company's behavior and policies and to make better decisions on asset (re)allocation accordingly.

2.2 A Closer Look at Impact Investing

There it follows that the debate on asset allocation strategies has been fueled by how ESG issues drive investors' interests and choices, thus paving the way to impact investing as a widening operating area: it refers to channeling money to companies, organizations and funds with the intention to generate a measurable, beneficial social or environmental impact alongside a financial return; no surprise that impact investors actively seek to place capital in businesses, nonprofits and funds in industries such as renewable energy, housing, healthcare, education, microfinance and sustainable agriculture [2]. Target markets can be in either emerging or developed countries, though beneficiaries tend to be those that often have difficulty attracting other private capital, which confirms the key role of these investors as catalysts for sustainable development.

To stress this point, it makes sense to compare impact and responsible investment: the former is oriented towards social and environmental objectives, in order to achieve competitive – or even below – market returns, and target companies are selected based upon their social and/or environmental impact; the latter is aimed at financial yields considering ESG criteria, with financial return being more relevant than in the impact investors' perspective. Anyway, the relevance of responsible investment should not be underestimated, provided that traditional investment is aimed at a financial return – possibly the highest expected yield that is deemed consistent with the implied risk – without any consideration being reserved to the criteria under investigation.

It is not a case that special attention has been paid to defining several Principles for Responsible Investment in an attempt at developing a more economically efficient and sustainable global financial system: they offer a menu of possible actions for incorporating ESG issues into investment practice, including appropriate disclosure (for instance, in terms of standardized reporting on ESG issues and of integrating them within annual financial reports); therefore, responsible investment is described as an approach to investment that explicitly acknowledges the relevance of ESG factors to the performance and profitability of investment and to the long-term health and stability of the market as a whole [3]. In line with these thoughts, impact investment can be comparatively classified as closer to philanthropy, which by the way does not rely on any expectation of financial performance.

2.3 Drawing upon AI

While ESG investing has been gaining traction, as implied by more and more money being put to work with companies that strive to make the world a better place, similar goals have been pursued by developing AI applications across all industries, with the realms of banking and financial services providing ground for accelerated progress. Actually, AI has proven able to generate a massive competitive advantage, though it does not seem enough to invest in cutting-edge technologies and algorithms, in finance as well as in other areas of business: for instance, the need to rewire decision making and operations has been highlighted, in an effort to support banks, financial institutions, as well as other companies and organizations, to extract additional value by making the most of their relationship with AI; harnessing its full potential should include investment in people and processes, in order to deploy whatever human-machine interaction situations call for, thus adapting as needed to ever changing circumstances and scenarios [4].

Nobody would question that AI is still revolutionizing how both consumers and companies access and manage their finances. This holds particularly true about institutions that operate in the financial arena, with those involved in the fintech industry spearheading the list of the actual and potential beneficiaries of AI applications: their benefits in finance include enabling frictionless, 24/7 customer interactions, reducing the need for repetitive work and lowering human error, to report just a few examples [5]; everything from chatbots and virtual assistants to fraud detection and task automation is set to take further advantage from AI, with special attention to be paid to machine learning which "sits at the intersection of a number of emergent and established disciplines including pattern recognition, financial econometrics, statistical computing, probabilistic programming and dynamic programming" [6].

Accordingly, AI applications can be said to entail "the core technology of new technological revolution and industrial transformation" and to transcend "the traditional means of simulating human intelligence by a computer, such as man–machine gaming, machine identification, and natural language processing" [7]: along this pathway, it makes sense to conceive financial intelligence "as one of the new intelligent needs in the AI 2.0 era", that features "the new generation of AI, based on the new information environment and new development goals" [8]; further steps to be envisaged draw upon the potential of financial intelligence to become a "financial brain", thanks to – among others – a fast and accurate machine learning capability to handle complex data. Areas that are likely to benefit from FinAI – or AI in finance – range from wealth management to financial identity authentication, from smart financial consulting to risk management, and from security in blockchain to bankruptcy prediction, to shed light only on most often quoted operating fields that can be fruitfully engaged in.

3 Evidence from the Financial Industry

Focusing on asset management, as consumers have increasingly devoted energy to reach their financial independence, providing the ability to manage one's financial health has surfaced as the driving force behind a wider and wider recourse to AI-powered applications in personal finance, which makes their adoption a must for any financial institution aiming to act as a top player in this area. The need for wealth management services to embrace technological capabilities has emerged stronger in the last few years, due to both demand and supply factors: examples include robo-advisors (RAs), that have been able to both enhance and disrupt the way in which clients are served by banks and other providers of the services at issue; actually, RAs' assets under management have posted a significant growth that is credited towards "competitiveness on pricing, transparency and services and better expected returns linked to the use of quantitative finance and technology with less subjective human intervention" [9].

A four step evolution is still being developed, which has already made digital wealth management advance from 1.0 to 4.0 [10]: at first, clients have been enabled to receive product or portfolio proposals after answering a questionnaire to single out suitable options, with product variety including not only stocks and bonds but also other financial assets, such as exchange-traded funds (ETFs); later on, profiling questionnaires have been combined with more sophisticated risk management techniques in order to allow for direct investments via self-learning AI investment algorithms. They monitor and adjust single client portfolios in real time by shifting between different asset classes based on changing market conditions and individual investment needs – such as profit, risk appetite and liquidity preferences – thus keeping on track with selected investment strategies.

Since more and more often they reflect investors' attitude towards putting their money where their values are and steering away from initiatives that are viewed as polluting, endangering or sinful (such as oil companies, gun manufacturers or casinos), technological progress is key to boost impact investing. Supporting arguments lead to emphasize that many RAs now seek out ETFs that specialize in ESG investments and optimize portfolios in such a way as to approximate what modern portfolio theory would otherwise dictate: for instance, RAs can support investors – even those who embrace do-it-yourself impact investing – by orienting them towards lower-cost ETFs and by preventing them to forgo the benefits of proper diversification against risk [11] while privileging ESG criteria in asset allocation and portfolio management; the other side of the coin leads to consider the unprecedented challenges that asset managers face, encompassing how to ensure that their sustainability efforts are not construed as "greenwashing", in the light of the ever-increasing focus by regulators on it, and how to best formulate and implement ESG policies and procedures, given the growing support of technological advances.

4 Illustrative Case Studies

Focusing on latest developments, it comes natural to stress that AI allows to collect and analyze more information than ever before, thus contributing to better account for ESG risks and opportunities in both investing and providing related services. At the crossroads of technology, innovation and sustainability, AI applications have contributed to modernize ESG investing and further positive effects are on the cards: here it suffices to evoke computer algorithms that have been trained to find and analyze tone and content and that can digest all of the information available about a company, which would imply a huge task for human beings to do at a reasonable speed; popularized programs aimed at assessing the tone of texts can automate functions that would be impossibly labor-intensive [12].

4.1 AI Software Options for complying with ESG criteria

In an attempt at bridging the gap between theory and practice, it seems appropriate to analyze some case studies that do not only confirm the role of AI for ESG investing but also provide ground for disseminating best practices. A non-exhaustive list of potential areas of interest include software options that are based upon AI and that can bring about advantages for investment professionals, besides benefits for investors: since ESG analysis has become a more and more important part of the investing process, the recourse to AI software has increased human understanding of big data, thus making it easier to identify and manage relevant information; as a result, ESG data, factors and metrics, have been made more manageable, thanks to AI applications set to identify correlations, trends and outliers that would otherwise be buried in a deluge of data [13].

The search for specific AI software options that allow to better assess ESG risks and opportunities as they surface leads to unveil the plans developed by France's Apiday: it was founded in 2021 "to help investors get on top of ESG reporting", as announced upon the successful bagging of € 3.2 million in seed funding for international expansion and the launch of specific AI-driven software [14]; an all-in-one platform is made available to manage extra-financial data through a simple, fast and reliable process that goes from sharing an ESG questionnaire to gathering the suggested documents and finally leads to approve answers and share the results, according to the institutional website (https://apiday.com/en/). In a few words, Apiday's goal can be identified with enabling "every company to be sustainability minded by helping them manage their ESG data as systematically and accurately as their financial data" [15].

By making recourse to the proposed tool, that falls within the scope of Software as a Service (known as SaaS), companies can reportedly "gain control of their sustainability agenda, answer ESG data requests in moments, reduce reputational risks and unlock new opportunities". As far as developments on the cards, after launching its first tool for companies in March 2022, Apiday is now focusing on a new product tailored for private equity and venture capital firms as well as for ESG consultants, in sight of providing them with the support needed to comply with the European Sustainable Finance Disclosure Regulation: it has been designed to lay down harmonized rules applicable to "financial market participants and financial advisecompany, according to its websiters on transparency with regard to the integration of sustainability risks and the consideration of adverse sustainability impacts in their processes and the provision of sustainability-related information with respect to financial products"; within this framework, "'sustainable investment' means an investment in an economic activity that contributes to an environmental objective... or an investment in an economic activity that contributes to a social objective... or an investment in human capital or economically or socially disadvantaged communities..." [16].

4.2 AI Trading Software for ESG Investing

Besides ESG reporting software, positive prospects can be associated with the market segment centered around AI trading software that has been specially designed to support ESG investing: benefits include the ability to trade 24/7, automated order execution and the ability to identify – and take advantage of – valuable opportunities that human traders would most likely miss; the features to look for in securing the software under investigation encompass ease of use, real time data, flexibility, customer support and pricing considerations [17]. With quite a number of specific programs available, each with its own features and capabilities, what really matters is that an accurate selection is made based on individual needs and preferences after narrowing down options at hands.

For instance, some AI-powered investing apps are popular with active traders but investors may rather choose among the alternative solutions that offer trading in the preferred types of assets, such as shares issued by better ESG performing companies. A case in point has to do with Ellevest, that is considered not only one of the top-ranked RAs but also "the top app for socially responsible investing", provided that the declared mission is designed "to help women investors make smart investing decisions" (though – needless to say – anybody can become a client) [18]; in other words, Ellevest aims at getting "more money in the hands of women" and proud itself of being built as a woman-first financial company, according to its website(https://www.ellevest.com/#/).

The journey began in 2014 thanks to a team of industry experts with decades of experience on Wall Street, despite the prevailing belief that women don't need their own investing platform. Ellevest allows clients to invest up to 53% in ESG or social impact funds, that make up its Impact portfolios: they are designed to invest in companies that have more women leaders, that support affordable housing and community services, and that have higher standards for sustainability; the underlying motto is set to emphasize that "financial returns and social returns don't have to be mutually exclusive" and it sounds meaningful that the Impact portfolio is estimated to only imply a little higher expenses to investors than Ellevest's low-cost core portfolio.

4.3 AI-Based ESG Ratings for Financial Institutions

AI-powered tools have also provided stronger and stronger support to assets managers – and to financial institutions in more general terms, including insurance companies – within the framework of ESG investing: for instance, they increasingly use AI-based ESG ratings to make investment decisions, with advantages of doing so that encompass daily updates and deeper insight also involving sentiment analysis; it is not a case that "some of the biggest global investors including Japan's Government Pension Investment Fund are using computer algorithms to gain deeper insights into the ESG effectiveness of companies they invest in" [19]. All in all, asset managers tend to increasingly resort to ESG rating agencies that deploy AI and as such make their evaluations diverge from traditional evaluators in terms of who performs the assessment, the frequency of updates and the main sources of information.

Therefore, it sounds appealing to take a look at some distinguishing features of these agencies, starting with Switzerland-based RepRisk. It prides itself in running the world's largest database on ESG risks, that is daily updated, and warrants more than 15 years of unbroken data history: as announced through its website (https://www.reprisk.com/), this company combines AI and advanced machine learning with human intelligence to identify material ESG risks on companies, real assets, and countries; clients and partners can be provided with due-diligence grade data and metrics mapped to 101 ESG risk factors and to standards such as those stemming from the UN Global Compact, the UN 2030 Agenda and its SDGs and the Sustainability Accounting Standards Board metrics.

Another key player is FactSet's Truvalue that applies AI to uncover timely, objective ESG data to identify both risks and opportunities at the speed of current events by focusing on company ESG behavior from external sources and includes both positive and negative events that go beyond traditional sources of ESG risk data, as it can be learned from the institutional website (https://insight.factset.com/resources/at-a-glance-factset-truvalue-insight360-datafeed). Finally, Arabesque S-Ray GmbH depicts itself as a quantitative and data-driven technology that has been developed to assess the sustainability performance of more than 8,000 global publicly listed companies and that relies upon four distinct sustainability measurements: ESG Score is just one of them, together with GC Score that is based on the core principles of the UN Global Compact; in addition, Preferences Filter helps users to check company activities against their personal values and Temperature™ Score allows to quantify the extent to which corporations are contributing to the rise in global temperature through their greenhouse gas emissions [20].

5 Conclusion

To conclude, AI-based ESG ratings are unlikely to completely replace human analysts in the near future, but the recourse to AI can usefully complement traditional assessments and help to address biases while improving understanding of evaluation results. With ESG and AI being considered the two disruptors of modern finance, there is wide room for further combining them to deliver sustainable intelligence, not only in rating but also in other operating fields, including AI trading software for ESG investing and

AI software options for complying with ESG criteria: actually, they are growing in popularity, especially with younger investors being appealed by the technologically-focused investment platforms that RAs are based on; many more of them have begun to include ESG funds in order to meet the mounting demand of their users.

With the corona crisis stimulating unprecedented government action worldwide and people disproportionately hurt by jobless and poverty, not to mention the huge number of those who lost loved ones, it comes as no surprise that growing attention has been paid to these issues in both the real and the financial sphere of the economy. Actually, companies have been increasingly concerned of discovering what they really were and what they might become by addressing them: as they continue to gather momentum, there seem to be no choice but to cope with them more extensively and/or intensively; meanwhile, a belief has emerged among some development experts that is set to pave the way for attracting more private capital to address global challenges if these criteria are part of the investment equation.

Reliable surveys show that ESG money currently goes to Europe and North America, but the door has been recently opened to potential interest in other markets: marginalized communities and low-income countries might hopefully gain traction soon, which may help to fulfill the 2030 Agenda; the adoption of AI-powered technical solutions, such as those briefly overviewed, sounds rewarding. Therefore, a moral imperative can be devised to share lessons learned, replicate success stories and disseminate best practices as a contribution to sustainable development, to benefit not only those directly involved in applying ESG criteria and deploying AI, but our global village as a whole.

References

1. Niemoller, J.: Sustainability vs ESG: What's the difference, and why does it matter? (March 30, 2021). https://www.perillon.com/blog/sustainability-vs-esg, 18 Aug 2022
2. Impact Investment. UN Industrial Development Organization (UNIDO) Knowledge Hub, Vienna (2022). https://hub.unido.org/training-modules-investment, 13 Aug 2022
3. Sustainable development goals. UN sustainable development solutions network, New York, NY (n.d.). https://sustainabledevelopment.un.org/content/documents/7969EconSDPartnerships.pdf, 13 Aug 2022
4. Artificial Intelligence and AI at scale. Boston consulting group, Boston, MA (n.d.). https://www.bcg.com/capabilities/digital-technology-data/artificial-intelligence?utm_source=search&utm_medium=cpc&utm_campaign=digital&utm_description=none&utm_topic=ai&utm_geo=global&utm_content=ai-in-action&gclid=Cj0KCQjwl92XBhC7ARIsAHL9ak2hbYPpxVpUryIRjf3pevq6YWteBWEyZhcIeCFunYmr3NYUfkP_LwaAi1pEALw_wcB, 14 Aug 2022
5. Artificial intelligence in financial services: applications and benefits of AI in finance. Insider intelligence, https://www.insiderintelligence.com/insights/ai-in-finance/, 14 Aug 2022
6. Dixon, M., Halperin, I., Bilokon, P.: Machine Learning in Finance. Springer, Cham (2020). https://doi.org/10.1007/978-3-030-41068-1
7. Zheng, X., Zhu, M., Li, Q., Chen, C., Tan, Y.: FinBrain: when finance meets AI 2.0. Front. Inf. Technol. Electron. Eng. **20**, 914–924 (2019)
8. Pan Y.: Heading toward artificial intelligence 2.0. Engineering, 2(4), 409–413 (2016)
9. Phoon K. F., Koh C. C. F.: Robo-advisors and wealth management. (2018). Journal of Alternative Investments. 20, (3), 79–94. Research collection lee kong chian school of business. https://ink.library.smu.edu.sg/lkcsb_research/6502/, 16 Aug 2022

10. The expansion of Robo-Advisory in wealth management. Deloitte, Berlin (August 2016). https://www2.deloitte.com/content/dam/Deloitte/de/Documents/financialservices/Deloitte-Robo-safe.pdf, 16 Aug 2022
11. Hayes A.: Impact investing with a robo-advisor. Investopedia (November 7, 2021). https://www.investopedia.com/impact-investing-with-a-robo-advisor-4775379, 16 Aug 2022
12. S&P Global: How can AI help ESG investing? New York, NY (February 25, 2020). https://www.spglobal.com/en/research-insights/articles/how-can-ai-help-esg-investing, 16 Aug 2022
13. 15 Best ESG artificial intelligence software options for investing. The impact investor (June 7, 2022). https://theimpactinvestor.com/esg-artificial-intelligence/, 16 Aug 2022
14. Petzinger J.: France's apiday launches AI-powered tool to help investors get on top of ESG reporting. Tech.eu, Fores Media Ltd., London (June 8, 2022). https://tech.eu/2022/06/08/frances-apiday-is-out-to-help-investors-get-esg-reporting-right/, 16 Aug 2022
15. Apiday. Speedinvest, Vienna (n.d.). https://www.speedinvest.com/portfolio/apiday, 17 Aug 2022
16. Regulation (EU) 2019/2088 of the European Parliament and of the Council of 27 November 2019 on sustainability-related disclosures in the financial services sector, as amended by Regulation (EU) 2020/852 of the European Parliament and of the Council of 18 June 2020 on the establishment of a framework to facilitate sustainable investment. European Parliament, Brussels.
17. Best AI trading software. ESG - The Report. https://www.esgthereport.com/best-ai-trading-software/, 16 Aug 2022
18. Royal J.: 10 best investment apps in August 2022, Bankrate (August 16, 2022). https://www.bankrate.com/investing/best-investment-apps/, 18 Aug 2022
19. How artificial intelligence is reshaping ESG ratings. Nomura, Tokyo (n.d.). https://www.nomuraconnects.com/focused-thinking-posts/how-artificial-intelligence-is-reshaping-esg-ratings/, 18 Aug 2022
20. Arabesque S-Ray. Methodology (v.2.6), Frankfurt Am Main (September 2020). https://arabesque.com/docs/sray/S-Ray%20Methodology%20v260.pdf, 18 Aug 2022

Fuzzy Analysis Approaches to Intellectual Potential Modeling in Micro Business

Iryna Moiseienko[1], Ivanna Dronyuk[2], and Myroslava Klapchuk[3](✉)

[1] Prydniprovska State Academy of Civil Engineering and Architecture, 24a Chernyshevsky Str., Dnipro 4600, Ukraine
[2] Artificial Intelligence System Department, Lviv Polytechnic National University, Bandera Str. 12, Lviv 79013, Ukraine
[3] Department of Mathematics, Lviv Polytechnic National University, Bandera Str. 12, Lviv 79013, Ukraine
myroslava.i.klapchuk@lpnu.ua

Abstract. The article presents approaches to entrepreneurship structure intellectual potential. It is suggested to implement microbusiness intellectual potential analysis based on fuzzy studies. To investigate intellectual micro-business potential, it is suggested to use the fuzzy analyses apparatus. Using mentioned business-plan characteristics of entrepreneurs and businesses, one can define by the expert way the tightness of cause-and-effect ties and quantitative measurements (mutual influences). With the help of fuzzy analysis mathematic transformations, we can define the hidden powers of some elements upon the others (determined). Their quantitative dimensions will be risk measures of business plan carrying out.

Keywords: Entrepreneur-beginner · Intellectual potential · Entrepreneur key competencies · Fuzzy logics · Mutual influences matrix (incidences)

1 Introduction

Based on the rapid formation of a new, "intellectual," economy one can observe the necessity to systematize gathered experience in the intellectual capital investigation as one of the capital variety and its potential determination.

From the point of view of the theory of entrepreneurship, a specific structure is determined by the subject of entrepreneurship, which shows personal interests through the performance of entrepreneurial functions – initiative, management, and risk. The organizational form of a business entity is understood as "an integrated system of interdependent structural, cultural, spatial and technological components that provide the flow of events with a certain character, image, and order" [14].

Based on these definitions, modern business entities can be considered structures of different integration levels of internal and external information, knowledge, and material flows. That is, the functioning of business entities should be viewed through the prism of their intellectual capital, and the possibility of integration through the prism of intellectual potential [11].

Z. Hu et al. (Eds.): AIMEE 2022, LNDECT 159, pp. 25–34, 2023.
https://doi.org/10.1007/978-3-031-24468-1_3

Intellectual capital (from lat. Intellectualis capitalize – main, primary, or main knowledge) we can consider as the capital variety, which possesses relevant capital features and at the same time only for its typical specification and peculiarities. Intellectual capital in the context of technological, administrative, and market innovations is knowledge, which may provide profit and can be estimated.

Enterprise's intellectual potential must differ from its intellectual potential. Enterprise's intellectual potential – is knowledge, provided by the intellectual resources at present and in the future, and can be used to solve some task or achieve some particular aim. It means that intellectual potential is made out of consistent parts of the intellectual potential, but their combination and the ability to manage knowledge about their potential.

Taking into account the reflections mentioned above, it is suggested to consider intellectual potential as a system, with peculiar to it elements (intellectual capital and knowledge management system), ties and peculiarities innovations, competitiveness, energy-information) and provision system (information, functional and organizational) [11].

Research and analysis of the structure intellectual potential as to the element foresee the determination of, for example, resource potential on the stages of the product making and its implementation/ realization, intellectual resources management on these stages, and so on [11].

That is why the problem lies in the element analysis basis choosing and its aims and methods selection. Under the conditions of quantitative measurement impossibility of intellectual potential different characteristics, it is necessary to use expert methods and fuzzy analysis.

Summarizing the carried out before analyses of intellectual potential level analyses learning of the entrepreneurship structures potential, we can come to the following conclusions:

1) Information-intellectual matters influence upon separate economic subjects functioning results, and economic growth generally has a peculiar character and is in the constant change according to the level of efficient forces development and mechanisms complication of market representatives correlation;
2) The most developed direction in the theoretical-methodological scheme is research, which investigates separate information-intellectual matters influence upon the result of economic representatives functioning in the frames of human capital concept and economic potentials theory;
3) Intellectual firm actives problems formation as a combination of all information-intellectual matters development is actively discussed recently in scientific-practical literature. However, such actives' formation and usage mechanisms and their influence on firm effectiveness and economic growth generally require deeper theoretical and methodological formalization.

The aim of micro business intellectual potential research with the help of complex methods (fuzzy logics) is the definition of the distribution of the most effective costs according to business-plan and the micro-business intellectual characteristics level potential according to the appropriate borrower level risk.

2 The Current State of the Problem

Reviews of the latest investigation in open innovations are presented [4, 13]. Qualitative analysis for seven case studies of SMEs active in digital (high-tech) or social economy (low-tech) sectors is presented in [7]. Investigation of open innovation practices of SMEs and to explore the potential size and country specificities for Spain and Portugal is presented in [9]. The interesting result is obtained in [8]: most open innovation activities related to the pioneering behavior. The methodology for capturing, measuring, and benchmarking open innovation practices of SMEs is proposed in [10]. Such indicators are chosen: knowledge and technology sourcing activities; sources of knowledge; innovation networks; intellectual property protection; innovation expenditure; human capital.

That R & D cooperation between firms and universities is essentially needed for innovations that are shown in [17]. In [15] there is shown that cooperation is very important for innovation. Theoretical thought development under the conditions of economic development foresees the research of collective, individual, and institutional knowledge.

These researches anticipate economic, financial, sociologic, and information measures usage, as it is reflected in the research

1) Intellectual and human capital theories [16];
2) Intellectual capital measurement tools [2, 3, 5];
3) Theories of building intellectual potential based on critical competencies [6, 12];
4) Measurement of key competencies using fuzzy sets and interaction matrices [10].

Neural networks are used for innovation solving economic tasks [18–20]. Creative industries for increasing intellectual capital are proposed in [21–23]. The short review shows that creativity, innovation, and entrepreneurship have to be combined for creating intellectual potential.

3 Competencies of the Business Startup

The competencies of a novice businessman determine the success and development of his business. The essential competencies of the business startup in a creative economy include those listed in Table 1 [1].

Entrepreneurial Potential can be implemented based on the use of basic business skills. Essential skills can be used across all roles and industries, primarily designed to help your startup succeed.

General skills required for startup include soft skills. Search skills that come through via your available experiences and previous employment. It's always to expand and build upon these skills, as it will further Entrepreneurial potential for success.

The most important of them are as follows:

1) Innovation. New businesses thrive on ideas.
2) Communication. Communication and interpersonal skills at work, Effective wrote communication and communication should be a reflection of the startup.

Table 1. The essential competencies of the business startup.

Competencies	Business process
Manage financials and physical assets	QuickBooks, QuickBooks alternatives
Planning	Critical processes that can be split out later, including market opportunity, requirements, product definition, business model, sales process, and organization
Product process	Document requirements, features, metrics, and milestones. SEO marketing
Funding process	Networking and documents for investors and institutional investors
Human resources Leverage information technology Customer service and support	Recruiting, hiring, paying, and training applications required, databases design, and backups scheduled, coding for the web people want a contact, real expertise, and quick response

3) Personal qualities. Emotional endurance and resilience affect relationships and determine business success.
4) Project planning and strategizing. Starting a business will mean you have to manage multiple different projects simultaneously. Writing a business plan, you'll be able to transition to business strategy [3, 15].

Another list of the competencies at their most basic level:

1) Being flexible;
2) Being a team player;
3) Demonstrating an entrepreneurial spirit;
4) Using a collaborative approach;
5) Having a customer focus;
6) Being able to "get things done";
7) Being innovative;
8) Being an analytical or conceptual thinker;
9) Possessing strong business acumen;
10) Having advanced negotiating skills [5].

Thus, essential competencies allow you to implement business innovations using personal and entrepreneurial skills.

Competencies are employee behaviors, skills, or attributes that directly affect performance. The more complex the job, the higher the impact competencies have on business startups.

4 Methods and Researches Analyses

For the intellectual potential analysis of entrepreneur structure at a micro level, it is suggested to use a complex method formed by integrating expert and formalized approaches based on fuzzy logic.

Let's consider separate elements of fuzzy analyses theory, which are used for the algorithm of hidden influences upon discovering business-plan effectiveness realization on micro-business level (entrepreneur-starter). Fuzzy analysis of the researched elements A

$$A\{(x, \mu_a(x))\} \tag{1}$$

Mathematically is identified as a combination of regulated couples, which consist of elements x of the universal analysis X and appropriate levels of belonging, $\mu_a(x)$

$$\mu_a(x) : X \to [0; 1] \tag{2}$$

Fuzzy relation R between analyses X and Y will be called a function

$$R : X \times Y \to L, \tag{3}$$

where L is a complete distribution grating.

If analyses X and Y are completed, fuzzy relations R between X and Y can be presented with the help of matrix relations. Analysis's elements X and Y correspond to tapes and columns of such a matrix, and on the tape's cross x and column y, the element $R(x, y)$ is located.

Business intellectual potential diagnostics realization based on entrepreneur's intellectual potential and business intellectual potential indexes and their dependences in formalized view can be presented like this:

$$S = P_i, V_j, \tag{4}$$

where P_i is the list of researched features: entrepreneur's management competencies (K), business intellectual potential indexes (B), business matters effectiveness realization (E);

$$P_i = \{K, B, E\}, \tag{5}$$

V_j is dependent between the 1st and 2nd levels.

Expert determination influences of some researched features upon the others:

$$K \to B, B \to E, K \to E. \tag{6}$$

Intellectual potential features analyses are described in the following way.

$$\begin{aligned} K &= \{k_1, \ldots, k_i, \ldots, k_n\} \\ B &= \{b_1, \ldots, b_j, \ldots, b_m\} \\ E &= \{e_1, \ldots, e_l, \ldots, e_d\} \end{aligned} \tag{7}$$

The features are interchangeable, which means that "k" features can be determined for a particular element "b" features; the meaning "b" is choosing for parts "e".

5 Results and Discussion: The Methodology for Assessing Intellectual Potential

Summarizing foreign and Ukrainian scientists' experience in the various activity results assessment methods usage and intellectual assets, the assessment methods of entrepreneur-beginner intellectual potential are suggested, which combine quantitative and qualitative assessment parameters and reflect the requirements to the choice of its elements.

Combining quantitative parameters of estimation with qualitative parameters of element entrepreneurship structures, intellectual potential evaluation is offered a stage intellectual possible analyses methodology.

Stage 1. Analyses based on outer data of public accounting with the help of quantitative and financial indices of enterprise activity.

Stage 2. Analyses and estimates quantitative intellectual capital parameters based on questionnaire personnel and e data.

Stage 3. To establish ties among the indices and characteristics based on matrix methods of measurement and analyses.

Stage 4. Calculations of integral estimation indices and econometric measures of intellectual potential influence on the results.

Microbusiness intellectual potential analysis foresees the specifics of this potential element consistency, which has the following structure in the shortened/simplified version. Microbusiness intellectual potential (IP_m) consists of entrepreneur-beginner intellectual potential (IP_p) and business idea intellectual potential, which is realized on the determined territory (IP_t).

$$IP_m = IP_p + IP_t. \tag{8}$$

The intellectual potential of an entrepreneur-beginner is formed using his key competencies. The term "key competencies" is suggested to consider in the following ways:

1. to establish the availability of skills and knowledge in some particular activity sphere (or the personnel competence the way it is);

2. to establish the relevance to the strategies requirements development of the enterprise according to the market requirements.

So, let us consider the three multiplications of the researches features of the enterprise intellectual potential R: $K, B, E \subset R$ (micro business intellectual potential).

Entrepreneur competencies "k" includes:

1) Activity prediction and planning;
2) Personnel management skills;
3) Creative innovation thinking;
4) Skills to form aims and tasks as to their carrying out;
5) Skills to obtain and form new knowledge;
6) To organize knowledge implementation;
7) Knowledge level about product competitiveness as to its characteristics;
8) Knowledge level about business competitiveness as to its location place;

9) Knowledge about consumer's needs;
10) Knowledge level about market channels.

Business characteristics are presented in Table 2.

Table 2. Business Characteristics List.

Intellectual potential Business "b"	Business plan "e" matters
Business state support	Production activity planning "expenses-profits."
Partnership relations in staff and business	Determination of payability risks system
Competitiveness level on the territory	Competitive advantages of the business location and suppliers determination
High consumption level	Competitive advantages of selling
Entrepreneurship activity strategic character	Competitive advantages of production
Personnel professionalism level Licenses usage strategy (franchising)	Personnel management (rewards, salary, qualification raise, etc.) Selling organization (discounts, merchandising, etc.)
Optimal small size business choice strategy	Fast profitability, minimal personnel quantity engagement)
Participation strategy in large business product	Supplies and market organization
Developed Sales network	Marketing Calculation carrying out and consumers' needs research; long-term supply agreements;

Mutual influences of these features can be presented with the help of matrix (relations) of incidences.

$$A = \{a_{ij}\}, a_{ij} = {}^{k_i}/_{b_j},$$
$$D = \{d_{jl}\}, d_{jl} = {}^{b_j}/_{e_l}, \quad (9)$$
$$C = \{c_{il}\}, c_{il} = {}^{k_i}/_{e_l}.$$

If one feature influences the other one, it is added as one if it is absent zero [11]. The expert estimated how key management competencies influence the indices of intellectual potential and intellectual potential level upon business-plan realization efficiency. With the help of fuzzy logic transformations, the hidden influences of management competencies on business realization are revealed.

Expert analysis tasks method with the help of fuzzy analysis methods are the following:

1. To establish the entrepreneurs' intellectual characteristics and business ideas.
2. Establish the list of business plan matters which provide business success.

3. To assess with the help of expert dimensions strategic business characteristics, influence management competencies, and entrepreneur-beginner knowledge level upon activity results to complete business-plan matters.

Implementation algorithm foresees following dependences definition.

1. Correlation between intellectual potential indices and business – plan (numeral correlation matrix 10/10).

2. Correlation establishment among entrepreneur's characteristics and business plan matters (matrix 10/10);

3. Corresponding matrixes construction:

– Experts' competencies assessment (divergence assessment);

– Characteristics' determination, which the most influence the activity success;

– The most effective costs division establishment among business plan matters.

The expert questionnaire should be done to implement tasks.

1. About the influence level of each entrepreneur characteristic (entrepreneur competencies) for each business-plan matter.

2. The division of the practical costs among business plan matters is about a business plan. Each matter level influences activity effectiveness.

Using algorithm [11] about operations over matrixes, the hidden matrix in fluences Δ_{il} is defined.

6 Conclusions

Features of activity in the conditions of the knowledge economy require a modification of methods for solving complex problems of the functioning and development of business structures the emergence of new approaches to the analysis of business practices. Research methods are methods, techniques for conducting research. The choice of research methods integration of different methods in the conduct of study allows for an adequate diagnosis of the state of functioning of a complex economic system and to detect directions of change.

It is impossible to assess and describe the relationship between critical competencies, elements of intellectual potential, and business plan activities using the mathematical apparatus of statistics or other numerical methods due to the lack of quantitative measurements of these influences.

Fuzzy logic models that use the concepts of incidents (influence) and matrices of incidents that display quantitative expert assessments of the impact of some factors on others make it possible to numerically describe the studied forces and detect hidden (undetected) effects. It is proposed to use the fuzzy logic apparatus and, based on the concepts of incidence (influence) and incidence matrices, which display quantitative expert assessments of the impact of some factors on others, determine the level of their power and detect hidden (undetected) effects.

Quantifying these influences will measure the risk of running the business plan. The presented model is of scientific and practical interest and can be used to develop practical methods for analyzing the intellectual potential of a business and assessing the risk of lending or financing it. Suggested entrepreneur structures intellectual potential analysis methods practical usage requires considering the specifics and activity directions, territory potential and experts' assessment trustworthy.

References

1. Eight core competencies for all startups. https://www.caycon.com/blog/8-core-competencies-for-startups
2. Andriesson, D.: Implementing the KPMG value explorer. critical success factors for applying IC measurement tools. J. Intell. Capital 6(4), 474–488 (2005). https://doi.org/10.1108/14691930510628771
3. Andriopoulos, C., Lowe, A.: Enhancing organizational creativity: the process of perpetual challenging. Manag. Decis. **38**, 734–742 (2000). https://doi.org/10.1108/00251740010379768
4. Bogers, M., Chesbrough, H., Moedas, C.: Open innovation: research, practices, and policies. Calif. Manage. Rev. **60**(2), 5–16 (2018). https://doi.org/10.1177/0008125617745086
5. Bontis, N.: Assessing knowledge assets: a review of the models used to measure intellectual capital. Int. J. Manag. Rev. **3**(1), 41–60 (2001). https://doi.org/10.1111/1468-2370.00053
6. Dronyuk, I., Moyseyenko, I., Kuck, J., Horina, O.: Adaptive models of information management in crisis. In: 2020 IEEE 15th International Conference on Computer Sciences and Information Technologies (CSIT), pp. 357–360 (2020). https://doi.org/10.1109/CSIT49958.2020.9321972
7. Dubouloz, S., Bocquet, R., Equey Balzli, C., Gardet, E., Gandia, R.: Smes' open innovation: applying a barrier approach. Calif. Manage. Rev. **64**(1), 113–137 (2021). https://doi.org/10.1177/00081256211052679
8. Hochleitner, F., Arbussa, A., Coenders, G.: Inbound open innovation in SMEs: Indicators, non-financial outcomes and entry-timing. Technol. Anal. Strat. Manag. **29**(2), 204–218 (2017). https://doi.org/10.1080/09537325.2016.1211264
9. Hochleitner, F., Arbussa, A., Coenders, G.: Evolution of inbound openness profiles in the innovation practices of small and medium-sized enterprises in Spain and Portugal. Int. J. Entrepreneurship Innov. Manag. 24(1), 73–96 (2020). https://doi.org/10.1504/IJEIM.2020.105281
10. Jones-Evans, D., Gkikas, A., Rhisiart, M., MacKenzie, N.: Measuring open innovation in SMEs. In: Researching open innovation in SMEs. pp. 399–427 (2018). https://doi.org/10.1142/9789813230972.0013
11. Moiseienko, I.: Intellectual potential management: Monograf., p. 304. Avers, Lviv (2007)
12. Moiseienko, I., Demchyshyn, M.: Regression analysis of intellectual potential. Actual Prob. Econ. **10**, 142–148 (2008)
13. Santoro, G., Ferraris, A., Giacosa, E., Giovando, G.: How SMEs engage in open innovation: a survey. J. Knowl. Econ. **9**(2), 561–574 (2016). https://doi.org/10.1007/s13132-015-0350-8
14. Shalley, C., Hitt, M., Zhou, J.: The Oxford Handbook of Creativity, Innovation, and Entrepreneurship, Chap. 3: Entrepreneurship, p. 540. Oxford University Press (2015)
15. Stefan, I., Bengtsson, L.: Appropriability: a key to opening innovation internationally? Int. J. Technol. Manag. **71**(3–4), 232–252 (2016). https://doi.org/10.1504/IJTM.2016.078570
16. Stewart, T.: Intellectual capital: the new wealth of organizations, p. 320. Crown Business (1998)
17. Veer, T., Lorenz, A., Blind, K.: How open is too open? The mitigating role of appropriation mechanisms in R&D cooperation settings. In: R & D Management, pp. 1113–1128 (2016). https://doi.org/10.1111/radm.12232
18. Peleshko, D., Rak, T., Izonin, I.: Image superresolution via divergence matrix and automatic detection of crossover. IJISA **8**(12), 1–8 (2016). https://doi.org/10.5815/ijisa.2016.12.01
19. Izonin, I., Trostianchyn, A., Duriagina, Z., Tkachenko, R., Tepla, T., Lotoshynska, N.: The combined use of the wiener polynomial and SVM for material classification task in medical implants production. IJISA **10**(9), 40–47 (2018). https://doi.org/10.5815/ijisa.2018.09.05

20. Shakhovska, N., Montenegro, S., Kryvenchuk, Y., Zakharchuk, M.: The neurocontroller for satellite rotation. IJISA **11**(3), 1 (2019). https://doi.org/10.5815/ijisa.2019.03.01
21. Moiseienko, I., Dronyuk, I., Vasylchak, S., Kravchenko, M., Petrynyak, U.: Information technologies implementation for creative entrepreneurship and innovations development in Ukraine. Lecture Notes Data Eng. Commun. Tech. **135**, 624–633 (2022). https://doi.org/10.1007/978-3-031-04809-8_57
22. Javed, Z., Mahmood, W.: A survey based study on fog computing awareness. IJITCS **13**(2), 49–62 (2021). https://doi.org/10.5815/ijitcs.2021.02.05
23. Faryad, S., Batool, H., Asif, M., Yasin, A.: Impact of Internet of Things (IoT) as persuasive technology. IJITCS **13**(6), 16–28 (2021). https://doi.org/10.5815/ijitcs.2021.06.02

A Model for Weather Forecasting Based on Parallel Calculations

Lesia Mochurad(✉), Oleh Kotsiumbas, and Iryna Protsyk

Lviv Polytechnic National University, S. Bandera Str., 12, Lviv 79013, Ukraine
lesiamochurad@gmail.com

Abstract. With the rapid accumulation of processed data in the modeling of a problem, it is necessary to apply the parallelization of the computational process, in order to be able to make decisions in real-time. In particular, parallel calculations are considered a tool for modeling complex phenomena and processes, as well as in solving cumbersome scientific and engineering problems: in the study of atmospheric phenomena, weather forecasting, environmental pollution, geology, seismology; applied and nuclear physics, electrical engineering, particle physics, materials science, nanotechnology, microelectronics; research in the fields of biotechnology, ecology, population biology, virology, genetics, pharmacology, and molecular chemistry. The paper constructs a model for weather forecasting using a regression equation based on parallel calculations. The parallelism of the algorithm for finding the coefficients of the curvilinear regression equation is performed. At the same time, the advantage of the possibility of calculating the coefficients of curves of any order is obtained. For concurrent calculations, we use the built-in concurrent library in Python, as well as properties such as multithreading. To verify the reliability of the obtained results on the basis of numerical experiments, a theoretical estimation of parallel acceleration according to Gustafson's law was performed. The dependence of the execution time of the proposed algorithm on the curve order for the different numbers of processes is investigated. A comparison of predicted and actual temperature data is made, and the coefficient of determination is calculated. In order to obtain real benefits from the implementation of parallel calculations, the current date was expanded and indicators of acceleration and efficiency of the proposed algorithm were found. The advantages obtained indicate the significance and value of the article.

Keywords: Parallel computing · Machine learning · Efficiency indicators · Concurrent library · Gustafson's law

1 Introduction

Weather forecasting is an integral part of everyone's life [1] because much of our life depends on weather conditions. For this purpose, whole meteorological stations with a large number of equipment and workers are used [2]. So the question is, how can such a high cost of resources be replaced by computer and machine learning methods alone?

At work [3], an analysis of the 500 most relevant scientific articles published since 2018 on the methods of machine learning in the field of climate and weather forecasting.

Z. Hu et al. (Eds.): AIMEE 2022, LNDECT 159, pp. 35–46, 2023.
https://doi.org/10.1007/978-3-031-24468-1_4

As a result, the authors conclude that machine learning methods will be a key feature in future weather forecasts.

As you know [4, 5], weather forecasting is one of the most difficult scientific and technical problems in the world. The accuracy of the model and its timely release are important.

One of the methods for predicting data is the regression equation [6]. It takes all the data into account and tries to draw the line as close and accurate as possible through our data. But this is not so simple, because there are many types of regressions [7], and different types are used for different tasks.

The aim of this study is to build a model for weather forecasting using regression equations based on parallel calculations. The dataset "Hourly energy demand generation and weather" was chosen as data [8]. The dataset contains information on the state of the weather every hour from January 1, 2015, to December 31, 2018, in the cities of Spain. However, in the course of a study to optimize the computational process based on parallel computing, this data set was expanded to 2021. The aim of the study is to accelerate the operation of the algorithm for finding the regression equation using parallel computational methods for the problem of weather forecasting.

2 Related Works

Parallel computing [9] is not something radically new in the world of technology, they are applied to all possible algorithms and approaches.

The regression algorithm has been known and used since the 1800s [10], especially in the field of numerical data processing – from financial statements to mechanical calculations. With the rise of machine learning, neural networks, and big data, the need to calculate regression has increased exponentially, as virtually any attempt to predict the value of a variable begins with an attempt to use linear, curvilinear, polynomial, or logistic regression.

This has become the foundation, and no one is thinking about improving the same linear regression, because there is nothing to improve. A single formula allows you to tell about the data after more than ten minutes of looking at the numbers.

The main activities to improve the efficiency of regression are two approaches. The first is the use of multiple cores (and processes and threads, respectively), and the second is the use of multiple data sources. In one case we divide the calculation into several parts, in another, we divide the data into several parts. Both cases use the theory that several smaller tasks are easier to handle than one large one.

GenStat [11] has made some progress in this, developing a program for entering data, setting parameters, and obtaining results through cloud computing, where parallel calculations are used, in terms of dividing the sample into smaller ones and finding coefficients on many cores of their server. However, not everyone will be able to appreciate this achievement, because the details of the algorithm used are not disclosed, and to use the program you need to purchase a key.

There are many third-party frameworks for many languages, the authors of which are also trying to implement a parallel calculation of linear regression. An example is RegParallel [12] for the R language from kevinblighe. This is a fairly powerful library for working with linear and logistic regressions in terms of big data processing with the ability to specify the number of cores, enable nested parallelism and change the confidence interval. Functionality is cumbersome, gives a significant advantage only for big data, while with small samples it works even slower.

Another promising area is the use of GPUs to speed up calculations [13], because the complexity of calculations is small, only a large number. Experiments are currently underway with the implementation of these modifications, and how you can see the acceleration of work in this way can be felt.

In [14], the authors show how reference vector machines can be used for the prediction of the linear and nonlinear nature of data. However, most of the existing systems are implemented using statistical approaches for the machine of reference vectors, which can not give an accurate forecast because they can not catch sudden changes in weather conditions. The method proposed in [14] uses the concept of multilinear regression, which can give better results than existing methods. However, the authors do not analyze the time spent on such a forecast and the possibility of using parallel calculations.

Paper [15] presents a neural network approach for weather forecast improvement. The method proposed by the authors allowed us to obtain a forecasting accuracy of 0.7689. Our novel algorithm is based on Parallel Calculations and Uses a Curvilinear Regression Equation capable of increasing forecast accuracy to 0.8587.

In this paper, the parallelization of the algorithm for finding the coefficients of the regression equation is performed. An important advantage is the ability to calculate the coefficients of curves of any order. The solution is also implemented in Python, which supports many libraries and tools for working with data and using machine learning. This means that the results can be visualized, used, or further processed in the widest range of ways.

3 Materials and Methods

The study of the form of connection sometimes necessitates the use of nonlinear (curvilinear) regression equations [6]. This is due to the fact that the interaction between the features that characterize individual processes is often more complex than just proportional (linear) relationships.

A parabolic regression equation [16], or in other words a second-order regression equation, is usually used to determine a nonlinear relationship. Finding the coefficients for regression second order looks (1).

$$\begin{cases} \sum y = a * n + b * \sum x + c * \sum x^2, \\ \sum x * y = a * \sum x + b * \sum x^2 + c * \sum x^3, \\ \sum x^2 * y = a * \sum x^2 + b * \sum x^3 + c * \sum x^4, \end{cases} \tag{1}$$

where x is the independent variable, y is the dependent variable, n is the sample size, a, b, c are the coefficients of the equation.

And it is quite logical that with increasing order the system of equations will have more parameters and require more cumbersome calculations, despite the fact that the curve itself will acquire more bends [17] (which will be shown below).

In the software implementation of the proposed algorithm, it is necessary to write a function that will make up the regression equation for the specified samples x and y, as well as the order of the curve. To begin, we need to make a matrix A of size $((n + 1) * (n + 1))$, where n is the order of the curve. The matrix will look like this: $\mathrm{a_{ij}} = \sum x^{i+j}$, where i and j – matrix row and column number, respectively. Next, we need to compose a vector B of size $n + 1$ so that: $b_{\mathrm{i}} = \sum y \cdot x^i$, where i – the ordinal number of the element in the vector.

Thus we obtain a system of linear algebraic equations of the following form: $A \cdot coffs = B$, where *coffs* – the required coefficients of our regression Eq. (2).

$$\mathrm{y} = \sum\nolimits_{\mathrm{i}=0}^{\mathrm{n}+1} \mathrm{coffs_i} * \mathrm{x}^{\mathrm{i}}, \tag{2}$$

where x is the independent variable, y is the dependent variable, *coffs* are the coefficients of the equation, and n is the order of the curve.

Finding the coefficients of the curvilinear regression equation consists of two stages: finding the elements of the matrix A and the vector B and finding the solutions of the system of linear algebraic equations. If the second stage of the algorithm does not make sense to parallelize, because there will not be much data and the creation of flows (processes) will take much more time than their work, then the right decision will accelerate only the first stage of the algorithm.

Parallelization will be implemented as follows:

1) Create a list L of size $n * 2 + 1 + n + 1$, filled with zeros, where n is the order of the curve;
2) Create a function that will be performed by parallel streams (processes), which will fill the list with results; the function takes a list of tuples, each of which contains the identifier of the operation (1- the sum of the elements of list X to a given degree, 2 – the sum of the products of elements of list Y to elements of list X to a given degree), list X containing independent variable, list Y, contains the dependent variable, degree, reference to the list created in step 1;
3) Form a list of arguments for a parallel function and divide it into blocks equal to the number of flows (processes);
4) Run streams with each block;
5) Fill in the matrix A, where each element $a_{ij} = L_{i+j}$, and the vector B, where each element $b_i = L_{-(i+1)}$;
6) Solve system of linear equations (or linear system): $A \cdot coffs = B$, where *coffs* are the required coefficients of the regression equation.

Given that we used the Python programming language to work, the built-in concurrent.

library is ideal for parallelization in this case. This library allows you to create parallel independent processes and control the number of processes running simultaneously.

4 Research Results

4.1 Experiments with the Initial Dataset

To begin, calculate the theoretical acceleration. Our algorithm consists of two parts: finding the elements of the matrix and the vector and solving the SLAR. To calculate the theoretical acceleration, calculate the execution time of each of the parts for different orders of the curve (see Table 1).

Table 1. Comparison of execution time of each part of the algorithm

Order	Search for items		Solving linear system		The whole algorithm	
	Time, s	%	Time, s	%	Time, s	%
1	0.044017	97.78	0.000998	2.22	0.045015	100
2	0.092351	98.89	0.001025	1.11	0.093376	100
5	0.464041	99.78	0.000994	0.22	0.465036	100
10	1.8433	99.95	0.000967	0.05	1.844276	100
15	4.24452	99.96	0.001698	0.04	4.2462178	100
25	22.51112	99.99	0.002	0.01	22.513119	100

Let's move directly to the calculation of theoretical acceleration. To do this, we use Gustafson's law (3).

$$S(P) = P - \alpha(P - 1), \tag{3}$$

where P is the number of processors, S is the acceleration, and α not the parallel part of the process.

Table 2. Dependence of theoretical acceleration on the order of the curve for different numbers of processes

The order of the curve	Number of parallel processes		
	1	2	4
1	1	1.9778	3.9334
2	1	1.9889	3.9667
5	1	1.9978	3.9934
10	1	1.9995	3.9985
15	1	1.9996	3.9988
25	1	1.9999	3.9997

We will perform our experiments on 1, 2, and 4 processors, so we will calculate the theoretical acceleration on these processors for different curve orders. The results of the calculations are given in Table 2.

Therefore, the theoretical acceleration approximately equal to the number of processors of the corresponding computer system was obtained.

Since the parallel part takes up almost 100% of the time of the algorithm, the theoretical acceleration is always close to the number of processes. This assessment allows us to conclude that for more processes we will get a similar trend in the results.

Now let's move directly to comparing the execution speed of parallel and serial algorithms for our data. We will compare the time to find the average coefficients of our curve equation by year for curves of a different order and see which curve of which order will visually give the best result (see Table 3 and Table 4).

Table 3. Dependence of algorithm execution time on curve order for different number of processes

The order of the curve	Consequence	Number of parallel processes				
		1	2	4	8	16
2	0.083	0.35	0.25	0.31	0.44	0.64
5	0.208	0.59	0.37	0.41	0.55	0.75
10	0.43	0.92	0.76	0.58	0.63	0.75
15	0.69	1.35	0.95	0.85	0.89	1.02
25	1.45	2.11	1.61	1.33	1.32	1.52

Table 4. Dependence of acceleration on the order of the curve for different number of processes

The order of the curve	Number of parallel processes				
	1	2	4	8	16
2	0.23	0.33	0.26	0.18	0.13
5	0.35	0.56	0.5	0.37	0.28
10	0.46	0.56	0.74	0.68	0.57
15	0.51	0.72	0.81	0.77	0.67
25	0.68	0.9	1.09	1.09	0.95

The obtained results indicate the need to expand the input data in order to obtain clear advantages from the parallelization of the computing process.

4.2 Experiments with Advanced Dataset

With this amount of data, the acceleration is very small, so you need to somehow increase the amount of data. We have time data in the form of year, month, day, and hour. The best option is to add another column - minutes. However, filling in the gaps, in this case, will be quite simple – using linear interpolation and evenly distributing the weather every minute between hours.

Thus, we increase the amount of our data by 60 times and now we have 365*24*60 = 525600 data for regression, which may well be enough to speed up our algorithm.

Once again, it makes no sense to look for theoretical acceleration, because it was already very close to the number of processes, and in the extended dataset x and y samples will increase, but the size of the matrix and vector will remain the same, so the theoretical acceleration will be even closer to the number of processes. Because the percentage of the sequential part in the proposed algorithm will be very close to zero.

Now let's do the same research with the extended dataset as with the non-extended one (see Table 5, 6, 7 and Figs. 1, 2).

Table 5. Dependence of algorithm execution time on the number of processes for different curve order for extended data

The order of the curve	Consequence	Number of parallel processes				
		1	2	4	8	16
2	3.98	10.98	10.22	10.37	13.68	20.42
5	11.17	18.28	14.54	13.34	14.85	22.36
10	26.17	33.85	23.29	19.04	19.73	27.01
15	46.41	56.02	38.04	26.76	26.24	33.61
25	103.80	105.73	67.14	43.35	36.46	37.07

Table 6. Dependence of acceleration on the number of processes for different curve order for extended data

The order of the curve	Number of parallel processes				
	1	2	4	8	16
2	0.36	0.38	0.38	0.29	0.19
5	0.61	0.76	0.83	0.75	0.44
10	0.77	1.12	1.37	1.32	0.96
15	0.82	1.22	1.73	1.76	1.38
25	0.98	1.54	2.39	2.84	2.80

So, as a result of using the proposed algorithm, an acceleration of 2.8 was obtained and an efficiency of 0.98 was achieved. These results can be significantly improved by varying the number of threads and choosing an appropriate computing system.

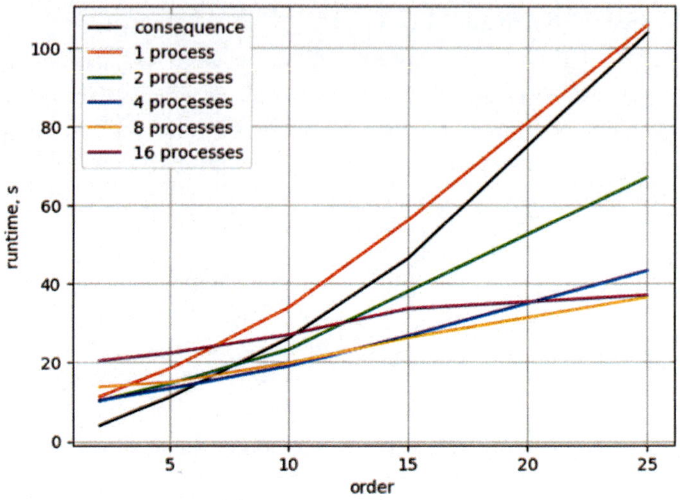

Fig. 1. Dependence of operating time on the curve order for a different number of processes for extended data

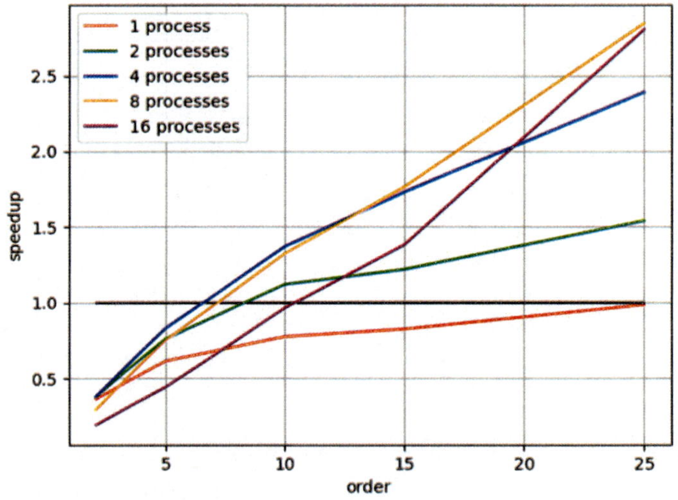

Fig. 2. Dependence of acceleration on the order of the curve for different number of processes for extended data

4.3 Weather Forecasting Using Regression

To begin with, we need to assign a special identifier to each minute of the year to form sample *X* (independent variable). We will do this with the help of a dictionary, where the key will be a train (month, day, hour, minute), and the value – the identifier itself. Let's create this dictionary using the following functions. Let's take 2016 as a basis, because it is high, and it has February 29.

Table 7. Dependence of efficiency on the number of processes for different curve order for extended data

The order of the curve	Number of parallel processes				
	1	2	4	8	16
2	0.36	0.19	0.095	0.036	0.004
5	0.61	0.38	0.207	0.094	0.027
10	0.77	0.56	0.34	0.165	0.06
15	0.82	0.61	0.43	0.22	0.08
25	0.98	0.77	0.59	0.35	0.175

Table 8. Comparison of projected data and actual temperature data of Valencia for 2019–2021

Date (dd.mm)	Predicted temperature	2019	2020	2021
15.01	10.6	9.5	11	10.5
15.02	11.09	9.5	12	13.5
15.03	12.87	15	14.5	12
15.04	15.41	11	15	14
15.05	18.73	13	11.5	23
15.06	23.29	25.5	22.5	–
15.07	26.66	27.5	24.5	–
15.08	26.22	26.5	28.5	–
15.09	22.98	25	26.5	–
15.10	18.49	19.5	17.5	–
15.11	13.98	11.5	14	–
15.12	11.56	14	14	–
Coefficient of determination	–	0.8587	0.8012	0.8015 (for 5 months)

As already analyzed, with increasing order of the regression curve, the accuracy of the model increases, so we choose the largest tested order – the 25th.

As described earlier, we will look for the average coefficient of our equation based on each year and use the resulting equation to look for the temperature for the specified day.

As a result of averaging the obtained coefficients for each year, we obtain the following function (see Fig. 3).

Using this equation, which is graphically represented in Fig. 3 we will predict the temperature. Let's compare the real data for 2019–2021. To do this, we found on the Internet weather data for these years in the city of Valencia.

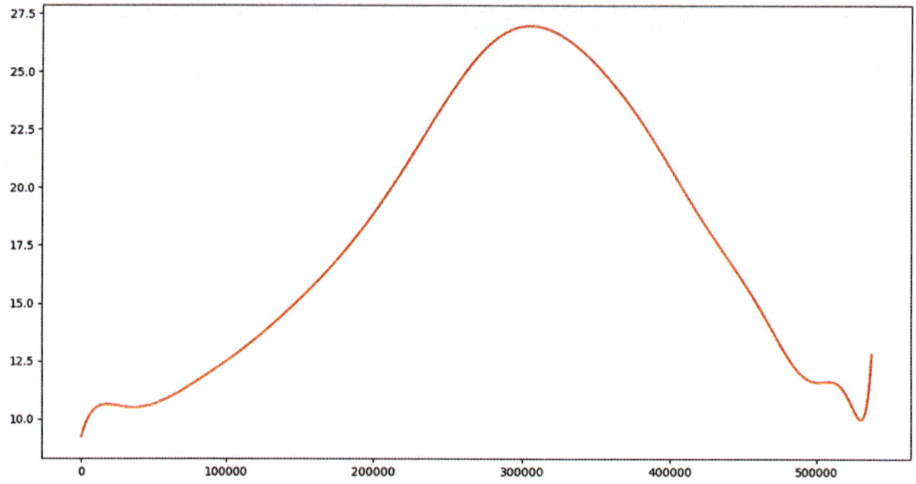

Fig. 3. Graph of the received regression

5 Conclusions

In this paper, we aimed to accelerate the algorithm for building a model for weather forecasting using curvilinear regression based on parallel calculations. As a basis for parallelization was chosen such a property as a lot of fluidity. The Python concurrent library was used for the software implementation of the proposed algorithm. Theoretical estimates of acceleration were obtained, which were analyzed for reliability on the basis of Gustafson's law.

As you can see from the Sect. 2 the built model adequately builds the regression equation, as well as provides an accuracy of more than 80%. Also in the same section, in the Subsect.4.2, we were able to achieve a maximum acceleration of 2.84 times for the regression curve of order 25 on 8 parallel processes. Obviously, these results can be significantly improved by varying the number of parallel processes and choosing the appropriate multicore computing system.

This study can be a starting point for further experiments and optimizations of the developed algorithm [18]. Given that weather forecasting is still relevant to mankind, this method can be convenient for weather forecasting in a short time, without the need for cumbersome meteorological centers, as it analyzes data from previous years. Of course, meteorological centers are indispensable, as they are the main provider of information for us, but if the task is to predict the approximate (probable) state of temperature, then this method will be very convenient.

Further research will be carried out in three main areas for increasing the accuracy of solving the predicting the weather forecasting:

1) The use of different machine learning methods [19, 20];
2) Use the weather forecasting model in BP/GA hybrid technology [21] in combination with modern parallel computing technologies;

3) To investigate the method using the inverse propagation neural network to predict the weather based on parallel calculations [22].

References

1. Davenport, F.V., Diffenbaugh, N.S.: Using machine learning to analyze physical causes of climate change: a case study of U.S. Midwest extreme precipitation. Geophys. Res. Lett. **48**, 1–11 (2021)
2. Chavan, G., Momin, B.: An integrated approach for weather forecasting over the Internet of Things: a brief review. In: 2017 International Conference on I-SMAC (IoT in Social, Mobile, Analytics, and Cloud) (I-SMAC), pp. 83–88 (2017)
3. Bochenek, B., Ustrnul, Z.: Machine learning in weather prediction and climate analyses. applications and perspectives. Atmosphere 13(2), 180 (2022)
4. Patil, T., Shah, K.: Weather forecasting analysis using linear and logistic regression algorithm. Int. Res. J. Eng. Technol. (IRJET) **8**(6), 2557–2564 (2021)
5. Nanda, S.K., Tripathy, D.P., Nayak, S.K., Mohapatra, S.: Prediction of rainfall in India using artificial neural network (ANN) models. Int. J. Intell. Syst. Appl. (IJISA) **5**(12), 1–22 (2013)
6. Mochurad, L.: Optimization of regression analysis by conducting parallel calculations. In: COLINS-2021: 5th International Conference on Computational Linguistics and Intelligent Systems, pp. 982–996 (2021)
7. Yin, Z., Hadzimustafic, J., Kann, A., Wang, Y.: On statistical nowcasting of road surface temperature. Meteorol Appl. **26**, 1–13 (2019)
8. Data sampling [Electronic resource]: https://www.kaggle.com/nicholasjhana/energy-consumption-generation-prices-and-weather?select=weather_features.csv
9. D'Agostino, D., Leporati, F., Torquati, M., Xue, J.: Guest editorial: special section on new trends in parallel and distributed computing for human sensible applications. IEEE Trans. Emerg. Top. Comput. **9**(4), 1640–1641 (2021)
10. Miguel-Hurtado, O., Guest, R., Stevenage, S.V., Neil, G.J., Black, S.: Comparing machine learning classifiers and linear/logistic regression to explore the relationship between hand dimensions and demographic characteristics. PLoS ONE **11**(11), e0165521 (2016)
11. [Electronic resource]: https://www.vsni.co.uk/software/genstat
12. [Electronic resource]: https://statisticsbyjim.com/regression/difference-between-linear-nonlinear-regression-models/
13. Mochurad, L., Dereviannyi, A., Antoniv, U.: Classification of X-Ray images of the chest using convolutional neural networks. IDDM 2021 Informatics & Data-Driven Medicine. In: Proceedings of the 4th International Conference on Informatics & Data-Driven Medicine. Valencia, Spain, November 19 - 21, pp. 269–282 (2021)
14. Chavan, G., Momin, B.: A novel approach for forecasting the linear and nonlinear weather data using support vector regression. In: Sa, P., Bakshi, S., Hatzilygeroudis, I., Sahoo, M. (eds) Recent Findings in Intelligent Computing Techniques. Advances in Intelligent Systems and Computing, p. 707 (2019)
15. Anusha, N., Chaithanya, M.S., Reddy, G.J.: Weather prediction using multi linear regression algorithm. IOP Conf. Ser.: Mater. Sci. Eng. 590, 012034 (2019)
16. Chen, H.-Y., Chen, C.: Evaluation of calibration equations by using regression analysis: an example of chemical analysis. Sensors **22**(2), 447 (2022)
17. Meloun, M., Militký, J.: 8-Nonlinear regression models. statistical data analysis. A practical guide, pp. 667–762 (2011)

18. Peleshko, D., Rak, T., Izonin, I.: Image superresolution via divergence matrix and automatic detection of crossover. Int. J. Int. Syst. Appl. (IJISA) **8**(12), 1–8 (2016)
19. Izonin, I., Trostianchyn, A., Duriagina, Z., Tkachenko, R., Tepla, T., Lotoshynska, N.: The combined use of the wiener polynomial and svm for material classification task in medical implants production. Int. J. Intell. Syst. Appl. (IJISA) **10**(9), 40–47 (2018)
20. Khan, M.Z.: Hybrid ensemble learning technique for software defect prediction. Int. J. Modern Educ. Comput. Sci. (IJMECS) **12**(1), 1–10 (2020)
21. Singh, Sh., Gill, J.: Temporal weather prediction using back propagation based genetic algorithm technique. Int. J. Intell. Syst. Appl. (IJISA) **6**(12), 55–61 (2014)
22. Barman, D., Singha, R.K., Chowdhury, N.: Prediction of possible business of a newly launched film using ordinal values of film-genres. Int. J. Intell. Syst. Appl. (IJISA) **5**(6), 53–60 (2013)

Steganography in TCP/IP Networks

Oleksandr Tymchenko[1](✉) and Bohdana Havrysh[2](✉)

[1] Faculty of Mathematics and Computer Science, University of Warmia and Mazury, 10-719 Olsztyn, Poland
olexandr.tymchenko@uwm.edu.pl

[2] Department of Publishing Information Technologies, Lviv Polytechnic National University, Lviv 79013, Ukraine
dana.havrysh@gmail.com

Abstract. Due to the large amount of data that is transmitted over computer networks, web traffic is an attractive environment for steganographic techniques. The question is, can data that is sent over the Internet be used as a medium, perhaps without the knowledge of the sender and recipient? In this work, methods of hiding and data transmission are considered, using some shortcomings of data transmission protocols of computer networks. The IP and TCP protocols, existing possibilities of embedding in the headers of data that are invisible and do not affect the transmission of packets are considered in detail. Steganographic transmission was simulated under conditions of a loaded network (packet loss), estimates of the possibility of reliable transmission of a steganographic message were obtained to detect the compromise of fields that appeared to be safe.

Keywords: Computer networks · Steganography · Protocol · Identification fields · Fragmentation · Uniqueness · Coding

1 Introduction

All objects that require access to the WWW must protect their systems and data, isolate themselves from the Internet using a firewall, proxy server, etc. On the other hand, they create problems for steganography – how to sew data that should not get inside the network in normal and "innocent" network traffic (for example, HTTP transmission from a public web server)?

However, data transmission protocols have some imperfections that can be exploited for covert data transmission. For example, the analysis showed that not all free or unused fields of TCP/IP packets can be used for this. In addition, the question of the possibility of steganographic transmission in conditions of packet loss, which is common in the Internet, remains open. The answer to this can be obtained on the basis of detailed analysis of data transmission protocols and simulation of steganographic transmission over the network.

The aim of the work is to analyze the use of steganography in computer networks, in particular the most frequently used TCP / IP networks. The conclusions presented in the paper apply to both TCP/IP local networks and the global network – the Internet.

Z. Hu et al. (Eds.): AIMEE 2022, LNDECT 159, pp. 47–56, 2023.
https://doi.org/10.1007/978-3-031-24468-1_5

Network steganography uses network protocols as carriers of secret data. In general, this is a family of methods for modifying data in the headers of network protocols and in the payload fields of packets, changing the structure of packet transmission. A common feature of all network steganography methods is the creation of hidden transmission channels with their help in any open channel in which there is redundancy [1].

Network steganography methods can be divided into three groups:un

1) Package modification methods:

   - methods that change data in header fields of network protocols.
   - methods that change data in the payload fields of packets – these are well-known algorithms for watermarks introduction, language codecs and other steganographic techniques for hiding data;
   - methods combining two previous classes;

2) Steganography methods that change the structure and parameters of packet transmission:

   - methods in which the passing packets order changes;
   - methods that change the delay between packets;
   - methods that introduce intentional packet loss by omitting sender sequence numbers;

3) Mixed (hybrid) methods change the content and terms of packages delivery, and the order of their transmission. At the same time, two approaches are used: rare delays of audio packets LACK (Lost Audio Packets Steganography) and retransmission of packets RSTEG (Retransmission Steganography) [2].

The Transcoding Steganography (TranSteg), the modifying network packets method, which changes the payload of a VoIP packet, is successful due to the popularity of programs that provide voice and video communication over the Internet. The method consists in compressing the payload of a network packet due to transcoding and can be used wherever data compression is possible [2].

As an example of the second class of methods, the use of SCTP mechanisms (Stream control transport protocol) [5] – transport protocol with packet control can be indicated. SCTP steganography uses the characteristic features of this protocol, such as multi-threading and the use of multiple interfaces (multi-homing). Changing the contents of SCTP packets methods are based on the fact that each part of a TCP packet can have variable parameters.

The RSTEG (Retransmission Steganography) method is based on the mechanism of packet resending. The sender sends the packet, but the receiver does not respond to the packet with an acknowledgment flag. The packet resending mechanism is activated, and now a packet with a steganogram inside is sent, which also does not receive an acknowledgment. Next, the original package without hidden attachments is sent, with a confirmation of successful receipt [3].

The possibilities of steganographic transmission in the conditions of packet loss, which is common via the Internet are solved only by retransmission methods.

Therefore, analysis and simulation of steganographic transmission through lossy networks remains relevant.

2 TCP/IP Protocol Analysis

Elements that can be used to create hidden channels may be specified at each level of the ISO/OSI model. The methods used for such purpose are different and depend on the characteristics of the layer models, and more specifically – the characteristics used at this level of the network protocol, hardware and software [1, 3].

Let's analyze the features of IP and TCP protocols. The answer to the question of data transmission possibility in secure network protocols will be obtained by studying the packet structure of these protocols (Fig. 1). An analysis of their suitability for data transmission other than developers, in the elements of IP and TCP headers, which can be used to create hidden channels, identified the actual suitability, advantages and disadvantages, detection capabilities [2].

Bits 0-3	4-7	8-15	16-18	19-23	24-31
Versions	HLEN	Service Type	General Length		
Identification			Flags	Fragmentation Bias	
Life Time		Protocol	Header Checksum		
Sender IP Adress					
ReceiverIP Adress					
Options					Addition
Data (65535 withoutTitle)					
...					

Fig. 1. The structure of the IP package

2.1 Identification of Hidden Channels

2.1.1 IP Protocol

A closer look at the structure of the IP package, which is the main transport unit at the IP network layer, can identify several elements that may be a potential target for a hidden channel. Not all of these elements are equally attractive and appropriate.

2.1.2 Type of Service Field

The first field of the IP datagram header is the service type field (Fig. 1). This 8-bit field does not have a clearly defined structure and can be interpreted differently in existing IP implementations. This uncertainty can be used to create a hidden channel [4–6].

This field in the IP network can be used or not, depending on the hardware (routers) and software (operating systems, applications). This decision depends on network creators and administrators. In addition, due to the mentioned uncertainty, the value contained in the TOS field is used to control the operation of the network. In the context of networks that use *Differentiated Services QoS, the ToS field is called Differentiated Services Codepoint (DSCP).*

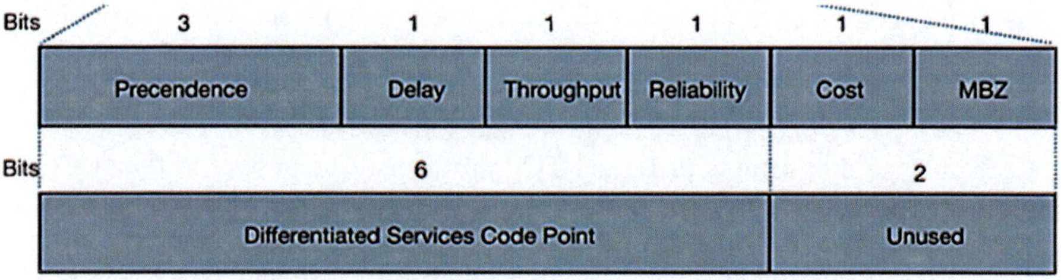

Fig. 2. Two interpretations of the ToS/DSCP field

The ToS/DSCP field can be used to send hidden data in several ways.
For Example:

1) use of the entire ToS field (ie all eight bits) for data transmission;
2) use one unused (youngest) bit, not to affect the values, that are already entered in the field (and not to interfere with the transmission method);
3) use one bit that is a comme one – ie – a bit indicating a small delay (bit 4);
4) use of the three left (oldest) bits of the ToS field, which determine the priority and in practice are ignored in most networks;
5) use of the two youngest, unused bits of the DSCP field (in networks using DifServic technology).

These methods differ in some basic functions. First, the most obvious difference in the amount (bandwidth) generated by the hidden channel – from a whole byte to one bit per IP packet (ie from 20 to 1500 bytes transported in the packet – the maximum size of the transport unit (MTU) in Ethernet networks usually set up to 1500 bytes). Secondly, the degree of concealment of the created channel is different [7, 9].

Monitoring of transmitted and modified IP packages by the person supervising the operation of the network and their analysis will reveal some irregularities in the ToS field in the first case, because under no circumstances is it possible to use the entire field in a "normal" working IP network. In addition, the use of unused bits may be grounds for suspicion. Using bits that are allowed but not used on a particular network will be less suspicious and may be caused by software or configuration errors. In the case of detection of data transmission using the ToS field, the interruption of data transmission is simply to save another value in this field [8].

2.1.3 Identification Field

The next object is a 16-bit identification field. It is used to uniquely identify each package and must be unique, but importantly – only within one transmission for the transmitter-receiver pair (host-receiver-transmitter host). In other words, as long as we provide different ID numbers for different packages, we are free to choose them. This feature allows you to encode seemingly independent, "random" values [10, 11].

If the identification numbers created independently of the TCP/IP stack algorithms are unique, the connection will not be affected. Analysis of IP packages may suggest that the identification has been slightly modified, but we cannot say this with certainty without knowing the operating system from which the data is transmitted and the exact information about the IP transmission through it. As mentioned above, the method of generating ID numbers for each subsequent packages depends on the implementation of the TCP/IP stack in the operating system.

If in the next stage of generating consecutive values of the identification field care is taken to ensure that they do not depend on each other, and there is no statistical analysis to detect hidden data – the channel formed in this way can be considered relatively secure [12].

2.1.4 Fragmentation Sign Field

There are three bits of the fragmentation attribute in the header of the IP package, after the identification field. They determine the processing of the datagram by the router in case of need for defragmentation and serve as information when fragmentation occurred. Two of these three bits are used for this. The oldest bit of the fragmentation feature field is not used (it is reserved and is often called RF – Reserved Flag), and is always zero [13, 14]. However, tests show that setting this bit does not affect how router datagrams are processed (the value of this bit is not checked). Thus, the unused fragmentation bit could be used for data transmission.

2.1.5 Option Field

The option field in the IP package can be variable in length and is not used in most cases. The accepted method of division and numbering options leaves the possibility to determine their own IP parameters, as the existence of options is given outside the RFC document describing IP [13, 15]. The option allows to create previously unused numbers and send data using them. The power of such a hidden channel varies from one bit per option (signaling the presence or absence of this option) to several bytes in the case of an option with parameters. If the "own" option of the IP variant is applied, this volume is limited only by the size of the data transferred by the datagram and the maximum transport unit (in Ethernet networks 1500 bytes).

Sent data stored in IP settings is not at risk of deleting properly configured routers. Parameters outside the RFC will not be interpreted or changed, the data processor will leave them unchanged.

3 TCP Protocol Analysis

3.1 Initial Sequence Number

An integral part of the transport protocol is the concept of three-way traffic, the mechanism used to start a TCP connection. The Initial Sequence Number (ISN), selected at the beginning of the connection, is placed in the first segment sent by the party that imposes the TCP connection. The value of ISN is random, it must be changed regardless of the existing (imposed) connections [16] in such a way that they cannot be predicted statistically. In existing implementations of TCP/IP stacks, the main emphasis is on the latter to counter attacks aimed at capturing the session and forgery.

Apart from the starting number, there is no need for ISN uniqueness in other connections. Therefore, given the complete randomness of this field, you can use it to send hidden data once. The ISN field in the TCP segment header is 32 bits, so you can send up to 4 bytes of data at a time.

Detecting a hidden transmission through the original serial number is difficult. All TCP connections must be monitored and ISN analyzed for each connection. Preventing covert transmissions without interrupting the corresponding TCP connection requires the use of proxy technology for all connections that may be hidden.

3.2 Reserved Field

The 5-bit field reserved in the TCP segment header, according to the RFC document defining the protocol [17–19], is always zero. As with unused ToS field bits and reserved IP datagram header fragmentation bits, it can be used for transmission in the case of a reserved field.

It should be noted that in some cases, there is a risk that routers will reject TCP packages with a non-zero backup field. This behavior is not in accordance with RFC 793 [18, 20], but sometimes it can happen.

3.3 Indication of Importance

The TCP segment header has a 16-bit field called the Urgent Pointer. According to the protocol definition [19], this field is used only when the URG bit is set in the TCP segment header. The concept of using these 16 bits (usually unused) to transmit hidden data is to write the transmitted value to the importance field while leaving the URG bit zero. Then the proof of importance will not be interpreted. Packages with such a modified header (incorrect in terms of RFC compliance) – as shown in the experiments, arrive without hindrance.

Due to the size limit of the IP datagram to the size specified by the MTU, the index cannot be too large. In the case of an Ethernet-based network, the MTU value is usually 1,500 octets. Thus, for the value entered in the field to be plausible and reliable, it must be less than 1460 (1500 minus the minimum IP and TCP header sizes).

3.4 Analysis of the Options Field

However, unlike IP options, TCP options are applicable and may contain at least one of three options: Maximum Segment Size (MSS), SACK-Permitted, No Operation, which are usually sent during TCP installation union.

The RFC authors for TCP have left the option to create their own option parameters. Using this fact, as well as the fact that many TCP settings are not used, you can try to use the TCP option field for data transmission. Also, in this case, the size limit is only the maximum length of the TCP segment, but you should consider the risk of interest in packages containing "exotic" options.

4 Steganographic Data Transmission Simulation

Among the above, to simulate steganographic data transfer the potential hidden channels that exist in the IP and TCP headers, the authors have selected two, the most promising and attractive for software implementation.

A general idea of how the simulator works (transmitter and receiver are on routers, and the client and server are on separate machines) is given in Fig. 3. Thicker lines indicate those network segments through which hidden data is transmitted.

The sender and receiver analyze each packet and check whether:

1) the packet is not damaged and has sufficient length to accommodate IP and TCP headers;
2) the TCP segment contained within carries data.

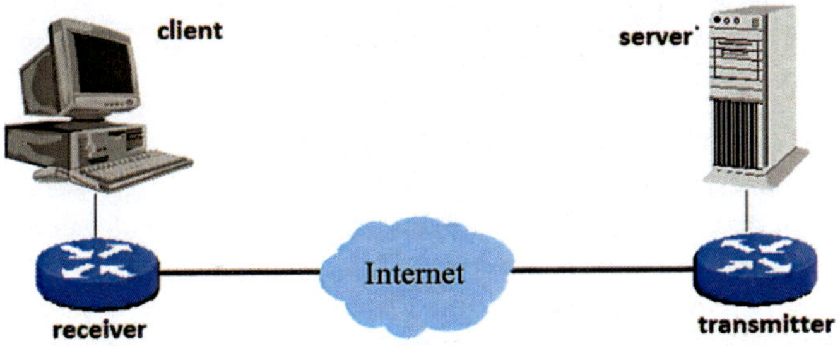

Fig. 3. Steganographic data transfer simulator

It is not advisable to use channels that offer one or more bits (less than 8). This is due to the difficulty of data reading sent to the destination. Because the purpose of the stego system is to send short text messages, we need to convert the resulting data into 8-bit characters. Because IP is a protocol without a connection, the loss of one or more bits (in this case, the network is very often loaded) makes it impossible to correctly read the character – one bit from the datagram.

4.1 The Identification Field in the IP Head

The field that offers the hidden channel with the largest capacity, the 16-bit authentication field, was selected from the IP datagram header. In order to ensure the uniqueness of the identification number and its uniqueness, the data is used only for the high byte (eight high bits) of the identification field, while the lower bytes are random.

Two different approaches have been used within this channel:

1) transmission of 8 bits of encoded character in one datagram;
2) divide the 8-bit value into two parts of 4 bits and send them to two consecutive datagrams.

The first method is simple and consists in encoding the transmitted character and placing the eight received bits in the high byte of the identification field.

The second method uses only the four oldest bits to transmit characters. The remaining four bits are used to send the following data (module 16), so that when you receive them you can restore the correct order. Thus, protection against reading data from datagrams that reached their destination in a different order than they left the source of transmission was introduced.

4.2 Results Discussion

Hiding Data in the IP Header Identification Field. The message was not received in its entirety. Four characters out of the 21-character text were lost. Since the received symbols consist of four bit "halves", after one part is lost, the whole symbol cannot be reproduced correctly – non-alphanumeric characters appear in the receiver's text. However, thanks to the numbering of the characters, the rest of the text was displayed in the correct order. 6 out of 52 transmitted datagrams were lost, i.e. 11% of the total number – with a high load on the network, greater losses occur. Therefore, the transfer will have to be repeated several times.

Hiding Data in the TCP Segment Header. This time the data was hidden in the TCP segment header, in the importance field. The transmitted message was corrupted. 4 of the 16 characters that make up the message were lost, i.e. 25% of the total. Due to the lack of character numbering in this method and the inability of the program to determine which characters have been lost, reading the message is more difficult. On the other hand, due to the hiding of one symbol in the datagram and, as a result, the higher bandwidth of the created hidden channel, the transmission was shorter than in the previous case and required the transmission of less data containing the hidden message.

5 Summary and Conclusion

Recently, methods have gained popularity in which hidden information is transmitted through computer networks using the peculiarities of data transmission protocols. However, the search for opportunities to use IP and TCP protocols to transmit hidden data should not be limited to just one hidden channel.

After finding several different methods, there is a possibility to choose and use them depending on the conditions in a particular network. Therefore, you should not use a hybrid combination of several header fields to create a larger channel – even if not all, some can be reset in certain circumstances, which makes it completely impossible to use such a channel. It should be noted that that the simulation of covert transmission in conditions of packet loss showed low capabilities of methods with modification of packet headers.

Acknowledgment. The authors are appreciative to colleagues for their support and appropriate suggestions, which allowed improving the materials of the article.

References

1. Frączek, W.: Krzysztof Szczypiorski Perfect undetectability of network steganography Perfect undetectability of network steganography. Security Comm. Networks **9**, 2998–3010 (2016)
2. Mazurczyk, W., Szczypiorski K. Steganography of VoIP streams.: In: MeersmanR, TariZ (eds) Springer-Verlag,2009. Wydziału Elektroniki and Technik Informacyjnych
3. Kyryk, M.I., Pleskanka, N.M., Tymchenko, O.V.: Metody ta modeli upravlinnia trafikom v rozpodilenykh infokomunikatsiinykh systemakh: monohr. — Lviv, Ukr. akad. drukarstva, 2017. — 264 s. (ISBN 978-966-322-473-2)
4. Kovalskyi, B., Dubnevych, M., Holubnyk, T., Pysanchyn, N., Havrysh, B.: Development of a technology for eliminating color rendering imperfections in digital photographic images. Eastern-Europ. J. Enterprise Technol. **1**(2), 40–47 (2019). https://doi.org/10.15587/1729-4061.2019.154512
5. Kurada, R.R., Kanadam, K.P.: A novel evolutionary automatic data clustering algorithm using teaching-learning-based optimization. Int. J. Intell. Syst. Appl. (IJISA) **10**(5), 61–70 (2018). https://doi.org/10.5815/ijisa.2018.05.07
6. Sharifzadeh, M., Aloraini, M., Schonfeld, D.: Adaptive batch size image merging steganography and quantized gaussian image steganography. IEEE Trans. Inf. Forensics Secur. **15**, 867–879 (2020). https://doi.org/10.1109/TIFS.2019.2929441
7. Bisikalo, O.V., Kovtun, V.V., Kovtun, O.V.: Modeling of the estimation of the time to failure of the information system for critical use. In: 10th International Conference on Advanced Computer Information Technologies (ACIT), Deggendorf, Germany, 2020, pp. 140–143 (2020). https://doi.org/10.1109/ACIT49673.2020.9208883
8. Koulla Moulla, D., Abran, A., Kolyang: Duration estimation models for open source software projects. Int. J. Inf. Technol. Comput. Sci. (IJITCS) **13**(1), 1–17 (2021). DOI: https://doi.org/10.5815/ijitcs.2021.01.01
9. Onyezewe, A., Kana, A.F., Abdullahi, F.B., Abdulsalami, A.O.: An enhanced adaptive k-nearest neighbor classifier using simulated annealing. Int. J. Intell. Syst. Appl. (IJISA) **13**(1), 34–44 (2021). Doi: https://doi.org/10.5815/ijisa.2021.01.03
10. Baziyad, M., Rabie, T., Kamel, I.: Directional pixogram: a new approach for video steganography. In: 2020 Advances in Science and Engineering Technology International Conferences (ASET), 2020, pp. 1–5. doi: https://doi.org/10.1109/ASET48392.2020.9118180
11. Bykov, M., Kovtun, V., Ivasyuk, I., Kotyra, A., Mussabekova, A.: The automated speaker recognition system of critical use. In: Proceedings of Photonics Applications in Astronomy, Communications, Industry, and High Energy Physics Experiments, vol. 10808, 108082V (2018). https://doi.org/10.1117/12.2501688

12. Peleshko, D., Rak, T., Izonin, I.: Image superresolution via divergence matrix and automatic detection of crossover. Int. J. Intell. Syst. Appl. (IJISA) **8**(12), 1–8 (2016). https://doi.org/10.5815/ijisa.2016.12.01
13. Watni, D., Chawla, S.: A comparative evaluation of jpeg steganography. In: 2019 5th International Conference on Signal Processing, Computing and Control (ISPCC), pp. 36–40 (2019). https://doi.org/10.1109/ISPCC48220.2019.8988383
14. Tymchenko, O.O., Havrysh, B., Khamula, O., Lysenko, S., & Havrysh, K.: Risks of Loss of Personal Data in the Process of Sending and Printing Documents. CITRisk, CEUR Workshop Proceedings, vol. 2805, pp. 373–384 (2020) 1st International Workshop on Computational and Information Technologies for Risk-Informed Systems, CITRisk 2020 Virtual, Kherson15 October 2020 through 16 October 2020Code 166724 (2020)
15. Hajduk, V., Levicky, D.: Cover selection steganography with intra-image scanning. In: 2018 28th International Conference Radioelektronika (RADIOELEKTRONIKA), pp. 1–4 (2018). https://doi.org/10.1109/RADIOELEK.2018.8376370
16. Pashayev, F.H., Pashayeva, S.E., Jafarov, J.M., Farhad, P.I.: Determination method of some operational characteristics of information search system in directive document database. Int. J. Math. Sci. Comput. (IJMSC) **6**(3), 1–11 (2020). https://doi.org/10.5815/ijmsc.2020.03.01
17. Audhi, S., Mascarenhas, M.: Secure mechanism for communication using image steganography. In: 2019 2nd International Conference on Intelligent Computing, Instrumentation and Control Technologies (ICICICT), pp. 729–732 (2019). https://doi.org/10.1109/ICICICT46008.2019.8993337
18. Auzinger, W., Obelovska, K., Dronyuk, I., Pelekh, K., Stolyarchuk, R.: A continuous model for states in CSMA/CA-based wireless local networks derived from state transition diagrams. In: Saraswat, M., Roy, S., Chowdhury, C., Gandomi, A.H. (eds.) Proceedings of International Conference on Data Science and Applications. LNNS, vol. 287, pp. 571–579. Springer, Singapore (2022). https://doi.org/10.1007/978-981-16-5348-3_45
19. Arora, H., Bansal, C., Dagar, S.: Comparative study of image steganography techniques. In: 2018 International Conference on Advances in Computing, Communication Control and Networking (ICACCCN), pp. 982–985 (2018). https://doi.org/10.1109/ICACCCN.2018.8748451.Dronyuk, Ivanna, O.F., Kryvinska, N.: Constructing of digital watermark based on generalized fourier transform. Electronics **9**(7), 1108 (2020). https://doi.org/10.3390/electronics9071108
20. Rashmi, N., Jyothi, K.: An improved method for reversible data hiding steganography combined with cryptography. In: 2018 2nd International Conference on Inventive Systems and Control (ICISC), 2018, pp. 81–84 (2018). https://doi.org/10.1109/ICISC.2018.8398946

Design of Intake Pressure Measurement and Control System Based on LabVIEW

Geng E. Zhang[1,2], Yi Wei[1] (✉), and Zhishi Wang[1]

[1] Faculty of Intelligent Manufacturing, Nanning University, Nanning 530000, China
78704108@qq.com

[2] Faculty of Engineering, University Malaysia Sabah, Sabah, Malaysia

Abstract. Air intake pressure sensor as a kind of automobile and aerospace components, are widely used in the field of automobile. The engine intake pressure sensor has a very important influence on the engine to achieve the best working efficiency, in order to improve the dynamic performance of the engine and economic characteristics, it is necessary to accurately and timely monitor the intake pressure of the engine. This paper designed a set of to the engine intake pressure measurement and control system, and the experiment was carried out on a Weizhi small car. The experimental results show that the designed intake pressure measurement and control system can accurately measure and control the engine intake pressure, realized the engine air intake pressure sensor signal detection and monitoring, real-time data display, automatic storage. The system has the advantages of simple operation, easy programming, strong expansion and good measurement and control effect.

Keywords: Intake pressure sensor · LabVIEW · System design

1 Introduction

The main function of the engine intake pressure sensor is to sense the change of vacuum degree in the intake manifold with different working speed and load of the engine, and then convert the change of internal resistance into voltage signal for the ECU to modify its fuel injection amount and ignition timing angle [1]. It makes important contribution to improve the power and economy of the engine. So, the engine air intake pressure sensor performance measurement and control research and innovation breakthrough are especially important. In order to improve the economic characteristics and the dynamic characteristics of engine, it is necessary to detect and monitor the engine intake pressure sensor signal. Building a monitoring system is the premise of engine intake pressure measurement and control, how to timely monitor the engine intake pressure is also the research content of this topic.

Professor Junke Hu of school of Mechanical and Electrical Engineering, Central South University and his academic research team have developed a set of automatic control system for engine intake pressure. The automatic control pressure system technology can effectively control and improve the actual intake pressure of the engine,

Z. Hu et al. (Eds.): AIMEE 2022, LNDECT 159, pp. 57–66, 2023.
https://doi.org/10.1007/978-3-031-24468-1_6

realize automatic control of the actual intake pressure of the engine, and improve the dynamic performance of the engine under ideal conditions, but not applicable to all altitudes due to the constraints of conditions [2].

Virtual instrument is a kind of instrument based on computer. The close combination of computer and instrument, based on a common computer hardware and software and operating system equipment, realizes the characteristics of various instruments which is also the main trend of the development of computer technology at present [3, 5].

2 Hardware System Design

The core hardware of the intake pressure measurement and control system is mainly composed of the intake pressure sensors, data acquisition cards, actuators, computers, keyboards and monitors [6].

2.1 Data Acquisition Card

The main indicators of choosing data acquisition and processing card are the highest data bus resolution and data accuracy, the highest precision sampling data speed, the number of bits of the channel, the hardware type of the data bus connection line and port, the need for data isolation, the need for other calibration hardware functions, support for software drivers and their software platforms, etc. [7, 9].

The intake pressure measurement and control system mainly has the following characteristics: less measurement parameters, high sampling rate, strong compatibility requirements, digital input, etc. [10]. After deeply studying the related board design foundation of NI Company, and combined with the above factors, USB 6216 model data acquisition card was selected. The main performance indicators of this acquisition card are described as follows:

(1) Analog input resolution: 16 bits;
(2) Maximum sampling rate: 400 KS/s;
(3) Analog output: 2 channels;
(4) Digital I/O port: 24;
(5) Counter/timer: 2;
(6) Number of analog input channels: 8 differential channels or 16 single terminals;
(7) Analog output resolution: 16 bits.

2.2 Intake Air Pressure Sensor

In this test, there are four power wires connecting the intake pressure sensor with the ECU. The ports are shown in Fig. 1, including the power supply power line with the working voltage of 4.8 V–5.1 V (which may vary slightly according to different models used), the intake pressure signal line, the intake temperature signal line and the ground line. When the engine is in idle state, the vacuum degree of the intake manifold is high (the absolute vacuum pressure is low), the resistance value of the sensor is large, and the sensor will automatically output a low voltage signal of 1.5 V–2.1 V (which may vary slightly according to the use of different models). When the throttle valve is fully opened, the vacuum degree of the intake manifold is low (the absolute pressure is high), and the resistance value of the sensor is small, and the sensor will output a high voltage signal of 3.6 V–4.8 V (which may vary slightly according to the use of different models). The connection between the intake pressure sensor of the test vehicle and ECU is shown in Fig. 2.

Fig. 1. Intake pressure sensor port of the test vehicle

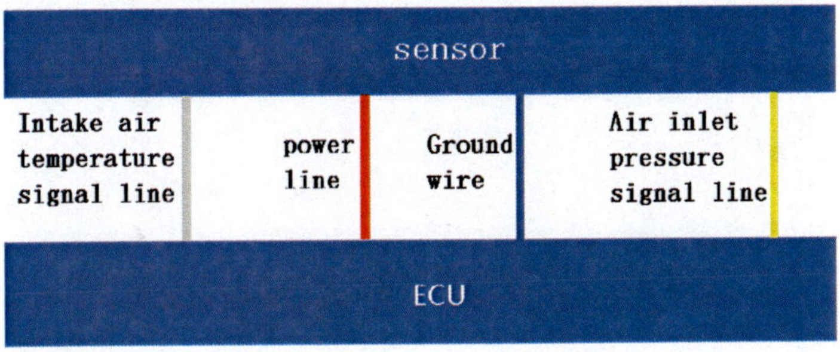

Fig. 2. Schematic diagram of the connection between the intake pressure sensor and ECU of the test vehicle

3 Software System Design

In the software design mainly includes the following basic aspects: the LabVIEW program block diagram, the front panel design, the establishment of the acquisition connection between data acquisition card and measurement and control system [11, 13, 14].

3.1 LabVIEW Program Block Diagram Design

The functional structures mainly used in the design of the program block diagram in Fig. 3 are sequential structure and While loop. The sequential structure is to ensure that the program can be executed step by step. When other functions or controls are added to the sequential structure, these functions can be executed in sequence, and the program will not be dereliction during the runtime, resulting in the termination of the operation. The While loop ensures that data can be continuously collected and transmitted in real time for display on the front panel. In the design of this paper, some operational functions are specially added: numerical value to small string transformation (operation function): can make the number transformation into a decimal (including fractions) format; Write text (voltage function): Write the voltage value into TXT text.

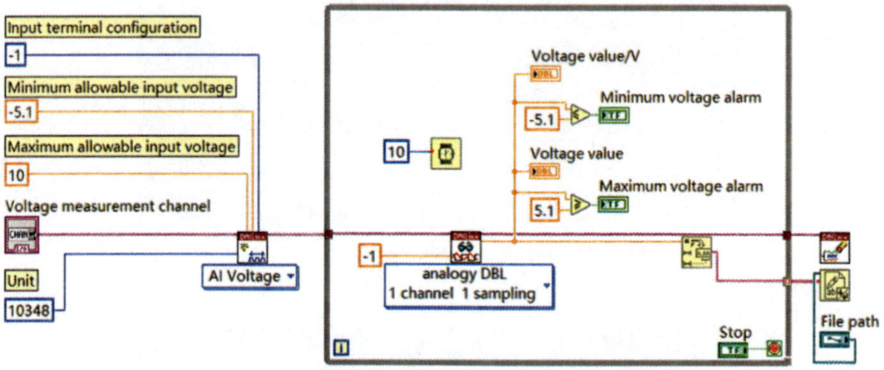

Fig. 3. LabVIEW program block diagram design

(1) Data Acquisition Module

Data acquisition is the first step of the measurement and control system. Without the acquisition of voltage signals, it is impossible to monitor the signals of the intake pressure sensor. The data acquisition module consists of virtual channel (VI) created by DAQmx, which creates the acquisition channel configuration of the sensor voltage signal. DAQmx can set input parameters, such as maximum permissible input voltage, minimum permissible input voltage, input wiring configuration, voltage measurement channel, etc. [15]. Continuous acquisition can be achieved using the While loop [16], as shown in Fig. 4.

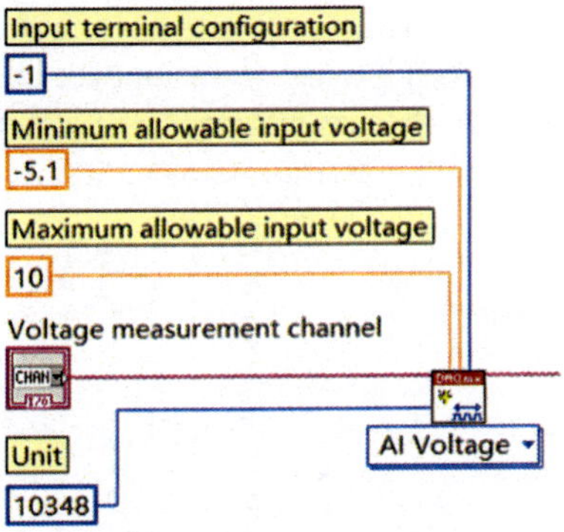

Fig. 4. Data acquisition module

(2) Measurement and display module

Through DAQmx read (function) to read the voltage signal collected in the virtual channel created by DAQmx, and real-time display of its data through waveform charts and numerical display controls. In addition, two alarm lamp plates are added to the system to compare the voltage signal read with the threshold value. When the voltage signal is greater than or less than the set voltage threshold value, two alarm lamps are lit respectively. The specific program block diagram is shown in Fig. 5.

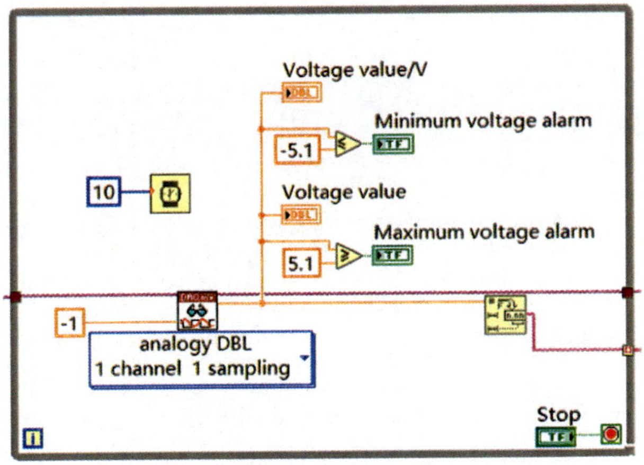

Fig. 5. Measurement and display module

(3) Data storage module

In this module is mainly to choose three controls to read the voltage signal to save, the first step is to use the numeric to decimal string conversion (function), so that the digital signal into decimal (fraction) format, this can improve the accuracy and accuracy of the data. The next step is to write a string or array to a file in line order by writing a text file (function). Finally, create a path on the written text (function) that can be stored in the file, this total process is shown in Fig. 6.

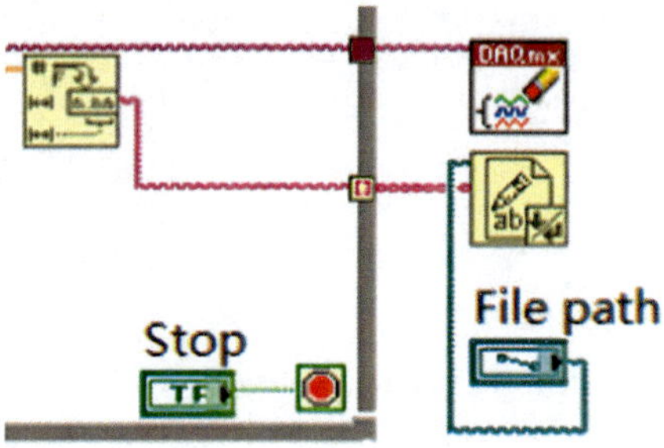

Fig. 6. Data storage module

3.2 LabVIEW Front Panel Design

The front panel of the system is mainly composed of waveform display window, threshold alarm lamp, voltage measurement channel, file storage path, data display window and function control button, as shown in Fig. 7.

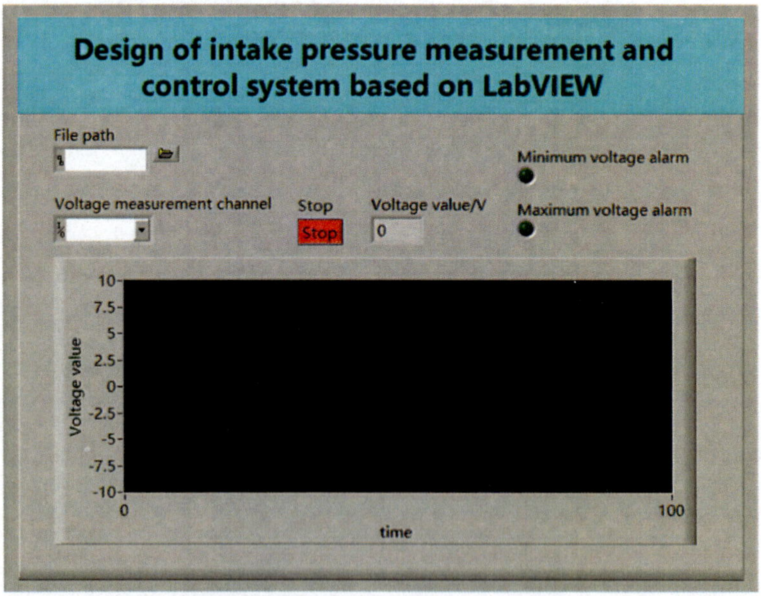

Fig. 7. Front panel design of LabVIEW

In the design of the front panel, the main selected components are two Boolean indicator light and stop button, waveform chart, numeric indicator. The purpose of placing the numeric indicator is to display the current voltage value of the intake pressure sensor in real time. The purpose of selecting the waveform display control is to monitor the voltage change curve of the intake pressure sensor more intuitively. When the voltage signal of the intake pressure sensor reaches the highest and lowest threshold during the acquisition process, the Boolean indicator can give an alarm. In the front panel, we can see the voltage measurement channel frame, in which to select the measurement channel to use, and then confirm that the acquisition card connected successfully. If the test result is not shown, can choose measurement point open channel drop-down box to refresh, then open the drop-down box to select the correct voltage measurement channels, finally select data storage location from the file path box.

4 Experimental Testing

4.1 Experimental Conditions

A Weizhi 1.5 L car was selected for this test. In the test process, the data acquisition card, the measurement and control system and the intake pressure sensor are firstly connected to measure the voltage signal of the intake pressure sensor of the engine, and then the operation is carried out. In the process of the engine completely relaxing to completely stepping on the accelerator pedal, the changes of the intake pressure and voltage signal can be accurately recorded and saved.

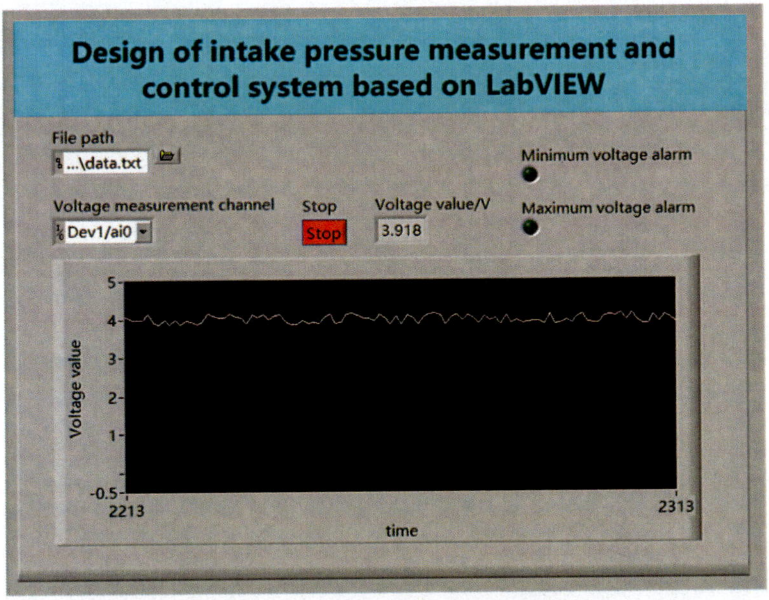

Fig. 8. Software operating voltage signal display in the test

4.2 Experimental Procedure

1) Open the computer and start the test program;
2) Connect the data acquisition card and select the appropriate measurement channel;
3) One of the acquisition ends of the data acquisition card is connected with the intake pressure signal line of the intake pressure sensor through the probe; and the other acquisition end is connected with the ground lead;
4) Create a new TXT text, select it as the file storage path, run the test program;
5) Press the accelerator pedal slowly until it is completely crushed;
6) Stop the test, release the accelerator pedal, and click the "Stop" button to stop the data collection.

The software operating voltage signal in the test is shown in Fig. 8.

4.3 Experiment Result Analysis

In order to more intuitively show the voltage signal output changes of the pressure sensor, the output characteristic curve is made after the test data is analyzed and sorted out, as shown in Fig. 9.

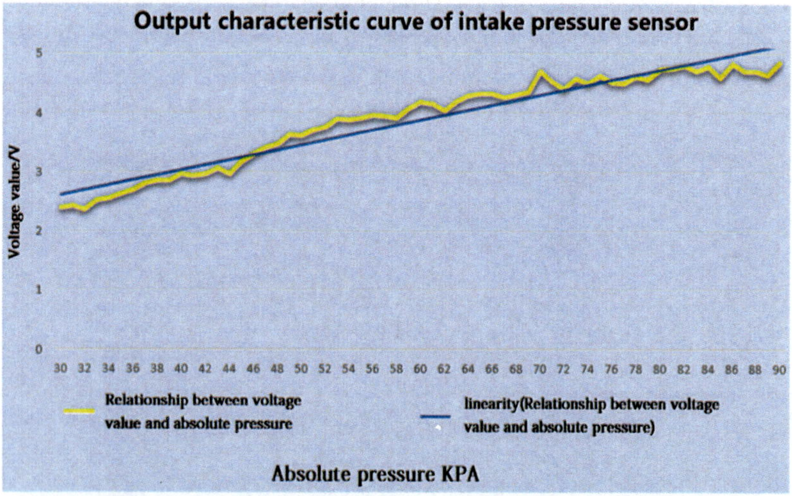

Fig. 9. Output characteristic curve of intake pressure sensor

It can be seen from the Fig. 9. That the output voltage value of the engine intake pressure sensor rises linearly with the change of absolute pressure. The greater the voltage value, the greater the absolute pressure, which conforms to the conclusion of its working principle.

In this chapter, using the measurement and control system designed in this paper to experiment, the engine air intake pressure sensor output voltage signal can be measured for detection and monitoring. Then the output characteristic curve between the output voltage signal of the engine intake pressure sensor and the absolute pressure is obtained, it is concluded that the higher engine air intake pressure sensor output voltage, the higher the absolute pressure. The linear relationship between the two is consistent with the working principle and the expected goal of the measurement and control system is achieved.

5 Conclusion

In this paper, on the platform of LabVIEW, designed the intake pressure measurement and control system based on LabVIEW, realized the automatic monitoring of the performance of the intake pressure sensor. The measurement and control system designed has several advantages such as intuitive user interface, convenient monitoring, simple maintenance, accurate detection, easy extension and reliable operation. It has made a certain contribution to improve the power and economy of the engine.

Acknowledgment. This papert is supported by: (1) Sub-project of Construction of China-ASEAN International Joint Laboratory for Comprehensive Transportation (Phase I), No. GuiKeAA21077011-7.

References

1. Wei, L.: Principle and Fault Detection of New Automotive Sensors and Actuators. China Machine Press, Beijing (2013)
2. Hu, J., Zhao, C., Niu, Q., Xu, K.: Development and research of automatic control system for engine intake pressure. Manufacturing Autom. **40**(08), 53–55+58 (2018)
3. Song, R., Zhang, D., Cheng, Y., Lan, Z.: Gas pressure testing system based on LabVIEW. Chinese Hydraulics & Pneumatics (12), 28–31 (2008)
4. Xu, T.: Pressure sensor test and evaluation system based on LabVIEW. Instrument Technol. (06), 4–5+48 (2012)
5. Dan, L.: Design of Gas Flow Measurement System Based on LabVIEW. Nanjing University of Science and Technology, Jiangsu (2010)
6. Lan, W.Y.: Research on multi-channel pressure monitoring system based on virtual instrument. Harbin Engineering University (2005)
7. Haosai, L., Gao, W.: Design of pressure test system based on LabVIEW. Electron. World **19**, 161 (2014)
8. Hui, Y., Qiang, Z.: Pressure data acquisition and processing system based on LabVIEW and AVR Microcontroller. Control Instruments Chemical Indust. **37**(11), 92–94 (2010)
9. Bo, L., Liu, X., He, X.: Measurement system for wind turbines noises assessment based on LabVIEW. Measurement **44**(2), 36–38 (2010)
10. Li, H., Wang, G.: Design of pressure test system based on LabVIEW. Electronic World, [EB/OL]. https://www.doc88.com/p-302517798560.html, 2021-05
11. Nan, J.: Principle and application of intake pressure sensor for vehicle. Light Vehicle Technol. **10**, 29–32 (2013)
12. Juan, Y., Qingliang, G.: Analysis of the principle and application of intake pressure sensor. Agricultural Mach. **8**, 73–74 (2009)
13. Duan, Z., Liu, J., Yang, Y.: Design of pneumatic proportional pressure control system based on LabVIEW. Comput. Digital Eng. **43**(03), 442–445+461 (2015)

14. Cao, C.: Pressure sensor testing system based on LabVIEW. Nanjing University (2014)
15. Jun, H.: Development of Pressure Reducing Valve Test Equipment based on LabVIEW. Ningbo University, Zhejiang (2017)
16. Wang, X.: Design of hydraulic support pressure monitoring system based on LabVIEW. Coal Mine Mach. **36**(12), 39–41 (2015)

A New Method of Search Engine Optimization Based on Semantic Kernel Idea

Sergey Orekhov(✉), Michail Godlevsky, Hennadiy Malyhon, and Tetiana Goncharenko

Computer Science and Software Engineering Department, National Technical University Kharkiv Polytechnical Institute, Kharkiv 61002, Ukraine
sergey_orekhov@yahoo.com

Abstract. The results of more than 30 projects in the field of search engine optimization on the Internet show that the effectiveness of existing approaches is catastrophically reduced. Paid promotion methods no longer guarantee a banal increase in the traffic of a given web resource, and even more so the number of leads. The reason is that competition has moved to the Internet as much as possible. So, we need to look for new alternative methods. Therefore, the paper proposes to consider a new alternative way to promote goods on the Internet. This approach is a new information technology – virtual promotion. It is based on the creation of a virtual image of a product on the Internet. We called this virtual image the semantic kernel. Such a kernel is placed on the Internet in order to attract a potential buyer. It is placed in the Internet nodes according to a special online sales scheme, which is based on a map of the user's travels on the Internet. This scheme is called the virtual promotion map. The map must guarantee maximum traffic to a given web resource. However, it will not be possible to immediately build a map. Therefore, its creation is iterative. We are, as it were, setting ourselves up to find effective ways between the semantic core and the user who is ready to buy our product online. The article offers a description of the virtual promotion technology. The main idea of virtual promotion is that promotion is a logistical channel between a company and a client. We put a special message (semantic core) into the channel. He moves through the logistics channel and tries to attract as many new buyers of the goods as possible. This article presents the results of the work of special software that implements the synthesis, quality assessment and management of the semantic kernel of web content. The program generates and manages two channels: the first is intended for the distribution of the semantic kernel, the second is for assessing the quality of such distribution. The first channel combines the actions of transporting, storing and searching for semantic kernel s, depending on the needs of a potential buyer. The second channel generates a list of nodes where the core should be placed in order to be noticed by real buyers. If the kernel is visible, then the user is buying the product. The proposed virtual promotion software was tested on the basis of a real WEB resource. This WEB resource provides a web service to the US market. The introduction of a new method of search engine optimization made it possible to increase the traffic of the WEB resource tenfold.

Keywords: Virtual promotion · Semantic kernel · Search engine optimization · Customer journey map

Z. Hu et al. (Eds.): AIMEE 2022, LNDECT 159, pp. 67–77, 2023.
https://doi.org/10.1007/978-3-031-24468-1_7

1 Introduction and Related Works

The semantic kernel (SK) is one of the elements of the new technology of virtual promotion of goods and services on the Internet (Fig. 1) [1]. From the point of view of the classical theory of marketing [2, 3], semantic kernel is a message that the seller of goods sends through channels that are formed on the Internet. The purpose of this message is to attract potential customers to purchase a given product or service. That is, SK also becomes a management tool that affects the whole complex of virtual promotion. Then the SK can be analyzed as follows – Fig. 2.

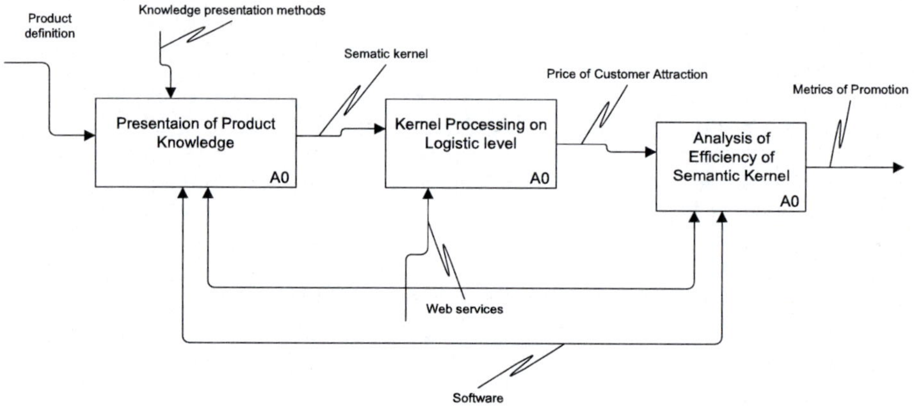

Fig. 1. Virtual promotion process

Analyzing the above conclusions regarding the two-tier system of virtual promotion, we can say that SK is an element of two levels: marketing and distribution (Fig. 2). At the marketing level, SK plays the role of communicating the product and the need it meets. The message should contain information about the time parameters and geography of the product or service.

An online message is presented as a set of keywords that a potential buyer as a user can see online during his/her trip. That is, the user's travel map online should include the point where the SK is contained in the form of a set of keywords, or rather HTML content. This travel map is opened by a search engine and other mobile applications using the HTTP protocol. This is the second level – the logistics of information and knowledge about a product or service on the Internet [4, 5].

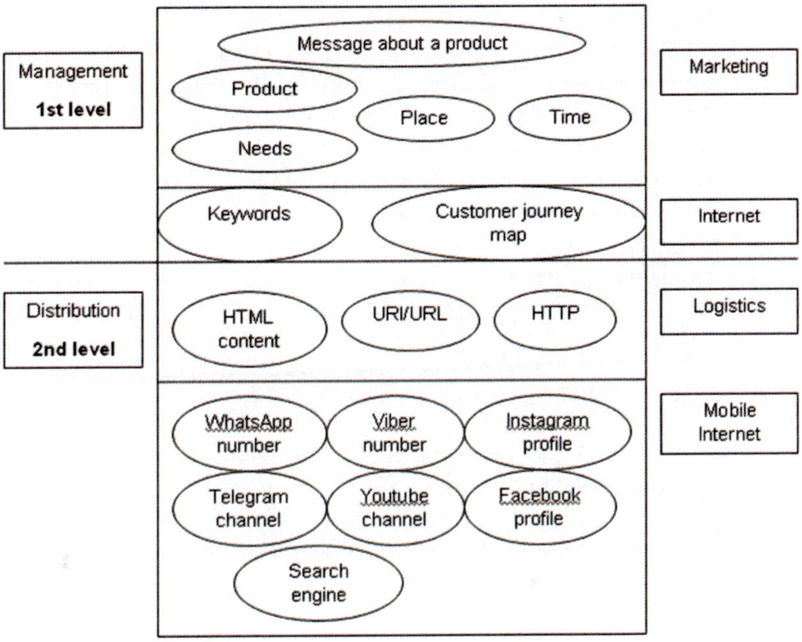

Fig. 2. The structure of virtual promotion

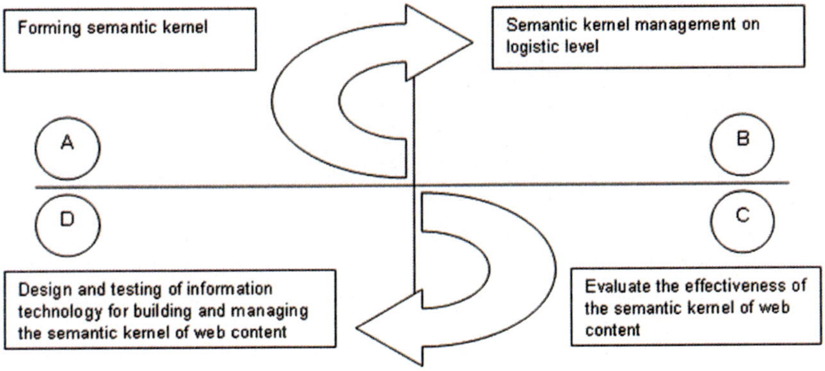

Fig. 3. Task tree

Figure 3 shows the task tree of virtual promotion technology in the terms of solving the problem of building and managing the semantic kernel of web content. Thus, the first main task is the structural-parametric identification of the semantic kernel of web content (HTML format). Based on the nature of SK, we can say that it is a reflection (annotation) of the product, which is offered for sale. In other words, SK is the representation of knowledge about a product, which is formed on the basis of a description of a product or service.

Then the problem of SK formation is transformed into the problem of presenting knowledge about the product by artificial intelligence methods [6]. That is, you need to choose a form of knowledge presentation that meets the requirements of the Internet technologies and conditions in terms of marketing.

When solving the problems of virtual promotion, SK plays the role of control action to activate the level of knowledge distribution, then it plays the role of presenting knowledge about the product, moving the message about the product to the end user (potential buyer).

That is, you need to manage the SK to correctly choose the channel of promotion and information technology to support such promotion. Therefore, the paper proposes to consider a new problem of management of SK of web content at the logistic level.

2 Problem Statement

Studies [5] show that SK for the real web resource is changing, moreover, it is necessary to change SK to have an effective promotion in terms of profitability. In other words, due to the actions of competitors, the market situation is formed on the Internet, which corresponds to the current version of the Internet. That is, there is a problem of SK management in different market situations [6].

The next new topical task of the dissertation is formation of metrics and criteria for evaluating the effectiveness of the semantic kernel from the standpoint of profitability of product promotion on the Internet. This fact was recorded in [1].

The latest new task is dedicated to the creation of information technology that supports the solution of the above three problems. The variants of software components of such technology were presented in [7].

Thus, the paper will consider the design and testing of software for management of the semantic kernel of web content through the solution of four main tasks:

1) Synthesis of the semantic kernel based on methods of knowledge representation and processing of large data sets (data mining, big data). The purpose of solving the problem is to form the semantic kernel of web content as a control impulse for the logistical level of the marketing channel. Given the fact the semantic kernel is a message that is formed in the channel in order to motivate a potential customer to buy a given product or service.
2) Management of the semantic kernel of web content based on artificial intelligence methods to set up an effective marketing level.
3) Formulation of metrics and criteria for the effectiveness of the semantic kernel of web content for the analysis of the profitability of product promotion on the Internet.
4) Verification of the built software on the basis of a number of web projects in Ukraine and the USA.

The purpose of forming the semantic kernel of web content is to increase the effectiveness of virtual promotion of goods or services [7]. The semantic kernel is a message in the virtual promotion marketing channel. For operation with the semantic kernel, it is solved by a set of research tasks, which is presented in Fig. 3. The analysis of this task tree allows us to say the following:

1) The first task is to form the message itself, adapted to the structure and properties of the promotion channel.
2) The next task describes the situation of changing the message in the channel in the case of low efficiency of promotion or "aging" of web content and the kernel due to changes in the preferences of potential buyers of goods or services. The change of the semantic kernel is also due to the action of competitors who introduce new forms of semantic kernels to describe the needs of a potential buyer.
3) The change of the semantic kernel should ensure that the effectiveness increases. To do this, it is necessary to form a criterion for the optimal change of the semantic kernel or metrics that confirms this fact.
4) The very variants of the semantic kernel, and most importantly the web content on the basis of which the kernel is formed, have the characteristics of large amount of data. Therefore, it is necessary to solve these three problems in automatic mode using modern cloud solutions and technologies for processing large amount of textual information in HTML format.

Thus, the task of the research is formulated as follows: to find such a rational variant of the semantic kernel that delivers the maximum function of the efficiency of the promotion channel.

3 Proposed Approach

According to the formulation of the problem, it is necessary to solve several tasks. The first task is to synthesize a set of semantic kernels. Next, you need to split this set into clusters. Each cluster describes a class of need. If we promote a given product, then the buyer wants to know what need this product satisfies. Therefore, the entire set of semantic kernels is divided into these classes of needs. Having found out what classes we have, we need to find the centers of mass of these classes. Centers of mass will be our semantic kernels that need uploading to promotion channels on the Internet.

By loading the first kernel, we activate the kernel management loop. Each such download will start a stage of the cycle. The completion of the control cycle stage is possible for two reasons. The first is when the process of promotion is no longer needed and must be stopped. The second reason is when the efficiency of the loaded kernel drops according to web statistics. If we have the second reason, then we need to change the kernel. If our product must cover the same need as it was in the previous management cycle, then we load the kernel into the promotion channels, which are close in the class of needs to the center of mass. Another strategy for changing a kernel in a channel is to augment an existing kernel with new kernels. That is, we create a composite kernel by adding new elements to an existing one.

Thus, we form a semantic kernel to change a management cycle in the promotion channel. However, to implement this cycle, it is required to introduce a metric and criterion for the effectiveness of the semantic kernel in the promotion channel.

The paper proposes to use the following metric to select the best elements for semantic kernel [8–11]:

$$m = \begin{cases} log_2|a|f(a), notincluded \\ log_2|a|(f(a) - \frac{1}{P(T_a)}\sum_{b\in T_a} f(b)), included \end{cases} \tag{1}$$

where a– kernel candidate, $f(a)$ – frequency of appearance of the candidate, T_a- set of candidates that include the keyword, $P(T_a)$ - number of candidates for T_a, $\sum f(b)$- sum of candidate frequencies $b \in T_a$.

To evaluate the effectiveness of the semantic kernel in the channel, the following non-linear programming problem is proposed:

$$\sum_i \sum_j \frac{w_j a_{ij}}{x_{ij}} \to min \tag{2}$$

$$\sum_i \sum_j \frac{D}{\frac{a_{ij}}{x_{ij}} + C} \geq \sum_i R_i \tag{3}$$

$$\sum_j w_j = 1 \tag{4}$$

$$x_{ij} \geq 0 \tag{5}$$

where x_{ij} - traffic in j th node of promotion channel done during i th time period, w_j - weight of a node, R_i - rate of return predefined for i th time period.

Thus, the physical meaning of problem (2)–(5) is that it is required to find out the minimum traffic in the nodes of the promotion channel, which guarantees this semantic kernel. If the given kernel does not allow reaching the given level of traffic, then it is required to change it. That is, if web statistics shows that the semantic kernel in the promotion channel guarantees the required level of traffic at the stage of the control cycle, then it does not need to be replaced. If the traffic drops, then at the next stage of the cycle we change the kernel to a new one or expand what we have.

Figure 4 shows the algorithm of the proposed approach to managing semantic kernels. First, a set of semantic kernels is synthesized in the form of semantic networks [6]. The vertices of these networks form clusters – classes of needs. We choose the centers of these classes as candidates for semantic kernels. Using the metric (1), we select 5–10 candidates for the semantic kernel of the web resource. This is a set of K-applicants. When the K-applicants are selected, we start the management cycle. The first kernel (the first K-applicant) in the list is placed in the promotion channel and the value of the function (3) is calculated. Knowing the required level of traffic, we control web statistics. If the traffic value grows, then the kernel performs its function. If the traffic.

value drops, then this means that the kernel requires correction. We can either expand the kernel with the second position on our list of candidates, or replace the existing kernel with a new one. The management process continues as long as the company is interested in promoting a given product, which corresponds to the list of semantic kernels obtained at the first step.

4 Experiment and Its Results

Consider the implementation of the algorithm (Fig. 4) on a real example. As an example, the web content of the American site of web services in the field of building a psychological portrait of a person was taken.

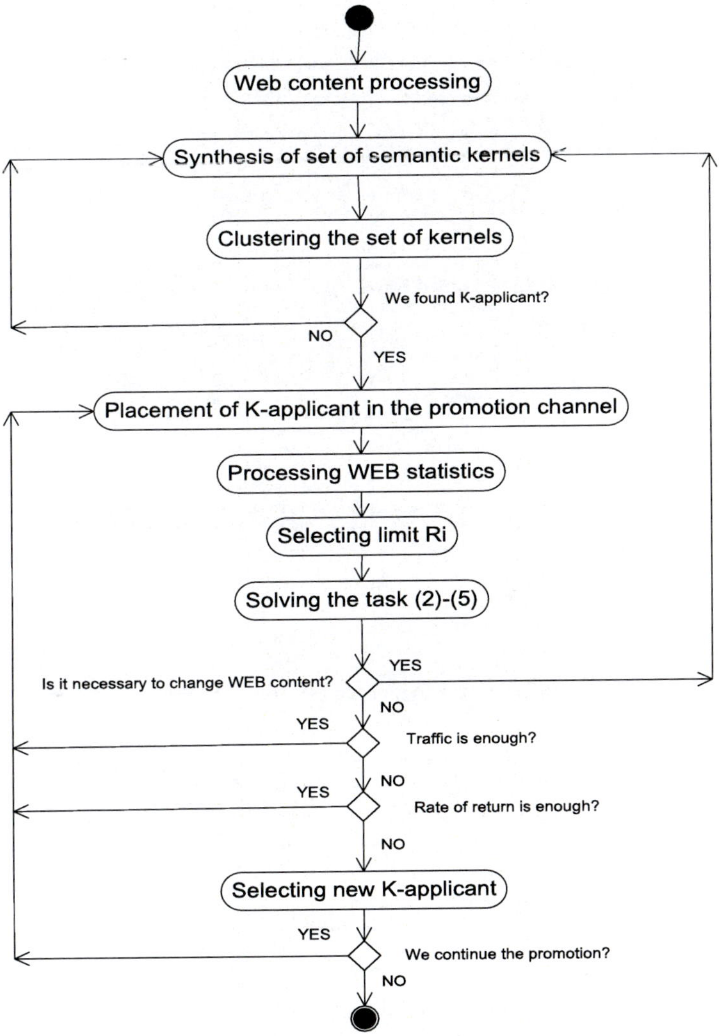

Fig. 4. Activity diagram – algorithm

Figure 5 shows an example of web content from the specified web project. Figure 6 shows a fragment of the semantic web that describes this web content.

1	The combination of the scales listed below provides a dynamic psychological portrait of a person (mood, state of mind, etc.) at any point in time. The descriptions of the scales can help you make informed decisions about actions and events. 1. Self-esteem. This scale deals with a person's level of self-esteem, enthusiasm, and leadership ability. It illustrates how the person feels about him or herself, and how confident he or she is in facing the day. It shows how much energy the person has to deal with whatever comes along and whether he or she would rather lead or take a back seat on any particular day...

Fig. 5. Example of web content

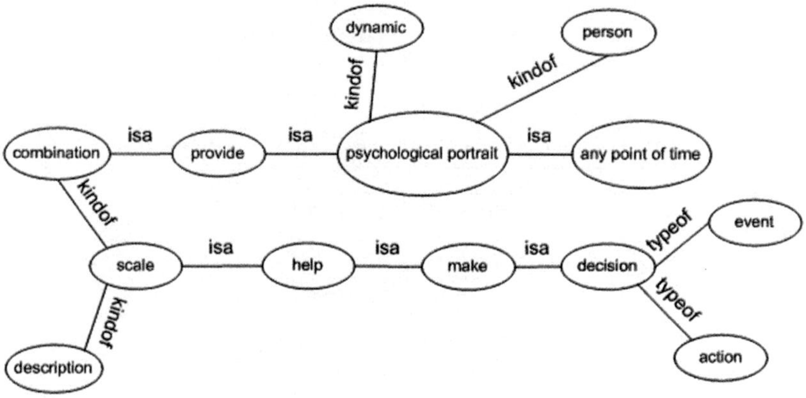

Fig. 6. The fragment of semantic network

№	K-applicants	Metric value
1	Psychological portrait of person	9.51
2	Forecast	9
3	At any point in time	7.92
4	Success	5.61
5	Time	5

Fig. 7. The set of K-applicants

Then, using the "maximum flow" clustering algorithm and metric (1), a list of candidates (K-applicants) for the semantic kernel of this web resource was obtained (Fig. 7).

We take the first kernel from the list of candidates (Fig. 7) and start the control cycle. To do this, this kernel was placed in thirty nodes of the Internet. Further, during the first stage of the cycle, 9 months long, the level of traffic was tested (Fig. 8).

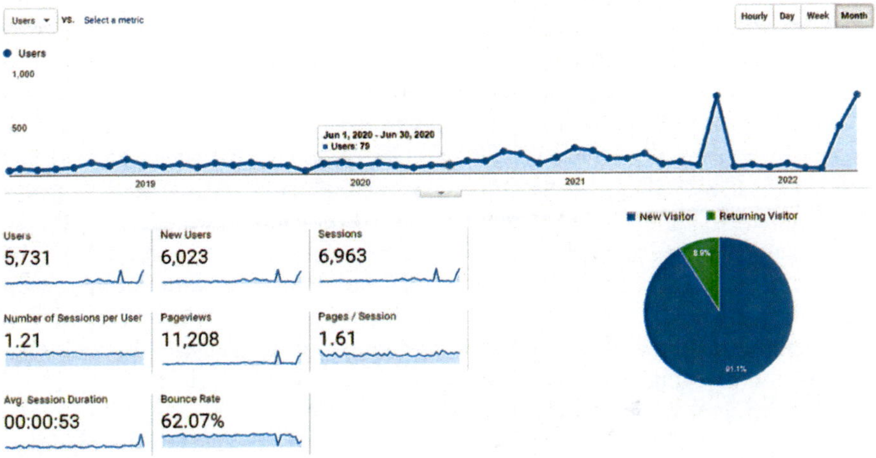

Fig. 8. Testing WEB statistics

The first step of the cycle lasted from June 1, 2020. As can be seen from Fig. 8, at first, traffic was growing gradually, but at the end of the stage it began to fall. Therefore, the solution of problem (2)–(5) was performed, which showed that a sufficient level of traffic is approximately 100 visits per month to each node of the promotion channel. In this case, it is sufficient to have only four such nodes.

Then, at the second stage, a second candidate (Fig. 7) was added to the existing kernel. As we can see from the web statistics (Fig. 8), there was an increase in traffic again. However, after 9 months, traffic began to fall again. Therefore, at the third stage, the kernel was again expanded at the expense of the third K-applicant. It was the last expansion that gave the desired effect and traffic increased 10 times.

Thus, the management cycle was 27 months until the web resource received the required level of traffic. At the same time, the classic search engine optimization problem was solved in an alternative way.

Consider the main theoretical and practical results that were obtained in the work.

5 Summary and Conclusion

First, it is necessary to determine the initial conditions that were before the cycle of situational management and the result obtained briefly.

The beginning of the cycle started when the level of web resource traffic was at the level of 70–80 units per month. At the end of the cycle, which worked for almost two years, the traffic was already 700 units. Thus, the overall assessment of the use of information technology is positive (Fig. 8). However, it is necessary to analyze the results from the standpoint of using certain methods and algorithms, which are proposed in the paper.

A new effect called "aging" was discovered in the work on semantic kernel [7]. This phenomenon is that the semantic kernel, like the real product on the market, lives its life cycle, so to speak, separately in cyberspace. It, similar to the life cycle of a real product on the market, has four stages (Fig. 8). Accordingly, this effect coincides with the classical theory of marketing.

Then there are short cycles. We believe that the reason for this behavior is the effect of a "snowball", when adding new elements to the semantic kernel leads to a sharp emergence of new users of the site. This fact was caused by the use in the work of the strategy of constant building of the semantic kernel in this WEB project.

The problem of nonlinear programming (2)–(5) becomes a new method of estimating the efficiency of a channel with a semantic kernel. Analysis of the process of solving this problem on a real test example shows that the increase in traffic in the nodes of the promotion channel can not be unlimited. There is a certain threshold after which a further increase in traffic does not lead to an increase in profitability. In addition, the introduction of the weights of the objective function (2) corresponds to the fact that the nodes have different weights in the promotion channel.

Defining this threshold is an important point that determines the upper limit to which you must strive for success.

The next practical result should be considered as to the confirmation of the empirical rules that were identified before the implementation of this work. The first rule that was known earlier was the existence of the semantic kernel itself and the need to define it [6, 7].

The second rule, which was described in the literature on search engine optimization on the Internet [12–15], describes the fact that approximately 5–7% of keywords in web content affect the final result. According to Fig. 7, the maximum number of words in the semantic kernel to achieve the desired result is from five to seven words.

Thus, the following results are reflected in the work:

1) For the first time the software for managing the semantic kernel of web content of a real WEB resource that provides services on the US market was used.
2) The practical results of using the technology were confirmed by positive statistics on the growth of traffic and profitability of this test web resource.
3) The results of the use of software indicate an increase in the value of the basic metrics of the effectiveness of the search engine optimization process on the Internet tenfold.
4) The obtained results are completely confirmed by the classical theory of marketing and product promotion.
5) The connection between the number of nodes in the promotion channel, where the semantic kernel is placed, and the level of traffic and profitability of the web resource was proved.
6) We have confirmed the accuracy of the developed strategy of step-by-step increase of the semantic kernel in the cycle of situational management.
7) We have confirmed the implementation of four empirical rules that were previously known from the literature. The implementation of these rules confirms the efficiency of the proposed models, algorithms, and programs.

References

1. Orekhov, S., Malyhon, H.: Virtual promotion knowledge management technology. Bulletin of the National Technical University "KhPI". Series: System analysis, control and information technology. — Kharkiv: NTU «KhPI» **1**(3), 74–79 (2020)
2. Kotler, P., Keller, K.: Marketing Management, 812 p. Prentice Hall. (2012)
3. Kotler, P., Armstrong, G., Saunders, J., Wong, V.: Principles of marketing. Prentice Hall Europe, 1036 p. (1999)
4. Benotmane, Z., Belalem, G., Neki, A.: A cost measurement system of logistics process. I.J. Inf. Eng. Electron. Bus. **5**, 23–29 (2018)
5. Tang, P., Tian, H.: Research on the layout of national economic mobilization logistics centers. I.J. Intell. Syst. Appl. **1**, 44–50 (2010)
6. Godlevsky, M., Orekhov, S., Orekhova, E.: Theoretical fundamentals of search engine optimization based on machine learning. CEUR-WS **1844**, 23–32 (2017)
7. Orekhov, S., Malyhon, H., Stratienko, N., Goncharenko, T.: Software development for semantic kernel forming. CEUR-WS **2870**, 1312–1322 (2021)
8. Bendle, N., Farris, P., Pfeifer, P., Reibstein, D.: Marketing metrics. The manager's guide to measuring marketing performance, 456 p. Pearson Education, Inc., USA (2016)
9. Hashimova, K.K.: Development of an effective method of data collection for advertising and marketing on the internet. I. J. Math. Sci. Comput. **3**, 1–11 (2021)
10. Hashimova, K.K.: Analysis Method of Internet Advertising-Marketing Information's Dynamic Changes. I.J. Inf. Eng. Electron. Bus. **5**, 28–33 (2017)
11. Khan, A., Baharudin, B., Lee, L., Khairullah, K.: A review of machine learning algorithms for text-documents classification. J. Adv. Inf. Technol. **1**(1), 4–20 (2010)
12. Veglis, A., Giomelakis, D.: Search Engine Optimization. Future Internet, 104 p. (2021)
13. Kelsey, T.: Introduction to Search Engine Optimization. Apress, 126 p. (2017)
14. Pawade, D.Y.: Analyzing the Impact of Search Engine Optimization Techniques on Web Development Using Experiential and Collaborative Learning Techniques. I.J. Modern Educ. Comput. Sci. **2**, 1–10 (2021)
15. Kumar, V., Tripathi, A.K., Chandra, N.: An Efficient and Optimized Sematic Web Enabled Framework (EOSWEF) for Google Search Engine Using Ontology. I.J. Inf. Eng. Electron. Bus. **5**, 40–45 (2019)

Analysis of Road Capacity Based on Vehicle Energy Consumption Formula in Single Speed Limit Zone Under Periodic Boundary Conditions

Fanglei Luo[1], Lanxiang Wei[1(✉)], Chunmei Lu[2], Bingjie Gu[1], Zhiyu Luo[1], and Deye Huang[1]

[1] School of Intelligent Manufacturing, Nanning University, Nanning 530299, China
wlx071102@163.com

[2] School of Physics and Mechanical and Electrical Engineering, Hechi University, Hechi 546300, China

Abstract. Vehicle energy consumption is the kinetic energy loss of vehicle. In order to explore the impact of road speed limit on vehicle energy consumption, based on the cellular automata NaSch traffic flow model, a formula for vehicle energy consumption is proposed for a one-way single-lane road segment with a single speed limit zone, and under the periodic boundary condition numerically simulate the influence of the length of the speed limit zone and the maximum speed allowed in the speed limit zone. When performing numerical simulation, one of the two quantities of the length of the speed-limiting zone and the maximum speed allowed in the speed-limiting zone is arbitrarily selected, and the corresponding energy-density diagram and basic diagram can be obtained by changing the size of the other quantity, the influence of these two quantities on vehicle energy consumption and road traffic can be known by analyzing the graph. After research, it is found that the longer the length of the speed limit zone and the smaller the maximum speed allowed in the speed limit zone, the smaller the energy consumption of the vehicle, the smaller the traffic flow and the average speed of the vehicle on the road, indicating that the road traffic is more prone to congestion phenomenon.

Keywords: Speed limit zone · Cellular automata · Energy consumption · Traffic flow · Numerical simulation

1 Introduction

If the vehicle encounters road maintenance, curves, tunnels and other road conditions when driving on the road, it must slow down appropriately to ensure driving safety. Therefore, it is generally necessary to limit the maximum speed of the vehicle on this road section, that is, to set up a speed limit zone [1]. Traffic jams are prone to occur when vehicles passing through the speed limit zone, causing various traffic problems. Therefore, it is necessary to study how to solve the traffic problem in the speed limit zone

Z. Hu et al. (Eds.): AIMEE 2022, LNDECT 159, pp. 78–87, 2023.
https://doi.org/10.1007/978-3-031-24468-1_8

to reduce the occurrence of traffic congestion and traffic accidents and reduce energy consumption.

The cellular automata traffic flow model is defined in a cellular space composed of cells with discrete and finite states, and is a dynamic system that evolves in discrete time dimensions according to certain local rules [2]. Its powerful complex computing capabilities, inherent parallel computing capabilities, high dynamic characteristics, and high flexibility make it powerful in simulating the spatiotemporal dynamic evolution of complex transportation systems [3]. For this reason, this model is widely applied in traffic flow research. For example: references 7–8 applied the model to the research on traffic capacity and speed prediction of road intersections; references 9–10 applied the model to the research on issues related to lane speed limit; references 11–13 applied the model to the research of on-street parking related problems.

When a vehicle travels on a speed-limited road section, kinetic energy loss occurs due to deceleration, referred to as energy consumption. Based on the cellular automata traffic flow model, references 14–17 proposes a vehicle energy consumption formula, and analyzes the impact of tunnels, highways, and on-street parking behaviors on road traffic from the perspective of vehicle energy consumption. References 18–20 use artificial intelligence to study the related problems of dynamic vehicles. It can be seen that in the existing references, there are few theoretical studies on analyzing the capacity of speed-limited sections from the perspective of vehicle energy consumption. Based on the cellular automata NaSch traffic flow model, this paper proposes a vehicle energy consumption formula for a one-way single-lane road section with a speed limit zone, and numerically simulates and analyzes the influence to the vehicle energy consumption by the length of the speed limit zone and the maximum speed allowed in the speed limit zone under the periodic boundary conditions, in order to clarify the microscopic principles of traffic congestion and dissipation, and to provide some references for further research on measures to improve the capacity of speed-limited sections.

2 Establishment of Model Rules for Road Including Speed Limit Zones

A one-dimensional lattice is used to represent a single lane, that is, the lane is composed of 1000 cells, and the actual road length corresponding to each cell is 7.5 m. L_b represents the length of the normal driving zone, and L_a represents the length of the speed limit zone, so the total length of the road is $L = L_b + L_a = 7.5$ km. Each cell is in the state of having a car or not having a car at each moment. The system adopts periodic boundary conditions, that is, the vehicle travels from left to right, and when it leaves the right boundary, it starts again from the left boundary [4].

2.1 Determination of Vehicle Maximum Speed and Delay Probability

Assuming that each vehicle on the road occupies a cell, the state of each vehicle is represented by the vehicle's speed v, $v \in [0, v_{\max}]$, that is, the speed can take any value between 0 and the maximum speed. When the vehicle is driving in the speed limit zone, its maximum speed takes a smaller value, represented by $v_{\max 1}$; the random.

delay probability takes a larger value, represented by P_0, $P_0 = 0.4$; when the vehicle is driving in the normal driving zone, its maximum speed takes the larger value, represented by $v_{\max2}$, $v_{\max2} = 5$ (moving 5 cells per second, the corresponding actual speed is 135 km/h); the random delay probability takes the smaller value, represented by P, $P = 0.25$. In this paper, the units of length are all dimensionless in units of cells.

2.2 Evolution Rules of the NaSch Model

The evolution rules of NaSch model are as follows [6]:

1) Deterministic acceleration process:

$$v_n(t) \rightarrow min(v_n(t) + 1, v_{max};$$

2) Deterministic deceleration process:

$$v_n(t) \rightarrow min(v_n(t), gap_n(t));$$

3) Randomly decelerate the process with probability P or P_0:

$$v_n(t) \rightarrow max(v_n(t) - 1, 0);$$

4) Location update process:

$$x_n(t) \rightarrow x_n(t) + v_n(t).$$

Among them, $v_n(t)$ and $x_n(t)$ are the speed and position of the n th vehicle at time t respectively, $x_{n+1}(t)$ is the position of the $n + 1$th vehicle at time t, $gap_n(t) = x_{n+1}(t) - x_n(t) - 1$ is the number of cells between the nth vehicle and the immediately preceding vehicle at time t. At each time step $t \rightarrow t + 1$ (one time step is 1 s), all vehicles update their velocities and positions according to the evolution rules described above.

2.3 Definition of Vehicle Energy Consumption

At time t-1, the kinetic energy of the vehicle with mass m and speed v is mv^2 /2 when moving. If the vehicle decelerates, the kinetic energy at time t is less than the kinetic energy at time t-1, that is, the kinetic energy decreases, indicating that the vehicle has generated energy consumption. The energy consumption of the nth vehicle from t-1 to t is [5]:

$$e(n, t)\begin{cases} \frac{m}{2}\left[v^2(n, t-1) - v^2(n, t)\right], & v(n, t) < v(n, t-1) \\ 0, & v(n, t) \geq v(n, t-1) \end{cases}$$

The average total energy consumption of each vehicle per unit time is [5]:

$$E = \frac{1}{T}\frac{1}{N}\sum_{t=t_0}^{t_0+T-1}\sum_{n=1}^{N} e(n, t)$$

In formula N is the total number of vehicles, T is the statistical time, and t_0 is the relaxation time.

According to the evolution rules of the NaSch model, the vehicle has deterministic deceleration behavior and random deceleration behavior. The energy consumption generated by the deterministic deceleration behavior of the vehicle is represented by E_i, which is called deterministic deceleration energy consumption. E_r represents the energy consumption generated by the random deceleration behavior of the vehicle, which is called random deceleration energy consumption. And E represents the sum of the two energy consumption, which is called the total energy consumption.

3 Numerical Simulation and Analysis

Assuming that there are N vehicles on a road with a length of L, the calculation formulas of vehicle density, average traffic flow and average vehicle speed are as follows:

Vehicle density:

$$\rho = \frac{N}{L} \tag{1}$$

Average traffic flow:

$$J = \rho V \tag{2}$$

Average speed:

$$V = \frac{1}{T}\frac{1}{N}\sum_{t=t_0}^{t_0+T-1}\sum_{n=1}^{N} v_n(t) \tag{3}$$

At the beginning of timing, vehicles are randomly distributed on a one-dimensional 1000-cell chain L according to a certain vehicle density ρ. 40,000 time steps are taken for each operation. In order to eliminate the influence of the transient state, the first 20,000 time steps are not counted, and the speed $v_n(t)$ is accumulated in each time step of the subsequent 20,000 time steps. The average speed of the run is the average of the speed over time over 20,000 time steps. Taking into account the existence of random problems, in order to reduce the random error, a total of 10 runs were performed, and then the results of the 10 runs were averaged.

3.1 Influence of Speed Limit Zone Length L_a on Vehicle Energy Consumption

Figure 1 is the result of numerical simulation under the condition that the maximum allowable speed $v_{\max 1} = 2$ in the speed-limiting zone and the length L_a of the speed-limiting zone are taken as 100, 200, 300 and 400 respectively.

Figure 1(a) is a deterministic deceleration energy consumption-density diagram, namely the E_i-ρ diagram. It can be seen from this figure that when ρ is small, the number of vehicles on the roadway is small, the distance between the two cars is large, the driver can drive the vehicle at high speed, and the deceleration behavior is less, so the kinetic energy loss of the vehicle is small; with the increase of ρ, the number of vehicles on the road increases, the distance between the two vehicles decreases, the vehicle deceleration behavior increases, and E_i gradually increases to the maximum value; when ρ is larger, the distance between the two vehicles becomes very small, and the vehicles on the road appear low speed. In the behavior of following the car, the kinetic energy loss of the vehicle becomes smaller and smaller, and E_i gradually decreases to zero with the increase of ρ. In addition, Fig. 1(a) also reflects: when L_a is larger, the corresponding maximum value of E_i is smaller; in the case of the same ρ, the larger L_a is, the smaller E_i is.

Figure 1(b) is the energy consumption-density diagram of random deceleration, namely the E_r-ρ diagram. This figure shows that E_r has a large value when ρ is small, and as ρ increases, E_r gradually decreases to zero. When the density is low, the number of vehicles on the roadway is small, but the vehicles have random deceleration behavior, and the energy consumption of random deceleration has a non-zero initial value; when the number of vehicles increases, the vehicles on the roadway are crowded, and the speed difference between the front and rear of the vehicle is small. As a result, the kinetic energy loss is small, and the energy consumption of random deceleration is reduced to zero. In addition, Fig. 1(b) also reflects: when $\rho < 0.15$, the larger L_a is, the smaller the corresponding maximum value of E_r is, and in the case of the same ρ, the larger L_a is, the smaller E_r is; Within the range of 0.15 to 1.0, the four curves in the figure overlap,

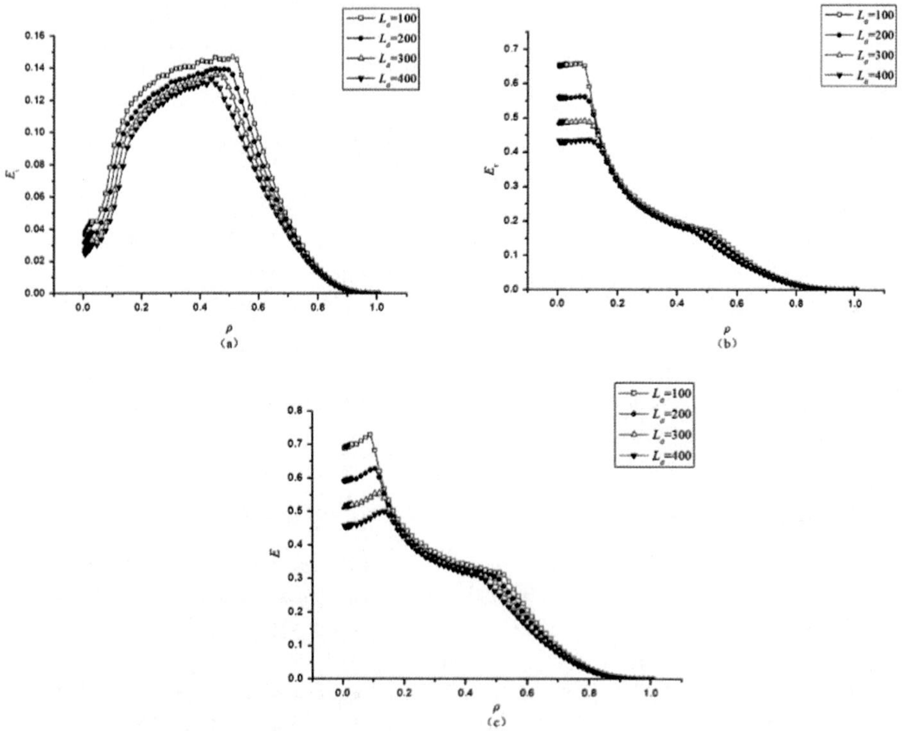

Fig. 1. Energy consumption-density diagrams under different L_a ($v_{\max 1} = 2$)

indicating that different L_a have the same effect on the energy consumption of random deceleration.

Figure 1(c) is the total energy consumption-density diagram, namely the E-ρ diagram. In this figure, as ρ increases, E first increases and then decreases to zero. In addition, when ρ is less than about 0.15, the larger L_a is, the smaller the corresponding maximum value of E is, and in the case of the same ρ, the larger L_a is, the smaller E is; when ρ is about 0.15–1.0, the four curves in the figure overlap, indicating that different L_a have the same effect on the total energy consumption.

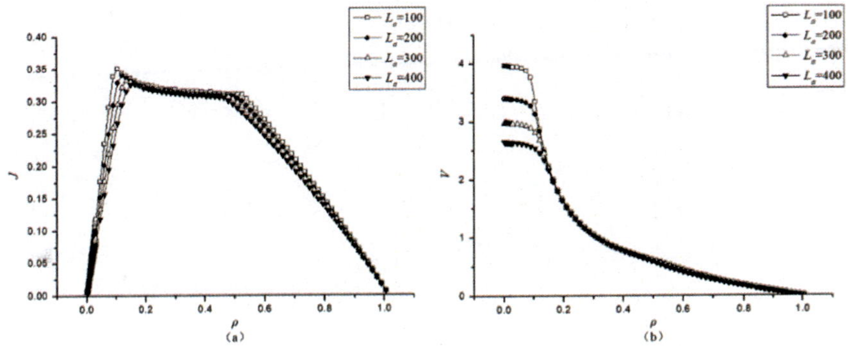

Fig. 2. Basic diagrams under different L_a ($v_{\max 1} = 2$)

When $v_{\max 1} = 2$, and L_a takes 100, 200, 300, and 400, the flow-density and speed-density basic diagrams are shown in Fig. 2(a), (b). Figure 2(a), (b) show that when the vehicle density is low, due to the small number of vehicles on the road, the distance between the vehicles is large, and there is no mutual interference with each other, the driver can keep driving the vehicle at high speed, so the average speed of the vehicle is large; in the medium density range, due to the large number of vehicles on the roadway, the degree of mutual interference between each other increases, and the vehicle speed decreases, so the flow rate increases with the density. In the high density range, because the number of vehicles on the roadway is large, the mutual interference between each other is very large, and the average speed of the vehicles decreases greatly, so the traffic flow quickly drops to zero. In addition, Fig. 2(a), (b).

also shows that the larger L_a is, the smaller the maximum average speed and the maximum flow; when the density is the same, the larger L_a is, the smaller the average speed and traffic flow.

Comparing the energy consumption-density diagrams with the flow-density and speed-density basic diagrams, it can be seen that: keep $v_{\max 1}$ unchanged, the larger L_a is, the smaller the vehicle energy consumption, the smaller the traffic flow and the average vehicle speed, and the easier it is for the traffic lane to appear traffic jam phenomenon.

3.2 The Influence of the Maximum Speed vmax1 Allowed in the Speed Limit Zone on the Vehicle Energy Consumption

Figure 3 is the result of numerical simulation when the length of the speed-limiting zone is set $L_a = 300$, and the maximum speed $v_{\max 1}$ allowed in the speed-limiting zone is 1, 2, 3, and 4 respectively.

Figure 3(a) is a deterministic deceleration energy consumption-density diagram, namely the E_i-ρ diagram. It can be seen from this figure that when ρ is small, the number of vehicles on the roadway is small, the distance between the vehicles is large, the driver can drive the vehicle at high speed, and the deceleration behavior is small, so the kinetic energy loss of the vehicle is small; If ρ is large, the number of vehicles on the driving lane increases, the distance between the two vehicles becomes smaller, the vehicle deceleration behavior increases, and E_i gradually increases to the maximum value; when ρ is larger, the distance between vehicles becomes very small, the behavior of following the car at low speed appears, and E_i gradually decreases to zero with the increase of ρ. In addition, Fig. 3(a) also reflects: when $v_{\max 1}$ is larger, the corresponding maximum value of E_i is larger; in the case of the same ρ, the larger $v_{\max 1}$ is, the larger E_i is.

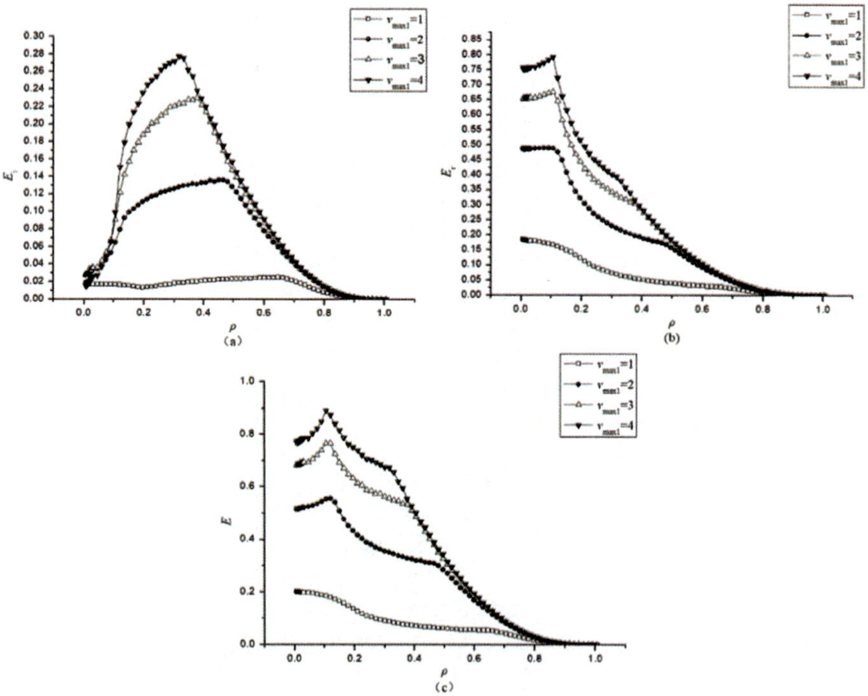

Fig. 3. Energy consumption-density diagrams under different $v_{\max 1}$ ($La = 300$)

Figure 3(b) is the energy consumption-density diagram of random deceleration, namely the E_r-ρ diagram. The figure shows that E_r has a large value when ρ is small, while E_r gradually decreases to zero as ρ increases. When the density is low, the number of vehicles on the roadway is small, but the vehicles decelerate randomly, and E_r has a non-zero initial value; when the number of vehicles increases, the vehicles on the roadway are crowded, and the speed difference between the front and rear of the vehicle is small, resulting in kinetic energy loss small, E_r decreases to zero. In addition, Fig. 3(b) also reflects that the larger $v_{\max1}$ is, the larger the corresponding maximum value of E_r is, and in the case of the same ρ, the larger $v_{\max1}$ is, the larger the E_r is.

Figure 3(c) is the total energy consumption-density diagram, namely the E-ρ diagram. Analyzing this figure, it is found that with the increase of ρ, the value of E first increases and then decreases to zero. When $v_{\max1}$ is larger, the corresponding maximum value of E is larger, and when ρ is the same, the larger $v_{\max1}$ is, the larger E is.

When $L_a = 300$ and $v_{\max1}$ takes 1, 2, 3, and 4 respectively, the flow-density and speed-density basic diagrams are shown in Fig. 4(a), (b). Figure 4(a) shows that when the vehicle density is low, due to the small number of vehicles on the roadway, the distance between vehicles is large and there is no mutual interference between them, therefore, the average speed of vehicles is large, and the traffic flow increases with the increase of density. In the medium density range, due to the large number of vehicles on the roadway, the degree of mutual interference between them increases, and the speed of the vehicles decreases, so the flow remains basically unchanged with the increase of density; in the high density range, because the number of vehicles on the roadway is large, the degree of mutual interference between them is large, and the average speed of the vehicles decreases greatly, so the traffic flow quickly decreases to zero. In addition, Fig. 4(a), (b) also shows that the larger the $v_{\max1}$, the larger the maximum average speed and the maximum flow; when the density is the same, the larger the $v_{\max1}$, the larger the average speed and traffic flow.

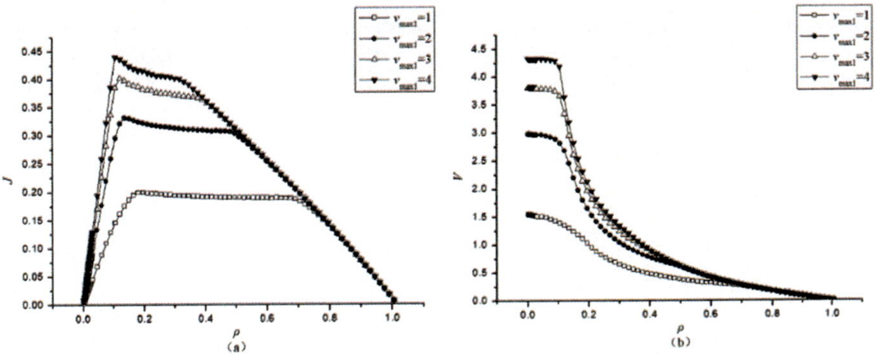

Fig. 4. Basic diagrams under different $v_{\max1}$ ($L_a = 300$)

Comparing the energy consumption-density diagrams with the flow-density and speed-density basic diagrams, we can see that: keeping L_a constant, the larger $v_{\max1}$, the greater the vehicle energy consumption, the greater the traffic flow and the average vehicle speed, and the smoother the road.

4 Conclusion

In this paper, based on the cellular automata NaSch traffic flow model, a numerical simulation is carried out for a one-way single-lane road section with a speed limit zone, and the pair of the length L_a of the speed limit zone and the maximum speed $v_{\max 1}$ allowed in the speed limit zone under the periodic boundary condition which could impact on vehicle energy consumption is studied. The results of the study are as follows:

1) The larger the length L_a of the speed limit zone, the smaller the maximum value of the vehicle deterministic deceleration energy consumption E_i, the random deceleration energy consumption E_r and the total energy consumption E, and the smaller the corresponding maximum value of the traffic flow and the average vehicle speed;
2) The smaller the maximum allowable speed $v_{\max 1}$ in the speed limit zone is, the smaller the maximum value of the vehicle deterministic deceleration energy consumption E_i, the random deceleration energy consumption E_r and the total energy consumption E is, and the corresponding maximum value of traffic flow and vehicle average speed is also smaller;
3) Under the same vehicle density, the larger L_a and the smaller $v_{\max 1}$, the smaller the three energy consumption values of the vehicle, and the smaller the corresponding flow and average speed.

To sum up, the larger the length L_a of the speed-limiting zone and the smaller the. maximum allowable speed $v_{\max 1}$ of the speed-limiting zone, the easier the speed-limited road section is to enter the congestion state. The above conclusions can provide some valuable references for the study of measures to further improve the traffic capacity of speed-limited sections.

Acknowledgment. This project is supported by Autonomous Region-level College Students Innovation and Entrepreneurship Training Program Project (Project No.: S202211549081) and Nanning University Professor Cultivation Engineering Project (Project No.: 2020JSGC01).

References

1. Xianqing, Y., Wei, Z., Kang, Q., et al.: Influence of speed limit on traffic flow. Comput. Phys. **24**(4), 499–504 (2007). (in Chinese)
2. Siling, G.: Study on the phase transition characteristics of cellular automata traffic flow model and traffic measured analysis. Nanning: Guangxi University, pp. 15–22 (2006) (in Chinese)
3. Xue, Y., Liyun, D., Shiqiang, D.: An improved one-dimensional cellular automata traffic flow model and the influence of deceleration probability. Acta Physica Sinica **50**(3), 445–449 (2001)
4. Xiaoming, L., Li, W.: Cellular automata traffic flow model considering on-street parking. J. Jilin Univ. (Eng. Sci. Edn) **42**(2), 327–333 (2012). (in Chinese)
5. Huanhuan, T., Xue, Y., Sanjun, K., et al.: Research on energy consumption of cellular automata mixed traffic flow model. Acta Physica Sinica **58**(7), 4506–4513 (2009)
6. Chowdhury, D., Santen, L., Schreckenberg, A.: Statistical physics of vehicular traffic and some related systems. Phys. Rept. **329**, 199–329 (2000)

7. Jiaxiu, P., Xue, Y., Yujuan, L., et al.: Effect of road structure on the capacity of a signalized road intersection. Chin. Phys. B **18**(10), 4169–4176 (2009)
8. Tian, J.F., Jia, B., Li, X.G., Gao, Z.Y.: A new car-following model considering velocity anticipation. Chin. Phys. B **19**(1), 010511–010518 (2010)
9. Kaixian, C., Junwei, Z., Yongsheng, Q., et al.: Research on the characteristics of traffic flow speed limit control in zones affected by expressway traffic accidents. Highway Eng. **43**(2), 91–95 (2018). (in Chinese)
10. Jun, Y., Jingyi, S.: Speed limit by lane research on related issues[J]. Science Technology and Engineering **12**(15), 3780–3782 (2012). (in Chinese)
11. Lanxiang, W., Yujuan, L., Youkun, Z.: Research on the impact of on-street parking belts on traffic flow. Western Transp. Sci. Technol. **10**, 93–98 (2015). (in Chinese)
12. Lanxiang, W., Yujuan, L., Youkun, Z.: Analysis of the impact of on-street parking on traffic flow. J. Hechi Univ. **35**(5), 55–61 (2015). (in Chinese)
13. Lanxiang, W., Yujuan, L.: The influence of delay probability and on-street parking strip length on road traffic. Southwest Normal Univ. J. (Nat. Sci. Edn) **41**(5), 115–121 (2016). (in Chinese)
14. Lanxiang, W., Yujuan, L., Jinsong, P., et al.: Research on the influence of tunnels on the energy consumption of highway traffic flow. Highway Automobile Transp. **2**, 66–69 (2016). (in Chinese)
15. Yanhong, F.: Traffic Flow Modeling of Some Complex Road Sections and Research on Traffic Energy Consumption and Traffic Accidents. Guilin: College of Physical Science and Technology, Guangxi Normal University, pp. 13–22 (2013). (in Chinese)
16. Lanxiang, W.: A cellular automaton traffic flow energy consumption model considering on-street parking. Sci. Technol. Bull. **33**(1), 216–220 (2017). (in Chinese)
17. Yujuan, L., Guixue, L.: Looking at the traffic capacity of the curved road section from the kinetic energy loss. Sichuan Normal Univ. J. (Nat. Sci. Edn) **34**(3), 355–359 (2011). (in Chinese)
18. Suprapto, K., Makhrus, F.: Point based forecasting model of vehicle queue with extreme learning machine method and correlation analysis. Int. J. Intell. Syst. Appl. (IJISA) **13**(3), 11–22 (2021)
19. Afshany, A.J., Tourani, A., Shahbahrami, A., et al.: Parallel implementation of a video-based vehicle speed measurement system for municipal roadways. Int. J. Intell. Syst. Appl. (IJISA) **11**(11), 25–37 (2019)
20. Dhanya, K.M., Kanmani, S.: Dynamic vehicle routing problem: solution by ant colony optimization with hybrid immigrant schemes. Int. J. Intell. Syst. Appl. (IJISA) **9**(7), 52–60 (2017)

Research and Design of Four Rotor UAV Based on Cascade PID

Deshen Lv, Chengquan Liang, and Ying Zhang(✉)

School of Intelligent Manufacturing, Nanning University, Nanning 530200, China
876357930@qq.com

Abstract. Four rotor UAV is usually greatly affected by wind and wind direction. The stability of UAV flight cannot be guaranteed due to weather changes, so it has not been widely used in industrial occasions. Aiming at this problem, this paper presents a design scheme of four rotor UAV using cascade PID control. Stm32f103c8t6 single chip microcomputer is used for data processing and motion control, and mpu6050 sensor is used to measure the angle and radian of the UAV. The flight power is provided by four 720 hollow cup motors. The single chip microcomputer uses cascade PID to control the motor, and then communicates with the remote controller through nRF24L01 module. The PC terminal displays the attitude of the UAV and debugs PID parameters. The test results show that the designed four rotor UAV can fly smoothly in some harsh environments and can carry certain heavy objects. The research of this paper provides a reference for the application of UAV.

Keywords: Stm32f103c8t6single chip · Cascade PID · MPU6050

1 Introduction

With the rapid development of society and the continuous progress of science and technology, people's life is also moving towards automation. Now, unmanned aerial vehicles can be used in various fields, such as military strike, police pursuit, disaster relief, transmission line inspection, aerial photography, etc. [1]. At present, the types of UAVs include: single axle helicopter and fixed wing aircraft. Fixed wing aircraft has good operability, flexibility and stable flight, and can take off, land and hover vertically [2, 3].

After the 1990s, the research of MEMS inertial navigation system was mature, but the data noise was large, and various digital algorithms were needed to denoise. At that time, the operation speed of single chip microcomputer was not fast, and the flight stability was still poor [4]. It was not until around 2005 that a stable multi rotor UAV was produced [5, 6]. Later, with the development of technology, a variety of UAVs appeared, such as aerial photography UAV, electric power line patrol UAV, plant protection UAV, etc. [7].

Many scholars and industry insiders at home and abroad have carried out research on UAV. Literature [8] takes STM32 microcontroller as the core, uses mpu9250 to collect the attitude data of four rotor UAV, and cooperates with Kalman filter algorithm to fuse and

Z. Hu et al. (Eds.): AIMEE 2022, LNDECT 159, pp. 88–99, 2023.
https://doi.org/10.1007/978-3-031-24468-1_9

filter the data collected by the sensor, so as to achieve accurate attitude measurement. Literature [9] uses mpu6050 sensor to collect the attitude information of the aircraft, and solves three attitude angles according to the collected attitude information. Finally, only Simulink simulation is done. Literature [10] takes the four rotor helicopter as the research object. The attitude adopts the back-stepping sliding mode technology with adaptive gain. Finally, the system is relatively stable and robust.

The four rotor UAV system involves aerodynamics, MEMS inertial attitude detection, data fusion, data filtering, attitude resolution and other technologies [11–13]. The design of four rotor UAV mainly solves the problems of aerodynamics, MEMS device attitude detection, data fusion, data filtering, solution and so on [14, 15]. By using the cascade PID algorithm, the inner loop controls the angular velocity and the outer loop controls the angle, so that the UAV can fly smoothly, and has certain anti-interference ability and the ability to restore the initial attitude.

2 System Structure Scheme

The whole system consists of two parts: the remote control part and the UAV. The first is remote control system which mainly sends the read rocker value to the UAV through nRF24L01, so that the remote control can control the UAV, and 0.96 inch OLED and serial port display parameters; The second part is the UAV system which mainly drives MPU6050 to read the original data through a single chip microcomputer, then calculates the attitude angle through a series of operations, and controls the motor speed in combination with the data of the remote controller received by nRF24L01 and the PID control algorithm [16]. The design scheme of the system mainly includes two parts: the design of the remote controller and the design of the UAV. This part includes the analysis of the working principle of the four rotor UAV.

2.1 Design of Remote Controller

The remote controller uses stm32F103C8t6 single chip microcomputer, 360° rocker module, 0.96-inch OLED screen, TP4059 lithium battery charging chip, dual 20K potentiometer, buzzer and LED. Functions to be completed by the remote control:

(1) Use the timer update interrupt to provide a system clock for the remote control system. After each time, execute the corresponding function to poll and switch the task.
(2) MCU uses adc + dma peripherals to read out the original values of the rocker and potentiometer. After the amplitude limiting function, the amplitude is limited in the range of 1000–2000. Finally, the final data is obtained through mean filtering. The ADC is also used to read the battery voltage entering the resistance divider.
(3) The MCU drives the NRF24L01P module through SPI peripherals, which is configured to send the processed data to the UAV. When the UAV is about to send it to the remote controller, the NRF24L01 module is set to the receiving mode [17].
(4) The 0.96- inch OLED screen is driven by IIC software to display parameters, which is convenient for debugging.

(5) Use USART1 peripherals to transmit the data of UAV and remote controller to the upper computer through CH340 module. The upper computer can also send commands to the remote controller, and then send them to the UAV through the wireless module of the remote controller.
(6) The charger is connected to USB and recharges the lithium battery through the TP4059 lithium battery charging chip. When the battery is charging, the indicator light is on; when the charging is completed, the battery full indicator lights up.
(7) When the voltage of remote controller and UAV is lower than a certain value, the buzzer will alarm.

2.2 Design of UAV

The UAV needs to use STM32F103C8T6 MCU, 720 hollow cup motor, MPU6050 gyroscope, TP4059 lithium battery charging chip and other hardware. UAV should complete the following functions.

(1) The timer update interrupt is used to provide the clock for the UAV system. After each time, the corresponding function is executed to poll and switch the task.
(2) Configure ADC to read the battery voltage after resistance voltage division.
(3) Use the PWM function of the timer to give an input signal to the 720 hollow cup drive circuit to control the motor speed.
(4) The acceleration and gyroscope data of MPU6050 sensor are read through IIC, and filtered. Finally, the attitude angle is calculated by using the attitude solution complementary filtering algorithm based on quaternion [18].
(5) The attitude of UAV is controlled by cascade PID. The inner loop PID controller manages the diagonal speed and outputs the corresponding motor PWM value; The outer loop PID controller calculates the angle error, and the output is the input of the inner loop PID [19].
(6) NRF24L01 module is used to receive the data sent by the remote controller. When idle, it will be configured as the sending mode to send the UAV data to the remote controller.

2.3 Working Principle of Four Rotor UAV

The principle of the four rotor UAV is to control the rotation of four 720 hollow cup motors to drive the blades to produce an upward lift. If each motor is controlled separately, it will produce different lift, so as to control the attitude and position of the UAV [20]. UAV can realize eight kinds of movements in the air, which are vertical rise, vertical fall, forward movement, backward movement, left movement, backward movement, clockwise change of course, counterclockwise change of course. The structural situation of UAV can be divided into two types: (1) X mode; (2) Cross mode. The attitude measurement and control of the two structural modes are not very different. In order to facilitate understanding, this design studies x mode.

2.3.1 Vertical Rise and Fall

When the UAV rises to a certain height and is stable, the vertical upward movement of the UAV can be realized by adjusting the speed of the four hollow cup motors of the UAV at the same time. Conversely, turning down the rotation speed of the four hollow cup motors of the UAV at the same time will realize the vertical descending movement of the UAV. As shown in Fig. 1.

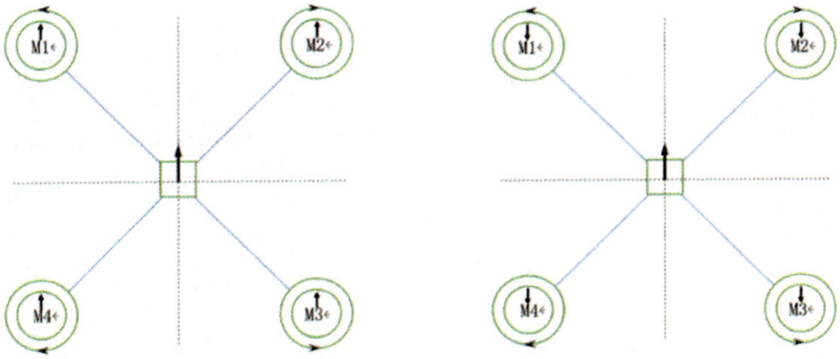

Fig. 1. Vertical rise (left), vertical fall (right)

2.3.2 Forward and Backward Movement

When the UAV rises to a certain height and is stable, at the same time, the UAV can move forward by increasing the speed of M3 and M4 hollow cup motors and decreasing or maintaining the speed of M1 and M2. On the contrary, at the same time, the UAV can move backward by increasing the rotation speed of M1 and M2 hollow cup motors and decreasing or maintaining the rotation speed of M3 and M4. See Fig. 2.

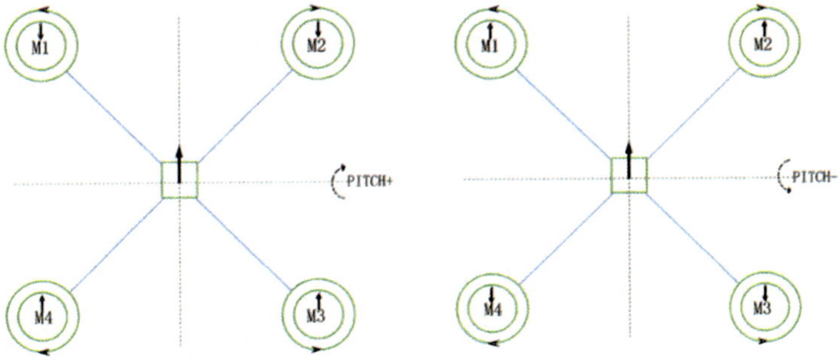

Fig. 2. Forward movement (left), backward movement (right)

2.3.3 Left and Right Movement

When the UAV rises to a certain height and is stable, at the same time, the UAV can move to the left by increasing the rotation speed of M2 and M3 hollow cup motors and decreasing or maintaining the rotation speed of M1 and M4. On the contrary, at the same time, the UAV can move to the right by increasing the rotation speed of M1 and M4 hollow cup motors and decreasing or maintaining the rotation speed of M2 and M3. As presented in Fig. 3.

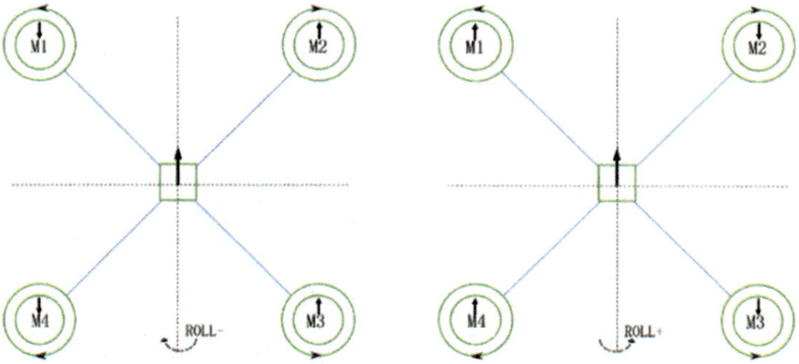

Fig. 3. Left movement (left figure), right movement (right figure)

2.3.4 Change Course Counterclockwise and Clockwise

When the UAV rises to a certain height and is stable, at the same time, the rotation speed of M1 and M3 hollow cup motors of the UAV is adjusted up, and the rotation speed of M2 and M4 is adjusted down or the rotation speed is unchanged, the course of the UAV can be changed counterclockwise. On the contrary, the UAV can change its course clockwise by increasing the M2 and M4 speed, decreasing the M1 and M3 speed or keeping the same speed, which is shown in Fig. 4.

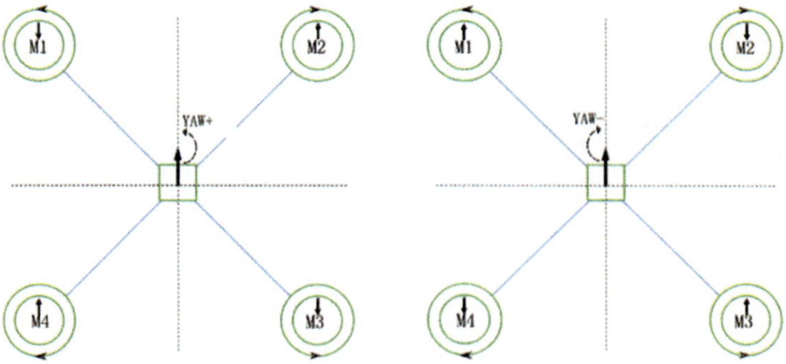

Fig. 4. Change the course counterclockwise (left), and change the course clockwise (right)

3 Four Rotor UAV Algorithm

The four rotor UAV is vulnerable to external environmental factors. It processes the acceleration and gyroscope data collected by mpu6050 and solves the attitude. At this time, it is necessary to add some software algorithms to the software design to reduce the error impact.

3.1 Complementary Filtering Algorithm for Attitude Solution

Complementary filtering algorithm can filter low-frequency and high-frequency interference at the same time, which is convenient to realize sensor data fusion. On mpu6050, IIR low-pass filter is used for accelerometer [21], and arithmetic average filter is used for gyroscope. The two filters form a complementary filtering algorithm before attitude angle calculation. Finally, the gyroscope data obtained by arithmetic mean filtering is converted into radians / s and integrated relative to the gyroscope. The complementary filter model is shown in Fig. 5.

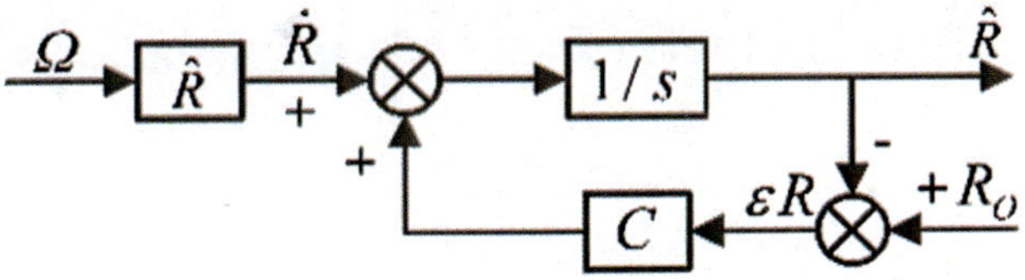

Fig. 5. Complementary filter model

Expression of complementary filter in time domain is:

$$R^\wedge = k\_p(R\_0 - R^\wedge) + R^\wedge\Omega \tag{1}$$

The expression of complementary filter in time domain:

$$R^\wedge = k_{\_p}\ (R_{\_0} - R^\wedge) + R^\wedge\Omega \tag{2}$$

For the above acceleration original value of MPU6050, for the IIR parameter, take every 1ms = 0.001 s to read MPU6050 data, mpu6050 reads the maximum operating frequency of the data function 10 Hz, and the acceleration original waveform and filter waveform are shown in Fig. 6. The filtered data smoother than the original data can be obtained by performing the arithmetic mean filtering algorithm on the gyroscope. The original waveform and filtered waveform of the gyroscope are displayed on MultiWiiConf as shown in Fig. 7. Finally, the three relatively important attitude angles of UAV can be obtained by attitude calculation. Both Fig. 6 and Fig. 7 are the waveforms generated by the UAV on the table and gently patted by hand. It can be seen that the original waveform changes greatly, and some of them are errors caused by the industrial defects of the gyroscope chip itself. Although the errors can be reduced by the filtering algorithm, they cannot be completely eliminated. If MPU6050 filters only once, the error of gyroscope will continue to increase after a period of time from the start of operation, which will cause serious danger to the UAV system. In the program, after reading mpu6050 data register every time, convert the data, and then calibrate MPU6050.

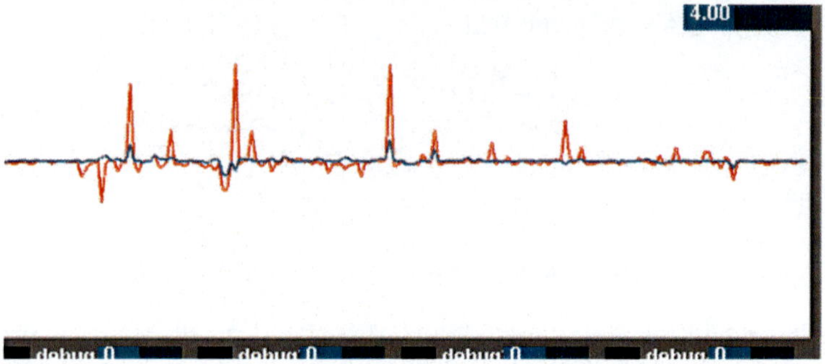

Fig. 6. Original acceleration waveform (red) and filtered waveform (blue)

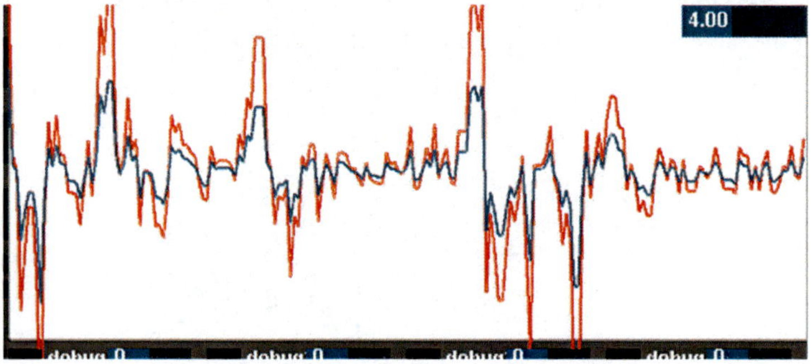

Fig. 7. Gyro original waveform (red) and filtered waveform (blue). (Color figure online)

3.2 Attitude Angle Calculation

Attitude solution involves the position of mpu6050 gyroscope on the circuit board. The coordinate axis of UAV at different positions will also be different, and the attitude angle solved by quaternion complementary filtering algorithm will also be different. At this time, trigonometric function solution is required. For trigonometric function solution, this design adopts the method in literature [22], which is not described here repeatedly.

3.3 Cascade PID Controller

PID controller is the most common feedback loop algorithm in industry. The elements of feedback include three parts: measurement, comparison and execution. The actual value of the controlled variable is compared with the expected value, and the deviation is used to correct the response of the system for adjustment and control. Basic indicators of PID controller: (1) stability (P and I reduce system stability, D improve system stability); (2) Accuracy (P and I improve the steady-state accuracy, D has no effect); (3) Rapidity (P and D improve response speed, I reduce response speed). Cascade PID refers to two PID algorithms, which are connected in series. It can enhance the anti-interference of the system, because it controls more variables than a single PID controller, making the UAV system more stable [23].

The discrete formula of cascade PID is shown in (3):

$$u(k) = K_p * e(k) + K_I * \sum_{(i=0)}^{k} k * e(i) + K_D * [e(k) - e(k-1)] \quad (3)$$

The block diagram is shown in Fig. 8.

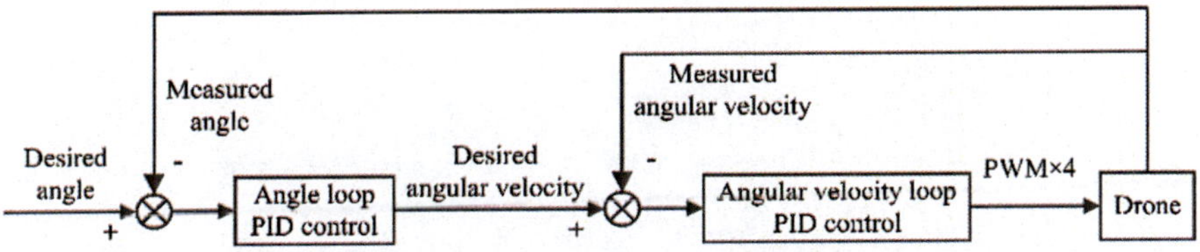

Fig. 8. Cascade PID block diagram

The expected angle comes from the remote control data, and the feedback measurement angle comes from MPU6050. The error obtained by subtracting the actual measured value from the expected value is used as the input of the angle loop PID control. The angle loop PID calculates the angle error, and finally the calculation result is used as the input of the angle loop. The angular velocity loop PID control is a positional PID, which takes the angular velocity error calculated by subtracting mpu6050 from the angular velocity loop output as the input of the angular velocity loop PID control, and the angular velocity loop PID calculates and outputs the angular velocity, and distributes the output value to four motors.

In the angle loop PID input, two attitude angle controls of PITCH and ROLL are added, and the control value of the corresponding potentiometer is added, which is convenient for potentiometer debugging. The outer loop PID input is the value after the roll angle (or pitch angle) is reduced by 13 times less the roll angle (or pitch angle) given on the remote control, that is, the expected angle is about -35° to 35°, and the value of the fine-tuning potentiometer is reduced by 100 times before and after the remote control potentiometer is added (or left and right). Because the potentiometer is used for fine-tuning, the reduction ratio will be relatively large in the program, and the rocker value is to control the flight direction of the UAV, so it cannot be reduced by 100 times like the fine-tuning potentiometer, which will cause the remote control to not control the motion of the UAV well.

The output of the angle loop and angular velocity loop PID controller is obtained from the cascade PID discrete formula of formula (2), roll_ PID and pitch_ PID is the outer loop PID parameter structure, gyro_ roll_ PID and gyro_ pitch_ PID is the inner loop PID parameter structure. Because the input of the outer ring is the expected value of the front, rear, left and right directions given by the remote control, roll_ PID and pitch_ The parameters of PID are the same. The output of the outer ring is the input of the inner ring, and the yaw rocker value on the remote control is also the input. However, in the lock tail and non-lock tail modes, it can be seen that the rocker value yaw is added to the non-lock tail, but not to the non lock tail. So the inner ring gyro_ roll_ PID and gyro_ pitch_ The parameters of PID are the same.

4 Four Rotor UAV Commissioning

Altium designer 17 is used for the circuit design of the four rotor UAV to design the schematic diagram and PCB. After welding, the physical object is shown in Fig. 9.

Fig. 9. Entity of four rotor UAV

After completing the physical welding, the main work is to debug the PID parameters of the UAV.

In order to facilitate the PID parameter adjustment of UAV, this design uses serial port to display the parameters to the upper computer, and can set the parameters of UAV at the same time. The upper computer displays PID parameters as can be seen in Fig. 10.

Fig. 10. PID parameters displayed by upper computer

In Fig. 10, pitch represents the inner ring parameters of pitch angle and roll angle, yaw represents the inner ring parameters of yaw angle, and roll represents the outer ring parameters of pitch angle and roll angle.

The last work is to adjust the PID parameters of UAV. The steps are generally as follows:

(1) Pitch and roll inner ring P. Mainly for the control of the angular speed of the UAV, the smaller the inner ring P is, the smaller the UAV's resilience is, and the slower the system response is. The larger the inner ring P is, the greater the resilience is, and the faster the system response is. When the inner ring P is too large, the stability of the system will become poor and vibration will occur.
(2) Pitch and roll inner ring I. It mainly eliminates the static error of MPU6050, which is generally set small or not. When the inner ring I is too large, the recovery force of the UAV will become smaller and the response speed will become lower.
(3) Pitch and roll inner ring D. It is used to suppress oscillation, assist inner loop P, improve system stability and eliminate UAV oscillation.
(4) Yawinner ring I. When the pitch and roll parameters of the inner ring are determined, if the four axes spin, adjust the inner ring I of YAW to eliminate the static error of yaws.
(5) Pitch and roll outer ring P. It is mainly used to control the value of the remote control. The larger "P" is, the less time it takes for the UAV to reach the angle when flying.
(6) Pitch and roll outer ring P. Improve system stability.
(7) Pitch and roll outer ring I. Like the determination of pitch and roll inner ring I parameters, it is mainly used to eliminate the angular error of mpu6050.
(8) Pitch and roll outer ring D. Auxiliary pitch and roll outer ring P.
(9) YAW inner ring P. The UAV is controlled by the remote controller to turn forcefully.

5 Conclusion

This paper introduces the control principle of four hollow cup motors of four rotor UAV in detail, explains the scheme of remote controller and UAV, gives the method of data filtering and the application of cascade PID controller in this design, and finally designs a physical entity to verify its function.

When the external interference is not great, the UAV can fly smoothly until the battery is exhausted. The UAV can control the UAV to move forward and backward, left and right, up and down, left and right through the remote control. The remote controller can also send data to the upper computer through the serial port module, and the upper computer can also send the adjusted parameters to the UAV. Both the UAV and the remote controller can charge the lithium battery and have a fully charged display. Current problems are: when the UAV runs for a long time, the MPU6050 data will drift, and the readings will deviate. In general, this design is available and successful and can offer a salutary lesson for the future UAV design and improvement.

References

1. Chaoqiong, Y., Faqing, W.: Design of STM32 four axis aircraft. Electronic world **01**, 187–188 (2022). (in Chinese)
2. Xuanbing, L., Xuanbing, L., Shaolei, Z., Zhicai, X., et al.: Track tracking wind resistance of fixed wing UAV. Mod. Defense Technol. **50**(01), 41–47 (2022). (in Chinese)
3. Zhdanov, S., Kadet, N., Silkov, V., et al.: The mathematical model for research of the UAV longitudinal moving. Int. J. Eng. Manufact. (IJEM), **13**(5), 29–39 (2021)
4. Fei, D.: Algorithm and software design and implementation of personal navigation system based on MEMS inertial devices. Dongnan University, Nanjing (2017). (in Chinese)
5. Hou, J.: Research on attitude calculation and control system of four axis aircraft. Taiyuan University of Technology, Taiyuan (2018) (in Chinese)
6. Adeleke, O.A., Abbe, G.E., Jemitola, P.O., Thomas, S.: Design of the wing of a medium altitude long endurance UAV. Int. J. Eng. Manufact. (IJEM), **12**(1), 37–47 (2022)
7. Ooi, K.R., et al.: Design of hoeckens linkage based walking robot with MPU6050 IMU as navigation sensor. J. Phys. Conf. Ser., **1969**(1) (2021)
8. Xiaoyin, W., Zhang, X., Jiaxiang, L., et al.: Control system design of four rotor UAV. Microcomput. Appli, **38**(01), 20–22 (2022). (in Chinese)
9. Wang, K., Li, P., Tang, H.: Design of attitude angle algorithm for four rotor UAV. Electr. Test (23), 60–61+57 (2021) (in Chinese)
10. Derafa, L., Ouldali, A., Madani, T., Benallegue, A.: Non-linear control algorithm for the four rotors UAV attitude tracking problem. Aeronaut. J. **115**(1165) (2011)
11. Diankuan, D., Chenhao, Z., Tianguang, J.: Design of four rotor UAV control system based on STM32. Mod. Electron. Technol. **19**, 19–23 (2021). (in Chinese)
12. Chen, L., Li, T.: UAV control system based on back-stepping and active disturbance rejection control. In: Proceedings of the 3rd Joint International Conference on Information Technology, Mechanical and Electronic Engineering, pp. 142–146 (2018) (in Chinese)
13. Gokhale, S., Dalvi, A., Siddavatam, I.: Industrial control systems honeypot: a formal analysis of Conpot. Int. J. Comput. Netw. Inf. Sec. (IJCNIS), **12**(6), 44–56 (2020)
14. Malleswara Rao, J., Rao, C.V., Senthil Kumar, A., Gopala Krishna, B., Dadhwal, V.K.: Spatiotemporal data fusion using dictionary learning and temporal edge primitives. Int. J. Image Graph. Signal Process. (IJIGSP), **7**(10), 47–59 (2015)
15. Mohammadloo, S., Rashidi, A.J.: Fusion-based sensor selection for optimal state estimation and minimum cost (intelligent optimization approach). Int. J. Intell. Syst. Appli. (IJISA), **4**(4), 1–8 (2012)
16. Ling, Y.: Study on motion mathematical model of four rotor UAV. In: Proceedings of 2017 3rd International Conference on Environment, Biology, Medicine and Computer Applications (ICEBMCA 2017), pp. 96–105 (2017)
17. Zhang, Y., Jiang, Z., Wu, Y., et al.: Research on aircraft condition monitoring system based on nrf24l01+. J. Astronau. Metrology a Measur., 2021(4) (in Chinese)
18. Can, K., BAŞÇİ, A.: Real time trajectory tracking control of a four rotor UAV with backstepping controller. Iğdır Üniversitesi Fen Bilimleri Enstitüsü Dergisi (3) (2016)
19. Can, K., BAŞÇI, A.: Real time trajectory tracking control of a four rotor UAV with backstepping controller. J. Inst. Sci. Technol. **6**(3) (2016)
20. Derafa, L., Benallegue, A., Fridman, L.: Super twisting control algorithm for the attitude tracking of a four rotors UAV. J. Franklin Inst. **349**(2) (2012)
21. Min, J.: Binocular stereo vision control method for landing position of four rotor UAV. In: 2020 3rd International Conference on Computer Science, Electronics and Engineering Management (2020)

22. Sun, Z., Liu, C., Li, M., Zhao, J., Xu, Y.: Application of anti integral saturation cascade PID control algorithm in superheated steam temperature control. J. Phys. Conf. Ser. **2087**(1) (2021)
23. Bao, N., Ran, X., Wu, Z., Xue, Y., Wang, K.: Research on attitude controller of quadcopter based on cascade PID control algorithm. In: Proceedings of 2017 IEEE 2nd Information Technology, Networking, Electronic and Automation Control Conference (ITNEC 2017), pp. 1528–1532 (2017)

Diagnosis on Wrinkle in Film Transmission System and Study on Roller Traction Characteristics

Xichen Lin, Weihong Yang, Yan Bi, Wenhui Li, Xinjian Zhang, and Min Kong(✉)

Department of Mechanical and Electrical Engineering, Wenhua College, Wuhan 430073, China
westlifelxc@163.com

Abstract. The roller is a common traction and guiding mechanism in the film transmission system, which directly affects the transmission stability and surface quality of the film. In this paper, the wrinkle phenomenon of a production line during film transmission is studied. The modeling and finite element analysis of the system consisting of traction roll and working groove are carried out. The solution roller is regarded as a liquid film lubrication rotor system, and the simulation analysis is carried out. The liquid film bearing capacity and roller frictions are calculated for different roller deformation and liquid film thickness. The calculation results show that the thickness of the liquid film has a significant impact on the traction capacity of the roller, which is the main reason for the film folding. Some measures need to be taken to avoid the formation of the liquid film.

Keywords: Roller · Film · Traction characteristics

1 Introduction

In the film transmission system, the roller is often used as the traction and guidance mechanism. Through the friction between the film and the surface of the roller, the film plays the role of direction guidance and support movement. However, due to the influence of working environment, device parameters and other factors, sometimes it will lead to tension changes at the local contact position, make the film appear abnormal phenomena such as surface wrinkle deformation and lateral offset, and affect the production efficiency and product quality. The surface characteristics between the film and the roller and the traction of the roller are one of the important factors affecting the stability of film transmission. As a relatively complex system composed of many factors, there are also many influencing factors, which need to be comprehensively analyzed from the perspective of the system [1–3].

Jeenkour et al. Derived the film thickness equation by using Reynolds equation, convex guide roll shape function and film deflection equation [4]. Using the finite difference method and Newton iterative method, the numerical analysis shows that the central film thickness and the minimum film thickness of the convex guide roll are reduced, and the structural parameters of the guide roll affect the position of the minimum film thickness.

Z. Hu et al. (Eds.): AIMEE 2022, LNDECT 159, pp. 100–114, 2023.
https://doi.org/10.1007/978-3-031-24468-1_10

Ma et al. Studied the air entrainment between the moving film and the guide roller, and analyzed the velocity distribution and pressure characteristics of the entrained gas by using computational fluid dynamics [5]. Mario et al. Established the traction model of the guide roll and analyzed the effects of friction coefficient, wrap angle and tension on the speed difference between the film and the guide roll [6]. Chen et al. Studied an aerostatic bearing device, which can form an air film between the film and the air guide roller to realize non-contact support and transportation [7]. The whole flow field and the pressure distribution of a single throttle at different positions were numerically simulated [8, 9]. As a relatively complex system composed of many factors, there are also many influencing factors, which need to be comprehensively analyzed from the perspective of the system [10–13]. This paper focuses on the analysis of the influence of traction roller on the stability of film transmission in view of the wrinkles in a film production line.

2 Film Production Equipment and Transmission System

2.1 Overall Structure of Film Production Line

The general layout of the film production line is shown in Fig. 1. The traction rollers used in the production line adopt an up-down staggered layout. Driven by the motor, the roller drives the film to move along the specified path. It can be seen from the figure that this production line is equipped with six working tanks with the same configuration, which are closely connected one by one. Each working tank is filled with a chemical solution customized according to the production requirements. The film on the production line and each traction roller are immersed in the chemical solution, and the whole transmission process is completed below the liquid level.

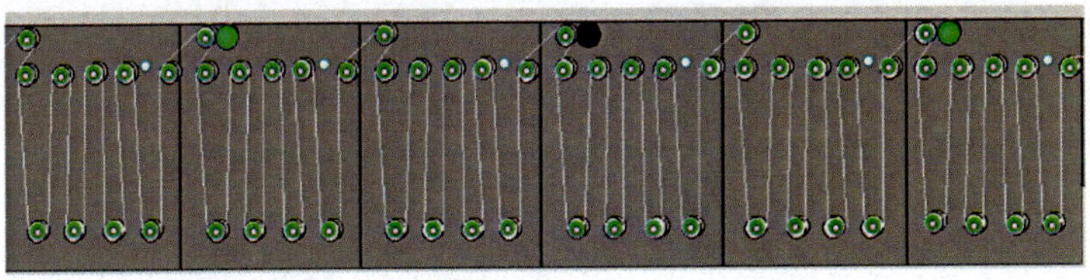

Fig. 1. General layout of production line

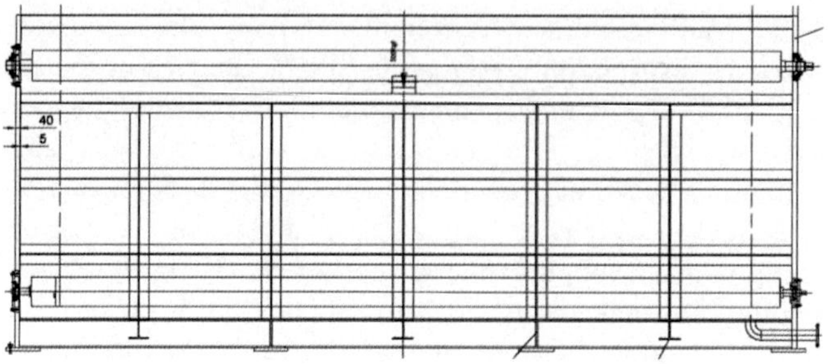

Fig. 2. Side view of working groove of film production line

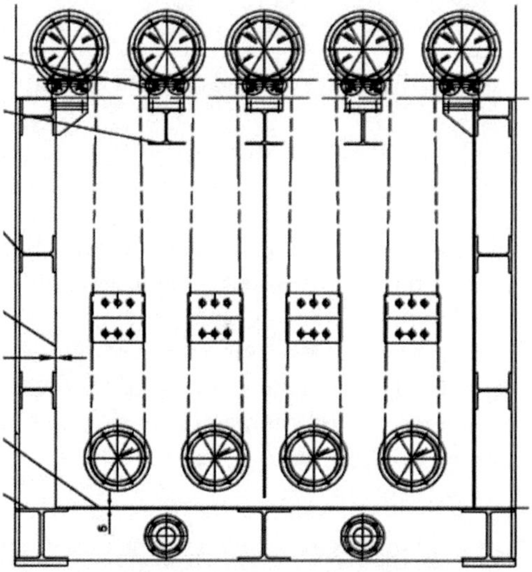

Fig. 3. Main view of working slot of film production line

Figure 1 and 2 show the assembly drawing of the working groove of the film production line. For the need of enterprise confidentiality, the geometric parameters of the tank are not marked in the figure.

2.2 Problems in the Production Line

After the production line is started, it can operate normally, stably and efficiently in the initial stage. However, when the production line continues to work for about 5 to 8 days, the transmission roller begins to slip, and there are slight wrinkles on the film surface, which has a certain impact on the product quality. After the workers stop the machine, disassemble the rollers and other parts in the production line, calibrate and install them, and restart the machine. The production line can resume normal operation, and the film surface is smooth, but after several days, the roller slipping and film folding phenomenon appear again. Observing another production line with the same layout scheme, it is found that this phenomenon also exists. After the above shutdown, disassembly, calibration and installation, the film wrinkles cannot be completely eliminated. Therefore, considering the impact of other factors on the production line, it is mainly reflected in the following two aspects.

The traction roll has a long span, and there is no support in the middle. Due to its self weight, it sinks slightly in the central area, showing the characteristics of convex roll. There is a certain amount of displacement along the length of the whole roll, causing the film to have a velocity difference in the width direction and wrinkle;

Due to the viscosity effect of the solution, the solution is drawn into the gap between the roller and the film, forming a layer of lubricating liquid film in the gap, reducing the surface friction, resulting in the loss of traction of the roller and wrinkles.

3 Finite Element Analysis of Working Tank

According to the assembly drawing of the working groove provided by the enterprise, a three-dimensional model is established, as shown in Fig. 4. The groove is equipped with 5 upper rollers and 4 lower rollers, the two sides of the traction roller are bearings, and the bearing end seat is connected with the load-bearing steel plate; In the horizontal direction and at the bottom of the tank, a cross I-beam structure is used to bear the load, and the bottom of the tank is equipped with a solution inlet and outlet.

Fig. 4. Three dimensional assembly drawing of working groove

In order to conduct ANSYS simulation analysis, the following treatments are made for the working slot:

Exert a vertical downward gravity on the tank, and impose a fixed constraint on the bottom surface of the 8 foot pads;

The tank is filled with solution, hydrostatic pressure is applied to its interior, and the inner wall is selected as the action surface.

After the restraint and load are applied, as shown in Fig. 5, the deformation at the center of the staggered bottom of the I-beam reaches the maximum, which is 0.0083558 m.

Considering the displacement caused by the self weight of the roller, the upper and lower rollers are treated as follows:

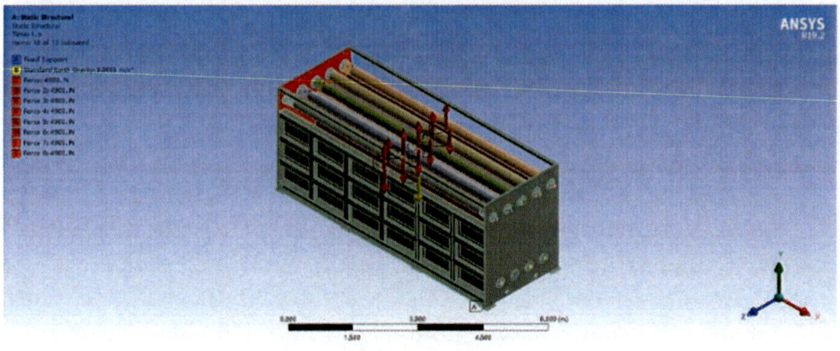

Fig. 5. Cloud diagram of constraints and stresses on the tank

Apply vertical downward gravity to the roller, one side of the roller is fixed constraint, and the other side is remote displacement constraint;

The roller is immersed in the solution, and the density of the roller is changed to the difference between the density of stainless steel and that of the solution.

After the load and constraint are applied, the displacement of the upper and lower rollers is shown in Figs. 6, 7, 8 and 9.

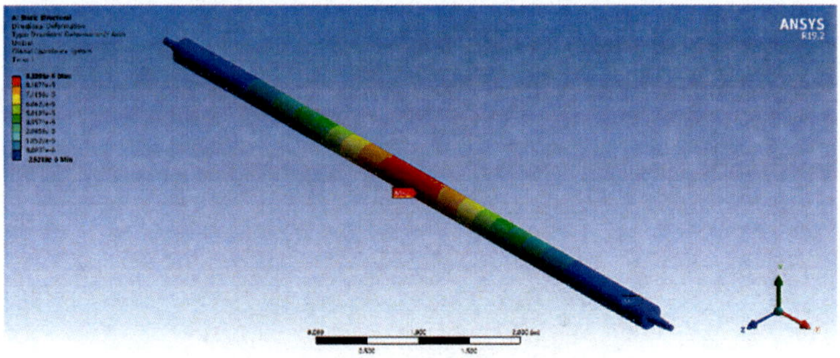

Fig. 6. Nephogram of y-axis displacement of upper roll

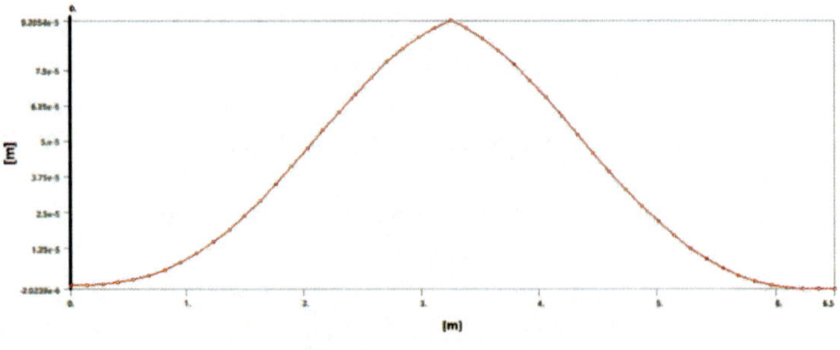

Fig. 7. Bending deformation of upper roll bottom

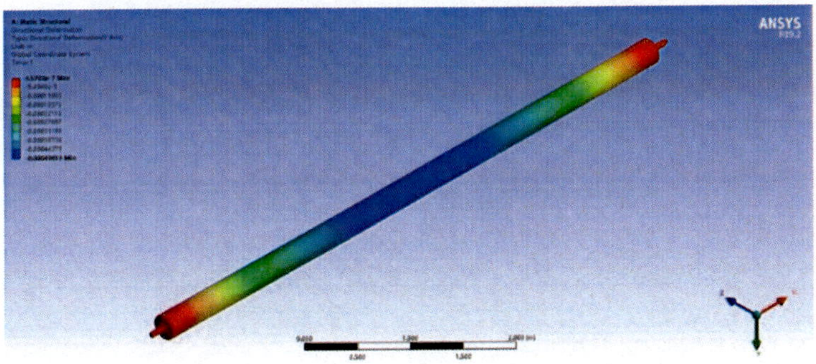

Fig. 8. Nephogram of lower roll displacement along Y axis

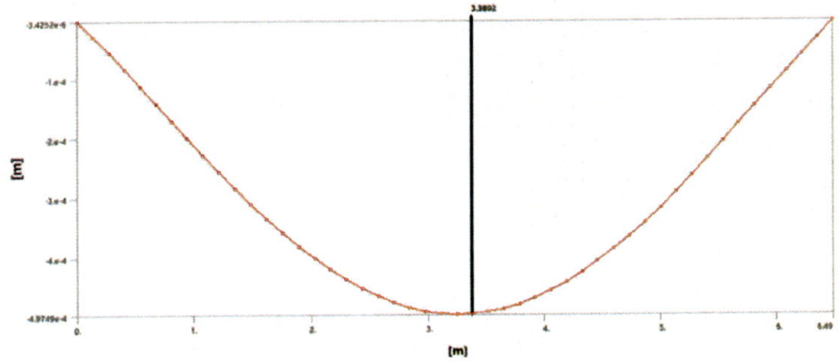

Fig. 9. Bending deformation of lower roll bottom

The results show that the maximum deformation value of the upper roll is 0.092 mm and the maximum deformation value of the lower roll is 0.498 mm, which are located in the center of the roller. From the center to the left and right sides, the deformation value gradually decreases and shows good symmetry. The random local slip and wrinkling phenomenon should not be considered as the main reason, and subsequent analysis is needed.

4 Analysis of Roller Traction Characteristics in the Working Groove

Consider the influence of roller traction characteristics. Since both the roller and the film are immersed in the solution, during long-term operation, due to the viscosity effect, a small amount of solution is sucked into the gap between the film and the roller, gradually filling the entire gap, forming a relatively uniform liquid film, as shown in Fig. 10. The liquid film gradually turns the film into wet friction with the roller. The surface friction changes, and its value is proportional to the bearing capacity of the liquid film, which affects its traction characteristics.

The characteristics of the production solution provided by the enterprise were tested and its parameters were measured. The test objects include the unused pure solution

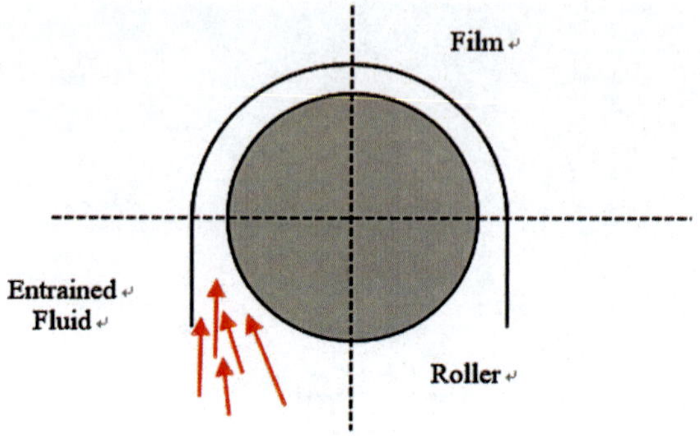

Fig. 10. Entrainment of solution by roller film gap

and the mixed solution extracted from working tank 1 to working tank 6 (six working tanks as shown in Fig. 3) after production. The test is carried out at standard atmospheric pressure, room temperature 16 °C, test the data for 10 times, and take the average value. Parameters are shown in Table 1:

Table 1. Parameters of solutions in the production line

Solution	Pure	Tank1	Tank2	Tank3	Tank4	Tank5	Tank6
Stiffness g/ml	1.2717	1.2821	1.3209	1.3292	1.3326	1.3346	1.3337
Viscosity $mPa \cdot s$	1.108	1.124	1.172	1.134	1.292	1.316	1.226
Internal friction coefficient mm^2/s	0.8713	0.8767	0.8873	0.8531	0.9695	0.9861	0.9192

4.1 Mathematical Model of Roller Film System

Comparing the solution parameters with water, under the same conditions, the density of water is 0.9989 g/ml, the viscosity is 1.111 mPa · s, and the internal friction coefficient of water is about 1.112 mm^2/s. The comparison shows that the viscosity of the solution is similar to that of water, and the internal friction coefficient is lower than that of water. It can be predicted that when the solution is drawn into the gap between the roller and the film to form a liquid film, its friction is also low.

Assuming that the solution is incompressible, laminar flow, ignoring the weight of the liquid layer and the inertia force during movement, and considering the liquid film when the roller rotates as an invisible variable rigid body, the Reynolds equation of the liquid film pressure distribution shown in Fig. 3, 4, 5, 6, 7, 8, 9, 10 and 11 is:

$$\frac{\partial}{\partial X}\left(\frac{h^3}{\mu}\frac{\partial \rho}{\partial X}\right)+\frac{\partial}{\partial Z}\left(\frac{h^3}{\mu}\frac{\partial \rho}{\partial Z}\right)=6\frac{\partial}{\partial X'}(Uh)+12\frac{\partial h}{\partial t} \tag{1}$$

where h is the liquid film thickness, U is the linear velocity of the outer surface of the roller, ω Is the angular speed at which the roller rotates around its center, R_b is the radius

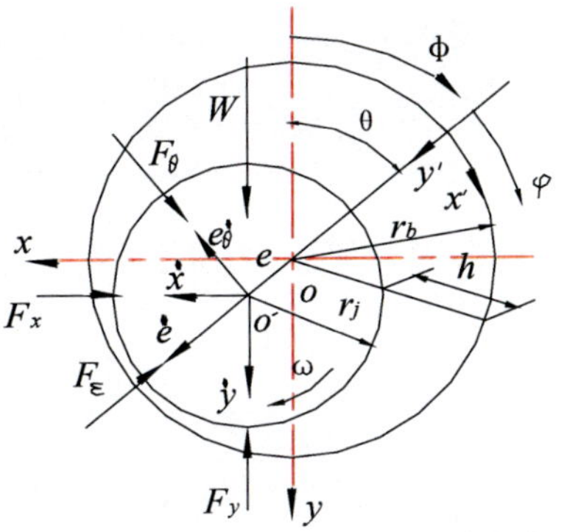

Fig. 11. Geometric relationship between roller and film

of roller with film, R_i is the roller radius, $c = R_b - R_i$ is the gap dimension between the film and the roller, e is its eccentricity, ε is its eccentricity. The following dimensionless variables are introduced:

$$\varphi = \frac{x^{'}}{R_b}, \overline{z} = \frac{2z}{L}, \psi = \frac{c}{R_b}, \overline{h} = 1 + \varepsilon \cos\varphi,$$

$$\overline{p}^{'} = \frac{\psi^2}{6\mu\omega}p, \lambda = \frac{L}{2R_b}, q = \frac{2}{\omega^*}\frac{d\varepsilon}{dt}, \overline{p} = \frac{\psi^2}{6\mu\omega^*}p$$

Then, the Reynolds equation can be reduced to a general dimensionless form:

$$\frac{\partial}{\partial\varphi}\left(\overline{h}^3\frac{\partial p}{\partial\varphi}\right) + \frac{1}{\lambda^2}\frac{\partial}{\partial Z}\left(\overline{h}^3\frac{\partial\overline{p}}{\partial\overline{z}}\right) = -\varepsilon\sin\varphi + q\cos\varphi \tag{2}$$

Then the Reynolds equation can be reduced to a general dimensionless form:

$$\overline{p}|_{\overline{z}=\mp1} = 0, \overline{p}(\varphi, \overline{z}) = \overline{p}(\varphi + 2\pi, \overline{z}) \tag{3}$$

According to the variational principle, an equivalent variational problem can be derived:

$$\left\{J(\overline{p}) = \int\int_\Omega\left\{\frac{\overline{h}^3}{2}\left[\frac{\partial\overline{p}}{\partial\varphi}\right]^2 + \frac{1}{\lambda^2}\left[\frac{\partial\overline{p}}{\partial\varphi}\right]^2 - (\varepsilon\ \sin\ \varphi - q\ \cos\ \varphi)\overline{p}\right\}d\Omega = \min\right. \tag{4}$$

By dividing the liquid film into triangular elements and performing linear interpolation, the variational equation can be obtained:

$$J(\overline{p}) = \sum_1^{2n} J^\varepsilon(\overline{p}) = \sum_1^{2n}\iint_\Omega\left\{\frac{\overline{h}^3}{2}\left[\frac{\partial\overline{p}}{\partial\varphi}\right]^2 + \left[\frac{\partial\overline{p}}{\partial z}\right]^2 - (\varepsilon\sin\varphi - q\cos\varphi)\right\}d\Omega \tag{5}$$

The total stiffness matrix and total linear equations can be obtained by using the over relaxation iteration method

$$K\overline{p} = \varepsilon W_{\sin\varphi} \tag{6}$$

where the stiffness matrix and load vector are respectively:

$$K = \sum\nolimits_{\varepsilon=1}^{N} K^{\varepsilon}, K^{\varepsilon} = \iint_{\Omega}\left\{\overline{h}^{3}\left(\left[\frac{\partial N}{\partial\varphi}\right]^{T}\left[\frac{\partial N}{\partial\varphi}\right] + \frac{1}{\lambda^{2}}\left[\frac{\partial N}{\partial\overline{z}}\right]^{T}\left[\frac{\partial N}{\partial\overline{z}}\right]\right)\right\}d\Omega \tag{7}$$

$$W_{\sin\varphi} = \sum\nolimits_{e=1}^{N} W_{\sin\varphi}^{e}, W_{\sin\varphi}^{e} = \iint_{\Omega_e}\left[\frac{\partial N}{\partial\varphi}\right]^{T}\sin\varphi d\Omega_e \tag{8}$$

where N is a shape function, and the pressure distribution $\overline{p}$ of the liquid film can be obtained by substituting the boundary conditions into the above formula-, The dimensionless bearing capacity $\overline{W}$ can be obtained by integrating on the solution domain-Horizontal component $\overline{W}_H$ and vertical component $\overline{W}_V$:

$$\overline{W}_H = \int_{-1}^{1}\int_{0}^{2\pi}\overline{p}\sin\theta d\theta d\overline{z}, \overline{W}_V = \int_{-1}^{1}\int_{0}^{2\pi}\overline{p}\cos\theta d\theta d\overline{z} \tag{9}$$

Then, the dimensionless bearing capacity of liquid film $\overline{w}$ And its action angle ϕ is:

$$\overline{W} = (\overline{W}_H + \overline{W}_V)^{1/2}, \phi = arctan(\overline{W}_H/\overline{W}_V) \tag{10}$$

The liquid film bearing capacity W is

$$W = \frac{6LD_b\mu\omega}{\psi^2}\overline{W} \tag{11}$$

4.2 Simulation Analysis of Roller Characteristics

According to the finite element simulation results in 2, take the maximum deformation of 0.5 mm in the upper and lower rollers as the eccentricity of the roller, the liquid film gap is taken as c = 2 mm, and the rotating speed Ω = 35 r/min. Set 51 * 21 nodes on the half cylindrical surface where the roller contacts the film, and the number of grids is 50 * 20 = 1000. Use MATLAB software for simulation calculation to solve the distribution of liquid film bearing capacity in the contact area between the roller and the film. Where X-axis is the length of the roller, Y-axis is the half cylindrical surface of the roller where the liquid film is located (0–180°), and Z-axis is the oil film bearing capacity, in N (Fig. 12).

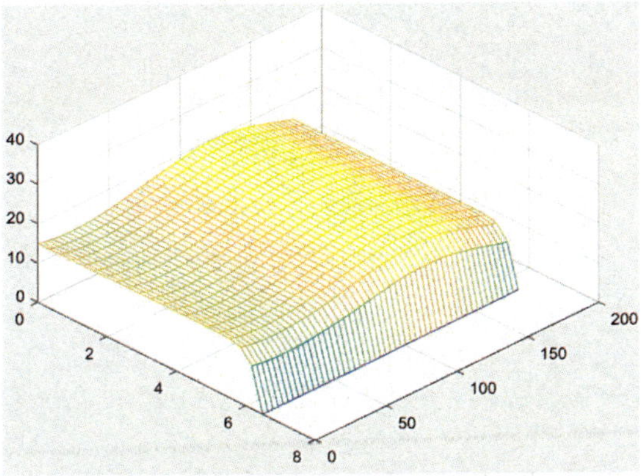

Fig. 12. Distribution of liquid film bearing capacity

The total bearing capacity of the liquid film is calculated. The horizontal component is 53.5619N, and the vertical component is 80.3860N, which is about *1%* of the self weight of the roller. Therefore, it is considered that the bearing capacity of the oil film between the roller and the film has little effect on the deformation of the roller, mainly on its friction.

Now, considering the influence of the vertical deformation of the roller, it gradually increases from 0.5 mm to 3 mm, so that the thickness of the liquid film also increases, and the value is equal to the deformation of the roller. The simulation calculation of its liquid film bearing capacity is carried out, and the results are shown in Figs. 13 and 14.

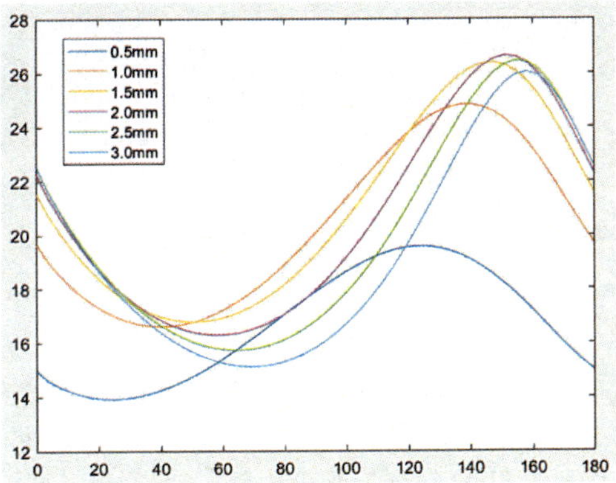

Fig. 13. Circumferential bearing capacity curve of liquid film in the middle section of roller

It is presented that as the vertical deformation of the roller increases, the bearing capacity of the liquid film begins to increase, but it begins to decrease after exceeding a certain degree (deformation of 2.0 mm). In the circumferential direction of the middle section, the distribution of the liquid film force also began to change, and the larger value shifted towards both sides of the roller.

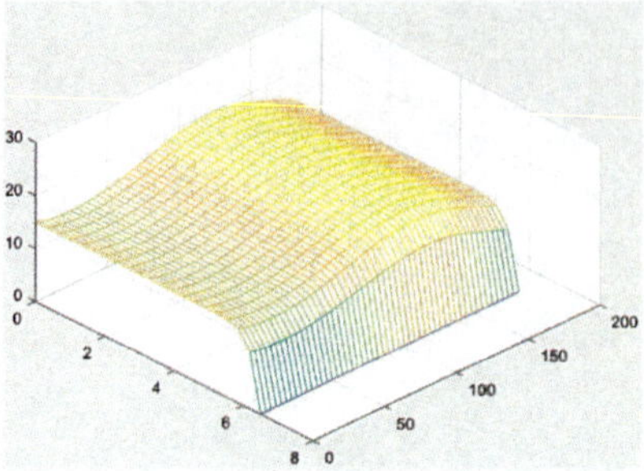

Deformation 0.5mm

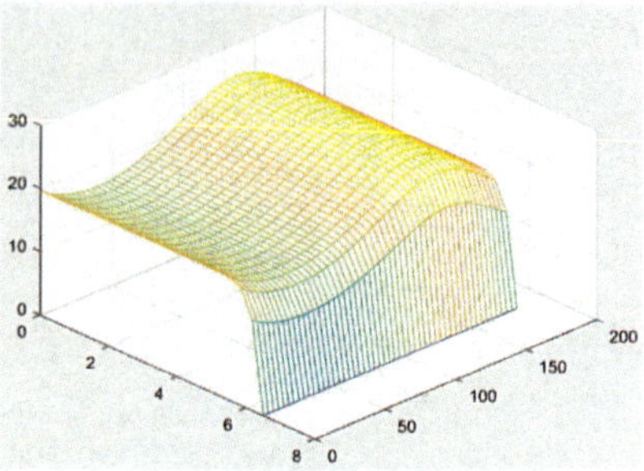

Deformation 1.0mm

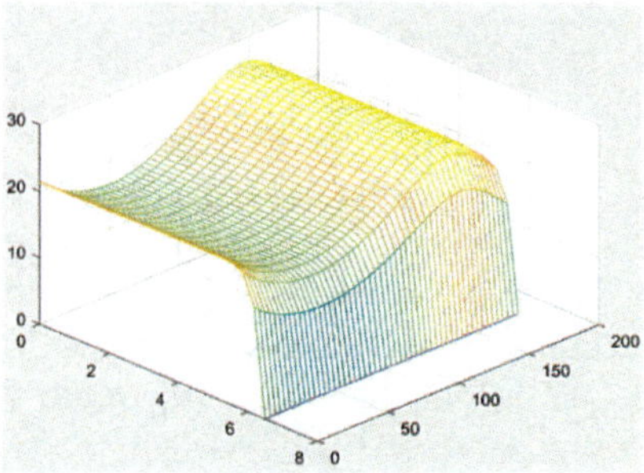

Deformation 1.5mm

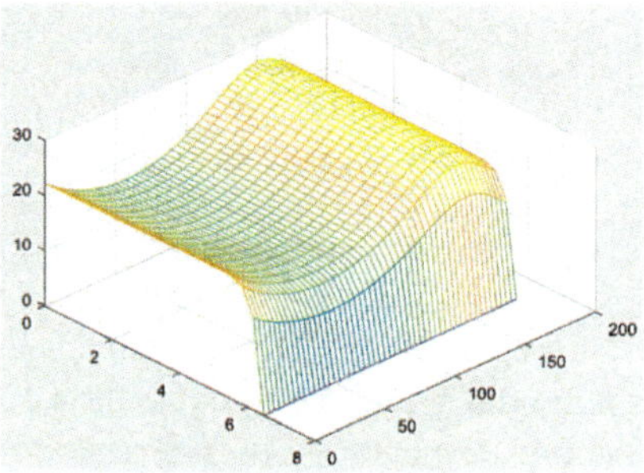

Deformation 2.0mm

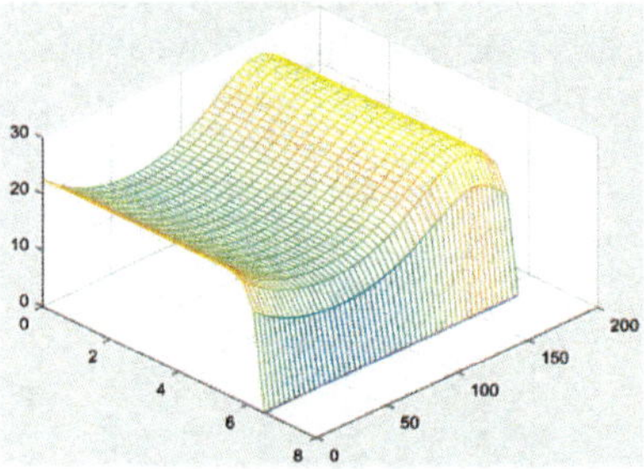

Deformation 2.5mm

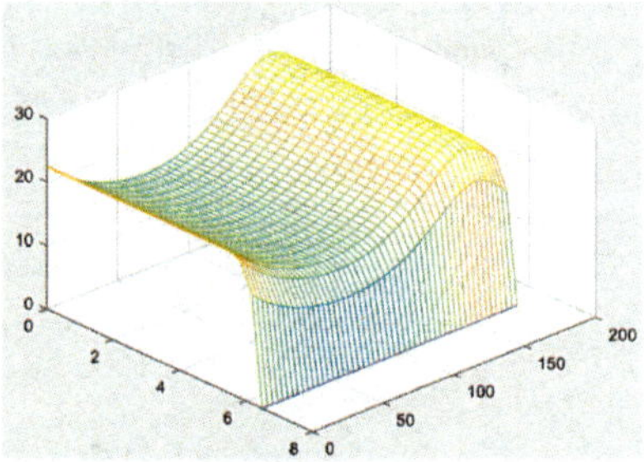

Deformation 3.0mm

Fig. 14. Distribution of liquid film bearing capacity

In order to further study the influence of the liquid film on the traction characteristics of the system, the vertical deformation of the roller is now maintained in the same numerical variation range, and the thickness of the liquid film is increased, but the incremental value is changed to twice the deformation of the roller, and the bearing capacity is simulated and calculated, as shown in Fig. 15 and Fig. 16.

According to the calculation results, when the thickness of the liquid film further increases, the peak value of its bearing capacity shifts while the value decreases significantly, and the maximum value is only 3.56n. It can be considered that the increase of the thickness of the liquid film will significantly reduce the bearing capacity of the liquid film.

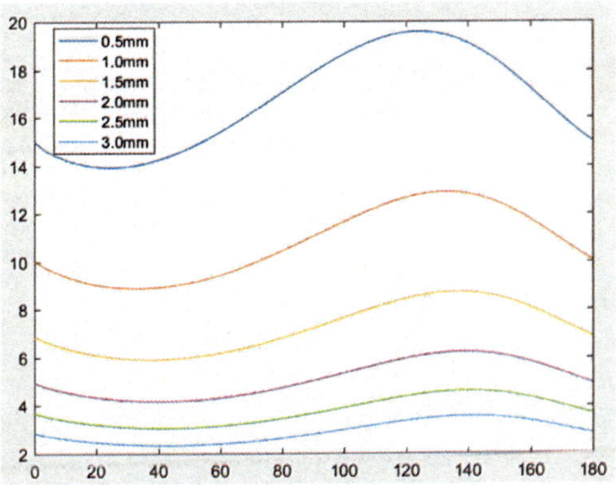

Fig. 15. Circumferential bearing capacity curve of liquid film in the middle section of roller

Since the system size and solution composition cannot be changed, only the roller speed is left for the adjustable parameters. Now increase the rotating speed from the initial 35 r/min to 60 r/min, and calculate the friction force on the contact surface between the film and the roller based on the above two deformations and the bearing capacity distribution under the condition of liquid film thickness, as shown in Fig. 17 and 18. Axis *X* in the figure is the deformation of the roller, in mm; *Y*-axis is the speed, in r/min; The *Z* axis is the friction force, in *N*.

It can be seen from the figure that when the thickness of the liquid film is small, the total friction of the roller changes little even if it is vertically deformed, and the friction increases gradually with the increase of the rotating speed. However, when the thickness of the liquid film becomes larger, the friction decreases rapidly, and when the deformation is 3.0 mm, it is only 10% of the first case, resulting in obvious slippage.

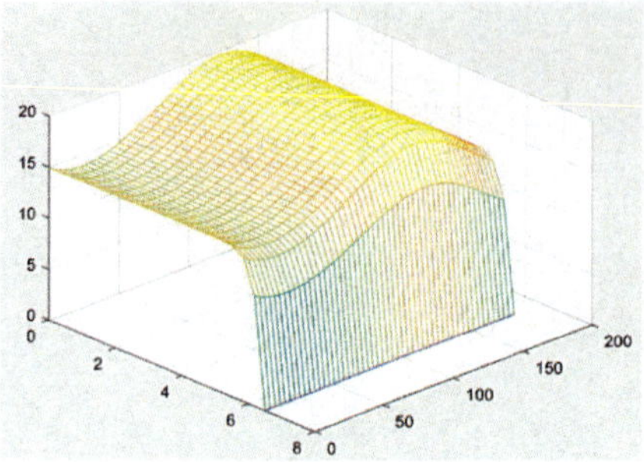

Deformation 0.5mm

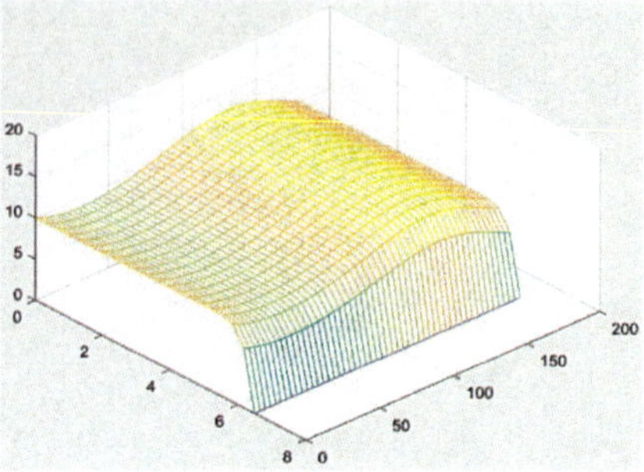

Deformation 1.0mm

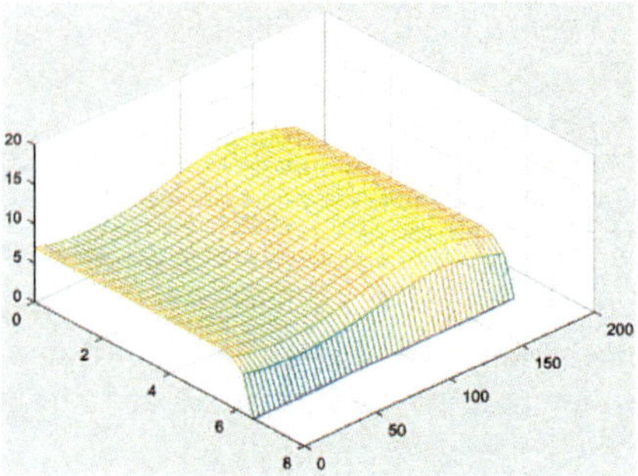

Deformation 1.5mm

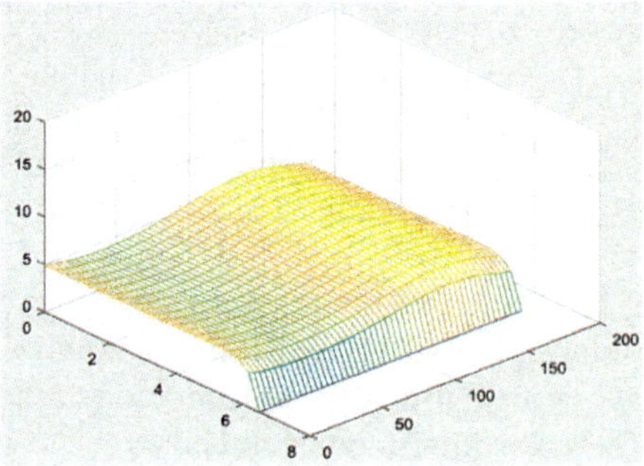

Deformation 2.0mm

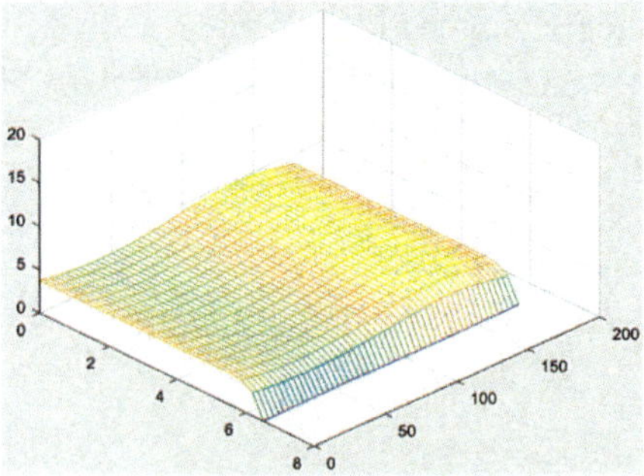

Deformation 2.5mm

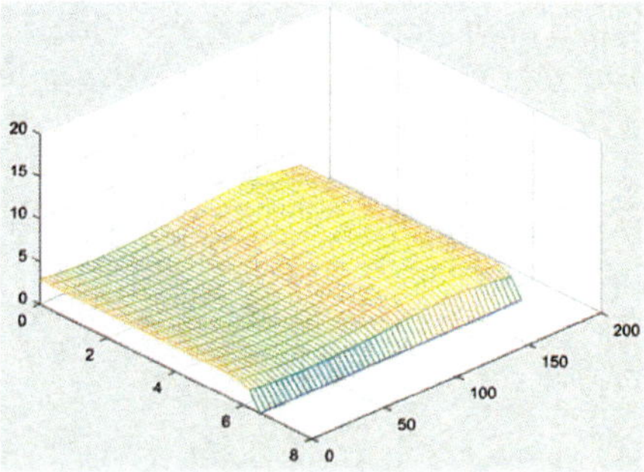

Deformation 3.0mm

Fig. 16. Distribution of liquid film bearing capacity

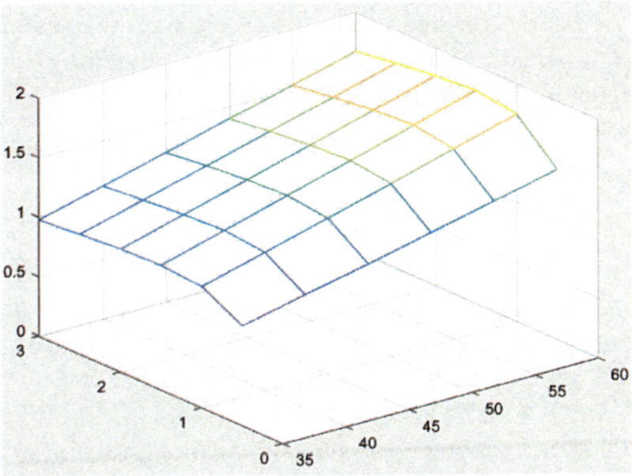

Fig. 17. Liquid film friction (the thickness increment is equal to the deformation of the roller)

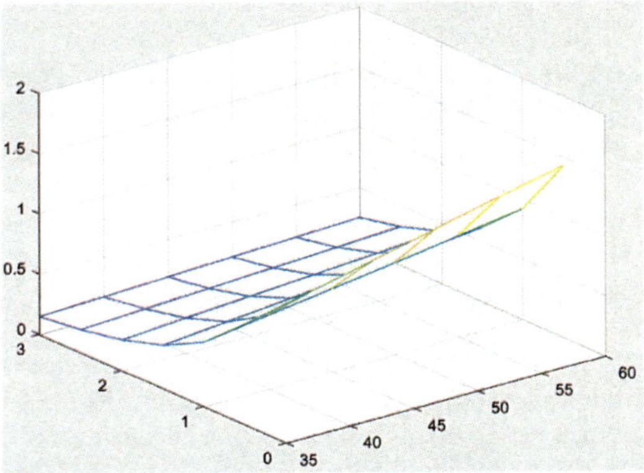

Fig. 18. Liquid film friction (the thickness increment is twice the deformation of the roller)

5 Conclusion

In this paper, the target film conveying system and roller are simulated and analyzed. Combined with the calculation results and considering the actual production situation, the following conclusions are drawn:

After the film conveying system works for a period of time, the solution gradually enters between the film and the roller, and fills the entire gap, forming a layer of liquid film, resulting in the reduction of friction;

The roller has certain vertical deformation when working, and the gap with the film foundation may be further increased during rotation, which will increase the thickness of the liquid film and significantly reduce its bearing capacity, resulting in minimal friction; The liquid film bearing capacity in the contact area between the roller and the film varies according to the position and the thickness of the liquid film in the gap, which makes the friction force in each area of the film different, resulting in a potential slip trend; The shape or support of the roller shall be adjusted to prevent the solution from forming a thick liquid film as far as possible.

Acknowledgment. This paper is supported by the project of outstanding young and middle-aged scientific and technological innovation team of colleges and universities in Hubei Province: "digital twins of intelligent logistics equipment" (T2021043).

References

1. Omri, M.N., Ben Aonne, H.: Towards an intelligent approach to workflow integration in a quality management system. Int. J. Intell. Syst. Appl. (IJISA) **14**(3), 54–73 (2022)
2. Tikito, I., Souissi, N.: Meta-analysis of systematic literature review methods. Int. J. Intell. Syst. Appl. (IJISA) **4**(8), 46–52 (2012)
3. Cao, M., Wang, G., Sun, L.: Research on key technology of new concept tyre building production line. Int. J. Eng. Manuf. (IJEM) **10**(6), 11–19 (2020)
4. Jeenkour, P.: A numerical study on air-entrainment between a web and a convex guide roller in a web transportation process. Appl. Mech. Mater. **39**(2), 110–115 (2013)
5. Ma, L.E., Wu, Z.Y., Wu, J.M., et al.: Air entrainment between moving membrane and the guide roller. ICSV2014 (6), 4597–4603 (2014)
6. Ma, L.E., Shao, M.Y., Wu, J.M., et al.: Research on traction characteristics of the guide roller in the web transfer system. J. Xi'an Univ. Technol. **32**(4), 462–467 (2016)
7. Chen, S.S., Chen, W.H., et al.: Development of an aerostatic bearing system for roll-to-roll printed electronics. J. Micromech. Microeng. **28**(6), 065002 (2018)
8. Janssens, S.D., et al.: Boundary curvature effect on the wrinkling of thin suspended films. Appl. Phys. Lett. **116**(19), 193702 (2020)
9. Jiang, S., Wang, Y.: Modeling of static stiffness for linear motion roller guide. J. Tribol. **141**(11), 1 (2019)
10. Ma, L., Mei, X., Li, Y., et al.: Analysis of the mechanical characteristics of the guide roller of the film transmission system. J. Xi'an Jiaotong Univ. **48**(11), 6 (2014). (in Chinese)
11. Ma, L., Shao, M., Wu, J., et al.: Study on the traction characteristics of guide rollers in film transmission system. J. Xi'an Univ. Technol. **32**(4), 6 (2016). (in Chinese)
12. Shao, M.: Study on the traction characteristics and follow-up of guide rollers in film transmission system. Xi'an University of Technology, Xi'an (2016). (in Chinese)
13. Kong, F.: Study on the flow field characteristics of the gas entrained between the moving film and the guide roller. Xi'an University of Technology, Xi'an (2016). (in Chinese)

Developing MFCC-CNN Based Voice Recognition System with Data Augmentation and Overfitting Solving Techniques

Ali Aswad[1], Essa Alghannam[1,2,3(✉)], and Qingying Zhang[3,4]

[1] Department of Mechatronics Engineering, Tishreen University, Lattakia, Syria
essaalghannam@tishreen.edu.sy
[2] School of Engineering, Manara University, Lattakia, Syria
[3] Wuhan University of Technology, Wuhan, China
[4] Royal Institute of Technology, Stockholm, Sweden

Abstract. The ever-increasing need for intelligent systems technology capable of enhancing human-machine interaction in different areas of life has become a very important topic. However, since speech is the main used method of communication, deep Learning-based speech recognition systems still have poor performance when trained on small datasets. To solve this problem, an automated speech recognition system based on a convolutional neural network and MFCC algorithm, is proposed in this paper. Firstly, the data augmentation technique is applied to the dataset in order to double it. Then, MFCCs are extracted from all the audio files. After that, CNN model is trained on the MFCCs with the help of early stopping and dropout techniques to prevent the overfitting problem. Finally, the best model with the highest performance is selected. Simulation results on Kaggle TensorFlow Speech Recognition Challenge dataset, which contains 20 basic command words, showed that the selected model achieved 88.35% accuracy.

Keywords: Voice recognition · MFCC · Convolutional neural networks

1 Introduction

Speech is simply a series of pressure waves that are produced by the vibration of the vocal cords, which in turn causes the air around them to vibrate. A transducer, such as a microphone, records these sound waves and converted them into an electrical signal, which is processed then using advanced signal processing techniques so that computers can learn to understand speech from prior experience over time.

Automatic speech recognition ASR systems recognize human spoken words and convert them into a machine-readable format such as text or command. This technology allows users to control their digital devices using their voice instead of using typical input tools, and help people who have difficulty dealing with a normal interface like keyboard, mouse, touchpad, etc. The main applications of these systems are: call centers, smart homes, dictation systems, assistive applications, mobile and embedded devices.

Z. Hu et al. (Eds.): AIMEE 2022, LNDECT 159, pp. 115–125, 2023.
https://doi.org/10.1007/978-3-031-24468-1_11

An enormous amount of research has been done on the use of machine learning for speed processing applications during the past decades which has achieved much better results when compared to other used techniques [1, 2]. Deep learning algorithms are used by modern speech recognition systems [3, 4], where better results are achieved by the use of a convolution neural network [5].

1.1 Before Deep Neural Networks DNNs

Decades ago, deep learning emerged as a new sub-field of machine learning, and since that time, it has been tested and used in a range of different research topics. Its algorithms have been used to enhance the capabilities of computers so that they understand what humans can do, including speech recognition which has received great attention over the past five decades [6, 7].

First attempts to develop interactive acoustic machines started in the 1950s; they were based on the direct conversion of an audio signal into a series of certain sound-like units [8]. The first speech recognition systems were built and tried at Bell Laboratories, where a single-speaker number recognition system was developed [9].

The conventional speech recognition systems were representing the speech signals using Gaussian Mixture Models (GMMs) and hidden Markov models (HMMs), which are considered statistically inefficient when used to model non-linear or near non-linear functions [10].

1.2 After Deep Neural Networks DNNs

IBM was the first to implement HMMs in speech recognition in the 1980s and since then HMM-GMM became the dominant phonemic model until 2006 when the use of deep learning in speech recognition was proposed, which later transformed from basic artificial neural networks to DNNs. DNN-HMM became the main phonemic model which showed strong recognition ability even with a large vocabulary, and with the rapid development of speech technology, more neural networks are used in the field of speech recognition, such as CNN and RNN.

Several examples have proven that using DNN leads to better results than classical models as in 2012 when Microsoft released its Deep Learning-based speech system, MAVIS, which clearly showed that the word error rate was reduced by 30% compared to Gaussian mixtures-based models [11].

A comparison was made between convolutional neural network and basic neural network to determine the best speech recognition model with the best performance [12]. It was found that the convolutional neural network performed better in speech recognition compared to other neural networks when the audio data was converted into an image and fed into a CNN.

Recurrent neural networks RNNs have been trained with the use of Hessian free optimization in [13], and they have proven their ability to create sequential text characters.

A review in the field of speech recognition has been conducted in [14], which concluded that even though deep processing architectures can provide improvements in this area, the choice of features and the architecture with which they are combined, including layer width, can also be important factors.

A summary of the work that has been done by Microsoft in the area of speech using deep learning was conducted in [15]. The paper showed the different capabilities and limitations of Deep Learning-based speech recognition systems. The experimental results have shown that the spectrogram features of speech are more advanced to Mel Frequency Cepstral Coefficients MFCC with deep neural networks compared to the traditional models using GMMs-HMMs.

In this paper, a voice recognition system based on MFCC for acoustic features extraction and CNN for patterns extraction from MFCC images was proposed with the use of a data augmentation technique for increasing the size of the dataset, and early stopping in addition to dropout techniques for solving the overfitting problem. As the proposed system was trained on a set of command words, such as: stop, off, right, ..., and numbers 0–9, it can be helpful in many fields such as industrial, domestic, and medical fields where humans can control their assistant robots using their voice directly.

2 Dataset, Pre-processing, and Data Augmentation

The dataset used in this paper is the Speech Commands Dataset released by Google on August 3, 2017. It contains 47349 one-second audio clips of 20 short words (yes, no, down, up, right, left, off, on, go, stop, zero, one, two, three, four, five, six, seven, eight, nine). The audio files were crowdsourced by Google to collect single-word commands (rather than words as said and used in conversation). This group of 20 core words audio files was recorded and repeated 5 times by most speakers.

The dataset was divided into 80% training data and 20% validation data after choosing 800 files to be used in model testing. The training data is used to train the network, where the loss function values are computed via feed forward and learnable parameters are updated via backpropagation. The validation data is used to evaluate the model during the training process, fine-tune the hyperparameters, and perform model selection. The test set is used only once when training is completed to evaluate the performance of the final model.

Since the duration of some recordings in the dataset is less than 1 s and the sampling rate is very high, pre-processing steps have been used. Therefore, after reading the audio files, the following pre-processing steps have been done to deal with these two problems:

1) Padding short clips: short clips are not compatible with the system so a function that adds zero values to those files has been used to make the duration of all files 1 s.
2) Resampling: The sampling rate of the files in the dataset was 16000 Hz so they have been resampled at 8000 Hz since most speech-related frequencies are at 8000 Hz.

Data augmentation was used in this paper by shifting the pitch of each audio file in the dataset, which led to double the size of this dataset. Pitch shifting is a method for making audio thinner or thicker. As the source of the audio moves closer or away from the receiver, the successive signals reach the receiver at either smaller or larger intervals of time, respectively. This time difference causes a change in the frequency of the audio being heard by the receiver. We can perform a similar pitch-shifting exercise by splitting an audio signal into smaller frames, time-stretching these audio frames, and then resampling the frames to generate an approximation of the original audio sequence. The algorithm is summarized as: "Stretch in time, then resample". Figure 1 shows an example of pitch shifting by 2 octaves.

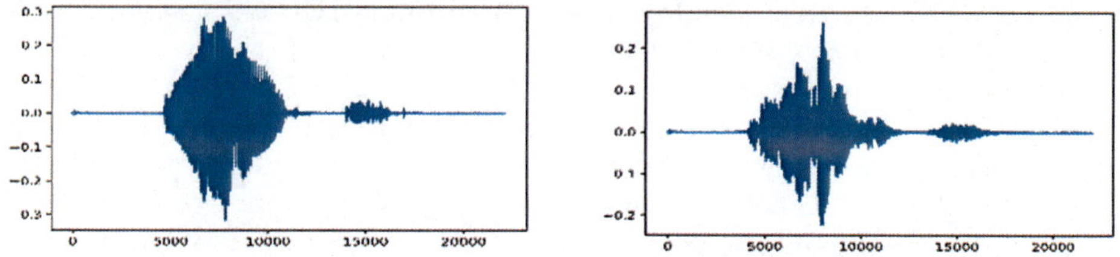

Fig. 1. Pitch shifting by 2 octaves

The next step is to convert audio files into images through the MFCC algorithm that gives 13 coefficients for each frame of an audio file.

3 Mel Frequency Cepstral Coefficients

Most automatic speech recognition systems extract features from the acoustic signal instead of using the full speech signal due to the large variations in the speech signal. Before the introduction of the MFCC algorithm, Linear Prediction Coefficients (LPC) and Linear Prediction Cepstral Coefficients (LPCCs) [17] were the most used methods at that time and were used together with Hidden Markov models [18] in speech recognition systems.

Recently, MFCCs is the most popular and used feature extraction algorithm in the field of automatic speech recognition. The block diagram of the MFCC algorithm is shown in Fig. 2.

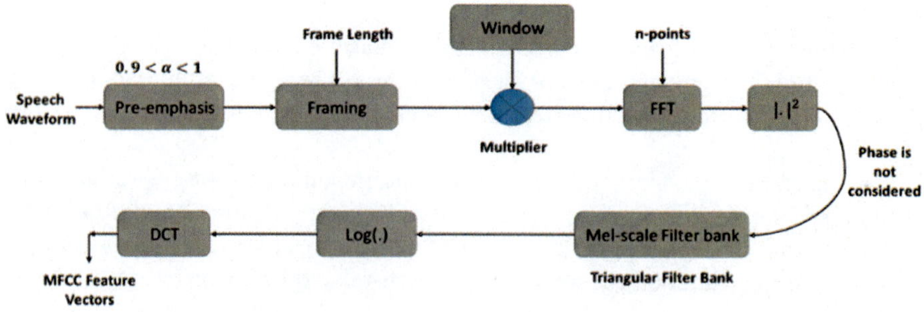

Fig. 2. Block diagram of MFCC algorithm

The pre-emphasis purpose is to amplify the power and increase the magnitude within the spectrum of the higher frequencies in the input speech signal because the high frequencies tend to have a small magnitude compared to the lower ones. The pre-emphasis equation is:

$$Y[n] = X[n] - \alpha X[n-1] \tag{1}$$

Y is the output speech signal, ***X*** is the input signal and α is the coefficient which is typically between 0.95 – 0.97.

Framing divides the acoustic signal into sub-frames (20–40 ms), which are then multiplied by hamming function. The Hamming function W is:

$$W(n) = \begin{Bmatrix} 0.54 - 0.46\cos(\frac{2\pi n}{N-1}) & 0 \le n \le N-1 \\ 0 & otherwise \end{Bmatrix} \tag{2}$$

N is the frame length. After that, the Fast Fourier transform is applied to calculate the frequency spectrum on each frame with k = 512:

$$X[k] = \sum_{n=0}^{N-1} x[n]e^{-j2\frac{\pi}{N}kn} \tag{3}$$

As phase information is not helpful in speech recognition, the power of the spectrum is calculated by square *X[k].*

The next step is applying triangular band-pass filters to the signal in mel-scale, where there is one filter for each mel-frequency component. The output of each filter represents the amount of spectral energy present in the frequency components covered by the filter:

$$Y_t[m] = \sum_{k=1}^{N} H_m[k]|X_t[k]|^2 \tag{4}$$

where, $H_m[k]$ are the filter functions. Since the human response to signal level is logarithmic, because humans are less sensitive to slight differences in amplitude at high amplitudes than at low amplitudes, we use logarithm which compresses the dynamic range of values. Finally, Discrete Cosine Transform (DCT) is applied to the output of the mel-scale filter banks:

$$y_t[k] = \sum_{m=1}^{M} \log(|Y_t(m)|)\cos(k(m-0.5)\frac{\pi}{M}) \ , \ k = 0, ..., J \tag{5}$$

The result of the conversion is called Mel Frequency Cepstrum Coefficient Fig. 3. The set of the coefficients is called acoustic vectors. Therefore, each input utterance is transformed into a sequence of acoustic vector.

4 Convolution Neural Network

The main application of CNN is image processing but it has been used in many other fields like voice and video processing. It was proposed in 1998 [19]. Figure 4 shows the architecture of CNN, which consists of the following layers: a convolution layer whose

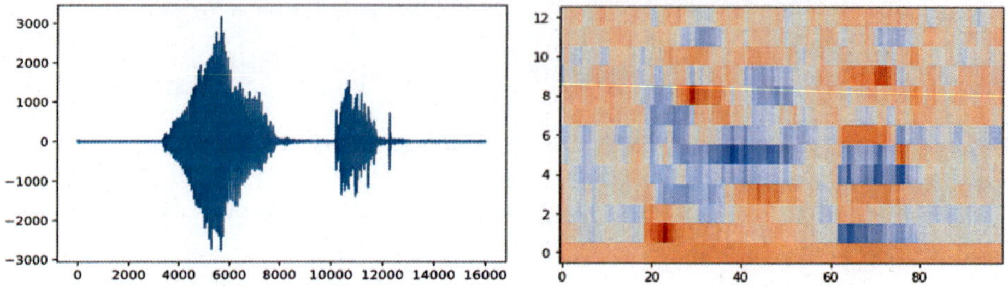

Fig. 3. Speech signal (left), and the corresponding MFCC (right)

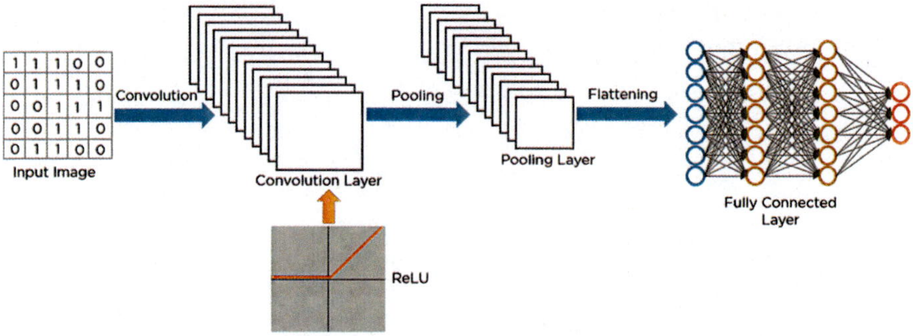

Fig. 4. The architecture of CNN

purpose is to extract features from input data, a pooling layer that is used to reduce the dimension of the data, fully connected layer that is similar to the multi-layer perceptron.

In this study, we have designed a CNN network in order to classify speech signals based on its MFCC features. Since the problem to be solved in this paper is kind of a multi-classification problem so cross-entropy loss function shown in Eq. 6 has been used:

$$L = -\sum_{c=1}^{M} y_{o,c} \log(p_{o,c}) \tag{6}$$

M is the number of classes. $y_{o,c}$ is a binary indicator; it takes "1" when class label c is a correct classification for observation o, and "0" in case of incorrect classification. $p_{o,c}$ is predicted probability for recognizing an observation o as a class c.

The backpropagation algorithm has been used for training process whilst for optimization process, gradient descent algorithm has been used and the learnable parameter (w) has been updated as follows:

$$w := w - \alpha * \frac{\partial L}{\partial w} \tag{7}$$

α is learning rate, $\frac{\partial L}{\partial w}$ is the gradient of the loss function with respect to the learnable parameter (w). Table 1 shows the architecture of this model.

Table 1. Convolutional neural network architecture

Layer	Input shape	Output shape	Weights	Activation function
Input layer	49 × 13 × 1	49 × 13 × 1	0	None
Convolutional layer	49 × 13 × 1	49 × 13 × 16	800	ReLU
Dropout layer	49 × 13 × 16	49 × 13 × 16	0	None
Convolutional layer	49 × 13 × 16	49 × 13 × 32	12832	ReLU
Dropout layer	*49 × 13 × 32*	*49 × 13 × 32*	*0*	*None*
Convolutional layer	49 × 13 × 32	49 × 13 × 64	18496	ReLU
Maxpooling layer	49 × 13 × 64	24 × 6 × 64	0	None
Dropout layer	49 × 13 × 64	24 × 6 × 64	0	None
Flatten layer	49 × 13 × 64	9216 × 1	0	None
Fully connected layer	9216 × 1	256 × 1	2359552	ReLU
Dropout layer	256 × 1	256 × 1	0	None
Fully connected layer	256 × 1	128 × 1	32896	ReLU
Dropout layer	128 × 1	128 × 1	0	None
Output layer	128 × 1	20 × 1	2580	Softmax

5 Experiments and Results

The purpose of this study is to build an effective voice recognition system which is able to identify a group of audio commands, and to minimize as well the number of layers in the proposed convolutional neural network architecture. This will definitely make the proposed deep learning model simpler and also save time. It is worth mentioning that all experiments in this paper were conducted on hardware: core i5 7th Gen with 8 GB RAM.

5.1 Problems and Solutions

Overfitting Problem: one of the biggest challenges when training deep learning models is choosing training time or a number of epochs. When choosing a large number of iterations, the overfitting problem appears, this is because the model is built too complex with too many parameters. The overfitted model is not generalizable to never-seen-before data. However, if the training time is short, the underfitting problem appears, which means that the model has not trained enough on the training data, so the performance on the training data and test data is low and the error rate is high.

To solve this problem, the following techniques have been used:

(1) Early stopping technique, which stops training the model before Overfitting occurs. The model tries to chase the loss function crazily on the training data, by tuning the parameters. It keeps another set of data as the validation set and as it goes on

training, it keeps a record of the loss function on the validation data, and when it sees that there is no improvement on the validation set, it stops, rather than going all the epochs.

It is built on the fact that the weight updates in gradient descent become significantly smaller as the model approaches minima. The training is stopped when the update becomes as small as 0.001, as stopping at this point minimizes loss and saves computing power by preventing any unnecessary epochs.

(2) Regularization through dropout, which is a recently introduced regularization technique in which randomly selected activations are given a value of 0 during training, so that the model becomes less sensitive to specific weights in the network. This in turn results a network that have more ability to be generalized. The dropout value used in this paper is 0.3.

Small Dataset Size Problem: Due to the small size of the available dataset, a data augmentation technique has been used, which increases the amount of data. To obtain more data, even though we have no more data, slight transformations or adjustments are made to the current data set. The neural network will think that the new data is distinct from the existing data. Having a large data set is critical to the good performance of the network. Modern neural networks usually have millions of parameters. Of course, if we have a lot of parameters, we need to train the machine learning model on a proportional number of examples to get good performance. Data augmentation is also effective in reducing overfitting.

5.2 Model Evaluation

Since we are dealing with a multi-classification problem, the categorical cross-entropy function is used as a loss function. The training was done with batch size 32. Three metrics were monitored to review model performance: Accuracy, Loss, and Confusion Matrix because tracking the accuracy only gives an incomplete picture of the model's performance.

Model accuracy is shown in Fig. 5, the accuracy was low at the beginning of training because the network is initialized with random weights, as the training process and iterations progress the network begins to learn the features in the audio clips so the recognition accuracy becomes greater and greater. It also appears from the figure that the accuracy of verification data recognition is greater than the accuracy of training data recognition except at the fortieth iteration. The training process stops at epochs 48 due to the early stopping technique. The highest accuracy value the best model has reached was 82.33% on the training set and 88.35% on the validation set.

Model loss is shown in Fig. 6, which monitors the values of the categorical cross-entropy function during the training process. The value of the loss function given by the best model was 0.5731 on the training set and 0.5572 on the validation set.

Most of the time, accuracy increases as loss decreases, but this is not always the case.

Accuracy and loss have different definitions and measure different things. They often appear inversely proportional but there is no mathematical relationship between them.

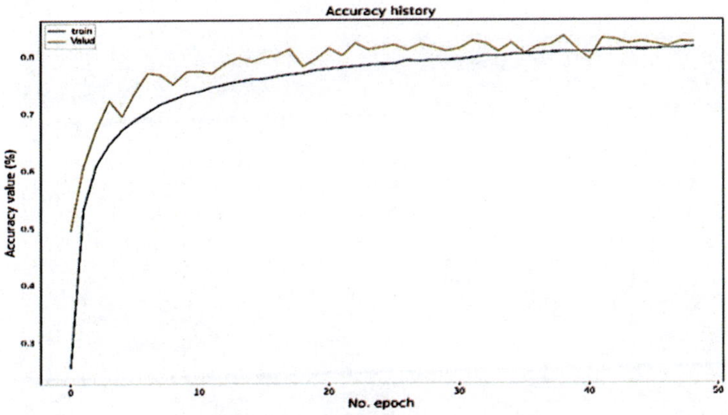

Fig. 5. Model accuracy vs epoch

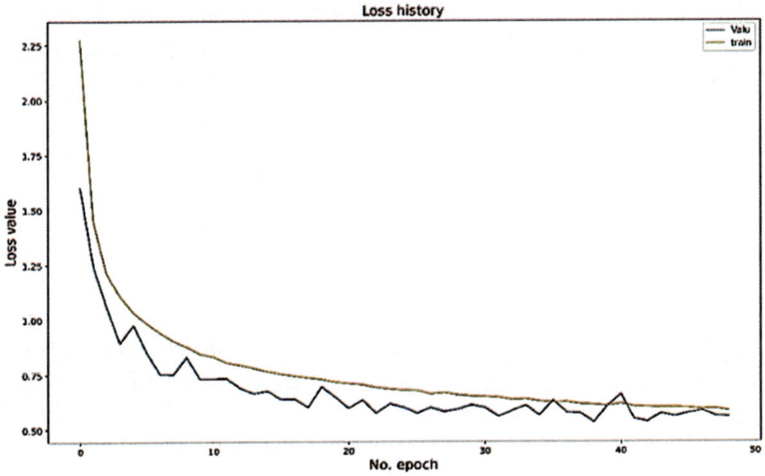

Fig. 6. Model loss vs epoch

The resulting confusion matrix is shown in Fig. 7, which is a well-known matrix in the field of machine learning. It has been used to test the performance of a classifier and measure the effectiveness of a model by displaying the percentage of correct and incorrect classification. It allows us to see the results of using the classifier and simply show the analytical results in a powerful visual model.

The last line in the matrix, for example, represents the results of classifying 945 Zero words. The model recognizes 820 words as zero, 9 as yes, 11 as up, 57 as two, 2 as stop, and 8 as six, 2 as seven, 4 as no, 1 as nine, 23 as go, 1 as four, 3 as eight, and 4 as down. That means 86.77% accuracy.

It has been seen that the CNN model has performed well on a used dataset. The model has achieved 88.35% accuracy. An early stopping technique has been implemented during training the model. The training has stopped the execution when the model tried

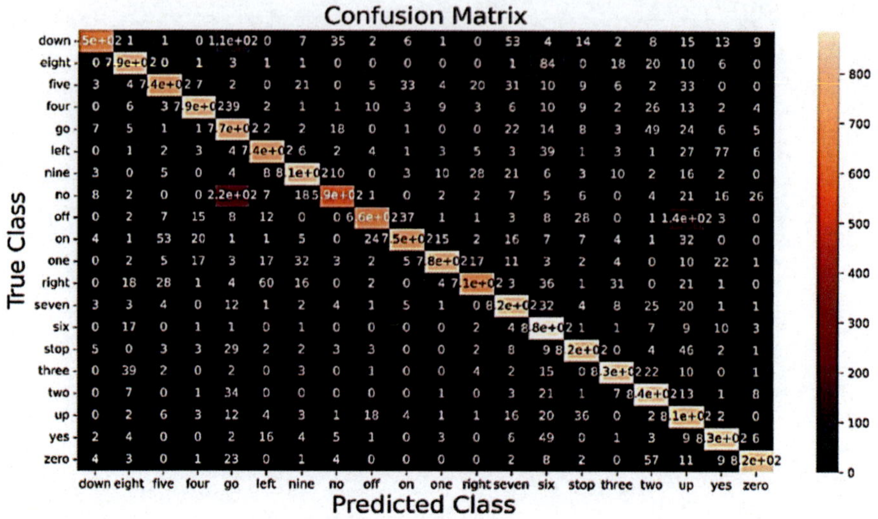

Fig. 7. Confusion matrix

to overfit. The CNN model has been executed through 47 whole epochs before training stopping at the end of the 48th iteration.

6 Conclusion

This research paper has proposed an automatic speech recognition system based on two algorithms for features extraction, firstly the unique features of the speech acoustic signal has been extracted using MFCCs, then these features were fed to the CNN algorithm for further features learning and classification.

The study showed that using techniques such as early stopping, dropout, and data augmentation have helped to solve the overfitting problem and make the model more able to be generalized which in turn enhanced the model performance. Moreover, this paper showed that the use of MFCC following by CNN model as a double-phase feature extraction has improved the accuracy, reduced the complexity of the model and saved time. The future work will be on increasing the model recognition capability in noisy environments.

References

1. Anusuya, M.A., Katti, S.K.: Speech recognition by machine, a review. Int. J. Comput. Sci. Inf. Secur. IJCSIS **6**(3), 181–205 (2010)
2. Nassif, A.B., Shahin, I., Attili, I., Azzeh, M., Shaalan, K.: Speech recognition using deep neural networks: a systematic review. IEEE Access, 1 (2019)
3. Mohamed, A.R., Dahl, G.E., Hinton, G.E.: Acoustic modeling using deep belief networks. IEEE Trans. Audio Speech Lang. Process. **20**, 14–22 (2012)

4. Yu, D., Seide, F., Li, G.: Conversational speech transcription using context-dependent deep neural network. In: Proceedings of the 12th Annual Conference of the International; Speech Communication Association, pp. 27–31 (2011)
5. Sainath, T.N., Mohamed, A.R., Kingsbury, B., Ramabhadran, B.: Deep convolutional neural networks for LVCSR. In: Proceedings of the 2013 IEEE International Conference on Acoustics, Speech and Signal Processing, Vancouver B, pp. 8614–8618 (2013)
6. Singh, H., Bathla, A.K.: A survey on speech recognition. Int. J. Adv. Res. Comput. Eng. Technol. **2**(6), 2186–2189 (2013)
7. Anusuya, M.A., Katti, S.K.: Speech recognition by machine: a review. Int. J. Comput. Sci. Inf. Secur. **6**(3), 181–205 (2009)
8. Davis, K.H., Biddulph, R., Balashek, S.: Automatic recognition of spoken digits. Acoust. Soc. Am. **24**(6), 627–642 (1952)
9. Klevans, R., Rodman, R.: Voice Recognition. Artech House, Boston (1997)
10. Singh, H., Bathla, A.K.: A survey on speech recognition. Int. J. Adv. Res. Comput. Eng. Technol **2**(6), 2186–2189 (2013)
11. Xie, Y., Le, L., Zhou, Y., Raghavan, V.V.: Deep learning for natural language processing. In: Handbook of Statistics. Elsevier, Amsterdam (2018)
12. Bansal, M., Thivakaran, T.K.: Analysis of speech recognition using convolutional neural network **11**(1) (2020)
13. Martens, J., Sutskever, I.: Learning recurrent neural networks with Hessian-free optimization. In: ICML, pp. 1033–1040 (2011)
14. Morgan, N.: Deep and wide: multiple layers in automatic speech recognition. IEEE Trans. Audio Speech Lang. Process **20**(1), 7–13 (2012)
15. Deng, L., et al.: Recent advances in deep learning for speech research at Microsoft. IEEE Acoust Speech Signal Process., 8604–8608 (2013)
16. Davis, S., Mermelstein, P.: Comparison of parametric representations for monosyllabic word recognition in continuously spoken sentences. IEEE Trans. Acoust. Speech Signal Process. **28**(4), 357–366 (1980)
17. Gupta, H., Gupta, D.: LPC and LPCC method of feature extraction in Speech Recognition System. In: Cloud System and Big Data Engineering (Confluence). IEEE (2016)
18. Mashao, D.J., Gotoh, Y., Silverman, H.F.: Analysis of LPC/DFT features for an HMM-based alphadigit recognizer. IEEE Signal Process. Lett. **3**(4), 103–106 (1996)
19. Lecun, Y., et al.: Gradient-based learning applied to document recognition. IEEE **86**(11), 2278–2324 (1998)

A Review of Research on Parking Planning and Site Selection

Yuan Zheng(✉), Qi Zeng, Chen Zhang, Bo Jiang, Hui Zhang, and Rui Xiao

Central Southern China Electric Power Design Institute Co., Ltd., of China Power Engineering Consulting Group, Wuhan 430063, China
zy5698@csepdi.com

Abstract. With the vigorous development of economy and the acceleration of urbanization, the number of cars in the city continues to grow exponentially, and the parking problem is becoming increasingly serious. This paper focuses on urban parking planning and location, analyzes the research status at home and abroad, and summarizes the relevant literature such as parking demand forecasting and parking location methods. Literature analysis shows that parking demand forecasting is rarely able to accurately predict the supply of parking spaces, and only consider the scale of parking supply, ignoring the constraints of parking supply on parking demand. For parking location, it is necessary to further study the parking planning and location problems based on the direction of national policies and the increasingly mature wisdom means, comprehensively considering both sides of parking supply and demand, so as to solve the problems of unreasonable parking planning and low utilization rate of parking space, and provide reference for the development of urban smart parking.

Keywords: Public car park · ArcGIS · Parking demand forecast · Site selection

1 Instruction

With the vigorous development of the economy and the continuous acceleration of urbanization, traffic contradictions have become increasingly prominent. However, the limited ground land and environmental constraints make the development of road traffic face difficulties. The continuous increase of vehicles and the increasing congestion of roads form Contradiction has become the primary problem in the development of transportation today. Urban car ownership continues to grow at a blowout. As of 2020, the national car ownership will reach 280 million, with more than 3 million in 13 cities across the country. However, while the number of cars continues to grow, the gap of parking spaces is also increasing, and the total number of parking gaps nationwide has risen to 80 million.

Due to a series of problems such as insufficient number of parking spaces, unbalanced parking supply and demand, unreasonable parking structure, unscientific ratio, and unbalanced parking layout, local traffic congestion in the city, road disorder, air pollution, and reduced public satisfaction have been caused. A series of derivative phenomena have

Z. Hu et al. (Eds.): AIMEE 2022, LNDECT 159, pp. 126–135, 2023.
https://doi.org/10.1007/978-3-031-24468-1_12

seriously affected the image of the city and the quality of life of citizens. In recent years, "difficulty in parking" has become one of the livelihood problems that urban residents have responded strongly to. The imbalance between the supply and demand of urban parking has caused static traffic to have a greater impact on dynamic traffic. Parking has become one of the main causes of urban traffic congestion, restricting the sustainable development of cities, and has also become the focus of government attention.

Therefore, it is imperative to strengthen the construction of urban public parking lots. In order to improve the efficiency of urban space utilization, promote the composite utilization of land, improve the level of urban traffic operation, and increase the carrying capacity of urban roads, parking lots play an extremely important role. To match the growing demand for parking.

At the same time, under the background of new infrastructure, the state has continuously strengthened the construction of parking lots in recent years, accelerated the pace of construction of new intelligent infrastructure, and put forward higher quality and service requirements for the planning and design of urban parking garages. Issues related to the planning and design of parking facilities will be the key research content in the field of transportation during the "14th Five-Year Plan" period.

The author summarizes and summarizes the latest research progress in the main research fields such as parking impact indicators, demand forecasting methods, and site selection methods.

2 Research on Parking Demand Forecast

Foreign scholars have studied parking demand forecast earlier. W. Bowman Cutter et al. [1]. Analyzed whether the growth of parking demand will significantly increase the demand for parking lots through the investigation of parking demand in some cities and towns of Los Angeles, and proposed different land types and Parking space control standards for parking lots in different areas. Jospeh C. Oppnelnader proposed a single-objective site selection model with the parking convenience as the goal and the parking lot service radius and maximum walking distance as constraints, so as to carry out planning site selection and berth allocation.

A TSI research report shows that off-street parking and on-street parking are significantly different in terms of service objects and service functions. The report points out that almost all on-street parking facilities mainly meet the parking needs that are not long-term parking needs. But short-term parking demand.

William H.K.Lam found through research that parking demand, walking distance, parking lot capacity and parking charges will significantly affect parking behavior [2]. Passengers' parking decisions in travel are hierarchical selection structures, and based on this, the establishment of a time-dependent road network equilibrium model. Austin, TerenceW studied the parking demand allocation in CBD, by investigating the driver's final parking location, and using the correlation between walking time and parking cost, to simulate the driver's parking choice, and then allocate different parking behaviors according to the parking behavior. Parking needs of the plot.

Domestic scholars also gradually began to study related content.

Zhan Xiaolan analyzed the supply and demand relationship between parking facilities and road network capacity for the central area of the city, found the relationship between the two, and proposed effective measures to maintain the balance between the two [3]. Wei Lianyu and others proposed a regional quantification model of parking demand by introducing the degree of location dominance [4]. On this basis, they improved the parking generation rate model and applied it in practice. Bai Yu et al. [5]. Proposed an effective prediction method based on road network capacity by analyzing the influence of travel distribution and road network capacity on forecasting of parking demand, and taking regional road network capacity as a constraint. Wang Fengyuan et al. analyzed the basic principles of land use and demand prediction models, used the land use characteristic function to describe the relationship between the land use nature and the parking generation rate, and used the traffic impact function to analyze the impact of the growth rate of road network traffic and parking rate on parking [6]. On this basis, a parking demand model based on land use and traffic characteristics is established. Wu Jiayou and Liu Shuhong7 analyzed the relationship between traffic demand and surrounding land use and traffic flow based on the technology of a large number of data surveys, and proposed a parking demand prediction method based on location characteristics and induced traffic [7].

Several commonly used parking prediction model methods are introduced below.

1) Parking Incidence Model

   The parking occurrence rate model is a model that combines the parking demand forecasting method with the land property for analysis and forecasting. The textual definition of the parking generation rate is the number of parked vehicles generated per unit land area of a certain nature throughout the day. The mathematical formula is expressed as follows.

$$P_j = f\left(L_{ij}\right) = \sum{}_i \alpha_i \cdot L_{ij} \tag{1}$$

   Among them, the basic daily parking demand (standard number of cars or parking spaces) in the jth subdivision is represented by P_j; the i-type land use index (land use indicator) in the jth subdivision is represented by L_{ij}; the parking generation rate index of the i-class land is represented by α_i Standard number of trains/(land-use unit index·day).

   This method needs to investigate several sections and buildings with different land use characteristics in the research area, and calculate the parking demand of motor vehicles in this area according to the parking demand and traffic influence function generated by different land use characteristics. The peak parking demand predicted by this method is closely related to the characteristics of land use, and has high reliability in spatial distribution; but the disadvantage is that the workload is large, and the longer the forecast period, the worse the accuracy of the traffic impact function, and the long-term parking The error in the required total scale accuracy will be larger.

2) Correlation Analysis Method

   The correlation analysis method means that the parking generation rate of different types of land is often the result of the combined effect of factors such as land use, population, and traffic volume. The correlation analysis method needs to consider multiple influencing factors, and the mathematical model is as follows.

$$P_{di} = K_0 + K_1(EP_{di}) + K_2(PO_{di}) + K_3(FA_{di}) + K_4(DV_{di}) + K_5(RS_{di}) \quad (2)$$

Among them, P_{di} is the parking demand (parking space) in area i during peak hours in year d; EP_{di} is the number of jobs in area i in year d; PO_{di} is the population of area i in year d; FA_{di} is the floor area of buildings in area i in year d; DV_{di} is the number of units (enterprises) in district i in year d; RS_{di} is the retail service industry in district i in year d; K_i $(i = 0, 1, 2, \ldots)$ refers to is the regression coefficient.

The correlation analysis method makes predictions by establishing the functional relationship between parking demand and urban economic activities and land use variables. It requires detailed planning-related data such as population, employment, and urban economic activities. This method considers many relevant factors. The prediction method is relatively strict, but during modeling, multiple parameters need to be estimated. The workload is large, the method is more responsible but the practicability is not wide, and the forecast time is short. Over time, the influencing factors of various factors will also change, resulting in errors in the prediction results.

3) Motor Vehicle OD Prediction Method

The vehicle OD prediction method, also known as the travel attraction model, refers to the direct relationship between parking demand and regional travel attraction. The regional travel attraction (person-times/day) can be converted into the actual number of vehicles arriving according to the proportion of travel modes, and then according to the peak hour coefficient and the average parking rate of motor vehicles, the demand for motor vehicle parking during peak hours can be obtained. The key to the vehicle OD prediction method is to determine the sharing proportion of the traffic mode and the load capacity of the vehicle. The mathematical model can be expressed as follows.

$$P_i = (A_i + B_i + C_i + D_i + E_i) \times \alpha \quad (3)$$

Among them, P_i represents the daily parking demand of the i community; A_i and B represent the daily attraction of buses, minibuses, taxis, large trucks and small trucks in the i community; β is the average parking rate of motor vehicles (0.8–0.9).

The motor vehicle OD prediction method calculates the demand for motor vehicle parking according to the motor vehicle OD data predicted in the near and long term. It is necessary to have complete OD data before forecasting. The peak parking demand predicted on the basis of the traffic development strategy is the demand for macro control and has a guiding role in the formation of the urban dynamic and static traffic system. However, this method has a strong dependence on OD and a weak sense of spatial distribution.

In summary, there are many and complex factors affecting parking demand. In addition, the traditional parking demand forecasting method cannot consider the time-varying trend of parking demand, the actual operability is not strong, and the accuracy of the prediction result is difficult to guarantee. As technology further develops, the availability of parking data increases. From the perspective of quantitative analysis, it is more and more practical to use time series analysis methods to predict and analyze the time-varying situation of parking demand.

Based on the existing research results of parking demand forecasting at home and abroad, the existing forecasting methods are summarized, as shown in the Table 1.

Table 1. Comparison of parking demand forecasting methods

Model	Input data	Calculation method	Advantage	Shortcoming
Total forecast method	Population, expected travel distance, expected time, land area, etc.	Establish travel demand, travel accessibility, urban development scale constraints, passenger traffic constraints, transportation energy consumption, and road resource constraints	The model is simple and the required data is easier to collect	Not many variables, low precision
Parking Generation Rate Model	Land use index, parking generation rate index, current parking generation rate	Calculate the parking demand according to the parking generation rate generated by different types of land	The predicted value is closely related to the land use information. Strong reliability and high accuracy in spatial distribution	The longer the number of years, the worse the accuracy of the parking generation rate, and the error of the long-term parking demand scale
Correlation analysis	Population, economy, traffic demand, number of jobs, land use status, etc	Find the impact mechanism of land use and social and economic development on parking demand, so as to forecast the demand	Considering many factors, the calculation is simple	The multiple linear regression model has many undetermined coefficients, the investigation is time-consuming and laborious, and the data collection is difficult

(*continued*)

Table 1. (*continued*)

Model	Input data	Calculation method	Advantage	Shortcoming
Motor vehicle O D prediction method	Percentage of stop-and-go trips, traffic attraction of various types of vehicles, peak parking correction factor, parking space turnover rate	Calculate the OD scene of vehicle travel in the community according to the travel attraction and mode sharing of the traffic area, and then calculate the demand for parking spaces in the area during peak hours according to the conversion factor during peak hours	Apply the data of urban transportation planning, with better connection with other planning and data sharing	The basic data requirements are relatively high, and a relatively complete future motor vehicle OD data is required
Traffic Volume-Parking Demand Prediction Model	In the future, the amount of parking at night in various regions and the amount of traffic attracted by each time period	Calculate the demand for motor vehicle parking based on the regional traffic volume	Good data sharing is relatively simple and practical when the research is small	The data will change as the scope of the study becomes larger, reducing the accuracy of the predictions
Land use model	The nature of the land, the number of employed persons, the number of parking spaces, etc	/	The data requirements are simple, and the prediction cost is low	The model requires high accuracy of building area and employment, so the prediction accuracy of large urban areas is poor.

3 Research on the Method of Parking Location Selection

Foreign research on infrastructure planning has achieved certain results.

Siegel and Koller and others developed a comprehensive index on the spatial accessibility of the medical and health sector in order to solve the shortage of outpatient services in the underserved areas of Germany, which helps to help the underserved areas between various departments to determine the accessibility Sexuality and planning of medical facility layout [8]. Ismail Onden et al. integrated the powerful spatial analysis capabilities of GIS with network analysis functions so as to be fully applied in the process of location selection of retail service facilities [9]. Network analysis tools such as GIS can simplify location problems, coordinate data, and build network models to determine all coverage areas suitable for retail store locations; spatial analysis can identify spatial distribution, connect specific locations, and determine hierarchical spatial models. And it can realize visualization functions such as map overlay and comparison. Ritsema et al. used GIS technology to assist the location selection and layout of public service facilities, used accessibility analysis to determine the service scope of public service facilities, and used the competition degree estimation method and the gravity surface method for optimization [10]. It can also be applied to the site selection of urban public parking lots. Monika Mangla et al. proposed two functions of parking guidance system. With these functions, the system central server will obtain dynamic information from the external network, and use the same parking suggestion to the driver to achieve automatic parking [11]. M.O. Reza and other intelligent parking systems can obtain information about available parking spaces and process them, and then the car is placed in a certain position. Based on this, a parking assistant system is constructed [12]. Khalid Mahboob et al. proposed a self-parking system. In a narrow crowded system, the environment and actuator are read through sensors, and the car can park itself [13].

Domestic research on parking lot site selection has gradually achieved certain results. In order to break the bottleneck of GIS analysis and processing capability and speed, Wang Tuodi proposed a parallel GIS spatial location analysis scheme, mainly through the parallel buffer algorithm based on arc division and the parallel stacking algorithm based on horizontal scanning strip division [14]. Two algorithms to achieve efficient parallelization of fast analysis of GIS spatial location. Zhan Changgen took the location selection of public parking lots in the main urban area of Lanzhou City as an example, combined the powerful spatial analysis ability of GIS with the analytic hierarchy process, and comprehensively considered the five influencing factors of the location selection of public parking lots in Lanzhou urban area from multiple perspectives [15]. The Analytic Hierarchy Process (AHP) model is constructed, the weights of each factor are determined, and GIS is used for spatial processing and analysis, and the final comprehensive site selection suitability map of the public parking lot is obtained. Zhou Huazhen proposed a method for selecting the location of urban public parking lot by combining GIS spatial analysis technology and genetic algorithm [16]. This paper introduces the steps of selecting the location of public parking lot by combining genetic algorithm and GIS technology. First, determine the goals to be achieved and the various influencing factors to be considered in the location selection problem, and then use the powerful spatial analysis technology of GIS to process and analyze the data. Get the location of parking demand points in the study area. Yi Hongjie based on the limitations of traditional site selection methods and the advantages of GIS applied to spatial site selection, proposed to combine the site selection problem of urban public parking lots with GIS multi-dimensional decision-making analysis, and proposed a combination of GIS multi-quasi-decision decision [17]. Analytical and Analytic Hierarchy Process (AHP) site selection model, and propose a minimum distance summation site selection model. Zhang Feifei et al. used spatial analysis methods such as buffer analysis, overlay analysis, and Kriging interpolation analysis commonly used in GIS in the process of site selection for bus stops [18]. By analyzing the buffer area of the bus station with the corresponding service radius, the location range of adjacent stations is determined according to the distance of the buffer zone, and the comprehensive influence degree layer of each influence factor on the location of the bus station is calculated. Tang Shao-jun put forward a new optimal location model for multi-criteria decision analysis based on GIS on the basis of analyzing the theories and methods of common public service facility layout and location selection, and established the basic theory, algorithm design and analysis process of the model [19]. Be explained. Mo Weiliang and others first listed and analyzed various factors affecting urban public parking lots, pointed out the shortcomings of the single-objective optimization model, and then proposed a genetic algorithm that can consider multiple objective conditions to solve the multi-objective location selection of urban public parking lots [20]. Model, using MATLAB to solve the location selection of the model, and taking the urban public parking lot site in a certain area as an example to verify, indicating that the use of this model can better solve the problem of public parking lot location. Guo Tao et al. proposed a new model of genetic multi-objective optimization algorithm under the background of existing and at least one new parking lot, and analyzed the advantages and disadvantages of the model [21]. It has generalization and practicability, and provides theoretical methods and basis for parking lot space location selection.

In summary, the current research on the layout planning of parking facilities at home and abroad has achieved certain results, and it has been verified that the GIS system has practical significance in infrastructure location selection. However, most of the existing research results are difficult to give specific planning methods from the overall level, and due to the large differences in traffic characteristics, parking behavior characteristics and road traffic systems between different typical areas, the existing parking planning methods are not suitable for different areas. Regional applicability and design errors are large.

4 Conclusion

By summarizing the research status at home and abroad, it is found that when domestic and foreign scholars study the problem of urban parking demand forecasting, most of the parking forecasts are considered from the perspective of parking demand. Generally, a parking demand model is established around factors such as traffic demand and land use properties to predict the number of parking needs. Few predict the parking scale from the perspective of parking supply, ignoring the anti-constraint effect of parking supply on parking demand, let alone parking demand. From the perspective of the combination of dynamic traffic and static traffic, this paper deeply analyzes the influencing factors of urban parking supply.

From the perspective of parking site selection methods, most urban public parking lot planning studies use traditional mathematical models for site selection research. However, the traditional mathematical site selection models are complex and difficult to solve, and the solution results are quite different from the actual site selection environment. It is only theoretical and lacks practicality. Therefore, the GIS system has strong practical significance in site selection, but it is also necessary to combine the traffic characteristics, parking behavior characteristics and road traffic system of each typical area to match the model with the demand.

References

1. Bowman Cutter, W., Fanco, S.F.: Optill supply of parking at Los Angeles country. Transp. Res. Part A **46**(6), 34–36 (2012)
2. Lam, W.H.K., Li, Z.C., Huang, H.J., et al.: Modeling time-dependent travel choice problems in road networks with multiple user classes and multiple parking facilities. Transp. Res. Part B Methodol. **40**(5), 368–395 (2006)
3. Zhan, X.: Research on the relationship between the supply of parking facilities and the capacity of the road network in the city center. Traffic Transp. **85**(1), 77–79 (2005)
4. Wei, L., Zhou, Y., Zhang, C., Liu, W.: Research on parking demand prediction model with improved parking generation rate. J. Hebei Univ. Technol. **8**(5), 101–106 (2007). (in Chinese)
5. Yu, B., Xue, K., Yang, X.: Prediction method of parking demand based on road network capacity. Chinese J. Transp. Eng. **1**(4), 49–52 (2004). (in Chinese)
6. Wang, F., Zou, S., Yan, Y., et al.: Prediction method of parking demand based on land use and traffic characteristics. Chinese J. Transp. Eng. **7**(2), 84–88 (2004). (in Chinese)
7. Wu, J., Liu, S.: Research on the forecasting model of parking demand based on location analysis and induced traffic. J. Chongqing Jiaotong Univ. **8**(3), 104–106, 114 (2004). (in Chinese)
8. Siegel, M., Koller, D., Vogt, V., et al.: Developing a composite index of spatial accessibility across different health care sectors: a German example. Health Policy **120**(2), 205–212 (2016)
9. Nden, I., Sen, C.G., Sen, A.: Integration of integer programming with GIS analyzing abilities for determining the convenience levels of retail stores. Procedia Soc. Behav. Sci. **62**(1), 1144–1149 (2012)
10. Tsai, J., Miller, M.: Integrated planning for school and community. Transp. Res. Rec. J. Transp. Res. Board **1922**(1), 111–117 (2005)
11. Mangla, M., Garg, D.: An automated parking guidance system for megacities. Inf. Technol. Comput. Sci. **68**(1), 52–58 (2018)
12. Reza, M.O., et al: Smart parking system with image processing facility. Intell. Syst. Appl. **18**(3), 41–47 (2012)
13. Arhab, N., Jahan, M.S., Oussalah, M.: Car parking user's behavior using news articles mining based approach. Transp. Res. Procedia **55**(3), 26–33 (2021)
14. Wang, T.: Parallel Research and Implementation of GIS Spatial Location Algorithm. Yanshan University, Qinhuangdao (2016). (in Chinese)
15. Zhan, C., Wang, L., Wu, Y.: Location of public parking lots in the main urban area of Lanzhou City based on the combination of GIS and AHP. Surv. Map. Spat. Geogr. Inf. **76**(8), 1–4, 8 (2016). (in Chinese)
16. Zhou, H.: Research on site selection optimization of urban public parking lots. Lanzhou Jiaotong University, Lanzhou (2015). (in Chinese)
17. Yi, H.: GIS-based public parking lot location analysis. Southwest Jiaotong University, Chengdu (2013). (in Chinese)
18. Zhang, F., Xu, J., Xie, X.: The application of GIS spatial analysis method in bus stop location selection. Geospat. Inf. **60**(1), 118–120, 14 (2011)
19. Xu, W.: Research on the application of GIS spatial processing modeling in urban planning and design. Jiangxi Normal University, Nanchang (2008). (in Chinese)

20. Mo, W., Li, J., Li, H., Han, Y.: Using genetic algorithm to solve multi-objective location model of public parking lot. J. Huazhong Univ. Sci. Technol. (Urban Sci. Edn.) **11**(6), 67–69 (2006). (in Chinese)
21. Tao, G., Tao, Y.: Research on the location model of public parking lot based on GA. J. Transp. Eng. Inf. **3**(1), 95–98+115 (2006). (in Chinese)

Application of Improved SFM Adaptive Threshold Algorithm in Automatic 3D Reconstruction of Remote Sensing Images

Meiyan Li[1] and Shiling Huang[2](✉)

[1] Global Software College, Nanning University, Nanning 530000, China
[2] Intelligent Manufacturing College, Nanning University, Nanning 530000, China
2625240992@qq.com

Abstract. Motion recovery structure method has become a research hotspot of 3D reconstruction because of its advantages of low equipment cost, easy to carry, wide range of applicable scenes and so on. This paper proposes to use RANSAC adaptive threshold method to improve the SFM model estimation process, and proposes to combine the feature-based matching (FBM) method and the area based matching (ABM) method with NCC (normalized cross correlation) and LSM (least squares image matching) to perform FBM at the top of the image pyramid, Then, the location of matching points is adjusted layer by layer by using the region matching method with improved NCC and LSM, and finally a high-precision matching location is obtained. The modified matching points are used for basic matrix f estimation, so that the threshold can be adaptive to various scenes and data sets, and improve the accuracy of model estimation and final three dimensional reconstructions.

Keywords: SFM · Model estimation · RANSAC · Adaptive threshold

1 Introduction

3D reconstruction technology, which is widely used in medicine, driverless, geographic information measurement and other fields, is a research hotspot in computer vision, image processing, and machine vision. It depicts the real scene into a mathematical model that conforms to the logical expression of the computer through the processes of deep data acquisition, preprocessing, point cloud registration and fusion, surface generation and so on [1]. The main methods of 3D reconstruction are software modeling, 3D scanning and SFM. Software modeling method uses ready-made 3D modeling software to build 3D models. This method has low equipment cost but requires a lot of time investment. Scanning method is to detect the target structure by emitting laser or infrared ray through three-dimensional scanning equipment. It has the advantages of high precision, fast scanning speed and so on, but the cost is high and the equipment volume is large [2]. SFM method is a process of 3D reconstruction from multiple views, which uses a set of 2D image scenes taken from different angles to estimate the 3D structure [3]. Its

Z. Hu et al. (Eds.): AIMEE 2022, LNDECT 159, pp. 136–149, 2023.
https://doi.org/10.1007/978-3-031-24468-1_13

main equipment is the camera (mobile phone) with RGB camera. Compared with the three-dimensional scanning method, it has the advantages of low cost, small size, easy to carry and easy to embed [4]. SFM for 3D reconstruction is not constrained by the site environment. Large scale ground 3D reconstruction can be carried out by using the aerial image taken by UAV, or the moving camera can be placed on the moving object to obtain the moving image to complete the 3D reconstruction of street view or object [5]. However, at present, in the process of large-scale 3D reconstruction using SFM, most people pay more attention to the efficiency of reconstruction and ignore the accuracy of reconstruction.

With the development of stereo vision 3D reconstruction technology, people put forward higher requirements for measurement accuracy, measurement range, operation complexity and automation [4]. The massive data obtained from aerial photography has expanded the measurement range. Many studies have also tried to introduce SFM process into aerial photogrammetry for aerial triangulation, so as to improve the automation of aerial photogrammetry. At present, the most widely used traditional SFM method is bundler [6]. Incremental binding adjustment (BA) successfully completes the three-dimensional reconstruction of unstructured images on the Internet. However, when restoring the 3D points of UAV aerial images, the cumulative error generated by Ba leads to the re projection error of the calculated 3D points is very uneven, the middle point error is small, and the peripheral point error is very large, the accuracy is difficult to meet the requirements of photogrammetry specifications [4]. In the process of combining computer vision with photogrammetry, people focus on the realization of automation; However, accuracy is also an important research content in the field of photogrammetry. Therefore, improving the reconstruction accuracy has become an important goal of SFM method in the field of photogrammetry. This paper will study an improved SFM method based on RANSAC algorithm, in order to ensure the reconstruction efficiency and improve the reconstruction accuracy at the same time [7, 8].

2 SFM Algorithm to Realize Three-Dimensional Reconstruction Principle

2.1 SFM Theoretical Basis

2.1.1 Coordinate System

In the field of computer vision, various coordinate systems can be used to represent the specified position in the image. Coordinate system is used to place interrelated elements, pixels and coordinates in the spatial coordinate system are related to the position in the image [9]. The coordinates in the 3-D coordinate system describe the 3-D positioning and origin of the system. The coordinate system is divided into pixel coordinate system, image coordinate system, camera coordinate system and world coordinate system.

Pixel coordinate system: in the pixel coordinate system, images are regarded as grids of discrete elements, sorted from top to bottom and left to right. For example, the coordinates (U, V) of a pixel point represent the U column and V row in the image array respectively.

Image coordinate system: it represents the actual size of the actual target on the photosensitive element of the camera, expressed in two-dimensional plane coordinate system (x, y). In the pixel coordinate system, the unit is pixels, there is no actual physical unit, but the physical unit (usually mm) is established in the image coordinate system. The origin of the image coordinate system is the intersection of the camera optical axis and the image plane. The distance between the image plane and the camera optical center (small hole) is f, where f is the focal length. The pixel coordinate system and the image coordinate system are on the same plane, and they are the relationship of translation transformation. The image coordinate systems x and y are parallel to the U and V rows of the pixel coordinate system respectively.

Camera coordinate system: the point represented in the camera coordinate system is described by the origin located in the optical center of the camera, and the observation target is described by the position and angle of view of the camera [9]. In this system, the X axis of the camera coordinate system points to the right, the Y axis points down, and the Z axis points away from the camera, parallel to the image coordinate systems x and Y respectively.

World coordinate system: used to describe the positional relationship between different cameras. Generally, the camera coordinate system does not coincide with the world coordinate system. In reality, the target point and the position of the camera are different physical positions in the world coordinate system.

2.1.2 Discrimination of Camera Imaging Geometric Relationship

The geometric relationship of camera imaging is shown in Fig. 1.

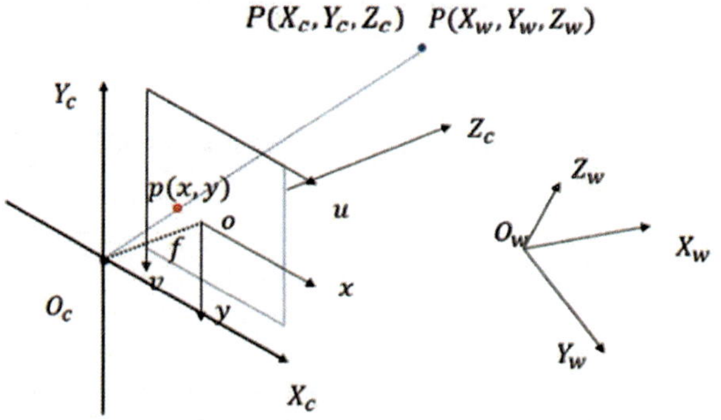

Fig. 1. Geometric relation diagram of camera imaging

Qw-XwYwZw is the world coordinate system, which describes the position of the camera, in *m*. *Qc-XcYcZc* is the camera coordinate system, with the optical center as the origin, and the unit is m. *O-XY* is the image coordinate system, and the intersection of the optical axis *ZC* and the image is the origin 0, in *mm*. UV pixel coordinate system, the origin is the upper left corner of the image, and the unit is pixel. Point *P (XW, YW, ZW)* is the real point in the world coordinate system. Point *P (XC, YC, ZC)* is the point observed in the camera coordinate system (the same point as point *P* in the out of bounds coordinate system). Point *P (x, y)* is the imaging point of point *P* in the image, and the coordinates in the pixel coordinate system are *(U, V)*.

Suppose that in the camera coordinate system, the projection of the observation point *P(Xc, Yc, Zc)* in the image coordinate system is point *p(x, y)*, and their projection homogeneous matrix relationship is shown in Eq. 1.

$$p = K[I\ 0]P_C \tag{1}$$

In Eq. 1, *P* is the image coordinate, P_C is the camera coordinate, *K* is the camera internal parameter matrix, which determines the mapping of space points to image points. This matrix is completely determined by the camera production process.

Different camera position observation points *P(Xw, Yw, Zw)* get different P *P(Xc, Yc, Zc)*, under each camera coordinate system. In order to objectively describe the realistic objective position of the target in the world coordinate system, it is necessary to rotate *(R)* and shift *(T)* the camera coordinate system and the world coordinate system. Their homogeneous matrix relationship is shown in Eq. 2.

$$P_C = [\begin{matrix} R & T \\ 0 & 1 \end{matrix}]P_w \tag{2}$$

In Eq. 2, *R* is the rotation matrix, *T* is the translation matrix, P_c is the camera coordinate, P_w is the world coordinate.

Substitute Eq. 2 into Eq. 1 to obtain the projection relationship between the point *P(Xw, Yw, Zw)* in the world coordinate system and the point *P (x, y)* in the image coordinate system. Its homogeneous matrix is shown in Eq. 3.

$$p = K[R\ T]P_W \tag{3}$$

$$p = MP_W \tag{4}$$

In Eq. 3, $\left[R\ T\right]$ is the external parameter matrix of the camera, which reflects the relationship between the camera position and the world coordinate system. Combine the internal parameter matrix *K* and $\left[R\ T\right]$ in Eq. 3 to obtain Eq. 4, M is called the projection matrix.

Use external parameters to convert the world points into camera coordinates, and use internal parameters to map camera coordinates to the image plane.

2.1.3 Camera Calibration and Requirements

Camera calibration, also known as geometric camera calibration, is to estimate the parameters of the lens and image sensor of an image or video camera. These parameters can be used to correct lens distortion, measure the size of the object in world units, or determine the position of the camera in the scene [10]. These tasks are used to detect and measure objects in machine vision and other applications, as well as robotics, navigation systems and 3-D scene reconstruction. Camera parameters include intrinsic parameters, extrinsic parameters and distortion coefficient. To estimate camera parameters, 3-D world points and their corresponding 2-D image points are needed. These correspondences can be obtained by using multiple images of a calibration pattern (such as a checkerboard). Using the correspondences, camera parameters can be solved. After calibrating the camera, the accuracy of estimated parameters can be evaluated by drawing the relative position and calibration pattern of the camera, calculating the re projection error and calculating the parameter estimation error.

The accuracy of camera calibration has a direct impact on the later 3D reconstruction calculation. The camera standard methods are divided into traditional calibration method and self-calibration method [10]. The traditional calibration method uses the correspondence between the actual three-dimensional coordinate points of the calibration pattern (usually black-and-white checkerboard) and its corresponding image points to estimate camera parameters. This method is suitable for applications requiring high accuracy. Self-calibration method: it does not need to calibrate patterns, but estimates camera parameters through the geometric constraint relationship between multiple images. This method is suitable for large-scale modeling occasions with low accuracy requirements [11–14].

2.2 SFM Algorithm and Its Application in 3D Reconstruction

According to the theoretical basis of SFM, the problem of SFM can be transformed into the following mathematical problems. It is known that the coordinates of 3D points in the world coordinate system in M images are:

$$p_{ij}(i = 1, \cdots, m, j = 1, \cdots, n) \tag{5}$$

According to the projection relationship of Eq. (5), the following relationship is obtained:

$$p_{ij} = M_i P_j \tag{6}$$

M_i is the projection matrix of the camera corresponding to the picture.

The solution process is as follows:

The projection matrix M_i ($i = 1, \ldots, m$) of m moving cameras and the coordinates of N three-dimensional points is P_j ($j = 1, \ldots, n$).

The three-dimensional points in the world coordinate system M_i and P_j can be calculated according to the camera calibration method and formula (6).

The basic steps of SFM algorithm to realize 3D reconstruction are: firstly, detect, extract and match the feature points of the image; Secondly, the internal and external parameter matrix is estimated, and the three-dimensional point coordinates are obtained by triangulation; Finally, the point cloud optimization, paste texture and other operations.

3 Improved RANSAC Adaptive Threshold Method

3.1 RANSAC Adaptive Threshold Method and Parameter Analysis

RANSAC was first proposed by Fischler and Bolles in 1981. It is a random sampling process, which can reliably estimate the parameters of the model. The idea of RANSAC algorithm is to randomly select a certain number of samples for many times, extract as few samples as possible but enough to estimate the parameters of the model each time, and then classify all the data according to the parameters of the model. One class is called interior points, that is, the data is within the allowable error range of the parameter model [15, 16]; One is called outliers. The data deviates far from the normal range. Outside the allowable error range of the parameter model, these outliers are often considered to be the noise of the data set [17]. Iterative calculation, find out the number of interior points of each random sampling data within the allowable error range of the parameter model, and take the estimation with the largest number of interior points as the optimal estimation of the parameters for re estimating the model.

RANSAC algorithm depends on a key parameter: threshold t. If t is too small, the number of interior points will be small, which leads to inaccurate models. Sometimes, the number of interior points is too small to even estimate a model. If t is too large, the outer point damages the inner point, which also leads to modeling errors. In the actual setting, because of the different elements in the data set, the estimation of each model should be solved independently. Therefore, different data sets should call a specific threshold to adapt to the corresponding data noise.

People usually rely on experience to judge the selection of threshold, but this empirical value is often not applicable to all data sets. This paper proposes an improved RANSAC algorithm, which sets an adaptive dynamic threshold for the input data estimated by each model, and finds the threshold corresponding to the input data. This threshold can not only obtain enough interior points to ensure that the model can be estimated, but also ensure that it will not accept too many exterior points, Get the best possible interior point quality, so as to improve the accuracy of model parameters.

This paper improves the threshold change process of RANSAC algorithm as follows:

1) Set the variation range of the threshold and determine the minimum and maximum values;
2) Using a method similar to binary search, the number of interior points is calculated from the middle value of the threshold to judge whether the maximum interior point set is enough to estimate a model;
3) If the number of inner points is enough to estimate the model, the part whose threshold value is less than the middle value is searched circularly until the threshold value to which the camera can be added just happens.
4) If the number of inner points is not enough to estimate a model, the part whose threshold is greater than the middle value is searched circularly until the camera can be added.

In the process of SFM camera pose estimation, it first selects the two most stable views to generate 3D point clouds, and then compares the new camera with the previous model. RANSAC repeats random sampling. If the re projection error is less than the threshold, it is added to the previous model, optimizes the R and t of the camera, and calculates the position and 3D points of the camera [18]. Repeat this process, adding one camera at a time until all camera pose estimation and 3D point coordinate recovery are completed. Generally, only a part of the images are used for 3D reconstruction, and those image sets that cannot be used for reconstruction are not selected in advance, but are automatically selected during the operation of the algorithm. At this time, if the threshold is set too small, it may make many cameras unable to join. If the threshold is set too large, it may accept views with large re projection errors, which will affect the accuracy of the whole reconstruction result [19, 20]. In this paper, the improved SFM method introduces RANSAC adaptive threshold algorithm in both image matching and model estimation. The specific process of the improved SFM method is shown in Fig. 2.

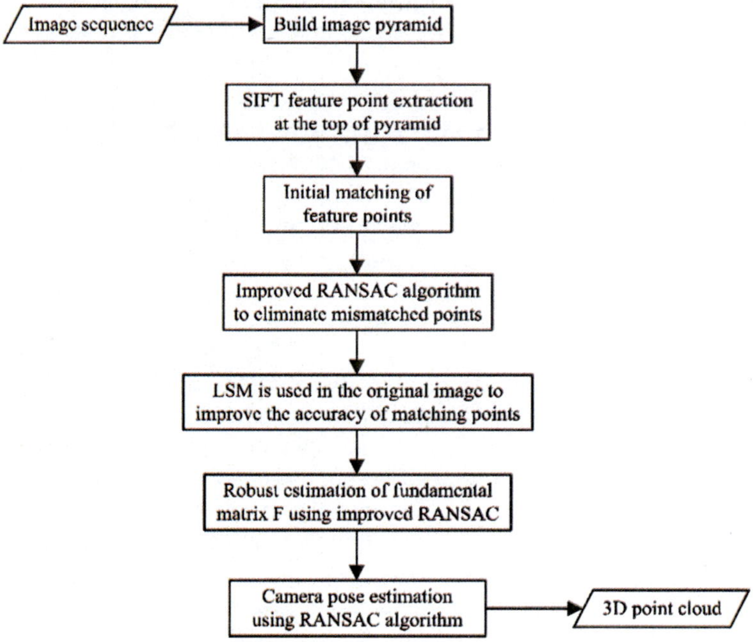

Fig. 2. Improved SFM flow chart

3.2 Algorithm Implementation Process

The general feature point matching and SFM processes are described in the following Algorithms 1 and 2. The output of the first algorithm is a pair of matching points, which contains the estimated basic matrix. Single response estimation is used to select the initial image pair from a large number of matches while maintaining a sufficiently wide baseline. The second algorithm takes these matching pairs as input, generates 3D point clouds and calculates the camera pose.

The asterisk (*) in the algorithm is the improvement of this paper, that is, using the improved RANSAC method to automatically adjust the threshold to adapt to the corresponding image, so as to generate a more accurate SFM.

Algorithm 1: Calculation of matching pairs corresponding to geometric consistency
Requirements: image sets
Confirm: match the geometric consistent correspondence between pairs of points
Calculate assumed matching pairs
Detect feature points and generate description operators in each image
Match description operator (force or approximate nearest neighbor)
*Filter geometric consistent matching points:
Estimation basis matrix f
Estimate homography matrix H

Algorithm 2: iterative SFM
Demand: camera internal calibration parameters (matrix K, possibly from EXIF data)
Corresponding points with geometric consistency in pairs
Confirm: 3D point cloud
Camera pose
Calculate the corresponding tracks
Calculate the connected graph G (the point represents the image, and if there is enough matching, an edge e is generated)
Compare the homography matrix and select an edge as the initial image pair
Generate the initial reconstruction result and shrink the e-edge
When G contains e, loop
*Robust estimation of camera pose and 3D points
Triangulate new tracks, shrink e
Execute Ba
End cycle

4 Algorithm Test and Experimental Result Analysis

Figure 3 is a comparison diagram of matching points obtained by Algorithm 1 under different thresholds. A blue line represents a pair of matching points, and the occurrence of cross lines indicates the occurrence of mismatching points, that is, the so-called "outer points".

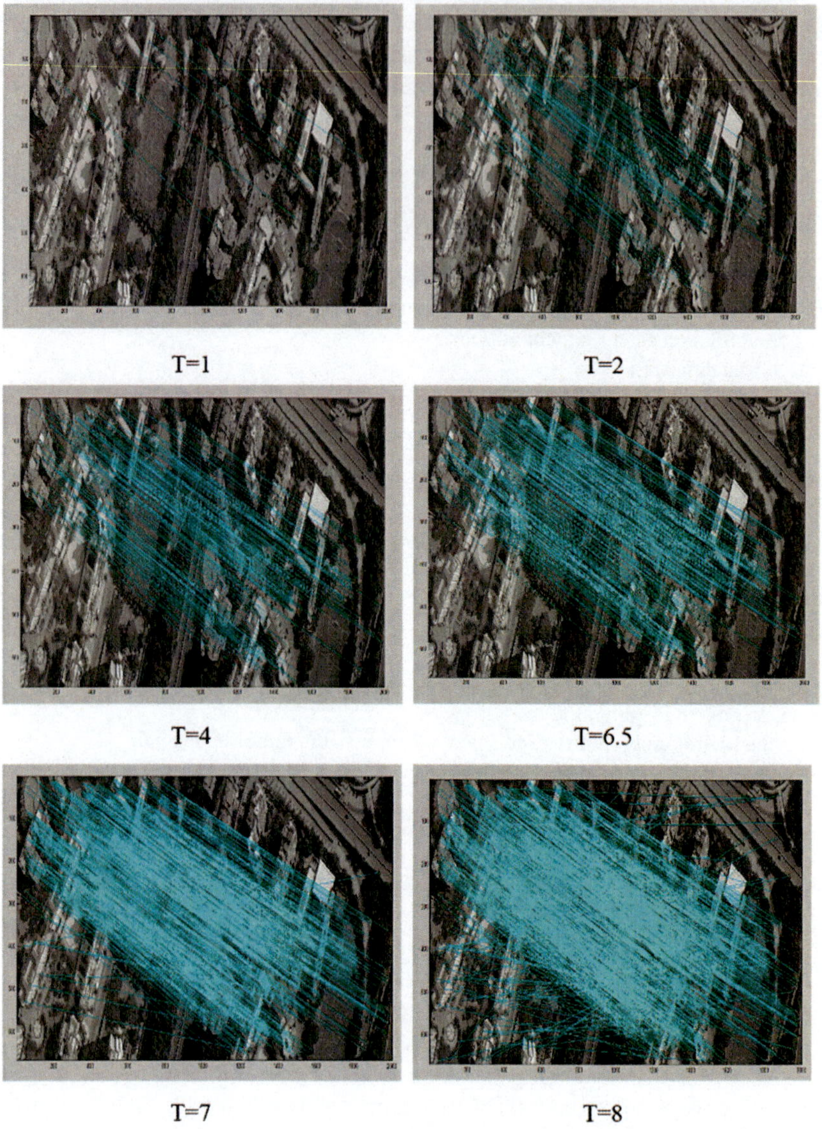

Fig. 3. Matching points under different thresholds

In order to verify the accuracy of the improved algorithm in this paper, the experiment uses 10 images as input (some data sources are shown in Fig. 4) to verify the estimated camera pose accuracy.

Fig. 4. Image sequence used in the experiment

The experimental results are shown in (a) and (b) of Fig. 5. The former is the translation baseline error, and the latter is the angle error. The results show that when using the improved RANSAC algorithm to estimate the camera pose, the angle error and baseline error are smaller than those of the traditional SFM algorithm, and the error value of the improved algorithm is relatively uniform. Although the angle error difference is not large, with an average difference of about 0.2°, assuming an error of 0.2° in the 10 m ray direction, the radian difference is about $(0.2\pi/180) \times 10\text{m} = 3.5\text{cm}$. However, the aerial ray of UAV is more than 100 m, so reducing the angle error of 0.2° can also improve the accuracy of back calculation 3D reconstruction. Therefore, using adaptive threshold can get more accurate camera pose.

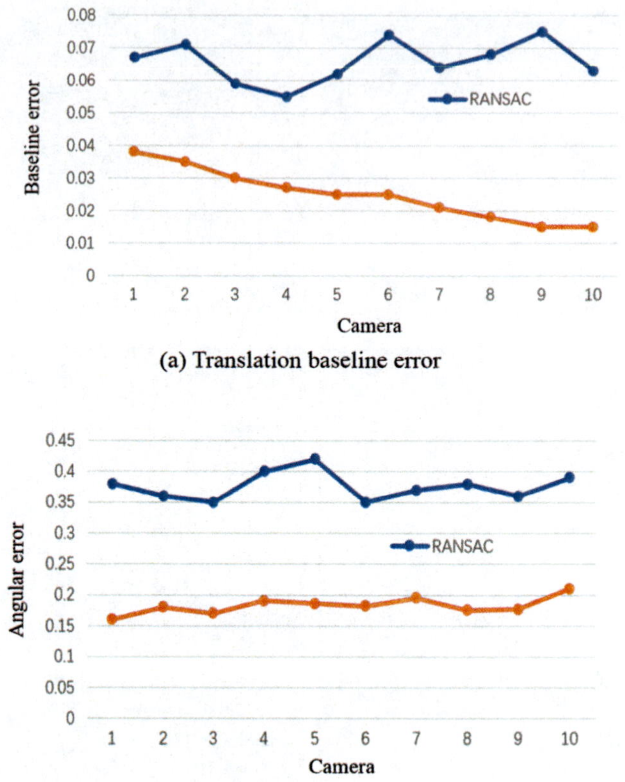

(a) Translation baseline error

(b) Angular error

Fig. 5. Estimation error of improved SFM model

The UAV aerial 3D reconstruction point cloud and camera pose estimation obtained by using the traditional SFM method and the improved SFM method in this paper are shown in Fig. 6. It can be seen from the figure that there are many error points below the horizon when the traditional SFM method (left figure) recovers the point cloud coordinates, while the error points recovered by the improved method (right figure) are significantly less than the former.

(a) aditional SFM method

(b) Improved SFM method

Fig. 6. Point cloud and camera pose recovered by SFM method

5 Conclusion

Based on the theory of computer stereo vision, this paper studies the traditional SFM method. On this basis, it studies the main steps of reconstruction, and looks for a 3D reconstruction method that can obtain higher accuracy. The imaging principle of camera, camera model, camera calibration and the principle of image stereo matching are studied. According to the accuracy requirements of photogrammetry, an accurate camera calibration method based on ABM and FBM is proposed. The goal of this method is to find accurate matching points, estimate accurate basic matrix, and then use the method based on essential matrix or basic matrix for camera self-calibration to obtain accurate camera parameters. This calibration result provides high-precision dense point clouds for subsequent 3D reconstruction.

Through the study of SFM method, we know that RANSAC, a robust estimation algorithm, is used in the traditional SFM method for basic matrix estimation and camera pose estimation. However, in the process of RANSAC estimation, a global fixed threshold is used, which makes it difficult for users to choose a threshold that can adapt to all input data sets and get accurate estimation results. Therefore, this paper improves the RANSAC algorithm by using adaptive dynamic threshold, so that in the process of model estimation, we can not only ensure to get enough internal points for model estimation, but also be affected by external points as little as possible. The experimental analysis shows that a new SFM process proposed in this paper can calculate the 3D point cloud with high accuracy.

Acknowledgment. This project is supported by the Guangxi University Young and middle-aged teachers' scientific research basic ability improvement project (2021ky1811).

References

1. Chen, Y.: Research on 3D Object Pose Estimation Based on Object Coordinates[D]. Beijing University of Posts and Telecommunications, Beijing (Jan 2021) (in Chinese)
2. Awati, A.S., Patil, M.R.: Inpainting of structural reconstruction of monuments using singular value decomposition refinement of patches. Int. J. Image Graph. Signal Process. (IJIGSP), **11**(5), 44–53 (2019)
3. Goklani, H.S., Sarvaiya, J.N., Abdul, F.: A review on image reconstruction using compressed sensing algorithms: OMP, CoSaMP and NIHT. Int. J. Image Graph. Signal Process. (IJIGSP), **9**(8), 30–41 (2017)
4. Kavitha, A.V., Srikrishna, A., Satyanarayana, Ch.: An efficient texture feature extraction algorithm for high resolution land cover remote sensing image classification. Int. J. Image Graph. Signal Process. (IJIGSP), **10**(112), 21–28 (2018)
5. Tianze, C., Baozhen, G., Qijun, L.: Reprojection-optimized free binocular camera pose estimation method[J]. China Optics **14**(06), 1400–1409 (2021) (in Chinese)
6. Zhu, H.: 3D Reconstruction and Pose Estimation of Objects. Zhejiang University, Hangzhou (Jun 2022) (in Chinese)
7. Li, J., Li, Y., Yang, Y.: Vehicle LiDAR ground point cloud extraction based on improved RANSAC algorithm. J. Henan Inst. Technol. (Nat. Sci. Ed.), **34**(01), 49–54 (2022) (in Chinese)
8. Gautam, P., Raj, K.: Edge detection based on ant colony optimization using adaptive thresholding technique. Int. J. Image Graph. Signal Process. (IJIGSP), **10**(7), 60–68 (2018)
9. Li, G., Dong, L., Pan, L., Henghai, F.: The calibration algorithm of a 3D color measurement system based on the line feature. Int. J. Image Graph. Signal Process. (IJIGSP), **1**(1), 17–24 (2009)
10. Guan, J., Pan, W.: Ground point cloud extraction algorithm based on multi-region RANSAC. Electr. Technol. Softw. Eng. 14176–177 (2020) (in Chinese)
11. Shaopeng, D., Rufei, L., Yongning, C., Peng, W.: A point cloud adaptive slope filtering method considering Terrain. Remote Sensing Information **04**, 108–113 (2019). (in Chinese)
12. Chunjian, H., Rui, P., Ying, C.: Binocular ranging method based on improved ORB-RANSAC. Advances in Lasers and Optoelectronics **58**(22), 366–373 (2021). (in Chinese)
13. Shengdao, L., Chengcong, X., Guohua, Z.: Measurement accuracy analysis of binocular vision system in long-distance 3D coordinate measurement. Adv. Lasers Optoelectron. **14**, 439–445 (2021). (in Chinese)
14. Qingzhe, Z., Yong, W.: Binocular stereo vision calibration accuracy evaluation method based on Epipolar constraint. Adv. Lasers Optoelectron. **23**, 240–246 (2019). (in Chinese)
15. Yanguo, F., Chai Jianglong, X., Mingming, W.B., Qiushi, H.: Improved image registration based on ORB and RANSAC fusion. Opt. Precision Eng. **03**, 702–717 (2019). (in Chinese)

16. Qin, W., Jin-e, Y., Zhi, L., Yueping, X., Guoquan, C.: Panoramic image stitching technology based on improved RANSAC algorithm. Optoelectron. Laser **12**, 1253–1261 (2021). (in Chinese)
17. Khot, T., Agrawal, S., Tulsiani, S., et al.: Learning Unsupervised Multi-View Stereopsis via Robust Photometric Consistency (2019)
18. Whelan, T., Salas-Moreno, R.F., Glocker, B., Davison, A.J., Leutenegger, S.L.: ElasticFusion: real-time dense SLAM and light source estimation. Int. J. Robot. Res., **35**(14), 1697–1716 (2016)
19. Mellado, N., Aiger, D., Mitra, N.J.: Super 4PCS fast global pointcloud registration via smart indexing. Comput. Graph. Forum, **33**(5) (2014)
20. Penate-Sanchez, A., Andrade-Cetto, J., Moreno-Noguer, F.: Exhaustive linearization for robust camera pose and focal length estimation. IEEE Trans. Pattern Anal. Mach. Intell. **35**(10), 2387–2400 (2013)

Explore the Application of Big Data Technology in Modern Enterprise Logistics Management

Mei E. Xie and Quanxing Ou(✉)

School of Business Administration, Wuhan Business University, Wuhan 430056, China
253637228@qq.com

Abstract. The rapid development of Internet and e-commerce has triggered the explosive growth of data volume, and big data is being integrated and developed with various industries, and it has an extremely important role in logistics management. This paper firstly introduces the basic characteristics and technical components of big data, related research results at home and abroad, data mining technology and data warehouse technology. Based on the investigation of enterprises, the specific application of big data technology in modern enterprise logistics management is studied, and a logistics center is selected as a case to apply time series model to forecast the demand of transportation freight volume. The research in this paper argues that intelligent logistics based on big data has great advantages in terms of efficiency, cost and user experience, and will also fundamentally change the current mode of logistics operation. The rational use of big data technology will play a positive role in the management and decision-making, customer relationship maintenance, and resource allocation of logistics enterprises.

Keywords: Big data technology · Artificial intelligence · Logistics management

1 Introduction

With the advent of the 5G era, the logistics industry has started to develop in the direction of intelligent logistics. Big data technology is the basis and core for intelligent logistics to play its important role. The logistics industry generates huge amount of data in various aspects such as goods flow, vehicle tracking and warehousing, etc. Analyzing these logistics big data will help people deeply understand the hidden laws behind logistics activities, optimize the logistics process and improve logistics efficiency. The existing research results at home and abroad mainly focus on the application of big data technology in modern logistics, data mining, data analysis and prediction of big data into various aspects of logistics enterprises, the innovation and prospect of logistics management in big data environment, the impact of big data on intelligent logistics, data science and supply chain management, the relationship between economic growth and big data, automation of big data application and personnel training. Use qualitative and quantitative methods to address relevant issues and predict outcomes by considering data quality.

Z. Hu et al. (Eds.): AIMEE 2022, LNDECT 159, pp. 150–161, 2023.
https://doi.org/10.1007/978-3-031-24468-1_14

This paper adopts literature research method, case study method, fieldwork method, modeling research method and other research methods to study the basic characteristics of big data, related research results at home and abroad, types of technologies of big data and specific applications of big data technology in modern enterprise logistics management. Discovering the valuable information hidden behind the massive data is the important business value of big data. Big data is a golden key to open the mysterious "black continent" in the logistics field, with these logistics big data, the so-called logistics "black continent" will no longer exist, people can fully understand the laws behind the logistics operation through data, with the help of big data With the help of big data technology, the data of each logistics link can be summarized, classified, integrated, analyzed and refined, providing important support and guidance for enterprise strategic planning, operation management and daily operation, thus effectively improving the overall service level of the logistics industry. The research on the application of big data in this paper will promote the transformation of logistics industry from rough service to personalized service [1].

2 Basic Features of Big Data

2.1 Large Amount of Data

From the perspective of data volume, Big Data refers to a huge collection of data that cannot be accessed, managed and processed within a tolerable time frame using traditional information technology and hardware and software tools, and requires a scalable computing architecture to support its storage, processing and analysis. According to estimates made by the international data corporation (IDC), a leading consulting firm, the amount of data generated by human society has been growing at a rate of 50% per year that is, doubling approximately every two years, which is known as Moore's Law of Big Data. This means that the amount of data generated by humans in the last two years is equal to the sum of all the data generated before.

2.2 Wide Range of Types

Big Data has many sources of data, with scientific research, enterprise applications, and Web applications all generating a constant stream of new and varied types of data. Biological big data, traffic big data, medical big data, telecommunication big data, electric power big data, financial big data, etc., all show "blowout" growth, involving a very large number, has jumped from terabyte (TB) level to beat byte (PB) level. Big data is very rich in data types, but it can be divided into two major categories, namely structured data and unstructured data of which the former accounts for about 10%, mainly referring to data stored in relational databases, and the latter accounts for about 90%, with a wide variety, mainly including email, audio, video, WeChat, Weibo, location information, link information, cell phone call information, web logs, etc. [2].

2.3 Fast Processing Speed

Many applications in the era of big data need to give real-time analysis results based on rapidly generated data for guiding production and life practices, so the speed of data processing and analysis usually has to reach second- or even millisecond-level response. Taking Google's big data engine Dremel as an example, it is a scalable and interactive real-time query system for read-only nested data analysis. By combining multi-level tree execution process and columnar data structure, it can do the aggregated query of trillions of tables in seconds, and the system can scale to thousands of CPUs to meet the demand of tens of thousands of Google users to operate pat-byte (petabyte) level data, and it can complete the query of pat-byte (petabyte) level data in 2 ~ 3 s.

2.4 Low Value Density

The value density of big data is much lower than the data already in traditional relational databases. In the era of big data, a lot of valuable information is scattered in the massive data. Take the district surveillance video as an example, if no unexpected events happen, the data continuously generated are of no value, and when unexpected situations such as theft occur, only the small segment of video that records the process of the event is valuable. However, in order to be able to obtain that precious video in case of theft and other accidents, we have to invest a lot of money to purchase surveillance equipment, network equipment, storage devices, and consume a lot of electricity and storage space to save the surveillance data coming from the cameras continuously [3].

3 Domestic and International Related Research

3.1 Review of Domestic Research

In terms of the application of big data technology in modern logistics, Liu Caifang (2017) summarized the important functions and impacts of big data technology in the application of logistics industry in a comprehensive manner from several aspects [4]. Li He (2018) effectively integrates technologies such as data mining, data analysis and prediction of big data into all aspects of logistics enterprises, establishes an information sharing platform for logistics collaboration, improves the economic efficiency of enterprises while realizing logistics information integration and intelligent logistics management, and promotes the healthy and sustainable development of China's modern logistics industry [5]. In terms of innovation and prospect of logistics management in the big data environment, Cai Yaxuan (2020) argues that the big data environment has attracted new development opportunities for logistics enterprises, but at the same time, it has also brought considerable challenges to enterprise management [6]. Cui Genhua (2020) sorted out the application of big data technology in logistics management, the impact of big data on intelligent logistics, proposed corresponding optimization measures and argued that big data is the way to effective management of enterprise logistics, transformation and upgrading, getting stronger and bigger, and achieving intelligent logistics [7].

3.2 Review of Foreign Research

In terms of theoretical research on the application of big data technologies, Anitha P et al. (2018) studied bridging the gap between the two intersecting fields of data science and supply chain management. Data can be analyzed for inventory management, forecasting and prediction in the form of reports, queries and forecasts. It is the application of qualitative and quantitative methods to solve relevant problems and predict outcomes by considering data quality. The research covers the investigation of how big data technologies can be used for the storage, processing, management, interpretation and visualization of data in the supply chain area [8]. Luis Emilio Alvarez-Dionisi (2017) presents the results of a study on the appropriate skills needed to use big data technologies, aiming to address the link between economic growth and big data [9].

In terms of data analytics research in big data technology, Ankur Jariwala et al. (2022) argued that every field generates a large amount of data. Therefore, in order to apply automation on any dataset, properly trained data plays an important role in achieving efficient and accurate results. Essaid EL HAJI et al. [10]. (2020) studied the modeling of a big data intelligent digital ecosystem for educational and career guidance for students and young people seeking their first job or retraining [11].

3.3 Review of Literature

In summary, the existing domestic research results mainly focus on the application of big data technology in modern logistics, data mining, data analysis and prediction of big data and other technologies into various aspects of logistics enterprises, the innovation and prospect of logistics management in the big data environment, and the impact of big data on intelligent logistics.

Foreign studies mainly focus on data science and supply chain management, the relationship between economic growth and big data, automation of big data applications and talent training, using qualitative and quantitative methods to solve related problems and predict the results by considering data quality. For this paper, the research on the application of big data technology in modern enterprise logistics management has certain reference and inspiration, which provides the theoretical basis and support basis for this paper.

4 Types of Big Data Technologies

Big data technologies are applied to all aspects of end-to-end big data systems, including data access, data pre-processing, data storage, data processing, data visualization, data governance, as well as security and privacy protection. Among them, the key technologies of big data are mainly as follows.

4.1 Distributed Database Technology

Distributed database is a logically unified database consisting of physically scattered multiple database units connected together. With the rising demand for database from big data applications in various industries, database technology faces rapid growth of data and dramatic expansion of system scale, and constantly puts forward higher requirements for system scalability and maintainability. At present, structured data is the main focus, combined with spatial, text, time series, graph and other unstructured data fusion data analysis has become an important demand direction for users [12].

4.2 Distributed Database Technology

Based on industry demand and technology development, distributed storage mainly presents 3 trends. First, the distributed storage technology based on hardware processing. At present, most of the storage still uses HDD (traditional hard disk), and a few use SSD (solid state drive), or SSD + HDD model. Second, the distributed storage technology based on converged storage. For the existing storage system for block storage, file storage, object storage, the basic needs of big data storage, to provide a system to support the convergence of a variety of protocols to reduce storage costs and improve the speed of online. Third is the convergence of artificial intelligence technology [13].

4.3 Stream Computing Technology

Streaming computing refers to the processing and analysis of data as it flows in, and is often used in large-scale computing scenarios that deal with high-speed concurrency and high timeliness requirements. The key to the stream computing system is the stream computing engine, and the current stream computing engine mainly has the following features: support for stream computing model, which can perform real-time computing on streaming data; support for incremental computing, which can perform incremental processing on local data; support for event triggering, which can provide timely response to changes in real time; support for flow control, which can avoid crashing or performance degradation due to high flow or.

4.4 Graph Database Technology

A graph database is a database that uses graph structure for semantic queries. Compared with the relational model, the graph data model has unique advantages. One is the ability to model data sets with complex or even arbitrary structures with the help of edge labels; second is that graph models can perform complex queries or analysis involving multi-hop relationships between data entities very efficiently. Third, it is more flexible and can easily create and dynamically convert data to reduce the cost of schema migration. Fourth, the graph database is good at handling complex relationships in a mesh, which has wider applications in financial big data, social network analysis, recommendation, security prevention and control, logistics and other fields.

4.5 Artificial Intelligence

Artificial intelligence is not a new technology, but it has proved to be practical over the years. In many ways, big data has played a role in promoting the development of artificial intelligence through two branches of artificial intelligence: machine learning and deep learning. As we all know, machine learning means that computers can learn without tedious programming. Applying this to big data analysis, machine learning enables the system to view historical data, identify patterns, build models, predict future results, and is mainly related to predictive analysis technology. On the other hand, deep learning is a kind of machine learning that imitates the working principle of human brain. It creates artificial neural network and uses multi-layer algorithm to analyze data. In big data technology, it enables analysis tools to recognize the content in images and videos, and then process them accordingly [14].

5 The Specific Application of Big Data Technology in Enterprise Logistics Management

5.1 Application in Corporate Logistics Distribution

With the intensive application of Internet, online shopping is becoming more and more common, and express delivery has become the most common phenomenon. For online shopping logistics delivery, in the mathematical model is a typical traveling saleman problem (TSP), the traveling saleman problem is also translated as traveling salesman problem, cargo stretcher problem, is the most basic route problem, the problem is seeking a single traveler from the starting point, through all the given demand points, and finally back to the origin of the minimum path Cost. The traveler problem in logistics is described as how to determine the shortest route for a logistics distribution company that wants to deliver all the orders of individual customers along the shortest route. The idea of solving the traveler's problem is not complicated, but when the number of online shoppers is large.

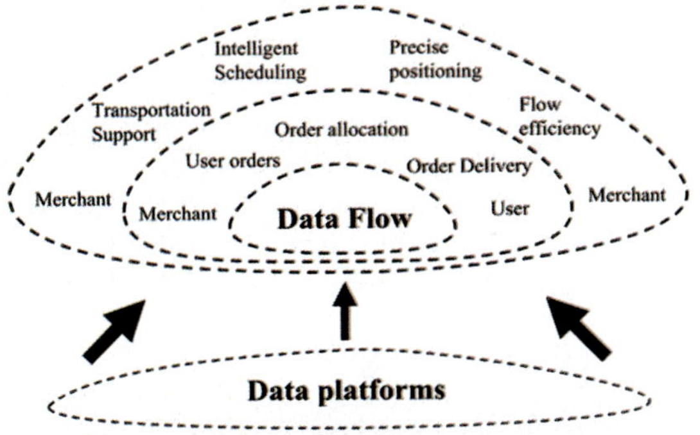

Fig. 1. Big Data Distribution Management Chart

It creates a very large number of possibilities, so big data technologies such as Hadoop must be used (Fig. 1). The following are the three main ideas for mathematically solving the traveler's problem: the collapsed route construction method, the collapsed route improvement method, and the collapsed synthesis heuristic method [15].

5.2 Application in Corporate Warehousing

The development of omni-channel shopping has brought about high requirements in various aspects such as order size, order structure and fulfillment timeliness, and the difficulty and complexity of logistics order processing are increasing day by day. For the warehouse side, it needs to face the uncertain market order changes and provide customers with high certainty of logistics service quality. At the same time, along with the in-depth application of big data technology in the logistics industry, it gradually plays a role in platform data, basic data, strategy management, receiving appointments, quality inspection, warehousing, shelving, wave, picking, inventory, value-added service management, etc. Industry-oriented, through scenario-based implantation, with scenario-based algorithms, policy rules and scheduling, refined prediction and behavior control, digging into the business characteristics of the industry to form a global optimization of cargo space and information system solutions that meet its characteristics.

5.3 Application in Enterprise Transportation Management: Time Series-Based Demand Forecasting of Transport Freight Volumes

In transportation management, the big data application covers the whole network and the whole chain of transportation services such as trunk transportation, feeder transportation, city distribution transportation, pick-up transportation and transfer transportation. It allows transportation management to increase efficiency, promote safety, control punctuality and provide more refined process control. Before transportation, we can prepare resources and load and unload efficiently; during transportation, we can visualize the process and correct deviations in time; after transportation, we can assess and settle and analyze data [16].

Based on big data analysis, applying time series model, by modeling and predicting long-term trends, seasonal changes, cyclic changes, etc., it can effectively predict the transportation volume of goods in the next stage, so as to quickly respond and reasonably plan and adjust the transportation capacity plan in the next stage. A logistics center is selected as the analysis case, and the actual freight volume of the logistics center from April 2021 to January 2022 is obtained through research collection, collation, analysis and aggregation as shown in Table 1 below. Now the gray forecasting method is used to forecast its cargo transportation volume from February 2022-May 2022.

Table 1. April 2021-January 2022 freight volume

Time	Cargo volume (kg)
April 2021	108341.73
May 2021	454117.32
June 2021	692469.48
July 2021	631172.28
August 2021	336489.39
September 2021	557620.47
October 2021	185857.80
November 2021	477058.62
December 2021	579161.28
January 2022	545744.22

5.3.1 Gray Prediction Model

Let the original data columns be

$x^{(0)} = (x^{(0)}(1), x^{(0)}(2), x^{(0)}(3), \ldots, x^{(0)}(n))$,

Calculate the series ratio,

$$(k) = \frac{x^{(0)}(k-1)}{x^{(0)}(k)}, k = 2, 3, 4, \ldots, n \tag{1}$$

If all the level ratios fall in the tolerable coverage interval $X = (e^{\frac{-2}{n+1}}, e^{\frac{2}{n+1}})$, then you can directly create GM (1,1) Model,and make gray predictions, otherwise, do a translation change to the data transform,

$$y^{(0)}(k) = x^{(0)}(k) + c, k = 1, 2, 3, \ldots, n \tag{2}$$

Establishment GM (1,1) Model, $x^{(0)}(k) + az^{(1)}(k) = b$, regression analysis was used to find estimated values of a,b, The corresponding whitening model is then

$$\frac{dx^{(1)}(t)}{dt} + ax^{(1)}(t) = b \tag{3}$$

Solve for

$$x^{(1)}(t) = (x^{(0)}(1) - \frac{b}{a})e^{-a(t-1)} + \frac{b}{a} \tag{4}$$

Thus the predicted value is obtained

$$\hat{x}^{(1)}(k+1) = (x^{(0)}(1) - \frac{b}{a})e^{-ak} + \frac{b}{a}, k = 1, 2, \ldots, n-1 \tag{5}$$

From this, the predicted values are obtained accordingly:

$$\hat{x}^{(0)}(k+1) = \hat{x}^{(1)}(k+1) - \hat{x}^{(1)}(k), k = 1, 2, \ldots, n-1 \tag{6}$$

Residual test, calculation of relative residuals

$$\varepsilon(\mathrm{k}) = \frac{\mathrm{x}^{(0)}(\mathrm{k}) - \hat{x}^{(0)}(k)}{\mathrm{x}^{(0)}(\mathrm{k})}, \mathrm{k} = 1, 2, \ldots, \mathrm{n} \tag{7}$$

If for all the $|\varepsilon(\mathrm{k})| < 0.1$, then it is considered to reach the higher requirements; Otherwise, if for all $|\varepsilon(\mathrm{k})| < 0.\ 2$, Then the general requirements are considered to be met.

5.3.2 Gray GM(1,1) Model Demand Forecast Analysis

Substituting the cargo transportation volume of the logistics center into the model and calculating the level ratios of the series, it is found that some of the level ratio values do not fall in the tolerable coverage interval [0.834 1.199], Therefore, a translation transformation of the data, Take 2077407, At this point, the level ratios all fall within the tolerable coverage interval, can be created directly GM(1,1) model and make gray predictions. The estimated values of parameters a, b were calculated to be 0.0028 and 2612270.45, respectively. as shown in Table 2.

Table 2. Model construction results

Development factor a	Gray role volume b	Posterior test difference ratio C value	Small error probability p-value
0.0028	2612270.4479	0.5799	0.700

Among them, the posterior difference ratio C value is used for the model accuracy level test, the smaller the value the better, generally the C value is less than 0.35 then the model accuracy level is good, the C value is less than 0.5 means the model accuracy is qualified, the C value is less than 0.65 means the model accuracy is basically qualified, if the C value is greater than 0.65, then the model accuracy level is not qualified. And the small error probability p-value is generally less than 0.7 means the model is not qualified, less than 0.8 means the model is barely qualified, less than 0.95 means the model is qualified, and more than 0.95 means the model accuracy is very good. From the above table, it can be seen that the posterior difference ratio C value $0.580 < 0.65$ means that the model accuracy level is basically qualified. In addition, the small error probability p-value is $0.700 < 0.8$, which means the model accuracy is barely qualified [17, 18].

After calculation, the forecasted values of freight volume from February to May 2022 are obtained as follows in Table 3.

Table 3. February 2022-May 2022 freight volume forecast

Time	Cargo volume (kg)	Predicted value
April 2021	108341.730	108341.730
May 2021	454117.320	524929.487
June 2021	692469.480	517527.489
July 2021	631172.280	510146.545
August 2021	336489.390	502786.595
September 2021	557620.470	495447.579
October 2021	185857.800	488129.438
November 2021	477058.620	480832.113
December 2021	579161.280	473555.544
January 2022	545744.220	466299.672
February 2022	–	459064.439
March 2022	–	451849.785
April 2022	–	444655.652
May 2022	–	437481.982

From the forecast results in Table 3, it is clear that the freight volume of this logistics center shows a significant decreasing trend after January 2022.

6 Conclusion

Through the above analysis, we can understand that big data technology has played an increasingly important role in the process of logistics management of enterprises. By using the data center built by big data technology, the hidden value of a variety of data can be better explored to improve the management and development benefits of logistics enterprises themselves. Therefore, enterprises should regard big data as a strategic resource in the process of their own development, give full play to the advantages of big data technology, and better carry out their own strategic planning and business model planning, so as to occupy a favorable position under the platform of big data and provide a smooth and efficient platform for enterprise development.

In the era of big data, the scale of having data and the ability to apply data have become the key to competition among enterprises, and the effective use of big data resources has become an important influencing factor of national competitiveness. Data, as an important production factor, is creating huge social, economic and scientific research value for business and consumers, and enterprises need to strengthen the organization of logistics big data research as well as resource support for big data research. In terms of public policy on big data, the state and industry should strengthen the cultivation of big data talent, create incentives to promote data sharing, develop policies that balance data use with data security protection, establish an intellectual property framework that effectively promotes innovation, overcome technological barriers and accelerate R&D of key technologies, and ensure investment in ICT infrastructure [19].

On the basis of the research on big data in enterprise logistics distribution, warehouse management and transportation management, the next step will be to deepen the research on the application of big data technology in intelligent storage. Intelligent storage is based on big data processing technology, the use of software systems to better integrate people and equipment, and constantly improve the level of intelligence, so that people and equipment can play their respective advantages, to achieve the best state of the system, and constantly evolve. Big data-based intelligent storage, in terms of efficiency, cost, user experience, etc. has great advantages, but also will fundamentally change the current mode of logistics operations.

Acknowledgment. This paper is supported by project of The National Social Science Fund of China (18BJY138) and 2022 China logistics society, China Federation of logistics and purchasing (2022CSLKT3–137).

References

1. Yang, L.: Logistics information mining and application in the era of big data. Logist. Technol. **33**(23), 414–416 (2014)
2. Zhang, J.: Exploration of logistics management informationization upgrade in the era of big data. Wealth Today **8**(11), 77–78 (2021)
3. Su, Z., Zhao, C.: Introduction to the construction of enterprise logistics management system in the context of big data. Bus. News **11**(12), 95–96 (2020)
4. Liu, C.F: Research on the application of big data technology in modern logistics. Intell. Comput. Appli, **7**(06), 169–170+177 (2017)
5. Li, H., Wang, L.: Application of big data technology in the development of modern logistics industry in China. Market Week (Theory Res.), **7**(03), 42–43+65 (2018)
6. Cai, Y.: Analysis of logistics management innovation under the big data environment. Times Econ. Trade **9**(23), 71–72 (2020)
7. Cui, G.: Application and prospect of big data in logistics management. China Storage Transport. **11**(07), 127–129 (2020)
8. Seuring, S., Müller, M.: From a literature review to a conceptual framework for sustainable supply chain management. J. Clean. Prod. **16**(15), 1699–1710 (2008)
9. Alvarez-Dionisi, L.E.: Envisioning skills for adopting, managing, and implementing big data technology in the 21st century. Int. J. Inf. Technol. Comput. Sci. **9**(1), 18–25 (2017)
10. Jariwala, A., Chaudhari, A., Bhatt, C., Le, D.-N.: Data quality for AI tool: exploratory data analysis on IBM API. Int. J. Intell. Syst. Appli. **14**(1), 42–56 (2022)
11. El Haji, E., Azmani, A.: Proposal of a digital ecosystem based on big data and artificial intelligence to support educational and vocational guidance. Int. J. Mod. Educ. Comput. Sci. (IJMECS), **12**(4), 1–11 (2020)
12. Silva, N., Barros, J., Santos, M.Y., Costa, C., Paulo Cortez, M.: Advancing logistics 4.0 with the implementation of a big data warehouse: a demonstration case for the automotive industry. Electronics **10**(18), 2221 (2021)

13. Babu, M.M., Rahman, M., Alam, A., et al.: Exploring big data-driven innovation in the manufacturing sector: evidence from UK firms. Ann. Oper. Res. **9**, 1–28 (2021)
14. Big, S.S.: Data analytics in manufacturing: a bibliometric analysis of research in the field of business management. Int. J. Prod. Res. **10**, 1–29 (2021)
15. Chen, J., Jin, C.Y.: A study on the collaborative inventory management of big data supply chain: case of China's beer industry. J. Korea Soc. Comput. Inf. **26**(3), 77–88 (2021)
16. Sundarakani, B., Ajaykumar, A., Gunasekaran, A.: Big data driven supply chain design and applications for blockchain: An action research using case study approach. Omega **102**(10), 24–52 (2021)
17. Chen, Y.T., Sun, E.W., Chang, M.F., et al.: Pragmatic real-time logistics management with traffic IoT infrastructure: Big data predictive analytics of freight travel time for Logistics 4.0. Int. J. Prod. Econ. **238**, 108157 (2021)
18. Benotmane, Z., Belalem, G., Neki, A.: A cost measurement system of logistics process. Int. J. Inf. Eng. Electron. Bus. **10**(5), 23–29 (2018)
19. Arora, M., Bhardwaj, I.: Artificial intelligence in collaborative information system. Int. J. Mod. Educ. Comput. Sci. **14**(1), 44–55 (2022)

Research on Financial Performance Evaluation of New Retail Listed Companies

Yeye Zhou(✉) and Xue Hu

School of Economics and Management, Wuchang Shouyi University, Wuhan 430064, China
610643847@qq.com

Abstract. With the advent of the era of big data, the development of retail industry also needs to constantly change to match the environment, traditional retail industry needs to reform in order to survive in the market. At the same time, the new retail industry transformation has become the general trend, and the economic changes brought by the transformation can be directly reflected through the financial performance. Therefore, this paper selects 52 new retail listed companies as research samples, using factor analysis method to evaluate the financial performance indicators, analyze the current situation of new retail companies in China. The research shows that the financial performance of the new retail listed companies is good, but the overall level still has some room for improvement. In view of the current situation of the financial performance of the new retail industry, suggestions are put forward to improve the financial performance from the four aspects of service platform, debt ratio, supply chain distribution, and economies of scale, so as to promote the good development of China's new retail market.

Keywords: New retail transformation · New retail listed company · Financial performance

1 Introduction

Due to the emergence of diversified sales channels with Electronic Commerce, the traditional retail industry has lost a large number of consumers in the case of competitive disadvantage, and it still has to pay higher labor costs and rents. At present, it is difficult to continue to operate. As far as e-commerce itself is concerned, because of its low technical content and low start-up funds, and uneven quality of goods on the Internet, and the situation of e-commerce has become precarious. Therefore, retail enterprises use high-tech platforms and data to achieve business model upgrading to occupy the industry market share, and the era of new retail transformation also follows. The impact of new retail transformation is most intuitively reflected in the financial performance of enterprises, and the financial performance evaluation of enterprises can comprehensively express the effect of enterprises in cost control, the effect of asset utilization management, the effect of allocation of capital sources and the composition of shareholders' equity return rate.

Z. Hu et al. (Eds.): AIMEE 2022, LNDECT 159, pp. 162–174, 2023.
https://doi.org/10.1007/978-3-031-24468-1_15

The new retail concept was first proposed by Alibaba founder Jack Ma in October 2016 at the Hangzhou Cloud Summit. The release of the theory provides a strong support for the germination and subsequent development of new retail. Because the development goal of early retail industry is not clear, the concept of new retailing has been defined with the development of the retail market. "New retail" refers to a new retail model that individuals or enterprises use, it rely on the Internet, and adopt new means such as the Network of Things, big data and artificial intelligence to reform and upgrade the production and circulation process of products, combine with online services, offline experience and modern logistics. In view of the fact that China's new retail industry has not yet developed to a mature period, this paper defines the new retail enterprise as the enterprise that has initially carried out the strategic layout of new retail when analyzing the financial performance of the retail industry in China.

The Department of Statistics and Evaluation of Ministry of Finance (1999) believes that enterprise performance includes two levels of concept, Operating performance and Operator's performance. Operating performance refers to the operating performance of an enterprise in a certain operating period, namely the final results of enterprise operation management, also called financial performance. Operator's performance refers to the extent that the managers of enterprises influence the formulation and operation of enterprise strategy, the Operating results and performance level in the process of enterprise operation, also called management performance. In this paper, the analysis of financial performance refers to the Operating performance, analyze the operation effect of the enterprise from the financial aspect.

2 Research Basis and Research Design

2.1 Literature Review

Domestic and foreign scholars have mainly conducted research in two aspects, one is the analysis of "new retail", and the other is the analysis of changes in corporate financial performance brought about by new retail transformation.

With the development of new retail transformation, scholars at home and abroad have expressed their views on the background, essence and development path of "new retail". Wang Xue (2018) pointed out that the era background of consumption upgrading and mobile shopping has created a diversified new retail model [1]. Yan Ningning and Li Tao (2019) pointed out that "new retail" is the use of big data as an analysis tool and means to sort out the needs of consumers, combined with a variety of online and offline marketing channels, while the artificial intelligence, cloud computing and other scientific and technological innovation achievements to be used, and the use of new forms of logistics and commercial sales activities [2]. Lan Hong and Zhao Jiawei (2020) believe that to promote the development of the new retail industry, we should establish a consumer-centered relationship-based marketing concept [3]. Li Ying, Ding Lu and He Jiajia (2020) show that new retail transformation is conducive to promoting the improvement of relationship performance and operational performance, which is the only way for the current.

development of the retail industry [4]. Bertoni Steven et al. (2020) believe that under the impact of the epidemic, enterprises must also carry out diversified reforms, and it is the trend of the times for the new retail industry to replace the traditional retail industry [5].

Regarding the evaluation of the financial performance of enterprises by new retail transformation, scholars believe that in addition to the transformation driven by national policies and industry competition, new retail transformation will improve the financial performance of enterprises. Mihail Nikolaevich Dudin et al. (2018) used EVA's evaluation method to analyze the economic profits of food retail companies, which shows that it is easy to judge the financial performance of retailers using this method, and the economic profits of enterprises have improved after the transformation [6]. Liu Yalan (2018) took Amazon, Alibaba and JD.com as examples to analyze the development of new retail, and the results showed that the development prospects of enterprises after the transformation of new retail are good [7]. Hou Dongyan (2019) conducted a comprehensive analysis of Suning and concluded that its financial performance after transformation is good [8]. Li Jing (2021) analyzed the financial performance development of New World Department Store, and the results showed that its operational conditions after the transformation were improving [9].

Looking at the existing literature, although the research of scholars at home and abroad has achieved certain results, there are still shortcomings: (1) the research on the impact of new retail transformation should not be limited to a single enterprise, but also combined with the entire retail industry in China for research and analysis; (2) Foreign scholars study the difference between the transformation of new retail and traditional enterprises, mainly focusing on the study of the economic profits of enterprises alone, while there is less literature on the operation, profitability, debt servicing and development capabilities of enterprises' financial performance. Therefore, this paper focuses on the study of the changes in financial performance before and after the transformation of the retail industry, focusing on the combined analysis of the entire retail industry and a single enterprise, and paying attention to the changes in each individual financial performance indicator.

2.2 Research Methods

The financial performance evaluation method includes financial indicator analysis method and non-financial indicator analysis method. In this paper, when evaluating the financial performance of the new retail industry, in view of the unquantifiability of non-financial indicators, only financial indicators are selected to analyze the financial performance.

According to the principles of comparability, accessibility, reliability, operability and integrity of the evaluation index system, the selection factor analysis method in this paper systematically integrates the financial performance indicators and establishes a financial performance evaluation index system that is compatible with this paper. The advantage of factor analysis is that it synthesizes more influential data from a large number of data to reflect more financial information of the enterprise.

2.3 Framework of Financial Performance Evaluation Indicator System

Table 1 is the financial evaluation index system.

Table 1. Financial evaluation index system

Financial performance	Index			
Profitability	Net profit ratio	Gross income ratio	Total profit cost ratio	Operating profit ratio
Solvency	Quick ratio	Current ratio	Debt asset ratio	Oper-cash into current debt
Operating capacity	Iventory turover ratio		Total asset turover ratio	
Development capacity	Total asset grow ratio		Net profit grow ratio	

In order to improve the accuracy of the conclusion, this paper based on the five basic principles of financial performance evaluation index system, combine with a large number of literature from domestic and foreign experts and scholars, based on the four capabilities of financial performance, finally confirm 12 key indicators, set up a new financial evaluation index system.

2.4 Sample Selection and Data Sources

According to the industry classification standard set up by China Securities Regulatory Commission, there are 93 "retail" listed companies under "wholesale and retail" in Shenzhen Stock Exchange and Shanghai Stock Exchange. To improve comparability of data, 93 listed companies based on 2019 financial data will be screened as follows:

1) Excluding retail companies listed after 2019;
2) Excluding the companies currently subject to ST and * ST;
3) Excluding the pharmaceutical industry, construction industry and other specific industries of retail companies;
4) Excluding the companies that use the traditional retail model.

After the above four screening, finally selected 52 new retail listed companies as samples for research. The data of the financial evaluation index system of the sample companies come from websites such as China Wealth Net, Orient Fortune Net and Juchao Information Net.

3 Empirical Analysis

3.1 Descriptive Statistical Analysis

After the establishment of the financial performance evaluation system, in order to further reflect the effect of the data after factor analysis, the financial index data of 52 sample listed companies in China's new retail industry in 2019 are analyzed by descriptive statistics. The minimum value, maximum value, average value and standard deviation of financial performance indicators are shown in Table 2.

Table 2. Descriptive statistics of financial indicator data

Financial performance	Index	N	Min (M)	Max (X)	Mean(E)	standard deviation
Profitability	Net Profit Ratio	52	−15.24	133.35	5.68	18.86
	Gross Income Ratio	52	3.52	58.18	22.69	8.69
	Total Profit Cost Ratio	52	−16.48	134.05	7.57	19.37
	Operating Profit Ratio Ratio	52	−16.61	122.24	6.78	17.61
Solvency	Quick Ratio	52	0.16	7.33	1.10	1.28
	Current Ratio	52	0.26	7.66	1.40	1.26
	Debt Asset Ratio	52	0.07	83.65	50.31	19.53
	Oper-cash into Current Debt	52	−0.19	50.02	1.05	6.93
Operating capacity	Iventory Turover Ratio	52	0.84	1105.38	35.61	152.95
	Total Asset Turover Ratio	52	0.05	2.36	1.08	0.48
Development capacity	Total Asset Grow Ratio	52	−35.02	67.24	3.90	14.52
	Net Profit Grow Ratio	52	−11355056.28	5215.74	−218202	1574687

Data Source: China Financial Net, East Fortune Net and Great Tide Information Network, etc.

In the profitability indexes, the average cost and expense profit margin is 7.57%, and the standard deviation is 19.37%, which is the indicator with the largest gap; The difference between the minimum and maximum values of sales net profit and operating profit margin is large, indicating that the profitability of listed companies in China's new retail industry is uneven and the average profitability is low.

In the solvency indexes, there is a certain gap between the maximum and minimum value of the asset-liability ratio, indicating that the overall solvency of companies is still at a good level, but the solvency between the companies is not balanced.

In the operating capacity indexes, the maximum and minimum inventory turnover rate are very different, but the standard value of the total asset turnover rate is 0.48 times, and the total asset turnover capacity is comparable. Overall, the operating capacity of listed companies in China's new retail industry is better.

In the development capability indexes, the minimum of Net Profit Grow Ratio is −11355056.28%, which is due to the poor management of the new Huadu company, which leads to a sharp goodwill impairment of subsidiaries acquired by enterprises,

and a large number of stores with no hope of turning losses closed. It is not because of objective reasons that the profits decline sharply, so this sample data is not removed. The difference between the minimum and the maximum of Net Profit Grow Ratio is huge, indicates that the net profit growth of the company is extremely out of sync.

3.2 Factor Analysis

1) Data Standardization

Because of the different measurement units and orders of magnitude of each data, this section will deal with each index dimensionless. Dimensionless processing was performed using the Z-Score normalization method. The formula is as follows:

$$Z = (X - \mu) \div \delta$$

Z is the normalized data, X is the raw data, μ is the mean and δ is the standard deviation.

2) KMO and Bartlett test

KMO and Bartlett tests are required for variables before principal component analysis is performed.

Table 3. KMO and Bartlett tests

The number of tangents for KMO sampling	0.762	
Bartlett's spherical degree test	The chi-square that was last read	865.419
	degree of freedom (df)	66
	Significance (sig)	0.000

Table 3 shows that the result of KMO test is $0.762 > 0.5$, the effect of factor analysis is better. Bartlett test results sig < 0.05, indicating that the correlation between variables is strong, can be used for principal component analysis.

3) Extracting Common Factors

By KMO and Bartlett test, the selected financial indicators are suitable for principal component analysis. This section will extract the common factor in accordance

Table 4. Total variance explained

Fac-tor	Initial eigenvalue			Extract the sum of squares of the loads			Sum of the squares of the rotating loads		
	Fea-ture root	Variance interp-retation rate %	Cumul-ation %	Feature root	Variance interp-retation rate %	Cumu-lation %	Feature root	Variance interp-reta-tion rate %	Cumul-ation %
1	5.504	45.869	45.869	5.504	45.869	45.869	4.195	34.958	34.958
2	1.5	12.499	58.369	1.5	12.499	58.369	2.614	21.785	56.742
3	1.123	9.36	67.729	1.123	9.36	67.729	1.201	10.01	66.752
4	1.137	9.472	77.201	1.137	9.472	77.201	1.254	10.449	77.201
5	0.987	8.228	85.429						
6	0.653	5.44	90.869						
7	0.5	4.164	95.033						
8	0.428	3.568	98.601						
9	0.149	1.242	99.843						
10	0.014	0.117	99.96						
11	0.004	0.035	99.995						
12	0.001	0.005	100						

with the factor characteristic root is greater than 1, the purpose is to integrate the original multiple indicators into a small number of indicators. In the process of extracting common factors, the cumulative variance interpretation rate can be used to express the effect of factor analysis. The higher the cumulative variance interpretation rate is, the better the extraction degree of the factor to the information is.

Table 4 shows that there are four common factors if the factor characteristic root is greater than 1. Four common factors were selected to explain the original 12 variables, and the interpretation ratio of the four common factor to the 12 original variables was 77.201%. The results showed that the extraction of common factors was ideal.

4) Rotation Component Matrix

In order to describe the structure of factor load more clearly, the maximum variance method is used to rotate the component matrix. The rotated component matrix is shown in Table 5.

Table 5. Composition matrix after rotation

Financial performance	Index	Factor			
		1	2	3	4
Profitability	Net profit ratio	0.898	0.331	0.108	0.069
	Gross income ratio	0.764	−0.012	−0.07	−0.075
	Total profit cost ratio	0.905	0.348	0.086	0.063
	Operating profit ratio	0.898	0.351	0.092	0.066
Solvency	Quick ratio	0.567	0.782	0.006	0.043
	Current ratio	0.587	0.76	0	0.047
	Debt asset ratio	−0.258	−0.715	0.444	−0.223
	Oper-cash into current debt	−0.077	0.149	−0.879	−0.018
Operating capacity	Iventory turover ratio	0.025	0.533	−0.06	−0.078
	Total asset turover ratio	−0.626	0.067	0.66	−0.166
Development capacity	Total asset grow ratio	−0.141	0.466	0.216	0.645
	Net profit grow ratio	0.17	−0.174	−0.137	0.855

In Factor 1, the absolute value of the coefficient is larger than that of the net selling rate 0.898, the gross selling profit rate 0.764, the cost and expense profit margin 0.905, the operating profit margin all four indicators reflect profitability, so factor 1 can be named the profitability factor.

In factor 2, the absolute value of the coefficient is greater than that of the quick ratio 0.782, the current ratio 0.76, the asset liability rate 0.715, the inventory turnover rate 0.533, Quick ratio, current ratio and asset liability ratio reflect the solvency, inventory turnover reflects the operating capacity, so the factor 2 is named solvency factor.

In the factor 3, the absolute value of the coefficient is larger than that of the cash flow debt 0.879, total asset turnover 0.66, cash flow liabilities reflect solvency and total asset turnover reflects operating capability, so factor 3 is named operating capability factor.

In the factor 4, the absolute value of the coefficient is relatively large total assets growth rate 0.645, net profit growth rate of 0.855, these two indicators reflect the development ability, so will factor 4 named development ability factor.

5) Factor Score Calculation

This section further carries on the processing to the data, obtains the factor score coefficient of each factor, then calculates each factor score, the factor scores coefficient matrix is shown in Table 6.

Table 6. Factor score coefficient matrix

Financial performance	Index	Component			
		1	2	3	4
Profitability	Net profit ratio	0.229	−0.026	0.108	0.001
	Gross income ratio	0.273	−0.192	−0.074	−0.09
	Total profit cost ratio	0.227	−0.019	0.09	−0.007
	Operating profit ratio	0.224	−0.016	0.096	−0.004
Solvency	Quick ratio	0.005	0.31	0.061	−0.06
	Current ratio	0.018	0.291	0.053	−0.056
	Debt asset ratio	0.086	−0.293	0.315	−0.089
	Oper-cash into current debt	−0.05	0.038	−0.735	−0.065
Operating capacity	Iventory turover ratio	−0.118	0.305	−0.013	−0.122
	Total asset turover ratio	−0.235	0.237	0.313	−0.112
Development capacity	Total asset grow ratio	−0.187	0.256	0.249	0.51
	Net profit grow ratio	0.076	−0.239	−0.096	0.726

According to the score coefficient matrix, F_1, F_2, F_3 and F_4 are designated as profitability factor, solvency factor, operating ability factor and development ability factor, respectively. By naming $X_1, X_2, X_3, X_4, X_5, X_6, X_7, X_8, X_9, X_{10}, X_{11}$ and X_{12} as sales net profit margin, sales gross profit margin, cost profit margin, operating profit margin, quick ratio, current ratio, asset-liability ratio, cash flow liability ratio, inventory turnover rate, total asset turnover rate, total asset growth rate and net profit growth rate, the score functions of the following factors can be obtained:

$$F_1 0.229X_1 + 0.273X_2 + 0.227X_3 + 0.224X_4 + 0.005X_5 + 0.018X_6 + 0.086X_7 \\ -0.05X_8 - 0.118X_9 - 0.235X_{10} - 0.187X_{11} + 0.076X_{12} \quad (1)$$

$$F_2 - 0.026X_1 - 0.192X_2 - 0.019X_3 - 0.016X_4 + 0.31X_5 + 0.291X_6 - 0.293X_7 \\ +0.038X_8 + 0.305X_9 + 0.237X_{10} + 0.256X_{11} - 0.239X_{12} \quad (2)$$

$$F_3 0.108X_1 - 0.074X_2 + 0.09X_3 + 0.096X_4 + 0.06X_5 + 0.053X_6 + 0.315X_7 \\ -0.735 - 0.013X_9 + 0.313X_{10} + 0.249X_{11} - 0.096X_{12} \quad (3)$$

$$F_4 0.001X_1 - 0.09X_2 - 0.007X_3 - 0.004X_4 0.061X_5 - 0.056X_6 - 0.089X_7 \\ -0.065X_8 - 0.122X_9 - 0.112X_{10} + 0.51X_{11} + 0.726X_{12} \quad (4)$$

Taking the ratio of the variance interpretation ratio and the cumulative variance interpretation rate of the four common factors in Table 4 as weight, the common factor function is weighted, and the composite score formula is obtained:

$$F_{synthesis} = 34.958 * F_1 + 21.785 * F_2 + 10.01 * F_3 + 10.449 * F_4/77.201 \\ = 0.4528F_1 + 0.2822F_2 + 0.1297F_3 + 0.1353F_4 \quad (5)$$

3.3 Analysis of Results

The former use factor analysis method to deal with the four financial indicators, the final score data can directly reflect the financial performance of enterprises. The final score is 0 as the standard, the greater than 0 indicates that the ability is good. Table 7 presents a summary of the capabilities of 52 new retail listings.

Table 7. Factor scores

Financial performance	Score			
	> 2	(0,2]	[-2,0]	< -2
Profitability	1	21	30	0
Solvency	2	20	30	0
Operational capabilities	0	32	19	1
Develop capacity	1	26	24	1
synthesis	1	18	33	0

From the data of the profitability factor, it can be seen that there are 22 new retail listed companies with factor data greater than 0, of which 21 companies have scores between 0 and 2; in addition, there are 30 companies with negative factor data. Overall, there are large differences in the profitability of listed companies in the new retail industry.

From the data of the solvency factor, it can be seen that 22 companies have solvency scores greater than 0, of which 20 companies have scores between 0 and 2; in addition, 30 companies have negative solvency scores. It can be seen that the solvency of listed companies in China's new retail industry is in a low state.

From the data of the operating capacity factor, it can be seen that 32 companies finally got more than 0, and 20 companies had negative operating capacity. Since the number of enterprises with operational capacity scores above 0 is the largest, the operational capacity of enterprises is the best among the four financial indicators, indicating that the overall operating conditions of companies are developing well.

From the data of the development ability factor, it can be seen that 27 companies have a final score of more than 0, of which 26 companies score between 0 and 2; in addition, 25 companies have a negative development ability. This shows that although the new retail listed companies are developing well, there is still a gap.

According to the comprehensive score formula, it can be seen that 19 companies finally scored more than 0, of which 18 companies scored between 0 and 2; in addition, 33 companies scored negatively. On the whole, at present, there is an imbalance in the development of China's new retail industry.

4 Suggestions for Improving the Financial Performance of Listed Companies in the New Retail Industry

4.1 Establish an Omni-channel Service Platform

Based to the analysis of profitability, the current profitability of China's new retail industry still needs to be improved; through the analysis of 52 companies, it is found that the new retail listed companies whose profits have been improved have mainly created an omni-channel integrated operation platform after transformation. Therefore, the establishment of an omni-channel service platform is conducive to improving the financial performance of enterprises.

For the establishment of the omni-channel service platform, it is necessary to maintain the expansion of physical stores offline. In the case of sufficient funds, enterprises can appropriately expand the scope of self-operated retail, further expand the consumer group. In addition, you can also build offline physical stores, create your own brand, enhance brand influence. At the same time, it is also necessary to strengthen the development of online retail platforms. By joining the existing online sales platform or opening an online platform on its own, we will make up for the shortcomings caused by the limitations of physical stores in terms of customers, resources, information, etc., improve the online operation mechanism, and form a unique value advantage.

4.2 Reasonable Allocation of the Proportion of Liabilities

According to the above analysis of solvency, it can be seen that the current solvency of China's new retail industry is low. Therefore, in order to improve the financial performance of enterprises, the proportion of liabilities should be reasonably allocated. In the process of operation, enterprises adjust the proportion of long-term and short-term liabilities in a timely manner according to the needs of development. When formulating long-term strategies, it is reasonable to use the form of long-term liabilities to reduce the financial pressure of enterprises. In the follow-up development, enterprises need to further optimize the capital structure, while using the promotion of accounts receivable collection efforts, the use of wealthy funds to develop scientific and reasonable investment strategies, improve the efficiency of the use of funds.

4.3 Improve the Efficiency of Supply Chain Distribution

Although the operating capacity of the new retail listed companies is better, there is still room for improvement in the operation status after transformation, and the distribution efficiency of the main supply chain still needs to be further improved. With the development of the information age, new retail enterprises to improve the turnover speed of inventory has not been limited to the use of human handling and the establishment of scattered warehouses, whether it is the use of self-operated logistics providers, or third-party logistics distributors, its purpose is to transport goods to consumers as soon as possible. New retail enterprises can use big data to track the user's purchase footprint in real time, guide enterprises to reasonably arrange production and warehouse management, online platform to meet the needs of consumers to shop at any time, offline stores to coordinate logistics and distribution services, to help consumers save shopping costs and improve distribution efficiency.

4.4 Accelerate the Realization of Economies of Scale

Based on the above analysis of developing capacity, the new retail listed companies have a good development capacity, but to maintain the sustainable development, it is necessary to accelerate economies of scales. New retail enterprises can pass through M & A, rapidly expand the scale in order to achieve the optimal allocation of resources, so that the business costs continue to decline, the business profits continue to increase. In addition, the most commonly used way for enterprises to achieve economies of scale is to expand the scope of chain operations, Chain management has the characteristics of high standardization and strong replication, which facilitates the unified management of enterprises and scientific division of labor, and finally achieves economies of scales.

5 Conclusions

Through the factor analysis, it can be seen that the transformation of new retail can improve the financial performance of enterprises. The innovation of this paper is that the factor analysis method is used to conduct an empirical analysis of the financial performance of listed companies in China's overall new retail industry, and the results obtained are combined with the financial performance of the new retail transformation of a single enterprise, and the final conclusion is more influential. The content written in this article is conducive to the improvement of the financial performance of China's new retail companies after the transformation, and at the same time can play a role in promoting the development of China's new retail market and national economy.

The transformation of new retail has obvious improvement effect on the financial performance of case companies, and the analysis of transformation means is conducive to providing transformation suggestions for the entire new retail enterprises at present, helping enterprises to reduce mistakes in the transformation process. With the development of China's economy, the status of the new retail industry in the national economy will continue to improve, the sales of the retail industry will continue to rise, its scale is growing rapidly, and it also provides immeasurable contributions to the economic and social development of our country.

References

1. Xue, W.: Research on the social spatial characteristics and spatial influence of new retail formats: a case study of Hema Fresh Life. In: Proceedings of China Urban Planning Annual Conference. China Architecture and Building Press, Beijing, pp. 1–12 (2018). (in Chinese)
2. Ningning, Y., Tao, L.: Discussion on the transformation of traditional retail enterprises in China based on "new retail." Res. Bus. Econ. **8**(14), 104–107 (2019). (in Chinese)
3. Hong, L., Jiawei, Z.: Opportunities, challenges and coping strategies for the development of new retail industry under the background of the new crown epidemic. Southwest Finan. **11**(7), 3–16 (2020). (in Chinese)
4. Ying, L., Ding, L., Jiajia, H.: Research on the mechanism of supply chain synergy on performance of new retail enterprises. Bus. Econ. Res. **11**(19), 116–119 (2020). (in Chinese)
5. Steven, B., Maggie, M., Brianne, G., Katie, J.: The great retail reinvention. Forbes **12**(5), 61–74 (2020)
6. Dudin, M.N., Lyasnikov, N.V., Reshetov, K.Y.: Economic profit as indicator of food retailing enterprises performance. Eur. Res. Stud. J. **35**(1), 468–479 (2018)
7. Yalan, L.: A brief analysis of the development status of China's new retail industry. Bus. Econ. **8**(9), 80–83 (2018). (in Chinese)
8. Dongyan, H.: Comprehensive performance evaluation of Suning Tesco. Cooper-ative Economy and Technology **12**(15), 133–135 (2019). (in Chinese)
9. Jing, L.: Research on financial performance evaluation of new world. Hebei Enterprise **65**(1), 118–119 (2021). (in Chinese)

Research and Implementation of Automatic Navigation and Driving of Tracked Robot in Tunnel Based on SLAM

Xiaozhe Yang(✉), Kangning Feng, and Huiting Lu

School of Intelligent Manufacturing, Nanning University, Nanning 530299, China
970423706@qq.com

Abstract. In view of the low accuracy of traditional GPS positioning and the inability to locate accurately in the tunnel effectively, this paper studies and designs an intelligent crawler robot based on SLAM technology. The research scheme is based on the ROS robot operating system of raspberry pie as the software framework, based on the STM32 core circuit as the execution system, using IMU technology to locate the position of the crawler robot, using Lidar Scanning to input environmental information, accurately drawing a map based on Hector algorithm, calculating the driving route based on global path planning algorithm, and adjusting the path in real time based on local path planning algorithm to avoid obstacles, Finally, the tracked robot based on SLAM can drive automatically in the complex tunnel environment. Finally, the feasibility, reliability and stability of the design are proved by experimental tests, which provides some reference and help for the research of intelligent mobile robot in complex tunnel environment.

Keywords: Automatic driving · Tunnel · SLAM · ROS

1 Introduction

The automatic navigation of autonomous robots is a major innovation and progress in the history of human science and technology [1, 2]. The concept of autonomous driving (also known as driverless driving) originated in the late 1950s, and gradually expanded from the initial application in the military field to the industrial and civil fields [3]. Driverless technology has been constantly innovated.

In 1966, Nielson of Stanford University and his team developed the "Shakey" smart car, which can track low-speed targets and actively avoid static obstacles. It meets the most basic functions of smart cars and is undoubtedly the first smart car in human history [4].

The "cart" car developed by Hans Moravec in 1978 can draw the information map of the surrounding environment with visual sensors, which lays the foundation for the autonomous obstacle avoidance of mobile cars [5, 6]. In 2003, Kiva's "intelligent logistics vehicle" can move autonomously between two predetermined points, reducing labor costs, improving mobile efficiency, and significantly increasing the company's revenue.

Z. Hu et al. (Eds.): AIMEE 2022, LNDECT 159, pp. 175–188, 2023.
https://doi.org/10.1007/978-3-031-24468-1_16

China officially launched the research on driverless technology in the late 1970s. In 2000, the first intelligent wheelchair using sensor fusion system (including infrared sensor) was successfully developed. It could accurately locate and calculate the optimal driving route in real time to avoid obstacles [7, 8]. The first humanoid intelligent crawler robot "forerunner" in the history of China also came out in the same year. Its function was very powerful. It could not only walk independently, but also express its own situation in language.

Intelligent wheelchair, "pioneer" and "THMR-V" developed by Tsinghua University in 2003 are China's landmark achievements in the field of driverless [9].

At present, the mainstream automatic driving technology often faces new challenges when driving in the tunnel because of the weak GPS signal and the complex environment in the tunnel [10]. The automatic navigation driving using SLAM technology can not only use laser radar to scan the surrounding environment in real time to locate the position of the vehicle in the unknown environment, but also draw a map through laser scanning the surrounding environment and calculate the optimal driving route, so as to provide an effective solution for vehicle positioning and navigation in the complex and unknown tunnel environment.

2 System Design Scheme of Crawler Robot

Automatic navigation and control is an important function of robot operation [11]. This paper takes the realization of automatic navigation and driving of tracked robot in tunnel based on SLAM technology as the research object. The crawler robot mainly includes brush motor, crawler, robot metal structure, battery, voltage monitoring screen, HD screen, Lidar, GY-85 nine axis gyroscope, STM32 single chip microcomputer, raspberry pie 4B motherboard. The research and development are carried out on the mainstream ROS system platform. The ROS robot operating system is used to compile the automatic navigation function of the crawler robot. The raspberry 4B master under raspberry school is used as the decision control system and the STM32 master control board is used as the executive element of the system, and the IMU accelerator gyroscope sensor is equipped. By scanning and inputting the surrounding environment information by lidar and drawing the map and path planning by SLAM algorithm, the tracked robot can automatically move and avoid obstacles in the unknown tunnel environment.

2.1 Analysis of Usage Scenarios of Tracked Robot

The crawler robot designed in this paper needs to pass through the tunnel independently and safely. The environment in the tunnel is not only complex and the GPS signal is weak. At the same time, the tunnel may also cause water leakage, seepage, collapse and other problems due to bad weather, as well as traffic accidents caused by human factors, tunnel maintenance and other problems that hinder the passage. The crawler robot not only needs to pass through the tunnel safely, but also needs to accurately avoid all kinds of obstacles in the tunnel. Because the environment in the tunnel is complex and unknown, the tracked robot designed in this paper needs to meet the following three conditions:

(1) It can recognize the tunnel environment in the unknown tunnel environment with weak GPS signal and accurately locate the pose information of the tracked robot in the tunnel;
(2) It can accurately draw the environmental map in the tunnel;
(3) Can independently plan the walking route and avoid obstacles in the tunnel.

2.2 Overall Design Scheme of Crawler Robot

The crawler robot that can realize self-driving in the tunnel is composed of two parts: hardware and software.

The hardware consists of machinery, bottom control hardware and raspberry school master hardware. The mechanical part includes metal robot structure, motor, crawler and other parts. The bottom control hardware includes nine axis gyroscope, STM32 development board and motor drive module. The main control part of raspberry pie is mainly ROS robot operating system and Lidar.

The software consists of the bottom driver of STM32 and ROS robot operating system. Among them, the tasks that STM32 bottom drive needs to complete are:

1) motor PWM control speed;
2) Motor speed measurement;
3) Guide angle acquisition;
4) STM32 communicates with ROS and sends robot data to ROS.

The tasks that ROS needs to complete include:

1) Installing the system and installing ROS;
2) Create a function node of ROS foundation. The task of this node is to control the crawler robot, calculate and publish the odometer of the crawler robot, publish the robot pose information, and communicate with STM32;
3) Installation drawing algorithm;
4) Install the navigation algorithm.

The overall structure is shown in Fig. 1.

2.3 Automatic Navigation and Driving Technology of Tracked Robot Based on SLAM

When the self-driving crawler robot designed in this paper turns on the power, the Lidar starts to work, scans its surrounding environment 360° continuously, and transmits the information obtained by scanning to raspberry pie 4b, which is processed by SLAM algorithm to generate the corresponding map information [12]. Then, by drawing a successful two-dimensional grid map, use the A*(A-star) algorithm to calculate the route with the highest efficiency from the starting point to the end point, and calculate the obstacles on the route in combination with the local path algorithm [13], and then send a signal to the STM32 development board to drive the crawler robot to travel. In the process of driving, the crawler robot uses IMU technology to detect its own pose information in real time. Thus, the tracked robot can accurately locate its position and posture in the tunnel, avoid obstacles and adjust the driving route in real time. Realize the automatic driving of the crawler robot in the tunnel.

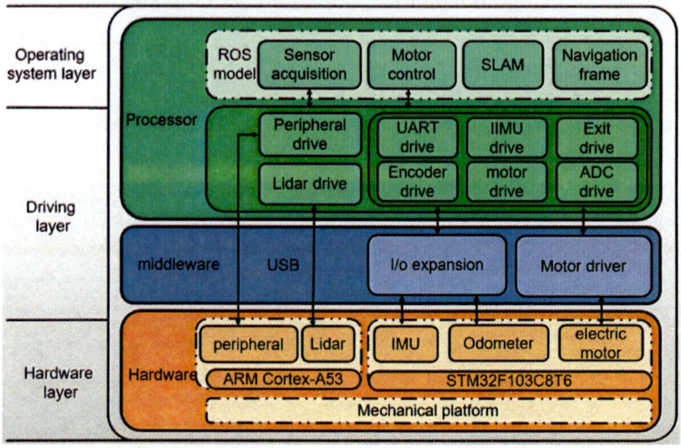

Fig. 1. Overall design scheme of crawler robot

2.3.1 SLAM Mapping Based on Hector Algorithm

Hector algorithm is a SLAM navigation technology based on two-dimensional space. It uses an inertial sensor system with good reliability and does not need to use odometer [14]. However, the standard of Lidar required for scanning environment is very high, which requires Lidar to have low noise and fast update rate. This system uses RPLIDAR A1 Lidar, which can scan and measure 360° environment, and adopts optical magnetic fusion technology. The maximum measurement radius is 12 m and the frequency is 8000 times/s.

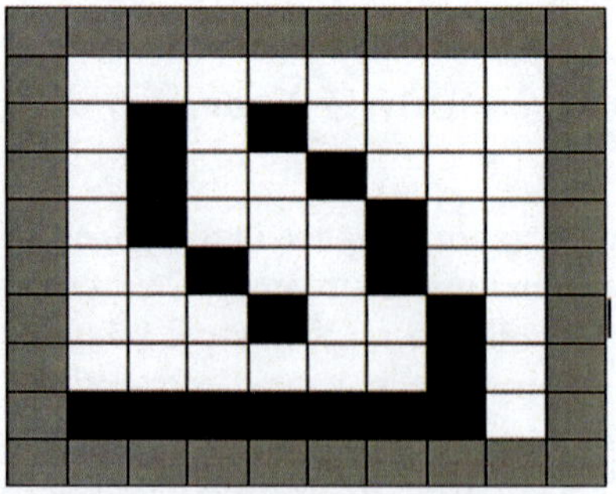

Fig. 2. Two dimensional grid map

Because the map drawn based on Hector algorithm is composed of countless grids in a two-dimensional plane, it is expressed as a two-dimensional grid map, which is called grid map for short. Grid map is a map created based on the probability of obstacles occupying the grid in the two-dimensional plane. Each grid can only be divided into three states: idle, unknown or occupied [15], as shown in Fig. 2.

Its advantage is that it is easy to build maps, occupies less memory, and can represent any environment. It consists of a SLAM front end and a SLAM back end. The front-end SLAM is used to calculate the real-time motion of the intelligent tracked robot, while the back-end SLAM is used to optimize the position of the intelligent tracked robot, and at the same time, the constraints with the pose information generated in the front-end system should be considered. The back-end system mainly includes map access, scan matching and multi-resolution map output.

(1) Map access: the intelligent crawler robot initializes the Lidar at the initial position and takes the generated laser data as the first processing frame [16]. At time t + 1, when the laser touches the barrier and is reflected, the Lidar will store these laser data and store them in the grid map at time t after calculation. If all laser data received at time t + 1 can be matched with the grid map at time t, it means that the grid map at time t + 1 and time t are successfully matched. In order to facilitate the processing of laser data and the extraction and discretization of environmental information obtained from the continuous scanning of laser radar during the continuous movement of intelligent tracked robot. However, the discreteness will produce inevitable errors to the accuracy of the data. In order to avoid these errors, bilinear interpolation method is used to estimate the grid occupancy probability of two-dimensional grid map (Fig. 3).

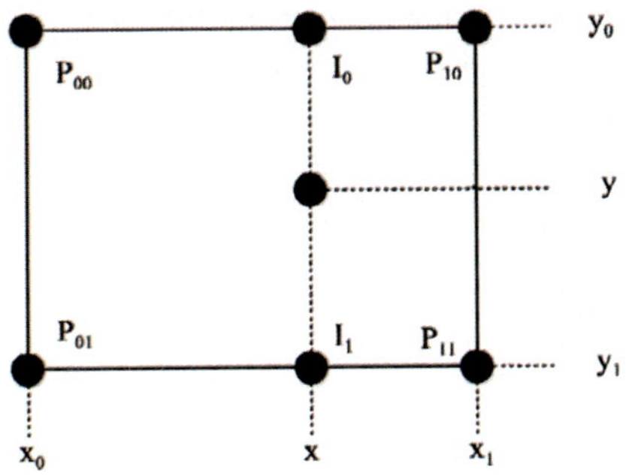

Fig. 3. Bilinear interpolation method

Suppose that the coordinate of a point is known as P_m, and the grid in the grid map represented with M(P_m). Use *(P00, P01, P10, P11)* to stand for the four grid coordinates around it, so as to simplify the calculation of the coordinate PM of the known point. Then calculate the bilinear interpolation of x, y axes first, which is expressed by formula

(1):

$$M(P_m) = \frac{y - y_0}{y_1 - y_1}\left(\frac{x - x_0}{x_1 - x_0}M(P_{11}) + \frac{(x_1 - x)}{(x_1 - x_0)}M(P_{01})\right) + \frac{y_1 - y}{y_1 - y_0}\left(\frac{x - x_0}{x_1 - x_0}M(P_{10}) + \frac{x_1 - x}{x_1 - x_0}M(P_{00})\right) \quad (1)$$

Calculate the partial derivative of *Pm* in formula (1) in the x-axis direction as shown in formula (2).

$$\frac{\partial \mathrm{M}}{\partial \mathrm{x}}(\mathrm{P_m}) \approx \frac{\mathrm{y} - \mathrm{y_0}}{\mathrm{y_1} - \mathrm{y_0}}(\mathrm{M}(\mathrm{P_{m11}}) - \mathrm{M}(\mathrm{P_m})) + \frac{\mathrm{y_1} - \mathrm{y}}{\mathrm{y_1} - \mathrm{y_0}}(\mathrm{M}(\mathrm{P_{10}}) - \mathrm{M}(\mathrm{P_{m00}}) \quad (2)$$

Get the partial derivative of *Pm* in formula (1) in the y-axis direction as shown in formula (3).

$$\frac{\partial \mathrm{M}}{\partial \mathrm{x}}(\mathrm{P_m}) \approx \frac{\mathrm{x} - \mathrm{x_0}}{\mathrm{x_1} - \mathrm{x_0}}(\mathrm{M}(\mathrm{P_{m11}}) - \mathrm{M}(\mathrm{P_m})) + \frac{\mathrm{x1} - \mathrm{x}}{\mathrm{x_1} - \mathrm{x_0}}(\mathrm{M}(\mathrm{P_{10}}) - \mathrm{M}(\mathrm{P_{00}})) \quad (3)$$

Combined by formula (1) (2) (3), the gradient value of coordinate *Pm* point is shown in formula (4)

$$\nabla M(P_m) = \left(\frac{\partial M}{\partial x}(Pm), \frac{\partial M}{\partial y}(P_m)\right) \quad (4)$$

(2) The next task after building the grid map is scanning matching, that is, the process of matching the scanned laser data with the given map. The closer the algorithm result of the laser occupation value m (*Pm*) is to 1, the better the matching effect. According to the laser data obtained at time t + 1, the position gain from time t to time t + 1 is calculated, and the best matching result is obtained through iterative matching at time t + 1. Complete the pose information estimation of intelligent mobile robot. In this process, the initial data collected by Lidar is used as the basis for map construction, and the data collected later is matched with the constructed map. Doing these tasks well requires a lot of complex and cumbersome calculations, and the best solution is to use the Gauss Newton gradient method, as shown in formula (5).

$$\eta^* = arg_\eta min \sum\nolimits_{j=1}^{n} [1 - M(S_J(\eta))]^2 \quad (5)$$

where: $\eta = (Px, Py, \beta)$ represents the optimal pose of intelligent tracked robot. And $Sj(\eta)$ represents the position coordinates of the Lidar in the grid map, as shown in formula (6).

$$S_j(x_j) = \begin{bmatrix} cos\beta & -sin\beta \\ sin\beta & cos\beta \end{bmatrix} \cdot \begin{bmatrix} S_{j,x} \\ S_{j,y} \end{bmatrix} + \begin{bmatrix} x \\ y \end{bmatrix} \quad (6)$$

In formula (6): $Sj(\eta)$ the two-dimensional grid map value of is $M(Sj(\eta))$. According to the idea of Gauss Newton gradient method, suppose an initialization value is

η, According to formula (7), the incremental value $\Delta\eta$ Carry out fuzzy calculation, and the result is shown in formula (7).

$$\sum_{j=1}^{n}\left[1 - M\left(S_j(\eta + \Delta\eta)\right)\right]^2 \to 0 \tag{7}$$

In Eq. (7), $M\left(S_j(\eta + \Delta\eta)\right)$ is difficult to solve to get its value, expand it with Taylor formula, and Incremental value $\Delta\eta$ of The result of η partial derivation is shown in formula (8).

$$\begin{aligned}&\nabla_\eta \sum_{j=1}^{n}\left[1 - M\left(S_j(\eta + \Delta\eta)\right)\right]^2 = 2 \cdot \sum_{j=1}^{n}\left[\nabla M\left(S_j(\eta)\right) \cdot \frac{\partial S_j(\eta)}{\partial\eta}\right]^T \\ &\left[1 - M(S_j(\eta)) - \nabla \mathrm{M}(S_j(\eta)) \cdot \frac{\partial S_j(\eta)}{\partial\eta}\Delta\eta\right]\end{aligned} \tag{8}$$

To sum up, $\Delta\eta$ is capable to be transformed into a function extremum problem.

When Eq. (8) is set to be 0, the incremental value can be obtained as $\Delta\eta$, shown in formula (9).

$$\Delta\eta = H^{-1}\sum_{j=1}^{n}\left[\nabla M\left(S_j(\eta)\right) \cdot \frac{\partial S_j(\eta)}{\partial\eta}\right]^T\left[1 - M\left(S_j(\eta)\right)\right] \tag{9}$$

In Eq. (9): the matrix of Hession is H, as presented in Eq. (10).

$$\mathrm{H} = \left[\nabla M\left(S_j(\eta)\right) \cdot \frac{\partial S_j(\eta)}{\partial\eta}\right]^T\left[\nabla M\left(S_j(\eta)\right) \cdot \frac{\partial S_j(\eta)}{\partial\eta}\right] \tag{10}$$

In formula (10): $\nabla M\ Sj\eta$ represents the gradient value of the map, which has been described in Eq. (3).

(3) Multi resolution map output: reduce the length and width of each grid in the two-dimensional grid map in proportion, so as to improve the definition of the two-dimensional grid map generated by Lidar scanning. And when using iterative calculation, the local extreme value can be reduced by this method.

2.3.2 Environmental Map and Path Planning

Route planning is to integrate environmental information from various sensors to create an environmental map. The automatic navigation of the crawler robot designed in this paper uses grid map, because its representation of obstacles is very simple and easy to implement. Grid map refers to dividing the environment map into square grids of equal size through a certain proportion. These square grids are divided into three different states according to the environmental information of the environmental map, namely, the dominant state, the unowned state and the unknown state. They are usually distinguished by black, white and gray. White represents the ownerless state, which means that the intelligent crawler robot can pass, while black represents the main state, which prohibits the intelligent crawler robot from passing, and gray represents the unrecognized environment. When an accurate and reliable grid map is established, the intelligent crawler robot can use the path planning algorithm for autonomous navigation, that is, it can independently move from the *U* point in the grid map to the *G* point according to the mastered grid map, and the *U* point and *G* point can be any two points in the grid map.

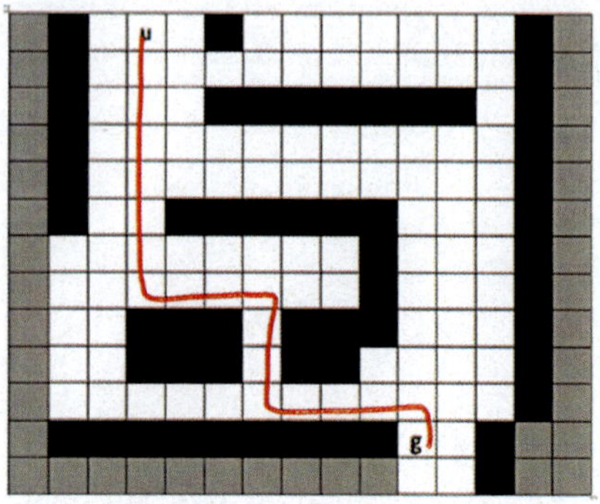

Fig. 4. Schematic diagram of path planning

The automatic navigation system includes global path planning and local path planning. The global path planning can calculate the best path for the intelligent crawler robot to reach the destination, while the local path planning can dynamically adjust the route calculated by the global path planning in real time to ensure that this route is the best route, and it can automatically avoid dynamic obstacles and rebuild the best route to the destination during the driving process of the intelligent crawler robot.

2.3.3 Global Path Planning

The problem of global path planning is to arbitrarily give a starting point and an ending point on the map that has obtained the map environment information. How can we find a safe route with the minimum distance between these two points, avoid obstacles on the route and calculate the most efficient route in real time and effectively? Simply put, on a known map, assuming that the intelligent tracked robot is at the starting point *U* and the target is point *G*, there are countless routes from point *U* to point *G*, but only one route is optimal, which is the work of the global path planning algorithm. In order to calculate the optimal route, the algorithm used by our intelligent crawler robot is A*(A-Star) algorithm, and its algorithm flow is shown in Fig. 5.

A* (A-star) algorithm is a heuristic path planning algorithm [17], which uses the estimation function to search the destination, and takes into account the actual cost from the starting point to the current location and the estimated cost from the current location to the destination at the same time. The expression of the function is shown in Eq. (11).

$$f(n) = g(n) + h(n) \tag{11}$$

In formula (11): $f(n)$ represents the estimated cost at the current point; $g(n)$ represents the actual cost from the starting point to the present point; $h(n)$ represents the estimated

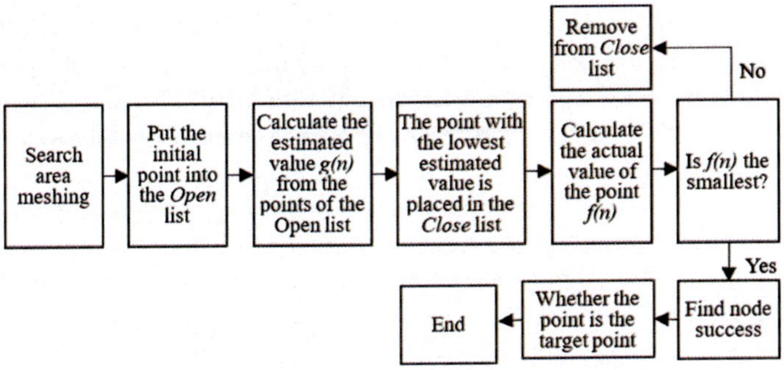

Fig. 5. A*(A-Star) algorithm flow chart

cost from the current point to the end point.

$$h(n)|x1 - x2| + |y1 - y2| \tag{12}$$

In formula (12): the coordinates of the current point are (x_1, y_1) and the coordinates of the target point are (X_2, Y_2).

When using A*(A-Star) algorithm to plan the route, use the close list and open list to store the points of the grid map respectively, and each point represents a grid. The points that have been scanned and matched and given the estimated cost are stored in open. The points with the lowest estimated cost are stored in close, and the route of each point is processed in close to create a mobile route. Figure 4 shows the path planning of the intelligent tracked robot using the A*(A-Star) algorithm to move from the starting point *U* to the ending point *G* in the map. The process steps are as follows:

(1) In the known map, as shown in Fig. 4, assuming that the *U* point is the starting point of the intelligent tracked robot, the *U* point is taken as the first calculation point, and the points around the *U* point are stored in open. The estimated cost *f(n)* of each point stored in open is calculated with formula (11).
(2) Compare the estimated cost of each node in *Open*, and then take the node with the lowest estimated cost as the current processing point, represented by *S*, and delete the *S* point from open and store it in *Close.*
(3) Assuming that the actual cost *g(n)* from an adjacent point K to the starting point *U* is smaller than the actual cost *g(n)* from the current processing point *S* to the starting point *U*, replace the adjacent point *K* with the current processing point *S*, store it in *Close*, and then delete the original point *S* from *Close*. Simply put, the point with the lowest actual cost is saved in *Close*.
(4) Repeat the above steps until all the optimal nodes to the target point *G* are calculated, then add these nodes to *Close*, and connect lines along the central point of each node, and the resulting line is the optimal path.

2.3.4 Local Path Planning

A*(A-star) algorithm can calculate the best path from the starting point to the end point, but it will be helpless in the face of dynamic obstacles on the environment map. Because the obstacles that appear when the intelligent tracked robot travels on its planned route are not always static. In order to solve this defect and avoid the collision of intelligent crawler robot in the process of driving, the local path planning dynamic window method (DWA) was born. This method can effectively avoid dynamic obstacles. This algorithm puts different speeds into the speed space according to the activity of intelligent crawler robot (V, ω) For comparison. Different speeds are used to simulate the driving route of the intelligent tracked robot in the specified time. Finally, the scoring rules are used to compare each driving route obtained from the simulation, and the route with the highest score is regarded as the route with the highest efficiency. The scoring rules are shown in (13).

$$G_{(v,\omega)} = \sigma(\alpha Xheading(v, \omega) + \beta Xdist(v, \omega) + \gamma Xvelocity(v, \omega)) \tag{13}$$

In the formula, *dist (V, ω)* represents the linear interval between the occupied grid in the grid map and the intelligent tracked robot traveling on the line. *Heading(v, ω)* indicates the score of the angle, and *velocity (v, ω)* is the velocity curvature change score.

3 Test Results and Analysis of Automatic Driving of Crawler Robot in Tunnel

3.1 Attitude Calibration of Crawler Robot

The tracked robot in this study is composed of hardware and software systems. ROS platform is run on raspberry pie 4B as the functional control of the tracked robot. IMU, linear velocity and angular velocity need to be calibrated before the experiment.

3.2 Map Construction Experiment of Crawler Robot

The requirement of the design is to realize automatic driving in the tunnel. This experiment will establish a tunnel environment with complex ground conditions, as shown in Fig. 6. The tunnel has a ground environment consistent with many actual tunnels, including factors with complex conditions such as mud pits, gravel, debris, vegetation and so on. Under this environment, the crawler robot will conduct map building experiments with Karto algorithm, Cartographer algorithm and Hector algorithm respectively. The effectiveness and accuracy of each algorithm have been verified by comparison.

The test process of map construction of crawler robot is described as follows: place the crawler robot at the entrance of the tunnel, mark the initial position of the crawler robot, open the Ubuntu system on the PC, connect and control the crawler robot with SSH instructions, and then open the control script with commands to control the crawler robot to move in the tunnel, scan the environment at the same time and generate grid maps, and save the maps drawn by Karto algorithm, cartographer algorithm and Hector algorithm separately, The test results are shown in Fig. 7(a), (b) and (c) respectively.

As can be seen from the mapping results, the constructed grid map is divided into three different color areas of black, white and gray [18]. The black is the barrier, which

Fig. 6. Experimental environment

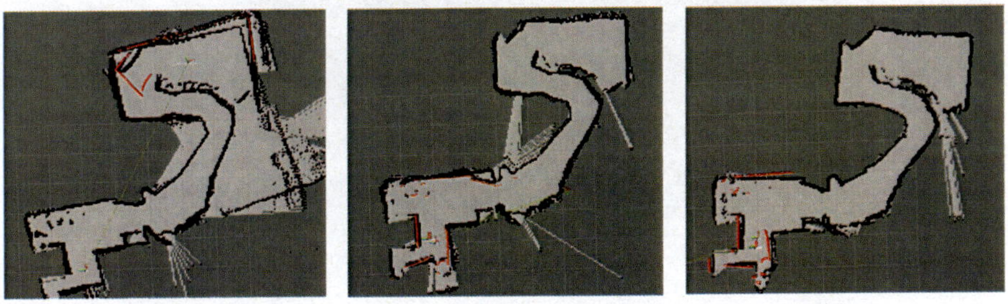

(a) (b) (c)

Fig. 7. Algorithm mapping of three algorithms, (a) Karto algorithm mapping (b) Cartographer algorithm mapping (c) Hector algorithm mapping

means it is impassable, the white is the open area, which means it is movable, and the gray area is where not scanned by the crawler robot [19].

By comparing the three SLAM mapping algorithms through experiments, the map constructed by Karto algorithm is the worst (as shown in Fig. 7(a)). The map constructed by cartographer algorithm has a little flaw (as shown in Fig. 7(b)). Although there are some small black spots in the map built by Hector algorithm, the black area can well describe the shape of the obstacles (as shown in Fig. 7(c)). Therefore, in an environment where various algorithms are used to build maps, Hector algorithm has the best effect, the second is Gmapping algorithm, and Karto SLAM algorithm is comparatively the worst.

3.3 Autonomous Navigation Test of Crawler Robot

The test process of autonomous navigation of crawler robot is as follows: in the tunnel environment when the map has been built, place the crawler robot outside the tunnel entrance as the starting point of navigation, and the tunnel exit as the end point, and several obstacles are put in the tunnel, as shown in Fig. 8, the left and right brick obstacles.

Place the crawler robot at the tunnel entrance, start the crawler robot, calculate the optimal path of the tunnel entrance and tunnel exit by the global path planning algorithm, and the Lidar scans the surrounding environment information in real time to detect whether there are obstacles. When the crawler robot drives into the tunnel, the Lidar scans two obstacles, at this time, the local path planning algorithm starts to calculate the position relationship between the obstacles and the crawler robot, and calculates a local path to avoid the obstacles to complete the passage, as shown in Fig. 8. Continue to the first turning point, and effectively complete the turning passage at the turning point, as shown in Fig. 9. The second turning point behind also passes safely and drives out of the tunnel, as shown in Fig. 10. Then, Complete the navigation task.

Fig. 8. Two obstacles at the tunnel entrance

Fig. 9. The first turning point in the tunnel

Through experiments, it can be seen that the crawler robot designed in this paper can avoid obstacles perfectly, and can maintain a certain distance between the travel path and obstacles, and can travel to the destination without error according to the planned road strength, which shows that it has the reliability of automatic driving. At the same time, the rotation angle of the crawler robot is relatively smooth when it turns on the route, and there is no left-right jitter during the driving process. It can drive smoothly, with basically the same speed and smooth driving, which shows that the crawler robot designed in this study has stability. It is proved that the crawler robot designed in this study can effectively realize automatic driving in complex environment in the tunnel.

Fig. 10. Tunnel second turning point (tunnel exit)

4 Conclusion

Through experiments, it is capable to be find that the crawler robot designed in this paper can avoid obstacles perfectly, and can maintain a certain distance between the travel path and obstacles, and can travel to the destination without error according to the planned road strength, which shows that it has the reliability of automatic driving.

At the same time, the rotation angle of the crawler robot is relatively smooth when it turns on the route, and there is no left-right jitter during the driving process. It can drive smoothly, with basically the same speed and smooth driving, which shows that the crawler robot designed in this study has stability. It is proved that the crawler robot designed in this study can effectively realize automatic driving in complex environment in the tunnel.

Acknowledgment. This project is supported by the project of scientific research and basic ability improvement for young and middle-aged teachers of Guangxi universities (No. 2022KY1781).

References

1. Tahir, N.M., Bature, U.I., Abubakar, K.A., Baba, M.A., Yarima, S.M.: Image recognition based autonomous driving: a deep learning approach. Int. J. Eng. Manuf. (IJEM) **10**(6), 11–19 (2020)
2. Omisore, M.O., Ofoegbu, E.O., Fayemiwo, M.A., Olokun, F.R.: Intelligent mobile application for route finding and transport cost analysis. Int. J. Eng. Manuf. (IJEM) **8**(9), 73–80 (2016)
3. Qi, H.: Research on path planning method of tracked unmanned vehicle. Wuhan University, Wuhan (2018). (in Chinese)
4. Yu, D.: Obstacle avoidance and trajectory planning of mobile robot. Zhejiang University, Hangzhou (2017). (in Chinese)
5. Rong, X.: Intelligent car based on ROS system. Nanjing University of Aeronautics and Astronautics, Nanjing (2018). (in Chinese)
6. Wei, T.: System design of autonomous navigation of ROS intelligent car. Anhui University of Engineering, Hefei (2020). (in Chinese)
7. Tiwari, A.K., Nadimpalli, S.V.: New fusion algorithm provides an alternative approach to robotic path planning. Int. J. Inf. Eng. Electron. Bus. (IJIEEB) **12**(3), 1–7 (2020)
8. Yousif, M., Salim, A., Jummar, W.K.: A robotic path planning by using crow swarm optimization algorithm. Int. J. Math. Sci. Comput. (IJMSC) **7**(1), 20–25 (2021)

9. Bhargav, S., Meeradevi, R., Mundada, M.R., Ravanavar, S.G.: An interactive cart with analytics recommendation and tracking-iCART. Int. J. Inf. Eng. Electron. Bus. (IJIEEB) **12**(2), 1–8 (2020)
10. Mohamed, S.I., Abdelhadi, M.: IoT bus navigation system with optimized routing using machine learning. Int. J. Inf. Technol. Comput. Sci. (IJITCS) **13**(3), 1–15 (2021)
11. Mohammed, R.H., Elnaghi, B.E., Bendary, F.A., Elserfi, K.: Trajectory tracking control and robustness analysis of a robotic manipulator using advanced control techniques. Int. J. Eng. Manuf. (IJEM) **8**(6), 42–54 (2018)
12. Rong, W., Yongjing, W.: Design of a multifunctional exploration robot based on slam. Mach. Electron. **37**(09), 51–53+58 (2019). (in Chinese)
13. Hart, P.E., Nilsson, N.J., Raphael, B.: A formal basis for the heuristic determination of minimum cost paths. IEEE Trans. Syst. Sci. Cybernet. **4**(2), 28–29 (1972)
14. Kohlbrecher, S., Stryk, O.V., Meyer, J., et al.: A flexible and scalable SLAM system with full 3D motion estimation. In: IEEE International Symposium on Safety. IEEE (2011)
15. Chen, J.: Design and implementation of autonomous navigation mcnaim wheel car based on laser radar. Heilongjiang University, Harbin (2021)
16. Wu, X., Chen, S.: Huang Yanwei Research on complex environment location based on lidar. Electromech. Technol. **01**, 31–34 (2021)
17. Stentz, A.: Optimal and efficient path planning for partially-known environments. In: Proceedings of IEEE International Conference on Robotics and Automation, pp. 3310–3317 (1994)
18. Qiu, C., Wang, L.: Design of full coverage path algorithm for disinfection robot. Autom. Instr. **37**(02), 39–43 (2022)
19. Hushe, G., Suryawanshi, R.: WiFi based indoor navigation system for closed buildings using smart phones. Int. J. Eng. Manuf. (IJEM) **11**(3), 35–44 (2021)

Research on Spaces of Public Parking Lots Based on Total Forecast Model-Taking Wuchang District as an Example

Yuan Zheng(✉), Dan He, Weiya Fu, Cong Zou, Bing Sun, and Qian Yu

Central Southern China Electric Power Design Institute Co., LTD. of China Power Engineering Consulting Group, Wuhan 430063, China
zy5698@csepdi.com

Abstract. With the continuous increase in the number of motor vehicles, the difficulty of parking has become one of the main reasons for traffic congestion. The construction of urban public parking lot is the focus of improving the relationship between parking supply and demand and thus improving the traffic capacity. Parking demand forecasting is the first step in the planning and construction of parking lot. Based on the total forecast model, combining with the idea of traffic district division, using the data of population and motor vehicle ownership, this paper studies the parking space of public parking lot, and takes Wuchang as an example for experimental analysis. The experimental results show that when the district division and related data are fine enough, the parking lot demand prediction can also preliminarily realize the parking lot location analysis, which provides a reference for public parking demand prediction and parking location planning.

Keywords: Parking demand forecast · Aggregate forecast model · Public Parking · GIS

1 Instruction

With the development of the urban economy, the number of motor vehicles continues to increase. By 2021, the number of vehicles in the country will reach 395 million, and there will be 27.5 million newly licensed drivers throughout the year [1]. The contradiction between the increasing number of motor vehicles and the limited number of parking spaces has become a major problem in the development of urban transportation. The problem of parking difficulties has become one of the main reasons for urban traffic congestion, which restricts the sustainable development of the city. Correspondingly, at present, most of the cities in China generally have outstanding problems such as insufficient number of parking spaces, unbalanced supply and demand, unreasonable parking structure, unscientific ratio, and unbalanced parking layout [2]. Therefore, it is imperative to strengthen the construction of urban public parking lots, especially the research on parking demand forecasting.

To this, the author studies a land use model that considers the relationship between land use and parking demand and adopts multiple linear regression analysis [3]; a model

Z. Hu et al. (Eds.): AIMEE 2022, LNDECT 159, pp. 189–198, 2023.
https://doi.org/10.1007/978-3-031-24468-1_17

for predicting regional parking spaces based on actual demand [4]; considers the relationship between parking facilities and road network capacity. The parking demand prediction model based on the constraints of road network capacity [5, 6]; an improved weighted parking generation rate model that considers traffic location, vehicle ownership level, travel level, and parking turnover rate [7]; an MRA model based on better fitting of parking demand and impact factors, beetle antennae search (BAS) algorithm and BP neural network combined demand forecasting algorithm model [8]. These models analyze the impact of single parking impact factors on parking demand, and are relatively independent in data acquisition, require a large amount of investigation, high threshold, low generality, and are not closely integrated with other planning, and the practicability of model parameters is not strong, difficult to apply theory to practice.

Therefore, this paper selects and improves the total parking demand forecasting model that is adapted to the planning data, and expands its application based on the idea of the OD survey of traffic districts for the division of traffic districts [9], using population distribution, vehicle ownership distribution, etc. Basic city data to realize demand forecasting of spaces in different regions.

2 Total Forecast Model

2.1 Model Overview

The total forecast model can predict the parking demand at the district level and above from the macro level, and comprehensively consider the population development and the number of motor vehicles in the planning year, and forecast the demand for parking spaces in the area [10]. In this paper, some improvements are made to the population distribution and motor vehicle distribution to achieve sub-regional prediction.

2.2 Technical Route

According to the principle of dividing traffic areas, the basic planning data such as population distribution and vehicle distribution are integrated, and then the demand for parking spaces is predicted based on the population and the number of motor vehicles in each traffic area. Finally, the weighted integration is the total demand for public parking spaces in each area, as shown in Fig. 1.

3 Specific Method

3.1 Traffic District Division

The purpose of dividing traffic cells is to fully understand the spatial distribution characteristics of traffic flow between traffic sources. Theoretically speaking, the smaller the traffic area is, the more precise the corresponding analysis can be. In this study, the parking demand forecast can be refined to the smaller area, which is more favorable for parking lot site selection. However, it will increase the workload of investigation and increase the cost of OD data investigation, which is not easy to obtain.

Generally speaking, the division only needs to comply with the following principles:

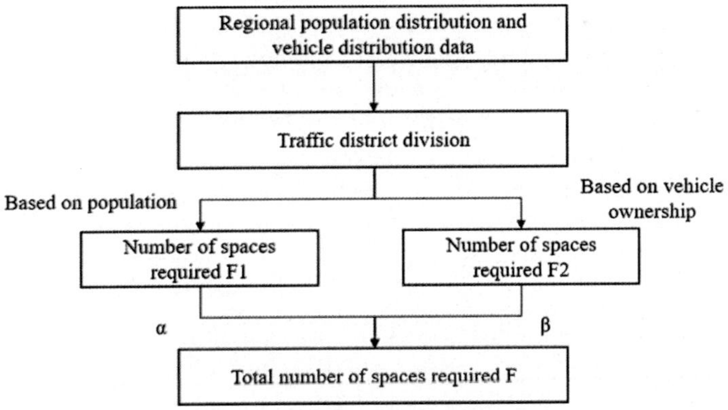

Fig. 1. Total forecast process

(1) Make the traffic survey reduce the workload as much as possible while meeting the accuracy requirements, enhance the practical operability of the survey, and reduce the difficulty of analysis and prediction;
(2) Homogeneity The nature of land use, economic level, and development level in the traffic area should be as consistent as possible;
(3) Consider historical survey data. It should be compatible or combined into the boundaries of previous surveys to facilitate the integration and comparison of new and old data, and reduce the workload of the survey. Similarly, data are generally stored in administrative divisions, and the division of traffic areas should generally not break administrative divisions to facilitate data acquisition;
(4) When there are natural or artificial divisions such as rivers and railways in the traffic division area, they should generally be used as the boundary of the traffic area, while avoiding the main road as the boundary of the traffic area;
(5) The special area should be divided into the traffic area separately. For example, large-scale tourist ecological scenic spots that have an important impact on traffic, etc.;
(6) With travel time as the standard, vehicles move slowly on congested roads. Therefore, the division of traffic areas in congested roads should be more detailed than in non-congested roads. Land use High-intensity areas are more granular than low-intensity areas, and densely populated areas are more granular than sparsely populated areas.

3.2 Parking Space Prediction Based on Population

Predicting urban public parking demand based on the population of the planned area. According to the requirements of "Urban Road Traffic Planning and Design Specification" (GB-50220–95), the total land area of urban social public parking lot can be calculated according to the planned urban population of 0.8–1.0 square meters per person, and the land for motor vehicle parking lot should be 80%–90%, and the land for non-motor vehicle parking lot should be 10%–20%.

The elastic coefficient of urban development is 1.0–1.3 according to the characteristics of future urban development. The area of the public parking lot for motor vehicles should be calculated according to the equivalent number of car parking spaces. For the ground public parking lot, each parking space should be 25–30 square meters; the construction area of the parking building and the underground parking lot, each parking space It should be 30–40 square meters. The specific calculation formula is as follows:

$$F_1 = \frac{PK_1K_2K_3}{S} \tag{1}$$

F_1 is the planned annual demand for parking spaces in urban public parking lots based on urban population. P is the population of the central urban area in the year of urban planning. K_1 is the ratio of the parking lot area that should be occupied by each planned population. K_2 is the elastic coefficient of urban development, according to the characteristics of the future development of the city. K_3 is the ratio of the parking lot area for motor vehicles to non-motor vehicles. S is the occupied area of a unit motor vehicle parking space.

3.3 Parking Space Prediction Based on Motor Vehicle Occupancy

Predicting urban public parking demand based on the number of motor vehicles in the planning area [11]. According to the "Urban Parking Planning Specification" GB_T51149–2016, the provisions for planning the number of motor vehicles and the total supply of motor vehicle parking spaces are as follows [12]. Among them, the parking spaces equipped with buildings account for more than 85% of the total demand, and the proportion of parking spaces in urban public parking lots is 10% to 15%. The specific calculation formula is as follows:

$$F_2 = Nt_1t_2 \tag{2}$$

F_2 is the number of parking spaces in urban public parking lots in the planning year based on motor vehicle ownership. N is the number of motor vehicles in the planning year. t_1 is the elastic coefficient of urban development. t_2 is the proportion of parking spaces in urban public parking lots.

3.4 Total Space Demand

According to the results of forecasting the demand of urban public parking lots by the above two methods, and comprehensively considering the future development of the city, the appropriate total demand for spaces F of urban public parking lots is obtained. The calculation formula is as follows:

$$F = \alpha F_1 + \beta F_2 \tag{3}$$

F is the demand for parking spaces in urban public parking lots in the planning year. α is the proportion coefficient based on the forecast of the planned population. β is the proportion coefficient based on the prediction of the number of motor vehicles.

4 Experiment and Analysis

4.1 Experimental Data

The basic geographic information data of Wuchang District in this study is obtained from AutoNavi Map using data crawling tools, and the coordinate system used by AutoNavi Map is the Mars coordinate system, which was developed by the National Bureau of Surveying and Mapping of China for the Internet map after offsetting. The coordinate system of a geographic information system. Therefore, it is necessary to use the corresponding coordinate conversion tool, and through the corresponding correction processing, the GCJO2 coordinates are converted into the WGS1984 geographic coordinate system. As shown in Fig. 2, the basic geographic data includes Wuchang district and county boundaries, waters, railways, provincial roads, national roads and other grades of roads. As shown in Fig. 3, Wuchang District data includes population distribution and vehicle distribution data.

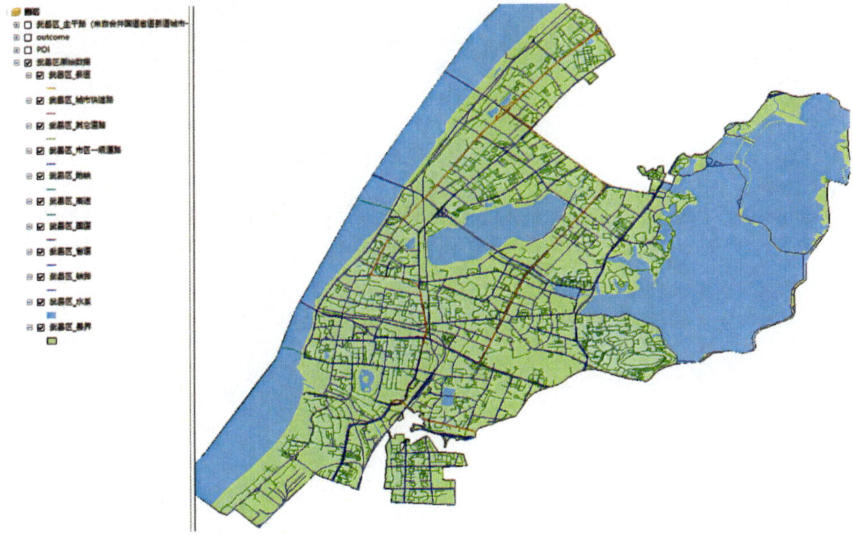

Fig. 2. Basic geographic information data of Wuchang District

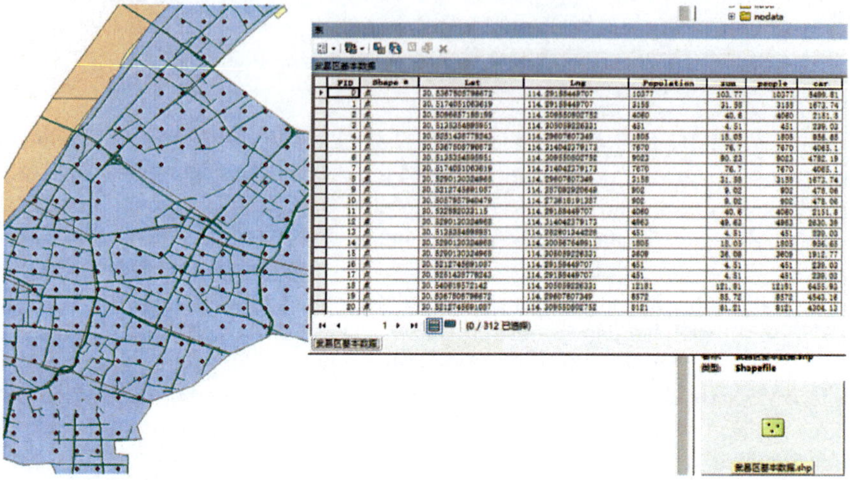

Fig. 3. Distribution data of population motor vehicles in Wuchang District

4.2 Traffic District Division

According to the principle of traffic district division, Wuchang district is divided into 13 traffic districts according to the characteristics of lakes, population density distribution and railways in Wuchang District, as shown in Fig. 4.

In terms of overall division, densely populated areas are divided into small traffic areas, planned to be small by the center, large by the periphery, large by lakes, and no other special areas. And correspondingly extract the population of each traffic district and the number of motor vehicles, and assign them to the attribute table.

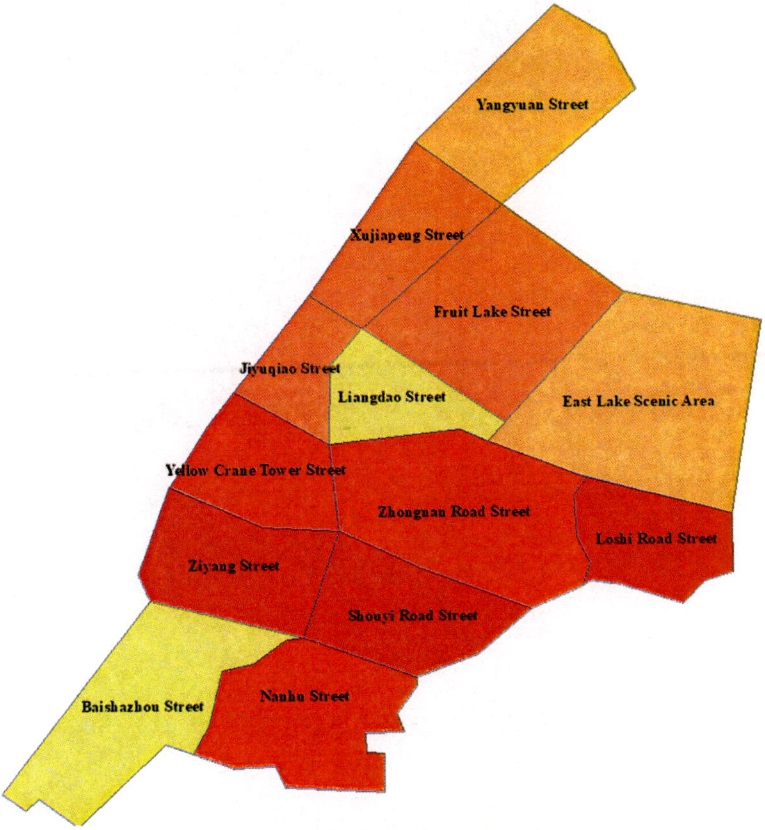

Fig. 4. Traffic district division

4.3 Demand Space Calculation

According to the standard value, the total land area of the urban social public parking lot is calculated as 1.0 square meters per person, and the land used for the motor vehicle parking lot is 90%. The area of the public parking lot for motor vehicles is 30 square meters per parking space based on the equivalent number of car parking spaces; based on the difference in population and motor vehicle ownership data, the demand for public parking spaces in each traffic district is calculated as shown in Fig. 5. The total demand for public parking spaces in Wuchang District is 123,000 vehicles.

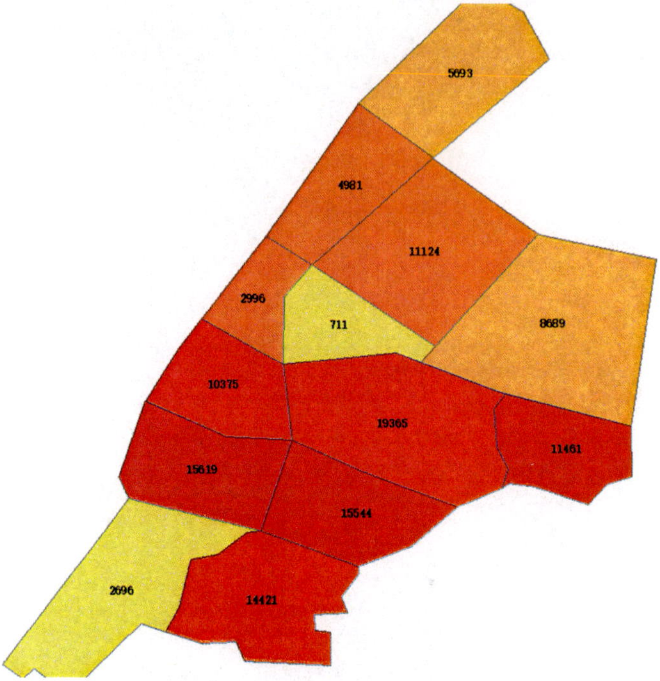

Fig. 5. Demand for parking spaces

4.4 Analysis of Results

The demanded spaces and density are shown in Table 1. The demanded spaces are predicted to have high demand for spaces in the areas of Yellow Crane Tower Street, Luoshi Road Street, Ziyang Street, and Shouyi Road Street, which are basically consistent with the parking reality in Wuchang District. For areas with high demand for parking spaces, traffic surveys, parking surveys, etc. can be carried out in a targeted manner to realize the planning and site selection of the parking lot.

The improved total demand forecasting model has the following advantages:

(1) The result has high precision. From the overall parking space demand forecast in the original area to the parking space demand forecast in each traffic area, the controllability is strong, and the obtained results are highly accurate.
(2) There is a certain site selection. When the population distribution data, vehicle distribution data, and traffic districts are sufficiently finely divided, the resulting output demand density of parking spaces narrows the research scope, so that the forecasting of parking spaces demand can also preliminarily realize the analysis of parking spaces.
(3) Associated with the basic information of the traffic area. The basic parameters of the total forecast model, the urban development coefficient, the proportion of public parking spaces, etc., can be combined with the information of specific traffic areas to obtain more targeted and accurate parking space forecast results.

Table 1. Demand for spaces and density of each traffic area

Name	Spaces F	Area (km^2)	Density (spaces/km^2)
Baishazhou street	2696	7.2518	371
Nanhu street	14421	6.2136	2320
Yellow crane tower street	10375	4.0443	2565
Zhongnan road street	19365	8.7093	2223
Loshi road street	11461	3.8159	3003
Yangyuan street	5693	5.7960	982
Fruit lake street	11124	7.6950	1445
East lake scenic area	8689	10.0887	861
Xujiapeng street	4981	4.2113	1182
Ziyang street	15619	4.9542	3152
Shouyi road street	15544	4.9227	3157
Jiyuqiao street	2996	2.4035	1246
Liangdao street	711	3.1189	227

5 Conclusion

This paper studies the total forecasting model of parking demand forecast, improves it combined with the idea of OD traffic district division, and takes Wuchang District as an example for experimental verification. The experimental conclusions are as follows.

(1) When the population distribution data, vehicle distribution data, and traffic districts are sufficiently finely divided, parking demand forecasting can also preliminarily realize parking lot location analysis, which provides new ideas and references for parking demand forecasting and parking planning.

(2) It can be used as a follow-up research direction to combine the specific urban development elasticity coefficient with information such as the nature of the land used in the traffic area.

References

1. Yan, F.: Review of China automotive coatings industry's development in 2021 and Outlook of 2022(II) China. Coatings **37**(02), 1–9 (2022). (in Chinese)
2. Shouhua, C.: A brief analysis of the current urban parking problems. Eng. Technol. **3**, 137 (2016). (in Chinese)
3. Xiliang, T., Di, H.: Parking demand prediction based on parking characteristic factors. Technol. Highway Transp. **34**(04), 117–122 (2018). (in Chinese)

4. Cutter, W.B., Fanco, S.F.: Optill supply of parking at los angeles country. Transp. Res. Part A **46**(06), 34–36 (2012)
5. Min, P., Li, W., Weilong, M.: The research on the forecasting method for the parking demand of the social and public parking area in the town CBD. Nat. Sci. J. Harbin Normal Univ. **28**(02), 78–81 (2012). (in Chinese)
6. Zhongyou, L.: Forecastin method of public parking demand based on the optimization model of the relationship between on road and off road parking. Traffic Transp. **36**(06), 24–28 (2020). (in Chinese)
7. Sun, J.: Research and practice of parking demand forcast model for urban mixed land utilization. Traffic Transp. **87**(01), 62–66 (2016). (in Chinese)
8. Janak, P., Pritikana, et al.: Study on demand and characteristics of parking system in urban areas: a review. J. Traffic Transp. Eng. (English Edition) **37**(01), 116–129 (2020)
9. Jinxia, S.: Study on parking demand forecasting method for center district of medium and small cities. Commun. Sci. Technol. Heilongjiang **39**(02), 133–135 (2016). (in Chinese)
10. You, W.: Study on site selection method for urban public parking lot based on GIS. Changsha University of Science & Technology 2019. (in Chinese)
11. Deng, H., Li, M.: Research on the gross parking demand based on urban vehicles polulation. Commun. Shipping **3**(03), 51–54+76 (2016). (in Chinese)
12. Fang, X., Jinghua, L., Zuokang, P.: Demand calculation of urban public parking lots based on population heat map—Taking haizhu district in Guangzhou as an example. Commun. Shipp. **7**(01), 69–73 (2020). (in Chinese)

Smart Supply Chain for Forest and Fruit Products Based on Judgment Matrix

Ruixuan Su[1], Chen Zhu[1], Jian Li[1], Wenjiang Xiang[2], and Dongyuan Ge[1](✉)

[1] School of Mechanical and Automotive Engineering, Guangxi University of Science and Technology, Liuzhou 545006, China
midu941qi@126.com

[2] School of Mechanical and Energy Engineering, Shaoyang University, Shaoyang 422004, China

Abstract. Taking the traditional supply chain operation model as a reference, the green supply chain operation model of forest fruit products is established on the basis of the investigation on the production and sales of general forest fruit products. At the same time, the green supply chain performance evaluation index system of forest and fruit products is constructed, and the analytic hierarchy process-fuzzy comprehensive evaluation method is used to obtain objective and scientific supply chain performance evaluation results based on the judgment matrix of each index, so as to achieve effective supply chain performance. Manage. In this paper, taking Wogan products in Liuzhou City, Guangxi Province as an example, the results of green supply chain performance evaluation are calculated, and the results are analyzed to optimize the operation of the supply chain.

Keywords: Green supply chain · Performance evaluation · Judgment matrix · Maximum eigenvalue · Eigenvectors

1 Introduction

In the "Notice of the State Council on Printing and Distributing the "14th Five-Year Plan" to Promote Agricultural and Rural Modernization", it is clearly stated that it is necessary to promote the development of the entire agricultural industrial chain, conform to the law of industrial development, improve the industrial chain, build the supply chain, upgrade the value chain, and improve the comprehensive benefits of agriculture. At the same time, with the improvement of national living standards, there are higher requirements for forest fruits, fresh and green, traceable, etc. [1]. So the study of the green supply chain of forest fruit products is indispensable.

2 Green Supply Chain Operation Model of Forest Fruit Products

2.1 Green Supply Chain Management

The definition of green supply chain is constantly enriching with the progress of the times. Chen Lei believes that in the low-carbon era, the management of green supply

Z. Hu et al. (Eds.): AIMEE 2022, LNDECT 159, pp. 199–210, 2023.
https://doi.org/10.1007/978-3-031-24468-1_18

chains should be continuously optimized with the goals of zero pollution, low energy consumption and zero emissions [2]. Yan Jie believes that green supply chain management is a consideration of the supply chain, resource utilization, and ecological environment impact [3]. Zhu et al. validated the cost, quality, and flexibility of environmental management to improve green supply chain operations [4]. Gimenez et al. argue that the goal of green supply chain management is to improve environmental efficiency and achieve zero pollutant emissions [5].

Supply chain management is the continuous optimization of the supply chain, so that environmental protection and resource utilization are in a dynamic balance. Obtain the optimal supply chain at the least cost.

2.2 Establish Green Supply Chain Operation Model of Forest Fruit Products

Forest fruit Green Supply Chain Operation Model including planning, procurement, production, logistics and supply chain support. Integrated supply chain planning is the most important link in the supply chain, procurement plan, production plan, sales plan, etc. to guide the "green" operation of the supply chain, determine the direction of the supply chain and its ability to resist market fluctuations. The main operation links of the supply chain such as green procurement and green production reflect the core of the green supply chain - green The circulation of information such as markets and production, the distribution of solid goods, the flow of returns and exchanges, and the reliable flow of funds are all supported by the smooth progress of the green supply chain. The operational model is showed as Fig. 1:

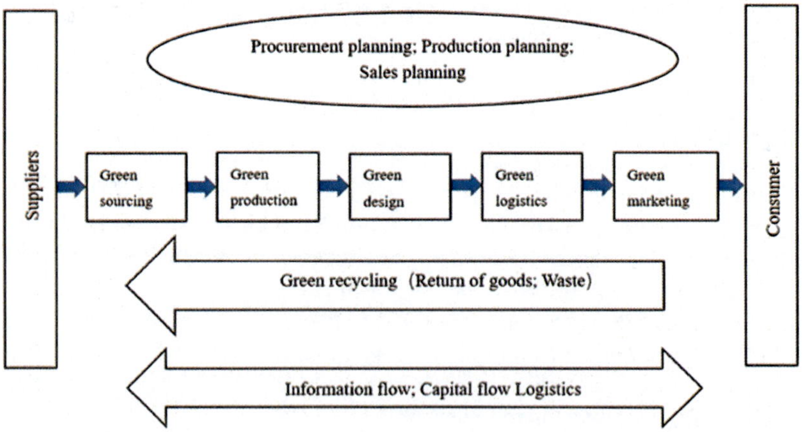

Fig. 1. Green supply chain operation model for forest fruit products

3 Construction of Green Supply Chain Performance Evaluation Index System

3.1 Construction Principles

To effectively manage the green supply chain, it is necessary to have a scientific, objective and reasonable evaluation index system for the supply chain as a whole [6]. Its construction principles are as follows:

I. The principle of comprehensiveness.
II. The principle of importance
III. The principle of combining sex and quantity.
IV. Scientific
V. Timeliness.

3.2 Construction of Performance Evaluation Index System

Combined with the operation mode of linguo green supply chain, the performance evaluation system is constructed, so it is divided into six first-level indicators of green procurement, green production, green design, green logistics, green marketing, and green consumption, on the basis of which 23 s-level indicators are subdivided in combination with relevant literature and expert opinions [7–10] (Table 1).

Table 1. Green supply chain performance evaluation system for forest fruit products

objective	Level 1 indicators	Level 2 indicators
Green supply chain performance evaluation system for forest fruit products	Green sourcing(A_1)	materials level of green certification (A_{11}); Ecological benefits of raw materials (A_{12}); Vendor Information(A_{13})
	Green production(A_2)	Policy support (A_{21}); Raw material utilization (A_{22}); Finished product conformity inspection (A_{23}); Green production technology and equipment (A_{24}); Waste disposal(A_{25})

(*continued*)

Table 1. (*continued*)

objective	Level 1 indicators	Level 2 indicators
	Green design(A_3)	Green packaging of the product (A_{31}); Energy efficient design (A_{32}); Green design input costs (A_{33})
	Green logistics(A_4)	Cargo damage rate (A_{41}); Green warehousing (A_{42}); Transport efficiency (A_{43}); Technology convergence (A_{44}); Green logistics packaging (A_{45}); Return response (A_{46})
	Green marketing(A_5)	Integration of green consumption concepts (A_{51}); Return on funds (A_{52}); Product trace-ability (A_{53})
	Green consumption(A_6)	Green consumption practice (A_{61}); Green environmental protection concept (A_{62}); Green consumer benefits (A_{63})

4 Green Supply Chain Performance Evaluation

The choice of analytic hierarchy and fuzzy comprehensive assessment is not only suitable for problems with qualitative and quantitative analysis, but also makes the analysis process methodical, easy to understand, and convincing.

4.1 Analytic Hierarchy

1) Building a hierarchical model (Fig. 2)

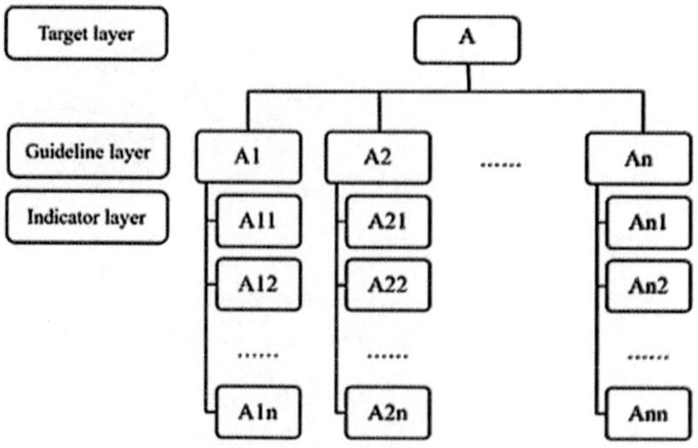

Fig. 2. Hierarchical Model Diagram

2) Construct Judgment Matrix

The basic idea is to compare the indicators in pairs, and use the scale of 1 to 9 to judge the importance of the two, and the matrix can be judged:

$$A = \begin{bmatrix} a_{11} & \cdots & a_{1n} \\ \vdots & \ddots & \vdots \\ a_{n1} & \cdots & a_{nn} \end{bmatrix} \tag{1}$$

3) Calculating The Index Weights Of Each layer (Square Root Method)

(1) Calculates the product of each row of elements of the matrix:

$$A_i = \prod_{j=1}^{n} a_{ij} \tag{2}$$

(2) Calculate the nth power root of A_i:

$$\overline{W_i} = \sqrt[n]{A_i} \tag{3}$$

(3) Normalize $\overline{w_i}$:

$$W_i = \frac{\overline{W_i}}{\sum_{i=1}^{n} \overline{W_i}} \tag{4}$$

4) Conformance Check

(1) Calculates the maximum feature root λ_{max}:

$$\lambda_{\max} = \sum_{i=1}^{n} \frac{(AW)_I}{nw_i} \tag{5}$$

(2) Find consistency metrics CI:

$$CI = \frac{\lambda_{\max} - n}{n-1} \tag{6}$$

An indicator of matrix consistency $CR = CI/RI$. When CR < 0.1, It can be determined that the consistency of matrix A is within the degree of satisfaction, its eigenvectors can be used as weight vectors. Or the matrix A needs to be updated.

4.2 Fuzzy Comprehensive Evaluation Method

Fuzzy comprehensive assessment method is to use its membership theory to transform qualitative evaluation (excellent, good, general, etc.) into quantitative evaluation to solve the uncertainty problem [10, 11]. The steps are as follows:

1) Establish a collection of evaluation indicators and a rating of comments

The criteria and target layers in the hierarchical model form multiple sets of evaluation metrics, denoted by $A = \{A|A_1, A_2, A_3 \ldots\ldots A_n\}$. The review rating is divided into $V = \{$Verygood, Better, Soso, Poor, Verypoor$\}$ $=\{90, 80, 70\ 60\ 50\}$.

2) Determine the affiliation of the evaluation index to the evaluation level

Its affiliation can be judged by multiple experts in the form of a questionnaire, and calculated by fuzzy statistical method:

$$r_{ij} = \frac{d_{\mathrm{ij}}}{d} \tag{7}$$

3) Construct a fuzzy evaluation matrix

According to the membership results of each index, a fuzzy evaluation matrix of each evaluation index set is constructed and represented by R:

$$R = \begin{bmatrix} \mathrm{r}_{11} & \cdots & \mathrm{r}_{1\mathrm{n}} \\ \vdots & \ddots & \vdots \\ \mathrm{r}_{\mathrm{m}1} & \cdots & \mathrm{r}_{\mathrm{mn}} \end{bmatrix} \tag{8}$$

4) Calculate the fuzzy comprehensive evaluation results

The product of the eigenvector corresponding to the maximum eigenvalue of the fuzzy evaluation matrix calculated by the analytic hierarchy method is the fuzzy comprehensive evaluation result B:

$$R = WR = (W_1, W_2, \ldots W_{\mathrm{m}}) \begin{bmatrix} \mathrm{r}_{11} & \cdots & \mathrm{r}_{1\mathrm{n}} \\ \vdots & \ddots & \vdots \\ \mathrm{r}_{\mathrm{m}1} & \cdots & \mathrm{r}_{\mathrm{mn}} \end{bmatrix} = (B_1, B_2, \ldots B_{\mathrm{n}}) \tag{9}$$

5) Calculate the fuzzy composite value

The fuzzy comprehensive value S is the product of the transposed matrix of the transposed matrix of $B_1^{'}$ and the number of scoring grades obtained by normalizing the fuzzy comprehensive evaluation result B_1 of the criterion layer index:

$$S = B_1^{'} V^T \tag{10}$$

5 Case Studies

5.1 Calculate the Weights of Each Metric

According to the green supply chain performance evaluation system of forest fruit products established in Table 2, the questionnaire was designed and scored by forest fruit industry experts. This article obtains an objective and true judgment matrix at the criterion layer and indicator layer from the 30 valid answer sheets of the reply. At the same time, with the help of Yaaph, the root square method is selected for calculation and consistency check. The judgment matrix of the target layer is shown in Table 2. The indicators of the criterion layer are shown in Table 3, 4, 5, 6, 7 and 8.

Table 2. The judgment matrix of the target layer

	A_1	A_2	A_3	A_4	A_5	A_6	The eigenvector corresponding to the largest eigenvalue
A_1	1	1/3	1/3	3	7	7	0.1744
A_2	3	1	1/2	5	5	8	0.2920
A_3	3	2	1	7	5	9	0.3849
A_4	1/3	1/5	1/7	1	3	5	0.0791
A_5	1/7	1/5	1/5	1/3	1	3	0.0449
A_6	1/7	1/8	1/9	1/5	1/3	1	0.0247
Consistency check: $\lambda_{max} = 6.4694$; $CR = 0.0745 < 0.1$; pass							

Table 3. Judgment matrix of green procurement

	A_{11}	A_{12}	A_{13}	The eigenvector corresponding to the largest eigenvalue
A_{11}	1	1/5	3	0.1947
A_{12}	5	1	6	0.7172
A_{13}	1/3	1/6	1	0.0881
Consistency check: $\lambda_{max} = 3.0940$; $CR = 0.0904 < 0.1$; pass				

Table 4. Judgment matrix of green production

	A_{21}	A_{22}	A_{23}	A_{24}	A_{25}	The eigenvector corresponding to the largest eigenvalue
A_{21}	1	1/6	1/7	1/5	1/5	0.0375

(continued)

Table 4. (*continued*)

	A_{21}	A_{22}	A_{23}	A_{24}	A_{25}	The eigenvector corresponding to the largest eigenvalue
A_{22}	6	1	1/4	1/4	2	0.1422
A_{23}	7	4	1	2	3	0.4197
A_{24}	5	4	1/2	1	2	0.2742
A_{25}	5	1/2	1/3	1/2	1	0.1264
Consistency check: $\lambda_{max} = 5.3642$; CR $= 0.0813 < 0.1$; pass						

Table 5. Judgment matrix of Green Design

	A_{31}	A_{32}	A_{33}	The eigenvector corresponding to the largest eigenvalue
A_{31}	1	1/5	1/2	0.1168
A_{32}	5	1	4	0.6833
A_{33}	2	1/4	1	0.1998
Consistency check: $\lambda_{max} = 3.0246$; CR $= 0.0236 < 0.1$; pass				

Table 6. Judgment matrix of green logistics

	A_{41}	A_{42}	A_{43}	A_{44}	A_{45}	A_{46}	The eigenvector corresponding to the largest eigenvalue
A_{41}	1	3	7	1/2	5	7	0.2980
A_{42}	1/3	1	2	1/5	2	4	0.1125
A_{43}	1/7	1/2	1	1/5	1/3	4	0.0575
A_{44}	2	5	5	1	8	7	0.4180
A_{45}	1/5	1/2	3	1/8	1	6	0.0868
A_{46}	1/7	1/4	1/4	1/7	1/6	1	0.0272
Consistency check: $\lambda_{\max} = 6.5402$; CR $= 0.0857 < 0.1$; pass							

Table 7. Judgment matrix of green marketing

	A_{51}	A_{52}	A_{53}	The eigenvector corresponding to the largest eigenvalue
A_{51}	1	1/3	3	0.2684
A_{52}	3	1	4	0.6144

(*continued*)

Table 7. *(continued)*

	A_{51}	A_{52}	A_{53}	The eigenvector corresponding to the largest eigenvalue
A_{53}	1/3	1/4	1	0.1172
Consistency check: $\lambda_{max} = 3.0858$; CR = 0.0825 < 0.1; pass				

Table 8. Judgment matrix of green consumption

	A_{61}	A_{62}	A_{63}	The eigenvector corresponding to the largest eigenvalue
A_{61}	1	2	5	0.5816
A_{62}	1/2	1	3	0.3090
A_{63}	1/5	1/3	1	0.1095
Consistency check: $\lambda_{max} = 3.0037$; CR = 0.0036 < 0.1; pass				

5.2 Overview

This article takes the wogan products in Liuzhou, Guangxi Province, as an example, and distributes electronic questionnaires to participants in the wo citrus industry, involving farmers, plantations, processing enterprises, courier waiters, etc. After the data statistics of 157 valid questionnaires, according to the calculation principle of the fuzzy comprehensive evaluation method, the affiliation of the indicators of the green supply chain of The Orange Industry in Liuzhou City was obtained [12] (Table 9).

Table 9. The green supply chain index of Liuzhou Wogan citrus industry is subordinate

Index		Membership				
		Verygood	Better	General	Poor	Very poor
A_1	A_{11}	0.1372	0.3621	0.2695	0.1968	0.0344
	A_{12}	0.1492	0.2976	0.2865	0.1527	0.1140
	A_{13}	0.1854	0.2962	0.2694	0.1063	0.1427
A_2	A_{21}	0.1297	0.1863	0.3065	0.1546	0.2229
	A_{22}	0.2163	0.2541	0.3097	0.1564	0.0635
	A_{23}	0.1637	0.3957	0.2783	0.1012	0.0611
	A_{24}	0.1786	0.3381	0.3215	0.1388	0.0230

(continued)

Table 9. *(continued)*

Index		Membership				
		Verygood	Better	General	Poor	Very poor
	A_{25}	0.2468	0.3078	0.2163	0.1864	0.0427
A_3	A_{31}	0.1378	0.3104	0.2164	0.1742	0.1612
	A_{32}	0.2261	0.3521	0.2375	0.1729	0.0114
	A_{33}	0.2114	0.2791	0.1378	0.1673	0.2044
A_4	A_{41}	0.0951	0.3683	0.2686	0.2093	0.0587
	A_{42}	0.1327	0.3085	0.2913	0.1764	0.0911
	A_{43}	0.1793	0.2993	0.2869	0.1836	0.0509
	A_{44}	0.1852	0.3105	0.2378	0.1308	0.1357
	A_{45}	0.1527	0.3741	0.2603	0.1219	0.0910
	A_{46}	0.1467	0.3061	0.2523	0.1421	0.1528
A_5	A_{51}	0.1638	0.2778	0.2574	0.1634	0.1376
	A_{52}	0.1604	0.2726	0.2681	0.1193	0.1796
	A_{53}	0.1931	0.2864	0.2739	0.1371	0.1095
A_6	A_{61}	0.1381	0.3678	0.2856	0.0947	0.1138
	A_{62}	0.1428	0.3992	0.2718	0.1131	0.0731
	A_{63}	0.1494	0.2916	0.2881	0.1419	1.1290

As can be obtained from the table above, the fuzzy evaluation matrix of the indicator layerR_1~R_6

Fuzzy comprehensive evaluation criterion layer result $B = WR$ can be obtained:

$$B_1 = (0.1947, 0.7172, 0.0881)\begin{bmatrix} X0.1372\ 0.3621\ 0.2695\ 0.1968\ 0.0344 \\ 0.1492\ \ 0.2976\ 0.2865\ 0.1527\ 0.1140 \\ 0.1854\ \ 0.2962\ 0.2694\ 0.1063\ 0.1427 \end{bmatrix}$$

$$= (0.1501, 0.3100, 0.2817, 0.1572, 0.1010)$$

The same can be obtained

$$B_2 = (0.1845, 0.3408, 0.2878, 0.1321, 0.0547)$$

$$B_3 = (0.2128, 0.3326, 0.2151, 0.1719, 0.0675)$$

$$B_4 = (0.1482, 0.3323, 0.2582, 0.1619, 0.0994)$$

$$B_5 = (0.1651, 0.2756, 0.2659, 0.1332, 0.1601)$$

$$B_6 = (0.1408, 0.3692, 0.2816, 0.1056, 0.2124)$$

Remember $R = (R_1, R_2, R_3, R_4, R_5R_6)^T$, Then the result of the target layer

$$B = WR = (0.1846, 0.3294, 0.2553, 0.1536, 0.0682)$$

Normalization

$$B' = (0.1863, 0.3324, 0.2576, 0.1550, 0.0867)$$

Comprehensive evaluation score

$B = B'V^T = (0.1863, 0.3324, 0.2576, 0.1550, 0.0867)(90, 80, 70, 60, 50)^T =$ 75.026.

That is to say, the green supply chain score of Wogan orange industry in Liuzhou is 75.026 points, which is at a good level. At the same time, scores of various indicators at the criterion level can be obtained: green procurement 72.51 points, green production 74.676 points, green design 74.506 points, green logistics 72.68 points, green marketing 71.517 points, and green consumption 78.876 points. Among them, the highest score is green consumption. With the continuous improvement of the living standards of the Chinese people, there are higher requirements for forest and fruit products, such as green, healthy and environmentally friendly. The lowest score is green marketing, indicating that the integration of green concepts and capital gains cannot achieve positive proportional growth in developing China, and the solution is to further improve people's green concepts.

6 Conclusion

This paper takes the Wogan orange products in Liuzhou as an example. The performance evaluation results indicate that green marketing and green procurement in the green supply chain have greater room for improvement than green consumption. This provides direction for the management of the green supply chain of Wogan products: in order to achieve a more efficient and green operation of the supply chain, it should be deeply optimized in terms of green marketing and green procurement. This provides ideas for the development of my country's green supply chain, obtains the performance evaluation results of the supply chain in a scientific and effective way, and optimizes the weaker links of the supply chain to improve the overall operation level of the supply chain.

Acknowledgement. This project is supported by National Natural Science Foundation of China (51765007) and (81960332), and by Logistics Education Reform and Teaching Research Project (JZW2018-037), and by Innovation Project of Guangxi Graduate Education, grant number YCSW2022438.

References

1. McKennon, L.: Research on the development of supply chain management theory based on Chinese context. Supply Chain Manag. **2**(09), 76–102 (2021). (in Chinese)
2. Lei, C.: Research on green supply chain management in the era of low-carbon economy. Chin. Foreign Entrepreneurs **2019**(30), 2 (2019). (in Chinese)
3. Jie, Y.: Under the low carbon economy- on green supply chain management. Bus. Observ. **05**, 75–77 (2022). (in Chinese)
4. Zhu, Q., Sarkis, J., Lai, K.H.: Confirmation of a measurements and model for green supply chain management practices Implementation. Int. J. Prod. Econ. **111**(2), 261–273 (2008)
5. Tachizawa, E.M., Gimenez, C., Sierra, V.: Green supply chain management approaches: Drivers and performance implications. Int. J. Oper. Prod. Manag. **35**(11), 1546–1566 (2015)
6. Jin, Y.: Supply Chain Strategy and Operation Design. China Fortune Press, Beijing (2013). (in Chinese)
7. Yang, L.: Information Analysis Methods and Practice. Northeast Normal University Press, Changchun (2017). (in Chinese)
8. Wang, Z., Yin, Q., Pan, Y., et al.: Progress on recycling of re-manufacturing closing-loop supply chain. Comput. Integr. Manuf. Syst. **27**(04), 1201–1210 (2021). (in Chinese)
9. Ge, D.Y., Zhang, Q.Y.: Supply Chain management and supplier's permeation and integration. In: Proceedings of the 4th International Conference on Material Handling & Logistics Systems, pp. 187–189+501 (2002)
10. Wei, Z., Xiong, X.: Evaluation and Modeling of Equipment Support Effectiveness. National Defense Industry Press, Beijing (2020). (in Chinese)
11. Wu, Y.: The operation mode and performance evaluation of the green supply chain of forest and fruit products in Hunan. Changsha: Central South University of Forestry Science and Technology Master's thesis (2020). (in Chinese)
12. Wang, Z., Zhao, F.: Research on comprehensive evaluation of agricultural cold chain green logistics based on fuzzy comprehensive evaluation method. J. Central China Normal Univ. (Natural Sci. Ed.) **49**(04): 546–550 (2015). (in Chinese)

Multistage SVR-RBF-Based Model for Heart Rate Prediction of Individuals

Ivan Izonin(✉), Roman Tkachenko, Rostyslav Holoven, Maksym Shavarskyi, Serhii Bukin, and Ihor Shevchuk

Lviv Polytechnic National University, S. Bandera Str., 12, Lviv 79013, Ukraine
ivanizonin@gmail.com

Abstract. The paper investigates the problem of improving the accuracy of solving the heart rate prediction of individual tasks, using machine learning tools. The authors developed a new multistage learning model based on the use of SVR with RBF kernel for the case of analysis of large datasets. The paper describes in detail the procedures of sequential learning of the developed model and its application. The simulation of the studied model was based on the author's software. Experimental studies were performed using a real-world dataset of large volume. The effectiveness of the developed model was evaluated using various performance indicators. The optimal parameters of the model operation were selected experimentally. A comparison of its accuracy with several known methods used to solve the stated problem is performed. The highest accuracy of the developed model compared to existing methods based on all six performance indicators is established. These advantages create several prerequisites for the practical use of the model in automated diagnostic systems in medicine, particularly in monitoring human stress.

Keywords: Cascading · Multistage learning · SVR · RBF functions · Precision model · Heart rate prediction · Individuals

1 Introduction

The task of the heart rate prediction of individuals is topical in many applied fields of medicine [1]. This indicator is used during diagnosis or monitoring and supports the patient's treatment process [2]. One of the possible options for applying an effective solution to this problem is also to determine the level of human stress [3].

It is known that stress negatively affects a person's emotional state and, as a consequence, his ability to work [4]. Therefore, timely determination of its level, especially for people involved in very responsible and risky areas of activity, can protect against adverse consequences when making decisions [5–7].

Existing machine learning methods do not provide sufficient prediction accuracy [8], leading to ensemble approaches [9]. Boosting, bagging, and stacking strategies are typical for developing high-precision prediction models [10, 11]. However, each of them has some significant disadvantages that reduce the effectiveness of their use, for

Z. Hu et al. (Eds.): AIMEE 2022, LNDECT 159, pp. 211–220, 2023.
https://doi.org/10.1007/978-3-031-24468-1_19

example, in the case of data analysis of large volumes [10–12]. Cascading as a minor fourth class of ensemble methods provides significantly higher accuracy than previous classes of ensemble methods [9]. However, selecting the correct training method for implementing this strategy, constructing the actual cascade algorithm, and its effective implementation is a non-trivial task [13–16].

That is why this work aims to develop a new multistage learning model to improve the accuracy of solving the heart rate prediction task of individuals in processing large datasets.

The main contribution of this paper can be summarized as follows: we have developed a new multistage learning model based on the use of SVR with RBF kernel for the case of large volume of data analysis; we have selected the optimal parameters of the studied model that provide the highest efficient of its operation; the highest accuracy of work of the developed model in comparison with existing methods for heart rate prediction of individuals is established experimentally based on six performance indicators.

2 Materials and Methods

2.1 Dataset Description

To solve the task of heart rate prediction of individuals in this paper, we used a real-world set of large amounts of data [3]. The challenge is to predict the heart rate of individuals based on 18 independent variables.

The dataset contains 369289 vectors, which were randomly divided into training (258502 vectors) and test (110787 vectors) samples. Based on them, experimental research was carried out in this work.

2.2 Multistage SVR-RBF-Based Model

The task of predicting the heart rate of individuals is not new [1]. Many researchers who have improved existing and developed new methods of processing use the dataset [3]. However, this paper presents a new, high-precision approach to solving the tasks. The authors have developed a new multistage learning model that significantly increases prediction accuracy in the case of analysis of large datasets.

The model is based on a cascading ensemble of machine learning methods [9]. The use of the classic SVR with RBF kernel does not provide sufficient accuracy for this method in automated systems monitoring human stress. However, SVR with RBF kernel provides the highest accuracy of solving the task among other kernels of the SVR method. Therefore it was chosen as the basic algorithm for training a new multistage model [17].

The training set is divided into N parts to implement the training procedure. Here N will determine the number of learning steps or equal to the number of stages of the proposed multistage model. In the first step of the training procedure, the first N sub-sample is used to train the first SVR1 of the proposed model. The next sub-sample is applied to SVR1, and the resulting output is added as an additional attribute. Further, extended in this way sub-sample is used to train SVR2. The third sub-sample is applied

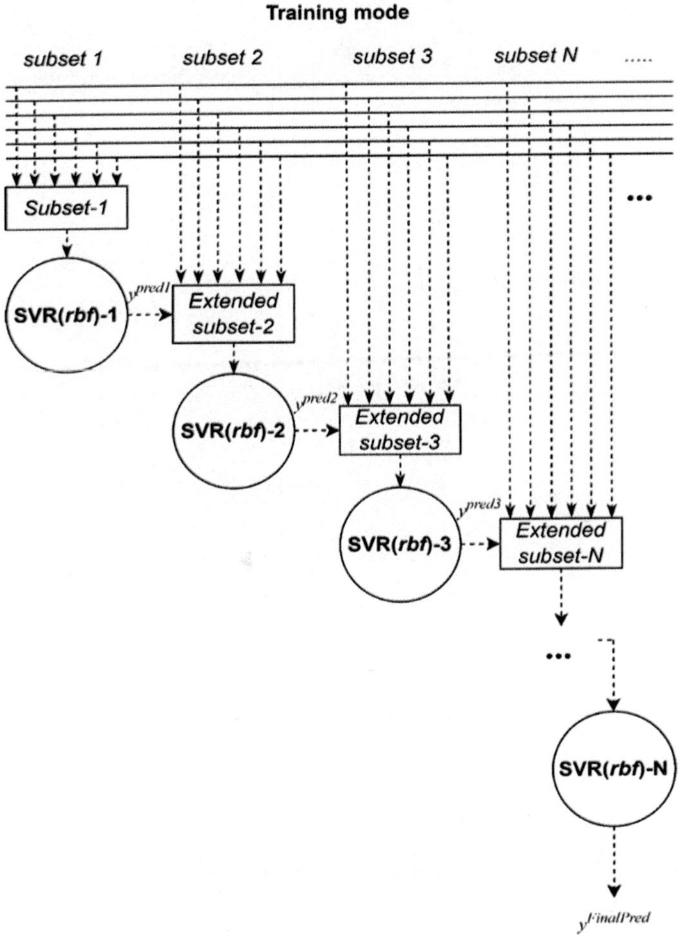

Fig. 1. Flowchart of the multistage SVR-RBF-based model in training mode

to SVR1, and its extended form (predicted outputs by SVR1 are added as an additional attribute) is applied to SVR2. In this case, obtained predictions (based on SVR2) are added to the third sub-sample as an additional attribute and, in this form, come for training by SVR3. The fourth and subsequent training sub-samples are also processed in a similar way. The last SVR from N returns the.

desired result. A flowchart of the multistage SVR-RBF-based model in training mode is shown in Fig. 1.

In test mode, we have one test sample. It is consistently applied to all N pre-trained SVRs, where the last SVR provides the desired result. A flowchart of the multistage SVR-RBF-based model in test mode is shown in Fig. 2.

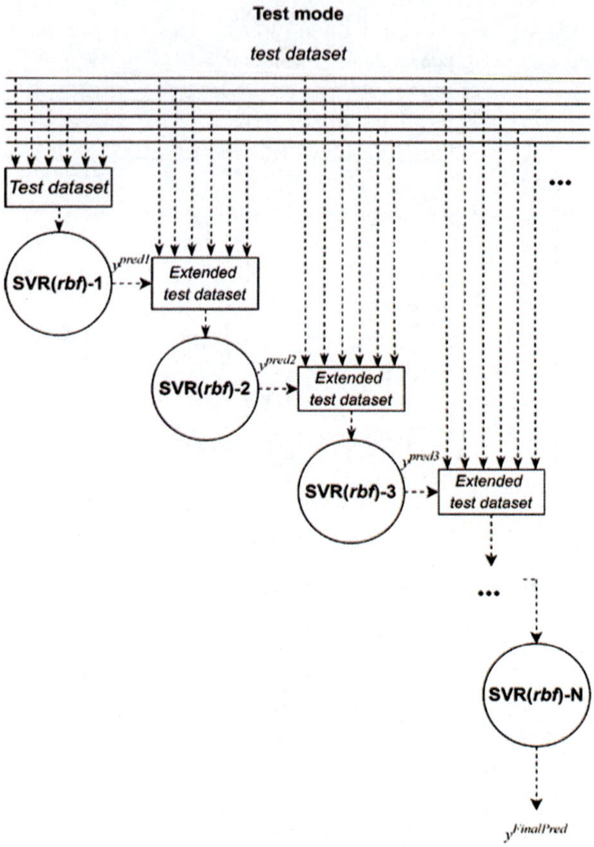

Fig. 2. Flowchart of the multistage SVR-RBF-based model in test mode

Thus, considering the outputs of each SVR preceding the current one ensures the implementation of the principle of linearization of the response surface [9], and nonlinear expansion of the inputs due to *rbf*-kernel [18] provides increased prediction accuracy. These principles became the basis of the developed multistage SVR-RBF-based model.

3 Modeling and Results

The proposed approach was modeled using the author's software. It is developed in Python. The experimental studies were performed on a computer DELL, intel CORE I7, 8 Gb RAM.

The training and test samples were normalized according to the min-max scaler scheme.

Because the multistage SVR-RBF-based model proposed in this paper may have different learning stages, the study was to find the optimal number of them that would provide the highest prediction accuracy. The experiment was conducted with a gradual increase in the number of stages of learning the model from 1 to 6.

The results obtained based on the Mean Squared Errors and Mean Absolute Errors are presented in Fig. 3

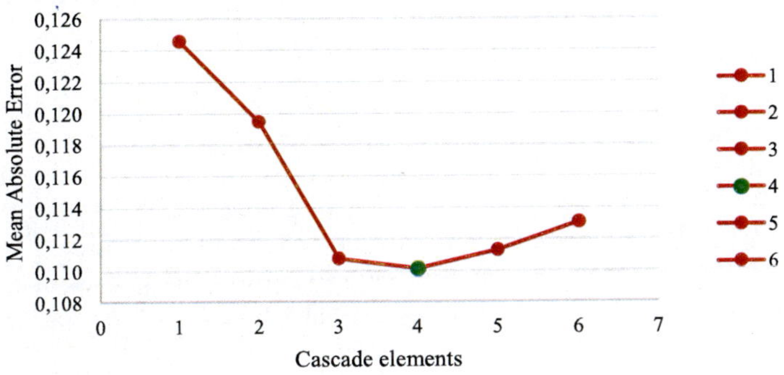

a)

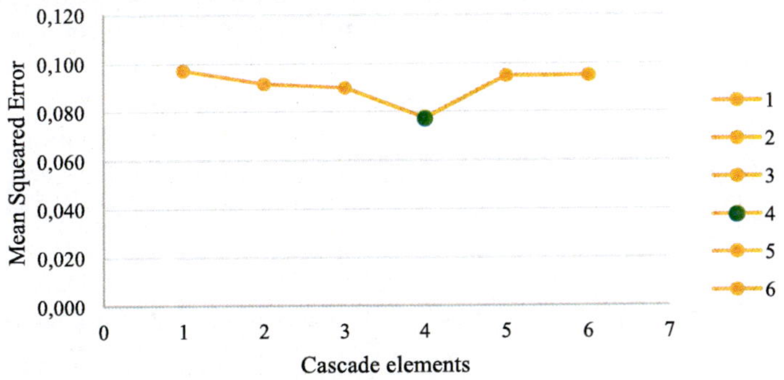

b)

Fig. 3. Errors of the proposed model (testing mode) when changing the number of stages of its training: a) MAE values; b) MSE values.

As can be seen from Fig. 3, both accuracy indicators with an increasing number of steps of the model learning procedure gradually increased. Stages of saturation of both accuracy indicators were acquired in the fourth consecutive step of training. Then both indicators began to decrease. This can be explained by the accumulation of errors at each previous level of the multistage SVR-RBF-based model, which led to an increase in the overall error of the model starting from the fifth step of sequential learning.

Based on this, Table 1 summarizes the values of seven different indicators of the accuracy of the multistage SVR-RBF-based model using four stages of sequential training. It should be noted that the table shows the errors in the model training and its application. In addition, the duration of the training procedure of the multistage SVR-RBF-based model is given in Table 1.

Table 1. Results

Performance indicator	Training mode	Test mode
Mean Absolute Error (MAE)	0.109	0.110
Mean Squared Error (MSE)	0.71	0.077
Root Mean Squared Error (RMSE)	0.275	0.278
Mean Absolute Percentage Error (MAPE)	0.001	0.001
Maximal Error (ME)	11.760	11.989
Median Absolute Error (MedAE)	0.062	0.062
Coefficient of determination (R2)	0.999	0.999
Training time, seconds	688.78	–

An essential element for evaluating machine learning models when processing large amounts of data is the ability to generalize the model and the risks of its overfitting.

As can be seen from Table 1, overfitting of the multistage SVR-RBF-based model proposed in this paper is not observed. The seven performance indicators show an increase in the model error in the application mode compared to the training mode. However, the difference between the corresponding errors of the two studied modes (training and test) is very insignificant. It indicates the high generalization properties of the proposed multistage SVR-RBF-based model in the case of processing a large dataset.

Because of all the above, the optimal value of the number of successive stages of learning of the proposed multistage SVR-RBF-based model to solve the problem of the heart rate prediction of individuals on a given large dataset will be equal to 4. This model was used later to assess its effectiveness compared to several existing methods.

4 Comparison and Discussion

The accuracy of the proposed multistage SVR-RBF-based model was compared using a number of existing methods used to solve the stated task [1]. In particular, the authors used:

1) LinearRegression;
2) RidgeRegression;
3) ARDRegression;
4) BayesianRidgeRegression;
5) AdaBoostRegressor;
6) GradientBoostingRegressor.

Their implementations are taken from Python language libraries. For ease of reproduction of the experiment, the parameters of all existing methods are handled by default.

The results of all the studied methods are shown in Fig. 4 and Fig. 5. In particular, Fig. 4 shows the MAE and MSE values of all studied methods for the application mode.

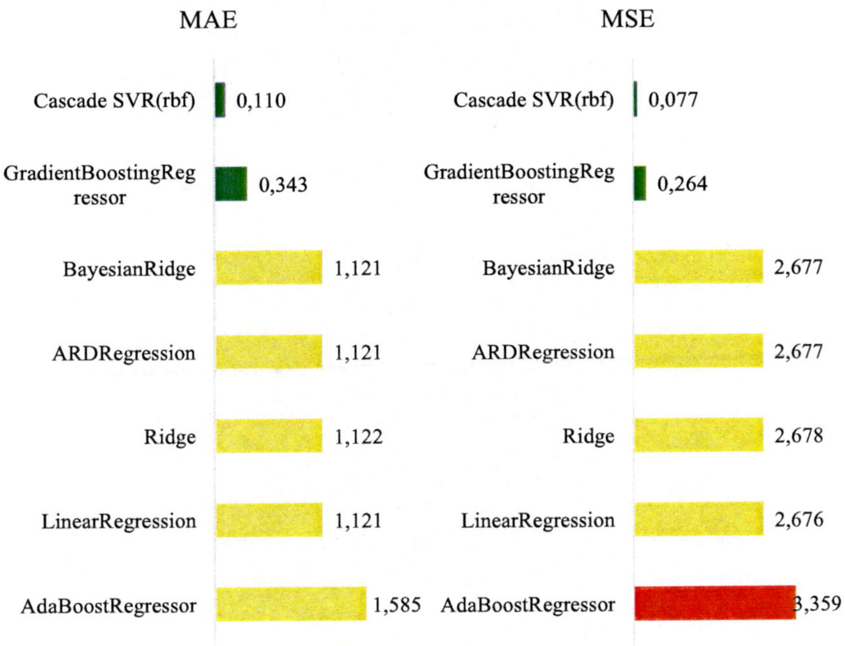

Fig. 4. Comparison of different ML-based models using: a)MAE; b)MSE

Figure 5 shows the values of MAPE and RMSE also for all studied methods in the test mode.

As can be seen from both figures, the lowest accuracy according to the four performance indicators is demonstrated by a boosted-based regressor - AdaBoost. On the contrary, the other method of this class - Gradient Boosting- shows good accuracy indicators for solving the stated task. Methods based on LinearRegression, RidgeRegression, ARDRegression, and BayesianRidgeRegression show satisfactory.

accuracy (highlighted in yellow). In addition, the values of each of the corresponding errors of these methods are almost indistinguishable and vary in a very small range.

The proposed multistage SVR-RBF-based model showed the best results in terms of accuracy based on all four indicators. In this case, the studied model provides at least 1.85 and at most 3.34 times higher accuracy (depending on the selected indicator) than the nearest regressor - Gradient Boosting. It indicates the possibility of using the proposed model to solve the problem of heart rate prediction of individuals in practice. Also, this approach can be used in many other areas [19–25].

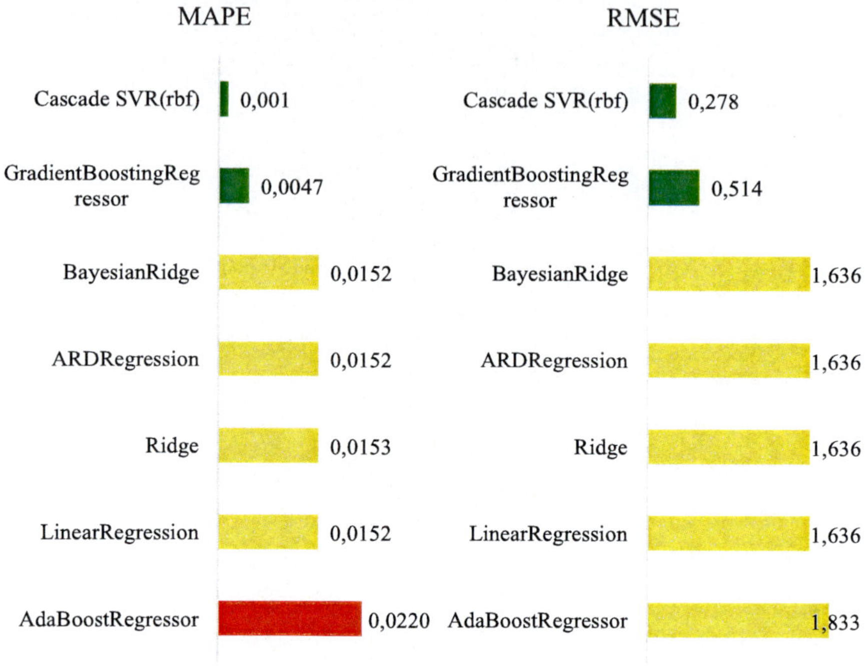

Fig. 5. Comparison of different ML-based models using: a)MAPE; b)RMSE

5 Conclusions

In this paper, the task of the heart rate prediction of individuals was considered. Its relevance is growing daily due to the different possibilities of applying such results. The paper develops a new multistage learning model to increase the prediction accuracy when processing large datasets. This model is based on the use of SVR with RBF kernel. The authors describe the training and application procedures of the developed model. Experimental research was conducted using proprietary software based on the use of a real-world dataset of a large volume. The authors found a high efficiency of using the developed model based on various performance indicators. By.

comparing with six other methods, the highest accuracy of solving the stated task is shown using the developed model. These advantages create several prerequisites for the practical use of the model in automated diagnostic systems in medicine, particularly in monitoring human stress.

Despite the high results obtained in terms of the accuracy of the proposed multistage SVR-RBF-based model, the duration of the procedure of its training through the sequential mode of operation is significant. That is why in the prospect of further research, we will consider the following potential opportunities:

1) the use of feature selection procedures to reduce the dimension of the input data space;

2) the use of non-iterative machine learning tools to replace the iterative SVR with *rbf*-kernel in the studied model;
3) development of other variants than the *rbf*-functions for nonlinear input extension;
4) the use of high-speed PCA, in particular by neural network tools, with the subsequent nonlinear expansion of the space of important main components in the proposed model.

Acknowledgment. The authors thank the reviewers for the correct and concise recommendations that helped present the materials better. The Krzysztof Skubiszewski Foundation support this research.

References

1. Matthew, O., et al.: A predictive analysis of heart rates using machine learning techniques. Int. J. Environ. Res. Public Health. **19**(4), 2417 (2022)
2. Babichev, S., et al.: Technology of gene expression profiles filtering based on wavelet analysis. IJISA. **10**(4), 1–7 (2018)
3. 'Heart Rate Prediction to Monitor Stress Level'. https://www.kaggle.com/vinayakshanawad/heart-rate-prediction-to-monitor-stress-level. Accessed 19 Jun 2022
4. Mochurad, L., Ya, H.: Modeling of psychomotor reactions of a person based on modification of the tapping test. Int. J. Comput. **20**(2), 190–200 (2021)
5. Bisikalo, O., et al.: Research of safety and survivability models of the information system for critical use. In: Proceedings of the 2020 IEEE 11th International Conference DESSERT, pp. 7–12. IEEE, Bahrain (2020)
6. Mashkov, V., Lytvynenko, V.: Method for unit self-diagnosis at system level. IJISA **11**(1), 1–12 (2019)
7. Teslyuk, V., et al.: Development and implementation of the technical accident prevention subsystem for the smart home system. IJISA **10**(1), 1–8 (2018)
8. Piletskiy, P., Chumachenko, D., Meniailov, I.: Development and analysis of intelligent recommendation system using machine learning approach. Adv. Intell. Syst. Comput. **1113**, 186–197 (2020)
9. Izonin, I., Tkachenko, R.: An approach towards the response surface linearization via ANN-based cascade scheme for regression modeling in Healthcare. Procedia Comput. Sci. **198**, 724–729 (2022)
10. Zhengbing, H., et al.: An evolving cascade system based on a set of neo - fuzzy nodes. IJISA **8**(9), 1–7 (2016)
11. Zhengbing, H., et al.: An ensemble of adaptive neuro-fuzzy kohonen networks for online data stream fuzzy clustering. IJMECS **8**(5), 12–18 (2016)
12. Mykhailyshyn, H., et al.: Designing network computing systems for intensive processing of information flows of data. Data-Centric Bus. Appl. **48**, 391–422 (2021)
13. Hovorushchenko, T., et al.: Ontology-based intelligent agent for determination of sufficiency of metric information in the software requirements. Adv. Intell. Syst. Comput. **1020**, 447–460 (2020)
14. Sheketa V., et al.: Empirical method of evaluating the numerical values of metrics in the process of medical software quality determination. In: Proceedings of the International Conference DASA, Bahrain, pp. 22–26. IEEE (2020)

15. Babichev, S., et al.: An effectiveness evaluation of information technology of gene expression profiles processing for gene networks reconstruction. IJISA **10**(7), 1–10 (2018)
16. Olszewski, S., et al.: Digital method of automated non-destructive diagnostics for high-power magnetron resonator blocks. IJIGSP **14**(1), 40–49 (2022)
17. Izonin, I.: The combined use of the wiener polynomial and SVM for material classification task in medical implants production. IJISA **10**(9), 40–47 (2018)
18. Kotsovsky, V., Batyuk, A.: On-line relaxation versus off-line spectral algorithm in the learning of polynomial neural units. In: Proceedings of the International Conference DSMP, Lviv, pp. 3–21. IEEE (2020)
19. Nehrey, M., Hnot, T.: Data Science Tools Application for Business Processes Modelling in Aviation. Cases on Modern Computer Systems in Aviation, pp. 617–631. IGI Global (2019)
20. Kaminskyi, A., et al.: Complex risk analysis of investing in agriculture ETFs. IJIEPR **31**(4), 579–586 (2020)
21. Dronyuk, I., Fedevych, O.: Traffic flows ateb-prediction method with fluctuation modeling using dirac functions. Commun. Comput. Inform. Sci. **718**, 3–13 (2017)
22. Oliinyk, A., et al.: Implementation of evolutionary methods of solving the travelling salesman problem in a robotic warehouse. Lect. Notes Data Eng. Commun. Technol. **48**, 263–292 (2021)
23. Mulyak, O., et al.: Influence of software reliability models on reliability measures of software and hardware systems. Eastern-Eur. J. Enterpr. Technol. **9**(76), 53–57 (2015)
24. Kaminskyi, A., Nehrey, M., Zomchak, L.: COVID-19: crisis or new opportunities time for the agricultural sector of Ukraine. IOP Conf. Ser. Earth Environ. Sci. **628**(1), 012031 (2021)
25. Zomchak, L., Nehrey, M.: Economic growth and capital investment: the empirical evidence. In: Hu, Z., Zhang, Q., Petoukhov, S., He, M. (eds.) Advances in Artificial Systems for Logistics Engineering. ICAILE 2022. Lecture Notes on Data Engineering and Communications Technologies, vol. 135, pp. 645–652. Springer, Cham (2022). https://doi.org/10.1007/978-3-031-04809-8_59

Differential Development of High-Quality Distribution Enterprises in the Middle Yangtze River Cities in the Context of Major Public Events

Yuze He[1], Mingfei Liu[1](✉), and Aiman Magde Abdalla Ahmed[2]

[1] School of Management, Wuhan University of Technology, Wuhan 430070, China
liumingfei5223@163.com

[2] Faisal AI Islam Bank, Khartoum 999129, Sudan

Abstract. This paper uses the entropy value method to calculate the indexes of high-quality development and shock resistance of commerce and distribution enterprises in cities along the middle reaches of Yangtze River, and explores the correlation between high-quality development and shock resistance using multiple linear regression. The results of the study show that: the level of high-quality development of commerce and distribution enterprises in the middle reaches of Yangtze River cities has gradually improved, showing a pattern of multi-center pulling and synergistic development of enterprises in cities along the Yangtze River; the structure of high-quality development of commerce and distribution enterprises in different cities is different, and the development structure of enterprises in central cities is better than those in peripheral cities, and the resistance to shocks is stronger; high-quality development and urbanization rate significantly improve the resistance to shocks of enterprises. Therefore, the trade circulation enterprises in the middle reaches of Yangtze River cities should optimize the multi-center networked structure, improve the comprehensive bearing capacity and play the role of spatial hub in the construction of the domestic grand cycle.

Keywords: Public emergencies · Circulation companies · Resilience

1 Introduction

Major public events usually occur quickly and have a wide range of impacts, which may break the balance of the social system in a short time and bring huge adverse effects. The rapid spread of novel coronavirus is such a major public event. It has had a great impact on the global supply chain, and also brought heavy losses to the circulation enterprises, which are the nodes connecting production and consumption. In this context, China has proposed to accelerate the construction of a new development pattern with the domestic grand cycle as the main body and the domestic and international double cycle promoting each other. The eleven provinces and cities along the Yangtze River Economic Belt, with huge population size and economic volume, should play a leading role in implementing the new development concept,

Z. Hu et al. (Eds.): AIMEE 2022, LNDECT 159, pp. 221–230, 2023.
https://doi.org/10.1007/978-3-031-24468-1_20

building a new development pattern of double circulation and promoting the national economy to achieve high-quality development. As an important part of economic operation, circulation enterprises promote the organic combination of the four flows of commercial flow, logistics, information flow and cash flow by connecting both sides of supply and demand, and jointly promote the efficient allocation of resources. In order to enhance the carrying capacity of the city cluster in the middle reaches of Yangtze River and become a hub of domestic large circulation, the high-quality development of circulation enterprises is essential.

In this paper, based on the existing researches on "distribution enterprises", "regional economic quality" and "economic development resilience", we review and sort out the existing researches in the context of major public emergencies. We review and sort out the existing research, summarize the shortcomings of current research, and propose the target and content of this paper. The current situation of high-quality development of trade and distribution enterprises in the middle reaches of Yangtze River urban agglomeration is studied, and its development process and current problems are explored from the current development situation. By collecting and processing the production and operation information of 29 cities in the middle reaches of Yangtze River, we construct the evaluation index system of high-quality development and study the basic situation of high-quality development.

2 Literature Review

2.1 High Quality Development

In the face of major public emergencies, the high-quality development of trade circulation enterprises in the middle reaches of Yangtze River cities must follow the new development concept. High-quality development is to focus on solving the basic contradictions of China's current society, and the path and method of solving the "new contradictions" of China through development [1]. To achieve high-quality development, we must meet the following six requirements: implementing the new development concept, insisting on quality first, prioritizing efficiency, focusing on structural reform on the supply side, moving the supply system and industrial structure to the middle and high end, significantly increasing the innovation and competitiveness of the national economy, and being able to meet the growing needs of the people for a better life [2]. High-quality development means that supply has high quality, demand has high quality, allocation has high quality, input and output has high quality, income distribution has high quality and other characteristics [3].

2.2 Commerce and Distribution Enterprises

At present, China is accelerating the construction of a new development pattern, and the city cluster in the middle reaches of Yangtze River is located in central China, which is of great strategic significance in the construction of the double cycle. The development of regional cities can be not only a structure of single-core pull development, but also a structure of multi-core common pull [4]. With the continuous development of the economic belt, the scale of the city is also expanding, and the multi-center regional city cluster is more conducive to economic development. The multi-center city cluster can break the spatial element calm of economic development, and the efficient flow of elements among multi-center, so that the regional multi-center structure can reach equilibrium again. At the current stage of China's economic development, polycentricity has become the goal of urban cluster development [5]. General Secretary Xi Jinping emphasized that "to build a new development pattern, the construction of a modern circulation system must be taken as an important strategic task" [6]. In September 2020, the eighth meeting of the Central Finance and Economics Commission pointed out that "Domestic circulation and international circulation are inseparable from an efficient modern circulation system, and the construction of a modern circulation system must be taken as an important strategic task in building a new development pattern" [7]. In building a new pattern of double circulation, it is especially important to coordinate development and security, to put security in a more prominent position, to consider more security factors in development, and to strive to achieve a dynamic balance between development and security [8].

2.3 Literature Summary

Therefore, distribution enterprises in the middle reaches of the Yangtze River should actively communicate to accelerate the completion of a polycentric development structure with the provincial capitals as the core nodes, enhance the carrying capacity of regional economic and trade activities, and achieve balanced regional development. Enterprises should give full play to their geographical advantages, actively communicate with national economic and trade activities, and promote the high-quality development of the domestic general circulation.

3 Research Methods

The entropy value method was used to evaluate the high-quality development of trade and distribution enterprises and their resilience to shocks in the middle reaches of Yangtze River cities. Population size and urbanization rate are used as the control variables, quality development index as the explanatory variables, and resilience index as the explanatory variables, and multiple linear regressions are conducted to find their correlation. Finally, Arcgis is used to map the high-quality development of trade circulation enterprises in the middle reaches of Yangtze River to study the regional development differences and the overall development evolution process.

4 Research Design

4.1 Construction of Evaluation Indicators

High-quality development and shock resistance are both the result of the joint action of the system, which cannot be evaluated by a single indicator, and a comprehensive multi-dimensional evaluation can more objectively and truly reflect the characteristics of high-quality development and shock resistance. High-quality development emphasizes following the new development concept. Referring to the research of Wang Lei and other scholars, this paper chooses to construct the evaluation system of high-quality development from four aspects: economic scale, input innovation, fruit sharing and balanced development. The specific indicators are shown in Table 1.

Table 1. Economic quality development evaluation index system

	Tier 1 Indicators	Secondary indicators	Directionality
High Quality Development Index	Economic Scale	Enterprise output	+
		Per capita consumption of social retail consumer goods	+
		Turnover of goods	+
	Investing in Innovation	Fixed Asset Investment Index	+
		Labor productivity	+
		Degree of chainage	+
	Sharing results	Employee's real average wage index	+
		Disposable income of employees	+
		Number of employed persons	+
	Coordinated Development	Asset-liability ratio index	–
		Increase enterprises above limit	+

Trade and circulation enterprises are the key to build a new development pattern. The high-quality development of circulation enterprises can open the key blockage point of the double cycle and guarantee the stability and security of our supply and production chains. This requires the distribution enterprises in the middle reaches of Yangtze River cities, on the one hand, to improve their carrying capacity and act as a hub to communicate the national economic and trade activities; on the other hand, in the face of major public events, the trade distribution enterprises should also be resilient enough to guarantee the security and stability of our economic and trade.

activities. This paper evaluates the shock resistance of distribution enterprises to major public events in five indicators from three levels: economic resilience, structural resilience and development resilience. See Table 2.

Table 2. Evaluation index system of shock resistance

	Tier 1 Indicators	Secondary indicators	Directionality
Shock resistance	Economic Resilience	Change rate of output share	+
		Rate of change of cargo turnover	+
	Structural Resilience	Change rate of chainization	+
		Change rate of investment and innovation scale	+
	Development Resilience	Change rate of active employees	+
		Change of average wage index	+

4.2 Research Model

Entropy method: In order to guarantee the objective and scientific nature of the evaluation index and avoid the influence of subjective factors in the process of determining the weights of each index, the data are processed by the entropy method. For each index, the greater the information entropy, the greater the role in the whole evaluation system.

Forward normalized processing formula:

$$X_{i,j}^{*} = \frac{X_{i,j} - X_{min}}{X_{max} - X_{min}} \tag{1}$$

Negative normalized processing formula:

$$X_{i,j}^{*} = \frac{X_{max} - X_{i,j}}{X_{max} - X_{min}} \tag{2}$$

Assign weights to each indicator and calculate the contribution of each indicator:

$$P_{i,j} = \frac{X_{i,j}^{*}}{\sum_{i} X_{i,j}^{*}} \tag{3}$$

Calculate the entropy value of each indicator j:

$$E_j = -\mathrm{K}\sum_{i} P_{i,j}, \ \mathrm{K} = \frac{1}{lnam} \tag{4}$$

Find the information utility value of each indicator j:

$$e_j = 1 - E_j \tag{5}$$

Calculate the weight of the jth indicator:

$$W_j = \frac{e_j}{\sum_{j}^{n} e_j} \tag{6}$$

Finally, the quality development index and shock resistance index are calculated:

$$Z_{i,j} = \sum_{i=1}^{n} W_j * X_{i,j}^* \tag{7}$$

Multiple linear regression: According to the existing research and the requirements of constructing a new development pattern in China under the new normal, high-quality development is to improve the shock resistance of trade and distribution enterprises. Meanwhile, due to the development differences of different cities, it also affects the anti-shock ability of enterprises. Therefore, this paper introduces urbanization rate and gross regional product as control variables to analyze the correlation between high-quality development and shock resistance:

$$res_i = \alpha + b_1 hq_i + b_2 urb_i + b_3 gdp_i + e \tag{8}$$

Among them: res_i, hq_i are the shock resistance index and high quality development index of each city, respectively; urb_i, gdp_i are the control variables, which denote the urbanization rate and gross regional product respectively. As a bridge between production and consumption, commercial and distribution enterprises can be affected by market changes at both ends. The higher the urbanization rate, the larger the consumer market and the larger the target customers of trade distribution enterprises.

4.3 Data Sources and Descriptions

In this paper, the panel data of 29 cities in the middle reaches of Yangtze River, including Wuhan, Huangshi, Ezhou, Huanggang, Changsha, Zhuzhou, Xiangtan, Nanchang, and Jiujiang, are used as samples from 2016–2020, and the data are obtained from China Statistical Yearbook, provincial and municipal statistical yearbooks and statistical bulletins, and the missing data are filled in by linear interpolation.

5 Empirical Analysis

The spatial pattern map (Fig. 1) and shock resistance map (Fig. 2) of the high-quality development level of trade and distribution enterprises in the middle reaches of Yangtze River cities were drawn using ArcGIS software by applying the high-quality development index and shock resistance index derived from the entropy value method above. The high-quality development index and shock resistance index are divided into five levels using the natural discontinuity method to reflect the spatial and temporal evolution characteristics and spatial distribution of the high-quality development of trade and distribution enterprises in the middle reaches of Yangtze River from 2016 to 2020, as well as their shock resistance in the face of major public events.

From the level of spatial development pattern, the high-quality development of trade circulation enterprises in the middle reaches of Yangtze River cities presents a polycentric development pattern, and the overall high quality development level is steadily improved from the center to the edge. Firstly, the enterprises located in the multi-center connection area are more strongly radiated by the central enterprises; secondly, for the commerce and distribution enterprises in the non-connection area of the neighboring cities, the core city enterprises have a certain siphoning effect on them, but the stability and structure of high-quality development is better than the edge enterprises. Finally, although the trade and distribution enterprises in the peripheral areas are weaker than their neighbors in terms of technological investment, they have an advantage in the scale of economic development and the overall speed of high-quality development is better than that of their neighbors.

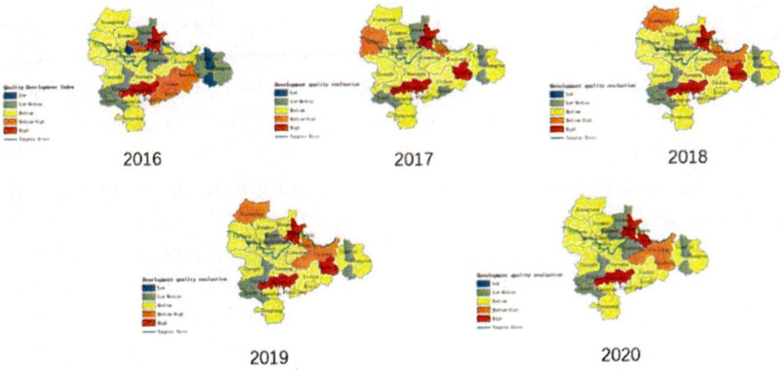

Fig. 1. Evolution of high-quality development of trade circulation enterprises in the middle reaches of Yangtze River cities

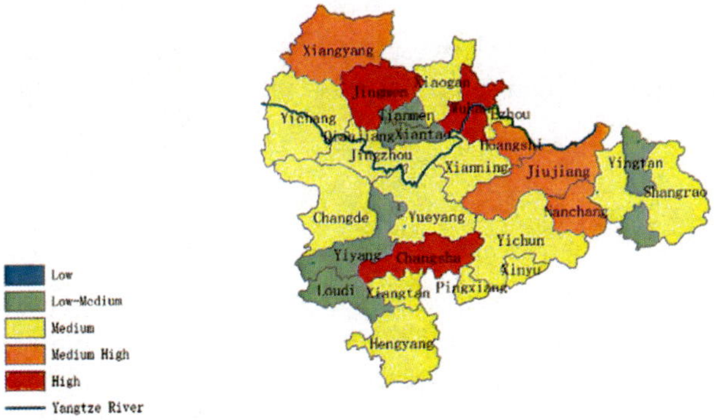

Fig. 2. Anti-shock capacity of trade circulation enterprises in the middle reaches of Yangtze River cities

The Yangtze River as the development link of the middle reaches of the Yangtze River city cluster also plays a key role in the high-quality development of enterprises. The overall high quality development level of trade and distribution enterprises in cities along the Yangtze River and the synergy of inter-city enterprise development are significantly better than those in other regions. A vertical analysis of the cities along the Yangtze River shows that the level of quality development and synergy in the downstream region is significantly better than that in the upstream region. With Wuhan and Nanchang as the economic centers and the Yangtze River as the backbone of shipping and transportation, the speed and level of high-quality development of commerce and distribution enterprises in the eastern city cluster are better than those in other regions, and the coordination of development is also more prominent.

In the risk resistance, excluding the special reason of Wuhan, the risk resistance of commerce and distribution enterprises in the middle reaches of Yangtze River cities.

remains multi-centered distribution. Combined with the structure of high-quality development, the business circulation enterprises near the center, although slower than the edge of the city, but benefit from better development structure and technical advantages, has a stronger anti-risk ability. At the same time, distribution enterprises in the east and inter-center enterprises have more stable risk resistance.

From the level of time development, the level of high-quality development has gradually increased, and the region has also gradually achieved synergistic development. With, the high-quality development was proposed in 2017, the same year, the middle reaches of the Yangtze River cities business circulation enterprises respond positively, medium development quality and above cities accounted for nearly 80%, up 17% from the previous year, while the multi-center radiation also began to play a role.

Combined with enterprise risk resilience, commerce and distribution enterprises in cities around the center outperform edge cities in terms of standardized production and operation, market stability and staff business level, and these cities perform better in terms of development and structural resilience. Economic and population size are the main advantages of peripheral cities.

From a comprehensive point of view, the regional enterprises in the middle reaches of the Yangtze River rely on the development pattern of the multi-center, while with the synergy of the Yangtze River Economic Belt, the level of high-quality development of regional enterprises has improved across the board. However, the high-quality development level still has the problem of unbalanced internal structure and external regional development.

6 Correlation Analysis

In this paper, a regression analysis is conducted on trade and distribution enterprises in 29 cities in the middle reaches of Yangtze River, and it is found that high-quality development has a significant positive effect on the shock resistance of enterprises. Multiple linear regression analysis was conducted with population size, urbanization rate and high-quality development index as independent variables and shock resistance capacity index as dependent variables. Firstly, the variance inflation factor VIF method was used for the diagnosis of multiple covariance, and the VIF values were all around 1, so there was no covariance among the independent variables. The correlation test was performed on the model (Table 3), and the p-value was <0.05, indicating that there was a linear relationship between the independent variables and the dependent variable see, and the regression results are shown in Table 4.

The regression results show that the level of high-quality development of trade circulation enterprises in the middle reaches of Yangtze River significantly affects the shock

Table 3. ANOVA[a]

Model		Sum of Squares	freedom	Mean Square	F	Significance
1	Regression	2.017	3	.672	18.621	.000[b]
	Residuals	.903	25	.036		
	Total	2.920	28			

Table 4. Regression results of the impact of high quality development on resilience to shocks

	B	Standard Error	Beta			Tolerances
Population Size	−.148	.122	−.143	−1.209	.238	.890
High Quality Development Index	.592	.095	.733	6.244	.000	.897
Urbanization Rate	.655	.203	.366	3.225	.003	.960

resistance of enterprises. As a bridge between production and consumption, the ultimate goal of high-quality development is to communicate and stabilize the two ends of the business, and high-quality development is to solve the current conflicts and pay more attention to the long-term interests of the future. Then the regression result also confirms the objectivity and reasonableness of the evaluation system of high-quality development.

For the two independent variables of population size and urbanization rate, the urbanization rate has a significant effect on the ability to resist shocks, and the regression coefficient is positive, indicating that the higher the urbanization rate, the larger the urban population size, the larger the consumer market size of trade and circulation enterprises, and the degree of enterprise chain will also increase, and the enterprises will increase their investment and innovation. This is conducive to the gradual standardization of labor services and goods, and the provision of better services for residents. Therefore, the city cluster in the middle reaches of Yangtze River should continue to promote regional integration, improve regional transportation facilitation, and better connect multi-centered trade activities; Efficient use of the Yangtze River's transportation and shipping advantages to enhance the region's carrying capacity in the domestic general circulation.

7 Conclusion

Through the study of trade circulation enterprises in 29 cities in the middle reaches of Yangtze River, the positive contribution of high-quality development to the anti-shock ability of enterprises is found, which confirms that high quality development of trade circulation enterprises can promote the construction of modern circulation system and domestic general circulation. This paper draws the following conclusions from the study:

1) The level of high-quality development of trade circulation enterprises in the middle reaches of Yangtze River cities has steadily increased, forming a multi-center development pattern of Wuhan, Changsha and Nanchang, with enterprises in the central cities jointly driving high-quality development of enterprises in neighboring cities. The Yangtze River plays a conduction role in the synergistic development of the region.
2) Among the cities in the middle reaches of the Yangtze River, there are differences in the internal structure of high-quality development of commerce and distribution enterprises, with the high quality of enterprises in the central city circle mainly driven by enterprise upgrading, and the enterprises in the peripheral cities mainly driven by the expansion of the consumer market scale.
3) By developing enterprises with high quality, the impact resistance can be improved, and the increase of urbanization rate in cities can also be improved by expanding consumption market and improve the efficiency of enterprises to enhance their resistance to shocks.

Acknowledgment. This project is supported by Institute of Distribution Research, IDR (IDR2021YB004).

References

1. Yang, W.: Implementing the spirit of the central economic work conference to promote high-quality development. Macroecon. Manage. **02**, 13–17 (2018). (in Chinese)
2. Qigui, Z.: Establishing an indicator system to promote high-quality development. Wen Wei Po, 2018-02-06(012). (in Chinese)
3. Wei, L.: Six connotations of high quality development. China Forestry Ind. **Z1**, 50–51 (2018). (in Chinese)
4. Lei, W., Qian, G.: Study on the impact of economic performance of the spatial structure of urban agglomerations in the middle reaches of Yangtze River[J]. Hum. Geogr. **33**(06), 96–102 (2018). (in Chinese)
5. Lei, W., Chengli, L.: Growth effects of the polycentric structure of urban agglomerations in central China[J]. Yangtze River Basin Resour. Environ. **27**(10), 2231–2240 (2018). (in Chinese)
6. Jinping, X.: Grasp the new development stage, implement the new development concept, and build a new development pattern. Seeking Truth **9**, 4–18 (2021). (in Chinese)
7. Huang, G., Hong, J., Dai, Q.: The High-quality development of China's economy: theory and applications special issue. J. Syst. Sci. Inform. **9**(6), 2 (2021). (in Chinese)
8. Ge, H., Wu, Q.: Research on the spatial influence of digital inclusive finance on the high-quality development of private economy. Manage. Sci. Res. **11**(2), 11 (2022). (in Chinese)
9. Ron, M.: Regional economic resilience, hysteresis and recessionary shocks. J. Econ. Geography **12**(1), 1–32 (2012)

Aesthetic Challenge and Reflection of AI Art in Film and Television Works from the Perspective of "Face Changing Technology"

Feng Bin[1,2](✉)

[1] Graduate School of Sejong University Cooperative Course of Performance, Film and Animation, Seoul 05006, South Korea
adrianbin@foxmail.com
[2] School of Journalism and Cultural Communication, Guangxi University of Finance and Economics, Nanning 530001, China

Abstract. With the rapid development of artificial intelligence technology, great changes have taken place in human life and aesthetic concepts. The "face changing" technology in the film and television cultural industry, with accurate capture and real-time technology, provides a new thinking and new aesthetic existence for the development of the film and television industry. Based on the conceptual characteristics, historical context and AI face changing technology of AI, this paper discusses the transcendence of AI art over traditional film and television art in film and television works, discusses the innovation of AI to traditional film and television art, the expansion of human creativity, and the challenge of human development mechanism, and briefly analyzes the curve of horror Valley theory - it is a hypothesis about human feelings about robots and non-human objects. On this basis, it reflects on the hidden dangers behind the wider application of artificial intelligence, with a view to clarifying the direction of the next step.

Keywords: Film and television works · Artificial intelligence art · Face change · Horror Valley theory

1 Introduction

The progress of science and technology has played an obvious role in promoting human civilization [1]. The change of material life has continuously improved human living conditions and quality of life. In the new century, modern science and technology continue to create miracles one after another [2]. Science is changing the spiritual and cultural life of mankind [3].

From the first appearance of computers in the 1940s to the 21st century, science and technology have developed by leaps and bounds. The "scientific and technological revolution" has deepened the connection and transformation between human beings and society. Technology has changed the traditional machinery and production mode, and

Z. Hu et al. (Eds.): AIMEE 2022, LNDECT 159, pp. 231–242, 2023.
https://doi.org/10.1007/978-3-031-24468-1_21

to a certain extent, has affected people's lives, including art appreciation and aesthetics [4–7].

The application of science and technology in the field of art has exposed people to scenes they have never seen or even imagined before, brought people a lot of visual impact, and expanded the public's vision. It will even change people's aesthetic outlook [8]. At the same time, it will also cause people to think about new technologies and innovative ideas [9–12]. Take film and television art for example. The application of "face changing" technology in the works, with the support of motion capture and real-time technology, has brought earth shaking changes to the development of the film industry [13–17].

1.1 Conceptual Analysis of Artificial Intelligence

Artificial intelligence, abbreviated as AI. It is a new technology discipline, which mainly studies and applies human intelligence theory and method technology to the new generation of robots.

As a branch of computer science, artificial intelligence can make a human like response by understanding the essence of intelligence and creating a new intelligent machine that is different from the traditional one. Robot technology, language recognition, image recognition, natural language processing and other professional systems are its main research directions. Since the establishment of the theoretical system of artificial intelligence technology, its application scope has gradually expanded, but artificial intelligence is not human intelligence after all, but can think like human beings.

1.2 Characteristic Analysis of Artificial Intelligence

Artificial intelligence is a technical science aiming at simulating, extending and extending the theories, methods, technologies and application systems of human intelligence [18–20]. Artificial intelligence in China started in the late 1970s. In recent years, with the continuous development of network technology and the continuous innovation of science and technology, artificial intelligence in China has also stepped into the fast lane.

1.2.1 Provide Services for Human Beings Through Calculation and Data

In essence, the essence of artificial intelligence is to serve human beings. Through special hardware technology, such as based on the set programming logic or software algorithm, through the collection, processing, analysis, extraction of data, the establishment of valuable information and knowledge model, so that some human "intelligent behavior" can be expanded and serve human beings. The essence of its technology is computer science.

1.2.2 Perceive the External Environment and Realize Human-Computer Interaction Complementarity

Artificial intelligence technology perceives the external environment through hearing, vision, smell, touch, etc., receives all kinds of information, and responds to this information, generating corresponding words, sounds, expressions and actions, which act on the environment and human beings. Or with the help of keyboard, mouse, gesture, posture, facial expression, force feedback or virtual scene, human and machine can interact, so that machines and devices can more and more "understand" human beings and achieve the effect of human-computer complementarity.

1.2.3 Self Adaptation and Evolutionary Iteration

AI technology has the ability to actively learn and adapt, and will automatically adjust parameters or update optimization models with the changes of environment, data or activities; In addition, on this basis, it can realize the evolution and iteration of machine objects and human subjects through in-depth connection and expansion with the cloud, people and things, so as to automatically adapt and expand to cope with the changing actual environment.

In film and television media, these three characteristics can be reflected, but the most important is that artificial intelligence is widely used instead of artificial in film and television post production.

1.3 Overview of AI Development

The development of artificial intelligence is revolutionary, inevitable and worldwide. Usually, people's first impression of AI comes from sci-fi films and television works. The creators create "Ai" on the screen through virtual imagination and film and television technology of AI. The "artificial intelligence" appearing in film and television works often has strong power, which is beyond the scope of human ability and has strong destructive and deterrent power. It is a kind of "strong artificial intelligence" or "super intelligence". The image of artificial intelligence presented in film and television works is determined by the author's subjective consciousness, which has a certain impact on the public's cognition.

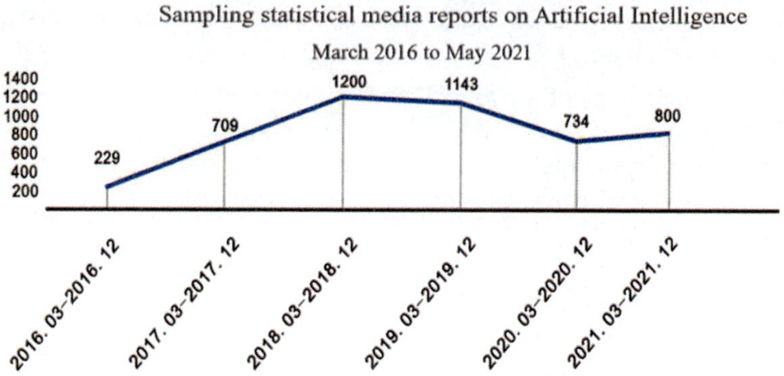

Fig. 1. Sampling statistical media reports on Artificial Intelligence

Since 2016, as AlphaGo defeated Li Shishi, the major media have paid more and more attention to and reported on AI (see Fig. 1 and Table 1), which has expanded the public's understanding of AI, let the public know AI from a single science fiction film and television work, broaden cognitive channels, let more people know AI, reduce the cognitive gap, and promote the development of AI.

Table 1. Sampling analysis of media reporting on the topic of AI (2016–2021)

	2016	2017	2018	2019	2020	2021
National Strategy/Policy System	4 (16.3%)	5 (5.5%)	15 (11%)	10 (9.3%)	8 (8.9%)	6 (8.3%)
Economic development	8 (33.7%)	12 (15.8%)	33 (24.6%)	33 (28.1%)	23 (25.9%)	20 (25.9%)
Culture/ Education	2 (8.2%)	12 (15.9%)	17 (12.5%)	16 (12.7%)	5 (5.2%)	4 (5.0%)
Society/People's Livelihood	1 (4.1%)	12 (15.3%)	28 (20.0%)	16 (12.7%)	12 (13%)	10 (1.3%)
Technology Innovation/ Technology	4 (16.3%)	20 (26.2%)	22 (15.4%)	22 (17.5%)	7 (7.9%)	5 (7.1%)
Academic/ Theory	1 (4.2%)			1 (0.8%)	2 (2.2%)	2 (2.3%)
Resources/	1 (4.2%)					
Medical		1 (1.2%)	8 (5.8%)	4 (3.3%)	2 (2.2%)	4 (3.3%)

(continued)

Table 1. (*continued*)

	2016	2017	2018	2019	2020	2021
Justice/Legal		6 (6.5%)	3 (2.2%)	3 (2.5%)	3 (3.4%)	2 (3.0%)
Political life		1 (1.1%)	2 (1.6%)	3 (2.5%)	1 (1.1%)	1 (1.2%)
Resources/ Ecology			1 (0.9%)			
Epidemic					14 (16.4%)	20 (20.1%)
Other	2 (7.6%)	5 (6.4%)	1 (0.9%)	7 (5%)	7 (7.9%)	7 (8.1%)
Total	23	74	130	115	84	83

As an infinite sustainable renewable resource, the research space in the field of artificial intelligence is huge, new development ways are constantly opened up, and the application prospect is broad.

2 Origin of "Face Changing Technology"

Face changing is to change others' faces into "own" faces through artificial intelligence technology. Face recognition is the premise of successful face change [21–23]. The early application of AI "face changing" technology appeared in Hollywood more than 20 years ago. Star Wars: Rogue one was released in 2016. Peter Cushing, the actor of the story character Governor Tarkin, died unexpectedly. The post production team of the crew used facial expression capture technology to capture the facial expression of an actor who looked like Peter, and finally revived it with CG technology, The effect is enough to confuse the fake with the real (see Fig. 3).

Fig. 2. Governor Tarkin "resurrected" in the movie

In recent years, as the main actors in some popular film and television works have gradually been exposed to negative news, they have been listed as "tainted artists" by the State Administration of radio, film and television, which has a direct impact on the ratings and benefits of film and television works. For example, in the great pediatrician, Tong Zhuo's face was replaced by another actor by AI. If it wasn't intentional, it really wouldn't be obvious. In addition to a certain color difference between the face and the side face, the whole looks natural and pressed (Fig. 2).

AI face changing operation is also used in "three thousand crows kill". Originally, Liu Lu played the role of "Qingqing" in the play, but she became a "bad actor" because she didn't cooperate with the security check. Since then, during post production, the drama company replaced the actor Zhang Dingding's face with "Qingqing" (see Fig. 4).

AI face changing technology has been widely used in film and television works, and has become an "emergency means" for dramas. With the widespread use of face changing technology in film and television dramas, the production, investment and production can stop losses in time, and some specific scenes can become more vivid through the blessing of special effects. It also further improves the film and television production technology, and also makes the performance forms of actors in some film and television works more technical and diversified.

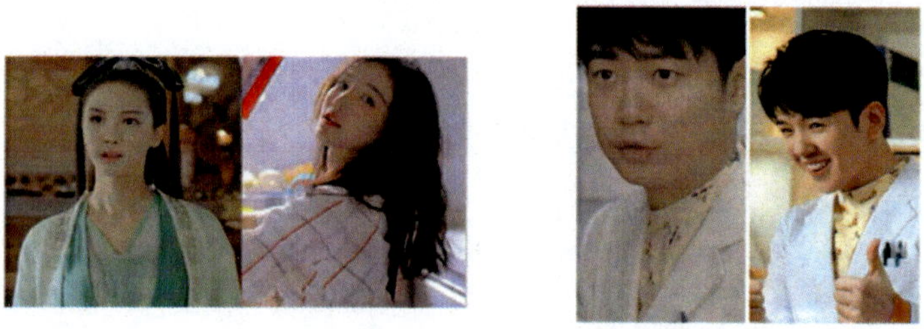

Fig. 3. AI face changing in TV series

2.1 AI Virtual Character "Hua Zhibing"

Hua Zhibing is an artificial intelligence virtual human based on "enlightenment 2.0". She has rich knowledge and theory, can interact friendly with humans, and can create music, poetry and painting (see Fig. 4). Hua Zhibing officially appeared in Beijing in 2021. She has a sweet voice and very realistic facial expressions. She not only has certain emotional communication and reasoning ability, but also can gradually "grow up" in continuous learning. Just as people can constantly correct mistakes through their own experience. Hua Zhibing organically integrates his knowledge gained in the new scene with his own model to make himself smarter.

According to data, by 2020, the market scale of AI core industries will exceed 150 billion yuan, and authoritative institutions predict that it will exceed 400 billion yuan by 2025. In the manufacturing industry, AI technology has significantly improved production efficiency and service level. Driven by the government, industry associations

Fig. 4. AI virtual character "Hua Zhibing"

and other departments, China's AI industry has ushered in explosive growth, and the scope of application scenarios in China's AI market will continue to expand to different fields.

2.2 Natural Person's Identification with AI Virtual Character Identity

In the film and television works created with the theme of science fiction, there are contradictions or cognitive differences between natural people and AI virtual people, but they can live in harmony overall, which is due to the recognition of AI virtual people by natural people to a certain extent. The "humanoid" appearance of AI virtual human makes people feel more intimate. In addition, robots are getting closer to humans in thinking, intelligence, behavior and other aspects, which greatly reduces the recognition ability of natural people to AI virtual human and weakens the sense of exclusion of "non-human".

In the film "artificial intelligence", the naive and lovely childlike image of AI virtual people in the slaughterhouse makes natural people no longer adhere to the original position of anti AI virtual characters, and gradually more people begin to have an attitude of acceptance and recognition.

However, the identity brought by this convergence of body language is actually the self-identity of natural people, but it is only temporary. However, the identity of "humanoid attribute" based on this is uncertain and limited. Robot scientist Mori Changhong's "uncanny valley" can illustrate this phenomenon.

Generally, the fear Valley phenomenon occurs because people make a "human" prediction when they see the humanoid AI virtual human, but at the same time, some "non-human" characteristics shown by the AI virtual human violate the original prediction. The conflict between this prediction and reality will produce additional brain activity, resulting in brain load, and human discomfort. This discomfort is called "disorder of empathy mechanism", which is presented by Mori Changhong with a graph (shown in Fig. 5).

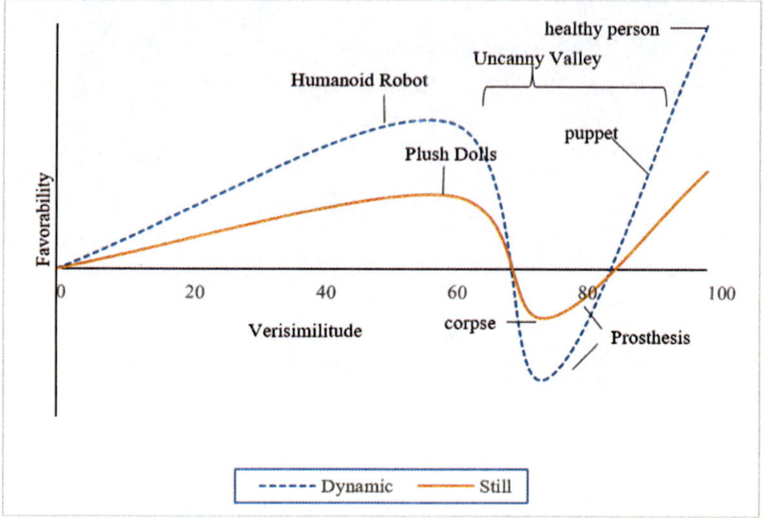

Fig. 5. Mori Changhong's "valley of terror" theory

3 Aesthetic Challenges of AI Art

3.1 AI Art Surpasses Traditional Film and Television Art

With the introduction of face changing technology and the emergence of AI virtual characters, the application of artificial intelligence art in the traditional film and television field will bring new artistic forms and aesthetic forms, and then expand the existing cultural space and artistic expression. Generally speaking, the closer the connection between film and television art and artificial intelligence, the stronger its technical limitations. In terms of image technology, technology directly changes the creative form and process of images. In the field of film production, not only science fiction works increasingly rely on post production, but also the types of realistic themes cannot be separated from the support of artificial intelligence technology. The visual painting effect synthesized by computer and live shooting has become the main way of film and television drama production. In many countries and regions with high AI technology, much video production is completely digital, which greatly exceeds our understanding of the original technology. In the creation of electronic music, it is difficult to distinguish between real musical instruments and synthetic music, and these music can also be put into the music list. With the wide application of 3D technology and simulation technology, the film can bring a new visual experience to the audience, and can have a benign interaction with the audience, so as to obtain an immersive on-site experience.

3.2 Innovation of Artificial Intelligence Art to Traditional Film and Television Art

While AI technology innovates traditional art, it breaks people's understanding of art space and changes art space and time to some extent. Using big data and AI software, we can comprehensively sort out and study human art history in a relatively short time, extract the essence and laws of art history, and imitate and create similar works of art. In the face of thousands or hundreds of years of literature and art development history, it is often difficult for us to understand all. In this process of art development, a special concept of art space-time and culture has been formed. Under the dual action of big data and artificial intelligence, the natural time of art works is compressed by artificial intelligence technology, which has greatly changed the concept of art space and space. Human artistic creation is the result of the interweaving and interaction between image and abstraction, rationality and irrationality, intuition and objectivity, consciousness and subconsciousness. The production of artificial intelligence is a kind of simulation of human psychological activities, and in the mode of thinking, it is a kind of technical thinking. However, the result of imitation has some perceptual thinking nature, which poses a challenge to the existing aesthetic understanding.

3.3 Challenges of Artificial Intelligence Art to Human Creativity

Although artificial intelligence technology is the crystallization of human wisdom, at the creative level, it must be transformed, such as machines, software and so on. However, in the deepest essence of human thoughts and feelings, AI technology still has a lot of room for technological improvement, because the closer it is to human itself, the more difficult it is to surpass. To some extent, the creation of artificial intelligence plays is a reflection of literature. Through the establishment of big data and models, after reaching a certain scale and experience, we will seek corresponding propositions to create. Such creation is very different from "human literature". Human literature itself takes language as the medium and artistic image expression as the object, so that aesthetics can be presented. This is a stable and unique value of literature, but also a kind of humanistic care, which can be shared by all humankind, far beyond the reach of technical rationality. Artificial intelligence may have the potential to surpass human beings in text combination, aesthetic extraction and so on. But the collision of beautiful words and the expression of artistic conception are unmatched by AI literature.

4 Reflection on the Art of Artificial Intelligence

In the creation of film art, artificial intelligence is based on scientific and technological progress. Artificial intelligence technology can increase people's imagination and exploration of film art. Artificial intelligence technology brings more real feelings to film art and visual perception. Artificial intelligence technology supports the development direction of film art. If we say that AI "face changing" technology in film and television works enables producers to avoid risks in time and completely fail to see the actors' previous appearance, so as to make up for some losses. Then the popular filter function in video software is also equivalent to the application of AI technology in the field of short video. For example, with the help of AI algorithm, Tiktok software generates virtual faces that differ greatly from the actual age. Due to the strong contrast between real and virtual images, it shows eye-catching contrast interest. But behind the wider application of artificial intelligence, some negative effects have also emerged.

4.1 How to Protect Human Privacy Data

Although "face changing technology" has brought convenience to the post production of film and television works, it has also caused many disputes. This is mainly due to the risk of personal privacy disclosure brought by artificial intelligence. It can disclose and intercept user data information and generate data portraits for individuals, to build a complete personal image, which not only involves privacy, but also comprehensively monitors individuals. Using big data and algorithms, artificial intelligence technology can "deeply imitate" human body, behavior and practical ability, and even bring hidden dangers to human security. With the advent of the data age and the increase of such computing environments and platforms, data abuse has become more and more frequent.

4.2 Can AI Robots Replace Humans

Taking the virtual human "Hua Zhibing" as an example, from the perspective of social ethics, Whether artificial intelligence such as "Hua Zhibing" is beneficial without harm? This will be an important topic in the future. How can the research team strictly limit "her" The content and form of interaction with the public need to be further studied and demonstrated. However, in many science fiction films, after robots have self-consciousness, their relationship with humans is at stake, and humans as tools no longer use them. Therefore, people should not only ask whether a scene in the film will be produced in the real world? Where is the bottom line of AI exploration? How does law regulate the AI industry?

5 Conclusion

In recent years, AI has made major breakthroughs in the film and television industry, promoting the prosperity and development of the film and television industry. However, the development of science and technology in every era needs to be balanced and constrained by law and ethics. China should establish a perfect, open and transparent artificial intelligence system, assume the responsibility for the design and application of artificial intelligence, and establish a dual regulatory mechanism. At the same time, we should strengthen the legislative work of artificial intelligence related laws, and establish and optimize the legal and ethical framework to ensure the healthy development of artificial intelligence.

In the context of technology, artificial intelligence will inevitably affect all aspects of the future development of the media industry, such as re understanding and interpretation of production, script, director and interpretation. However, unlike other industries, film and television works as works of art cannot be equated with general industrial products, whether in the media or in film and television post production. Although AI can greatly improve the production efficiency of these works of art, AI is just a tool, just like the pen in the painter's hand. From pencil, oil painting to the production of computer art works, tools can never replace the role of people, but tools can improve the efficiency of creation, and make art creation more diversified through new forms and spaces of art creation. Therefore, in the face of the challenge of artificial intelligence to the film and television industry, we must insist that artificial intelligence is always a tool, encourage the originality of human thinking in artistic creation, and avoid weakening the creativity of video production due to the abuse of artificial intelligence.

References

1. Paul, A.M.: The coding revolution. Sci. Am. **315**(2), 42–49 (2016)
2. Fang, Z., Yu, C., Chen, W.: Innovation and industrial development: meeting the challenges of the new technological revolution. Reg. Econ. Rev. **2**, 55–67 (2018). (in Chinese)
3. Lu, N.: The two facets of scientific and technological revolution and legal evolution. Contemp. Am. Rev. (1), 66–84 (2019). 122 (in Chinese)
4. Xuejun, L.: Art is reproduction: reflections on artificial intelligence art. Learn. Exp. **10**, 176–182 (2020). (in Chinese)
5. Zhen, Z.: Aesthetic challenges and reflections on artificial intelligence art. Shandong Soc. Sci. **10**, 181–185 (2019). (in Chinese)
6. Du, J., Du, X., Guan, S.: Research on the influence characteristics and management strategies of art aesthetic experience in the age of artificial intelligence. Art Baijia **35**(05), 24–29+63 (2019). (in Chinese)
7. Yong, L.: "Post-Humanity" conception of artificial intelligence and contemporary art production—also an aesthetic reflection on "Digital Humanities." Inner Mongolia Soc. Sci. **42**(02), 145–152 (2021). (in Chinese)
8. Rathi, J., Grewal, S.K.: Aesthetic QR: approaches for beautified, fast decoding, and secured QR codes. Int. J. Inf. Eng. Electron. Bus. (IJIEEB) **14**(3), 10–18 (2022)
9. Cao, M.: Aesthetic challenges of artificial intelligence art. Ethnic Art Res. **31**(06), 90–97 (2018). (in Chinese)
10. Wei, W., Yi, H.: Synergy, value and singularity: art reflection under the challenge of artificial intelligence. Art Acad. **02**, 1–12 (2018). (in Chinese)
11. Issak, A.: Artistic autonomy in AI art. arXiv preprint arXiv:2111.04437 1–2 (2021)
12. Andrews, C.: An embodied approach to AI art collaboration. Create. Cogn. 156–162 (2019)
13. Cetinic, E., She, J.: Understanding and creating art with AI: review and outlook. ACM Trans. Multimedia Comput. Commun. Appl. (TOMM) **18**(2), 1–22 (2022)
14. Ragot, M., Martin, N., Salomé, C.: AI-generated vs. Human Artworks. A Perception Bias Towards Artificial Intelligence?. Human Factors in Computing Systems, pp. 1–10 (2020)
15. Lu, D.: Al creates masterful art in just minutes. New Sci. **3228**, 16 (2019)
16. Lu, Y., Wang, Y., Tong, X., et al.: Face tracking in video sequences using particle filter based on skin color model and facial contour. In: 2008 Second International Symposium on Intelligent Information Technology Application, vol. 1, pp. 457–461. IEEE (2008)
17. Malciu, M., Preteux, F.J.: Tracking facial features in video sequences using a deformable-model-based approach. In: Mathematical Modeling, Estimation, and Imaging. SPIE, vol. 4121, pp. 51–62 (2000)
18. Singla, V., Thareja, R., Thareja, R.: Psychometric analysis using computational intelligence for smart choices. Int. J. Modern Educ. Comput. Sci. (IJMECS) **14**(2), 65–75 (2022)
19. Arora, M., Bhardwaj, I.: Artificial intelligence in collaborative information system. Int. J. Mod. Educ. Comput. Sci. (IJMECS) **14**(1), 44–55 (2022)
20. Jariwala, A., Chaudhari, A., Bhatt, C., Le, D.-N.: Data quality for AI tool: exploratory data analysis on IBM API. Int. J. Eng. Manuf. (IJEM), **14**(1), 42–56 (2022)
21. Ebenezer, O.: Graphic Design Principles and theories application in rendering aesthetic and functional installations for improved environmental sustainability and development. Int. J. Eng. Manuf. (IJEM) **9**(1), 21–37 (2019)

22. Kumar, J.A., Muniandy, B., Yahaya, W.A.J.W.: Emotional design in multimedia learning: how emotional intelligence moderates learning outcomes. Int. J. Mod. Educ. Comput. Sci. (IJMECS) **8**(5), 54–63 (2016)
23. Aliyu, I., Bomoi, M.A., Maishanu, M.: A comparative study of eigenface and fisherface algorithms based on OpenCV and Sci-kit libraries implementation. Int. J. Inform. Eng. Electron. Bus. (IJIEEB) **14**(3), 30–40 (2022)

Differentiation of Relevant Demands of Scientific and Technological Staff in Wuhan Based on PEST Model

Min Wang[1,2](✉)

[1] Department of Science and Technology, Wuhan Technology and Business University, Wuhan 430065, China
40754707@qq.com

[2] Survey Station of Science and Technology Staff in Hongshan District, Wuhan 430065, China

Abstract. Based on the relevant needs of science and technology staff in Wuhan, using literature research method, questionnaire survey method, quantitative and qualitative analysis method combined to analyze, sort out, review the relevant documents of the paper PEST model is used to analyze the factors affecting the enthusiasm of scientific and technological staff in universities from four aspects: politics, economy, society and technology. Compiled the widespread use of the questionnaire survey in Wuhan science and technology personnel demands, according to the internal relation between questionnaire items and developing trend, analyses the science and technology staff in the Wuhan 160 questionnaire results, suggested the government advocate science fair income distribution mechanism, perfect supporting policy support system, the government take the lead to build all kinds of scientific research platform, It is expected to promote the sustainable development of science and technology activities in Wuhan.

Keywords: Scientific and technological staff · PEST Model · Questionnaire survey

1 Introduction

The promotion of science and technology to the comprehensive national strength of all countries has reached an unprecedented height [1–3]. The way people think and the way society produces have been profoundly influenced by the progress of science and technology. For a long time, as the main body of scientific and technological innovation, scientific and technological personnel are the source power of scientific and technological innovation and social development. Respecting the subjectivity of science and technology workers and creating a good working environment for them is the fundamental guarantee for the development of science and technology. Scientific and technological innovation cannot be separated from the sustained investment of time, energy and material resources of scientific and technological personnel. To strengthen the protection of scientific and technological personnel is to protect the innovation ability. The living

Z. Hu et al. (Eds.): AIMEE 2022, LNDECT 159, pp. 243–252, 2023.
https://doi.org/10.1007/978-3-031-24468-1_22

conditions and working requirements of scientific and technical personnel are one of the key points of concern of the whole society [4–6].

Therefore, starting from the relevant needs of science and technology personnel in Wuhan, this study analyzed, sorted out and summarized the relevant data and literature of the paper by combining quantitative and qualitative analysis with literature research method, questionnaire survey method. PEST model is used to analyze the factors affecting the enthusiasm of scientific and technological workers in colleges and universities from four aspects: politics, economy, society and technology.

2 Scale Design Based on PEST Model

2.1 Theoretical Basis

Harvard economics professor Francis J.Aguilar first proposed the ETPS model in 1967 as a theoretical framework for judging the external macroeconomic environment of an enterprise. Gerry Johnson and Kevkevan Scholes in the United States put forward PEST model in 1999. ETPS and PETS are just in a different order. PEST model includes Political, Economic, social and Technological factors. PEST model is a data analysis model, which can provide a basis for judging trends, opportunities and challenges when enterprises formulate strategies. In the annual business planning, according to the conclusion of PEST analysis, the enterprise can comprehensively identify opportunities and threats in the macro environment, and make adjustments to the strategic planning and business strategy accordingly. PEST model analysis is a kind of thinking framework, and its value lies more in providing enterprise strategy makers with a judgment basis for trends, opportunities and challenges rather than conclusions [7]. Based on PEST model and combining the political, economic, social and technological aspects of Wuhan, this study designs a survey scale for scientific and technological workers in colleges and universities to find out the differences in scientific and technological work and needs, in order to provide a basis for the government to help scientific and technological staff.

From the political environment analysis, including national system, government policies, laws and regulations, to understand industry changes. At present, the government has issued relevant policy about research projects and achievements, such as excellent young, outstanding youth, national science and technology award, national science and technology progress prize such as policy, implement the development strategy of the national innovation drive compendium "deepening the reform of science and technology system implementation plan" action plan for promoting transformation of scientific and technological achievements transfer and other documents.

From the macro and micro aspects of economic environment analysis, such as gross national product, consumer preference, consumer income level. Understand the market demand, hobby and consumption behavior of science and technology workers. In 2021, Wuhan's GDP ranked ninth in China, and the total retail sales of consumer goods rose sharply, with the growth rate lower than the average of the country and the new first-tier cities. The export value of goods was far lower than the average of the new first-tier cities, but the growth rate was fast. In 2020, Wuhan's tax revenue was slightly lower than the average of new first-tier cities. The total consumption of Wuhan is 679.5 billion yuan, ranking the ninth in China. The per capita consumption of Wuhan is also not low, reaching 55,100 yuan. Among the 30 big cities, the per capita consumption of Wuhan also happens to rank the ninth.

In combination with the social environment, this study analyzes factors including population structure and trend, education level and national culture, and designs the demographic factors, social structure, mentality, social culture and living conditions of scientific and technological workers in the questionnaire to provide theoretical basis. Wuhan, the capital of Hubei Province, is a mega-city in the Yangtze River Basin and one of the four centers of science and education in China. It has 28 state key laboratories and 19 international engineering research centers of 2,158 scientific research institutions. There are 72 academicians of the Chinese Academy of Sciences and the Chinese Academy of Sciences. By the end of 2017, there were 88 institutions of higher learning in the city, with 115,000 postgraduates and 956,800 undergraduate and junior college students. There are 7 national key universities, 25 key laboratories directly under the Ministry of Education and hundreds of key laboratories directly under the Ministry of Education. The total number of undergraduates and postgraduates is 1.0726006 [8, 9].

Evaluate the technology environment, technology-related policies, current new product development capabilities and technology development trends, etc. Wuhan supports key laboratories, engineering technology research centers, technology innovation centers, industrial technology research institutes, and manufacturing innovation centers independently or jointly with universities and institutes, and municipal finance provides corresponding support to major national science and technology projects and key national R&D programs undertaken by enterprises. Major science and technology projects and enterprise technology innovation projects have been set up in the municipal science and technology planning system to support enterprises in tackling technological problems. Enterprises are encouraged to acquire intellectual property rights. Enterprises are encouraged to timely apply for and obtain invention patents and other intellectual property rights for their independent research and development achievements.

2.2 Scale Design

In order to ensure the objectivity and accuracy of the survey data, the research questionnaire was set up on the basis of closely combining the research background, and underwent scientific and standardized planning and strict implementation under the guidance and supervision of Hongshan Association for Science and Technology. Questionnaires were distributed in the form of questionnaires, 170 of which were from Hongshan Scientific and technological workers, 160 of which were qualified and invalid samples were removed. The effective recovery rate of the questionnaire was 94.1%, and a list of scientific and technological workers in colleges and universities was established, and relevant sampling data were saved for follow-up investigation. The questionnaire classifies and summarizes the 18 questions in the questionnaire and analyzes them from four dimensions: society, economy, policy and science and technology (Table 1).

Table 1. Dimension classification of the PEST model-based questionnaire for scientific and technological workers in colleges and university

Dimension	Category	Questionnaire the source term
Social	gender, age, education background, Title and discipline	multi-item 1、2、3、4、5
Economic	work intensity, income level, social status, reward mechanism	multi-item 6、12、13、14
Technological	types of projects hosted and participated, number of government projects participated, number of social science popularization activities participated	multi-item 7, 9, 10, 11, 15, 17, 18
Political	Policy orientation demands in scientific research and funding demands for the construction of research workers' investigation platform	multi-item 8, 16

3 Investigation and Analysis of the Development Status of Science and Technology Staff in Universities

3.1 Young and Middle-Aged Scientists Account for a Large Proportion, with Long Working Hours and Heavy Tasks, and Low Personal Satisfaction with Life

As the main force to promote social development and social progress [10, 11], the existing empirical research results have proved that the evaluation of overall life satisfaction of scientists has a significant impact on the efficiency of scientific research. The satisfaction degree of physical and mental health of university scientific and technological Staff affects the quality of life and the output of scientific research. The satisfaction degree of physical and mental health of university scientific and technological Staff affects the quality of life and the output of scientific research [12]. At present, the age of science and technology Staff in colleges and universities is between 30 and 50 years old, and the age difference is 20 years. Each age group pays different attention to health conditions, but intellectuals in colleges and universities have higher requirements for their own health and life satisfaction than their peers in other professions. There are 66 male and 94 female scientific and technological Staff in the effective questionnaire of the investigated population. The age structure ranges from under 30 to over 60 (Fig. 1). Degree distribution structure, master or above accounted for the largest proportion. The distribution of professional titles from low-level to senior, intermediate and vice high above the science and technology Staff accounted for a large proportion, total accounted for 71%.

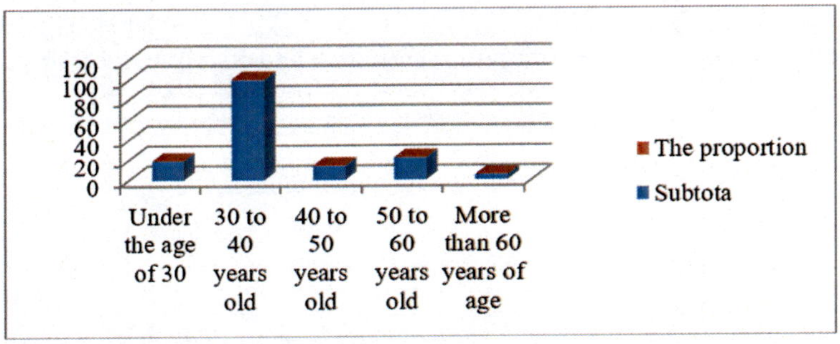

Fig. 1. Age structure of scientific and technological Staff in colleges and universities

The investigated data show that the working intensity and length of young scientific and technological Staff in colleges and universities are large. Although they have advantages in energy and physical strength compared with the elderly scientific and technological Staff, they are relatively weak in terms of experience, steadiness and stability. This means that young scientific and technological Staff are still in a growth stage, which is also a key stage of gradual stereotyping. There is a great space for development and a large factor of flexibility. However, long hours and high intensity work will have adverse effects on the physical and mental health of young scientific and technological Staff. Figure 2visually shows the working hours and working intensity of the investigated population.

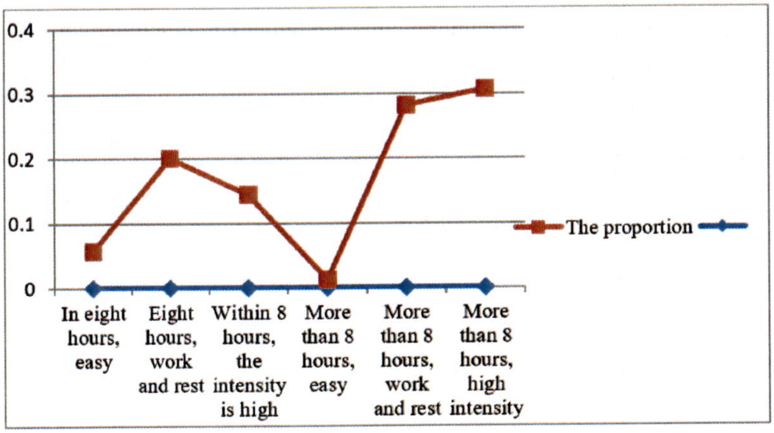

Fig. 2. Working hours and working intensity of scientific and technological Staff in colleges and universities

Figure 3 shows the life satisfaction of technicians. Job income is the most direct factor that affects the satisfaction of work and life of science and technology personnel. The survey reflects the income level of scientific and technological personnel in universities in Wuhan, among which the income of scientific and technological personnel in private universities is lower than that in public universities in Wuhan, and there is a certain gap with the income level of enterprise R&D departments and scientific research institutions.

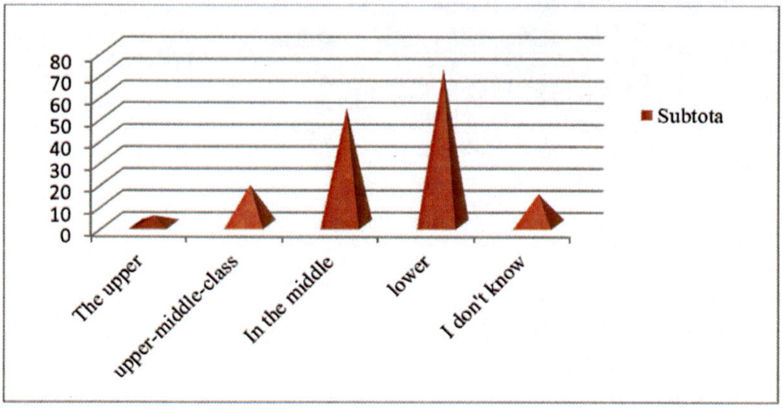

Fig. 3. Income of scientific and technological Staff in colleges and universities

3.2 Scientific and Technological Staff in Colleges and Universities Are Lack of Timely Policy Fund Platform Support in Participating in Scientific and Technological Activities

Figure 4 shows the research results of research projects involving researchers and social science popularization activities. Among them, the relevant data of the respondents show the willingness of scientific researchers to participate in social science popularization. One important data shows the willingness of scientific researchers to establish cross-school teams with the same research direction, accounting for about 72%.

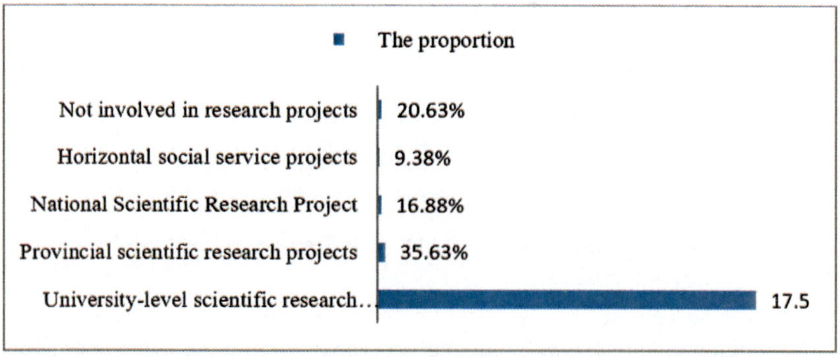

Fig. 4. Types of projects involving scientific and technological Staff in colleges and universities

The Fig. 4 shows the research results of research projects involving researchers and social science popularization activities. Among them, the relevant data of the respondents show the willingness of scientific researchers to participate in social science popularization.

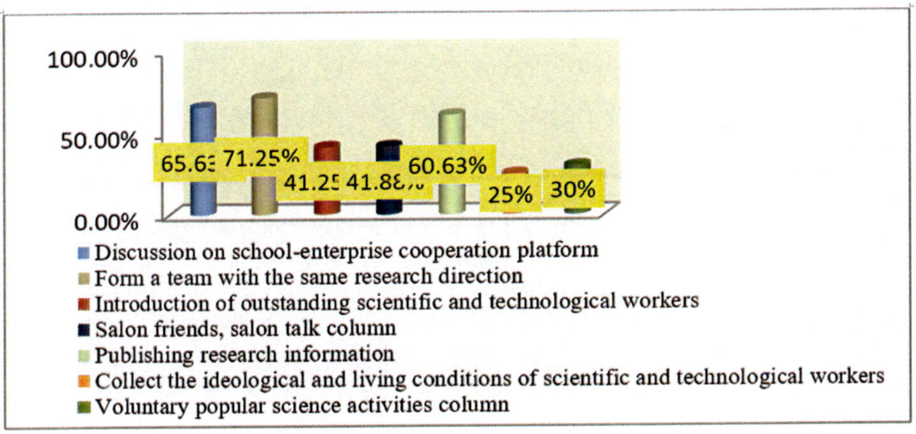

Fig. 5. Demands of the establishment of the investigation platform for scientific research Staff

Figure 5 shows the willingness of researchers to establish cross-school teams with the same research direction, accounting for about 72%.

4 Suggestions on Promoting the Development of Scientific and Technological Talents in Colleges and Universities

4.1 Advocating a Fair Income Distribution Mechanism for Scientific Research

University of science and technology staff of both teaching and scientific research work, the task is much, miscellaneous, the manager of the income and pay cost ratio, a blow to their scientific research enthusiasm, because, in the affirmation of the work to the science and technology staff need to establish a reasonable distribution mechanism, reduce the appraisal of scientific research workers teaching points, increase their scientific research evaluation score proportion, First, it can help science and technology staff devote themselves to their work. Secondly, the assessment score is proportional to the proportion of all kinds of bonuses. In particular, the scientific and technological workers with significant scientific research achievements should be rewarded with funds and materials in all aspects, and scientific research funds should be set up to protect and mobilize the enthusiasm of scientific and technological workers. Therefore, we should advocate schools to establish a fair research income distribution mechanism.

Departments at all levels need to provide opportunities for continuing education for young scientists and scientists to guide their requirements and directions for training and learning. In view of the personal development space of science and technology

Staff, comprehensive consideration, precise support for the development of science and technology personnel of different disciplines, effectively solve all kinds of problems affecting the development of young science and technology Staff. While appropriately increasing the basic income of scientific and technological Staff, the superior administrative authorities should plan long-term talent strategic deployment, establish files of scientific and technological Staff, regularly track and interview the living status and work demands of various kinds of scientific and technological Staff, and carry out various services centering on the physical and mental health of scientific and technological Staff in a planned way.

4.2 Improve Supporting Policies

Attach importance to basic research, first of all, in personnel, salary, reward policies to improve the management of science and technology personnel related policies and methods. Especially in view of the objective situation of young age, junior qualifications, poor pay and great pressure of highly educated talents who just entered the profession, relevant preferential policies should be formulated in terms of income distribution and professional title evaluation The economic development of Wuhan will promote the overall progress of Wuhan society, so the policies for scientific and technological activities of scientific and technological staff need to be adjusted in accordance with the actual situation. Especially around the project declaration, transformation and transfer evaluation system. The system needs to be further improved, and the achievement incubation department needs to further closely combine with the market form, fully protect the rights and interests of scientific and technological work in universities, mobilize the enthusiasm of scientific and technological workers, ensure the good cooperation of universities, government and enterprises in Wuhan, and produce high-quality scientific and technological achievements.

4.3 The Government Takes the Lead in Building Various Platforms for Scientific Research

The government builds an innovation and entrepreneurship incubation platform and provides favorable support to scientific and technological staff in terms of policies, funds, hardware laboratory facilities and experimental instruments. Real protection should be given to enterprises that acquire intellectual property rights of scientific and technological staff, so that enterprises with intellectual property rights can obtain multiple protections. Build an industry-university-research cooperation platform for enterprises to connect with researchers. Guiding the advantages of scientific and technological resources of universities and local strategic emerging industries can better combine to realize the common progress of universities and enterprises in Wuhan. The government builds its own public service platform to serve scientific and technological workers. Centering on the specific demands of scientific and technological staff, the government improves their deep needs, protects and transforms scientific and technological achievements, and at the same time solves some concerns of scientific and technological workers in their scientific research work.

5 Conclusion

Based on PEST analysis method and combining with national Political, Economic, Society and Technology, this study analyzes the macro environment of the work of science and Technology staff in Wuhan, and sets up a questionnaire widely used in the appeals of science and Technology staff in Wuhan. According to the internal relationship and development trend among the questionnaire projects, the accuracy of statistical research is analyzed and evaluated comprehensively in the form of scientific calculation and measurement, so as to help the government association to optimize the science and technology policy. Study found that the future Hongshan district in Wuhan city science and technology service as the space is large, suggest the local government should publish relevant policy, make good use of talent resources, encourage scientific and technological personnel out of the campus, development and applied research results to the local practice, in Wuhan city industry transformation and upgrading, promote scientific and technological innovation drive, contributing to national and local economic construction of science and technology intelligence, Further promote the development of Wuhan's science and technology service industry, improve the level of government organization and service capacity, and promote the sustainable development of Wuhan's science and technology activities.

Acknowledgment. This project is supported by Wuhan Association of Science and Technology. (200205).

References

1. Kocaleva, M., Stojanovic, I., Zdravev, Z.: Model of e-learning acceptance and use for teaching staff in higher education Institutions. Int. J. Mod. Educ. Comput. Sci. (IJMECS) **4**, 23–31 (2015)
2. Shi, L., Yao, Y., Xia, X., et al.: Practical teaching staff construction under new situation. Int. J. Educ. Manage. Eng. (IJEME) **2**, 57–61 (2011)
3. Ong, M.Y., Muniandy, B., et al.: Comparing the acceptance of key performance indicators management systems on perceived usefulness and perceived ease of use in a higher education institution in Malaysia. Int. J. Mod. Educ. Comput. Sci. (IJMECS), (10), 9–16 (2012)
4. Hong, Z., Hai, Y.: Human nature management in scientific research innovation: an analysis based on Maslow's Hierarchy of Needs. Sci. Technol. Manage. Res. **22**, 148–151 (2013). (in Chinese)
5. Zhou, G., Liu, W.: Problems and countermeasures in the layout of scientific and technological talents in China. Sci. Technol. Manage. 999–102 (2006). (in Chinese)
6. Li, K., Zhang, M., Yu, Q., Deng, D., Shi, H.: Research and analysis of the status of science and technology Staff in China. Today, Science Yuan, (11), 72–77 (2018). (in Chinese)
7. https://www.zhihu.com/question/386101603/answer/2362167111
8. Hongming, Y.: A new perspective of knowledge worker management: Job demand-resource model. Sci. Sci. Manage. Sci. Technol. **10**, 171–176 (2009). (in Chinese)
9. Wuhan Overview [EB/OL]. http://www.Wuhan.gov.cn/zjwh/
10. Wuhan Technology and Business University [EB/OL]. https://www.wtbu.edu.cn/

11. Kang, L., Chen, H.: Empirical analysis of influencing factors of job satisfaction of scientific researchers in China based on the Fourth National Survey report of science and technology staff. Sci. Technol. Rev. **10**, 99–108 (2021). (in Chinese)
12. Minli, S.: An empirical study on the relationship between psychological contract, job satisfaction and turnover intention of young scientific and technical personnel. Mod. Bus. **03**, 52–53 (2017). (in Chinese)

Application Possibilities of Data Science Tools in Agriculture: A Review

Maryna Nehrey(✉), Tatiana Koval, Nataliy Rogoza, and Liudmyla Galaieva

National University of Life and Environment Science of Ukraine, Kyiv 03041, Ukraine
marina.nehrey@gmail.com

Abstract. Digitization of agricultural production processes is a prerequisite for a successful strategy for the development of the agricultural sector. The process of implementing Data Science methods and algorithms into business processes has already begun and is gradually accelerating. However, it is far from clear how Data Science could make agricultural more effective and profitable. Availability of data on each agricultural object combined with well-developed algorithms and methods. Data Science allows to build mathematical models of the agricultural sector and farms, accurately calculate the algorithm of actions and predict the result. AgroTech, AgroFinTech, FoodTech and other businesses are actively developing in the technologically underdeveloped agricultural sector. This makes the agricultural sector attractive for investment and opens up new prospects for attracting financial resources. We show that Data Science tools in agricultural does not simply replace analogue technologies used in traditional agriculture. It offers new options for agriculture, including opportunities for more effective tools for agriculture producers and management. It allows for in-depth analysis and understanding of business processes, facilitates the structuring of problems, systematization of the agricultural sector as whole and individual farms. However, the use of Data Science tools is accompanied by certain risks, which also discussed in the paper.

Keywords: Data Science · Agriculture · Digitalization · Machine learning · AgroTech

1 Introduction

The Data Science toolkit opens up significant opportunities for all areas of business. For example, its algorithms can be used to create a selection of recommended products, music, movies, and the type of advertising that will work best for each consumer individually. Or you can determine the area of future crime and disarm the intruder. Identify suspicious user behavior online, detect a fraudulent scheme, and prevent cybercrime. Create a facial recognition application, and develop a voice assistant or a bot. Identify user psycho types, develop a marketing strategy for each psycho type and determine the best type of advertising. Predict the demand for ordering food in a restaurant, create an optimal menu, create an efficient delivery system and offer differentiated prices to

Z. Hu et al. (Eds.): AIMEE 2022, LNDECT 159, pp. 253–263, 2023.
https://doi.org/10.1007/978-3-031-24468-1_23

make the most efficient use of resources and maximize restaurant profits. Determine the probability of repayment of the loan by the customer, and calculate the individual interest rate. Evaluate the probability of occurrence of an insured event, and determine the optimal insurance package for each insured. Develop a weather forecast. The applications of Data Science are very broad. Industries, companies and individuals who use Data Science algorithms as a result of their application gain more efficient business, reduce costs, reduce the use of resources, increase efficiency and maximize profits.

The agricultural sector is a vital sector for all humanity. After all, the world's population is constantly increasing, and so is the need for food. In 2050, the agricultural sector must produce 70% more products than it does now. Therefore, it is necessary to develop the agricultural sector not only by increasing the number of animals and land, but also to a greater extent through more efficient technologies. Added to the problems of overpopulation are the consequences of the COVID-19 pandemic and the impact of the Russian-Ukrainian war on the agricultural sector. All of this raises significant concerns about the ability to maintain global food security at an appropriate level.

One of the possible ways to ensure the effectiveness of the agricultural sector is the application of Data Science technologies. Humankind has already experienced a wave of agricultural efficiency gains when mechanization was applied. The next wave of increased efficiency is possible precisely because of Data Science. Technology is evolving, becoming more accessible to any field of activity. Now it is possible to obtain data about each agricultural object, its living conditions, build mathematical models of its functioning, accurately calculate the algorithm of actions and assume the result.

Digitalization of agricultural production processes is a prerequisite for a successful strategy for the development of the agricultural sector. A new technological cluster AgroTech has emerged in the global agricultural space, causing considerable interest among investors and agricultural producers. The main players in the AgroTech cluster today are the United States, China, Canada, Israel, India, and Ukraine.

In this paper, we first identify the scope of our analysis and define our understanding of the potential uses of the Data Science toolkit. We then develop our approach by discussing the directions for the use of Data Science in agriculture. Next, we present the implications of the use of digital technologies for different aspects of agriculture policy, before moving on to a discussion of the main findings and implications of agriculture policy and research.

2 Literature Review

The agricultural sector is an industry with great potential for the use of Data Science. Agricultural production has a long production cycle, which with the help of Data Science tools can be improved, and made less resource-intensive and more efficient and profitable. Agriculture is subject to natural risks, and significant crop losses during cultivation, harvesting, and storage. Therefore, the application of Data Science methods and algorithms is a prerequisite for improving the efficiency of agriculture. At the heart of the transition to Agriculture 4.0 is the application of Data Science tools. Farmers are already using Data Science algorithms to monitor field conditions and crop quality, monitor livestock, fertilizer use, etc. However, this is only the beginning, and it is important to continue and intensify this process.

A study of the application of certain Data Science tools for precision agriculture was carried out in [1–3]. In the papers [4–7], the authors showed the possibilities of applying Data Science methods and algorithms in the agricultural sector. The increase in the possibilities of ensuring food security with the use of business analytics tools is highlighted in [8–11]. The analysis of ways to increase the efficiency of the agricultural sector using Big Data was carried out in [12–17]. Also, this approach can be used in many other areas [18–21].

3 Data and Methodology

Data Science includes a large number of methods and algorithms. The basic algorithms of Data Science are presented in Table 1.

Table 1. Data science algorithms

Algorithm class	Algorithm subclass	Algorithm
Supervised learning	Regression	Linear regression; Polynomial regression; Ridge/Lasso regression
	Classification	Logistic regression; Decision tree; Method of support vectors; Naive Bayesian classifier; k-nearest neighbors
Unsupervised learning	Segmentation	k-means; Agglomerative (hierarchical) segmentation; DBSCAN
	Frequency algorithms Dimension reduction	Associative rules; A priori Analysis of main components; Singular expansion of the matrix; Linear discriminant analysis; TSNE
Reinforcement leaning	Q-learning Of/On-policy algorithms Algorithms based on the model	Genetic algorithms; Tabular Q-learning; DQN; DDQN SAS; PPO Reinforcement learning using probabilistic models
Ensemble models	Bagging Boosting Stacking	Random forest AdaBoost, Gradient boosting

(continued)

Table 1. (*continued*)

Algorithm class	Algorithm subclass	Algorithm
Time series algorithms	Forecasting based on time series Decomposition of time series	ARIMA; Prophet Fourier distribution; Mel-spectograms (MEL); Mel-cepstral cofficients (MFCC)
Anomaly detection	Statistical algorithms One-Class Classification Based on Negative sample Autoencoders	Moving average Isolation Forest, Extended Isolation Forest, 1st class SVM MADI
Text processing algorithms	Frequency approaches Program Vector Representations	n-gram;TF-IDF W2V, Doc2Vec, Bert, Transformers, Universal encoders
Deep Neural Networks	Perceptron Convolutional neural network	Multilayer perceptron DCNN

The objectives of the study are to investigate a methodology of Data Science for application to the agriculture sector and future opportunities which could be open with new approaches.

4 Results and Discussion

4.1 Data Science Application

Data Science is the science of the 21st century. The rapid development of Data Science and the results of using its methods and algorithms contribute to the rapid spread of science and its active use in practical activities: in business, public administration, and other sciences. The basic task of Data Science is to process analytical data, build and verify mathematical models, search for patterns and relationships, and develop predictions.

The application of Data Science in agriculture creates the conditions for the rapid evolution of agricultural technology. The arrival in the agricultural sector of technology companies that improve agricultural production, creating, for example, a full-cycle production control system, has shown the need to move to a new level of implementation of digital technologies. The use of intelligent devices that transmit and process the current parameters of each element and its environment (temperature, air humidity, soil parameters, plant characteristics, animal condition, etc.) allows for managing the entire production system in both crop and livestock production. The use of Internet of Things technologies can automate most agricultural processes by creating a digital model of the entire production cycle and the interrelated elements of the value chain. The application of Data Science makes it possible to plan work schedules with fairly high accuracy, take emergency measures to prevent losses in the case of a detected threat, and calculate possible yields, production costs, income, and profits.

Agricultural production is characterized by a long value chain. Accordingly, there are a large number of unresolved agricultural problems that can be solved through digitalization, including the application of Data Science tools. A standard agricultural process, such as crop treatment (irrigation, fertilization, chemicalization) must take into account local features and natural variability and increase yields, saving resources and anticipating possible problems. Data Science algorithms must anticipate possible droughts or excess moisture, adjust fertilizer rates, record the appearance of weeds and insects, and provide a rapid response to any problems. Predicting possible disease outbreaks and preventing them is also possible thanks to Data Science.

During a season, farmers make about 50 different decisions: what crop to plant, when to plant, how to cultivate it, how to treat diseased plants, when to harvest, where to store it, to whom and at what price to sell it, etc. The lack of information to make decisions leads to a significant number of crops being lost in the production cycle. About 40% is lost during harvesting, storage and transportation. However, scientists have found that in addition to weather, 2/3 of the loss factors can now be controlled by automated control systems.

The use of Data Science in the agricultural sector allows for maximum digitalization of all stages of the production cycle. This reduces losses, increases business productivity, and ensures optimal management of resources. Integration of the received data with various Data Science tools that process them in the current time mode performs an evolutionary shift in decision-making for the farmer, providing the results of analysis of multiple factors and justification for further actions. The more sensors, sensors and field controllers are networked and share data, the smarter the agricultural business process management system becomes and the more useful information it can provide to the manager.

Based on Data Science algorithms, an agricultural management system is able to create recommendations for plant treatment and care or instructions for the automatic execution of robotic machinery. For example, a predictive analytical model helps determine that a 2-degree increase in temperature contributes to insect kill, or an increase in humidity above the limit can lead to a disease outbreak. Controlling these factors creates real value in modeling microclimatic conditions: if it's a greenhouse, you can keep the temperature from rising, and if it's a field, you can prudently monitor the area and act on the chemical in case of vermin. In this way, farmers can control natural factors, simulate precise business processes, and predict the outcome with high accuracy.

The basic components of Data Science are Business Itelligence, Traditional methods and Machine learning. Each of them can be used in agriculture to solve various tasks. The main tasks that can be solved with the help of Data Science, the Data Science tools used and the application area are presented in Table 2.

Table 2. Data Science in agriculture business

Data Science components	Business telligence	Traditional methods	Machine learning
Background	Data has been gathered and organized	Created and discussed BI reports	Created and discussed BI reports
Basic tasks	Using data to create reports and Dashboards to gain business insights	Assessing potential future scenarios by using advanced statistical methods	Utilizing artificial intelligence to predict behavior in unprecedented ways
Data Science tools	Analyzing the data: Extract info to the next forms: • metrics, • KPIs, • reports, • dashboards	Regression, Logistic regression, Clustering, Factor analysis, Time series	Supervised learning: SVMs, NNs, deep learning, random forests, Bayesian networks Unsupervised learning: k-means, deep learning Reinforcement learning: similar to supervised learning, but instead of minimizing the loss, one maximizes reward
Application area	Price optimization; Inventory management	User experience; Sales forecasting	Fraud detection; Client retention
Required experts	BI analyst, BI consultant, BI developer	Data Scientist Data analyst	Data Scientist Machine learning engineer

4.2 Implementation of Data Science in the Agriculture Sector

Data Science technologies are still new technologies for agriculture. Therefore, their implementation can sometimes cause farmers. Experience shows that the introduction of modern technologies takes place primarily in large agricultural enterprises and at the state level. And only after implementation at the macro level, has it become possible to use the Data Science toolkit at the level of individual farms.

For the state, the main process of improving agriculture is ensuring the intensification of the digitalization process. In the agricultural sector at the state level, digitalization can be carried out in many directions. Thus, digital registers of the country's agro-industrial complex can be created, which will allow to combine of disparate information, simplify analytical work in the industry and quickly make the necessary decisions. The creation of a system for monitoring and forecasting food security will also contribute to increasing the efficiency of agriculture. An electronic system for the analysis of subsidies to support the agro-industrial complex, an information system about agricultural land and land used or provided for agriculture as part of the land of other categories - such systems are also a necessary prerequisite for the further development of the agricultural sector. The main directions of the digitization process at the state level are presented in Fig. 1.

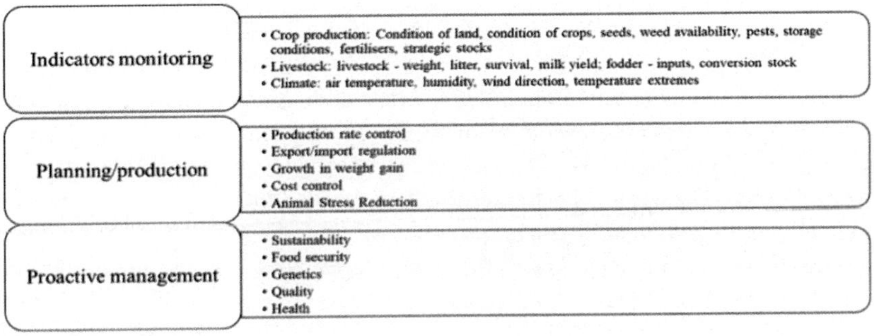

Fig. 1. Digitalization of public administration in the agricultural sector

When implementing Data Science technology in the agricultural sector, it can be a challenge to find the right specialists. After all, Data Scientists are needed in many fields: IT, creation, marketing, media, medicine, politics, science, medicine, public administration, etc. However, this challenge may also create new opportunities for the agricultural sector: by adopting a specialist from another area, the agricultural sector may also learn from his or her experience and effective management practices. For example, from the field of retail, wholesale and Internet sales Data scientists can be experienced in forecasting the demand for goods regarding seasonality, developing of personalized marketing or promotional offers, etc. The manufacturing sector can give the agricultural sector a specialist to predict production and consumption volumes, possible equipment failures, and product defects. A Data Scientist can take from the logistics sector the ability to construct optimal routes, determine delivery deadlines and plan warehouse utilization. In Finance, we need experience in risk analysis, calculating interest rates, assessing the creditworthiness of borrowers, and calculating the optimal insurance package. In IT, skills to develop software, web and mobile applications such as bots, search engines, big networks, social networks, and online services with content for different purposes.

4.3 Current Trends in the Development of the Agricultural Sector Using Information Technology

AgroFinTech is a field of innovative business in which companies that develops and provides financial services for agricultural enterprises work. Their main services are:

- improvement of credit opportunities for farmers and agricultural enterprises,
- development and implementation of a system for analyzing the real activity of farms, which allow business owners to reduce costs and increase profits,
- development of marketing tools that increase sales channels and client audience,
- finding alternative sources of monetization for farms, for example, through the provision of carbon credits.

Agricultural marketplaces. The creation of digital platforms for trade in agricultural products is a promising development for the global market of agricultural products. The tasks of companies that provide agricultural enterprises with online platforms are helping farmers purchase raw products and equipment from manufacturers, finding agricultural product sales channels and logistics partners for shipments. Successful agromarketplaces already exist today, such as the largest US digital grain trading platform Bushel and Chinese homegrown fresh produce giants Furong XingSheng and Dingdong Maicai.

Farm management systems. Data Science algorithms and models are used to develop digital systems and software in order to make the operational process on the farm more "intelligent", due to the analysis of data from many sensors, sensors and even satellites. Such tools help the farmer to determine how fertile the plot of land is, whether drastic climatic changes are expected in the near future, whether the plant is healthy or not? The technologies of "precision agriculture" allow you to calculate the amount and frequency of watering plants, the amount of necessary fertilizers or feed, the cost of maintaining a plot of land on the square. All these numerous tools are aimed at preserving and increasing productivity, reducing capacity, optimizing capacity and resources.

Farming-as-a-Service. Vertical farms, which are rapidly developing as a trend, are high-tech agro-industrial complexes for growing agricultural crops, the basis of which is the management of Data Science tools. Basically, microgreens, salad and a few vegetables and berries. Given their high yield, unpretentiousness, economy both from the point of view and from the point of view of the area resource, vertical farms allow solving the issue of availability of fresh vegetables, berries and greens in the conditions of a metropolis.

Autonomous machinery and farm work have several advantages over traditional machinery and personnel. The equipment can work independently at any time of the day and under different climatic conditions, avoids the risk of human injury or death when working in difficult conditions, for example, in mountainous areas, and only one specialist will be needed to control the entire farm complex. All this is possible thanks to the use of Data Science tools. Modern drones, robots and unmanned vehicles are equipped with sensitive sensors, cameras and computer vision technologies, which allow them to skillfully navigate in space, accurately harvest crops, carry out irrigation and processing of the plot without regard to the human factor.

The creation of alternative food products is mainly related to the production of alternatives for traditional products of animal origin: meat and dairy products. These are animal products made with vegetable and mushroom protein, and meat or ready-made meat products grown from scratch from cells. It is important to support this trend, as the demand for livestock products is constantly growing. This approach can be used to meet the corresponding demand using plant protein and insect protein. In addition, meat obtained by the cellular method, or its plant counterparts have a much smaller "carbon footprint" than its animal counterpart. Moreover, cell-based meat production does not involve the killing of an animal, and plant-based protein products are critical for vegans and vegetarians. Also, alternative foods can solve the problem of an allergic reaction to lactose. As for economic advantages, they are unconditional: the cost price of a vegetable protein product is usually lower than that of an animal-derived counterpart.

Synthetic biology is the use of an object's genetic material as a structure that can be programmed, like a computer. In this way, new or enhanced DNA properties of a certain agricultural crop or livestock breed are obtained to improve such things as yield and fertility, growth rates, resistance to harsh climatic conditions, and many others. Similar results can be achieved if genetic sections of other objects are "glued" to the culture's DNA, or if certain sections are "disabled" in the original DNA. Using Data Science tools, scientists and innovative companies create genetic code or molecules of inorganic material.

5 Summary and Conclusion

We highlight theoretical implications rather than a thorough analysis of empirical evidence for improving Data Science in agriculture. Our analysis shows that Data Science in general improves the efficiency of tools, mainly because it can improve computational accuracy, account for large amounts of data and improve business processes. Our study follows the following ideas for implementing Data Science in agriculture.

The process of introducing Data Science methods and algorithms into business processes has already begun and is gradually accelerating. The agricultural sector has already felt the first results of this process. The sale of farm products and their delivery to the end customer has become much faster and more efficient due to the development of digital channels for interaction between the producer and the customer through marketplaces and delivery services.

However, this process is also accompanied by certain challenges. The use of Data Science tools poses certain risks. For example, the development of cyber security systems in the agricultural sector occurs because of cyber threats to farm management systems. Agrarians should take this factor into account and apply preventive measures. A data-driven approach is being actively used to manage agriculture, improve yields and implement business resilience measures. In addition, the agricultural sector is becoming an active participant in the ESG concept of sustainability in the carbon footprint reduction segment. Farmers are actively partnering with large industrial and technology companies to reduce their environmental footprint and offset greenhouse gas emissions.

The application of Data Science algorithms enables in-depth analysis and understanding of business processes, facilitates structuring of problems, and provides systematization of the agricultural sector as a whole and individual farms. Agricultural sector management, based on Data Science algorithms, allows justifying decisions and automating business decision-making processes. Application of modern approaches to modeling and Data Science tools in the agricultural sector allows us to define a set of components of the agricultural sector management system, to move to a new level of interaction with stakeholders, to improve production efficiency, which as a result contributes to improving the competitiveness of farms and the agricultural sector.

References

1. Angon, P.B., Salehin, I., Khan, M.M.R., Mondal, S.: Cropland mapping expansion for production forecast: rainfall, relative humidity and temperature estimation. Int. J. Eng. Manuf. (IJEM) **11**(5), 25–40 (2021). https://doi.org/10.5815/ijem.2021.05.03

2. Kumar, K.A., Aju, D.: An Internet of Thing based Agribot (IOT- Agribot) for precision agriculture and farm monitoring. Int. J. Educ. Manage. Eng. (IJEME) **10**(4), 33–39 (2020). https://doi.org/10.5815/ijeme.2020.04.04
3. Arakeri, M.P., Arun, M., Padmini, R.K.: Analysis of late blight disease in tomato leaf using image processing techniques. Int. J. Eng. Manuf. (IJEM) **5**(4), 12–22 (2015). https://doi.org/10.5815/ijem.2015.04.02
4. Bhagawati, K., Bhagawati, R., Jini, D.: Intelligence and its application in agriculture: techniques to deal with variations and uncertainties. Int. J. Intell. Syst. Appl. (IJISA) **8**(9), 56–61 (2016). https://doi.org/10.5815/ijisa.2016.09.07
5. Klerkx, L., Jakku, E., Labarthe, P.: A review of social science on digital agriculture, smart farming and agriculture 4.0: new contributions and a future research agenda. NJAS: Wageningen J. Life Sci. **90–91**(1), 1–16 (2019). https://doi.org/10.1016/j.njas.2019.100315
6. Izonin, I., et al.: An extended-input GRNN and its application. Procedia Comput. Sci. **160**, 578–583 (2019)
7. Yang, J., Guo, X., Li, Y., Marinello, F., Ercisli, S., Zhang, Z.: A survey of few-shot learning in smart agriculture: developments, applications, and challenges. Plant Methods **18**(1), 1–12 (2022)
8. Starychenko, Y., Skrypnyk, A., Babenko, V., Klymenko, N., Tuzhyk, K.: Food security indices in Ukraine: forecast methods and trends. Estudios de Economia Aplicada, **38**(4), 1–14 (2020)
9. Davydenko N., Dibrova, A., Onishko, S., Fedoryshyna, L.: The influence of the gross regional product on the formation of the financial potential of the region. J. Optim. Industr. **14**(Special Issue), 153–157 (2021)
10. Dibrova, A., Davydenko, N., Nehoda, Y., Titenko, Z.: Financial provision for competitiveness of agricultural enterprises. Intellect. Econ. **14**(1), 67–75 (2020)
11. Kaminskyi, A., Nehrey, M., Rizun, N.: The impact of COVID-induced shock on the risk-return correspondence of agricultural ETFs. In: 2020 Special Edition of International Conference on Monitoring, Modeling and Management of Emergent Economy, M3E2-MLPEED 2020, pp. 204–218 (2020)
12. Morota, G., Ventura, R.V., Silva, F.F., Koyama, M., Fernando, S.C.: Big data analytics and precision animal agriculture symposium: machine learning and data mining advance predictive big data analysis in precision animal agriculture1. J. Animal Sci. **96**(4), 1540–1550 (2018). https://doi.org/10.1093/jas/sky014
13. Himesh, S., Prakasa Rao, E.V.S., Gouda, K.C., Ramesh, K.V., Rakesh, V., Mohapatra, G.N.: Digital revolution and Big Data: a new revolution in agriculture. CAB Rev. **13**(21), 1–7 (2018)
14. Coble, K.H., Mishra, A.K., Ferrell, S., Griffin, T.: Big data in agriculture: a challenge for the future. Appl. Econ. Perspect. Policy **40**(1), 79–96 (2018)
15. MayaGopal, P.S., Chintala, B.R.: Big data challenges and opportunities in agriculture. Int. J. Agric. Environ. Inf. Syst. **11**(1), 48–66 (2020). https://doi.org/10.4018/IJAEIS.2020010103
16. Sadiku, M.N., Ashaolu, T.J., Musa, S.M.: Big data in agriculture. Int. J. Sci. Adv. **1**(1), 1–10 (2020)
17. Matviychuk, A., Novoseletskyy, O., Vashchaiev, S., Velykoivanenko, H., Zubenko, I.: Fractal analysis of the economic sustainability of industrial enterprise. In: CEUR Workshop Proceedings, pp. 2422: 455–466 (2019)
18. Oliskevych, M., Beregova, G., Tokarchuk, V.: Fuel consumption in Ukraine: evidence from vector error correction model. Int. J. Energy Econ. Policy **8**(5), 58–63 (2018)
19. Hryhoruk, P., Khrushch, N., Grygoruk, S.: Assessing the investment capacity of the agricultural sector: case of Ukraine. In: 2020 10th International Conference on Advanced Computer Information Technologies, ACIT 2020 - Proceedings, pp. 183–187, 9208927 (2020)

20. Guryanova, L., Bolotova, O., Gvozdytskyi, V., Olena, S.: Long-term financial sustainability: an evaluation methodology with threats considerations. Rivista di Studi sulla Sostenibilita **1**, 47–69 (2020)
21. Shakhovska, N., Montenegro, S., Kryvenchuk, Y., Zakharchuk, M.: The neurocontroller for satellite rotation. Int. J. Intell. Syst. App. **11**(3), 1–10 (2019)

Clustering Stocks by ESG Score Values, Risks and Returns: Case of Expanded German Index DAX

Andrii Kaminskyi[1] and Maryna Nehrey[2(✉)]

[1] Taras Shevchenko National University of Kyiv, Kyiv 01033, Ukraine
[2] National University of Life and Environment Science of Ukraine, Kyiv 03041, Ukraine
marina.nehrey@gmail.com

Abstract. The paper is devoted to the application of the cluster approach to portfolio management, based on the inclusion of sustainability component in the clustering. ESG-score provided by S&P Global Corporate Sustainability Assessment was used as an integral numerical indicator of sustainability. The clustering was based on the ESG-score and its components (E-, S- and G-scorings), as well as such traditional investment indicators as expected return and risk. Each cluster corresponds to its own investment strategy in the form of creating a naive diversified portfolio based on stocks from that cluster. K-ratio was used as a numerical measure of strategy performance. The implementation of this approach was developed for the expanded German market DAX index (extended from 30 to 40 constituents at the end of 2021). The application of clustering to the index refers to the so-called "Index investing strategy". The classical version of this strategy does not consider the ESG prioritization and, accordingly, the related investors demand. Based on the developed clustering, it was grounded the opportunity to modify index strategies correspondingly to investors' sustainability attitude. Results were obtained from both conventional and fuzzy clustering tools. Each clustering identified 3 clusters that corresponded to the development of an investment strategy. However, clustering based on fuzzy tools reduces their number to 2. A comparative analysis of investment strategies corresponding to the clusters was conducted.

Keywords: Sustainable finance · ESG-score · Investment risk · K-means clustering · Fuzzy clustering · Portfolio management · Index investing strategy

1 Introduction

Today ESG investing is one of the most dynamic segments in the investment environment. This applies to both types of investments: direct and portfolio investments. For example, about 25% of all global assets under management (AUM) are now invested with an analysis of environmental, social, and governance factors (ESG) [1]. It is estimated that in 2020 the global volume of ESG AUM will be more than $23 trillion. The concept of ESG investing originated in the mid-1980s, when the Forum on Sustainable and Responsible Investment was organized [2]. The forum focused on promoting social,

Z. Hu et al. (Eds.): AIMEE 2022, LNDECT 159, pp. 264–276, 2023.
https://doi.org/10.1007/978-3-031-24468-1_24

moral, and ethical values in investment decision-making. Investments that meet these criteria are understood as "investment that "combine investors' financial objectives with their concerns about social, environmental and ethical (SEE) issues" [3]. On the other hand, the development of ESG design was also based on the concepts of EHS (Environment, Health, and Safety) and CSR (Corporate Social Responsibility), developing strong in 1990–2000. As a result, the design of ESG has evolved to date. Today this is holistic concept related to competitive advantage and risk-return management in short and long periods.

It can now be said with certainty that environmental, social and management criteria have been formed for their completeness. From today's perspective they are a set of standards for a company's behavior used by socially conscious investors to screen potential investments.

Namely,

1) The Environmental criteria consider how a company safeguards the environment, including corporate policies addressing climate change, carbon balance, ecosystem, waste management, water scarcity and its pollution and other similar issues.
2) The Social criteria cover the company's relations with employees, customers and the community in which it operates. DE&I (Diversity, Equity, and Inclusion) [4] is one of the basic principles that underlie this criteria.
3) The Governance criteria encompass, inter alia, issues of audits, internal controls, and shareholder rights.

At the same time, investors (especially portfolio investors), who attach great importance to ESG factors, usually do not have the ability to make a full-fledged assessment of a company's compliance with these criteria. Because it is a reality complex task. As a result, ESG score emerged. It is integrated numerical form allows expressing the level of compliance of the company with ESG criteria. The first such ESG scoring was built in 1999 by Innovest (e.g. [5]). Today, ESG scoring builds and offers to investors by several companies, such as MSCI, Refinitiv, Sustainalytics, S&P Global, and others. The idea of our research lies in the incorporating of ESG scores into the investment portfolio management objectives. Considering numerical representation of ESG score led us to the idea to include it into the approach of optimizing portfolio decisions.

From our point of view, the inclusion of the ESG score can be seen as an extension of the traditional approaches of modern portfolio management. Really, such approaches grounded on different forms of risk-return correspondence. As example, classical H.Markowitz model involve standard deviation as risk measure. At the same time investors have a propensity to invest with ESG factors considering.

The idea of our research is following. We involve ESG factors by means clustering together with traditional factors risk and return. After that we apply clustering approach for portfolio management. Considering homogeneity of stocks characteristics in concrete cluster we construct investment strategy in the form of creation naïve diversified portfolios though stocks in this cluster. The main logic позволяет инвесторам управлять портфелям с учетом их attitude for triple ESG-score – risk –expected return.

The practical implementation of our research was carried out for "index investing strategy". The logic includes changing the investments into whole index to the investments in clusters. The input for clustering embraced the stocks that are constituents of German stock index DAX (Deutscher Aktienindex). The actuality of such application connected with index basket expansion occurred in the 3rd quarter of 2021. The interconnection between parameters inside received clusters should clarify the specifics of DAX index investing strategy. In addition, the choice of DAX index was since the companies included in it have fairly accurate estimates of the S&P Global ESG score used by us. This increases reliability and ground-level of clustering.

Verification of performance investment strategies which correspond to clusters was done by investment indicators, one of which K-ratio.

2 Literature Review

Strong development of ESG investing last 10 years derives increasing number of publications. So, we think that the first step is logically to specify the paper [6]. This paper involves bibliometric systematization and visualization researches in ESG (more precisely this study searched Scopus for "ESG"). Authors demonstrated interesting dynamics of "Annual Scientific Production" through 1990 to 2020. It shows an exponential growth in scientific publications. This, together with practical publications, shows the high relevance of this topic.

Turning to the starting point for integrating ESG factors into finance and investment, a fundamental study should be noted, published in 2004 "Who Care Wins" [7]. This research and recommendations were prepared at the frameworks of a unified initiative of financial institutions to develop guidelines and recommendations on how to better integrate ESG factors in asset management, securities brokerage services and associated research functions. For today, considering from various angles it is possible to specify that it lay the groundwork for understanding ties financial markets with sustainable development.

The questions about up-to-date understanding meaning of ESG evaluation system are widely discussed in scientific literature. Thus, authors of article [8] note that environmental component mainly refers to the belief that the company can improve its environmental performance in production and operation and diminish the environmental cost per unit of output. At the same time, social responsibility refers to the companies' movement to higher business ethics, social ethics, legal standards and focus more to the internal relationship with the external society, including human rights, and interests of relevant parties [9].

Corporate governance responsibility can be considered as such mechanism which provides rationally divide the power of shareholders and employed managers around those who can be entrusted with that responsibility from development strategies to concrete actions [10–13].

One of the modern trends in the analysis of ESG factors and there applied to investing is ESG scoring. It is the presentation of ESG level in the numerical form. The logic and basic properties of ESG scoring is described from different points of views in [5]. The academic point of view to ESG scoring also is presented in many current publications.

In particular, we find the article fairly complete from an academic point of view [14]. In this study, the authors deconstruct ESG scoring at the E, S, and G factors and analyze affecting their levels to investment risk and returns. They have indicated that length of used time horizon has an important bearing on the indicators' significance. In the short term, they find that "G" is the dominant pillar because it strongly reflects event risks, such as fraud. In the long term, however, "E" and "S" indicators are more significant.

The questions about interrelations between ESG scoring and portfolio performance are considered in [15]. The logic of applying cluster approach for portfolio construction is described, in [16, 17]. Namely, paper [17] provides consideration with other risk measuring approach then presenting in our research. We used standard deviation as risk measure, but authors of [17] applied risk measuring based on mean-value at risk. The applying classification in other research areas is described, in [16–26]. The explanation of effectiveness of K-ratio is described in [27]. Also, this approach can be used in many other areas [28–32].

3 Data and Methodology

As a practical implementation of the considered approach, we have chosen clustering of companies that are included in the extended German stock index DAX. Index basket DAX has been expanded (since September 2021) from 30 companies to 40. The components of the index basket were chosen for analysis.

Three data sets were used for the basic cluster analysis. The first data set was for ESG score values for companies. ESG scoring (or similar titles "metric", "rating") are provided by different providers (some of them we denoted in Introduction). We grounded choice of SP Global scores following transparency methodology, which it has from our point of view. This scoring includes 3 subcomponents: "E-", "S-", and "G-" score values and integral ESG score. ESG score values, which were used in research, correspond to 2021 year.

The second group of data includes expected returns for considered companies. Period of analysis corresponds to 2017–2019 years in weekly time framing. One of the grounding of this time interval choosing raised from "quiet period" on the stock market, because in 2020 the pandemic of COVID-19 stipulated waves of turbulences. Therefore, it was implicated 157 values of weekly returns. The resource Investing.com was used for information about prices. Expected returns were calculated on this base.

Third group of data was constructed by applying the standard deviation operator to abovementioned series of weekly return.

Thus, initial data for analysis includes 6 indicators: 1) ESG score; 2) E score; 3) S score; 4) G score; 5) expected return; 6) risk.

For verification procedures we used weekly data from Jan 2020 to May 2022.

It is necessary to note the correctness of correspondences between data. Thus, we used companies' ESG score values for 2021 year. At the same time expected return and risk were estimated for 2017–2019 and verification period used as time horizon Jan 2020–May 2022. The consistency of data comparing arise from the fact that ESG scoring reflects results of company's long-term efforts towards E, S, and G implementation. It grounds the correctness to combine risk-return data from 2017–2019 and ESG score values from 2021. Because these score values are largely determined by the processes in the company in the period under consideration. Moreover, such logic corresponds to classical estimations by historical data using in modern portfolio theory.

There are 4 crucial methodological points was applied in our research and working out data. First methodological point is transforming classical risk-return correspondence into triple ESG score – risk – return. ESG score is indicator which correspond to implemented into company's practice environmental impact, social responsibility and governance rules. Investors looking for ESG score estimate how this company adequate their demands for sustainability. And here there is an objective question that this indicator reflects the risks that can manifest themselves when low ESG score. And also, the question concerns the long-term profitability of the company at different scoring values. There is a question about "intersections" between score values and other components from this triple. Does it methodologically correctly? Our answer "yes". Because ESG score focus on long-term returns and risks which raised from strategical implementation of E, S, G today. Second and third components of triple focus exclusive to financial performance in short and medium time periods. After that we think that compare these three indicators is methodologically correct. Because ESG score, in the essence, reflect qualitive attribute. And, in some extent, correspond to expectation of investor for putting money in this company.

Second methodological point of our research supposes to apply two approaches for clustering. First approach is "classical" which based on K-means. Second approach raised from fuzzy conception of clustering. The background of applying two approaches is explained by considering of investment strategies aligned to clusters. Other factors, including the investor's attitude to ESG scores and risk-return correspondence, are considering through investment strategies parameters. Classical clusters with precise frontiers may be very strong for investor's decisions. Fuzzy clustering helps to consider such questions.

Third methodological point concerns to cluster approach for portfolio constructing. One of the cornerstones of modern portfolio management is Markowitz portfolio theory which embraced optimization risk-return correspondence by use two-criterial nonlinear tasks. Generally, it is possible to add additionally linear constraints with ESG scores of the company (see as example, [1] for such approach). But applying such approach leads to specific structure of optimal portfolio (or, more generally, to analog of efficient frontier). This can generate in portfolios from an effective frontier "interweaving" of companies with high and low ESG levels. At the same time proposed cluster approach reflects some homogeneous set of stocks. This homogeneous includes also ESG context which investors involve in portfolio construction procedures. We find this to be an advantage in using the cluster approach to possibly reflect investor preferences regarding ESG. Hence, we adopted creation of naïve diversification portfolios (through clusters) for reflecting homogeneity based on clusters.

Fourth methodological point encompasses verification's procedure. The logic of verification concerns comparative analysis of investment results for naïve diversified portfolios corresponded clusters. One specificity in our case is that verification was realized at the time interval covered period of COVID-19 pandemic. It is well known, that pandemic has hit stock markets hard, creating more turbulence. Our approach included using pair indicators for verification: risk (standard deviation of returns) and K-ratio. These two indicators can estimate as volatility as maintenance of investment value and can be correctly applied despite market turbulence.

4 Results and Discussion

The application of the elaborated methodology to the corresponding data we received as first result rows with estimated characteristics for companies from the index DAX basket (Table 1). Two companies from the index basket were removed from consideration because it was lack of data (non-availability to access indicated resources). They are companies Covestro and Daimler Truck which are constituents of DAX basket. Thus, 38 companies were involved in cluster analysis. Table 1 also includes results of applying both approaches to clustering: classical (K-means) and fuzzy. The optimal number of clusters is 3 in both cases. Clusters named as A, B, C corresponds to results of classical clustering. Clusters named as F1, F2, F3 corresponds to results of fuzzy clustering.

A correlation analysis was applied to the received estimates. Its results are shown in Table 2. The results show that the expected return is not correlated with the ESG score values and its components. At the same time, the risk expressed by the standard deviation has an average (about 0.4) level of correlation for other indicators. The correlation is negative, indicating that the higher ESG score values corresponds to the lower average risk.

After that procedures of "conventional" clustering were applied. It was done by involving 5 parameters (E score value, S score value, G score value, ER, SDR). The results of clustering are pictured in Fig. 1. The analysis of clustering demonstrates some (not strong, but essential) tendency to correspond to ESG ordering. It is possible to see visually that the distribution of companies across clusters is not uniform. So, 50% of the companies belongs to cluster A, 29% - to cluster B, 21% - to cluster C.

The second step of applying clustering procedures included applying fuzzy clustering tools. The result of clustering using fuzzy tools is shown in Fig. 2. So, 50% of the companies came to cluster F1, 21% - to cluster F2, and 29% - to cluster F3.

Using fuzzy clustering point out one effect, which is represented in Fig. 2. Namely, F1 and F3 clusters cover in fuzzy sense cluster F2. So, it affects to the construction the set of investment strategies. Namely, to decrease their number to 2.

Verification procedures we provided by applying K-ratio. The K-ratio is a special valuation metric that test the consistency of asset's return in time. A higher K-Ratio is indicative of better performance than that of a low K-Ratio. K-Ratio base on considering the following cumulative value:

$$CValue = 1000\prod_{t=1}^{T}(1 + R_t)$$

where of R_t denotes return for period of t (weekly time frame).

K-ratio presents estimation of effectiveness through some form of risk-return correspondence. The description of interpretation is considered in [27].

Table 1. Clustering results

Companies	Tikers	ESG	E	S	G	ER	SDR	Clusters k-means	Clusters fuzzy
Allianz SE VNA O.N.	ALVG	93	97	95	90	0,0022	0,0219	A	F1
Deutsche Telekom AG	DTEGn	90	97	96	84	−0,0006	0,0215	A	F1
Munich Re	MUVGn	86	94	82	86	0,0027	0,0204	A	F1
Infineon Technologies AG	IFXGn	83	85	79	82	0,0024	0,0438	A	F1
Adidas AG (ADSGn)	ADSGn	82	98	74	81	0,0050	0,0318	A	F1
Linde plc	LINI	82	91	77	78	0,0041	0,0333	A	F1
Siemens	SIEGn	81	91	76	78	0,0004	0,0277	A	F1
SAP	SAPG	79	89	83	72	0,0027	0,0280	A	F1
PUMA SE	PUMG	78	89	73	77	0,0073	0,0386	A	F1
Bayerische Motoren Werke Aktiengesellschaft	BMWG	74	83	76	66	−0,0009	0,0290	A	F1
Deutsche Börse AG	DB1Gn	73	85	73	71	0,0042	0,0232	A	F1
Heidelberg Cement AG	HEIG	73	82	63	75	−0,0015	0,0310	A	F1
Deutsche Post AG	DPWGn	70	79	70	63	0,0007	0,0272	A	F1
RWE	RWEG	69	72	62	72	0,0065	0,0425	A	F1
Vonovia	VNAn	68	64	72	68	0,0032	0,0253	A	F1
Fresenius Medical Care AG & Co. KGaA	FMEG	65	75	62	67	−0,0008	0,0332	A	F1
Volkswagen group	VOWG_p	62	76	59	53	0,0022	0,0369	A	F1
Fresenius SE & Co. KGaA	FREG	61	67	52	68	−0,0019	0,0375	A	F1
Deutsche Bank Aktiengesellschaft	DBKGn	60	82	67	51	−0,0042	0,0450	A	F1
BASF SE	BASFn	49	62	42	43	−0,0013	0,0293	B	F2
Hannover RE	HNRGn	46	56	41	46	0,0036	0,0219	B	F2
MERCK Kommanditgesellschaft auf Aktien	MRCG	41	76	36	39	0,0007	0,0261	B	F2
Airbus SE	AIRG	39	51	39	30	0,0051	0,0310	B	F2
Brenntag	BNRGn	38	47	39	32	−0,0001	0,0278	B	F3
Henkel	HNKG_p	37	62	38	27	−0,0010	0,0258	B	F2

(continued)

Table 1. (*continued*)

Companies	Tikers	ESG	E	S	G	ER	SDR	Clusters k-means	Clusters fuzzy
Continental Aktiengesellschaft	CONG	36	58	21	28	−0,0025	0,0377	B	F2
Zalando SE	ZALG	36	64	30	25	0,0026	0,0516	B	F2
Bayer Aktiengesellschaft	BAYGn	35	69	37	27	−0,0013	0,0386	B	F2
E.ON	EONGn	34	36	31	34	0,0028	0,0322	C	F3
Mercedes Benz Group AG	MBGn	33	51	35	17	−0,0019	0,0341	B	F3
MTU Aero Engines AG	MTXGn	32	39	31	28	0,0057	0,0296	C	F3
Symrise AG	SY1G	32	39	23	34	0,0035	0,0249	C	F3
Sartorius	SATG_p	31	60	31	26	0,0072	0,0429	B	F2
Delivery Hero SE	DHER	28	28	28	27	0,0090	0,0526	C	F3
Siemens Healthineers	SHLG	25	47	19	26	0,0037	0,0315	C	F3
Qiagen	QIA	24	37	20	24	0,0015	0,0388	C	F3
HelloFresh SE	HFGG	23	27	18	23	0,0090	0,0679	C	F3
Porsche	PSHG_p	20	27	16	19	0,0021	0,0373	C	F3

Table 2. Correlation analysis

	ESG	E	S	G
Expected returns	−0,12	−0,26	−0,13	−0,05
St dev of returns	−0,40	−0,39	−0,41	−0,38

We constructed naïve diversified portfolio for clusters A, B, C and F1, F2, F3. They derive investment strategies which as a matter of fact are represented attitude to ESG level. After that analysis of time dynamic of *CValue* was realized and correspondence of K-ratio was estimated.

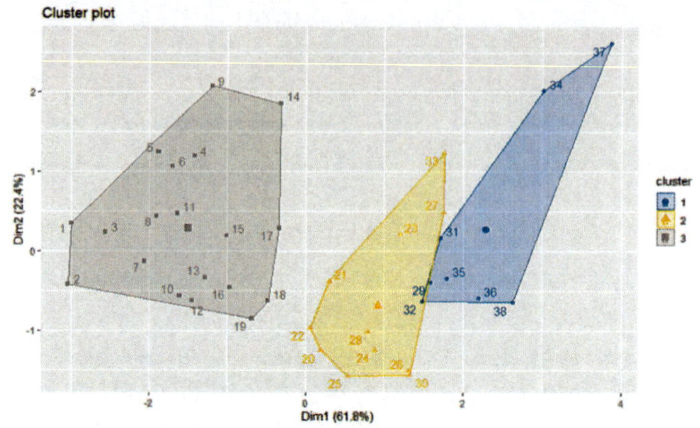

Fig. 1. K-means clustering results

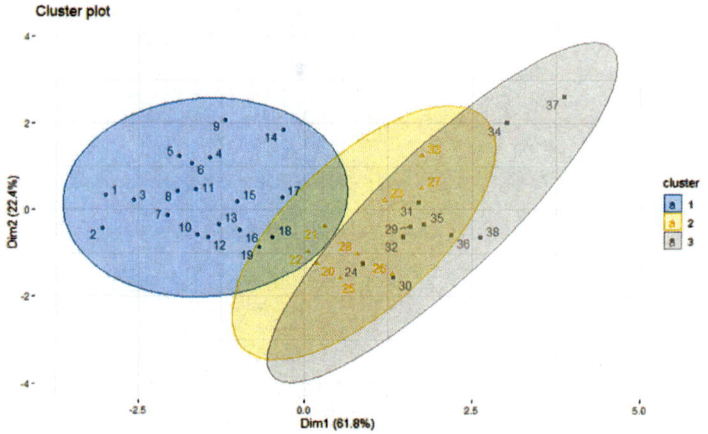

Fig. 2. Fuzzy clustering results

The dynamics of *CValue* for investment portfolios corresponding to clusters A, B, C are figured below at Fig. 3.

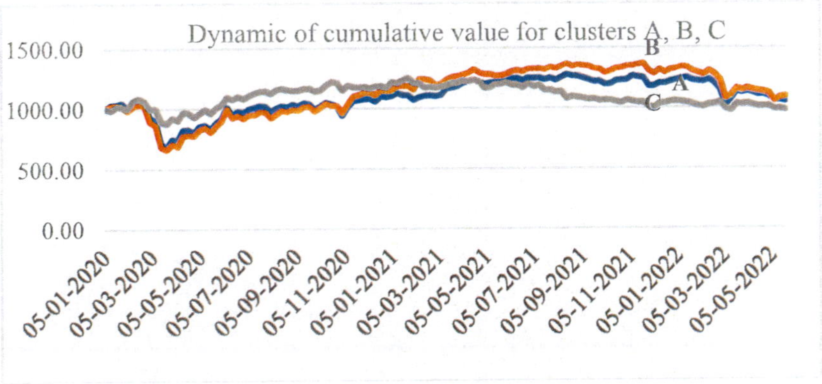

Fig. 3. Dynamic of *CValue* for K-means clusters A, B, C

Dynamics of *CValue* for clusters F1, F2, F3 (Fig. 4):

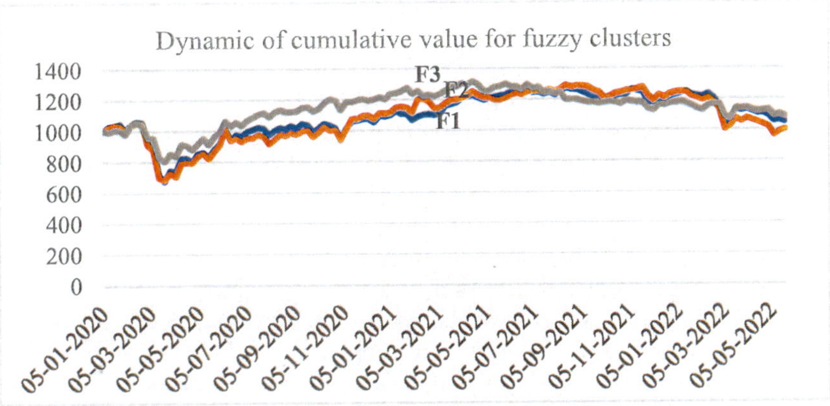

Fig. 4. Dynamic of *CValue* for fuzzy clusters F1, F2, F3

To apply operators of K-ratio and risk (measuring as volatility of *CValue*) is formed two-dimension estimation of investment strategies. Visualization of such estimation is done at the Fig. 5.

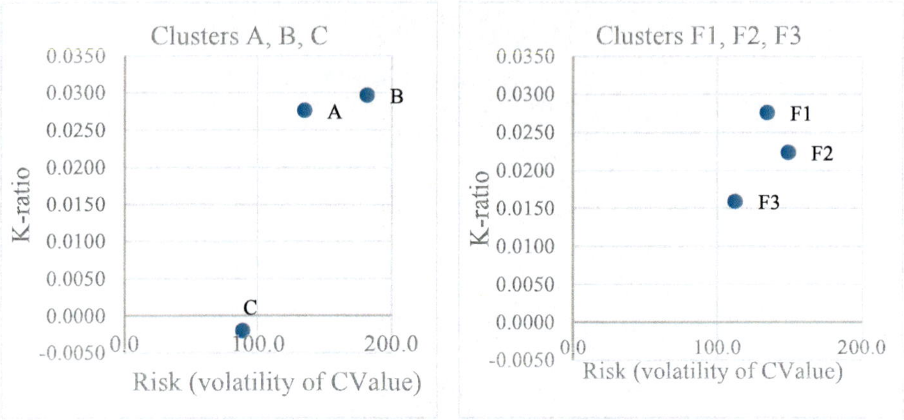

Fig. 5. K-ratio – risk (*CValue* volatility) correspondence

5 Summary and Conclusions

Considering investments from the point of view of factors E, S, and G is a rapidly developing trend in the market. More and more investors want to consider the level of ESG of companies in their investment decisions. At the same time, of course, taking into account the criteria of minimization of risk and maximization of profitability also remained under consideration. The integral indicator of ESG-score and its components reflected the "demand" of investors for sustainability indicators by a number of specialized companies. Thus, for many investors, the assessment of sustanability component is a good supplement to investment attractiveness analysis.

The problem we considered in the study is to use cluster analysis based on ESG score values, risk, and return to design investment portfolios. In our view, such cluster approach has a number of advantages over the extension approach of the classic Markowitz model. The advantage is that clustering generates relatively homogeneous groups of stocks. Including homogeneity by ESG score values. At the same time, Markowitz's approach is based on the correlation of returns and a clear vision of ESG for portfolio constituents is absent. This results in the portfolio-weighted ESG level being acceptable, but the components may have both high and very low ESG score scores. Which is not attractive for investors, because it negates the essence of sustainable investments. The cluster approach chosen for the study is therefore more appropriate in this context.

This paper presents the application of the cluster approach to the construction of investment strategies for the expanded German index DAX. The scope of the study included an additional specification. Namely, modification of the index investment strategy. The modification is that clustering is applied to the shares included in the index, and the clusters will contain only parts of the index basket. The application of two clustering techniques (K-means and fuzzy) allowed the formation of three clusters in each case. These clusters differ in the level of ESG scores of involving companies. Our research conducts assessment of investment strategies consisting in formation of naive diversified by clusters portfolios.

The results showed two interesting facts. The first is that a cluster with low ESG values generates a low K-ratio value for the cluster-specific investment strategy (in both clustering approaches). The second fact shows that the CValue for a cluster with low ESG scores increased during the COVID-19 pandemic, but after stabilizing it began to decrease. CValue for strategies corresponding to clusters with higher ESG scores showed the opposite dynamic. Falling during a pandemic and growing in a stable environment.

Thus, the cluster approach, which includes ESG scoring in addition to risks and returns, is a good tool for constructing investment strategies that take into account sustainability attitudes of investors.

References

1. Hill, J.: Environmental, social, and governance (ESG) investing: a balanced review of theoretical backgrounds and practical implications (2020)
2. US SIF: The Forum for Sustainable and Responsible Investment (2022). https://www.ussif.org/
3. Eurosif[S] (2022). https://www.eurosif.org/about-us/
4. Beyond the DE&I Acronym: What are Diversity, Equity, and Inclusion? (2022). https://www.ywboston.org/2019/03/beyond-the-acronym-dei/
5. What is an ESG Score? ESG The Report (2022). https://www.esgthereport.com/what-is-an-esg-score/
6. Gao, S., Meng, F., Gu, Z., Liu, Z., Farrukh, M.: Mapping and clustering analysis on environmental, social and governance field a bibliometric analysis using Scopus. Sustainability **13**(13), 7304 (2021). https://doi.org/10.3390/su13137304
7. Mashkov, V., Lytvynenko, V.: Method for unit self-diagnosis at system level. IJISA **11**(1), 1–12 (2019)
8. Brooks, C., Oikonomou, I.: The effects of environmental, social and governance disclosures and performance on firm value: a review of the literature in accounting and finance. Br. Account. Rev. **50**(1), 1–5 (2018)
9. Lee, K.H., Cin, B.C., Lee, E.Y.: Environmental responsibility and firm performance: the application of an environmental, social and governance model. Bus. Strateg. Environ. **25**(1), 40–53 (2016)
10. Aguilera, R.V., Rupp, D.E., Williams, C.A., Ganapathi, J.: Putting the S back in corporate social responsibility: a multilevel theory of social change in organizations. Acad. Manag. Rev. **32**(3), 836–863 (2007)
11. Poluektova, N., Klebanova, T., Guryanova, L.: Risk assessment of corporate infocommunication systems projects using Bayesian networks. In: 2018 International Scientific-Practical Conference on Problems of Infocommunications Science and Technology, PIC S and T 2018 - Proceedings, pp. 31–34, 8632150 (2019)
12. Dimitrov, I., Davydenko, N., Lotko, A., Dimitrova, A.: Comparative study of environmental determinants of entrepreneurship intentions of business students. In: 2019 International Conference on Creative Business for Smart and Sustainable Growth (CREBUS), p. 8840056. IEEE (2019)
13. Dimitrov, I., Davydenko, N., Lotko, A., Dimitrova, A.: Comparative study of main determinants of entrepreneurship intentions of business students. In: International Conference on Creative Business for Smart and Sustainable Growth, CreBUS 2019, p. 8840050 (2019)
14. Giese, G., Nagy, Z., Lee, L.E.: Deconstructing ESG ratings performance: risk and return for E, S, and G by time horizon, sector, and weighting. J. Portfolio Manage. **47**(3), 94–111 (2021). https://doi.org/10.3905/jpm.2020.1.198

15. Zehir, E., Aybars, A.: Is there any effect of ESG scores on portfolio performance? Evidence from Europe and Turkey. J. Capital Mark. Stud. **4**(2), 129–143 (2020). https://doi.org/10.1108/JCMS-09-2020-0034
16. Tola, V., Lillo, F., Gallegati, M., Mantegna, R.N.: Cluster analysis for portfolio optimization. J. Econ. Dyn. Control **32**(1), 235–258 (2008)
17. Kumari, S.K., Kumar, P., Priya, J., Surya, S., Bhurjee, A.K.: Mean-value at risk portfolio selection problem using clustering technique: a case study. In: AIP Conference Proceedings. AIP Publishing LLC, vol. 2112(1), p. 020178 (2019)
18. Izonin, I., et. al.: Stacking-based GRNN-SGTM Ensemble Model for Prediction Tasks. In: Proceedings of the 2020 International Conference DASA. Bahrain, pp. 326–330 IEEE (2020)
19. Chaikovska, I.I., Hryhoruk, P.M., Chaikovskyi, M.Y.: Fuzzy model for complex risk assessment of an enterprise investment project. In: CEUR Workshop Proceedings, vol. 3048, pp. 163–179 (2021)
20. Abdullah-Al, N., Niloy, S., Mahmudul Hasan, A., Rafia, N.T., Iffat, A.T., Lasker, E.A.: Home occupancy classification using machine learning techniques along with feature selection. Int. J. Eng. Manuf. (IJEM). **12**(3), 38–50 (2022). https://doi.org/10.5815/ijem.2022.03.04
21. Prashengit, D., Sunanda, G.: Skin lesion detection using fuzzy approach and classification with CNN. Int. J. Eng. Manuf. (IJEM) **11**(1), 11–18 (2021). https://doi.org/10.5815/ijem.2021.01.02
22. Ojugo, A.A., Nwankwo, O.: Tree-classification algorithm to ease user detection of predatory hijacked journals: empirical analysis of journal metrics rankings. Int. J. Eng. Manuf. (IJEM) **11**(4), 1–9 (2021)
23. Shakhovska, N., Montenegro, S., Kryvenchuk, Y., Zakharchuk, M.: The neurocontroller for satellite rotation. Int. J. Intell. Syst. App. **11**(3), 1–10 (2019)
24. Hagshenas, N., Mojarad, M., Arfaeinia, H.: A fuzzy approach to fault tolerant in cloud using the checkpoint migration technique. Int. J. Intell. Syst. App. **14**(3), 18–26 (2022). https://doi.org/10.5815/ijisa.2022.03.02
25. Rizqyawan, M.I., et al.: Comparing performance of supervised learning classifiers by tuning the hyperparameter on face recognition. Int. J. Intell. Syst. Appl. (IJISA) **13**(5), 1–13 (2021)
26. Matviychuk, A., Lukianenko, O., Miroshnychenko, I.: Neuro-fuzzy model of country's investment potential assessment. Fuzzy Econ. Rev. **24**(2), 65–68 (2019)
27. Kestner, L.N.: (Re) Introducing the K-Ratio (2013). https://doi.org/10.2139/ssrn.2230949. (SSRN 2230949)
28. Petrunenko, I., Grabchuk, I., Vlasenko, T., Petrova, E., Strikha, L.: Ensuring food security of Eu countries in the context of sustainable development. J. Manage. Inf. Decis. Sci. **24**(3), 1–12 (2021)
29. Hryhoruk, P., Grygoruk, S., Khrushch, N., Hovorushchenko, T.: Using non-metric multidimensional scaling for assessment of regions' economy in the context of their sustainable development. In: CEUR Workshop Proceedings, vol. 2713, pp. 315–333 (2020)
30. Oliskevych, M., Lukianenko, I.: European unemployment nonlinear dynamics over the business cycles: Markov switching approach. Global Bus. Econ. Rev. **22**(4), 375–401 (2020)
31. Klymenko, N., Nosovets, O., Sokolenko, L., Hryshchenko, O., Pisochenko, T.: Off-balance accounting in the modern information system of an enterprise. Acad. Account. Financ. Stud. J. **23**(2), 1–6 (2019)
32. Vlasenko, T.O., Chernysh, R.F., Dergach, A.V., Lobunets, T.V., Kurylo, O.B.: Investment security management in transition economies: legal and organizational aspects. Int. J. Econ. Bus. Adm. **8**(2), 200–209 (2020)

Energy Saving Optimization Scheme Design of Multi Train Cooperative Control in Urban Rail Transit

Yanzhi Pang, Jianqiu Chen(✉), and Yusheng Liu

School of Transportation, Nanning University, Nanning 530200, Guangxi, China
pangyanzhi@126.com

Abstract. The huge energy loss of subway trains cannot be ignored. According to the space-time distribution characteristics of rail transit passenger flow, this paper discusses the different energy consumption during peak and peak hours, and arranges multiple trains to absorb braking energy in time according to their respective working conditions, so as to improve the utilization efficiency of braking energy. By changing the acceleration of the train, it is in the energy consumption state of braking energy. During peak hours, by changing the dwell time of the train platform or changing the traction acceleration; On the premise that the total time of the second constant acceleration of the train remains unchanged, change the first and second traction accelerations of the train to make the train coasting in advance, so as to ensure that the train is in the state of energy consumption at the specified time. According to the working condition data of Nanning Metro Line 2, this paper verifies the designed multi vehicle cooperation scheme. The results show that the utilization rate of regenerative braking energy can be increased from about 85% to 100%. Some useful suggestions are put forward for improving the energy utilization rate of subway system.

Keywords: Regenerative braking · Multi train coordination · Train working condition · Spatial and temporal distribution of passenger flow

1 Introduction

According to statistics from relevant departments, by the end of 2021, nearly 50 cities in Mainland China have urban rail transit systems, and the total mileage is expected to exceed 7, 253 km, with 571.9 km of new lines in service [1]. Although rail transit has a low per capita energy consumption, it has a very high total electricity consumption, exceeding billions of kWh every year. In 2020, the electricity consumption of rapid transit in Beijing alone was up to 1.399 billion kWh, accounting for about 40% of the operating costs. The energy consumption of the rapid transit is mainly traction power consumption and auxiliary power consumption, with traction power consumption for train operations accounting for 70% of the total energy consumption. Therefore, reducing the traction power consumption of trains has great potential. Among the four modes of operation of rapid transit trains (traction, cruise, idle running, and braking), traction consumes

Z. Hu et al. (Eds.): AIMEE 2022, LNDECT 159, pp. 277–291, 2023.
https://doi.org/10.1007/978-3-031-24468-1_25

the most energy and accounts for 50% of the total traction power consumption. When the rapid transit train stops, it can generate regenerative braking energy equivalent to 30% of the traction energy, but 40% of such enormous regenerative braking energy is not utilized effectively. Therefore, it has gradually become the general trend of energy conservation for rapid transit trains to reasonably coordinate multiple trains and apply the energy generated by the stopped train to the towed train [2].

As of 2021, Nanning has five operating urban rail transit lines. However, economically speaking, not only does the total operating cost of Nanning Rail Group not achieve the economies of scale, but it increased accordingly. In addition, the traction power consumption of the rapid transit trains accounts for 50% to 60% of the total cost of rail transit transportation operation [3]. The braking energy equivalent to 30% of the traction power consumption is consumed in the form of thermal energy, which not only results in a waste of energy, but also increases the energy consumption of the tunnel fans due to the heat generated by the train. Therefore, by reducing the energy consumption of train traction and increasing the utilization rate of regenerative braking energy, we can effectively reduce the operating costs of the rail transit. Through the study of the four operating conditions of the train, we can match the train to be towed to the interval when the power supply generates braking energy so that the braking energy can be effectively utilized.

Regarding the energy-saving optimization of the rapid transit trains, overseas scholars believe that we can reduce the energy consumption of the train by preparing a suitable timetable. By adjusting the operation time and dwell time in the timetable, the direct utilization rate of regenerative braking energy can be improved. Nasri et al. used a genetic algorithm to match the most suitable dwell time for trains and the regenerative braking energy used by multiple trains together to successfully reduce 14% of the energy consumption. In addition, his team also studied the utilization rate of regenerative braking under different tracking intervals, and pointed out that different tracking intervals had different effects on the energy consumption of trains [4].

Scholars at home and abroad have various research directions on the energy conservation of rapid transit trains. Through the macroscopic readjustment and control of the scheduling system, some scholars optimized the number of trains, operating conditions, operation time in the interval, and other aspects, and prepared an energy-saving operation diagram. Whereas some other scholars studied on the four operating conditions of the train and found that trains adopt different operating conditions at different times in the interval to reduce the energy consumption of the train and improve the utilization rate of regenerative braking energy.

Zhang Deming and Liang Liang believe that the scheduling system, as the core of operation command, can prepare corresponding train operation diagrams through the technical application of trains, tractions, signals, and other aspects. The scheduling system adjusts the operation in the schedule according to the passenger flow and the actual operating conditions of the train, which can be divided into a flat-peak operation diagram and a peak operation diagram. The flat-peak operation diagram is relatively loose, and can achieve transportation capacity and energy by reducing the number of trains in service [5–7].

Without changing the timetable, Tang Zhaoquan proposed a study on multi-train collaborative energy-saving optimization based on the adjustment of the speed curve and discovered that the operating conditions of the trains driven by the ATO (Automatic Train Opeation) in the interval can be predicted. The system will match the optimal train when there will be train stopping in the interval, which will be inert in advance. When the braking energy is present in the power supply network, the optimal train will absorb the braking energy and change from the inert operating condition to the traction operating condition [8].

According to the principle of ATO, the force analysis of trains, and the optimization of the train operation curve based on dynamic planning and genetic algorithm, Tang Licheng proposed the optimization of the recommended speed curve of the train based on genetic algorithm so that the traction energy consumption can be reduced as much as possible when the train is consistent with the actual operation time [9, 10].

For the ATO automatic driving system of rail transit, Lu Wen proposed an automatic driving operation curve and controller to optimize the energy-saving driving curve of the train. He chose to solve the multiple targets of the train with the NPSO algorithm, and then obtained the optimal driving speed curve of the train combined with the controller algorithm based on the gray prediction fuzzy PID(Proportional Integral Derivative) control [11].

2 Subway Energy Saving Knowledge

2.1 Operation Conditions of Metro Train

The train is mainly affected by traction force F, total running resistance W, braking force B and gravity g during driving. Analyze the running status of the train, as shown in Table 1. Comparison of the nature of each working condition of the train.

The train operation status is analyzed by the formula, as shown in Table 1. Comparison of properties of trains under various working conditions.

Table 1. Comparison of train properties under various working conditions

Train acceleration	Train resultant force	Train operation condition
>0	>0	Traction condition
<0	<0	Coasting or Braking conditions
=0	=0	Cruise condition

(1) Traction Condition: the traction force F provided by the traction motor of the train is greater than the total running resistance, the traction motor consumes a lot of energy, and the train is in acceleration state. The train energy consumption formula is as follows:

$$E = \int F(t)v(t)dt \tag{1}$$

Where:

E is traction energy consumption;

t is the running time;

$F(t)$ is the train traction force at time t;

$v(t)$ is the running speed of the train at time t.

(2) Coasting condition: the train traction device does not provide traction force, and the train moves forward by the inertia of the train, and the traction motor of the train does not consume energy.
(3) Cruise condition: the resultant force of the train is 0. The traction or braking of the train depends on the total resistance of the train. The energy consumed by the traction motor in this stage is less than that in the traction stage, and the train is in the constant speed stage.
(4) Braking condition: the braking device of the train outputs braking force, so that the acceleration direction of the train is opposite to the driving direction, and the train is in deceleration state. The energy consumption under this working condition depends on the braking mode selected by the train. If the train adopts energy consumption braking, the train braking device will consume energy; If the train adopts regenerative braking, the kinetic energy before train braking can be converted into electric energy and fed back to the power supply network for other trains in the same power supply section.

Based on this, the rail operation company adjusts the train operation scheme in the section from the perspective of energy saving. If the distance between two stations is hours, the train adopts the working condition adjustment strategy of "traction coasting braking"; If the distance between stations is too large, the train will increase the speed to close to the limited speed, and use coasting, cruising and traction alternately until the train stops at the station.

2.2 Regenerative Braking

Regenerative braking is to convert the traction motor of the train from energy consumption to capacity, convert the kinetic energy of the train into electric energy, and feedback the electric energy to the power supply network through the traction transformer. The voltage of the power supply network of rail transit is saturated. When the regenerative braking energy generated by the train is fed back to the power supply network, the power supply network is in an oversaturated state. If it is in an oversaturated state for a long time, it may cause serious damage to the power supply network in the section. Therefore, how to quickly consume regenerative braking energy is also a problem. At present, the following utilization methods are adopted [12].

(1) Braking Resistance Consumption Type

The energy dissipation resistor is installed in the tunnel section. When the voltage of the power supply network rises rapidly, the energy dissipation resistor will absorb electric energy and consume it in the form of heat energy. The utilization method is simple and the equipment is expensive. However, this utilization method not only wastes energy, but also increases the energy consumption of tunnel fan due to the increase of interval heat.

(2) Capacitive Energy Storage Type

A huge capacitor is installed in the power supply section. By connecting with a two-way DC inverter and the power supply network, the excess energy of the power grid can be quickly stored. When the section train needs traction, the energy will be compensated to the power supply network. However, the cost of capacitor, the key component of this utilization mode, is expensive.

(3) Multi Train Cooperative utilization Type

Adjust the operating conditions of the train through technical means. If the train generates regenerative braking energy, it can just make the train requiring traction consume regenerative braking energy. However, this utilization method requires frequent adjustment of train diagram, and it is difficult to use in peak hours. Braking energy analysis: it is assumed that the change of mechanical energy in the braking process of the braking train is E_b. Resistance work during braking is E_r. The recovery rate of energy is generally 95%, then the recovered regenerative braking energy is E_c:

$$E_c = (E_b - E_r) \times 95\% \tag{2}$$

Among them, the energy used by the towed train is E_p:

$$E_p = E_c \times t_i / t_j \tag{3}$$

where: *ti* is the overlapping time of braking time of braking train and acceleration time of traction train; *tj* is the braking time of the train.

3 Energy Saving Model

Before you begin to format your paper, first write, and save the content as a separate text file. Keep your text and graphic files separate until after the text has been formatted and styled. Do not use hard tabs, and limit use of hard returns to only one return at the end of a paragraph. Do not add any kind of pagination anywhere in the paper. Do not number text heads-the template will do that for you.

Finally, complete content and organizational editing before formatting. Please take note of the following items when proofreading spelling and grammar:

3.1 Temporal and Spatial Distribution Characteristics of Passenger Flow

The passenger flow of rail transit is distributed according to the temporal and spatial characteristics, which is obviously imbalance and usually show a double-hump pattern, meaning heavy passenger flow during the morning and evening peak hours, while the passenger flow at other times is relatively sparse (flat hours). The flat hours with a light passenger flow makes it relatively easy to manage passenger flow in the station and adjust the operation schedule of trains. By changing the operation time or dwell time of the train in the interval, when braking energy is present, it can be used for the traction of the train. During the morning and evening peak hours, the passenger flow is heavy, and the rapid transit trains have a tight schedule. If we still adjust the operation schedule of trains, such as the dwell time and departure interval of trains, using the ATS method, we may not be able to mitigate the peak passenger flow in a timely manner, thereby reducing passengers' comfort and satisfaction with the rail transit system. On this basis, this paper puts forward an energy-saving proposal for trains based on braking time [13].

3.2 Model Hypothesis

For the convenience of the establishment of the model and analysis of energy consumption, the following hypothesis should be met:

(1) In this paper, only the conversion of traction energy, regenerative braking energy, train kinetic energy, and the total resistance of the train are considered, excluding the mass of passengers on the train;
(2) When the regenerative braking energy cannot be fully utilized, the excess braking energy can be consumed through the braking resistor; and
(3) ATO can calculate a new energy-saving speed curve after changing the operating conditions of the train.

3.3 Analysis of the Model

The train operating in the interval sends information such as the position, speed, and operating conditions of the train to ATS through wireless transmission. ATS compares the information collected and screens the train that stops and the matched train. Then, it simulates and calculates the braking time and braking energy of the train that stops, and filters the matched train. When the matched train cannot fully utilize the braking energy, the re-matched train will absorb the braking energy until the braking energy dissipates or there is no new matched train.

The physical quantities of the train that are known are the acceleration of the train, the position of the train, the generation of braking energy, the ending time and period, and the recommended speed of the train. Through the analysis of the single mass point model of the train, the train is subject to the train traction force, braking force, and total operation resistance in the driving direction.

When the braking energy is present in the traction power supply network, the following situations could occur.

Case I: The train is under the traction or cruise operating condition when the braking energy is present, which indicates that the train has the potential to absorb energy. The formula in this process is as follows:

$$t = \frac{m(v_t - v_0)}{F_t - F_c} \tag{4}$$

where

v_t is the speed of the train at time t;
v_0 is the initial speed;
t is the acceleration time of the train.

(1) A single train is under the operating condition of traction in the interval
If this period is in the flat hours, by changing the traction output of the train, reducing the acceleration of the train, and increasing the operation time of the train, the traction time is close to the time of generation of the regenerative braking energy (considering the punctuality of the train, if there is the traction of a new train during the time when the braking energy is present, we only need to adjust the time of the previous train to the time of traction of the new train, which is also applicable in a situation with multiple trains). If this period is during the peak hours, at this time, if the acceleration is reduced and the traction time of the train is prolonged, the subsequent trains will be delayed, which will not be able to mitigate passenger flows in time and may cause other emergencies and reduce passenger satisfaction. This also does not fall in line with the purposes of rapid transit services. Therefore, the rapid transit train operates according to the original speed curve to ensure the punctuality rate of the train.
(2) Multiple trains are under the operating condition of traction in the interval
The speed-time curve of operating conditions of the train is analyzed, and the trains are arranged to utilize the regenerative braking energy according to the sequence of the use of braking energy by the train. The trains are matched based on the principle of the maximum utilization of regenerative braking energy. If it is still not fully utilized, during the flat hours, the time of the use of regenerative braking energy by the train can be prolonged according to a single train is under the operating condition of traction in the interval and extend the time of the train on the platform or the operating time of the train in the interval.

Case 2: The train is idle running when the braking energy is present, which indicates that the train does not have the conditions to absorb the braking energy. The train should be under the idle running operating condition in advance so that the train is under the traction operating condition when the braking energy is present because the operating condition of idle running of the train cannot be changed and the train has to be idle running in advance if the secondary constant acceleration is used during the acceleration of the train. Figure 1 shows the comparison diagram after changing the operating conditions of the train.

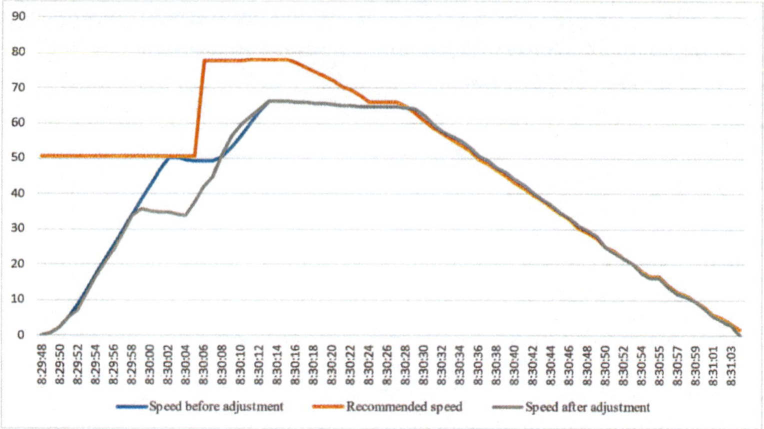

Fig. 1. Comparison diagram after changing train working condition

The train is under the operating condition of idle running when the braking energy is present is the maximum utilization of the regenerative braking energy. Through the acceleration of the two tractions of the train, the train is under the operating condition of idle running in advance, and the train can be under the operating condition of traction when the braking energy is present to achieve an energy-saving train operation. The known quantities include departure acceleration of the train a_1, speed at the time of idle running vL, second traction acceleration of the train a_2, cruise speed of the train vC, time consumed under different working conditions t_1, t_2, t_3, and time when the braking energy is present tb. Time of advance idle running of the train ta:

$$t_a = t_2 - t_b \tag{5}$$

Time of the first traction t_{f}:

$$t_r = t_1 - t_a \tag{6}$$

Time of the second traction t_{s}:

$$t_s = t_3 + t_a \tag{7}$$

After comparing with the acceleration range of the train, if it is consistent with the range, the acceleration of the second traction is calculated. During this process, the operation time of the train under different operating conditions and the punctuality of the train do not change. If the adjusted traction acceleration is not within the range, the time of idle running and acceleration can be reduced. The time of idle running should be 3 or 4 s.

In the flat hours, the time of idle running of the train is prolonged so that the train is under the operating condition of traction when the braking energy is present, and then the regenerative braking energy is used according to the time of braking energy and the time of operating conditions of each train. The analysis of energy consumption is as follows. Because the traction energy consumption of the train consists of regenerative braking energy and electric energy directly obtained by the traction power supply network, the train should also obtain the following energy from the power supply network Es:

$$E_s = (E_r + ma) \cdot t - E_c \tag{8}$$

If multiple trains are using the model together, the braking energy used should be distributed according to the ratio of time required by traction to total time. When the acceleration of the train remains unchanged, the above energy consumption formula can be directly applied for analysis. When the acceleration of the train changes, the energy consumption before and after the change should be compared and analyzed.

Before the change,

$$E_s = (E_r + ma_o) \cdot t \tag{9}$$

After the change,

$$E_s = (E_r + ma_c) \cdot t - E_c \tag{10}$$

where

a_o is original acceleration;

a_c is acceleration after change.

When the adjusted single train achieves the energy-saving performance, other trains should be reasonably matched to absorb the remaining energy until the regenerative braking energy is completely consumed or there is no matched train to realize the energy-saving operation of multiple trains.

4 Verification of the Proposal

4.1 Metro Line 2 of Nanning City

Nanning Metro Line 2 is the vertical line in the cross-shaped rapid transit planning network in Nanning, which is the main traffic line connecting the north and the south. The line starts from Xijin Station in the north and ends at Yudong Station in the south. The east extension line starts from Yudong Station, extends eastward along Liangyu Avenue, and ends at Tanze Station. It spans four urban areas including Xixiangtang District, Xingning District, Jiangnan District, and Liangqing District. The total length of Line 2 is 27.5 km, all of which are underground with a total of 23 stations.

4.2 Energy-saving Optimization Design of Nanning Metro Line 2

Through the analysis of the collected data, the timetable of Metro Line 2 is simulated with Excel. If Train 3 completes the regenerative braking between Jianshe Road Station and Shizitang Station, it can be seen from Table 2 and Table 3 that the regenerative braking energy is present at 8:29:27 and the braking energy dissipates at 8:30:01, existing for 34 s. Table 4 is then updated.

Table 2. Train time consumption of Nanning Metro Line 2

Running Direction	Xijin——Tanze		Running Direction	Tanze——Xijin	
Station	Platform dwell time	Interval operation time	Station	Platform dwell time	Interval operation time
Train Station (1)	58 s	1 min and 3 s	Shizitang Station	36 s	1 min and 43 s
ChaoYang square station (2)	59 s	2 min and 6 s	Jianshe Road Station	36 s	1 min and 16 s
Nanning theater station (3)	37 s	1 min and 27 s	Dashatian Station	39 s	1 min and 14 s
Fujian Garden station (4)	45 s	1 min and 7 s	Jiangnan Passenger Station	40 s	2 min and 1 s
Tinghong road station (5)	41 s	1 min and 16 s	Shizhuling Station	39 s	1 min and 11 s
Shizhuling station (6)	36 s	2 min and 1 s	Tinghong Road Station	38 s	1 min and 13 s
Jiangnan passenger station (7)	40 s	1 min and 18 s	Fujian Garden Station	38 s	1 min and 18 s
Dashatian station (8)	35 s	1 min and 16 s	Nanning Theater Station	39 s	2 min and 5 s
Jianshe road station (9)	35 s	1 min and 43 s	ChaoYang Square Station	59 s	1 min and 2 s
Shizitang station (10)	36 s	——	Train Station	60 s	——

Table 3. Time consumption of Nanning Metro Line 2 Train under various working conditions

Name between Stations	Interval time (seconds)	Traction 1 takes time (seconds)	Coasting 1 time consuming (seconds)	Traction 2 takes time (seconds)	Cruise time (seconds)	Braking time (seconds)
(7) to (8)	78	14	4	NO	13	36
(5) to (6)	66	14	5	6	15	43
(2) to (3)	126	13	3	17	55	35
(2) to (1)	62	14	6	6	NO	36

Table 4. List of matching trains for Nanning Metro Line 2

Train	Departure time	Inter station operation time (seconds)	End time of traction 1	Coasting end time	End time of traction 2	Cruise end time	Braking end time
4	8:29:23	78	8:29:37	8:29:41	8:29:54	8:29:58	8:30:41
5	8:29:48	76	8:30:02	8:30:07	8:30:13	8:30:28	8:31:04
6	8:28:02	126	8:28:15	8:28:18	8:28:35	8:29:30	8:30:05
13	8:29:35	62	8:29:49	8:29:55	8:30:01	8:30:01	8:30:37
14	8:29:34	78	8:29:48	8:29:52	8:30:05	8:30:09	8:30:52
15	8:28:14	121	8:28:27	8:28:30	8:28:47	8:29:37	8:30:12
16	8:29:25	76	8:29:39	8:29:44	8:29:50	8:30:05	8:30:41

Through Table 4, the time consumption of each working condition of the matched train is analyzed and sorted out, and the time of using regenerative braking energy of the matched train in Table 5 is summarized.

Table 5. Match the time when the train uses regenerative braking energy

Train	Start utilization time	End utilization time	Coasting period	Unused time period (seconds)	Braking energy utilization rate of single train
4	8:29:27	8:29:58	8:29:37–8:29:41	8	76.47%
5	8:29:48	8:30:01	NO	21	38.23%
6	8:29:27	8:29:30	NO	31	8.82%
13	8:29:35	8:30:01	8:29:49–8:29:55	14	58.82%
14	8:29:34	8:30:01	8:29:48–8:29:52	11	67.65%
15	8:29:27	8:29:37	NO	24	29.41%
16	8:29:27	8:30:01	8:29:39–8:29:44	5	85.29%

If the single train utilization model is used, the braking energy is allocated to Train 16, because its utilization rate of braking energy is the largest. However, if multiple trains jointly utilize the regenerative braking energy, the regenerative braking energy should be allocated according to the time. The Table5 can be used to match the train. Train 16 uses the energy for 8 s from 8:29:27 to 8:29:39. And then, matched Train 14 uses the braking energy for 7 s from 8:29:39 to 8:29:48, and matched Train 16 uses braking energy from 8:29:48 to 8:30:01. Through the coordinated utilization of multiple trains, the regenerative braking energy can be fully utilized, and the utilization rate of braking energy is improved from 85.29% to 100%. This proposal does not change the operating conditions of the train, so 100% energy of the train is saved, and we do not have to analyze the comparison of energy consumption of the train.

If Train 16 stops between Jianshe Road and Dashatian District, we can see from Table 5 that the braking energy is present at 8:30:05 and dissipates at 8:30:41. In the same way, the matched train is obtained. See Table 6 for the time consumption of operating conditions of the matched train.

Table 6. Time consuming for matching train working conditions

Train	Departure Time	Inter Station Operation Time (seconds)	End Time of Traction 1	Coasting End Time	End Time of Traction 2	Cruise End Time	Braking End Time
5	8:29:48	76	8:30:02	8:30:07	8:30:13	8:30:28	8:31:04
4	8:30:09	78	8:30:23	8:30:27	8:30:40	8:30:44	8:31:27

Through Table 6, the time consumption under each operating condition of matched Trains 4 and 5 is analyzed and summarized, and the time of using regenerative braking energy by matched trains is summarized in Table 7. The data are further analyzed on this basis.

Table 7. Match the time when the train uses regenerative braking energy

Train	Start Utilization Time	End Utilization Time	Coasting Period	Unused Time Period (seconds)	Braking Energy Utilization Rate of Single Train
5	8:30:07	8:30:28	8:30:02–8:30:07	15	58.33%
4	8:30:13	8:30:41	8:30:27–8:30:31	11	69.45%

By observing the time in the table, we can know that Trains 4 and 5 can both use the regenerative braking energy, and Train 4 has the highest utilization rate of regenerative braking. Considering the braking energy used by a single train, by selecting Train 4, the utilization of regenerative braking energy can be maximized.

When Trains 4 and 5 jointly utilize the regenerative braking energy, the utilization rate of energy can be higher, reaching up to 100%. According to the time sequence of the use of braking energy by trains, firstly, Train 5 begins to use the energy at 8:30:07 until the train stops and pulls in, and no longer absorbs the braking energy. Then, Train 4 begins to use braking energy at 8:30:28. The two trains use 94.4% of the braking energy. At this time, if the operating condition of the train is changed, it is possible to utilize all the regenerative braking energy.

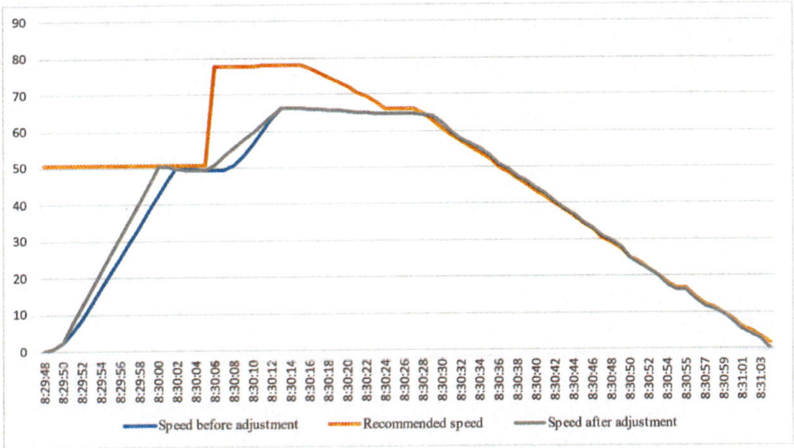

Fig. 2. Speed time curve before and after train adjustment

If Train 5 is under the operating condition of idle running when the braking energy is present, the operating condition of the train should be adjusted so that it is under the operating condition of traction.

In addition, the secondary constant acceleration is used for the speed limit of the line when the train leaves the station, so the operating condition of idle running should be added to ensure the comfort of passengers. The acceleration of Traction 1 is increased so that the train will reach the speed of the time of idle running at 8:30:00. And then, the train will be under the condition of idle running for 5 s so that the train can match the braking energy at the time of traction. Figure 2 shows the broken line diagram of speed and time before and after the train is adjusted.

Because the train should be under the operating condition of traction when braking energy is present, Train 5 leaves the station at an acceleration of 1.331 m/s2. The train reaches the speed at the time of idle running on the original speed curve at 8:30:00. And then, the train can absorb the regenerative braking energy for traction acceleration (extending the time of Traction 2), where acceleration is 0.61 m/s2. When the train's departure acceleration is changed, the energy consumption before and after the change of operating conditions of the train should be analyzed and verified.

Before the operating condition of the train is adjusted, by integrating the speed (the running distance is calculated with the graphic area method in this paper), the running distance of Traction 1 is 97.93 m and acceleration is 1.109 m/s2; the running distance of Traction 2 is 96.36 m and Acceleration 2 is 0.862 m/s2. Assuming that the resistance of the train is the physical quantity increasing with the displacement (W), because W is composed of the component force of gravity and the centrifugal force of the curve, the limit value is taken in this paper, which in this paper is assumed to be half of the gravity.

After adjustment, similarly, we can obtain that the acceleration of Traction 1 of the train is 1.331 m/ s2, its running distance is 83.94 m, the acceleration of Traction 2 is 0.61 m/ s2, and its running distance is 128.48 m. The braking distance of the train is 322.2 m, the speed at the time of braking is 64.44 km/h, and the braking time is 36 s.

By comparing the formula before and after adjusting the operating condition, we can obtain (the energy consumption after the change minus the energy consumption before change):

The calculation result is a number smaller than 0, which indicates that after adjusting the operating condition of Train 5, the energy before the change is greater than that after the change, which achieves energy-saving operation. On this basis, Train 4 is arranged to absorb the remaining regenerative braking energy to fully utilize the regenerative braking energy.

5 Conclusion

According to the temporal and spatial distribution characteristics of passengers, it is pointed out that trains need to adopt different schemes during flat peak and peak periods, so that the braking energy can be used by multiple vehicles and improve the utilization rate of regenerative braking energy. If the train is in energy consumption condition at the time of braking energy, the energy shall be absorbed according to the consumption time of train working condition, and the principle of maximum energy utilization shall be taken. If there is no energy consuming train currently, it is necessary to change the train state to make it under the energy consuming condition. For the peak period, the train can be in the energy consumption condition at the specified time by changing the train platform dwell time or changing the traction acceleration; For the peak period, the train needs to change the first and second traction acceleration of the train on the premise that the total time of the second constant acceleration remains unchanged, so as to make the train coasting in advance and ensure that the train is in the energy consumption condition at the specified time. This optimization method can increase the utilization rate of regenerative braking to the maximum.

Acknowledgment. This paper is supported by: (1) Sub-project of Construction of China-ASEAN International Joint Laboratory for Comprehensive Transportation (Phase I), No. GuiKeAA21077011-7; (2) 2022 District Level Undergraduate Teaching Reform Project: Research and Practice of Curriculum Teaching Quality Evaluation System in Applied Universities from the Perspective of OBE (2022JGA390); (3) School Level. Teaching Team Cultivation Project of Nanning University Rail Transit Teaching Team Project (2021JXTDPY04); (4) "Curriculum Ideological and Political Education" of Nanning University Demonstration Courses: Automotive electrical appliances: (2020SZSFK19).

References

1. Xu, Y.: Thoughts and prospects on energy conservation of urban rail transit. Sci. Technol. Wind. **23**, 232–233 (2019)
2. Gao, C.: Communication based Train Control System (CBTC). China Railway Press (2018)
3. Haichuan, T.: Energy saving optimization control of urban rail transit multi train operation. Southwest JiaoTong University, Chengdu (2015)
4. Nasri, A., Moghadam, M.F., Mokhtari, H.: Timetable optimization for maximum usage of regenerative energy of braking in electrical railway systems. In: International Symposium on Power Electronics Electrical Drives Automation and Motion, pp. 1218–1221. IEEE (2010)

5. Zhang, D., Liang, L.: Research on energy saving strategy of urban rail transit dispatching system. Railway Commun. Signal. **51**(9), 73–75+85 (2015)
6. Yuanyuan, L.: Operation scheduling integrated optimization method considering mixed operation of trains. Zhejiang University, Zhejiang (2018)
7. Xin, Y.: Research on optimization of train diagram for energy-saving urban rail transit. Beijing Jiaotong University, Beijing (2016)
8. Zhaoquan, T.: Research on multi train collaborative energy saving optimization based on speed curve adjustment. Beijing Jiaotong University, Beijing (2018)
9. Licheng, T.: Optimization of recommended speed curve of urban rail trains considering ATO control strategy. Beijing Jiaotong University, Beijing (2018)
10. Fengchu, L.: Research on optimization of optimal idling point of Train Based on Genetic Algorithm. Beijing Jiaotong University, Beijing (2014)
11. Wen, L.: Research on multi-objective optimization of Automatic Train Operation Process. Lanzhou Jiaotong University, Lanzhou (2017)
12. Liu, F., Wu, J., Zhang, Z., Lu, H., Li, K.: Operation optimisation of urban rail transit train base on energy saving. Int. J. Mater. Struct. Integr. **12**(1/2/3), 1 (2018). https://doi.org/10.1504/IJMSI.2018.10014930
13. Marius, W.T., et al.: Energy saving potentials of modern powertrains utilizing predictive driving algorithms in different traffic scenarios. Proc. Inst. Mech. Eng. Part D J. Automob. Eng. 234(4), 992–1005 (2019)

Advances in Medical and Technological Approaches

Research on Staff Training and Education Strategy of Highway Passenger Transport Enterprises with the High-Quality Development

Yan Chen(✉)

College of Traffic and Transportation, Nanning University, Guangxi 530200, China
2676501161@qq.com

Abstract. The highway passenger transport business is facing the competitive challenge of declining passenger volume and transportation proportion, and is also facing in a good opportunity for the high-quality development of building a strong transportation country. By analyzing the characteristics of high-quality transformation and development of highway passenger transport and the characteristics of talent demand, and analyzing the important role of compound talents, innovative talents and collaborative talents in the transformation and development of highway passenger transport, combined with the theory of enterprise human resources education and training, this paper puts forward the Trinity training and education content system of "knowledge + skills + literacy" for highway passenger transport enterprise talents, and makes it clear that knowledge, skills and quality are divided by training platform and industry experts and internal trainers are three different subjects to undertake. Finally, the implementation path of implementing and cultivating the education system is elaborated in detail.

Keywords: High-quality development · Highway passenger transport · Enterprise staff training and education

1 Introduction

Since 2020, China's road passenger transport has been facing more challenges. On the one hand, the overall passenger volume is declining, highway passenger volume decreased more. In 2020, the national commercial passenger volume was 9.665 billion, down 45.1% over the previous year, and completed passenger turnover 1925.143 billion km, down 45.5%; highway passenger volume was 6.894 billion, down 47.0% over the previous year, and completed passenger turnover was 464.101 billion kilometers, down 47.6%. From the perspective of transportation structure, highway passenger transport shared 71.3% of the total social commercial passenger transport volume in 2020 [1]. In 2021, the commercial passenger volume was 8.303 billion, down 14. 1% over the previous year, and the passenger turnover was 1975.815 billion kilometers, up 2.6%; the highway passenger volume was 5.087 billion, down 26.2% over the previous year, and the passenger turnover was 362.754 billion kilometers, down 21.8%. From the perspective of transportation structure, highway passenger transport undertook 61.3% of the

Z. Hu et al. (Eds.): AIMEE 2022, LNDECT 159, pp. 295–304, 2023.
https://doi.org/10.1007/978-3-031-24468-1_26

total social commercial passenger transport volume in 2021 [2]. The above data show that highway passenger transport is facing "double decline": the absolute value of social commercial passenger transport volume declines and the proportion in the overall passenger transport structure declines, therefore, the development trend is not optimistic. On the other hand, with the further development of transportation power construction, such as "national 123 travel traffic circle", passenger transport convenient smooth, intercity road passenger transport, urban and rural passenger service integration with construction goals, that determines the current and future a period of time, highway traffic must revolve around "first-class facilities, first-class technology, first-class management, first-class service", with high quality development to promote the policy decisions work, be a good first [3]. In supply and demand "bilateral" reform, driven by high quality transformation, highway passenger transport enterprises through high quality transformation development, meeting the demand of transport is still the basis of highway passenger transport enterprises, optimizing customer service and providing wisdom service is the main direction of highway passenger transport service quality, reforming the traditional business process model and promoting the service process autonomy is an important means of highway passenger transport service effect.

Most scholars' research on the high-quality development of highway passenger transport focuses on the innovation of new infrastructure construction and operation mode. For example, Yao Zhonghua, Gao Bo and Yang Yuanyi [4] proposed measures to improve equipment energy consumption, improve transportation structure and strengthen systems to promote the high-quality development of Zhejiang transportation industry; Taking Huai'an automobile group as an example, Chong Dou [5] pointed out that the main factors affecting the safety of highway passenger transport are driver factors and vehicle factors. However, the high-quality development of highway passenger transport needs a strong professional team to undertake its construction and operation. Taking the high-quality development of highway passenger transport as the research background, starting from the enterprise human resource management and education strategy, this paper analyzes the core backbone talents and talent cultivation strategy needed to promote the high-quality development of highway passenger transport, which is of great significance to improve the service level of highway passenger transport.

2 Analysis of the High-Quality Development Characteristics of Highway Passenger Transport

Under the guidance of the "four first-class" construction and operation direction, the high-quality development of highway passenger transport will start from both hardware facilities and soft services, and comprehensively conduct the transformation and upgrading of infrastructure, the promotion and application of new technologies, the deepening reform of management system and the mode innovation of travel services, so as to meet people's needs for a better life.

2.1 Taking into Account Both Social and Economic Benefits

As a quasi-public product, highway passenger transport provides people with safe, smooth, convenient and economic travel services, which is the social responsibility of enterprises and the guarantee of enterprise service quality. Due to its public service attribute, highway passenger transport has non-competitive characteristics and a wide range of service objects. Once the service quality problems, it will not only affect the reputation of the enterprise, but also may bring negative impact to the society. From this point of view, the highway passenger transport service should first fulfill its social responsibilities and ensure good social benefits [6]. At the same time, as a market-oriented business, passenger transport service should also ensure a certain profit return, in order to ensure the survival and development of enterprises, and its continuous operation. Therefore, high quality development requires highway passenger transport enterprises should provide high quality, high efficiency travel service, and improve the utilization rate of factor resources, maintain high capital efficiency, labor efficiency, environmental efficiency, resource efficiency, etc. With less cost, forming more effective output, to achieve a win-win of social benefit and economic benefit, is also one way of the highway passenger transport enterprises crack passenger volume drop problem.

2.2 Improve Energy and Efficiency in Parallel

The quality evaluation of highway passenger transport is mainly based on the satisfaction degree of passenger demand. Passenger travel needs are diversified, including safety, convenience, comfort, economy, shortening of the travel time, etc. From the perspective of demand management, by meeting the diversified needs of passengers and improving the perceived service quality of passengers, passenger satisfaction can be improved, so as to obtain a better quality evaluation. From the perspective of supply management, in order to meet the various demand characteristics of passengers, the supply of service can start from improving the level of service and service efficiency, and create a good service quality image through improving equipment and facilities, service means and operation and organization technology.

In the process of travel, the longer the travel time, the more need to the passenger travel, the more critical to the service. Passenger travel time combined with the arrive time (departure to the station, leave the station to the destination), waiting time, bus time, ride time, etc., what raise the service level should not only pay attention to passenger station-station in transit time, but also pay attention to the passengers what service get at the station and the convenience to arrive the station, that is, passenger service should cover passengers "travel chain", do well "travel chain" key link service, shorten the "travel chain" total time, will be able to make passengers get a good travel experience, improve the quality evaluation.

2.3 Seek Progress in Maintaining Stability, Driven by Innovation

The stability, sustainability and risk of business operation are also important signs to measure the quality of development. Since the beginning of the 14th Five-Year Plan, highway passenger transport is in the critical period of transforming the development mode, optimizing the supply structure and transforming the growth power. For high-quality development, we must make progress while maintaining stability and achieve sustained and steady growth. Highway passenger transport itself needs to maintain balanced the growth rate with efficiency, cost, social service, etc. and seamless connection with other transportation modes, such as railway, air and water transport to realize smooth passenger transportation. Only by optimizing the structure inside and outside of highway passenger transport can we maintain effective supply and stable growth.

Innovation-driven development. The key of improving the quality of supply and services is in innovation-driven development. The application and high-quality development of scientific and technological innovation aim to meet the needs of people and take the obstacles to the use of business process transformation and new technology and equipment, for example, some passengers have not formed the habit of face recognition and automatic ticket checking, which affect the gate efficiency. Innovation should take practical, applicable and easy to use as the basic principle. Highway passenger transport enterprises should focus on effectively solving prominent bottlenecks and deep-seated problems, accelerate the establishment of technological innovation institutions, accelerate the development of passenger transport business to digital, networked and intelligent, enhance the value chain, optimize the travel chain, and improve the quality of the supply system.

To sum up, the high-quality transformation and development of highway passenger transport enterprises is guided by policy planning, with the purpose of fully meeting the travel needs of passengers, and with the improvement of the quality of employees as the guarantee, to create a quality service supply system.

3 High-Quality Development of Passenger Transport Needs High-Quality Practitioners

Passenger participation in highway passenger service is very high, basically involved in the whole process of transportation production and operation. In this process, whether employees who directly contact passengers or do not directly, the production process is consistent with the passenger transportation process. To maintain real-time online working state, high quality service also puts forward higher requirements for the quality of employees [7].

3.1 Demand for Compound Talents

With the application of Internet technology in the field of highway transportation, the construction and operation of new infrastructure, and the development of new business models such as intelligent passenger transport and customized passenger transport [8], the passenger transport business talents needed should not only have outstanding experience in professional skills, but also have high information technology skills to do a good job in the relevant business planning, design, implementation and operation. In business planning and design, compound talents can combine the application of big data technology, network technology, information service technology with traditional business to realize intelligent passenger transport; In the process of business planning project implementation, compound talents can better grasp the project and do the project implementation management; In daily operation, compound talents can initially solve the technical problems on site, quickly resume production, and improve service efficiency.

3.2 Demand for Innovative Talents

In the transformation and development stage of highway passenger transport, innovative talents are essential. Innovative talents [9] with innovative entrepreneurial ability and creative spirit, can undertake hard work. Innovative talents are also pioneering, have extensive and solid knowledge in the field of highway passenger transport, have a high professional level. And their attention can highly concentrated, so that they can focus on the wisdom in the field of research and exploration, can create a new situation, make creative contributions to the development of passenger transport. Highway passenger transport through innovation driven development, must carry out business process reengineering, service equipment technology update or transformation, service gear upgrade, operation management mode reform and a series of wisdom passenger transport project. Due to the project innovation, risk and pioneering characteristics, project planning and operation is need innovative talents to host and participate in, to better implement, achieve the requirements of high-quality development.

3.3 Demand for Collaborative Talents

According to the characteristics of quasi-public products of passenger transport products and the requirements of passenger transport service quality standards, the passenger transport service work must be equipped with a certain scale of team to complete together, while the market operation requires the team to achieve efficient operation, in order to achieve the economic goal of passenger transport scale. If innovative talents and compound talents are the leaders of passenger transport business, collaborative talents are the executors of passenger transport business service operation system. Collaborative talents can quickly understand the objectives and requirements of work, be good at communication and cooperation with others, and can guide colleagues and passengers to act in accordance with the rules, which is conducive to the smooth achievement of work goals. In the current situation of poor service awareness, low service efficiency and insufficient work skills in highway passenger transport service, training or introducing cooperative talents can quickly reverse the now bad state, so as to improve the overall service quality.

4 To Build a Training and Education System

No matter what type of talent, whether it is the introduction or the enterprise itself training, to build a talent team that can complete the high-quality development goals of highway passenger transport business, we must establish an enterprise training system, through training, training, teaching, education and other ways, to build the own high-quality talent system.

The cultivation of high-quality talent team should first be carried out in the core backbone team that promotes the high-quality development of enterprises. Highway passenger transport enterprises are labor-intensive and technology intensive enterprises, and the scale of enterprises is generally more than 1000 people. In the talent structure of highway passenger transport enterprises, managers and engineering technicians are allocated according to 5% to 10% of the total number of enterprises, so enterprises with 1000 people need to allocate 50 to 100 managers and engineering technicians, which will be the first focus of the training and education system of highway passenger transport enterprises.

4.1 Build a High-Quality Training and Education Content System

The traditional human resource training of passenger transport enterprises mainly carries out pre-job training and post training around the post requirements. On this basis, career planning and management talent selection and training are carried out. The content and form are relatively single, and they cannot respond to the talent needs of high-quality transformation and development.

Highway passenger transport enterprises should fully rely on the Internet + education technology, choose the right training platform for cooperation, build "knowledge + skills + quality" trinity training education content system. Knowledge training delegated to the training platform. For skills training, senior technical personnel in the field shall be employed for hands-on teaching. Quality training, corporate culture training and industry competition training are given to the enterprise's own internal trainers.

4.1.1 Knowledge

Including job knowledge, service knowledge and industry knowledge, to provide intelligent passenger transport services for enterprise employees for knowledge reserve, to improve the service level of employees.

4.1.2 Skills

Including work skills, professional skills and comprehensive skills, it is an important training project to improve the quality and efficiency of passenger transport services. It can not only meet the daily work and operation requirements, but also cultivate a group of professional and technical talents for high-quality transformation and development.

4.1.3 Quality Class

The cultivation of quality needs a relatively long process. First of all, we should make use of appropriate opportunities and events to explore the characteristics of talents. Secondly, we should teach students in accordance with their aptitude, combine the personality characteristics and professional and technical ability, and guide them into corresponding compound talents, innovative talents and collaborative talents.

4.1.4 Corporate Culture Training

Including enterprise business philosophy, enterprise system, code of conduct and brand image and other contents. Enterprise management philosophy is the core enterprise culture, is the enterprise for years concise spirit of enterprise, is also the core part of enterprise culture training, especially in the transformation development stage, the enterprise concept into every employee work faith, let employees in the face of difficulties and setbacks, still can indomitable, complete the task. The publicity and implementation of the enterprise system is very important, the system is the expression of the enterprise business philosophy, do something or not do, every employee should know and follow the constraints of the enterprise system. The code of conduct also needs to be strengthened through training to build team cohesion. Corporate brand image plays a role as a bridge of external communication. In this aspect, enterprises often only pay attention to customer training, but ignore employee training. In fact, corporate employees are familiar with the corporate brand image, and can consciously maintain and publicize the corporate brand.

4.1.5 Industry Competition Structure Analysis and Training

Knowing oneself and knowing the enemy will not defeat a hundred battles. Business training must analyze the industry competition structure, so that employees can get familiar with the enterprise's suppliers, customers, similar competitive businesses and potential alternative businesses, as well as better understand the enterprise itself, which will help employees to strengthen their faith, rationally analyze the competitive problems encountered in the business work process, and take measures conducive to the enterprise.

4.2 Formulate a Scientific and Reasonable Implementation Path

In order to achieve the goal of high-quality talent team construction, implement the enterprise training content system, reserve human resources for the high-quality development of highway passenger transport, focus on the following four talent training implementation paths to achieve the goal.

4.2.1 Make a Good Human Resources Planning and Implement the Training Plan

Plan should determine the quality and quantity of all kinds of talents needed, determine the annual training objectives and contents of all kinds of talents, and determine the ways, forms and evaluation system of staff training (Fig. 1).

4.2.2 Build an Enterprise Internal Trainer Team

The training of quality, enterprise culture and industry competition to build the core competitiveness of enterprise talents should be completed by the enterprise's own internal trainer team. The internal trainer team is composed of senior executives, technical professionals, industry experts, etc., and is comprehensively managed by the human resources department. It is the most core and basic training resource of the enterprise.

4.2.3 Training List

The human resource management department can list the annual training list, encourage employees to actively participate in the training in the way of points system, complete the training content in various forms of self-study, teaching, mutual learning and other forms, and get points through the assessment. The training list should clearly list the

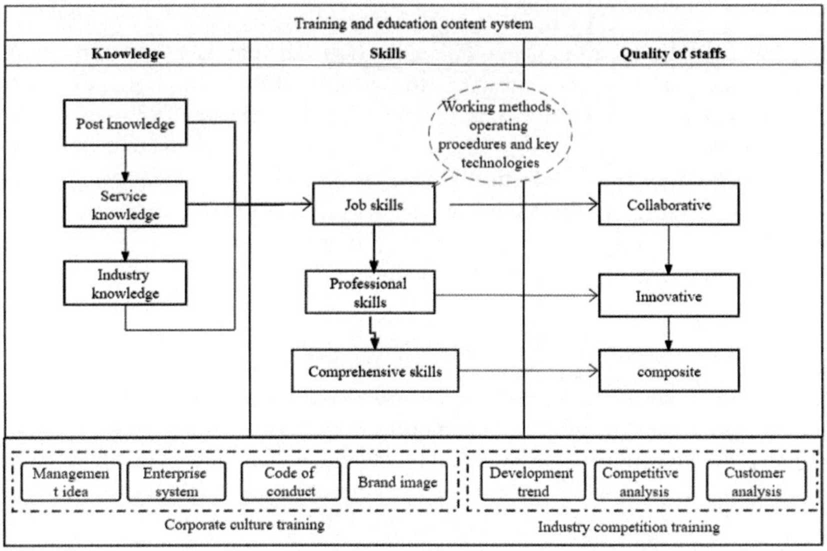

Fig. 1. Employee training content system of highway passenger transport enterprises under the background of high-quality development

required projects and elective projects, as well as the learning content and learning channels of each training program. It should encourage employees to choose and learn the elective projects based on personal development and enterprise development on the basis of completing the compulsory projects, and become characteristic talents in the field of passenger transport business.

4.2.4 Talent Description

All kinds of talent specifications, skills, knowledge and quality requirements required by the high-quality development of the enterprise, are compiled in the talent manual, which will encourage the enterprise employees to ask themselves according to the standards of the talent specification, exercise themselves, and strive to achieve the requirements of the talent specification, so as to become the high-quality talents most needed by the enterprise.

5 Conclusion

The high-quality transformation and development of highway passenger transport needs the support of high-quality talents. In "first-class facilities, first-class management, first-class technology, first-class management, first-class service" goal, highway passenger transport enterprises should carefully analyze how to balance social and economic benefits, seek improvement in stability, how to cultivate high-quality talents, fully integrate enterprise resources, build "knowledge + skills + quality" trinity, "enterprise culture + industry analysis" internal and external training content system, and through effective implementation path to achieve training goals, optimize service supply, to meet the social commercial passenger absolute decline and the share in the overall passenger transport structure decline that means "double drop" challenges.

1) The paper analyzes the characteristics of the high-quality development of highway passenger transport, and points out that as a quasi-public product, highway passenger transport should take into account social and economic benefits. The quality evaluation of highway passenger transport first takes the satisfaction of passenger demand as the main judgment basis, and its development should be stable and innovation driven. These works need a high-quality talent team to implement.
2) This paper studies the content of the training and education system of highway passenger transport enterprises, and points out that the key groups that the training and education system should first pay attention to are managers and engineering technicians. It is suggested that enterprises should establish a trinity training and education content system of "knowledge + skills + quality", and integrate enterprise culture and industry competition on this basis. It also puts forward that the training platform, external senior technicians and internal trainers should focus on carrying out targeted training respectively.

This paper also puts forward the path of the training and education system of highway passenger transport enterprises under the background of high-quality development, and discusses it in detail from four aspects: planning, team, list and instructions, which provides a better operation method for highway passenger transport enterprises to implement the training system.

Acknowledgement. This research has been financed by the second batch of school-level teaching team in 2019 "Applied Effective Teaching Design Teaching Team" (2019XJJXTD10).

References

1. Yao, Z., Gao, B., Yang, Y.: High quality development of Zhejiang transportation industry based on low-carbon perspectiv. Zhejiang Econ. **11**, 35–37 (2019). (in Chinese)
2. Davis, S.C., Diegel, S.W., Boundy, R.G.: Transportation energy data book: edition 33. Off. Sci. Tech. Inf. Tech. Rep. **176**(3), 319–338 (2017)
3. Wang, S., Liu, X., Zhou, C., et al.: Examining the impacts of socioeconomic factors, urban form, and transportation networks on CO2 emissions in China's megacities. Appl. Energy **185**(1), 189–200 (2017)
4. Tan, Z.: Talent management is the core force of the high-quality enterprise development. Charming China **32**, 359 (2019). (in Chinese)
5. Interconnection With wisdom, China's road passenger transport seeks innovation. Driving Park (05), 40–41 (2010). (in Chinese)
6. Li, J., Zhou, Y.: The exploration and practice in innovative personnel training of computer science and technology. Int. J. Educ. Manag. Eng. (IJEME) **2**(6) (2012)
7. Goldshtein, A., Vavilova, G., Rumkin, A., et al.: Detection of insulation defects in automated in-process testing of electric wire during its extrusion. Mat. Sci. Forum **94**(2), 141–150 (2019)

8. Rabeh, M., Petersen, J., O'Hearn, S.: Insulation optimization improves safety, reduces corrosion risks. Offshore **78**(2), 32–33 (2018)
9. Wang, P., Liu, H., Yuan, W., et al.: Carbon nanotubes regulated hollow skeleton carbon aerogel for effective thermal insulation and favorable mechanical behavior. J. Sol-Gel. Sci. Technol. **101**(1), 193–204 (2022)

General Methodological Issues of Agrocyborgs: (From Human to Plant)

Oleg N. Gurov[1(✉)], Georgy K. Tolokonnikov[2], Vyacheslav I. Chernoivanov[2], and Andrei Yu. Alekseev[3]

[1] MMIPT University, MGIMO University, Moscow, Russian Federation
gurov-on@ranepa.ru
[2] Federal Scientific Agro-Engineering Center VIM, RASciences, Moscow, Russian Federation
[3] MIPT, GAUGN, Moscow, Russian Federation

Abstract. The choice of this topic is determined by the importance of the intellectualization of agriculture in the current conditions. "Agrocyborg" is a scientific term, which meaning is formed at the junction of biological-technical and cultural-philosophical concepts of building and functioning of a biomach system. In the original version, agrocyborg is an agricultural worker, a bearer of soil traditions, and, due to symbiosis with high-tech tools, is also an electronic personality, a representative of eHomo, an electronic human. In various conditions of ascribing vital, mental and personal private phenomena to the biomach of the agro-industrial complex (AIC), the cyborg self appears in the guise of a human cyborg, an animal cyborg and a plant cyborg. The agrocyborg project is included in the methodology for building and applying biomach systems and is funded by the triad "human-machine-living". The authors suggest specific ways of constructing and using agrocyborg in animal husbandry and crop production. This allows highlighting the unsolvable, i.e., philosophical aspects of agrocyborg project, including the problems of causal informational interactions of bio- and techno-subsystems; of trusted attribution of cognitive phenomena to various classes of agrocyborgs; and of interdisciplinary coordination.

Keywords: Cyborg · Agrocyborg · Self-agrocyborg · Agrocyborg-human · Agrocyborg-animal · Agrocyborg-plant · Agro-industrial complex · Biomach system

1 Introduction

The purpose of this work is to conceptualize the concept of an agrocyborg as a promising project of the AIC. Agrocyborg embodies the socio-humanitarian values of the traditional peasant and the physical abilities of a human, greatly enhanced by technology. In existing studies, this term is widely interpreted [1–4]. The article attempts to expand the understanding of agrocyborg as a biomach system of human, animal and plant types. The paper attempts to create a common methodological and technological basis for standardization and unification of various acrocyborg projects.

Z. Hu et al. (Eds.): AIMEE 2022, LNDECT 159, pp. 305–315, 2023.
https://doi.org/10.1007/978-3-031-24468-1_27

A number of agrocyborg projects is briefly considered. The Self-agrocyborg seems to be the most basic version of this project, because it allows to confidently connect the causal relationships of private cognitive phenomena with the functions of attached technical means. How, for example, does a shovel become my technically extended hand, acquiring the status of an object of my subjective reality? Confidential attribution of cognitive phenomena to other agrocyborgs-humans is carried out due to assessment of cognitive functions of Self-agrocyborg. The cognitive architectonics of agrocyborg-animal is even more hypothetical from the standpoint of reliably attributing real mental functions of animals to it. However, it is quite acceptable to extend the behaviorist methodology to virtual reality of the "as it were mental life" of an animal. For example, a cow with technical means of milking is considered. Even more problematic is the "as if mental life" of the plant. However, technical extensions are also possible for this version of the project, for example, in the form of an "electronic nose" that regulates the vital functions of plants. The methodological invariant for the indicated classes of agrocyborgs is considered in the theory of the biomach system and its applications.

2 Biomach Systems in the Context of the Functional Systems Theory

There are dozens of definitions for a system and a systematic approach. In our opinion, the most profound approach is the theory of functional systems developed by P.K. Anokhin and his school [5]. For the AIC, an adequate systematic approach is set by the theory of biomach systems, within which the mathematical foundations for the construction and application of biomach systems were developed within the framework of the categorical theory of systems: functional, relational, ergatic, and other types [6, 7]. In the theory of biomach system as a categorical theory of systems, the main factor is the system-forming factor that provides a holistic coverage of subsystems and elements and their constructive formation, which includes the production of all types of agricultural products, including food of animal and vegetable origin. A biomach system necessarily includes a productive living (plants, animals, biomasses). It turns out the conceptual triad "human-machine-living". The implemented technologies with AI directly affect the productive life. At least two positions should be reflected here: 1) specific patterns of systems functioning with a productive living, which are not in purely social, human-machine, socio-technical and other systems - the listed ones are too general for our purposes, they do not have an "agro-component"; 2) development of the concept of trust in these specific technologies with AI. It should be noted the applied AI technologies used in the AIC for the needs of animal husbandry and crop production, farm management, logistics and marketing. These include remote control using satellites, drones and sensors; big data collection and analysis tools; robotization; internet of things [8, 9]. As an illustrative example, let's take the products of the American company Carbon Robotics, which has released a series of farming robots that destroy weeds without harming the soil. One such robot, using the thermal energy of a laser, can destroy 100,000 weeds in an hour [10]. However, in the light of such a large-scale challenge, changing the agricultural system and becoming it in accordance with the logic of a smart factory based on the introduction of the technologies indicated above does not qualitatively distinguish the AIC from conventional production and is not enough to provide an effective basis for a breakthrough digital transformation of the economy [1]. The second important difference is the presence of solvers with elements of artificial intelligence in biomach system. The aforementioned is directly reflected in the scheme of the biomach system.

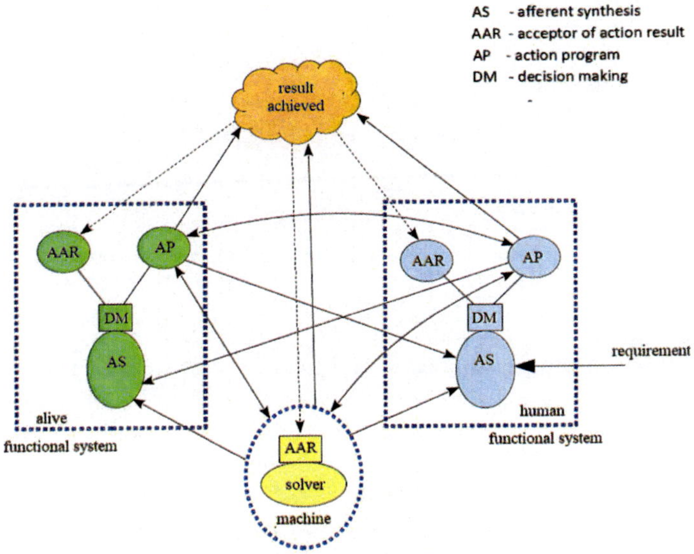

Fig. 1. Schematic diagram of biomach system

Solvers are built into the machine block of the biomach system (see Fig. 1), allowing the development of new algorithms for the machine's behavior, which are not included in it in advance by the designer (programmer). The demand for the inclusion of these components with solvers and productive living things in the agrarian system is updated by and large with the beginning and expanding in the AIC, the transition in technologies from average statistical criteria to the individual operation of machines with individual elements of living, in particular, individual animals and plants in precision farming, etc. Agrocyborg due to this model gets the following capacities:

1) to monitor in constant interaction with drones the state of the economy and collect data for subsequent analysis and prevent emergencies;
2) to participate directly in the processing and analysis of data within the framework of predictive analytics;
3) to get more qualitatively involved in "precision farming", the result of which is a multiple increase in yields [11];
4) to become a full-fledged participant in the marketplace (platform, ecosystem) for the sale and delivery of products to consumers for the purchase and sale of equipment and machinery with a break in the chain of intermediaries and their exclusion [5];
5) to ensure the breakthrough development of agricultural production with special regard for reserves, the assessment and justification [6];

6) to be a tool to achieve productive living conditions for the realization of the genetic potential laid down by breeders.

When designing agricultural machines and mechanisms, it is advisable to refer to the theory of the genome and to the methods of algebraic biology, as indicated in the diagram (Fig. 2).

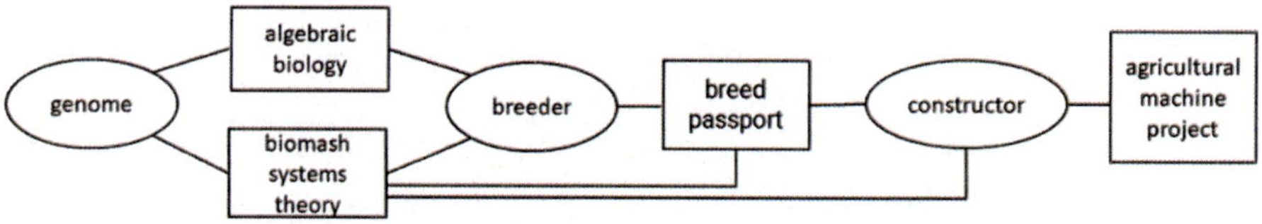

Fig. 2. Scheme from genome to agricultural machine project

In addition to such genetic foundation, the biomach system becomes a full-fledged agricultural machine when combined with plows, seeders, sprayers and other units for direct work with a productive living being. An example is sprayers operating on the principles of the "electronic nose" technology, which makes it possible to capture specific volatile substances emitted by individual plants in the event of a whitefly attack, process information and subsequently act on individual plants (insects located on them, distribution areas, etc.) to improving the health of plants. The electronic nose also successfully "sniffs out" tomato aphids, harmful insects of other crops. The information received from plants allows to pass from average spraying treatment to targeted treatment[12]. This saves the cost of the solution and prevents excessive soil contamination.

3 Agrocyborg Neural Interfaces and Chipization

As we noted above, the concept of cyborg has been actively used in scientific and mass discourse in recent decades. Cyborgization is a process of merging the bodily with the technological, combining the natural with the artificial. This is the convergence of the human and biological with the technological, the result of which already has a new nature, where the unity of consciousness with the body is supplemented by another component of this "trinity" - computer technology, which in turn supports three-dimensional (3D) semantics, linking a private cognitive phenomenon, a scientific description or an explanation of this phenomenon and a computer implementation of the phenomenon as an imitation, model or reproduction [5]. The concept of cyborg in the public sphere refers to people with artificial parts embedded (implanted) in them: dentures, pacemakers, heart valves, artificial kidneys, limb prostheses, exoskeletons and other mechanical and electronic implants built into the body. Since the last century, implants that are directly interfaced with the nervous system have been mostly common. The signals from the electrodes are amplified by the outer hair cells of the organ of Corti, transferred to the inner hair cells, from which, based on the sodium-potassium mechanism in the dendrites of the auditory nerve, action potentials are excited that propagate to the auditory part of the brain, where they are recognized [1].

The next level of cyborgization corresponds to invasive contact directly with brain neurons. In medicine (in connection with the treatment of epilepsy, Parkinson's disease, and other diseases), the technology of craniotomy has been developed; it is used for implanting an electrode grid, strip, or deep electrodes into the exposed surface of the cerebral cortex. In electrocorticography (ECoG), grids are used, as well as strips of round (plate) electrodes placed on the surface of one hemisphere. Electrode grids usually contain many current-collecting contacts that record impulses from a large surface area of the cerebral cortex. In stereotaxic EEG (SEEG), cylindrical electrodes are used, having up to a dozen current-collecting surfaces and penetrating through the skin, skull and all membranes of the brain to subcortical structures. This method provides monitoring of superficial and deep cortical structures [13]. As a result, direct neuronal electrocorticographic brain-computer interfaces arise, which, by picking up signals from brain neurons, provide control of electronic prosthetic and communication devices. They provide great opportunities for restoring functions lost due to neurological disorders in humans. In particular, ECoG signals recorded from the cerebral cortex are used as control signals for external devices (prostheses for paralysis, etc.). Work is underway to decode signals for imaginary speech, music, etc. [14]. The next stage of cyborgization technologies is the direction of the input signal to the cerebral cortex using ECoG electrodes for direct electrical stimulation. Electrical stimulation and simultaneous ECoG recording establish a bidirectional brain-computer interface. It becomes possible to control cyborgs and organize their interaction through direct neural interface communication, which may also exclude traditional sense organs.

In the Russian Federation, the "pilot" mass chipization began in 2019: ten volunteers from Tomsk State University implemented chip providing functions of a bank card, an electronic key for access to the facilities of the University, a transport card, etc.) [16]. Abroad, human chipping began in 1998 with the experiment of K. Warwick, who implanted an RFID implant under the skin - an implant with radio frequency identification. The chip allowed opening doors, turning on the light and giving voice commands [14]. The Swedish company Epicenter produces NFC chips implanted under the skin with covid-passports, NFC allows you to read data using a smartphone (the chip contains not only vaccination data, but also other documents). Tens of thousands of Swedes have implanted NFC to date [17]. From the simplest chips for reading information, a transition is already being made to the massive implantation of chips in the brain to pick up neuron signals and influence them. The technologies developed by I. Musk (Woke Studio), which use some versions of the ECoG and SEEG discussed above, are in the lead here. The chip size is 23 by 8 mm; it is implanted under the scalp and connected to the brain through holes in the skull with the thinnest filament electrodes [18]. A robot surgeon has been developed to implant the chip, the implantation operation under local anesthesia takes about an hour, and the client leaves the medical facility after the procedure on the same day [15].

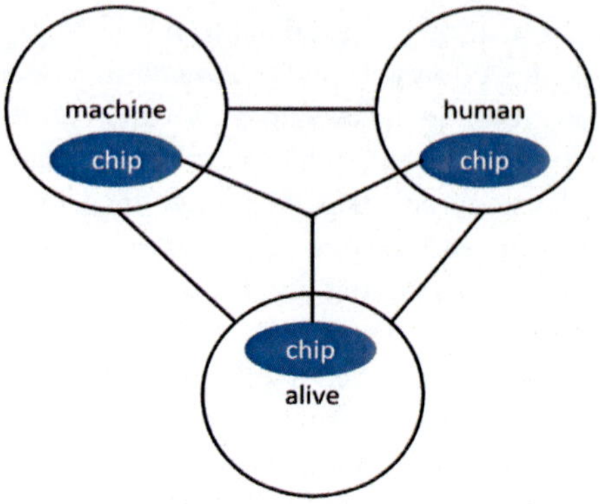

Fig. 3. Diagram of agrocyborg as a biomach system

4 Types of Agrocyborgs

The concept of agrocyborg, as noted above, is also the subject of philosophical research, shown in Fig. 3. It is unlikely that traditional questions of searching for causal relationships "psyche-brain" in the speculations of idealism, materialism, dualism, panpsychism, epiphenomenalism, etc., will be productive for cyborg building. These questions have a metaphysical status. Such a general conceptual format of the agrocyborg family is set by the functionalism paradigm of general artificial intelligence [19]. General artificial intelligence is a direction of scientific-theoretical and engineering-technological research focused on the construction and application of computer simulations, models and reproductions of cognitive phenomena of the widest range of life, mental, personal and social manifestations. In the formation of general functionalism, collective, definitive and observational approaches are distinguished. The collective approach is the collection, identification, coordination, formalization, systematization, unification, codification of all kinds of functionalist theories. The determinative approach is the analysis and identification of the main functionalist characteristics, relationships, patterns, causalities that are invariant with respect to the content of cognitive phenomena. The observational approach allows us to evaluate from the position of a human or a social community, immersed in the communicative "waves" of the virtual and real world, the various statuses of technological implementations of general functionalism. And paying tribute to the theory of systems, the role of a system-forming factor in the family of agrocyborgs with their various functions, characteristics and abilities, is Self-agrocyborg.

Self-agrocyborg: The concept of a human cyborg was voiced by L.V. Poskotinova (D.B.S., Ph.D.) in 2018. According to the speaker, the intellectualization of neurobiotechnologies contributes to software and hardware implementation of biofeedback by modeling the functions of the central nervous system, higher mental functions and the functions of visceral systems (cardiovascular, respiratory, gastroenteric and etc.). The praxeological question is valid. How ready is human to trust a machine to broadcast his own bodily, visceral sensations? If the machine is used only as a recording device, then human can trust the machine and follow the indicators that it presents to a human. But if the machine has its own goals for using human health indicators, then human risks receiving false feedback from indicators that may be far from his own. The question arises of testing equipment used for biofeedback for AI. At the heart of the methodology of such testing, it is preferable to combine the principles of introspection and self-examination (for example, analysis of frequency, rhythm, and fullness of the pulse). The Comprehensive Turing test [20] contains several introspective tests. They connect the phenomena of subjective reality, studied by the subject of this reality, with the results of computer simulation of these phenomena. The neurovisceral test concretizes these abstract tests in relation to the most important area, more important than the human mind - his health. There are about a hundred major versions of Turing test. Among them, Poskotinova's neurovisceral test seems to be very original and significant [21]. Self-cyborg is formed as follows. A comprehensive Turing test is being studied to select the necessary cognitive-computer competencies [20]. These competencies are formed from a list of values of a definite function: can the biomach system communicate; think; reason reasonably; function like a mentally healthy individual; understand; have a subconscious experience emotion, attribute mentality; create; self-formalize; self-organize; be indistinguishable from a human, but also not be a zombie; to have a different (foreign) consciousness; process ideas. In other words, can a biomach system do everything? This is what private Turing tests suit for.

Agrocyborg–Human (Not-self, Other, Alien): This is a biomach system, to which I attribute cognitive functions by analogy with my cognitive functions, and a cyborg, to which I attribute cognitive states and the content of these states, by analogy with my cognitive states and the content of these states. Since we have adopted the concept of general AI functionalism, which is based on the machine functionalism of H. Putnam, then all the arguments and objections to attributing my cognitive functions are fully consistent with the intersubjective methodology: to perceive the pain of another means to be in the same state of pain in which I find myself I am in conditions of certain pain stimulation and reactions, as well as an appropriate program of action with my biomach.

Attributing cognitive phenomena to an animal is much more problematic than the case of attributing them to another human. There are conflicting views on the attribution of cognitive phenomena to animals presented in the format of robots. Thus, J. Searle believes that "humans naturally attribute intentionality, for example, to primates or domestic animals, respectively, to monkeys or dogs. Roughly speaking, there are two reasons for this: without such an attribution, it is impossible to understand the intentions of animals, and it is obvious that animals are somewhat similar to humans (this is their eye, this is their nose, but this is skin, etc.). The coherence of animal behavior to human behavior, as well as the belief in the unity of nature, are justifications for a number of assumptions: the presence of mental states in animals; conditioning their behavior; producing the mental states of animals by the same mechanisms that generate the mental states of humans. Similar assumptions apply to the robot. However, we do not recognize them, because we know that the robot's behavior is the result of the work of a formal program and that the robot does not have that physical substance, on the basis of which real causal dependencies are generated. Therefore, the robot should be denied intentionality" [18]. Another cognitive philosopher, A. Solman, thinks differently. In the light of the analysis of the intersubjective phenomena of projective consciousness, he asks: "What is it like to be a stone, a sunflower, a bat, a human baby, an Alzheimer's patient, a down, a blind man, a female (for a male), a male (for a female)?" [22]. Let us emphasize that such questions directly correspond to the theme of attributing cognitive phenomena to Self-agrocyborg, agrocyborg-Other, agrocyborg-plant. In answering any of these questions, an incredible array of epistemological problems arises. And the general answer to the question "What is it like to be X?" the answer will be "I don't know". But to answer the question "What is it like to be a robot?" it is not difficult for a robot developer: with a completely known functional organization of the robot, what will happen to the robot in such and such conditions is known. But the developer thoroughly knows not only stimuli and reactions. After all, he is fully aware of the internal states of the robot: information structure, semantics, transformations, and ways of using elements [22]. Therefore, the answer to the question of what it is like to be a cyborg-animal or a cyborg-plant is very transparent: it is thoroughly known what it is like to be.

Agrocyborg-Animal (Agrocyborg Cow): This is a "milker-cow-milking machine" biomach system developed at VIM [23, 24]. For specialists, the problems with mastitis of cows milked by common milking machines are well known. One of the disadvantages of milking, which provokes mastitis, is that the vacuum of the milking machine acts on all parts of the udder, it does not take into account that in different parts the return of milk stops at different times. Consideration of physiology shows that the commonly used methods of milking using traditional milking machines based on average statistical data on animals are very far from the process of giving milk to the calf established by nature. This is influenced by both smells and sounds and functional systems formed in the body of a cow in the process of bearing a fetus. The number of hormones alone involved in the process of milk ejection reaches almost two dozen, among the hormones, oxytocin is the main one. According to the theory of lactation, as a result of mechanical and thermal effects on the udder receptors, nerve excitations occur that reach the spinal cord and pituitary gland, which releases oxytocin into the blood. The necessary concentration of oxytocin in the blood for milk flow lasts from 2 to 5 min. The destruction of oxytocin is accompanied by the cessation of the milk ejection reflex. One of the objectives of the milking agrocyborg is thus to maintain an appropriate level of oxytocin concentration in the animal's blood for milking. This requires an oxytocin sensor and a mechanism for influencing the cow's body. The sensor has not yet been developed, so concentration is determined by indirect signs, but stimulation, including the release of oxytocin, is carried out by the device developed at VIM (resonant frequency generator - RFG, see patent [25]). A special applicator has been developed for the cow to wear the RFG; if necessary, it is possible to pair the RFG with sensors and chips implanted with the cow. Traditional machine milking is perceived by animals as a stress factor, while the physiological reaction of the animal leads to an increased release of leukocytes, which are the main mass of somatic cells in milk, which worsens its quality. The specified mode of RFG allows reducing the release of leukocytes, which has a significant effect on improving the quality of milk. Agrocyborg-cow with RFG also mastered the function of suppressing the pathogens of the above-mentioned mastitis, which is already successfully used for the treatment and prevention of this disease [23–25].

Agrocyborg-plant: Agrocyborg plants as a living component in biomach system are also necessary and promising. In recent years, the concept of plants as intelligent biosystems has been developing. Unlike animals, the system properties of plants have not been sufficiently studied; in particular, the development of the theory of plant functional systems is just beginning [3]. Action potentials were found in plants (similar to electrical nerve signals in animals) [26–29], analogs of the nervous system, in which action potentials serve to transfer information, a new science of plant neurobiology, plant biosemiotics, has arisen [30]. The technology of the electronic nose as a sensor of plant state parameters was discussed above, but the electrical activity of plants plays no less a role in their functioning than in animals. Thus, an agrocyborg-plant does not fundamentally differ from the examples of agrocyborg animals considered above. It is obvious, however, that the degree of adequacy of attributing quasi-mental phenomena to plants is very low.

5 Conclusion

In connection with the development of cyborgs for humans, animals and plants, biomach systems arise in which the "machine" subsystem is built into the human, animal, and plant subsystems. As a result, an agrocyborg appears as an element of the biomach system. In this context, agrocyborg is a viable concept that can be put into practice [31]. It generates new opportunities due to the achievements of technological progress in such an important area for the survival and development of human civilization as the AIC. The study presents sections of the theory of biomach systems related to the concept of agrocyborg. The paper presents a philosophical understanding of cyborgs and agrocyborgs based on the dynamic boundaries of "human/technical", which makes it possible to give the scientific content of the term "agrocyborg" both biological-technical and philosophical sounding. Results can stimulate the development of the AIC, the electronics industry, artificial intelligence and informatics, and also contribute to the formation of a promising field of interdisciplinary scientific and technical, natural sciences and socio-humanitarian research. The unification and standardization of biomach system that develops Self-, Other, animal- and plant-agrocyborgs, is effective for the methodology of test functionalism. Unlike machine functionalism, which subordinates cognitive functions to algorithms of the Turing machine, the presented approach liberates the agricultural worker. "Agrohuman" is able to solve not only narrow technological problems (like increasing animal weight and milk yield), but also philosophical questions: whether a biomach as a cognitive computer system is able to exist, progress, be creative, behave morally, make friends, etc. We assume that the presented position is promising both for further interdisciplinary research and for the formation of functional models in real conditions, when the digital transformation and intellectualization of AIC is a crucial demand [32].

References

1. Gurov, O.N., Kon'kova, T.A.: Conceptual model of an agro-industrial cyborg. Artif. Soc. **16**(3) (2021)
2. Gurov, O.N.: Conceptual model of agro-industrial cyborg. In: Hu, Z., Wang, B., Petoukhov, S., He, M. (eds.) AIPE 2021. LNDECT, vol. 119, pp. 1-13. Springer, Cham (2022). https://doi.org/10.1007/978-3-030-97064-2_1

3. Chernoivanov, V.I., Tolokonnikov, G.K., Shogenov, Yu.Kh., Dorokhov, A.S.: Biomach systems, artificial intelligence and agrocyborgs. In: Trukhachev, V.I., Didmanidze, O.N. (eds.) Readings of Academician V.N. Boltinsky. Digest of articles. M.: Izdatel'stvo «Sam poligrafist», pp. 40–46 (2022)
4. Gurov, O.N.: Boundaries of "human" and "technical" (on the example of D. Cronenberg's work): dis. candidate of philosophical sciences: 09.00.13. - Moscow, 191 p. (2022)
5. Anokhin, P.K.: Principles of systemic organization of functions, M.: Nauka, 315 p. (1973)
6. Chernoivanov, V.I.: Biomashsystems: emergence, development and prospects. Biomashsystems **1**(1), 7–58 (2017)
7. Tolokonnikov, G.K.: Informal categorical theory of systems. Biomashsystems **2**(4), 41–144 (2018)
8. Saida, A., Yadav, R.K.: Review on: analysis of an IoT based blockchain technology. Int. J. Educ. Manag. Eng. (IJEME) **12**(2), 30–37 (2022). https://doi.org/10.5815/ijeme.2022.02.0
9. Kumar, K.A., Aju, D.: An internet of thing based Agribot (IOT-Agribot) for precision agriculture and farm monitoring. Int. J. Educ. Manag. Eng. (IJEME) **10**(4), 33–39 (2020). https://doi.org/10.5815/ijeme.2020.04.04
10. Cui, S., Cao, L., Acosta, N., Zhu, H., Ling, P.P.: Development of portable e-nose system for fast diagnosis of whitefly infestation in tomato plant in greenhouse. Chemosensors **9**: 297 (2021)
11. Babatunde, G., Emmanuel, A.A., Oluwaseun, O.R., Bunmi, O.B., Precious, A.E.: Impact of climatic change on agricultural product yield using K-means and multiple linear regressions. Int. J. Educ. Manag. Eng. (IJEME) (2019). https://doi.org/10.5815/ijeme.2019.03.02
12. Alekseyev, A.Yu.: Volumetric (3D) intensional semantics of the dictionary of an artificial society. Artif. Soc. **8**(1–4) (2013)
13. Shchekut'yev, G.A.: Neuromonitoring: general principles and applied methods. In: Zhukov, P.V. (ed.) Neurophysiological Studies in the Clinic. Collection of Materials of the Conference in Memory of Prof. G.A. Shchekutev, M.: Antidor, pp. 208–216 (2001)
14. Korzinov, N.: Microchip under the skin: human tuning. Popular Mech. **4**, 148–151 (2006)
15. Awadalla, M.H.A.: Spiking neural network and bull genetic algorithm for active vibration control. IJISA **10**(2), 17–26 (2018)
16. The vice-rector of a Tomsk university implants an electronic chip in himself as part of an experiment. https://nauka.tass.ru/nauka/5977697. Accessed 10 June 2022
17. In Sweden, they began to implant chips with COVID-passports into the human body. https://rg.ru/2021/12/19/v-shvecii-nachali-vzhivliat-v-telo-cheloveka-chipy-s-kovid-pasportami.html. Accessed 10 June 2022
18. Serl' Dzh.R.: Minds, brains, programs. In: Alekseev, A.Yu. (ed.) Turing Test. Robots. Zombie. M.: MIEM, pp. 6–20 (2006)
19. Alekseyev, A.Yu.: The general functionalist concept of artificial need as the basis of general artificial intelligence. Philos. Sci. **62**(11), 111–124 (2019). https://doi.org/10.30727/0235-1188-2019-62-11-111-124
20. Alekseyev, A.Yu.: Comprehensive Turing test: philosophical-methodological and socio-cultural aspects. M.: AInteLL, 304 p. (2013)
21. Poskotinova, L.V.: Neurovisceral Turing test: the need for development, methodological aspects. In: Reports at the Seminar "Neurophilosophy". Faculty of Philosophy, Moscow State University, 5 December 2018. https://scmai.ru/2018/12/05/. Accessed 10 June 2022
22. Slomen, A.: What does it mean to be a stone? In: Alekseev, A.Yu. (ed.) Turing Test. Robots. Zombie. M.: MIEM, pp. 86–102 (2006)
23. Lyubimov, V.T., Romanov, D.V., Tsoi, Yu.A., Ziganshin, B.G., Sitdikov, F.F.: Results of application of frequency resonance therapy for treatment of cow mastitis. In: BIO Web of Conferences, vol. 17, p. 00254 (2020). https://doi.org/10.1051/bioconf/20201700254

24. Lyubimov, V.Ye., Romanov, D.V.: Technical and technological justification for solving the problem of treating mastitis in cows using frequency-resonance therapy in a dairy farm. Biomashsystems **2**(1), 225–235 (2018)
25. Patent for invention: Romanov D.V., Lyubimov V.E. Method and electromechanical device for the prevention and treatment of udder diseases, stimulation of lactation in cows, Application № 2020127061/10(047622), 12 August 2020
26. Mironova, E.A., Shogenov, Y., Moiseenkova, V., Romanovsky, Y.: Bioelectric responses of plants to the low-intensive irradiation in the visible and infrared ranges. SPIE **3732**, 349–352 (1998)
27. Opritov, V.A., Pyatigin, S.S., Retivin, V.G.: Bioelectrogenesis in higher plants. M., Nauka, 213 p. (1991)
28. Vasilyev, V.A., Garkusha, I.V., Petrov, V.A., Romanovskii, Y., Shogenov, Y.: Light induced electrical activity of green plants. Biophysics **48**(4), 662–671 (2003)
29. Zimmermann, M.: Candidates for systemic signals in higher plants and the challenge of their identification. Plant Physiol. **170**(4), 407–419 (2016)
30. Stepanov, S.A.: Nervous system of plants: hypotheses and facts. Bul. Bot.garden Sarat. State University, **15**(4), 31–56 (2017)
31. Brenner, E.D.: Plant neurobiology: an integrated view of plant signaling. Trends Plant Sci. **11**(8), 413–419 (2006)
32. Vani, P.D., Rao, K.R.: Implementation of smart agriculture using CloudIoT and its geotagging on android platform. Int. J. Eng. Manuf. (IJEM) **9**(2), 43–53 (2019). https://doi.org/10.5815/ijem.2019.02.04

Biological Dualism Stochastics-Determinism and Its Modeling by Tensor-Unitary Transformations

Sergey V. Petoukhov[1,2(✉)]

[1] Mechanical Engineering Research Institute, Russian Academy of Sciences, Moscow, Russia
spetoukhov@gmail.com
[2] Moscow State Tchaikovsky Conservatory, Moscow, Russia

Abstract. His article is devoted to the algebraic modeling of interrelations between stochastic molecular processes and inherited deterministic structures in biological bodies. Using universal rules of stochastic organization of DNAs in genomes of higher and lower organisms, the author describes that structural feature of these genomic stochastics are connected with algebraic operations called “tensor-unitary transformations”. These operations allow algebraic modeling of the biological dualism “stochastics-and-determinism”, physiological Gestalt-phenomena, and some other bio-logical phenomena, which were not been algebraically modelled early. Some thoughts about features of tensor-unitary transformations and their possible profits for applications in algebraic biology and understanding of genetic intelligence are presented. It is supposed that they are useful for the development of quantum biomechanics to model biosimilar quantum systems capable of tensor self-development with the inheritance of memory of past states.

Keywords: Genomic DNA · Stochastic processes · Tensor product · Quantum informatics · Unitary transformation · Gestalt phenomena · Genetic intelligence

1 Introduction

Genetics as a science began with Mendel’s discovery of the stochastic rules of inheritance of traits in experiments on the crossing of organisms. Many processes in living bodies are of a stochastic nature. The well-known expressions “gene noise” or “cell noise” reflect the fact that even genetically identical cells within the same tissue exhibit different levels of protein expression, different sizes, and structures due to the stochastic nature of interactions of individual molecules in cells. Amazingly, this stochastics doesn’t prevent the genetic inheritance of parental traits to offspring. According to Mendel’s law of independent inheritance of traits, information from the level of DNA molecules dictates the macrostructure of living bodies through many independent channels, despite strong noises and stochastic interrelations of individual molecules. Thus, hair, eye, and skin colors are inherited independently of each other.

Z. Hu et al. (Eds.): AIMEE 2022, LNDECT 159, pp. 316–326, 2023.
https://doi.org/10.1007/978-3-031-24468-1_28

Accordingly, each organism is a machine of multichannel noise-immunity encoding in conditions of the biological dualism "stochastics-and-determinism". This dualism is a challenge for science and should be deeply studied. But, as the author can judge, no publications are known till now about algebraic modeling this dualism till now. This article presents results of such modeling in the important biological field by using new algebraic operations connected with traditional operations of quantum mechanics and informatics.

The presented topic is connected with inborn phenomena of Gestalt-biology (psychological, morphogenetic, motoric, homeostatic, sensory, etc.), in which stable holistic patterns are formed regardless of the wide variability of their constituent components. For example, Gestalt psychology studies the innate properties of the brain to form holistic images that are relatively independent of their particular components. This is an inherited fundamental property of the psyche: to seek in a disparate whole. One of many examples of the connection between stochastic and deterministic features in biology is given by the phenomenon of the biomechanics of movements by the classic of biomechanics N.A. Bernstein: the general target task of the movement is performed exactly regardless of the inaccuracies of its constituent motor subtasks [1, 2].

All physiological systems are genetically inherited through their genetic coding. Therefore, one should look for the origins of these inherited gestalt phenomena of physiology in the genetic system of DNA informatics. In the course of such searches, the author discovered that already in the information system of DNA molecules of the genomes of higher and lower organisms, universal gestalt phenomena of their stochastic organization are embedded [3, 4].

Let us briefly remind you of these rules. Genetic information is recorded in DNA molecules using long sequences of four nucleobases - adenine A, thymine T, cytosine C, and guanine G - and their combinations forming alphabets of 16 duplets, 64 triplets, 256 tetraplets, etc. These alphabets of 4 nucleobases, 16 duplets, etc. can be represented in the forms of members of the whole tensor family of matrices [5, 6].

Each of the long single-stranded DNAs in genomes of eukaryotes and prokaryotes can be represented as a sequence of single nucleobases (like as T-A-G-G-C-...), which contains certain percentages of each of the 4 nucleobases A, T, C, and G. The same DNA sequence can be represented as a sequence of paired nucleobases (like TA-GG-CA-...), which contains certain percentages of each of the members of the alphabet of 16 duplets. The same DNA sequence can be also represented as a sequence of triplets (like TAG-GCA-...), which contains certain percentages of each of the members of the alphabet of 64 triplets, etc. Calculation of the percentages of 4 nucleobases, 16 duplets, 64 triplets, etc. in the named representation of each concrete genomic DNA gives corresponding sets of percentages of members of these DNA alphabets. At first glance, these sets of probabilities of n-pelts are quite chaotic. But in this seeming chaos, there are many universal rules for n-pelt groupings that are valid for all studied genomes: special groupings of probabilities in these different sets are numerically interrelated by their sums although quantities of summands and percent values of separate summands in these sums are very different (it is a genomic type of gestalt phenomena since the equality of these sums exists under the wide variability of their summands). Moreover, in each of the studied genomic DNAs, these sets of probabilities are interrelated not only numerically but also algebraically using special systems of the tensor (Kronecker) products using stochastic components. In these phenomenological rules, it was revealed that nature for some reason uses in genomic stochastics an algebraic operation, which is called by the author a tensor-unitary transformation and whose properties are described and discussed in this article. It should be added that the author didn't meet such systems of algebraic operations, which are based on tensor-unitary transformations, in other fields of mathematical natural sciences.

Why universal stochastic rules of genomic DNAs are connected with tensor-unitary transformations? What mathematical features are provided by these transformations? Could they be used for modeling the biological gestalt phenomena and the dualism "stochastics-and-determinism"? Why living nature prefers the usage of tensor-unitary transformations in genetic systems?

The main aim of this article is to study these questions and present the initial author's results and thoughts about appropriate algebraic tools for modeling the important biological phenomena, which were not been algebraically modeled early but now are been including in the field of algebraic biology.

2 Research Methods

In this study, methods of tensor-matrix analysis and modeling of stochastic organization of genomic DNAs are used jointly with formalisms of quantum informatics and multidimensional hypercomplex numbers. In particular, this article relies on the terminology of the classic book [7] and uses the generally accepted Dirac notations:

- $|\psi>$ Means a vector, which is known also as a ket-vector;
- $<\psi|$ means a vector, which is dual to $|\psi>$ And known as a bra-vector;
- $<\varphi|\psi>$ means scalar (or inner) product between the vectors $<\varphi|$ and $|\psi>$;
- $|\varphi>\otimes|\psi>$ means the tensor product of $|\varphi>$ and $|\psi>$

In quantum informatics, a system of "n" qubits is considered. The computational basis states of this system are written in the form $|x_1x_2....x_n>$; a quantum state of such a system is specified by 2^n amplitudes [7, p. 17]. In our model approach, we in general interpret each of the above-mentioned n-plet sequences of genomic DNAs as a multi-qubit quantum system.

Correspondingly, the representation of any genomic DNA sequences as a sequence of 4 nucleobases C, A, T, and G, whose probabilities in this DNA are denoted by %C, %A, %T, and %G, gives the following 2-qubit expression (1):

$$\left|\psi_4\right>= (\%\ C)^{0.5}\left|00\right> +(\%\ A)^{0.5}\left|01\right> +(\%\ T)^{0.5}\left|10\right> +(\%\ G)^{0.5}\left|11\right> \quad (1)$$

The representation of any genomic DNA sequences as a sequence of 16 duplets CC, CA, CT, CG, AC, ..., whose probabilities in this DNA are denoted by %CC, %CA, %CT, %CG, %AC, ..., gives the following 4-qubit expression (2):

$$\begin{aligned}\left|\psi_{16}\right>= &(\%\ CC)^{0.5}\left|0000\right> +(\%\ CA)^{0.5}\left|0001\right> +(\%\ AC)^{0.5}\left|0010\right> +(\%\ AA)^{0.5}\left|0011\right> +\\ &+(\%\ CT)^{0.5}\left|0100\right> +(\%\ CG)^{0.5}\left|0101\right> +(\%\ AT)^{0.5}\left|0110\right> +(\%\ AG)^{0.5}\left|0111\right> +\\ &+(\%\ TC)^{0.5}\left|1000\right> +(\%\ TA)^{0.5}\left|1001\right> +(\%\ GC)^{0.5}\left|1010\right> +(\%\ GA)^{0.5}\left|1011\right> +\\ &+(\%\ TT)^{0.5}\left|1100\right> +(\%\ TG)^{0.5}\left[1101\right> +(\%\ GT)^{0.5}\left[1110\right> +(\%\ GG)^{0.5}\left|1111\right>\end{aligned} \quad (2)$$

Similar 2n-qubit expressions are possible for the representations of genomic DNAs as sequences of 64 triplets, 256 tetraplets, etc.

In the expressions (1) and (2), the correspondences between symbols of the computational basis states and probability amplitudes are not occasional but are based on the molecular binary-oppositional indicators of 4 nucleobases. Each of them is uniquely determined by its belonging to one or another part of the two molecular binary oppositions: "purine-or-pyrimidine" and "amino-or-keto". In this relation, nucleobase C is defined by the pair of indicators "amino and pyrimidine" and assigned to $|00>$ in (1); nucleobase A is defined by the pair of indicators "amino and purine" and assigned to $|01>$; the nucleobase T is defined by the pair of indicators "keto and pyrimidine" and assigned to $|10>$; the nucleobase G is defined by the pair of indicators "keto and purine" and assigned to $|11>$ (see more details in [3–6]).

In quantum mechanics and quantum informatics, unitary transformations have fundamental meaning: the evolution of closed quantum systems is described by unitary transformations, and all calculations in quantum informatics are based on unitary operators acting as quantum gates. Unitary transformations (they are orthogonal transformations in the case of real components) of vectors preserve their lengths and also preserve the inner products.

3 Results

The above-mentioned 2n-qubit expressions (1), (2), and their analogs for the cases of 64 triplets, 256 tetraplets, etc. for all of the wide set of genomic DNAs studied by the author are interconnected in a special algebraic manner. In the short article, these universal algebraic connections are described only briefly to attract attention to them using an example of probability data regarding the single-stranded DNA sequence of human chromosome #1, which contains about 250 million nucleotides and is presented in the GenBank: https://www.ncbi.nlm.nih.gov/nuccore/NC_000001.11 (more data are in the preprint [8]).

For human chromosome #1, in the expressions (1) and (2) have the following appearances (3) and (4) (here and below, the magnitudes of the probability amplitudes and the probabilities themselves are rounded to the fourth decimal place):

$$|\psi_4>= 0.2085^{0.5}|00> +0.2910^{0.5}|01> +0.2918^{0.5}|10> +0.2087^{0.5}11> \quad (3)$$

$$\begin{aligned}|\psi_{16}>&= 0.0541^{0.5}|0000> +0.0727^{0.5}|0001> +0.0503^{0.5}|0010> +0.0950^{0.5}|0011> +\\ &+0.0713^{0.5}|0100> +0.0103^{0.5}|0101> +0.0743^{0.5}|0110> +0.0714^{0.5}|0111> +\\ &+0.0601^{0.5}|1000> +0.0631^{0.5}|1001> +0.0440^{0.5}|1010> +0.0601^{0.5}|1011> +\\ &+0.0957^{0.5}|1100> +0.0729^{0.5}|1101> +0.0505^{0.5}|1110> +0.0542^{0.5}|1111>\end{aligned} \quad (4)$$

How phenomenological data in expressions (3) and (4) are interrelated? Is there a meaningful algebraic transformation capable of transforming the 2-qubit state $|\psi_4>$ (3) into the 4-qubit state $|\psi_{16}>$ (4)? Yes, there is and it is based on the tensor use of additional 2-qubit expressions (5–8):

$$|\psi_{BC}>= 0.2594^{0.5}|00> +0.3489^{0.5}|01> +0.3422^{0.5}|10> +0.0494^{0.5}|11>, \quad (5)$$

$$\left|\psi_{BA}\right> = 0.1729^{0.5}\left|00\right> + 0.3266^{0.5}\left|01\right> + 0.2553^{0.5}\left|10\right> + 0.2452^{0.5}\left|11\right>, \quad (6)$$

$$\left|\psi_{BT}\right> = 0.2059^{0.5}\left|00\right> + 0.2164^{0.5}\left|01\right> + 0.3279^{0.5}\left|10\right> + 0.2497^{0.5}\left|11\right>, \quad (7)$$

$$\left|\psi_{BG}\right> = 0.2109^{0.5}\left|00\right> + 0.2878^{0.5}\left|01\right> + 0.2417^{0.5}\left|10\right> + 0.2596^{0.5}\left|11\right>, \quad (8)$$

Really, let us take the 2-qubit $|\psi_4>$ (3) and tensor multiplicate each of its 4 computational basis states by a corresponding 2-qubit from (5–8); in the result, the 4-qubit state (4) arises:

$$\begin{aligned}&\left\{(0.2085)^{0.5}|00> \otimes|\psi_{BC}>\right\} + \left\{(0.2910)^{0.5}|01> \otimes|\psi_{BA}>\right\}+\\ &+\left\{(0.2918)^{0.5}|10> \otimes|\psi_{BT}>\right\} + \left\{(0.2087)^{0.5}|11> \otimes|\psi_{BG}>\right\} = |\psi_{16}>\end{aligned} \quad (9)$$

The described example with numeric data of the human chromosome #1 corresponds to the following general expression (10), which is appropriate for interconnections between $|\psi_4>$ (1) and $|\psi_{16}>$ (2) in genomic DNAs of different eukaryotes and prokaryotes:

$$\begin{aligned}
&(\%\,C)^{0.5}|00> \otimes\{(\%\,CC/\%\,C)^{0.5}|00> + (\%\,CA/\%\,C)^{0.5}|01> +\\
&\quad +(\%\,CT/\%\,C)^{0.5}|10> + (\%\,CG/\%\,C)^{0.5}|11>\}+\\
&+(\%\,A)^{0.5}|01> \otimes\{(\%\,AC/\%\,A)^{0.5}|00> + (\%\,AA/\%\,A)^{0.5}|01> +\\
&\quad +(\%\,AT/\%\,A)^{0.5}|10> + (\%\,AG/\%\,A)^{0.5}|11>\}+\\
&+(\%\,T)^{0.5}|10> \otimes\{(\%\,TC/\%\,T)^{0.5}|00> + (\%\,TA/\%\,T)^{0.5}|01> +\\
&\quad +(\%\,TT/\%\,T)^{0.5}|10> + (\%\,TG/\%\,T)^{0.5}|11>\}+\\
&+(\%\,G)^{0.5}|11> \otimes\{(\%\,GC/\%\,G)^{0.5}|00> + (\%\,GA/\%\,G)^{0.5}|01> +\\
&\quad +(\%\,GT/\%\,G)^{0.5}|10> + (\%\,GG/\%\,G)^{0.5}|11>\} =\\
&= (\%\,CC^{0.5}|0000> + \%\,CA^{0.5}|0001> + \%\,CT^{0.5}|0010> + \%\,CG^{0.5}|0011>)+\\
&+(\%\,AC^{0.5}|0100> + \%\,AA^{0.5}|0101> + \%\,AT^{0.5}|0110> + \%\,AG^{0.5}|0111>)+\\
&+\left(\%\,TC^{0.5}\left|1000\right> + \%\,TA^{0.5}\left|1001\right> + \%\,TT^{0.5}\left|1010\right> + \%\,TG^{0.5}\left|1011\right>\right)+\\
&+\left(\%\,GC^{0.5}\left|1100\right> + \%\,GA^{0.5}\left|1101\right> + \%\,GT^{0.5}\left|1110\right> + \%\,GG^{0.5}\left|1111\right>\right).
\end{aligned} \quad (10)$$

An unexpected and important phenomenological fact is that, in different genomic DNAs, the following sums of probabilities (11), which participate in (10), are equal to 1 with a high-level accuracy:

$$\begin{aligned}
\%\,CC/\%\,C + \%\,CA/\%\,C + \%\,CT/\%\,C + \%\,CG/\%\,C = 1\\
\%\,AC/\%\,A + \%\,AA/\%\,A + \%\,AT/\%\,A + \%\,AG/\%\,A = 1\\
\%\,TC/\%\,T + \%\,TA/\%\,T + \%\,TT/\%\,T + \%\,TG/\%\,T = 1\\
\%\,GC/\%\,G + \%\,GA/\%\,G + \%\,GT/\%\,G + \%\,GG/\%\,G = 1
\end{aligned} \quad (11)$$

It should be noted that, for genomic DNAs, similar tensor interconnections exist between quantum information representations of sets of probabilities of 16 duplets and 64 triplets, and also 64 triplets and 256 tetraplets, etc. (see the preprint [8]).

Multi-qubit expressions (1) and (2) can be matched with the appropriate vectors, whose coordinates are the probability amplitudes from these multi-qubits. For example, 2-qubit state (1) can be matched with the 4-dimensional vector

$$\mathbf{V_4} = \left[\%\,C^{0.5},\ \%\,A^{0.5},\ \%\,T^{0.5},\ \%\,G^{0.5}\right].$$

Besides this, the operation (9) with multi-qubits is matched - in the case of vectors - with the tensor product of each of the vector coordinates with an individual vector, whose sum of coordinates in the square is equal to 1 (qubit-like peculiarities). Such a coordinate-wise tensor multiplication of an arbitrary vector by qubit-like vectors generates a vector of tensorial increased dimension with the same length. Similar transformations are called "tensor-unitary transformations" by the author.

In general, tensor-unitary transformations are defined as transformations that preserve the norms (lengths) of vectors when they are tensor transformed into vectors of a space of increased dimension. Their above-described particular cases are presented in the universal rules of the stochastic organization of genomic DNAs.

Unlike unitary transformations, which transform the original n-dimensional vector into an n-dimensional vector of the same dimension, tensor-unitary transformations transform the original ("mother") n-dimensional vector into a "daughter" m-dimensional vector of increased dimension (m > n) with preserving the length of the mother vector. One can say that they provide the "inheritance" of the vector length into a vector space of tensorial increased dimension, which can be interpreted as the configuration space of a multiparameter system, which develops with an increasing number of parameters, by analogy with a biosystem under its ontogenetic growth.

Applying these tensor-unitary transformations to vectors uses a two-step process. First, the original n-dimensional vector, called the mother vector, is represented as the sum of all its basis vectors with their coordinate weights (for example, the vector [x, y] is represented as [x, 0] + [0, y]). Then each of these weight basis vectors is tensor-multiplied by the so-called "normvector" (qubit-like vector) ensuring the preservation of the length of the mother vector in the daughter vector of increased dimension. By definition, a normvector is a k-dimensional vector $[\alpha_0, \alpha_1, \alpha_2, \ldots, \alpha_{k\text{-}1}]$, whose sum of squares of coordinates is equal to one: $\alpha_0{}^2 + \alpha_1{}^2 + \alpha_2{}^2 + \ldots + \alpha_{k\text{-}1}{}^2 = 1$. Its coordinates can be considered as probability amplitudes.

This "qubit-like" property of normvectors introduces an element of probability (stochasticity) into the values of individual coordinates of the daughter multi-dimensional vector, whose special groupings of coordinates turn out to be deterministic memory carriers of the coordinates of the mother vector. Thus, tensor-unitary transformations, being stochastic-deterministic transformations, generate stochastic-deterministic vectors and make it possible to model the biological dualism "stochastics-determinism". Let us explain this with simple examples.

One can consider a simplest mother vector [x(t), y(t)] = [x(t), 0] + [0, y(t)] and tensor multiplicate its first weighted basis vector [x(t), 0] with the normvector $[\alpha_0, \alpha_1]$, and its second weighted basis vector [0, y(t)] with the normvector $[\beta_0, \beta_1]$ (here $\alpha_0^2 + \alpha_1^2 = 1$, and $\beta_0^2 + \beta_1^2 = 1$):

$$\mathbf{D}_1 = [x(t), 0] \otimes [\alpha_0, \alpha_1] + [0, y(t)] \otimes [\beta_0, \beta_1] = \left[x(t)\alpha_0, x(t)\alpha_1, y(t)\beta_0, y(t)\beta_1\right] \tag{12}$$

The values of the individual coordinates of the daughter vector $\mathbf{D_1}$ (12) are stochastic since each of them contains one of the probability amplitudes α_0, α_1, β_0, β_1 (there exist infinitely many options for the values of α_0, α_1, β_0, β_1, and their random selection changes the values of these individual coordinates but without changing the deterministic numerical relationships between special groupings of the coordinates of the daughter and mother vectors, which are described below).

Calculate the length $\|\mathbf{D_1}\|$ of the vector (12):

$$\|\mathbf{D}_1\| = \left\{x^2\alpha_0^2 + x^2\alpha_1^2 + y^2\beta_0^2 + y^2\beta_1^2\right\}^{0.5} = \left\{x^2\left(\alpha_0^2 + \alpha_1^2\right) + y^2\left(\beta_0^2 + \beta_1^2\right)\right\}^{0.5} = \left(x(t)^2 + y(t)^2\right)^{0.5} \tag{13}$$

This length (13) of the daughter 4-dimensional vector D1 is equal to the length of the mother 2-dimensional vector [x(t), y(t)] at any fixed time t. In other words, the daughter vector D1, having stochastic coordinates and increasing dimensionality, carries on an exact "memory" of the length of the mother vector at any fixed moment of time, regardless of the stochasticity of its individual coordinates, and in this respect is a deterministic entity. You now also note other deterministic features of this daughter's 4-dimensional vector, which are connected with its projections on coordinate planes.

A 4-dimensional vector space with a Cartesian system of numbered coordinates $[X_0, X_1, X_2, X_3]$ contains 6 coordinate planes:

$$(X_0, X_1), (X_2, X_3), (X_0, X_2), (X_0, X_3), (X_1, X_2), (X_1, X_3) \tag{14}$$

In the plane (X_0, X_1), the daughter vector $\mathbf{D_1}$ (12) is represented by its projection vector $\mathbf{M_{01}} = [x(t)\alpha_0, x(t)\alpha_1, 0, 0]$. Its length $\|\mathbf{M_{01}}\|$ (15) is equal to the first coordinate of the mother vector [x(t), y(t)] at any fixed time t:

$$\|\mathbf{M}_{01}\| = \left(a(t)^2\alpha_0^2 + a(t)^2\alpha_1^2\right)^{0.5} = \left\{a(t)^2\left(\alpha_0^2 + \alpha_1^2\right)\right\}^{0.5} = a(t) \tag{15}$$

In the plane (X_2, X_3), the daughter vector $\mathbf{D_1}$ (12) is represented by its projection vector $\mathbf{M_{23}} = [0, 0, y(t)\beta_0, y(t)\beta_1]$. Its length $\|\mathbf{M_{23}}\|$ (16) is equal to the second coordinate of the mother vector [x(t), y(t)] at any fixed time t:

$$\|\mathbf{M}_{23}\| = \left\{y(t)^2\beta_0^2 + y(t)^2\beta_1^2\right\}^{0.5} = \left\{y(t)^2\left(\beta_0^2 + \beta_1^2\right)\right\}^{0.5} = y(t) \tag{16}$$

Expressions (13, 15, 16) indicate that the daughter 4-dimensional vector $\mathbf{D_1}$, having stochastic coordinates, contains in the groupings of these coordinates the exact "memory" of all coordinates and length of the mother 2-dimensional vector and in this respect is a deterministic entity.

Thus, tensor-unitary transformations and families of daughter vectors, generated by them, make it possible to model the biological dualism "stochastics-and-determinism", as well as the phenomena of Gestalt biology, in which relations in groupings of elements are important and meaningful, but not individual elements in themselves. For example, we recognize a musical melody even when it is performed on different instruments and in different frequency ranges with a change in the sound frequencies of individual notes, which turn out to be insignificant, in contrast to the ratios in the groupings of musical frequencies. Since tensor-unitary transformations make it possible to model biological Gestalt phenomena, they can also be conditionally called Gestalt transformations, and the stochastic-deterministic vectors generated by them can be called Gestalt vectors.

It should be noted that the daughter 4-dimensional vector $\mathbf{D_1}$ (12) is not only the carrier of the exact memory of the coordinates of the mother 2-dimensional vector but also the carrier of new information in the planes (X_0, X_2), (X_0, X_3), (X_1, X_2), (X_1, X_3) of 4-dimensional space. In these planes, the daughter vector $\mathbf{D_1}$ is represented by its projection vectors $\mathbf{M_{02}} = [x(t)\alpha_0, 0, y(t)\beta_0, 0]$, $\mathbf{M_{03}} = [x(t)\alpha_0, 0, 0, y(t)\beta_1]$, $\mathbf{M_{12}} = [0, x(t)\alpha_1, y(t)\beta_0, 0]$, $\mathbf{M_{13}} = [0, x(t)\alpha_1, 0, y(t)\beta_1]$ whose lengths are not equal to the coordinates of the mother vector. The values of their lengths carry new information, which is entered into them by the stochastic coordinates of the normvectors α_0, α_1, β_0, β_1. These stochastic components can reflect, for example, the specifics of the stage of ontogenesis or the impacts on the organism from the external environment. Thus, the tensor-unitary transformation is a memory expansion operation (while maintaining all the previous memory) for a tensor extended multi-parameter system, which, along with the memory of the "ancestors" parameters, contains many new parameters or information.

The values of the inner products of two mother vectors - for example, the vectors [x, y] and [s, r] - and of two daughter vectors are equal only when both mother vectors are tensor multiplied by the same normvector, for example, $[\alpha_0, \alpha_1]$.

Tensor-unitary transformations can be repeatedly applied to generate or develop from the mother structure more and more complex multi-parameter daughter structures with a step-by-step increase in the dimension of their configuration spaces. In these growing daughter structures, the connections of parameters from mother structures will be preserved and new parameters and relationships will arise.

Tensor-unitary transformations also make it possible to simulate the replications (reproductions) of geometric configurations, which is illustrated by the following example. Let the variable coordinates of the mother vector [x(t), y(t)] correspond to the functions of the parametric definition of the curve, called the cardioid (we will call it the mother cardioid):

$$x(t) = \cos(t)^*(1 + \cos(t)); \quad y(t) = \sin(t)^*(1 + \cos(t)) \tag{17}$$

For simplicity, consider the tensor-unitary product in which both coordinates of this mother vector are tensor multiplied by the same normvector $\mathbf{W} = [0.5^{0.5}, 0.5^{0.5}]$, generating a daughter vector $\mathbf{R_4}$ (18):

$$\mathbf{R}_4 = [x(t), 0] \otimes \mathbf{W} + [0, y(t)] \otimes \mathbf{W} = \left[x(t)^*0.5^{0.5}, x(t)^*0.5^{0.5}, y(t)^*0.5^{0.5}, y(t)^*0.5^{0.5}\right] \tag{18}$$

In the four planes (X_0, X_2), (X_1, X_3), (X_0, X_3) and (X_1, X_2) of the Cartesian coordinate system, this daughter vector $\mathbf{R_4}$ is represented by the same projection vector $[x(t)^*0.5^{0.5}, y(t)^*0.5^{0.5}]$, whose functional coordinates correspond the parametric description of the cardioid. Thus, in these four different 2-dimensional subspaces of the

4-dimensional configuration space of the multi-parameter system, four identical daughter cardioids are generated, which are slightly reduced copies of the mother cardioid (17). This fact illustrates the general proposition that tensor-unitary transformations can serve as multiplication operators for images given by the mother vector.

One can turn now to the phenomena of coordinated biocycles. The organism is a huge chorus of coordinated cyclic processes, whose number increases as it develops ontogenetically from the embryonic to the mature state with step-by-step receiving new and new degrees of freedom, which are coordinated with already existing. For example, in an adult human body, this number reaches enormous values, since it contains approximately 100 trillion cells participating in these sets of coordinated cycles. Moreover, all these cyclic processes occur through stochastic interactions between individual molecules in cells.

But how to simulate such development of huge sets of coordinated cyclic processes under conditions of stochasticity of individual parameters of an organism and a huge step-by-step increase in the dimension of its configuration space? Tensor-unitary transformations allow us to offer a model answer to this question since they generate stochastic-deterministic vectors; in these vectors, with a purely stochastic nature of the values of individual coordinates, sets of coordinated stable cycles on certain groupings of coordinates can be represented, i.e. on certain groupings of subspaces of the configuration space of a developing multi-parametric organism. As it is known, in mathematics, the tensor product of matrices is used to increase the dimension of a vector space. The genetic coding system, together with the family of DNA alphabets, is structured specifically for the tensor product of matrices, which links, for example, matrix representations of different DNA alphabets into a single tensor family of matrices [5, 6].

One can add that tensor-unitary transformation can be applied not only in the case of the algebra of real numbers but also in cases of different systems of multidimensional numbers: complex numbers, hyperbolic numbers, quaternions by Hamilton, split-quaternions by Cockle, etc. Separate interest is applications of tensor-unitary transformations to block vectors and matrices. They can find interesting applications in many scientific problems, including, for example, processing biometric parameters, cryptography, optimization of information security systems and multidimensional time sequences, multilayer perceptron designing, the doctrine of evolution based on bio-antenna arrays, etc. [9–14].

4 Concluding Remarks

One of the founders of quantum mechanics, P. Jordan, argued that the laws of living organisms missed by science are the laws of probabilities [15]. In connection with this, the author discovered the universal rules for the stochastic organization of the DNA of the genomes of higher and lower organisms. An analysis of these universal rules revealed that structural relationships in the system of stochastic regularities of genomic DNA are associated with special tensor operations, which the author called "tensor-unitary". These operations make it possible to develop algebraic models of biological dualism "stochastics-and-determinism", physiological gestalt phenomena, and ontogenetic replications of structures in the language of matrix-vector algebra, quantum informatics, and quantum mechanics.

The name "tensor-unitary transformations" is determined by their functional analogies with unitary transformations, which play an important role in quantum mechanics and quantum computer science. The author believes that tensor-unitary transformations are useful for developing model ideas about living bodies as multi-parameter quantum biostructures capable of ontogenetic development with a tensor increase in the number of their degrees of freedom and a corresponding increase in the dimension of their configuration spaces. In other words, we are talking about the creation of quantum biomechanics (or quantum developmental biomechanics), which, unlike conventional quantum mechanics, is focused on modeling biosimilar quantum systems capable of tensor self-development with memory inheritance of past states through the use of tensor-unitary transformations and also of the dualism of "stochastics-and-determinism". The quantum developmental biomechanics can study the same problems (ontogenetic growth of body weight, cell differentiation, etc.), which are studied by developmental biology, but in the language of algebraic formalisms of quantum mechanics, quantum informatics, matrix analysis, etc. Special quantum informatics can arise in this modeling approach.

On this path, an in-depth understanding of genetic intelligence is possible, by which the author understands that part of the intellectual potentialities of living organisms, which allows the following: based on information in genetic molecules, build, for example, from a single fertilized cell a complex organism with trillions of cells in such a way that it possesses parental traits, which are reproduced in a noise-resistant multichannel way, despite strong noises and constantly changing of nutritional conditions and external influences in the course of life.

One can points out the major difference of the presented research from related research: other researches make no algebraical modeling of biological dualism stochastics-determinism. The described approach and new algebraic operations allow innovative combinations of biological modeling with quantum informatics and tensor-matrix theory of digital antenna arrays. It gives new possibilities for including of theoretical biology in the field of mathematical natural sciences.

References

1. Bernstein, N.A.: The Co-Ordination and Regulation Of Movements. New York, Pergamon Press, Oxford (1967)
2. Petoukhov, S.V., Svirin, V.I.: Stochastic rules in nucleotide sequences in genomes of higher and lower organisms. Int. J. Math. Sci. Comput. (IJMSC) **7**(2), 1–13 (2021)
3. Petoukhov, S.V.: Algebraic harmony and probabilities in genomes. long-range coherence in quantum code biology. Biosystems **209**, 104503 (2021). https://doi.org/10.1016/j.biosystems.2021.104503
4. Petoukhov, S.V.: Algebraic rules for the percentage composition of oligomers in genomes. Preprints **2021**, 2021010360 (2021)
5. Petoukhov S.V. Matrix genetics, algebras of genetic code, noise immunity. Moscow, RCD, 2008, 316 p. (in Russian). ISBN 978-5-93972-643-6
6. Petoukhov, S.V., He, M.: Symmetrical Analysis Techniques for Genetic Systems and Bioinformatics: Advanced Patterns and Applications. Hershey, USA, IGI Global (2010)
7. Nielsen, M.A., Chuang, I.L.: Quantum Computation and Quantum Information. Cambridge University Press, New York (2010). https://doi.org/10.1017/CBO9780511976667

8. Petoukhov, S.: The Stochastic Organization of Genomes and the Doctrine of Energy-Information Evolution Based on Bio-Antenna Arrays. Preprints, 2022030100 (2022)
9. Hu, Z.B., Tereikovskyi, I., Chernyshev, D., Tereikovska, L., Tereikovskyi, O., Dong Wang, D.: Procedure for processing biometric parameters based on wavelet trans-formations. Int. J. Mod. Educ. Comput. Sci. (IJMECS). **13**(2), 11–22 (2021)
10. Hu, Z.B., Gnatyuk, S., Okhrimenko, T., Tynymbayev, S., Iavich, M.: High-speed and secure PRNG for cryptographic applications. Int. J. Comput. Netw. Inf. Secur. (IJCNIS) **12**(3), 1–10 (2020)
11. Hu, Z.B., Yulia Khokhlachova, Y., Sydorenko, V., Opirskyy, I.: Method for optimiza-tion of information security systems behavior under conditions of influences. Int. J. Intell. Syst. App. (IJISA), **9**(12), 46–58 (2017)
12. Hu, Z.B., Mashtalir, S.V., Tyshchenko, O.K., Stolbovyi, M.I.: Video shots' matching via various length of multidimensional time sequences. Int. J. Intell. Syst. App. (IJISA) **9**(11), 10–16 (2017)
13. Hu, Z.B., Igor, A., Tereykovskiy, I.A., Tereykovska, L.O., Pogorelov, V.V.: Determination of structural parameters of multilayer perceptron designed to estimate parameters of technical systems. Int. J. Intell. Syst. App. (IJISA) **9**(10), 57–62 (2017)
14. Petoukhov, S.V.: The stochastic organization of genomes and the doctrine of energy-information evolution based on bio-antenna arrays. Biosystems, p. 104712 ISSN 0303–2647 (2022). https://doi.org/10.1016/j.biosystems.2022.104712
15. McFadden J., Al-Khalili, J.: The origins of quantum biology. Proc. R. Soc. A. **474**(2220), 1–13 (2018)

Design Automated Medical Information and Analytical Management System for Large and Medium-Sized Organizations

Nadiia Pasieka[1](✉), Yulia Romanyshyn[2], Svitlana Chupakhina[1], Mariia Oliinyk[3], Nataliia Kyrsta[1], and Andrii Pitulei[2]

[1] Vasyl Stefanyk Precarpathian National University, Ivano-Frankivsk 76016, Ukraine
pasyekanm@gmail.com
[2] National Technical University of Oil and Gas, Ivano-Frankivsk 76058, Ukraine
[3] Yuriy Fedkovych Chernivtsi National University, Chernivtsi 58012, Ukraine

Abstract. The development of an automated medical document verification system has implemented the automation of the clinic's business document management process, significantly speeding up and improving the interaction of medical staff in the workplace in order to provide better services. Efficient use of databases, and document-oriented data warehouses, including dual-use ones, allows certain operations to be performed efficiently and quickly, i.e. adding patient records, changing records, performing periodic or analytical data searches and operations to ensure quality decision management. The focus is on the interaction of distributed heterogeneous applications and independent institutional hardware, developing professional medical client applications; thus, improving the quantity and quality of information data flows required for efficient patient care. The designed software application allows automatic patient record keeping with the needs of the medical staff, especially in the management of patient visits (scheduling) and the creation of patient medical records in accordance with current international standards. This work is basic for a smoother integration of medical records and data and effective prevention, diagnosis and prognosis. As the appropriate approaches and standards used have improved health information technology functions as well as the quality and safety of patient care, we have created a global architecture of specific medical automation systems, coordinating internationally regulated HL7 medical services to ensure system scalability in the global future.

Keywords: Warehouses data · Electronic health record · Soft-ware applications · Health level seven international · Unified modelling language

1 Introduction

In studying the industry of creating and using medical information management systems, we can observe that it has been in existence for more than sixty years. How-ever, despite such a long existence and use of computerized medical management systems, they still retain significant time lags in the cost of creating and correcting departmental documentation. Because of the elementary lack of automated work-stations for practitioners

Z. Hu et al. (Eds.): AIMEE 2022, LNDECT 159, pp. 327–338, 2023.
https://doi.org/10.1007/978-3-031-24468-1_29

and registrars of medical automated management systems. Based on historical realities, medical information management systems today are the optimal and balanced solution for maintaining modern medical records. [1, 4] Which is intended for the process of automation of commercial activities of medical institutions and at the same time provides an effective maintenance of electronic patient records, simplify and automate statistical reporting. Such automated control systems also enable to manage financial flows of a medical institution, to keep effective records of material and technical resources. Carry out and perform optimization tasks for placement of patients with various degrees of diseases both in hospitals and in sanatoriums. [8, 10, 17] Also, the use of automated medical management systems ensures maximum protection of patients' personal data using distributed access rights for doctors to patients' personal rooms.

Thus, the relevance of the topic of scientific research is to consider the problem of effective use of existing and being developed information-analytical automated medical management systems for medical and economic activities of large medical institutions, namely hospitals, health centers, clinics. However, the use of automated medical management systems for small businesses (private dentistry, radiology offices, etc.) has a number of significant drawbacks, the main of which is a significant functional redundancy, and as a result, a long time to master because of the complexity of the internal architecture, as well as high initial cost and maintenance. [2]

The object of the study is the design process of an automated medical information-analytical system.

The subject of research is the effective automation of medical business processes in large and medium-sized medical organizations.

Novelty of scientific research is a systematic analysis of international experience in the application of medical information-analytical management systems of institutions.

The aim of the work is to design automated medical information and analytical management system for large and medium-sized organizations, the use of which will contribute to improving the efficiency of the polyclinic, the transition to a new level of care and treatment of patients, based on international experience in designing, re-search business processes occurring in a medical organization and the reception of the attending physician. [6, 13]

The aim of scientific work was to solve a number of interrelated problems: the analysis of the subject area, necessary for the formation of software requirements to create software, the actual choice of environment for the design of information-analytical management system of the medical organization, the design of management system in an object-oriented design language UML (Unified Modelling Language), development and detailing of functional modules based on methodology IDEF0 and DFD diagrams. [3, 5, 9]

Purpose of automated medical information-analytical system of institution management: to increase effectiveness of work of medical institution (reduction of time for documentation of medical information, standardization and unification), support for making medical decisions using medical practices, recorded in the system. Simplified maintenance of electronic patient records with automatic transfer of treatment information to the patient's personal cabinet, economic analysis of the effectiveness, use of medicines by means of information-analytical management system, formation of electronic schedule of doctors and medical auxiliary services. [18, 32]

2 Subject Area Research

2.1 Information-Analytical Management Systems and Business Processes of a Medical Institution

The purpose of creating and implementing an automated medical information and analytical management system is to organize the work of a large and medium-sized medical institution. At the same time, the effective use of modern information and communication technologies requires a significant revision of the principles and mechanisms of enterprise management. [7, 12]

Proceeding from the fact that various medical institutions provide medical ser-vices, so they can be regarded as effective production management systems. In turn, production management systems of medical institutions consist of groups of mutually independent components working together to achieve the end goal. [11]

These components of an information-analytical system are defined as processes. Thus, a production automated medical information-analytical management system consists of a group of interrelated processes that ensure the achievement of organizational goals. [19]

Business processes of automated medical information-analytical management system of the institution is a related set of repetitive actions (functions), which result in a final product - a medical service - in accordance with predetermined rules. [24]

Thus, a medical organization distinguishes between basic and auxiliary business processes. The main processes occurring in an automated medical information-analytical management system are those that add quality. In this case, these are the processes of providing the patient with quality medical care (quality medical ser-vices). [22]

Auxiliary processes form the infrastructure of an organization, in a medical institution - they create the conditions for performing the treatment and diagnostic process.

A medical information system is designed to provide information support for both core and auxiliary business processes of a medical institution. [15, 20] Therefore, the first stage of its design is a formalized description of these business processes, i.e. the construction of functional models of a medical organization, the description of sys-tem requirements specifications based on a comprehensive survey of the organization, data of the subject area entities, processes that use these entities and the links between them.

2.2 System Analysis and International Practice of Creating Automated Information-Analytical Systems of Medical Institution Management

The Electronic Health Record (EHR) is the focus of any automated information and analytical management system for medical institutions. Effective implementation of EHR began in the late 20th and early 21st centuries and is still underway in many countries, including Ukraine. Confidential medical information stored in the EHR is represented by a complex set of heterogeneous data, both textual and digital, graphic, which originally reside in different systems. For successful synchronization of these data in various automated information-analytical systems of medical institution management, it is necessary not only to transfer them through protocols, but also to per-form their transformation into the form accepted in other automated information-analytical systems, and vice versa. [16, 21, 23]

To solve the problem of MIS integration, the main international standards defining the rules of information exchange in medicine are described. This is exemplified by a large variety of international standards, often competing with each other (HL7, FHIR, OpenEHR, CEN TC/251, DICOM, ISO, TC 215, XDS). It is worth noting that the problem of electronic medical records is extremely closely connected with the peculiarities of the health care system in a particular country. Therefore, there is no point in talking about "direct" transfer of the experience of other countries. [25, 26, 29]

The Health Level Seven (HL7) is a model standard for the exchange, management and integration of electronic health information. The HL7 standard is used for the electronic exchange of documents in medical institutions, especially in those where the patient receives intensive care, such as hospitals. HL7 includes conceptual standards (HL7 RIM), application standards (HL7 CCOW), document standards (HL7 CDA), and messaging standards (HL7 v2, v3.0 and HL7 FHIR). These standards guide the creation of structured EHR documents in many countries, including Ukraine. [33]

The integrated OpenEHR environment is consistent with the new Standard for Electronic Case Record Management (EN 13606) in automated information and analytical management systems for health care facilities. It has been used in part in the United Kingdom's UK NHS Connecting for Health program and has been selected as the basis for a national program in Sweden. It is also being developed in a number of countries, such as Denmark, Slovakia, Chile, and Brazil. In addition, it is beginning to be used in commercial systems around the world.

The efforts of CEN, HL7, OpenEHR and other standards represent a collaborative effort to define common EHR standards. Harmonizing these different approaches will gradually help reduce the complexity of implementing common EHR-system internationally. According to Deloitte surveys of U.S. physicians, U.S. use of electronic health records in 2018 was as follows: three in four U.S. physicians believe that electronic health records EHR-system increase practice costs, outweighing any efficiency savings, and seven in ten believe that they reduce productivity.

Examining these trends in detail in a follow-up survey two years later, we note that they are essentially unchanged, and generally hold negative opinions about EHR-system. So, the study shows that the EHR transition process is very much about improvement, with 63% wanting it to be interoperable, while 57% were looking for simplification and streamlining to make it more efficient.

If we consider the research company Medscape conducted in 2017, where were interviewed more than 4000 specialists, including 2600 doctors and 800 medical workers on the feasibility of using EHR-systems in automated information-analytical systems of management of medical institutions. The results were as follows: large medical institutions used EHR-systems more often (more than 55%), medium medical practices only 29%, small medical practices only 4%. Summarizing the results of the survey on the use of EHR-systems, only 29.7% of respondents are using them, and another 20% are ready to implement these systems in the next year or two.

Most of the medical staff surveyed were satisfied with their EHR-system as part of a medical institution management information and analytical system, at 69.7%, and would recommend it to their colleagues. At the same time, however, almost one third of those surveyed, 30.3%, would not recommend the EHR-system. And of those health professionals who were dissatisfied with the existing EHR-system, only 16.3% reported an intention to upgrade or replace it.

3 Design of an Automated Information-Analytical Management System for a Medical Institution

When designing an automated information-analytical management system for a medical institution, it is necessary to perform a detailed analysis of the objectives of the system and identify the software and business requirements for individual users. Information for building the model of automated information-analytical system of medical institution management is taken on the basis of a comprehensive system survey of the organization for which the development is carried out, including the EHR module. Data collection begins with the study of the entities of the subject area, business processes that use these entities, and the relationships between them.

On the assumption that, the automated information-analytical system of medical institution management is designed to automate medical business processes and information support for employees of the Department of Defence.

Overload and short-circuit protection disconnection, and the protection system guarantees emergency shutdown. Electrical equipment standards must comply with fire safety requirements; no toxic fumes should be emitted in case of fire. Information security requirements:

In the event of the following emergencies, system information must be secured:

Power failure, Software failure, Hardware failure, Database crash.

In the event of a power outage, software or hardware failure on a workstation or database server, all recorded transactions must be terminated or unconfirmed transactions must be cancelled, and information from the last completed transaction must be saved. If the database is corrupted, you need to secure the information in the last data-base backup. A backup should be created at least once a day.

Registrar - manages the registration process, maintains (i.e., enters, changes, deletes) physician and patient data.

Patient - make an appointment with a doctor and receive medical care. Physicians make appointments and can schedule appointments.

Based on the needs of the participants, the systems analyst may suggest the following uses: "login", "make an appointment", keep a history, "make an appointment", "open enrolment", "close enrolment", "patient CRUD data", physician CRUD data. CRUD means create, read, update, delete (or create, receive, update, destroy). Based on the list of participants, construct a case map that includes a list of the functions and tasks of the system as a whole and each unit (subsystem), and the relationships between them, (see Fig. 1).

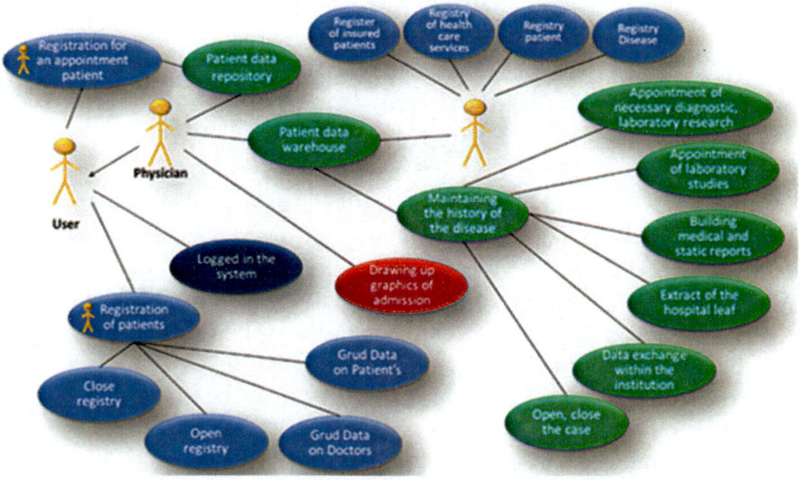

Fig. 1. Chart of precedents "Treatment of the patient"

The information system contains the following subsystems: "Login to the system", "Register for an appointment", "Making an appointment schedule", "Keeping a medical history".

Diagram of "Login to the system" activity.

(Short Description):

This use case describes the user logging in to the system (Fig. 2):

(Main Event Flow):

The system asks for a user name and password. The user enters a name and password.

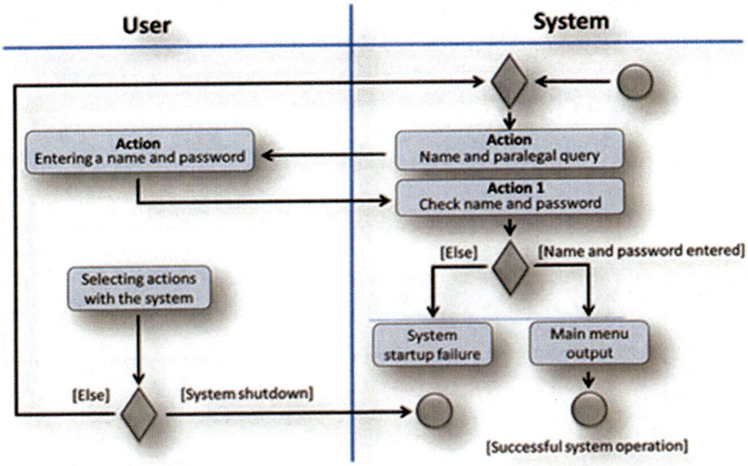

Fig. 2. Activity diagram "Login to the system"

The system validates the name and password, determines the user type (patient/physician/administrator) and displays the main menu giving access to system functions according to the user type.

(Alternate streams):

The system detects that the name and password combination is not correct.

The system reports the error and prompts the user to either re-enter the name and password, or refuse to log on.

The user informs the system of their choice.

According to the user's choice, either execution proceeds to the beginning of the main thread, or the use case is terminated.

(Preconditions): None. (Post conditions):

If the use case is successful, the system grants access to the main menu to the user who gave the correct username and password combination.

Otherwise, the system guarantees that the user who reported the wrong combination of name and password will not be granted access to the menu.

Diagram of the "Register for an appointment" activity.

As shown in Fig. 3:

(Brief Description):

This use case allows a patient to register for an appointment with a physician

(Main Event Flow):

The user communicates a desire to register for an appointment.

The system prompts the user to enter data. The system searches for the patient.

The system communicates the required action (i.e. it searches for free time in the schedule and prompts the patient to choose an appointment time).

The patient communicates their choice.

The system confirms the patient's choice.

The system ends the session. (Subordinate event streams):

Patient choice of time.
The user selects an available time in the doctor's schedule.
The system puts the patient on the schedule. (Alternate Event Streams):
The system reports an absent patient in the system.

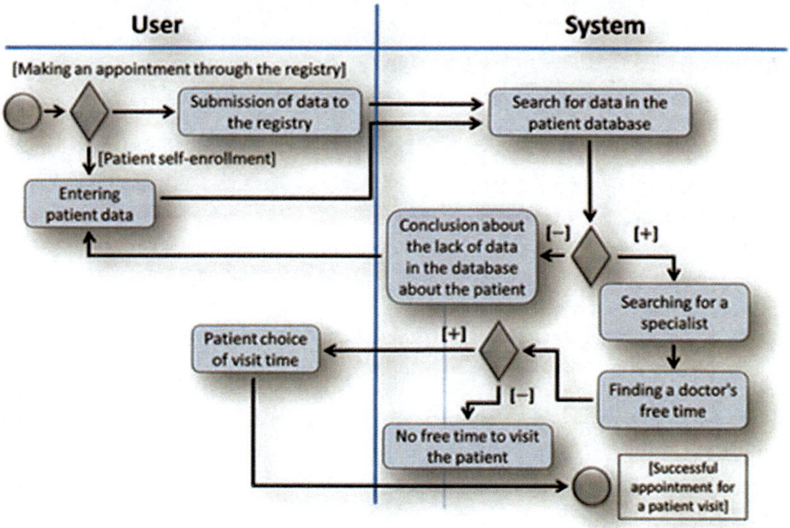

Fig. 3. Activity diagram "Registering for an appointment"

The system reports that there is no available appointment time. (Prerequisite): Absent. (Post condition):

If the use option completes successfully, the system will update the appointment schedule according to the user's selection.

Activity diagram "Making an appointment schedule".

(Brief description):

Allows the physician to create an appointment schedule. The system transmits information about the physician's staff, the created appointment schedules, and the patients who have made an appointment (Fig. 4).

(Main event stream): The user requests a physician search. The system confirms that the doctor is found. The system reconciles the appointment schedule. The system requests the appointment schedule settings for a specific physician. (Subordinate Event Streams): The system receives the schedule settings. The system prompts for schedule creation. The user adds the patient to the schedule. (Alternate event streams): The schedule for the period is created: The system prompts the user to change the schedule. Schedule modification. The user is prompted to search and add the patient to the schedule. (Prerequisites): The physician must log in to the system. (Post conditions): If successful, an appointment schedule is made.

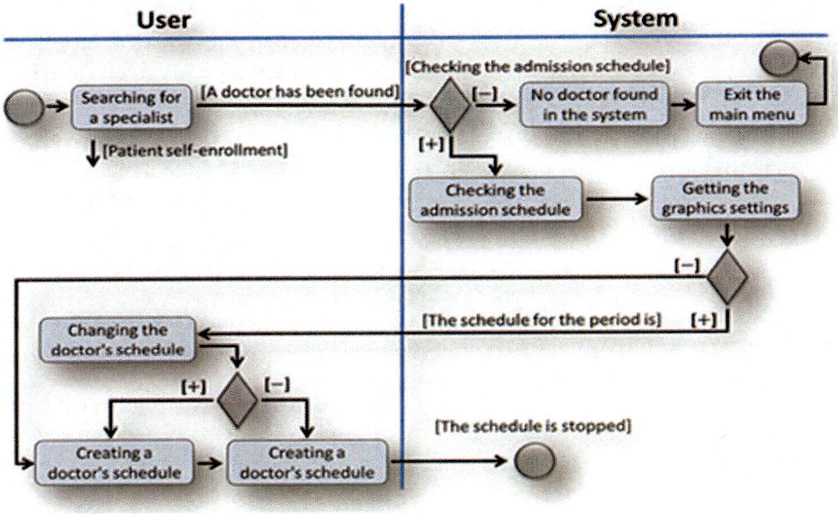

Fig. 4. Diagram of the activity "Making an admission schedule"

4 Conclusion

The development of the automated medical document review system automates the clinic's business document management process, which greatly speeds up and improves the interaction of medical staff in the workplace for better service delivery. Effective use of databases, databases and document-oriented data warehouses, including dual-use, allows you to efficiently and quickly perform certain operations, namely, adding records, making changes to them, performing regular or analytical data searches and operations to ensure high quality solution management. Focusing on the interaction between distributed heterogeneous applications and independent institutional hardware, professional medical client applications were developed; therefore, the quantity and quality of information data streams required for efficient patient care were improved. The developed application allows the automatic management of patient records, taking into account the needs of medical personnel, especially in the management of patient visits and the creation of patient medical records in accordance with applicable international standards. This work is the basis for a smoother integration of medical records and data for more effective prevention, diagnosis and prognosis. As relevant standards have improved health information technology functions as well as the quality and safety of patient care, we have completed a global architecture of specific medical automation systems, coordinating internationally regulated HL7 medical services to ensure system scalability on a global scale in the future. Thus, the following conclusions were formulated in the course of scientific research: defined the main medical processes taking place in a medical organization; regulatory and reference acts, which are of general obligatory character and serve as a normative basis for design of Electronic Health Record with basic functionality, were studied; international experience in the application of electronic health record was studied, the experience in the application of EHR-systems, conducting a survey of medical personnel, which raises the main issues related to the security of personal data, the cost of implementing EHR-systems. The study showed the advantages and disadvantages of applying EHR-systems for the specified period from 2018 to 2021. Thus, the tasks were solved in full, the goal was achieved – designed a medical information system for a medical organization, which will increase the efficiency of the use of human resources of the enterprise, which is relevant to the design of Electronic Health Record. The results of the work can be the basis for the development of the final product of the information system.

References

1. Koren, A., Jurčević, M., Huljenić, D.: Requirements and challenges in integration of aggregated personal health data for inclusion into formal electronic health records (EHR). In: The 2nd International Colloquium on Smart Grid Metrology (SMAGRIMET), pp. 1–5 (2019)
2. Opaliński, A., et al.: Medical data exploration based on the heterogeneous data sources aggregation system. In: Federated Conference on Computer Science and Information Systems (FedCSIS), Leipzig, Germany, pp. 591–597 (2019)
3. Rahman, A., Mitra, A., Rahman, F., Slepian, M.J.: Smart EHR - a big-data approach to automated collection and processing of multi-modal health signals in a doctor-patient encounter. In: IEEE International Conference on Big Data (Big Data), pp. 6198–6200 (2019)
4. Durnyak, B., Tymchenko, O., Tymchenko, O., Havrysh, B.: Applying the neuronetchic methodology to text images for their recognition. In: IEEE Second International Conference on Data Stream Mining and Processing (DSMP), pp. 584–589 (2018)
5. Sekar, B., et al.: Intelligent clinical decision support systems for patient-centered healthcare in breast cancer oncology. In: 20th International Conference on e-Health Networking, Applications and Services (Healthcom), Ostrava, pp. 1–6 (2018)
6. Xie, C., et al.: An efficient hierarchical key management scheme for access medical database. In: International Conference on Information, Communication and Engineering (ICICE), Xiamen, pp. 461–463 (2017)
7. Nguyen, D.C., Pathirana, P.N., Ding, M., Seneviratne, A.: Blockchain for secure EHRs sharing of mobile cloud based e-health systems. IEEE Access **7**, 66792–66806 (2019)
8. Barakah, D.M., Alwakeel, S.S., Shira, M.M.: The impact of advanced e-clinical systems and IT-based technologies on medical practice at Saudi tertiary hospitals. In: IEEE International Conference on Power, Control, Signals and Instrumentation Engineering (ICPCSI), Chennai, pp. 178–184 (2017)
9. Singh, D., Verma, S., Singla, J.: A comprehensive review of intelligent medical diagnostic systems. In: 4th International Conference on Trends in Electronics and Informatics (ICOEI), Tirunelveli, India, pp. 977–981 (2020)
10. Chebanenko, E., Denisova, L., Serobabov. A.: Intelligent processing of medical information for application in the expert system. In: Ural Symposium on Biomedical Engineering, Radioelectronics and Information Technology (USBEREIT), Yekaterinburg, Russia, pp. 0085–0088 (2020)
11. Frontoni, E., et al.: A decision support system for diabetes chronic care models based on general practitioner engagement and EHR data sharing. IEEE J. Transl. Eng. Health Med. **8**, 1–12 (2020)
12. Halim, E., Halim, P., Hebrard, M.: Indonesia medical knowledge management system: a proposal of medical KMS. In: International Conference on Information Management and Technology (ICIMTech), Yogyakarta, pp. 322–327 (2017)

13. Martins, J., Barbosa, R., Lourenço, N., Robin, J., Madeira, H.: Online verification through model checking of medical critical intelligent system. In: 50th Annual IEEE/IFIP International Conference on Dependable Systems and Networks Workshops (DSN-W), Valencia, Spain, pp. 32–37 (2020)
14. Schwertner, M.A., Rigo, S.J., Araújo, D.A., Silva, A.B., Eskofier, B.: Fostering natural language question answering over knowledge bases in oncology EHR. In: IEEE 32nd International Symposium on Computer-Based Medical Systems (CBMS), pp. 501–506 (2019)
15. Giannoulis, M., Marakakis, E., Kondylakis, H.: Developing a collaborative knowledge system for cancer diseases. In: IEEE 30th International Symposium on Computer-Based Medical Systems (CBMS), Thessaloniki, pp. 767–768 (2017)
16. Pasyeka, M., Sheketa, V., Pasieka, N., Chupakhina, S., Dronyuk, I.: System analysis of caching requests on network computing nodes. In: 3rd International Conference on Advanced Information and Communications Technologies (AICT), Lviv, Ukraine, pp. 216–222 (2019)
17. Sheketa, V., Poteriailo, L., Romanyshyn, Y., Pikh, V., Pasyeka, M., Chesanovskyy, M.: Case-based notations for technological problems solving in the knowledge-based environment. In: International Scientific and Technical Conference on Computer Sciences and Information Technologies, pp. 10–14 (2019)
18. Farooqui, N.A., Mehra, R.: Design of A data warehouse for medical information system using data mining techniques. In: Fifth International Conference on Parallel, Distributed and Grid Computing (PDGC), Solan Himachal Pradesh, India, pp. 199–203 (2018)
19. Farahani, N.Z., Sundaram, D.S.B., Enayati, M., Arunachalam, S.P., Pasupathy, K., Arruda-Olson, A.M.: Explanatory analysis of a machine learning model to identify hypertrophic cardiomyopathy patients from EHR using diagnostic codes. In: IEEE International Conference on Bioinformatics and Biomedicine (BIBM), pp. 1932–1937 (2020)
20. Ajayi, O., Abouali, M., Saadawi, T.: Secure Architecture for inter-healthcare electronic health records exchange. In: IEEE International IOT, Electronics and Mechatronics Conference (IEMTRONICS), pp. 1–6 (2020)
21. Pasieka, N., Sheketa, V., Romanyshyn, Y., Pasieka, M., Domska, U., Struk, A.: Models, methods and algorithms of web system architecture optimization. In: IEEE International Scientific-Practical Conference: Problems of Infocommunications Science and Technology, PIC S&T 2019, pp. 147–152 (2019)
22. Pasyeka, M., Sheketa, V., Pasieka, N., Chupakhina, S., Dronyuk, I.: System analysis of caching requests on network computing nodes. In: 3rd International Conference on Advanced Information and Communications Technologies, AICT 2019, pp. 216–222 (2019)
23. Romanyshyn, Y., Sheketa, V., Pikh, V., Poteriailo, L., Kalambet, Y., Pasieka, N.: Social-communication web technologies in the higher education as means of knowledge. In: 14th International Scientific and Technical Conference on Computer Sciences and Information Technologies, CSIT 2019, pp. 35–38 (2019)
24. Babichev, S., Durnyak, B., Zhydetskyy, V., Pikh, I., Senkivskyy, V.: Application of optics density-based clustering algorithm using inductive methods of complex system analysis. In: 14th International Conference on Computer Sciences and Information Technologies (CSIT), pp. 169–172 (2019a)
25. Chaganti, S., et al.: Electronic medical record context signatures improve diagnostic classification using medical image computing. IEEE J. Biomed. Health Inform. **23**(5), 2052–2062 (2019)
26. Hijazi, S., Obeid, N., Sabri, K.E.: On the logical foundation of a personalized medical prescription system. IEEE Access **8**, 6471–6483 (2020)
27. Jiang, S., Wu, H., Wang, L.: Patients-controlled secure and privacy-preserving EHRs sharing scheme based on consortium blockchain. In: IEEE Global Communications Conference (GLOBECOM), pp. 1–6 (2019)

28. Ramachandran, S., Obu Kiruthika, O., Ramasamy, A., Vanaja, R., Mukherjee, S.: A review on blockchain-based strategies for management of electronic health records (EHRs). In: International Conference on Smart Electronics and Communication (ICOSEC), pp. 341–346 (2020)
29. Sheketa, V., Zorin, V., Chupakhina, S., Kyrsta, N., Pasyeka, M., Pasieka, N.: Empirical method of evaluating the numerical values of metrics in the process of medical software quality determination. In: International Conference on Decision Aid Sciences and Application, DASA 2020, pp. 22–26 (2020)
30. Shkitsa, L., Kornuta, V., Kornuta, O., Bekish, I., Bui, V.: Information support of design innovation activity of the technical university. Manag. Syst. Prod. Eng. **28**(2), 127–132 (2020)
31. Shkitsa, L., Kornuta, V., Kornuta, O., Bui, V., Bekish, I.: In-campus way of the insight transfer technology. In: Ivanov, V., Trojanowska, J., Pavlenko, I., Zajac, J., Peraković, D. (eds.) DSMIE 2021. LNME, pp. 322–331. Springer, Cham (2021). https://doi.org/10.1007/978-3-030-77719-7_32
32. Tkachenko, R., Izonin, I., Kryvinska, N., Dronyuk, I., Zub, K.: An approach towards increasing prediction accuracy for the recovery of missing IoT data based on the GRNN-SGTM ensemble]. In: Sensors (Switzerland) (2019)
33. Bhatt, V., Li, J., Maharjan, B.: DocPal: a voice-based EHR assistant for health practitioners. In: IEEE International Conference on E-health Networking, Application and Services (HEALTHCOM), pp. 1–6 (2021)

Advances in Educational Approaches

The Influence of the «Abilympics» Championship on the System of Higher and Secondary Professional Education

Elena Fomina(✉) and Olga Gefele

Tver State Technical University, Tver 170026, Russian Federation
f-elena2008@yandex.ru

Abstract. The process of primary professionalization and formation of the competitiveness of students with disabilities and limited health opportunities in the conditions of fierce competition in the labor market is controversial and complex. The article is devoted to the consideration of the actualization of the «Abilympics» championship of professional skills within the framework of modern inclusion in higher education. «Abilympics» championships at various levels are designed to increase competitiveness, effective adaptation to the process of primary professionalization and the acquisition of primary professional experience by persons with disabilities and limited health opportunities. Based on the analysis of the development of the movement, the influence of the championship on the educational system is determined, and the question of the championship's assistance to the process of personal and professional growth of its participants is raised. The question is raised about the priority of technological, methodological and psychological and pedagogical support within the socio-cultural inclusion of persons with disabilities and limited health opportunities.

Keywords: Education · Professional excellence championship · Professional competencies · Inclusive education · Psychological and pedagogical support · Personal and professional realization

1 History of the «Abilympics» Movement

The «Abilympics» movement was born in 1971 in Japan. Since 1971, nine International Championships have been organized and successfully held with the participation of 50 countries/regions (Table 1, Figs. 1, 2 and 3) [1]. The geography of the participating countries changes from championship to championship; since 2000, the competition of professional skills was first held in Europe (Czech Republic), and in 2022 the Russian Federation will be the host [2].

The main goal of the championships is to draw attention to the problem of people with disabilities and people with limited health opportunities, change the stereotypes and prejudices that have developed in society about their professional abilities, the opportunity for successful self-realization in a particular profession and career growth, full participation in the socio-economic life of society.

Z. Hu et al. (Eds.): AIMEE 2022, LNDECT 159, pp. 341–350, 2023.
https://doi.org/10.1007/978-3-031-24468-1_30

Table 1. Statistics and geography of international championships

Number of the International Championship	Host country	Year	Number of participating countries/ regions	Number of partici-pants	Number of compe-tencies
I	Japan	1981	49	304	17
II	Colombia	1985	47	281	12
III	Hong Kong	1989	37	497	37
IV	Australia	1995	25	412	37
V	Czech	2000	21	375	46
VI	India	2003	23	424	44
VII	Japan	2007	23	365	27
VIII	South Korea	2011	36	447	36
IX	France	2016	31	465	50
X	Russia	2022			

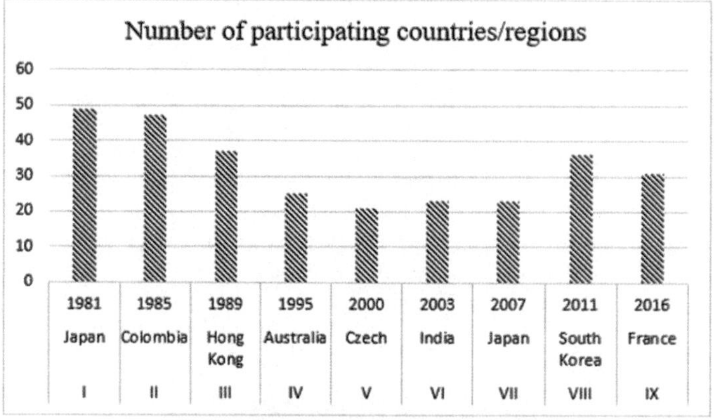

Fig. 1. Number of participating countries/regions

In addition, the championships are designed to promote the employment of disabled people, as they are held with the direct participation of employers, who get the opportunity to see and evaluate the professional competencies of the contestants in the reality.

Disabled people with any nosology have the opportunity to take part in the «Abilympics» championship: hearing and vision impairments, musculoskeletal disorders, neuropsychiatric and somatic disorders.

Competitions cover a wide range of professional competencies: IT technologies, working professions, professions related to the service sector, creative professions,

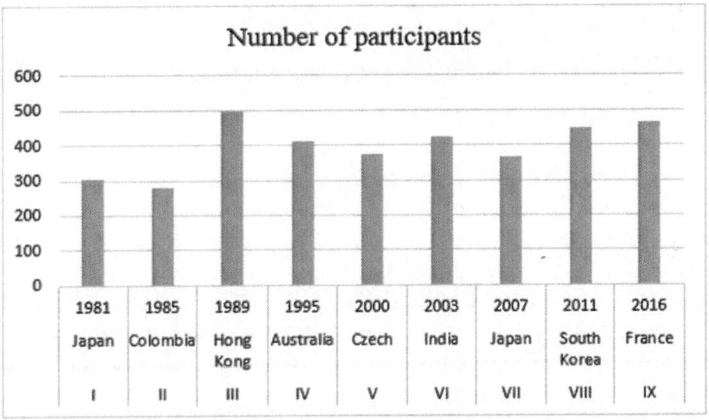

Fig. 2. Number of participants

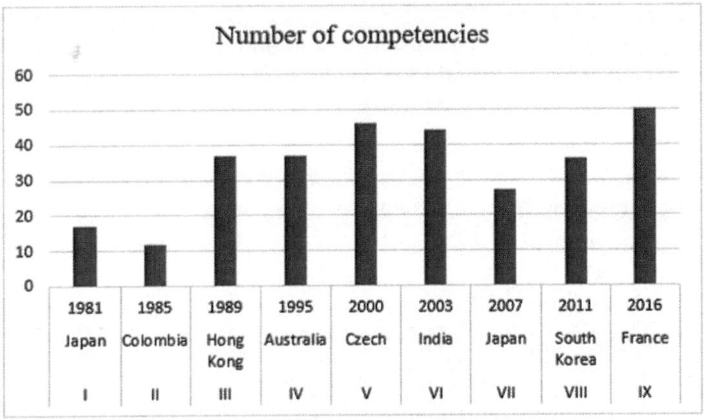

Fig. 3. Number of competencies

applied crafts. The list of competencies is constantly being adjusted depending on the socio-economic needs of the participating countries [3].

The International Championship is preceded by a series of qualifying events, which include a National Stage that determines the best participants in each country for each skill, who will subsequently become members of the national team and represent the country in international competitions. In turn, the National stage is preceded by the Regional qualifying stages, which allow to identify the strongest in each region of the country in terms of competencies and form a team for the National stage.

Competitors of three categories take part in the competitions - schoolchildren, students and specialists. Thus, the championship covers all age groups and provides an opportunity to prove themselves both to specialists with experience and work experience, and to novice participants, who often do not concede to their senior colleagues in professionalism, cause admiration and admiration for the diligence and dedication with which they go to victory, overcoming all difficulties.

2 The Influence of the «Abilympics» Championship on the Educational System

2.1 «Abilympics» and the Education System

Since 1971, the «Abilympics» movement has made a significant breakthrough in its development, expanding the geography of the participating countries, helping to draw attention to the problem of the disabled. In addition, the movement had a significant impact on the system of secondary vocational and higher education [4, 5].

Higher and secondary specialized educational institutions are training grounds for participants in the «Abilympics» championships. In addition, the championship participants are students of these educational institutions, which cannot but make their own adjustments to the educational process.

The main concept of inclusive education is to ensure the rights of every person to receive a quality education, regardless of the existing social, physical, mental, intellectual, cultural, ethnic, linguistic and other characteristics [6–8].

Thus, the education of people with disabilities and limited health opportunities is one of the urgent problems in modern education, both at the level of preschool and at the level of school, secondary vocational and higher education [6].

The following areas of impact of the championship on the educational system as a whole can be distinguished:

Case 1: Technological support, which is associated with the organization of the accessible environment of the educational institution;

Case 2: Methodological support associated with the introduction of new teaching methods, the development of adapted programs;

Case 3: Psychological and pedagogical support, which consists in changing stereotypes in relation to people with disabilities.

2.2 Technological Support as a Tool for Creating an Accessible Environment

The involvement of people with disabilities in the educational process dictates the creation of an accessible environment and the appropriate infrastructure of the educational institution. Undoubtedly, technological support is of paramount importance, because. The implementation of the educational process is impossible without the formation of the physical environment of the educational institution, which covers many stages from the access of a person with disabilities to an educational institution to the organization of educational space and recreation areas [6, 9].

Addressing the problem of the disabled has led to the introduction of special technological tools designed for disabled people with various nosologies, both in educational institutions and in all socially significant infrastructure facilities.

So for wheelchair users and people with impaired musculoskeletal system, the accessibility of entrance groups, classrooms and other premises is organized by removing barriers, installing ramps, lifting platforms, creating special jobs with additional space.

Jobs are being created for persons with hearing impairments, including the use of sound amplifying systems for both collective and individual use.

For the visually impaired and the blind, space is being adapted by installing tactile devices (using Braille), sound beacons and warning systems, visual devices and contrasting color marking of surfaces.

Undoubtedly, the development of an accessible environment will contribute to the fact that persons with disabilities will receive more opportunities for learning and obtaining a profession in demand.

2.3 Methodological Support

The education of people with disabilities and people with limited health opportunities dictates the introduction of specialized software into the educational process and the organization of a distance learning environment that helps to master the material if it is impossible to attend an educational institution for objective reasons. In this case, the question arises of the use of various remote technologies (chats, video lectures, video conferences, electronic textbooks, electronic software and methodological complexes, electronic reference books, etc.). As practice shows, the developed educational and methodological materials are useful for organizing the educational process with the entire mass of students, especially when switching to forced distance learning during a pandemic.

There is a retraining of teachers, an increase in the level of their professional competence, the introduction of new teaching methods.

One of the main advantages of the championship is the fact that employers are directly involved in the development of competitive tasks, who, like no one else, know what professional skills employees should have, what knowledge, skills and abilities are most in demand today. Thus, teachers of educational institutions, and in most cases they are the ones who prepare the participants of the championship, get a real picture of the needs of the labor market, which leads to the actualization of educational and methodological materials, lectures, practical tasks, and this, in turn, favorably contributes to in-depth formation of professional competencies.

2.4 Psychological and Pedagogical Support

The inclusive policy of the state prioritizes the involvement of people with disabilities in modern industries. Economic instability, tough competition in the labor market are becoming predetermining factors for the process of formation of the competitiveness of students with disabilities and limited health opportunities. Therefore, in the field of inclusive education at a university, it becomes significant in the framework of psychological and pedagogical support to diagnose the professional self-determination of students with disabilities and limited health opportunities in order to establish personal motivation for the chosen type of professional activity and the demand for the labor market.

The mission of both the championship itself and the educational institution within whose walls it is held is to remove the psychological barrier in the process of communication, learning and adaptation, both for a disabled student and those with whom he interacts both in the educational process (fellow students, teachers, methodologists), and during the championship.

The championship has a significant impact on the understanding of the main idea, which is that disabled people should participate equally in the life of society, thereby undergoing full socialization, and also have equal rights to receive education, i.e. has a positive impact on the formation of an inclusive culture in education and in society as a whole.

The ideology of inclusion is based on the idea of an inclusive society, changing social institutions in such a way that they contribute to a tolerant attitude towards people with disabilities and limited health opportunities [10, 11].

As practice shows, graduates with disabilities often experience difficulties in finding a job after graduation. There are many reasons for this. First of all, not all employers are ready to hire an employee with a disability, despite the fact that at present the state policy is aimed at quota jobs in almost all organizations, regardless of ownership. On the other hand, people with disabilities or limited health opportunities often have an emotional barrier, expressed in anxiety and self-doubt, in fear of not being able to compete in competitive selection. This fact leads to a decrease in self-esteem, underestimation of one's own capabilities, disappointment, and dependency.

The «Abilympics» Championship is designed to break these prejudices. The movement contributes to a change in mass consciousness already at the stage of preschool, and then school and higher education. Taking part in the championships of professional skills, which involve the performance of competitive tasks of a sufficiently high level, students with disabilities demonstrate their «limitless possibilities» to their peers, develop the ability to present themselves, gain experience in mobilizing their own volitional efforts, which ultimately contributes to successful socialization in society. This positive experience prevents the occurrence of discrimination and contributes to the creation of a comfortable educational environment.

One of the psychological and pedagogical conditions of inclusive higher education is qualified career guidance support, focused on «adequate formation of professional self-determination after the first stage of higher education (bachelor's) and successful choice of a profession at the second stage (master's) and its development, taking into account the health limitations of a student with a disability and with disabilities» [12].

On the other hand, for the productive entry of students of this category into society within the framework of inclusive education, it is necessary to use various mechanisms of socialization and professional training in order to develop the personal activity of each such student through the experience of professional and personal development, which can be acquired by participating in the competitive movement of professional skills «Abilympics».

The functionality of the championship is to provide psychological and pedagogical support aimed at improving or acquiring new skills and abilities, as well as drawing attention to the professional capabilities of the competition participants from potential employers [13, 14].

Thus, the goal of psychological and pedagogical support for students with disabilities and limited health opportunities in the process of preparing and holding a professional skills competition is to involve the participant in social interaction, to provide an opportunity to integrate into the educational and professional community.

The conditions for psychological and pedagogical support of students with disabilities and limited health opportunities, implemented in preparation for participation in the competition of professional skills, include:

- Differentiated conditions (optimal load mode in preparation for participation in a competency, especially if the competency for students of this category is new, for example, a student of the direction «Construction» masters the competency «IT-technologies. Text processing»);
- Psychological and pedagogical conditions (taking into account the individual psychological characteristics of the student, including his psychophysiological capabilities; correctional and developmental orientation of the training process; compliance with a comfortable psycho-emotional regime; in order to optimize the process of preparing and mastering the competitive task, the use of modern pedagogical, information, computer technologies in efficiency and accessibility plan);
- Specialized conditions (promotion of a set of special learning tasks focused on the special educational and professional needs of students with disabilities and limited health opportunities in preparation for the competitive task; differentiated and individualized training, taking into account the specifics of nosology and the direction of training students with disabilities and limited health opportunities; complex impact on the student, carried out in individual and group preparatory classes);
- Health-saving conditions (improving and security regime, prevention of physical, mental and psychological overload of students in the preparation and in the process of performing a competitive task during the championship of professional skills, compliance with sanitary and hygienic rules and norms);
- In the process of psychological and pedagogical support of students with disabilities and limited health opportunities in the preparation and during the championship of professional skills, the formation of professional maturity as a personal neoplasm takes place.

3 «Abilympics» as a Tool for Personal and Professional Growth

3.1 The Championship is a Launching Pad for the Growth Trajectory of Participants

Participation in the championship allows a student with a disability to determine the further trajectory of his personal and professional growth, which can subsequently determine his competitiveness.

«Abilimpics» can act as a tool for self-realization, on the other hand, it can be the basis for the formation of the competitiveness of students with disabilities and disabilities in the modern labor market.

Under the conditions of the competitive system, the competitiveness of students with disabilities and limited health opportunities is formed with the development of professional competence.

Professional competence (Dyachenko N.S., Karabanova L.B., Kuzmina A.A., Matveeva T.Yu.) is «a set of knowledge, skills and abilities that allow you to effectively and efficiently perform professional tasks in your subject area» [15].

Championships of professional excellence allow to analyze both the readiness of participants to perform professional activities, and allow to assess the level of development of professional competencies, which is a condition for determining the competitiveness of participants in the future during professional self-realization.

At present, traditionally, they began to take into account that such championships of professional excellence are not only a tool for determining the professional competence of participants, but also «a new technology for improving the skills of specialists, a method for training or identifying leaders in the professional field» [15].

3.2 The Prospect of Professional Realization and Building a Career of the Participants of the Championship

The main mission of the «Abilympics» championship is aimed at developing professional skills for people with disabilities and health limitations. In addition, participation in a professional competition is a start for further personal and professional implementation and a successful professional career. At the championship venues, participants may have the opportunity to receive professional assistance in finding a job.

«Abilympics», as an effective tool for career guidance, motivation, socialization and employment of people with all types of disabilities and health limitations, has a huge impact on the formation of a culture of attitude towards people with disabilities and health limitations in society.

Increasing requirements for universal personal competencies (communication skills, professional activities in a team, client-oriented approach) of a potential employee contribute to the improvement of skills and competencies, vocational training and retraining of a candidate with health limitations and disabilities, making him more in demand in the labor market and increasing his chances of employment.

As a result, the experience of holding a championship of professional excellence in the educational space shows that this event is more aimed not only at professional orientation on issues of further professional education in order to provide effective professional orientation and motivation to receive the next stage of professional education (master's degree) and promote sociocultural inclusion, but also on the implementation of personal and professional growth as predictors of the competitiveness of the championship participants.

4 Conclusion

Based on the foregoing, we can conclude that the «Abilympics» professional excellence championship has a significant impact on the modernization of the system of higher and secondary vocational education, changing it in three directions: technological, methodological, and psychological and pedagogical support.

The «Abilympics» movement, in the context of inclusive education, promotes successful sociocultural inclusion, socialization and integration of people with health limitations and disabilities into the labor sphere.

The championship as a competitive practice contributes to the development and implementation of personal and professional growth, thereby increasing the competitiveness of participants in the labor market.

Participation in the «Abilympics» professional excellence championship for students with health limitations and disabilities is an opportunity not only to show their professional skills and abilities, but also to acquire new knowledge, skills and abilities, which increases the competitiveness of participants during professional self-realization.

For organizations of higher or secondary vocational education, on the sites of which the championship is held, as well as the active participation of their students, it provides an opportunity to demonstrate that the environment of educational institutions, including the social one, is accessible not only for teaching people with health limitations and disabilities, but also for the implementation their developed professional skills and abilities.

References

1. Laden, T.: History of the international movement «Abilympics». Corporate Magazine of the «Abilympics» National Center, 1, 3–6 (2021)
2. Bikbulatova, A.A., Gribkova, N.B., Matraeva, L.V., Erokhin, S.G., Makeeva, D.R., Karplyuk, A.V.: The Russian historical aspect of the development of the international federation of «Abilympics». Res. J. Pharmaceutical, Biol. Chem. Sci. 9(5), 329–335 (2018)
3. Alekseevskikh D.Y.: «Abilympics» championships as a tool for career guidance and employment of people with disabilities. In the Collection: Actual Problems of Training and Promotion of Employment of Graduates of Engineering Universities from Among the Disabled. Collection of Articles, Reports and Materials of the All-Russian Conference, pp. 132–136 (2016)
4. Ahuja, S., Kaur, P., Panda, S.N.: Identification of influencing factors for enhancing online learning usage model: evidence from an Indian University. I.J. Educ. Manag. Eng. 2, 15–24 (2019)
5. Ghanyani, U.-S., Murad, M., Mahmood, W.: Crowd-based requirement engineering. I.J. Educ. Manag. Eng. 3, 43–53 (2018)
6. Gefele, O.F., Fomina, E.E.: Features of technological support as a predictor of successful implementation of inclusion in the system of higher education. Psychology and Pedagogy of Inclusive Education: Technology, Efficiency, Quality. Materials of the All-Russian Scientific-Practical Conference with International Participation. Lipetsk, pp. 74–77 (2021)
7. Kumar, M., Singh, A.J., Handa, D.: Literature survey on educational dropout prediction. I.J. Education and Management Engineering, 2, 8–19 (2017)
8. Xiu, L.P.: Based on Network moral education for teenagers path choices of dual influence. I.J. Education and Management Engineering, 10, 59–64 (2012)
9. Ostapenko, R.I., Bakulina, L.S., Barkalova, E.V.: Status and trends in the development of inclusion in Russia (based on the materials of the All-Russian meeting on inclusive vocational education). State Advisor **4**(24), 57–61 (2018)
10. Tyutyueva, I.A., Filyutina, T.N., Kostyleva, A.S.: Implementation of methodological approaches in preparing students with disabilities for professional competitions. Pedagogical Educ. Russia **5**, 162–169 (2020)
11. Podyablonskaya, O.V.: «Abilympics» is an Olympiad of opportunities. In the Collection: Strategic Directions for the Development of Education in the Orenburg Region. Scientific and Practical Conference with International Participation, pp. 831–835 (2017sss)
12. Evstifeeva, E.A., Filippchenkova, S.I., Gefele, O.F.: Features of the formation of the competitiveness of students with disabilities and with disabilities in the conditions of inclusion in the university. In: A Disabled Person in the Society of the XXI Century: a Collection of

Works of the II All-Russian Scientific and Practical Conference. Moscow, November 21–22, 2020. Moscow: Publisher: Moscow State University for the Humanities and Economics, 71–74 (2021)

13. Gribanova, E.M.: Competitions of professional skills «Abilympics» among persons with disabilities and disabilities. Pedagogical Search **8**, 27–30 (2021)
14. Isaeva, Y.N.: Accompanying persons with disabilities in preparation for participation in the Abilympics professional skill championships. Psychological and pedagogical support of the educational process: problems, prospects, technologies. In: Materials of the VII International Scientific and Practical Conference. Edited by A.I. Akhulkova, pp. 161–164 (2020)
15. Dyachenko, N.S., Karabanova, L.B., Khalimova, N.M.: Competitive movement as one of the conditions for the development of professional competencies of students of secondary vocational education. World Sci. Cult. Educ. **2**(81), 101–105 (2020)

Mobile Applications in Modern Social and Cultural Educational Practices

Yuliia Polovynchak(✉) and Viktoriia Bondarenko

National Law Library V.I. Vernadskyi National Library of Ukraine, Kyiv, Ukraine
polovynchak.j@nas.gov.ua

Abstract. The paper is devoted to the study of the development of educational mobile applications. By comparing the specifics of their distribution over the past five years, the authors found out the main trends in their distribution. Firstly, the formation of complex, multi-thematic and multi-format platforms for providing access to training courses has been defined as such. Secondly, the tendency to formalize self-education (the possibility of obtaining official certificates confirming the successful completion of the course). Thirdly, the growing role of various types of mobile applications as tools in the classical learning process. The principle of collaboration of different institutions in the creation of complex and high-tech mobile applications has also been noted as a trend. The connection between changes in the audience of educational services and the specifics of mobile applications as their tools has been substantiated. Emphasis has been placed on such characteristics of mobile applications as interactivity, virtuality, multimodality, personalization, and gamification.

Keywords: Mobile learning · Distance learning · Mobile applications · Self-education

1 Introduction

With Changes in the requirements for workable (effective) educational strategies have become characteristic of modern social and cultural processes. They are related, first of all, to the new features of the audience of educational services. Mosaicism, visuality, interactivity, and virtuality of perception have become typical of recipients of education of the information age. These characteristics obviously reduce the effectiveness of the classical linear teaching and learning process. At the same time, they can be a prerequisite for successful learning – in situations where educational methods not only consider these characteristics, but also work with them. This approach requires not only a revision of teaching and learning strategies, but also the selection of new teaching methods and tools.

Secondly, since the requirement of current modernity (according to Z. Bauman) is lifelong learning, consumers of educational services are much more likely than before to be professionals who either improve their professional level or acquire skills or abilities for self-fulfillment. Therefore, studying "on the run" became typical – in parallel with

Z. Hu et al. (Eds.): AIMEE 2022, LNDECT 159, pp. 351–358, 2023.
https://doi.org/10.1007/978-3-031-24468-1_31

professional activities or basic studying. Sometimes this metaphor takes on a literal meaning when dealing with self-education through listening to different types of audio materials during a run / a road trip. Children, adolescents, young people, as has been already mentioned – given the changing specifics of perception, also require a paradigm shift: from learning "behind the desk" to learning "in the game" and "in communication", as well as learning "in life", "in practice".

It should also be borne in mind that distance learning has become widespread today – as a forced response to epidemiological threats, but at the same time as a consciously chosen advantage. The growing share of distance learning, without a doubt, is also accompanied by a review of teaching methods and the search for information tools to ensure it.

It is obvious that educational and scientific communications undergo bilateral changes: from the demand for new teaching methods and acquiring of knowledge to the proposal of new information and communication technologies involved in the process of learning and personality formation. Mobile information and communication technologies can be considered as one of the effective means of educational communication in the knowledge society. This concept combines mobile devices with relevant software, content developed or adapted for use with such a device, as well as techniques and methods of working with them. The most popular mobile technologies include applications created by developers to implement a variety of tasks, including personality-centered education.

2 Research Methodology

The study of the specifics of the mobile application use in social and cultural processes requires a combination of methods of technical sciences and humanities. The use of social and cultural methods allows finding out the features of the audience of the educational process, the specifics of perception and assimilation of information by modern youth. Conclusions on these features, in particular, on the effectiveness of mobile applications, have been made based on the results of both the observation of student youth and the results of the questionnaire.

In the course of the research, the 50 most popular free and commercial applications focused on educational and developmental communication, according to the developers and sellers, have been identified and analyzed. The analysis of their content made it possible to substantiate the conclusion about the effectiveness of virtualization and gamification of the educational process. The comparative method and the method of classification have been used to determine the subject matter and orientation of popular educational applications.

3 Results of the Research

By Given the social significance of changes in the consumption and assimilation of information, as has been discussed above, both the scientific community and practitioners have quite actively discussed the issue of ensuring a new quality of education in view of state-of-the-art information and communication processes in recent decades.

H. Tkachuk, T. Yaroshenko, M.L. Bernackia, J.A. Greenea, H. Crompton, S.E. Elverici, H. Caldwell, K. Balliammanda considered the use of mobile information and communication technologies in education and new trends in the organisation of scientific and communication activities on the Internet in their research.

H. Tkachuk considered the introduction of mobile learning in terms of methodologies of the educational process: identifies the main components of mobile learning, analyses the types of content [1].

T. Yaroshenko clarifies the specifics of the impact of the Internet, cloud technologies, e-platforms and e-infrastructure, specialized services and resources, data search tools, mobile applications on the ways of scientific communication, their creation, openness and accessibility [2].

The scientific collection "Mobile technology, learning, and achievement: Advances in understanding and measuring the role of mobile technology in education", edited by M.L. Bernackia, J.A. Green, H. Crompton, studies the prospects and unstudied expectations for mobile learning. The materials of the collection consider mobile technologies as key components of empirical research, study their impact on the learning process and products, identify opportunities for collecting previously inaccessible data that improve understanding and simulation of the learning process [3]. The research "Switching Gears: Moving from e-Learning to m-Learning" explores challenges, opportunities and constraints reflected in efforts to transition from reliance on formal learning management systems for course development and delivery toward the infusion of media targeted to students' mobile devices [5].

A number of researchers, along with educational results, emphasize the effectiveness of communication technologies in socialization of students [6].

Caldwell H. studies the use of mobile technologies in education of teachers and quite rightly, in our opinion, defines them as a catalyst for new pedagogies based on a social constructivist model of learning in the teacher education programmed [7].

Traxler J. explores and articulates issues of mobile learning, in the context of develop the conceptualizations and evaluation of mobile learning. Researcher recognize that mobile learning is essentially personal, contextual, and situated [8].

A review and analysis of publications on the "mobilization" of the educational process, performed by one of the co-authors of this study, showed that modern mobile devices are used in the field of administration and organization of teaching for experts-practitioners, as well as technical support for learning for pupils and students, as well as young scientists. The earlier made conclusion that "mobile applications of higher education institutions, among other things, perform the following functions: administrative (informing about the schedule of classes, faculties, departments and teachers of the university); scientific and educational (informing about important scientific and educational events and news of the university); socio-cultural (the possibility of using the application as a "university social network")" remains relevant.

The observation made at one time about the transfer of educational materials (course plans, lecture notes, archives of video courses, assignments, exam questions) to leading universities for free access not only did not lose its relevance, but also.

gained further development. Thus, we can say that by creating publicly available mobile applications with educational materials, universities convert this content into social capital, demonstrating social responsibility and increasing the availability of knowledge and, consequently, the level of culture [9].

Special mobile applications run courses developed by universities, for example, Courser Online courses accumulates university developments to teach their courses online.

The above-mentioned predominance of visual perception confirms the popularity of platforms that accumulate video lectures, speeches and other educational, scientific or cognitive content, such as Khan Academy or TED.

Another type of mobile application that remains popular is multi-industry projects, which actually act as collections of a variety of online courses. These include, for example, Udemy Online Courses, which is positioned as a global market for online learning and teaching, offering tens of thousands of diverse courses for millions of students. The Ukrainian analogue (not so large-scale, but localized to the needs of its users) is the Prometheus project, which provides free high-quality educational opportunities, including through its mobile version, which is perfectly adapted to the user's device. The Ukrainian educational project Educational Era (EdEra) also creates full-fledged online courses.

In our opinion, educational platforms in the format of the "Meccano" of training courses, which include collections of forms and educational content with the ability for registered users to create their own projects using existing material or supplementing it with their own, can be considered promising. For example, Moodle (acronym for Modular Object-Oriented Dynamic Learning Environment) is a learning platform intended to create a personalized learning environment.

4 Discussion of Results

A new trend is the development and dissemination of applications, which are not so much complete educational or cognitive projects, as separate tools for in-depth work with sources. They can be considered an effective tool for gaining knowledge "in life" mentioned at the beginning of the paper. Such tools can be the applications based on the use of artificial intelligence for recognition of different objects. For example, Google Lens recognizes visual objects taken with a smartphone camera and automatically redirects them to the Internet resources with information about them; Shazam is a similar tool – an application that works with audio content. Thus, these programmed work on recognizing and providing encyclopedic help based on their own database and the Internet resources. The World History Timelines, Maps & Videos application based on an interactive timeline will help comparing events in space and time, detailing information with the help of video. These applications can be useful for the implementation of any project outside the classroom: student research on nature, human settlements, art. Similar approach is effective in organizing a variety of excursion programmed.

In fact, the diversification of practical learning is not only connected with the use of special online opportunities during traditional excursions, as mentioned above, but also through the transfer of study excursions and expeditions to virtuality. There are two promising applications in this regard: the interactive virtual world museum History pin, which combines materials from hundreds of cultural and historical institutions around the globe with user content (users only need to add their own photos to Google public map and accompany them with the text). History pin is characterized by high interactivity typical of this type of project – from user participation in filling the project with content to the possibility of creating scientific or artistic collages, forming collections or tours according to their own characteristics. A similar project is the Google Expeditions application, which offers 200 expeditions to different parts of the world. The user can either organize his/her own virtual expedition (choose a route, task, team), or join an already developed virtual trip.

The application-tools for the educational process may include narrower programmed, such as "Medieval Handwriting" for recognition of medieval manuscripts provided by the application. The possibilities of access to the source, reference encyclopedic material is growing, as, for instance, thanks to the Vkraina application, which presents digital versions of rare historical maps of the XVI-XVIII centuries.

At the same time, one of the largest in this field can be considered the "SCARLET" project (Special Collections using Augmented Reality to Enhance Learning and Teaching) developed by the University of Manchester and the Jisc Non-Profit Organization (focused on digital services and solutions in higher education institutions in the UK). The project is designed to develop effective tools to study the collections of rare books, manuscripts and archives and is based on the use of augmented reality. In essence, such information products operate on the principle of integrated interdisciplinary information products that allow the user to access distributed information from a single access point via a mobile phone, integrating not only information about the virtual space of the Internet environment, but also real, physical space [10].

It is also safe to say about the continuing trend towards self-education, in particular, the popularity of gamified (using gaming technology) and mobile (and therefore available at any time) self-training for various subjects. The Ukrainian material (statistics of downloads of popular applications) testifies to the active use by senior pupils of a variety of training to prepare for final and entrance exams (external independent evaluation).

Foreign language learning applications remain very popular and in demand. The specificity of the most popular ones is a rather narrow methodological base: the main component is the creation of a language environment and assistance in the implementation of language practices. Such applications connect millions of speakers from around the world who are interested in communication, thus making it accessible, differentiated by languages and the level of communication complexity, and an interactive educational space. At the same time, such methods allow intensifying the teaching and learning process. Thus, mobile applications focused on learning foreign languages can be part of a set of efforts aimed at language acquisition, or a way to maintain language practice and the appropriate level of proficiency in a particular language.

It is worth noting the importance of such practices for speakers of minority languages, for whom, especially outside the area of use of such a language, it was previously difficult to find a way to maintain a sufficient level of communication. Thus, even in the light of globalization processes, accompanied by the growing influence of global languages, especially English, we should not forget about additional opportunities for the implementation of much less common linguistic identities, their preservation and development when needed. We should note here the importance of implementing language practices through specialized mobile applications.

This trend of intensification of minority or even endangered languages (in Ukraine, they include Crimean Tatar) fits into the logic of current postcolonial processes, intensified by the spread of information technology. At the same time, the Ukrainian reality of these processes is manifested through rather specific, but at the same time demonstrative unnaturally high popularity of applications for the study or improvement of the Ukrainian language. Prolonged Russification processes that took place in the past, along with the functioning of a common information space with Russia until now, where the dominance of the Russian language is determined by a trite quantitative advantage, have led to the formation of a large number of Ukrainians aware of the need to increase Ukrainian language skills. Part of the audience of relevant applications, numbering in the tens, if not hundreds, may, of course, be foreigners in need of language practices. However, both statistics and surveys show that the share of Ukrainians among users is high enough to consider this issue in the context of postcolonial processes.

Multimodality, virtuality, interactivity, as we see, are the defining characteristics of the new generation of educational applications. It is worth noting that before that we were talking about a tool for education and self-education of adolescents, students or adults. It is obvious that these characteristics will only intensify for younger users, and in their case, it will be more of cognitive and entertaining applications. These include, for example, two thematic projects supported by the Ukrainian Cultural Foundation: "Charivnyi Svit.UA", dedicated to the Ukrainian mythological characters, and "Znai Nashykh" - a project by artist Yurii Zhuravel about the great Ukrainians. Competitive element and gamification are the basis of another cognitive application aimed at children and adolescents: "Moia Ukraina" – a mobile application that collected more than 600 questions from 10 different areas of knowledge related to Ukraine (history, culture, geography, science, sports, etc.). Therefore, a combination of visual material, video, game elements, test tasks make these applications effective tools for obtaining or updating knowledge of ethnography, folklore, history, geography for audiences of all ages, including children.

5 Conclusion

Thus, mobile applications designed for educational or developmental purposes have been studied in terms of their development over the past five years. Statistical analysis of the popularity of relevant developments in accordance with their content, technologies used, current methodological approaches allowed drawing a number of conclusions.

Firstly, along with the preservation of popularity of traditional platforms that accumulate video lectures – from individual speeches to entire courses, they become more sophisticated, turning into a kind of "meccanos" for the formation of educational material blocks according to individual requirements. The process of formalization has gained development with regard to them – the opportunity to not only take courses at the most prestigious universities, but also, obtain a certificate by passing online exams. Another feature is that such platforms become a tool not only for self-education, but also for teaching.

Secondly, with the sophistication of application-based technologies, such as augmented reality technologies or recognition algorithms using artificial intelligence, a cluster of applications is emerging that can be defined as highly effective tools for intensifying the learning process. This intensification focuses mainly on virtualization, interactivity, multimodality and multioperativeness (a combination of not only different media, but also different operations).

That is, the defining feature of educational processes is the gradual retreat from the linear construction of courses and their mostly textual representation, focusing on the multidimensionality of understanding the surrounding reality. Along with this, there has been a tendency towards the emergence and spread of applications – tools for independent practical research.

Thirdly, the spread and sophistication of mobile technologies in educational processes directly reflects the focus on the personalization of learning – from the variability of educational programmed to the individual selection of effective methods and techniques of work. An important aspect of personalization and interactivity of education and self-education is compliance with the needs and capabilities of the audience. Thus, the removability and interactivity of education provide compensatory functions – from the ability to access educational and scientific resources regardless of place of residence, health and social status, to overcoming postcolonial or any other destruction.

At the same time, the analysis is the basis for forecasting the further development of this area of education and self-education. Given the sophistication of the toolkit, the increase in its number to a situation where you need to navigate the existing resources, we can predict that in the future its effective use will be associated not so much with fully autonomous self-education, but with mediation, curatorial, methodological efforts of experts. That is, the use of new technologies as an element of successful acquisition of knowledge is the most promising in the context of systemic education.

Acknowledgment. Yuliia Polovynchak made conclusions regarding the socio-cultural impact of modern digital capabilities of mobile applications.

Viktoriia Bondarenko studied mobile applications as one of the tools of education and self-education.

References

1. Tkachuk, H.: Peculiarities of the implementation of mobile learning: prospects, advantages and disadvantages. Inf. Technol. Teach. Learn. Aids **64**(2), 13–22 (2018). (in Ukrainian)
2. Yaroshenko, T.: Digital support for research at a modern university: key challenges. In: Material of the Speech at the XII International Scientific-Practical Conference "Information, analysis, forecast - strategic levers of effective public administration", 2019 (Electronic resource), access mode: http://ekmair.ukma.edu.ua/handle/123456789/16527 (in Ukrainian)
3. Bernackia, M.L., Greenea, J.A., Crompton, H.: Mobile technology, learning, and achievement: advances in understanding and measuring the role of mobile technology in education. Contemp. Educ. Psychol. **60** (2020) (Electronic resource). https://mobt3ath.com/uplode/books/book-97227.pdf
4. Crow, R., et al.: Switching Gears: moving from e-learning to m-learning. MERLOT J. Online Learn. Teach. **6**(1) (2010) (Electronic resource). https://www.researchgate.net/publication/228821708_Switching_Gears_Moving_from_e-Learning_to_m-Learning

5. Elverici, S.E.: Mobile technologies for education: attitudes to social media. Eur. J. Foreign Lang. Teach. **5**(3), 11–12 (2021)
6. Ahmad, T.: Student perceptions on using cell phones as learning tools: Implications for mobile technology usage in Caribbean higher education institutions. PSU Res. Rev. **4**(1), 25–43 (2020)
7. Caldwell, H.: Mobile technologies as a catalyst for pedagogic innovation within teacher education. Int. J. Mobile Blended Learn. (IJMBL) **4**(2) (2018)
8. Traxler, J.: Defining, discussing, and evaluating mobile learning: the moving finger writes and having writ, Int. Rev. Res. Open Distance Learn. **8**(2) (2007)
9. Bondarenko, V.: Mobile applications as a means of communicating in a knowledge society: an educational aspect. Acad. Papers Vernadsky National Library Ukraine (48), 576–590 (2017) (in Ukrainian)
10. Bondarenko, V., Hranchak, T.: Augmented reality library projects (AR): foreign experience. Ukrainian J. Library Sci. Inf. Sci. (48), 100–114 (2021) (in Ukrainian)

Research and Practice of PLC Experiment Teaching Based on OBE and Digital Twin

Yuexia Tang and Guangping Li(✉)

College of Electromechanical and Quality Technology Engineering, Nanning University, Guangxi, Nanning 530200, China
118039434@qq.com

Abstract. Experimental teaching refers to an organizational form of practical teaching in which students observe, measure and analyze research objects under the guidance of teachers, using instruments and equipment, and under the condition of human control, so as to obtain knowledge and develop ability. It has been widely used in basic courses and professional courses in Colleges and universities. Its purpose is not only to verify book knowledge, but also to cultivate students' ability to correctly use instruments and equipment, test, adjust, analyze, synthesize, design experimental schemes, and prepare experimental reports. In view of a series of problems such as insufficient investment of experimental equipment and no hardware involved in virtual simulation in PLC experimental teaching, this paper discusses the basic idea and related problems of developing experimental teaching platform by using digital twin technology. Integrate the OBE education concept into the teaching, define the ability training objectives, establish the experimental course system, carry out various teaching reforms relying on the digital twin technology, and complete the teaching evaluation through the student ability training effect experiment, providing a new idea for the PLC Course Experimental Teaching Reform and the application of the digital twin technology in the teaching.

Keywords: Experimental teaching · PLC · OBE · Digital twin

1 Introduction

1.1 OBE Education Concept

Outcomes based education (OBE) was first proposed in the basic education reform in the United States and Australia [1, 2]. It is a reverse design based on the orientation of training objectives, which requires educators to have a clear understanding of the abilities that students should achieve when they graduate, and design the corresponding education system to provide guarantee. Emphasize the main role of students in the learning process, and pay attention to cultivating students' ability to transform theoretical knowledge into solving practical problems [3–5]. Compared with the traditional higher education idea, OBE education idea provides a clearer goal and more flexible method for the reform of experimental teaching.

Z. Hu et al. (Eds.): AIMEE 2022, LNDECT 159, pp. 359–368, 2023.
https://doi.org/10.1007/978-3-031-24468-1_32

1.2 Development Background of Digital Twin Experimental Teaching Platform

With the acceleration of informatization, in recent years, a large number of open and shared high-quality teaching resources have been popularized all over the world, which has brought great convenience to network theory teaching. However, it is not enough for practical teaching to learn only by watching videos. Students need to do more to ensure good learning results, otherwise it is difficult to achieve the ability and level that match the technical standards of professional posts.

Digital twin DT was first proposed in the training course for senior executives in Michigan, USA in 2003. It creates a virtual simulation model of physical entities in a digital way, simulates the behavior of physical entities with the help of data, promotes the interaction and integration of the physical world and the information world through virtual and real interactive feedback, data fusion analysis, decision iteration optimization and other means, and adds or expands new capabilities for physical entities. Since 2010, with the promotion of NASA, Siemens and other institutions and companies, digital twin technology has been applied more and more widely. Domestic scholars Tao Fei and others have also studied digital twin technology in many aspects, and put forward the standard system, five-dimensional model and ten application fields based on digital twin technology [8, 9]. Cuiyihui, jiangaiguo and others applied digital twin technology to the intelligent monitoring system of aeroengine and drilling platform [10, 11]. Chu LeYang et al. summarized the application of digital twin technology in the field of education [12, 13]. In general, the practical application in teaching is less involved. Digital twin technology has the technical characteristics of virtual reality symbiosis, high virtual simulation, high real-time interaction and deep insight. It has great application value for solving the practical problems in the experimental teaching of PLC Course [14, 15]. Based on the OBE teaching concept, this paper develops a PLC experimental platform based on digital twin, and puts forward feasible solutions to a series of problems existing in the experiment.

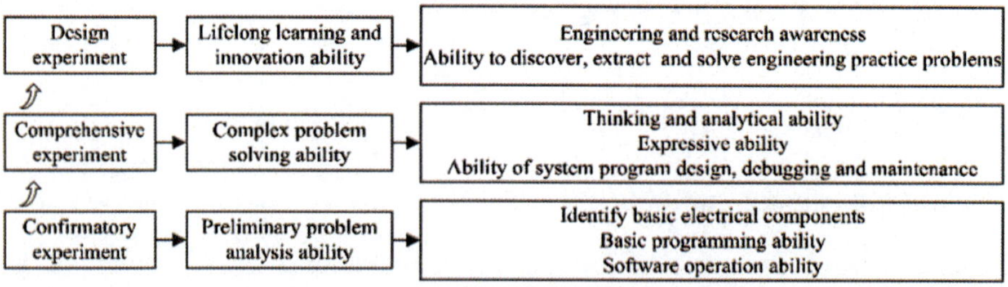

Fig. 1. Objectives of curriculum ability training

2 OBE Capability Objectives and PLC Course System

2.1 Capability Objectives Based on OBE Concept

The OBE education concept is result oriented. The results mentioned here are the final learning results of students and the maximum ability that students can achieve after learning at a certain stage. Systematically study the application of OBE concept in PLC course experiment teaching, and determine that students' learning achievements are three progressive abilities, such as preliminary problem analysis ability, complex problem solving ability, lifelong learning and innovation ability, as shown in Fig. 1. The level of confirmatory experiment is low, which is to let students verify the theoretical knowledge learned through experiments and cultivate students' ability to analyze preliminary problems. Comprehensive experiment is the flexible use of curriculum knowledge points to cultivate and improve students' ability to solve complex problems. The design experiment has the highest requirements. The students design and implement the experimental scheme to cultivate the students' lifelong learning and innovation ability.

2.2 PLC Course System

PLC (programmable logic controller) is a practical and comprehensive course. Students should complete hardware circuit design, programming, software and hardware debugging and other work in the course experiment. At present, there are three main PLC experiment methods and problems:

(1) The real PLC controller and experimental module are used to debug the program. When the program is running, the LED indicator on the experimental module is observed to verify the program. Because there is no actual controlled object, lack of hardware facilities, and no workplace environment, students' interest in learning is not high, and the learning effect is poor.
(2) Using the comprehensive experimental device developed by a professional teaching instrument company. This kind of experimental device has a wide variety of controlled objects. It can use real sensors to control various actuators. Through the disassembly, installation, programming and operation debugging of the equipment, students have high enthusiasm for participation, which can improve students' practical ability. However, due to the limitations of the number, variety and site of experimental devices, the experiment requires long hours, consumes a lot of experimental consumables, and the purchase and daily maintenance costs of experimental devices are high, resulting in the decline of experimental teaching effect.
(3) Using virtual simulation technology, create virtual environment or tools to simulate real cases for operation, verification, design, operation and other experiments. Through the visual virtual simulation of the experimental object, the structure, function, behavior and the dynamic simulation of the user's thinking process and behavior are completed. In this experimental mode, only the program software is debugged without involving hardware circuit design, and the course requirements cannot be fully met [16, 17].

The above problems are common in all colleges and universities. It needs to be solved urgently.

3 PLC Experimental Teaching Platform Based on OBE and Digital Twin

The construction of experimental teaching platform can provide powerful conditions for the improvement of students' practical ability [18]. In order to meet the requirements of the results oriented education of the OBE education concept, classroom teaching should achieve at least the following five changes: from the indoctrination classroom to the dialogue classroom, from the closed classroom to the open classroom, from the knowledge classroom to the ability classroom, from the emphasis on learning over thinking to the combination of learning and thinking, and from the emphasis on teaching over learning to the focus on learning. According to the characteristics of PLC classroom experiment and the classroom teaching requirements of OBE teaching concept, this paper establishes a PLC experimental teaching platform based on OBE and digital twins. The overall framework of the experimental platform is shown in Fig. 2. The experimental platform is mainly composed of three modules: Digital twin model, data acquisition plug-in and IO interface circuit board.

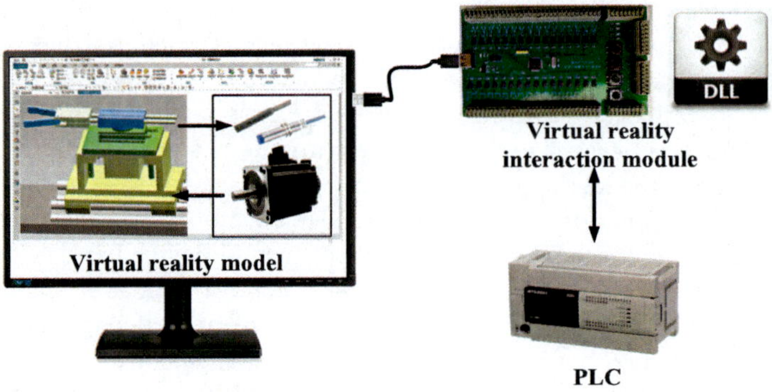

Fig. 2. Structure of PLC experimental platform based on digital twin

The digital twin model is established by NX-MCD software, including virtual mechanical structure, virtual sensor and actuator; Independently developed and produced IO interface circuit board to realize virtual sensor realization and actual actuator virtualization; The data acquisition plug-in is developed by using c\n\e program to realize the data sharing between the virtual simulation model and the IO interface circuit board. During the experiment, the PLC controller is introduced into the experimental platform. Through the wiring between PLC and IO interface circuit board, the virtual sensor and virtual actuator in the digital twin model are connected with PLC to realize virtual real interaction, complete hardware circuit design exercises, and solve the problems that do not involve hardware circuit design under the teaching mode of full virtual simulation experiment.

As the digital twin model takes the photos of various parts of the equipment collected on site, size and other related materials as the modeling standard, the effect is more realistic, and it has a more on-site experience than the LED lamp of the observation experiment module, so the students are highly motivated. Different production and experimental equipment can build the digital twin model. The cost of building, using and maintaining the digital twin model is far less than the actual equipment, which solves the problem of insufficient laboratory investment. The specific use process is shown in Fig. 3.

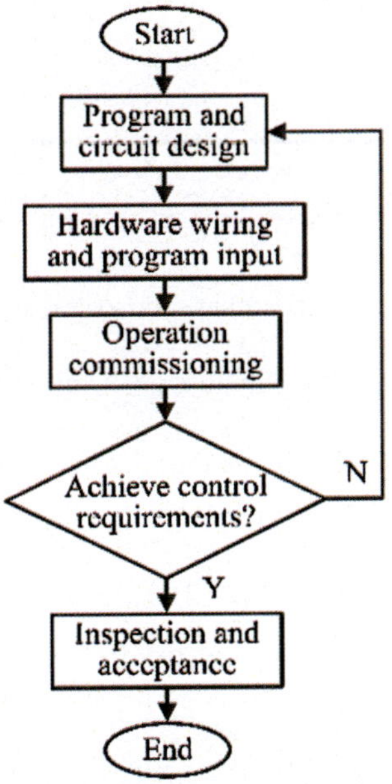

Fig. 3. Use process of experimental teaching platform

It is very necessary to strengthen the reform of teaching contents, teaching methods and examination methods [19].

In terms of teaching content, first of all, a set of digital twin models can be used at multiple computer terminals at the same time. The cost of IO terminal blocks is low, the number of experimental equipment is sufficient, students' per capita operation time is increased, their familiarity with the experiment is improved, the mechanical repetition related to operation is reduced, and more time is left for students to explore and communicate with teachers and students. Secondly, it is not limited to the existing hardware conditions, carry out school enterprise cooperation, use digital twin technology to build models of various industrial automation equipment, follow up the update of enterprise equipment at any time, broaden students' knowledge application background, draw inferences from one instance, stimulate students' interest in learning, and cultivate students' exploration spirit and practical ability. Finally, gradually reduce the proportion of confirmatory experiments in the experimental curriculum system, and increase the proportion of comprehensive and design experiments.

Table 1. Test scores of self-locking circuit

Assessment	Score	Criterion
functional testing	70	(1) when both conveyor belts are not open, the signal lamp will be on continuously (10 points) (2) press the start button 1, the upper conveyor belt runs (5 points), the signal light flashes as required (10 points); and press the stop button 1, the conveyor belt stops running (5 points) (3) press the start button 2, the lower drive belt runs (5 points), the signal light flashes as required (10 points); press the stop button 1, the conveyor belt stops running (5 points) (4) both conveyors are running and the signal light is not on (10 points) (5) after the experiment, turn off the computer and clean the test bench (10 points)
Innovation points	30	(1) 10 points will be given to the top 10 and 5 points will be given to the 11th-20th (2) add new functions on the basis of fundamental requirements (20 points) (3) points are given as appropriate in other cases
Deduct points	some	(1) 5 points is deducted for those who cannot answer the questions correctly during the experiment (2) if there is no self-inspection and the answer is wrong, 10 points is deducted (3) 10 points is deducted for live wiring or disconnection (4) if the equipment is damaged due to non-standard operation, 10 points is deducted (5) 20 points is deducted if incorrect operation causes short circuit and laboratory trip (6) deduct points according to other circumstances

As for the teaching methods, the PLC experimental platform based on digital twin adopts the interconnection of real PLC controller, virtual sensor and virtual actuator, so as to realize the deep virtual real fusion and provide greater fault tolerance. While providing experimental interactive experience, it also greatly improves the safety of the experiment, changing the traditional "closed" teaching situation when students.

conduct electrical operation experiments, Students boldly try various ideas and ideas, solve various problems encountered, and improve their comprehensive ability.

About assessment methods, according to the diversity of digital twin models, the traditional assessment method of "uniform test questions" is changed, and the open or semi open assessment method is adopted for comprehensive evaluation. Specific scoring standards are set for each experimental project: each experiment sets the scores for basic operation and requirements, adds challenging content or free play for students, and gives innovation bonus points, as shown in Table 1. The results of each experiment are included in the final grade to improve the students' attention to the experiment and enthusiasm for participation. Through the open or semi open experimental content and the whole process evaluation, stimulate students' interest and internal motivation in experimental exploration, guide students to become the main body of learning, and cultivate students' innovative consciousness.

Relying on the PLC experimental teaching platform based on digital twin and under the guidance of OBE concept, the above five changes have been basically realized through a series of teaching reforms in teaching contents, teaching methods and assessment methods.

4 Development, Application and Effect Evaluation of Experimental Teaching Platform

4.1 Development of Teaching Resources of Digital Twin Experimental Platform

Experimental teaching resources are the basis for all experimental teaching activities. The teaching resources of PLC experimental platform based on digital Twin are mainly the digital twin models of various industrial automation equipment. Taking the opportunity of the construction of the school level teaching team - the design practical teaching team of Nanning University, through the combination of introduction and training, a high-quality and reasonably structured practical teaching and development team has been formed. The teaching team introduced the digital twin technology into the PLC experiment teaching, established the PLC Experiment Platform Based on the digital twin, and trained the skills of insufficient number of equipment, unrepeatable, high loss, complicated maintenance, long time-consuming and high risk coefficient in the experiment through the digital twin model repeatedly, so as to achieve effective learning consistent with the operation of the production process. After two years of construction, the teaching team has built and accumulated a number of teaching resources. These teaching resources are all open to students. In addition to completing course experiments, students can also carry out various independent innovation practices. On the one hand, it can improve students' comprehensive ability of consulting materials, 3D modeling, programming and debugging. On the other hand, it can also quickly increase the number of models.

4.2 Teaching Evaluation

After the completion of the PLC experimental teaching platform based on the digital.

twin, it has been applied to the PLC experimental teaching of the electrical engineering and automation majors in 2015 and 2016, and achieved remarkable results.

(1) The teaching effect has been significantly improved

After introducing the OBE education concept and establishing a PLC experimental teaching platform based on digital twins, the practice was carried out for the PLC experimental teaching of the electrical engineering and Automation Specialty in 2015 and 2016. The course assessment score table is shown in Figs. 4, 5 and 6. From the analysis of assessment scores, compared with 2014 students who have not undergone teaching reform, 2015 students have significantly improved their mastery of theoretical knowledge and practical application ability. Subsequently, the teaching team continued to improve the teaching resource construction, teaching content, teaching method design and assessment technical indicators of the experimental platform according to the 2015 level implementation process and feedback. The 2016 level practice results were further improved compared with 2015 level. The teaching mode and experimental platform have been widely welcomed by students and praised by peers.

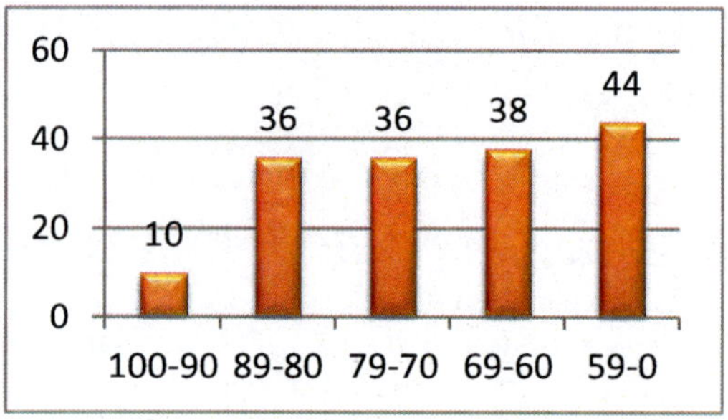

Fig. 4. Courses assessment results of Students enrolled in 2014

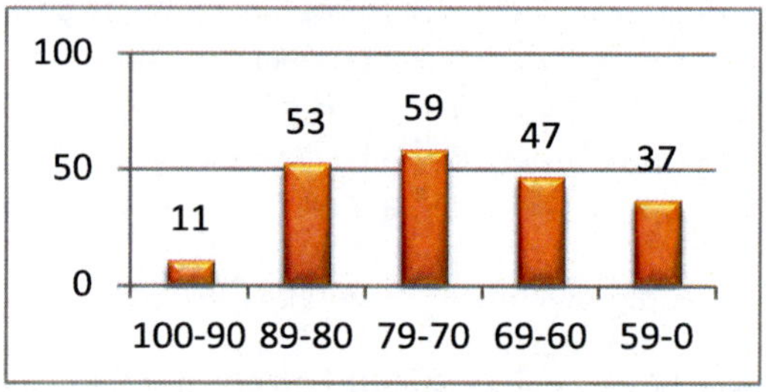

Fig. 5. Courses assessment results of Students enrolled in 2015

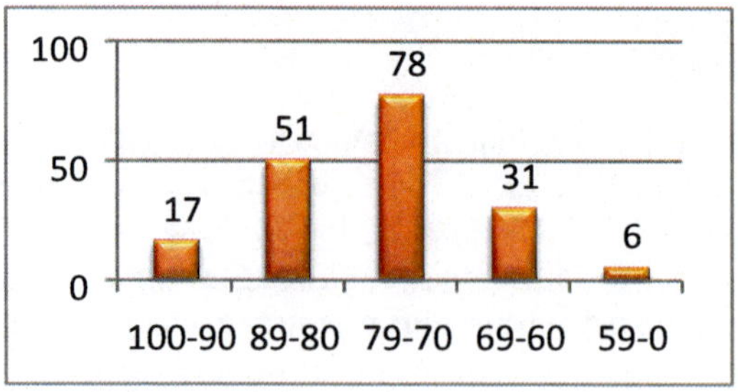

Fig. 6. Courses assessment results of Students enrolled in 2016

The teaching idea of taking ability training as the goal, the teaching method of carrying out ability training with actual industrial equipment as the prototype and a variety of digital twin models are of great help to expand students' knowledge, improve their ability of thinking and analysis, and cultivate their ability to solve practical engineering problems.

(2) The experimental efficiency is significantly improved

After the introduction of digital twin technology, the efficiency of PLC course experiment has been greatly improved. Taking the design experiment of step-by-step instruction as an example, in the traditional teaching mode, the school only has 3 sets of the actual equipment yl335b production line sorting system for the experiment. According to the national standard of Undergraduate Experimental Teaching: 2 persons of a group, a teaching class needs to be divided into 7–8 batches to complete the experimental project. According to students' knowledge and skills, the completion time of each batch is uncertain. By introducing the digital twin technology, a digital twin experimental platform is installed on each computer, and students can operate the model to complete the experiment, which can be conducted in multiple groups at the same time, significantly improving the efficiency of the experiment.

(3) The comprehensive ability has been significantly enhanced

Since the implementation of teaching reform, the effect of cultivating students' comprehensive ability has been obvious. In the past three years, students of this major have been granted 15 patents, won 1 national first prize, 2 s prizes, 4 first prizes, 6 s prizes and 4 third prizes at the autonomous region level in the National Undergraduate Electronic Design Competition; One national award in the National Undergraduate Mathematical Modeling Competition; One national Bronze Award and one gold award of Guangxi in the 'Internet +' innovation and entrepreneurship competition, and four papers were published.

5 Conclusion

In view of a series of problems existing in PLC experimental teaching, under the guidance of OBE education concept, the ability objectives required by the PLC course are clarified, a verification, comprehensive and design level progressive experimental course system is established, an experimental platform based on digital twin technology is developed, the teaching content, teaching methods and assessment methods are reformed, and the experimental platform resources are enriched through the construction of teaching team and the participation of students, Finally, through the evaluation of students' training effect and experiment development efficiency, it provides a new idea for students' ability training and a new case for the application of digital twin technology in teaching.

It can be popularized and applied in the experimental teaching of other related majors, and play a greater role in the cultivation of innovative and practical talents.

Acknowledgments. This project is supported by Guangxi Higher Education Undergraduate Teaching Reform Project-1 (2016jga100); Guangxi Higher Education Undergraduate Teaching Reform Project-2 (2018jgb376); Teaching team construction project of Nanning University.

References

1. Zhao, W., Liu, M., Liu, X., Han, L.: Practice of practical training teaching system based on OBE concept[J]. Exp. Technol. Manage. **35**(03), 185–189+232 (2018). (in Chinese)
2. Bo, W., Meiling, W., Wei, L., et al.: Electronic technology practice teaching reform based on OBE concept[J]. Lab. Res. Explor. **38**(10), 228–231 (2019). (in Chinese)
3. Yanhong, B.: Research and Practice on formative evaluation of curriculum objectives under the background of engineering education certification[J]. China Higher Educ. Res. **11**(12), 60–64 (2019). (in Chinese)
4. Jie, L.: Practical teaching construction of water treatment system based on OBE concept[J]. Lab. Res. Explor. **38**(03), 194–198 (2019). (in Chinese)
5. Shengfa, F., Jie, H., Zhang Xianmei, X., Ligong.: Research on undergraduate teaching and training system based on the concept of "output oriented" (OBE)[J]. Educ. Theo. Pract. **39**(24), 6–8 (2019). (in Chinese)
6. Uhlemann, H.J., Lehmann, C., Steinhilper, R.: The digital twin: realizing the cyber-physical production system for industry 4.0[J]. Procedia CIRP **13**(61), 335–340 (2017)
7. Negri, E., Fumagalli, L., Macchi, M.: A review of the roles of digital twin in CPS-based production systems[J]. Procedia Manuf. **11**, 939–948 (2017)
8. Fei, T., Weiran, L., Meng, Z., et al.: Digital twin five dimensional model and its application in ten fields[J]. Comput. Integrated Manuf. Syst. **25**(01), 1–18 (2019). (in Chinese)
9. Fei, T., Ying, C., Jiangfeng, C., et al.: Theory and technology of physical information fusion in digital twin workshop[J]. Comput. Integrated Manuf. Syst. **23**(08), 1603–1611 (2017). (in Chinese)
10. Yihui, C., Bintao, Y., Yi, F., Xin, X.: Application of digital twin technology in Aeroengine intelligent production line[J]. Aeroengine **45**(05), 93–96 (2019). (in Chinese)
11. Aiguo, J., Jinjiang, W., Ming, G., et al.: Application of intelligent technology of digital twin driven semi-submersible drilling platform[J]. Marine Eng. **48**(05), 49–52+55 (2019). (in Chinese)
12. Leyang, C., Weidong, C., Yue, T., et al.: Virtual reality symbiosis: the prospect of digital twin (DT) technology and its educational application - Also on the reconstruction of ubiquitous intelligent learning space[J]. J. Dist. Educ. **37**(05), 3–12 (2019). (in Chinese)
13. Leyang, C., Weidong, C., Yue, T., Sisi, Z.: Reshaping experience: the prospect of extended reality (XR) technology and its educational application -also on the trend of "integration of education and new technology"[J]. J. Dist. Educ. **37**(01), 17–31 (2019). (in Chinese)
14. Jianfeng, L.,Hongjun, L., Yumin, M.: A framework of network virtual experiments for control courses[J]. Int. J. Educ. Manage. Eng.(IJEME), **2**(9), 49–55 (2012)
15. Qiu, M., Huang, C.: Applied the experimental teaching model of three phases into the environmental monitoring experiment[J]. Int. J. Educ. Manage. Eng.(IJEME), **1**(1): 47–50 (2011)
16. Guangping, L., Yuexia, T.: Construction of 3D simulation platform for assembly unit of automatic production line[J]. Exp. Technol. Manage. **37**(01): 141–144+148 (2020). (in Chinese)
17. Wang, X., Wenwen, X., Zehong, G., Zhijun, W.: Can the teaching application of virtual reality technology improve the learning effect- Meta analysis of 38 experiments and quasi experiments based on the perspective of instructional design[J]. J. Dist. Educ. **37**(06), 61–71 (2019). (in Chinese)
18. Ning, G.: Construction and implementation of innovation computer network practical teaching system[J]. Int. J. Educ. Manage. Eng.(IJEME) **1**(2), 30–35 (2011)
19. Fatima, S., Abdullah, S.: Improving teaching methodology in system analysis and design using problem based learning for ABET[J]. International J. Modern Educ. Comput. Sci. **5**(7), 60–68 (2013)

Integration of Empowerment and Industrialization Based on Intelligent Manufacturing Education

Dongmei Liu[1,2] and Ievgen Konoplianchenko[2](✉)

[1] Sumy National Agrarian University, Sumy 40021, Ukraine
[2] Xinyang Technician College, Xinyang 464000, China
Ldmei1122100@126.com

Abstract. Intelligent manufacturing is an innovative mode of mechanical manufacturing formed by the integration of the new generation of information technology and manufacturing engineering. It is an important starting point to achieve the high-quality development of the manufacturing industry. It is an interdisciplinary, multi-objective and multi-agent field. Its concept and technology run through all links of manufacturing activities such as product design, manufacturing, management and service, involving many technical disciplines, knowledge fields, knowledge units and knowledge nodes. From the perspective of education, this paper combines the education of intelligent manufacturing discipline with the national development policy and the current situation of the region, systematically explores the technical guarantee system for the integration of intelligent manufacturing discipline into the development of industrialization and informatization, gives full play to the action ability and professional characteristics of intelligent manufacturing discipline, enhances students' professional knowledge, and helps enterprises to carry out the integration of industrialization and informatization.

Keywords: Intelligent manufacturing discipline · Collaborative · Practical teaching · Integration of industrialization and informatization · Enterprise digital transformation

1 Overview of Integration of Industrialization and Informatization

With the wide application of information technology in the world, the research on the new industrialization road closely related to the national economic development has also attracted more and more attention in China since the reform and opening up. Because it has a strong driving force for industrial technological innovation, informatization has gradually replaced the central position of labor and capital in the industrial era and become the core driving force for promoting development and reform. In this context, in order to further promote the transformation of industrial upgrading and economic upgrading mode, the Chinese government put forward the concept of integration of industrialization and informatization, and began to formulate relevant policies to promote the integration of industrialization and informatization of.

Z. Hu et al. (Eds.): AIMEE 2022, LNDECT 159, pp. 369–377, 2023.
https://doi.org/10.1007/978-3-031-24468-1_33

enterprises.

The integration of industrialization and informatization refers to driving industrialization with informatization, promoting informatization with industrialization, taking a new road to industrialization and pursuing a sustainable development model. It is a deep combination of industrialization and informatization at a high level. The integration of industrialization and informatization is both a systematic project and an innovative work. It is an inevitable choice for China's economic transformation and structural adjustment. It is an important support for the implementation of the scientific outlook on development and the realization of the Chinese dream. For enterprises, the integration of industrialization and informatization is an effective way to enhance the overall sustainable competitiveness of enterprises, an inevitable means to promote the informatization, systematization and automation of the operation and management process, and an important means to improve the operating efficiency and labor productivity of enterprises, reduce operating costs, change the development mode, improve the development quality and enhance the core competitiveness.

2 Talent Training Mode

With the development of artificial intelligence technology, intelligent manufacturing has become an inevitable trend. According to relevant data, in recent years, the intelligent manufacturing industry will maintain an average annual compound growth rate of about 15%. It is estimated that by 2023, the output value of global intelligent manufacturing will reach about US $2310.8 billion [1]. Against the background that the global manufacturing industry is facing major adjustments and domestic economic development has entered the new normal, China has taken the development of intelligent manufacturing technology and the promotion of comprehensive upgrading of the manufacturing industry as a major strategy, and has issued corresponding development plans. It can be seen that cultivating innovative engineering science and technology talents is the key to building an innovative country, implementing the strategy of rejuvenating the country through science and education and the strategy of strengthening the country through talents [2].

Made in China 2025 clearly points out that it is necessary to accelerate the integrated development of new generation information technology and manufacturing technology, and take intelligent manufacturing as the main direction of the deep integration of industrialization and informatization [3]. Compared with the traditional manufacturing process, intelligent manufacturing pays more attention to networked and intelligent production, and the collaborative supply chain is seamlessly connected. The ubiquitous connection at the demand end enables the whole process of user participation and integrated services to extend the enterprise value [4, 5]. It can be seen that intelligent manufacturing is not only the integration and innovation of manufacturing technology and information technology, but also will reshape the manufacturing industry from multiple dimensions such as product form, manufacturing mode, business philosophy, market form and industry management, resulting in profound changes in the demand for talents in the manufacturing industry. The shortage of talents in the intelligent manufacturing industry, the uneven distribution of talents at all levels, the shortage of teachers in

Colleges and universities, the unclear positioning of talent training, the imperfect curriculum system, the unreasonable professional curriculum and other problems seriously restrict the long-term development of China's intelligent manufacturing industry in the future [6–13]. Therefore, how to scientifically cultivate high-level intelligent manufacturing professionals who meet the requirements of different industries and positions is an important task for higher engineering educators, and also a difficult problem to be solved for enterprise development. Therefore, colleges and universities should carry out multi-dimensional training of intelligent manufacturing talents, adjust the talent training mode and methods, and improve the graduates' professional skills and ability to solve practical problems [14]. First, colleges and enterprises jointly run professional and technical qualification certification and training services for enterprises, introduce professional and technical personnel from enterprises for teaching, strengthen the teaching force of Vocational and technical skills, break the traditional school running mode of colleges and universities, or work closely with local governments to accelerate the transformation of local colleges and Universities to application-oriented through the construction of practical teaching bases and innovation and entrepreneurship bases [15]; Secondly, to change the traditional classroom teaching mode, we can cooperate with local enterprises or industry and information technology departments to carry out off campus social classes, and hire some experts, engineers, technicians with rich practical experience to teach [15, 16]. This not only solves the problem of the shortage of college teachers, but also can lead students to the enterprise scene, and promote the integration of industry and education to form a local characteristic and application-oriented teaching mode [17]; Finally, colleges and universities make use of the advantages of intelligent manufacturing theory to intervene in local enterprises to cultivate localized intelligent manufacturing talents. Through specific cooperation projects between local governments and colleges and universities, while adopting the teaching forms of "subject system" and "project system" [18] in the curriculum system, they organize social practice activities to cultivate the ability of college students to apply theoretical knowledge to practice [19, 20]. Talent training in colleges and universities is a huge systematic project. It is necessary to comply with the development needs of the times to promote education reform, build a multi-dimensional education model by using the "school enterprise local" collaborative talent training, carry out a teaching method focusing on the integration of production, teaching and research, and promote the orderly development of education reform through the establishment of research institutions and innovation and entrepreneurship institutions through multiple assistance, so as to provide professional talent training Technical input and other support.

3 Practical Teaching Path

Through the analysis and Reflection on the relationship between the intelligent manufacturing specialty and the integration of the two modernizations, according to the actual situation of the integration of the two modernizations, combined with the undergraduate teaching content and practical teaching system of the intelligent manufacturing discipline. Try to promote the organic combination of intelligent manufacturing discipline education and the integration of industrialization and informatization from the aspects of discipline construction, talent training and curriculum construction, and explore the discipline innovation method and implementation path for the integration of intelligent manufacturing discipline into the integration of industrialization and informatization in China. It is specifically expressed in three levels according to the hierarchical structure (as shown in Fig. 1). The first level is the target layer of this study, the second level is the criterion layer, which includes the main links involved in achieving the goals, and the third level is the scheme layer [21]. The teaching methods and measures set to achieve the expected goals are the concretization of the objects of the criterion layer.

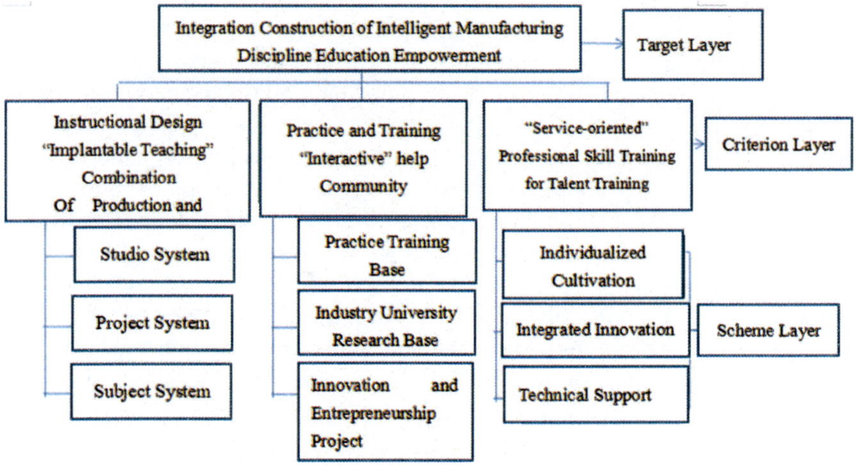

Fig. 1. Hierarchical model of practice teaching system structure

3.1 "Implantable Teaching" Industry Teaching Combination

In the course teaching, such teaching methods as "studio system", "project system" and "subject system" are used to implant the integration planning project into the teaching task, and the teaching content is set according to local conditions. In terms of design teaching, relying on enterprises and government departments and combining the actual project needs of enterprises, CIMS\MES\PLM\ERP and other implementation studios are formed to carry out the dialogue and group empowerment between enterprise projects and design teaching, and complete the comprehensive practical teaching design in the courses of enterprise digital transformation, enterprise integration and promotion, and enterprise digital factory design.

3.2 "Interactive" Help Community

According to the specific content of the integration planning and development of the two modernizations of the enterprise, the existing production equipment and industrial software application advantages of the enterprise shall be excavated, the project-based teaching mode shall be adopted in combination with the practical course content in the intelligent manufacturing professional course system, the teaching design of two-way flow, internal and external linkage, and the interaction of production and learning shall be explored, and the "Assistance Plan" shall be implemented for the enterprise through the practice group of "the integration assistance team of the two modernizations". At the same time, it has established internship and training bases, industry university research bases, school enterprise cooperation and transformation platforms with enterprises, combined with funding guarantee methods such as college students' innovation and entrepreneurship projects and teaching reform, and built a long-term effective linkage cooperation and assistance mechanism of normalization and institutionalization, so as to improve the talent education mechanism of intelligent manufacturing discipline while promoting the integration of enterprise industrialization and digitalization transformation and development.

3.3 "Service-Oriented" Professional Skill Training

Students majoring in intelligent manufacturing emphasize "service" in the process of integrating the two modernizations of enterprises with actions. They should fully consider the actual production situation of enterprises, lead students to conduct in-depth investigation on the manufacturing environment, production topics, upstream and downstream of the supply chain of enterprises in the mode of practical experience, carry out diversified personalized cultivation and integration innovation, and focus on training students' professional skills in engineering technology and implementation technology, Provide effective point-to-point technical support for industrial software and automation hardware facilities recommended for the integration of industrialization and informatization. For example, the digital factory construction is designed through simulation to improve the accuracy of the project, improve the production efficiency and product quality of the enterprise, give play to the role of education in serving the society, and realize the organic combination of the promotion and transformation of the educational achievements of intelligent manufacturing discipline in Colleges and Universities and the integration of the two.

4 Practice Teaching Docking with Enterprise Implementation

The specific content of practical teaching of Intelligent Manufacturing Specialty in Colleges and universities should be implemented by enterprises. Intelligent manufacturing specialty is an interdisciplinary specialty. The curriculum can cover mechanical engineering, industrial engineering, automation design, industrial software and other teaching contents. Through the creative reorganization of innovative design and intelligent design of digital chemical plants, we can open up the boundary between universities and enterprises, give full play to the rooted and.

productive nature of the discipline, activate the professional advantages of the discipline, and take the initiative to establish discipline education to connect with the practical problems of enterprise digital transformation, Through effective "implantable teaching", systematically explore the technical support system for the integration of intelligent manufacturing and industrialization, as shown in Fig. 2.

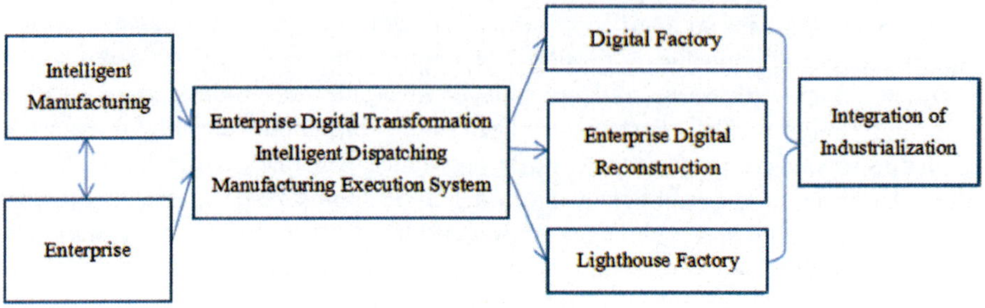

Fig. 2. Analysis of practice teaching docking with local services

4.1 Theoretical Basis

When the American scholar Bertalanffy first put forward the general system theory, he pointed out that the system is "a complex of elements in mutual interaction". American scholars Fred and Jackson hold the same view. They believe that the system is that some elements act in a special way to form an interdependent, complementary or restrictive relationship. Qian Xuesen, a Chinese scholar, believes that a system is an organic whole composed of components that both act on and depend on each other. At the same time, this organic whole is subordinate to a larger system. From these points of view, the academic community unanimously agreed that there are interrelated and interactive elements in the system, and these elements form a specific structure.

The purpose of intelligent manufacturing education is to cultivate talents with interdisciplinary knowledge, which needs to follow the growth law of engineering talents and the practical requirements of contemporary society. Therefore, intelligent manufacturing education should reflect the optimal link composition and element combination of interdisciplinary engineering talents training, and reflect the general characteristics and characteristics of the times of Engineering talents. The training of interdisciplinary talents in the education system of intelligent manufacturing specialty is not a simple process, but a large system with strong complexity. The large system also contains many subsystems, in which there are countless constituent elements. A complex network structure is formed among different systems, elements, and systems, which produces certain functions. At the same time, the training of intelligent manufacturing professionals based on interdisciplinary is also a highly open system. The system maintains continuous exchange with its own environment and provides continuous vitality to the system.

4.2 Empirical Research

An important node for the implementation of the integration of industrialization and informatization is to develop and expand enterprises, which are the support of the national economy. "Research on enterprise digital transformation in Xinyang City" finally completes the planning and design objectives of the project through full communication between the school and the enterprise, and in combination with the teaching content of intelligent manufacturing courses. First, in the early stage, a teaching team meeting the requirements of the project was organized to organize targeted teaching in terms of theoretical explanation and practical guidance, which was carried out in stages based on the curriculum system. The team teachers were responsible for the explanation of the theoretical courses, and set up a theoretical knowledge framework for the integration of the two modernizations; Secondly, the school organized students to go deep into enterprises to investigate the current situation of the integration of industrialization and informatization and intelligent manufacturing application in the region, improved students' sense of social responsibility through resource collation and analysis, organized students to conduct field interviews or questionnaires during the implementation stage of the course to master the common vision of the enterprise, and conducted in-depth investigation on the equipment automation degree and industrial software application degree in the digital transformation of enterprises in combination with the teaching of intelligent manufacturing course. In order to design and improve the efficiency and quality of the enterprise, make corresponding teaching plans, and clarify specific promotion objectives and tasks; Finally, take advantage of the opportunity of enterprise digital transformation to improve the enterprise digital transformation system in the overall urban area, and build a benchmark enterprise for the integration of two modernizations in Xinyang based on the enterprise digital transformation in Xinyang and the construction of Intelligent Manufacturing Specialty in the school: first, use the intelligent manufacturing professionals trained by the school to help the enterprise carry out hardware intelligent transformation; Secondly, using the advanced management concept of Lighthouse factory to help enterprises evaluate and guide the application of industrial software; Finally, considering the implementation path of enterprise digital transformation and lighthouse factory to improve quality and efficiency, help enterprises achieve the purpose of improving quality and efficiency.

5 Conclusion

The implementation of the integration strategy of enterprise industrialization and informatization has brought opportunities for the construction of intelligent manufacturing specialty and talent training in colleges and universities, and put forward higher requirements at the same time. The intelligent manufacturing specialty should seize the opportunity to build a talent training program and teaching system that adapt to the development of the times and the needs of local economic construction, pay attention to the cultivation of students' spirit of serving enterprises, cultivate applied professionals with high quality and high skills, and realize the benign interaction between talent training and serving the society. At the same time, colleges and universities should give full play to the advantages of scientific research resources, organize teachers' technical teams to go deep into enterprises, introduce enterprise digital transformation construction projects and enterprise quality improvement and efficiency enhancement projects into the classroom through cooperation with the government or enterprises, cultivate students' professional qualities in the process of classroom teaching in combination with actual cases, and build a bridge between the training of intelligent manufacturing professionals and the integration of enterprises, Starting with the enterprise hardware automation transformation and the application of enterprise industrial software and management software, this paper puts forward corresponding planning and design strategies, integrates enterprise resources, promotes enterprise revitalization through overall planning of regional enterprises, makes the course teaching content more practical, and realizes the goal of the integration of intelligent manufacturing education and enterprise.

References

1. Intelligent factory upgrading intelligent manufacturing continues to grow at a high speed http://www.sohu.com/a/226570332_793144, 2019/12/24–07:50 (in Chinese)
2. Tianyou, C.: Research and practice of innovative automation engineering technology talent training mode. Higher Eng. Educ. Res. **3**, 1–4 (2019). (in Chinese)
3. Notice of the state council on printing and distributing made in China 2025 [EB/OL] http://www.gov.cn/gongbao/content/2015/content_2873744.htm (in Chinese)
4. Lieyun, D.: Thinking on the cultivation of innovative engineering science and technology talents in intelligent construction. Res. Higher Eng. Educ. **5**, 1–4 (2019). (in Chinese)
5. Brockman, O., Manufacturing, I.: Subversion and reconstruction of future industrial models and formats. China Machine Press, Beijing (2015). (in Chinese)
6. Shuaibang, F., Yi, S., Mingjing, L., et al.: Qitian garden Research on spatial pattern and development countermeasures of intelligent manufacturing industry in Liaoning Province. Econ. Res. Guid. **11**, 26–28 (2022). (in Chinese)
7. Ling, D., Zhiliang, J.: Build a "city of craftsmen" and a "Jiaojiang model" for the construction and reform of industrial workers. China Human Resour. Soc. Secur. (04), 28–29 (2022) (in Chinese)
8. Yunyang, Y., Wentao, C.: Research on the construction of intelligent manufacturing talent training service platform under the background of industrial transformation and Upgrading -- Taking Wenzhou intelligent manufacturing talent training as an example. Sci. Technol. Econ. Market (03), 106–109 (2022) (in Chinese)
9. Qi, Z.: Summary of the sub forum on opportunities and challenges for the development of vocational education in the era of intelligent manufacturing. China education fiscal policy consultation report (2015–2019) (2021) (in Chinese)
10. Jiajun, Z.: Strategies for skills shortage of small and medium-sized enterprises in the post epidemic era. Mod. Enterp. (11), 45–46 (2021) (in Chinese)
11. Shuai, D.: Research on the shortage of highly skilled talents in HJB company and Its Countermeasures. Xiangtan University, Xiangtan (2020) (in Chinese)
12. Baohong, L., Kexue, L., Hong, L.: Research on the shortage of intelligent manufacturing technology talents in China. Heilongjiang Sci. 10(05), 110–111 (2019) (in Chinese)
13. Luciano, R.G., Salvatierra, M.J.: Stressors and stress-coping mechanisms of academic scholars in HEIs: a basis for stress management plan formulation. Int. J. Educ. Manag. Eng. (IJEME) 12(3), 1–12 (2022)
14. Jingjing, D.: Research on the impact of artificial intelligence development on China's manufacturing employment. Liaoning University (2021) (in Chinese)
15. Yong, C., Gaolian, S., Zhe, G., et al.: Research on the construction of productive training base in higher vocational colleges facing intelligent factory -- taking suzhou industrial vocational and technical college as an example. Light industry technology, 37(09), 167–168 (2021) (in Chinese)
16. Yan, G., Xinfa, S., Yuhang, S., et al.: Analysis of basic capabilities of industry digitalization enabled industries and suggestions on improvement methods. Electromech. Eng. Technol. 50(06), 34–39 (2021) (in Chinese)
17. Yongbo, L., Minzhi, L., Hui, Y.: Create online open learning resources to help improve the professional quality of workers in the intelligent manufacturing industry. Sci. Technol. Innov. (02), 69–71 (2020) (in Chinese)

18. Faizi, J., Umar, M.S.: A conceptual framework for software engineering education: project based learning approach integrated with industrial collaboration. Int. J. Educ. Manag. Eng. (IJEME), 11(5), 46–53 (2021)
19. Jiangling, X.: A preliminary study on the reform of industrial robot technology as an applied undergraduate. Mod. Enterp. **10**, 128–129 (2019). (in Chinese)
20. Zhaoxin, W.: On the construction of intelligent manufacturing technology specialty. Occupation **08**, 73 (2018). (in Chinese)
21. Qiu-yan, C.: Researh on the practical path of enabling rural revitalization by landscape architecture education in colleges and universities. J. Shaoguan Univ. 42(12) (2021) (in Chinese)

Reform and Practice of Engineering Surveying Practical Teaching System for Transportation Majors Under the Background of Engineering Education Professional Accreditation

Jie Shang[1], Lanfang Zhang[1](✉), Jianqiu Chen[1], and Yubin Lv[2]

[1] College of Transportation, Nanning College, Nanning 530000, China
2158655849@qq.com
[2] Disease Control and Prevention Center of Guangxi Zhuang Autonomous Region, Nanning 530000, China

Abstract. Engineering education certification is a kind of professional authentication based on international standards, which is the general trend of the development of new engineering. As a professional basic course for transportation majors, engineering survey has the characteristics of outstanding practical teaching and is also an important indicator of professional certification. This paper studies the problems existing in the current practical teaching of Engineering Surveying for transportation majors, such as the contradiction between the compression of class hours and the requirements for improving teaching objectives, the mismatch between teaching content and measuring equipment and post requirements, the fragmentation of assessment and evaluation indicators, and the lack of deep engineering background of teachers. Based on the concept of engineering education certification, this paper proposes to revise the syllabus by benchmarking 12 graduation requirements, It is suggested to adopt the online + offline mixed and task driven teaching mode, build a multi-dimensional evaluation system, and put forward the construction direction of the teaching staff, so as to achieve the continuous improvement of teaching quality.

Keywords: Engineering education certification · Engineering survey · Practical teaching system

1 Introduction

As an integral part of the training of senior engineering talents in the higher education system, engineering education is the fundamental guarantee to promote scientific and technological innovation and progress, and it is also an important foundation to promote the development of the national economy [1, 2]. As one of the important ways to improve the quality of engineering education, the professional certification system of engineering education, i.e. Engineering Education Professional.

Accreditation, abbreviated as EEPA at this paper, is an important means to ensure the quality of engineering education, which has been paid more and more attention [3].

Z. Hu et al. (Eds.): AIMEE 2022, LNDECT 159, pp. 378–390, 2023.
https://doi.org/10.1007/978-3-031-24468-1_34

Engineering education certification, also known as specialized/professional certification, or, EEPA, refers to the specialized certification implemented by professional certification organizations for vocational professional education (programmatic) offered by higher education institutions [4].

In June 2016, China officially joined the Washington Agreement, one of the most influential engineering education degree mutual recognition agreements in the world. A consensus of industry education is that the basic engineering courses in the curriculum system are required to have good support for the engineering ability in civil engineering and other related fields [5]. The objectives of engineering education certification in China are: to build a quality monitoring system for engineering education in China, promote the reform of Engineering Education in China, and further improve the quality of engineering education [6]. Establish an engineering education certification system connected with the engineer system, promote the connection between engineering education and the business community, and enhance the adaptability of engineering education talent training to industrial development; Promote international mutual recognition of China's engineering education and enhance international competitiveness.

Taking engineering education certification as the background, fully understanding the relationship between engineering certification and education reform, further clarifying the teaching objectives and constantly improving the teaching process will help to promote the continuous development of education [7–11]. This is also the original intention of this paper.

2 Three Core Concepts of Engineering Education Certification

The professional certification system of Engineering Education in Colleges and universities is one of the important ways to improve the quality of engineering education.

The three core concepts of engineering education certification are "student centered (SC), outcome based education (OBE), and continuous quality improvement (CQI)". They are also known as the troika to promote engineering education and improve the quality of engineering education.

(1) The notion of SC (-Student Center)

Student centered concept is the purpose of EEPA. The educational objectives focus on the cultivation of students, the design of teaching content focuses on the cultivation of students' ability, teachers and educational resources meet the students' learning effect, and the focus of evaluation is the evaluation of students' effect [10, 12–14]. The student-centered certification concept reflects the importance attached to the educational process.

(2) The notion of OBE (-Outcome Based Education)

Outcome/achievement/result based education is a requirement of EEPA. Reverse design from the final outcome to determine the appropriateness of all teaching towards the final outcome. The starting point of teaching is not what teachers want to teach, but what is needed to achieve the final results [15]. This concept reflects the concern for educational outcomes.

(3) The notion of CQI (-continuous quality improvement)

CQI is a mechanism that EEPA pays special attention to. Establish a regular evaluation mechanism and make continuous improvement [16, 17]. The training objectives, graduation requirements and teaching links should be evaluated. Every teacher is responsible for continuous improvement. The effect of continuous improvement is reflected by students' performance. It corresponds to the concern about the value-added of education.

The three concepts of engineering education professional certification meet the new needs of contemporary engineering education talent training, and strengthen the whole person training integrating engineering technology with humanities, society, law and environment. The certification standard of engineering education is also an educational quality assurance mechanism guided by the new quality concept of education.

Guided by these ideas, ensure the achievement of the abilities of all undergraduate graduates, and emphasize the "output" quality of education, that is, what abilities graduates have, what they can do and what they will do when they leave school. The engineering certification standard stipulates 12 graduation requirements, including general ability (including sense of responsibility and Literacy) and engineering professional ability. It also stipulates that schools must have a corresponding quality assurance system and be able to make continuous improvement, that is, to establish a continuous quality improvement cycle mechanism of "finding problems - timely feedback - agile response - effective improvement". The engineering certification standard also imposes strict requirements on the curriculum system, teaching staff, support conditions and other aspects from the general standards and professional supplementary standards. Curriculum is the carrier of talent training and professional construction. Curriculum construction is a crucial link in achieving the goal of engineering education certification [18–22].

3 Course Nature of Engineering Surveying Practice and Problems in Teaching

Only by clarifying the nature and requirements of the course and deeply analyzing the problems existing in the course teaching can we put forward targeted solutions, improve the course teaching and make it play a better role in talent training.

3.1 Nature of Engineering Survey Course

Engineering surveying is a basic course for transportation and civil engineering majors. The teaching goal is to teach students to master relevant theoretical knowledge, and then be able to complete the measurement work carried out in each stage of engineering construction survey, design, construction and management, so as to provide basic data and drawings for the project and ensure the quality of the project. With the continuous development of modern engineering construction and the application of new measurement technology, engineering measurement is of great significance in all stages of engineering construction.

Engineering surveying course has obvious practical characteristics. In the process of course teaching, we should not only teach students the basic theoretical knowledge of measurement, but also cultivate students' operating skills in operating and using measuring instruments, and improve students' practical operation and engineering adaptability. At the same time, measurement work runs through the construction of all kinds of transportation projects and plays a very important role in the project. It can be seen that engineering surveying has a significant impact on the training quality of transportation professionals [23]. Under the background of engineering education professional certification, with the continuous development of new surveying and mapping technology, it is worth discussing how to balance the teaching content of Engineering Surveying and innovate the teaching methods, which requires teachers to keep up with the forefront of science and technology and engineering practice, keep pace with the times, and constantly deepen the teaching reform of engineering surveying [24].

3.2 Problems in Engineering Survey Course

The problems existing in engineering surveying course mainly include the following four aspects.

(1) The amount of class hours runs counter to the teaching objectives

In the process of teaching reform in Colleges and universities, the total class hours are generally compressed. The total class hours required for graduation of transportation major in our school are 1156, and the engineering measurement courses and engineering measurement practice courses are 32 and 24 respectively, accounting for only 4.8% of the total class hours, which is less than that of similar colleges and universities, and may be further compressed. However, with the continuous development of electronic informatization, automation and intelligence of Surveying and mapping technology, new measuring instruments, new theories, new technologies and other new knowledge continue to appear, and the amount of knowledge is increasing. With the advancement of engineering education certification, the requirements for graduates' engineering ability are also increasing. The gradual compression of class hours and the increasing improvement of teaching objectives have formed an irreconcilable.

(2) The teaching content and measuring equipment do not match the job requirements

As the compilation, approval and publication of teaching materials require a certain amount of time, but the development of Surveying and mapping technology is very rapid, so it often appears that once the engineering surveying related teaching materials are published, some of their contents are far behind the forefront of the industry, and the two are difficult to keep pace. It directly leads to less introduction of new knowledge of Surveying and mapping in the theoretical teaching of engineering surveying, and some teaching contents lag behind the development of Surveying and mapping technology. Moreover, it is difficult for the renewal speed of the instruments and equipment in the school surveying and mapping laboratory to catch up with the development speed of new technology, so that the talent training cannot fully meet the needs of industry posts.

In addition, the content of surveying practice is mainly to conduct traverse survey and topographic map mapping in the campus, but in the actual job work, surveying is not only used in project setting out and control point layout, but also closely related to measurement in the detection and monitoring industry, which requires personnel to have high practical ability and theoretical analysis ability [25]. On the other hand, the campus terrain is relatively simple and students are familiar with it. It is difficult to stimulate students' initiative and enthusiasm to practice in the school.

(3) Assessment and evaluation indicators are not systematic

The existing engineering measurement assessment methods are mainly theoretical examination + instrument practice, and the evaluation indicators are mainly test paper scores, peacetime homework, practical performance, etc. The assessment content is relatively scattered, and the indicators are independent. The main assessment is the students' mastery of a single knowledge point or skill point, which cannot truly reflect the achievement of students' ability to solve complex problems in the engineering certification standard.

(4) Teachers' engineering background is not deep enough

The engineering certification standard requires that teachers should have sufficient teaching ability, professional level, engineering experience, communication ability and professional development ability, and be able to carry out research on engineering practice problems and participate in academic exchanges. Teachers' engineering background should meet the needs of professional teaching. However, among the teachers, the school's own young teachers account for the mainstream, and some teachers' professional experience is from "school to school", lacking working experience in engineering practice. It is difficult to combine engineering practice in teaching, and it is also difficult to provide sufficient guidance for students' career planning and vocational education.

4 Reform of Engineering Measurement Practice Teaching Based on the Concept of EEPA

Under the guidance of the concept of engineering education certification, the construction of practical teaching system needs to be carried out from the following aspects.

4.1 Revision and Improvement of Syllabus

The practical teaching system of engineering surveying consists of two parts: in class practice and centralized practice. In class practice is mainly a supplement to theoretical teaching. With the theoretical teaching, the familiarity and practical operation of a single measuring instrument are completed alternately, and the measurement principles and knowledge of theoretical teaching are simply applied in the practical operation. Centralized practice is mainly to comprehensively use the knowledge of engineering measurement and other professional courses, close to the actual engineering production, and use the means of engineering measurement to solve the professional problems of engineering practice with a certain degree of complexity. Combing the teaching contents and teaching objectives in the syllabus, integrating theoretical learning into practice and modularizing the learning contents according to the professional characteristics, as shown in Table 1.

Table 1. Correspondence between teaching content and teaching objectives

Category	Content	Knowledge and ability goals	Responsibility and attainment objectives
Practice in class	Elevation survey	Follow the principle of leveling, and use the automatic ANPING level to lay, measure and calculate the leveling route.	Gradually form a professional quality of consulting relevant norms in professional work, being serious, rigorous and hardworking
	Angle measurement	Follow the principle and measurement steps of the return method and the direction observation method, and use the angle measurement function of the total station to measure and calculate the angle.	
	Distance measurement	Follow the principle of distance measurement, and use the total station distance measurement function and steel ruler to measure the distance respectively.	
	Small area control measurement	According to the principle of small area control measurement, the total station will be used to encrypt the control points.	
Centralized practice	Setting out of highway middle pile	Use the total station to set out the coordinates of highway middle piles in combination with the relevant contents of road engineering in the course of introduction to transportation engineering.	
	Highway profile design elevation setting out	Coordinate setting out of highway middle piles is carried out with a level in combination with the relevant contents of road engineering in the course of introduction to transportation engineering.	
	Highway side pile setting out	GPS-RTK receiver is used to set out the coordinates of highway side piles in combination with the relevant contents of road engineering in the course of introduction to transportation engineering.	

4.2 Reform of Teaching Mode and Method Based on SC and OBE)

The reform of teaching mode and method in this course highlights SC and OBE in the above three core concepts.

(1) Online + offline hybrid teaching

The flexibility of online teaching allows students to learn anytime, anywhere, and ruminate knowledge according to their acceptance ability [26]. Through the construction of online teaching resources, the explanation of theoretical knowledge, the demonstration of experimental instruments and other contents are put into online courses, so as to achieve the pre-positioning of this part of teaching content, requiring students to achieve low-level goals before class, and make full preparations for teaching in the classroom. In the offline classroom, teachers can focus on the explanation of key and difficult points and the guidance of practical operation, make full use of teacher resources, and maximize the effectiveness of offline classroom. The contradiction between the compression of class hours and the improvement of teaching objectives can be well solved through online + offline hybrid teaching. Moreover, adopting this teaching mode can take into account the differences of each student's individual foundation and learning ability, and students can control the rhythm by watching repeatedly and at double speed. It has better met the requirements of "engineering education certification" that takes students as the center and makes the achievements of teaching reform benefit every student.

(2) Task driven Teaching

In order to solve the problem that the teaching content and measuring equipment are out of line with the post requirements, the teaching material content needs to be structured and reconstructed. As shown in Table 1, the teaching content is divided into seven specific tasks, and each specific task is divided into six links: the release of the task book, students' formulation and implementation plan, plan guidance and revision, preparation before task implementation, task implementation (guidance during implementation) and task summary. The design of task content is closely.linked to the teaching objectives and engineering practice. The difficulty of the seven tasks is from easy to difficult, and the training of students is gradually transitioned to meeting the job requirements.

4.3 Teaching Process Supervision and Evaluation (CQI)

The implementation of CQI is to take a series of effective measures in teaching to achieve continuous improvement and effectively improve teaching quality.

(1) Establish a whole process and multi-dimensional assessment and evaluation system

In the construction of engineering education certification major, the establishment of achievement evaluation system can evaluate whether the students' ability meets the graduation requirements and the degree of realizing the talent training objectives. The assessment result of a single course is an important element of the achievement evaluation matrix, so the scientificity and effectiveness of the course evaluation method is the guarantee of the quality of the achievement evaluation system. In order to make the course assessment results truly reflect the degree of achievement of students' ability to solve complex problems in the engineering certification standard, the whole process of evaluation should be carried out, and the evaluation process is shown in Fig. 1. And establish a multi-dimensional assessment and evaluation system, as shown in Tables 2, 3, 4 and 5.

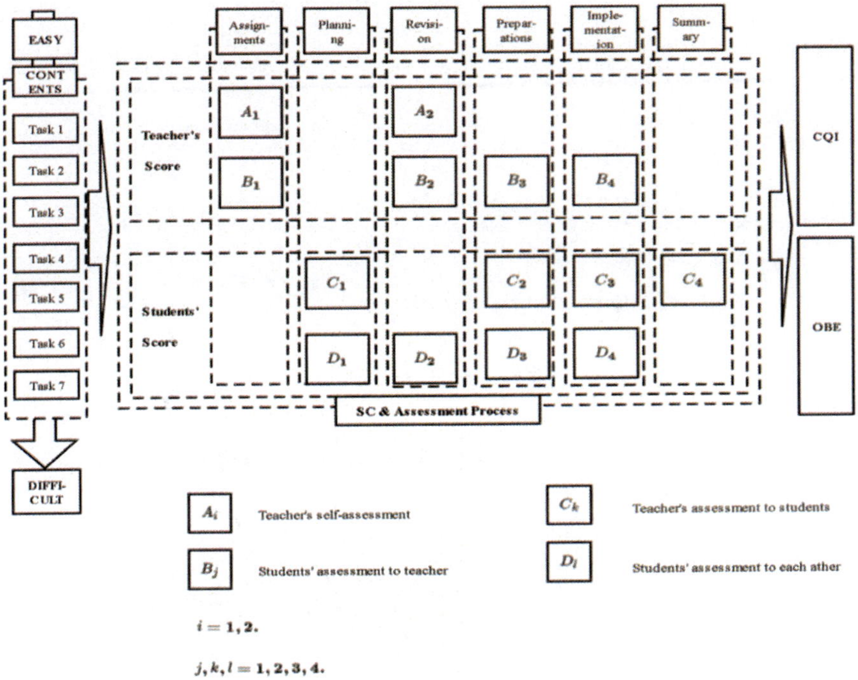

Fig. 1. Evaluation process

Table 2. Evaluation system of A_i

	Evaluation dimension(P_n)	Weight(W_n)
A_1	Teaching tasks are in line with the teaching objectives	0.5
	Moderate workload of tasks	0.3
	Tasks content has been improved from last time	0.2
A_2	Instruction was given to each group of students	0.3
	Students indicated that the guidance was useful	0.4
	Documentation of the mentoring process has been carried out	0.2

$$A_i = \sum_{n=1}^{3} P_n W_n \quad (i = 1,\ 2) \tag{1}$$

$$A = 0.05 \times A_1 + 0.05 \times A_2 \tag{2}$$

Table 3. Evaluation system of B_j

	Evaluation dimension(P_n)	Weight(W_n)
B_1	Moderate workload of task	0.3
	The task is interesting	0.3
	The task is useful	0.4
B_2	I got the teacher's guidance	0.2
	The guidance is useful	0.5
	The teacher's teaching method is very effective	0.3
B_3	I got the teacher's guidance	0.3
	The teacher's guidance is very important to me	0.4
	The teacher is patient and serious	0.3
B_4	The teacher's guidance enabled me to solve important difficulties	0.4
	The teacher is patient and serious	0.3
	The teacher has profound professional knowledge	0.3

$$B_J = \sum_{n=1}^{3} P_n W_n \quad (j = 1,\ 2,\ 3,\ 4) \tag{3}$$

$$B = 0.15 \times B_1 + 0.25 \times B_2 + 0.30 \times B_3 + 0.20 \times B_4 \tag{4}$$

Teacher's score:

$$S_1 = A + B \tag{5}$$

$$C_K = \sum_{n=1}^{3} P_n W_n \quad (k = 1,\ 2,\ 3,\ 4) \tag{6}$$

$$C = 0.10 \times C_1 + 0.25 \times B_2 + 0.30 \times B_3 + 0.20 \times B_4 \tag{7}$$

$$D_1 = \sum_{n=1}^{3} P_n W_n \quad (K = 1,\ 2,\ 3,\ 4) \tag{8}$$

Table 4. Evaluation system of C_k

	Evaluation dimension(D_i)	Weight(W_i)
C_1	The form of the plan is very standardized	0.3
	The content of the plan is complete	0.3
	The plan can strongly support task implementation	0.4
C_2	Actively participated in the preparatory work	0.3
	The questions raised have been carefully considered	0.3
	Adequate preparation	0.4
C_3	Actively participated in the task implementation.	0.3
	In the process of implementation, serious problems have been put forward.	0.4
	The task is successfully implemented	0.3
C_4	Actively participate in the summary	0.3
	Problems in task implementation are put forward in the summary	0.4
	The summary has clear logic	0.3

Table 5. Evaluation system of D_l

	Evaluation dimension(P_n)	Weight(W_n)
D_1/D_2/ D_3/D_4	Actively participated in the planning/revision/preparing/Implementation work	0.4
	He/she have important contribution to the objectives of this stage	0.4
	I feel like I learned a lot from him/Her	0.2

$$D = 0.10 \times D_1 + 0.10 \times D_2 + 0.20 \times D_3 + 0.20 \times D_4 \tag{9}$$

Students' score:

$$S_2 = C + D \tag{10}$$

It can be seen from Fig. 1 that in this evaluation process, the scores of each teaching link are decomposed. In each link, the evaluation objects are teachers and students, and teachers and students have the right to evaluate. Therefore, the evaluation forms include self-evaluation and other evaluation. In the evaluation system, there are 4 primary indicators a, B, C and D, 14 secondary indicators and 42 tertiary indicators. The index system is student-centered, showing the characteristics of the whole process and multi-dimensional. Through the assessment, the objective and comprehensive process evaluation of teachers and students can be achieved, which not only obtains the effective evaluation of students' engineering ability, but also obtains the evaluation and real-time feedback of teachers. The results can be used as a reliable basis for the follow-up teaching reform and continuous quality improvement.

(2) Continuous improvement of teachers' level

In terms of improving the level of teachers, we mainly adopt the methods of "please come in" and "send out". On the one hand, enterprise experts and post experts are invited to the school, and they are invited to participate in the revision of the syllabus, the compilation of practical training instructions that closely follow the industry frontier and learning situation, and the guidance of the practical teaching process. On the other hand, full-time teachers in the school will be sent to front-line production posts. Give full play to the resource advantages of school enterprise cooperation units, let the school teachers participate in the project implementation process of the cooperation units, and increase the school teachers' engineering practice experience and the ability to keep up with the forefront of the industry. And seek in-depth cooperation with enterprises, promote enterprises and teachers and students to jointly carry out scientific research, practice and training, and cultivate students' practical ability and innovation ability. In addition, professional teacher training should be organized regularly to continuously improve the level of teachers.

5 Conclusion

Applying for engineering education certification is a major opportunity and challenge in the process of specialty construction. The engineering education certification standard puts forward specific standards for specialty construction. The important way to achieve these standards is to strengthen the construction of curriculum system.

According to the 12 graduation requirements in the EEPA standard, it is decomposed into several index points that are convenient for evaluation. In the evaluation process of index points, the evaluation method combining direct evaluation and indirect evaluation is adopted to evaluate the achievement of graduation requirements from the three dimensions of course performance evaluation, employer evaluation and graduate self-evaluation, and the evaluation results are analyzed to use the evaluation results for continuous improvement.

As a basic course for transportation majors, we should revise the syllabus by benchmarking the engineering education certification standard in the construction of the practical teaching system of engineering survey, so as to clarify the teaching objectives and teaching contents. In the process of teaching implementation, we should innovate the teaching mode and teaching method of highlighting SC and OBE, explore the "Online + offline" hybrid teaching mode and make full use of task driven teaching, so that students can actively study and explore, and gradually make progress.

in practice, so as to meet the graduation requirements. In order to ensure the continuous improvement of teaching effect and talent training quality, it is necessary to use CQI concepts and measures to establish an all-round and multi-dimensional evaluation system to help teachers find the direction of improvement and give reliable judgments and suggestions for the achievement of students' abilities. The continuous improvement of teachers' level should be achieved through communication and training. The research results provide strong support for the realization of training objectives and graduation requirements in the training of transportation professionals .

Acknowledgments. The study is supported by: (1). 2021 Guangxi higher education undergraduate teaching reform project "Research on Innovative Teaching Method of Seamless Integration of Transportation Professional Courses and Ideological and Political Education Based on Semantic Analysis", (No. 2021JGB428) and (2). The second batch of teaching team projects of Nanning University in 2019 "Applied and Effective Teaching Design Teaching Team", (No. 2019XJJXTD10).

References

1. Keishing, V., Renukadevi, S.: A. review of knowledge management based career exploration system in engineering education. Int. J. Educ. Manage. Eng. (IJEME) **1**(8), 8–15 (2016)
2. Weiss Kathleen, A., McDermott Mark, A., Brian, H.: Characterising immersive argument-based inquiry learning environments in school-based education: a systematic literature review. Stud. Sci. Educ. **58**(1), 15–47 (2022)
3. Sun, W., Hong, J.C., Dong, Y., Huang, Y., Fu, Q.: Self-directed learning predicts online learning engagement in higher education mediated by perceived value of knowing learning goals. The Asia-Pacific Educ. Res. 1–10 (2022)
4. Sun, M.: Teaching reform of engineering surveying course under the background of engineering education professional certification. Sci. Technol. Horizon **20**, 19–20 (2021) (in Chinese)
5. Junfeng, S.: Research on the construction of practical teaching system of civil engineering surveying under the background of engineering education certification. Heilongjiang Education (theory and practice) **10**, 12–14 (2020) (in Chinese)
6. Cen, G., Luo, J.Y.: Design of a teaching process monitoring and management system. Int. J. Educ. Manage. Eng. (IJEME) **2**(4): 128–33 (2012)
7. Dewi, D.K., Ardhana, W., Irtadji, Chusniyah, T., Sulianti, A.: Inquiry-based learning implementation to improve critical thinking of prospective teachers. Int. J. Inform. Educ. Technol. **11**(12), 638–645 (2021)
8. Catherine, A., Nathan, B., Erin, M.: The positive influence of inquiry-based learning teacher professional learning and industry partnerships on student engagement with STEM. Front. Educ. **6**, 319 (2021)
9. ur Shani, Z., Tali, T.: We know our students: teacher views of inquiry-based learning implementation in light of their working context. Proc. Singap. Natl. Acad. Sci. **15**(02), 91–104 (2021)
10. Coulter Robert, W.S., Lindsay, O:. Using student-led discussions and snapshot lectures to stimulate active learning and accountability: a mixed methods study on teaching an implementation science course. Pedagogy in Health Promotion **8**(1), 30–40 (2022)
11. Linbo, L., Wang Yanli, W., Bing.: Discussion on online classroom teaching mode under the concept of active learning. J. Phys: Conf. Ser. **1976**(1), 1–6 (2021)
12. Syamsidah, R., Muhiddin, A., Lullulangi, M.: Effectiveness of learning devices with inquiry learning models to increase skills creative thinking students. Educ. Lang. Sociol. Res. **1**(1), 128–136 (2020)
13. Chiu, P.H.P., Im, S.W.T., Shek, C.H.: Disciplinary variations in student perceptions of active learning classrooms. Int. J. Educ. Res. Open **3**, 1–7 (2022)
14. Howell, R.A.: Engaging students in education for sustainable development: The benefits of active learning, reflective practices and flipped classroom pedagogies. J. Clean. Prod. **325**, 1–11 (2021)
15. Downes The project-based teaching reform of civil engineering survey in applied universities based on OBE concept. Homestead **20**, 241–242 (2018) (in Chinese)

16. Huang, N.: Analysis and design of university teaching evaluation system based on JSP Platform. Int. J. Educ. Manage. Eng. (IJEME) **7**(3), 43–50 (2017)
17. Shi, J., Dong, C., Jiuyang.: Research and design of teaching evaluation system based on fuzzy model. Int. J. Educ. Manage. Eng. (IJEME) **2**(10), 45–51 (2012)
18. Ning, G.: Construction and implementation of innovation computer network practical teaching system. Int. J. Educ. Manage. Eng. (IJEME) **1**(2), 30–35 (2011)
19. Fu, Y., Zhao, L., Ye, F.: Construction and application of virtual teaching platform of special english for applied chemistry. Int. J. Educ. Manage. Eng.(IJEME), , **2**(2), 54–58 (2012)
20. Shi, J., Dong, C., Hou, J.: Research and design of teaching evaluation system based on fuzzy model. Int. J. Educ. Manage. Eng. (IJEME) **2**(10), 45–51 (2012)
21. Liu'an, K., Xiaomei, W., Lin, Y.: The research of teaching quality appraisal model based on AHP. Int. J. Educ. Manage. Eng.(IJEME) **2**(9), 29–34 (2012)
22. Fangxiu, W.: The course design for algorithm analysis and research on improving the students' comprehensive ability. Int. J. Educ. Manage. Eng. (IJEME) **2**(2), 42–46 (2012)
23. Guangchen, W., Yan, L.: Design of student ability achievement evaluation system based on engineering education certification - Taking surveying and mapping engineering as an example. Exam. Week. **54**, 152+164 (2016) (in Chinese)
24. Lingjun, H., Wuhua, Z., Renliang, Z.: Research on the teaching reform of civil engineering surveying for Applied Undergraduate Based on IEET engineering education certification. J. Hubei Second Normal Univ. **35**(08), 106–109 (2018). (in Chinese)
25. China Engineering Education Professional Certification Association Engineering education certification standard (Exposure Draft): T/CEEAA 001—2022[S] (in Chinese)
26. Judi, R., Lyn, C.: Addressing student motivation and learning experiences when taking teaching online. Theor. Pract. **61**(1), 129–139 (2022)

Experimental Flipped Classroom Teaching Mode Based on Mobile Terminal

Huang Yu and Haoliang Zhu(✉)

College of Mechatronical and Quality Technology Engineering, Nanning University, Nanning 530200, China
412363055@qq.com

Abstract. Flipped classroom provides a new way for teaching. Here, the center of knowledge interaction and application is no longer the teacher, but the promoter of learning. This new teaching mode provides the conditions for personalized learning, and changes the situation of students' passive learning role in the traditional teaching process. In the teaching process of experimental courses, the role and performance of this teaching model is particularly prominent. Taking the analog circuit experiment course as an example, this paper discusses the teaching practice of flipped classroom based on mobile terminal platform and media, designs the experimental course process including a series of links such as data preparation, self-study before class, classroom discussion, teacher assessment, experimental activities, summary and analysis, and studies the implementation methods, advantages and disadvantages of flipped classroom teaching, Through the study of specific cases, this paper illustrates the role and effect of flipped classroom experimental teaching mode in improving the quality of practical teaching.

Keywords: Flipped classroom · Course experiment · Video teaching · Mobile terminal

1 Introduction

Flipped classroom, also known as flipped classroom, is a new teaching model [1, 2]. It advocates "student-centered", through the organic integration of online learning and face-to-face teaching, to achieve personalized and controlled learning based on students' aptitude, so as to "flip" the traditional teaching process, teaching philosophy, teachers' and students' roles, teaching mode, etc., and transfer the decision-making power of learning from teachers to students [3].

The popularization of Internet and the application of computer technology in the field of education make the "flipped classroom" teaching model feasible and realistic. From 2011 Salman Khan put forward the teaching concept of "flipped classroom" at the Ted conference. Since then, flipped teaching has quickly received a great response in the world [4, 5]. Many famous universities in the world, such as Harvard University and Singapore International University, have adopted and popularized this teaching model. Many successful cases have also emerged in domestic higher education, involving basic courses, professional courses, experimental courses, foreign language courses, etc. [6–9].

Z. Hu et al. (Eds.): AIMEE 2022, LNDECT 159, pp. 391–399, 2023.
https://doi.org/10.1007/978-3-031-24468-1_35

With the continuous change of teaching philosophy and the continuous progress of teaching technology, flipped classroom teaching plays an important role in the teaching activities of colleges and universities.

Online learning is the first step of flipped classroom learning [10]. Before class, teachers upload teaching related materials to the teaching platform and publish learning tasks. Students watch online materials and independently plan the content and methods of learning before class, and complete the tasks assigned by teachers. Through online discussions with other students, they can solve problems in the learning process, improve students' ability to learn independently, analyze and solve problems, and promote students' personalized learning, Improve teaching quality and teaching effect.

The teaching mode of flipped classroom has its own progressiveness, but it also has limitations in combination with China's national conditions and educational status [11, 12]. At this stage, for some poor areas in the west, colleges and universities are limited by human, material and other school running conditions and teaching resources, and the proportion of poor students in the student source structure is large, some students do not have personal computers, and the network coverage is not high, which may lead to obstacles in flipped classroom teaching.

The use of flipped classroom in the teaching of experimental courses has been proved to be effective [13–15]. Analog circuit experiment course is an important part of analog circuit course. In the traditional analog circuit experiment teaching process, it is generally divided into three stages: experiment preview, experiment operation and experiment summary. Before doing the experiment, students mainly preview through the experiment instruction and fill in the experiment preview report, and then enter the laboratory for practical operation. During the experiment, because the students' preview is not thorough, and the teachers' point-to-point guidance is not available in the process of explaining the use of the equipment, there are some problems as follows: wrong wiring leads to damage to the equipment; Potential safety hazards of power utilization; Data error caused by non-standard use of measuring instruments and equipment; The key links or sequence errors in the experiment process lead to the failure of the experiment.

In view of the above problems, this paper has carried out the research on the flipped classroom teaching of analog circuit experiment based on the mobile terminal platform, and proposed the implementation process and method of flipped classroom through the case of experimental teaching.

2 Teaching Platform and Teaching Resource Design

2.1 Design of Teaching Platform Based on Mobile Terminal

At present, MOOC teaching platforms are mainly PC based, and excellent flipped classroom teaching platforms such as Coursera, EDX, and Udacity have emerged. According to the actual investigation on the 18 level electricity related majors of Nanning University, 76% of the students have personal computers and 100% of the students have smart phones. Among these users, the proportion of smart phone operating systems is 92% for Android and 8% for Apple IOS. Basically, all smart phone screens are larger than 4.0 inches and are equipped with wireless WiFi, This makes it possible to carry out the flipped teaching mode with mobile phone as the teaching platform.

The slow speed of network is one of the negative factors that restrict the development of network teaching in many schools. Considering the problems of mobile phone traffic and network speed, this paper designs a teaching platform Nymk, which takes a PC as the host server and is placed in the analog circuit training room and connected to students' mobile phones through WiFi to form a LAN. The platform includes jsmk Exe and xsmk APP. Teachers can easily transmit videos and other learning materials to the server. Students can install xsmk software on their mobile phones. After logging in, they can easily watch learning videos, download materials, communicate online, assess online, leave messages and feedback. The interaction process is shown in Fig. 1.

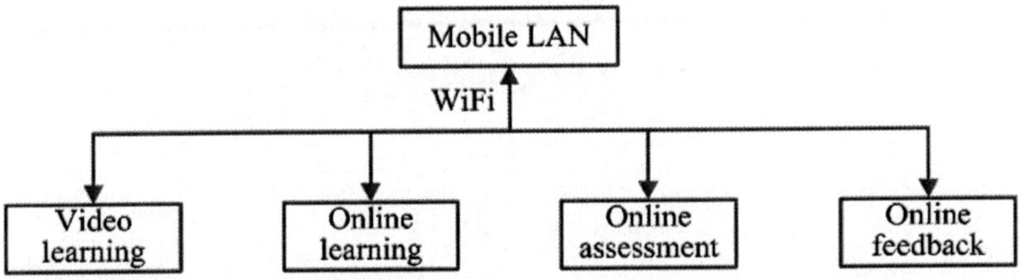

Fig. 1. Nymk teaching platform information interaction diagram

2.2 Construction of Learning Resources.

As this teaching platform is based on mobile terminals, the human and material resources that can be invested in the process of platform development are relatively limited. It is unable to have a variety of learning materials and high-definition video teaching materials like Coursera and EDX, and the software has limited functions. The platform mainly contains the following three types of learning resources:

(1) Video teaching resources. There are two types of video teaching resources. One is to use the camera to capture the experimental operation process, and then use the PPT to make a micro video with a length of 5 to 8 min after format conversion and video cutting. The main contents of the video explanation include: experimental purpose, experimental content, experimental instruments, experimental precautions, experimental operation process, experimental results and phenomenon demonstration; The second category of video resources is the explanation of Multisim software simulation examples. The simulation process of analog circuit experiment is completed by PC, so this part of video is recorded by screen video recording software, and the video time is 5 to 15 min;
(2) Text learning materials. Learning materials include electronic files of laboratory instructions, PPT, exercise explanation and other materials. Students can download the materials to mobile phones after logging in to the system;
(3) Online exercises. Each experimental unit contains four sets of test questions for students to learn. Exercises include the contents, knowledge, experimental precautions and experimental operation steps involved in each experiment, and are provided to

students in the form of multiple-choice questions and judgment questions for online assessment.

3 Implementation Process of Analog Circuit Experiment Flipping Classroom Teaching

Xuda proposed that the key links of flipped classroom of experimental courses are learning material preparation, autonomous learning before class, classroom activities, feedback and evaluation [16]. According to the characteristics of analog circuit experiment course, this paper divides the flipping experiment teaching process into four stages: Self-study before class, classroom discussion and teacher assessment, experimental classroom activities, summary and analysis. The flow chart is shown in Fig. 2.

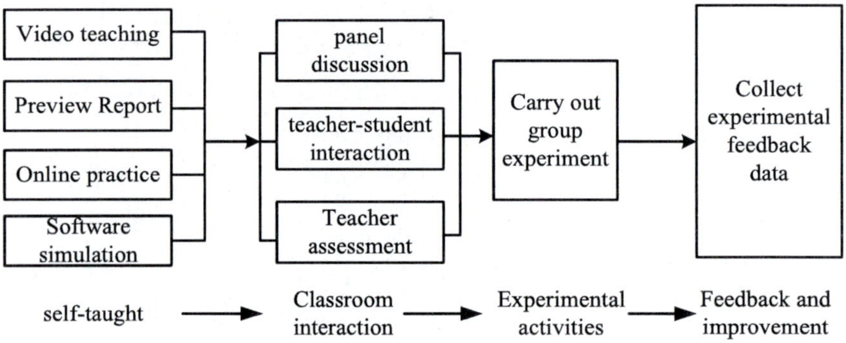

Fig. 2. Implementation process flow chart of analog circuit real flipping experiment class

3.1 Self-Study Before Class

(1) Video teaching. After students logging in to the NYMK teaching system with their mobile phones, they can watch (download) teaching videos online or browse electronic documents. They can learn by themselves through teaching videos, so that students can not only fully understand the experimental equipment, instruments, experimental precautions and experimental requirements before starting the experiment, but also conduct "simulation experiment" through video explanation, so as to improve the safety of experimental operation, the standardization of measuring instrument use The accuracy of measurement data can also reduce the accidental human error and the damage rate of experimental equipment [17].

(2) Online assessment. After video learning, students must log in to the online assessment end, and the system randomly assigns a set of assessment questions. Students must submit and participate in the assessment. This assessment can judge right and wrong, which can timely find their own problems in cognition, methods and operation methods, and enable students to have a deeper understanding of the experimental purpose, contents and instruments.

(3) Software simulation. The Circuit Simulation Software Multisim not only can highly simulate the circuit model, but also has the function of automatic error detection, so it is necessary to simulate before the experiment; After watching the video teaching and passing the online examination, students also require everyone to conduct software simulation, and send the simulation results to the teacher through the NNMC platform.

3.2 Class Discussion and Teacher Assessment

In order to improve students' enthusiasm and initiative in learning, the class was divided into several groups with 5 people, and the group leader was changed for each experiment. Before entering the laboratory, the team leader responsibility system is used to organize students to communicate and interact in the classroom. Then, the teacher assesses and communicates with each group in the form of students' experiment preview report and on-site questions. Only the learning groups that pass the assessment can enter the laboratory for experiments.

3.3 Experimental Activities

Before the experiment, the teacher briefly introduces the relevant matters needing attention in the experiment process and demonstrates the key links of the experiment before allowing students to enter the experiment. Through video learning before class, class discussion and assessment, students have a more intuitive understanding of the experimental content, experimental operation process, experimental results and experimental phenomena, because they can quickly enter the experimental state, significantly reducing the experimental preparation time. When students encounter problems during the experiment, they can turn on their mobile phones and solve them by themselves under the guidance of video explanation. At the end of the experiment, students are required to log in to the mobile terminal of the mobile phone and send the raw data and experimental phenomena measured during the experiment to the teacher in real time in the form of photos, so as to prevent experimental misconduct such as copying data.

3.4 Experiment Summary and Analysis

After the experiment, students can report the problems encountered in the experiment to teachers through the learning platform. Teachers summarize and analyze the problems encountered in the teaching process through students' feedback, sort out and research solutions, and record the solutions in the next experimental teaching video.

4 Case Study on Flipped Classroom of Experimental Course

There are three independent classes of electrical automation major of Nanning University in 2018, with a total of 140 students. After investigation, the student source structure, class number, class learning achievements and other data of the three classes are basically the same. In the spring of 2021, the author undertook the experimental teaching of analog circuit course in these three classes, implemented the flipped experimental teaching mode in class 1 of 18 Electric by using the method of case study, and selected the method of comparative research to analyze the implementation effect of flipped classroom. Assessment method: total experimental score = experimental preview report score × 20% + comprehensive test results × 30% + experimental report results × 50%. At the end of the period, the book was counted, and the specific data are shown in Table 1 and Table 2.

Table 1. Statistical table of experimental results

Class	Preview results	Test report results	Result of examination	Total score
Class 1	85.8	84.2	83.3	84.1
Class 2	77.4	82.1	77.2	78.7
Class 3	79.2	81.4	81.1	80.8

Table 2. Distribution statistics of total experimental scores

Class	Number of students	100–90	90–80	80–70	60–70	fail
Class 1	48	6	22	12	7	1
Class 2	47	4	13	21	6	3
Class 3	45	2	13	19	7	4

Statistics show that the average score of the final theoretical examination of class 1, class 2 and class 3 of electrical engineering in 2018 is 78.3, 78.7 and 77.6 respectively, with little difference in theoretical scores. Electrical class 2 got the best theoretical results; However, in terms of experimental scores, it can be concluded from the analysis in Table 1 and 2 and Fig. 1 and 2 that the average score of electrical class 1 is significantly higher than that of the other two classes, mainly because the proportion of students with good scores between 80 and 90 in the total experimental score of electrical class 1 in 2018 is significantly higher than that of the other two classes; After correcting the experimental preview report and the formal experimental report, it is found that the main reason for this phenomenon is that the students in class 1 of electrical engineering have higher scores in the experimental methods and experimental steps in the experimental preview report, and the accuracy of measuring the original experimental data during the experiment is significantly higher than that of the other two classes. In this way, the experimental results are closer to the ideal value. It can be seen that the use of flip mode teaching has played a certain role in improving the practical teaching of analog circuits.

During the observation of the students' experiment process, the author found that the students in class 1 of electrical engineering had significantly reduced their doubts during the experiment. When they encountered problems with the use of experimental instruments, their first reaction was to turn on their mobile phones to watch videos, while the students in class 2 and 3 of electrical engineering had significantly more questions than Class 1, and sometimes the teachers were even asked "too busy". The two classes had significantly more equipment operation errors in the experiment than Class 1; A very interesting phenomenon is that a few days before the last comprehensive experimental assessment, some students watched video teaching review and preparation with mobile phones in public places such as libraries, student canteens and stadiums, which also shows the flexibility and practicality of mobile terminal teaching.

5 Conclusion

It can be found from the research that due to the convenient carrying, flexible operation and strong interaction of mobile phones, the experimental flipping teaching mode based on the mobile terminal teaching platform has its unique and obvious advantages in improving the quality of students' experimental preview, the standardization of experimental equipment operation and the accuracy of measurement data.

5.1 Advantages of Mobile Terminal Experiment Flipping Teaching Mode

Applying this model to practical teaching can play the following three roles.

(1) It can obviously improve teachers' efficiency in class.
Under resource constraints, the ratio of students to teachers in Colleges and universities in Western China is generally high. Experimental courses are often carried out in a large class teaching mode. It is difficult for teachers to provide point-to-point guidance, and the use and operation of experimental equipment will also occupy a large amount of teaching time. Using this teaching mode can effectively solve this problem.

(2) The individualized level of teaching has been improved to a certain extent.
On the one hand, through mobile phones, students can reasonably arrange their learning progress according to their own learning situation; On the other hand, students can also make full use of the fragmented learning time, so that the personalized learning space has been effectively expanded.

(3) It can fully mobilize students' interest in learning and improve their ability of. autonomous learning.
It is applied to practical teaching in the form of flipped classroom. Students are familiar with the equipment in advance through self-study, and carry out simulation experiments, which can effectively improve the preview effect of students on the experimental purpose, experimental means, experimental process and experimental methods, and implement the educational concept of "student-centered learning" in engineering education certification; In addition, it also reduces the failure rate of the equipment, ensures the good operation of the experimental equipment, and shares the equipment maintenance workload of the experimental teachers.

Although this teaching model has been widely used in Colleges and universities all over the country and achieved some research results, it has also encountered many problems in the process of practice.

5.2 Disadvantages of Mobile Terminal Experiment Flipping Teaching Mode

(1) The network speed of mobile devices is slow, so they can only establish a LAN Platform Based on *WiFi* technology, and cannot connect to the Internet.

With the construction of China's network infrastructure project and the investment in western education, the western region has basically realized the full coverage of mobile 4G network, and most western universities have also realized free campus network services in the past two years. Therefore, this problem should be effectively solved in the future.

(2) The mobile device screen is relatively small, which is not conducive to the sensory experience of video teaching. This problem was prominent in the first two years. However, with the development of electronic technology, the cost of mobile electronic equipment is becoming lower and lower. Even in backward western universities, students can basically realize the burden free consumption of iPad and other large screen mobile learning equipment. This is also beneficial to the further teaching reform and implementation of the "this teaching mode" to a certain extent.

(3) The function of the teaching m is relatively simple (when the author carried out the research, he did not have the ability to develop professional teaching platforms, and there are not many mature products in the market). To solve this problem, there are many professional platforms in China now. As long as the model is effectively relocated.

(4) It increases the possibility for students to "play with mobile phones" in class. After investigation, it is found that many students are "deserting" in the process of using mobile phones for teaching, such as playing games, socializing, reading novels and so on. How to effectively solve this problem is the direction that the research group needs to work hard in the future teaching reform research.

Acknowledgment. This paper is funded: (1) Nanning University 2022 school level teaching reform project (2022xjjg21); (2) Provide support for the 2019 Guangxi Higher Education Undergraduate Teaching Reform Project (2019jga367);

References

1. Jinlei, Z., Ying, W., Baohui, Z.: Research on flipped classroom teaching model. J. Dist. Educ. **4**, 46–50 (2012). (**in Chinese**)
2. Qingsong, L.: Experimental curriculum design of "modern educational technology" based on flipped classroom. Lab. Res. Explor. **33**(1), 194–198 (2014). (**in Chinese**)
3. Akbar, B.: FonF practice model from theory to practice: CALL via focus on form approach and non-linear dynamic motivation to develop listening and speaking proficiency - ScienceDirect. Comput. Educ. **130**, 40–58 (2019)
4. Zhongliang, W., Lei, Z.: A preliminary study on flipped classroom teaching mode based on network learning space. China Audio Vis. Educ. **4**, 121–126 (2014). (**in Chinese**)
5. Yubin, H.: Reflection on flipped classroom teaching mode in colleges and universities based on teaching effectiveness. J. Higher Educ. **8**(05), 98–101 (2022) (in Chinese)

6. Xiang, W.: Application of flipped classroom in college english teaching in the 'Internet +' era. J. Hubei open vocat. coll. **35**(05), 191–192+195 (2022) (in Chinese)
7. Bogojevi, S., Zou, M.: Making infrastructure "Visible" in environmental law: the belt and road initiative and climate change friction. Transnational Environ. Law **10**(1), 35–56 (2021)
8. Hongwei, L., Changhong, Y., Qingjun, G.: Practical teaching of "operating system" course based on type training. Exp. Technol. Manag. **29**(1), 151–153 (2012). **(in Chinese)**
9. Qiliang, Z., Yongsheng, C.: Research on the application of flipped classroom in operating system experiment teaching. Exp. Technol. Manag. **31**(12), 173–176 (2014). **(in Chinese)**
10. Elizaphan, M., Robert, O.M., Oboko., Peter., Waiganjo, W.: Using machine learning techniques to support group formation in an online collaborative learning environment. Int. J. Intell. Syst. Appl. (IJISA) **9**(3), 26–33 (2017)
11. Xinglong, Z.: Progressiveness and limitations of flipped teaching. Chin. J. Educ. **20**(4), 65–68 (2013). **(in Chinese)**
12. Ahuja, S., Kaur, P., Panda, S.N.: Identification of influencing factors for enhancing online learning usage model: evidence from an Indian University. Int. J. Educ. Manag. Eng. (IJEME) **9**(2), 15–24 (2019)
13. Jianfeng, L., Hongjun, L., Yumin, M.: A framework of network virtual experiments for control courses. Int. J. Educ. Manag. Eng. (IJEME) **2**(9), 49–55 (2012)
14. Ning, G.: Construction and implementation of innovation computer network practical teaching system. Int. J. Educ. Manag. Eng. (IJEME) **1**(2), 30–35 (2011)
15. Qiu, M., Huang, C.: Applied the experimental teaching model of three phases into the environmental monitoring experiment. Int. J. Educ. Manag. Eng. (IJEME) **1**(1), 47–50 (2011)
16. Da, X.: Research on the support system of chemistry experiment teaching mode based on flipped classroom. J. Distance Educ. **31**(05), 107–112 (2013) (in Chinese)
17. Alkhafaji, S., Sriram, B.: Instructor's performance: a proposed model for online evaluation. Int. J. Inf. Eng. Electron. Bus. (IJEEB) **10**(1), 34–40 (2013)

Study on Insulation Detection of IT System Under AC-DC Mixed Injection

Zhaorui Lv(✉), Ying Yang, Xinjian Zhang, Haihua Li, Yanmin Lu, and Juan Fan

Mechanical and Electrical Department of Wuhan University, Wuhan University, Wuhan 430074, China
15347670@qq.com

Abstract. When the IT system occurs in a single-phase ground, the leakage protection on the line will cut off the power supply, causing an important load power off. Therefore, it is suggested in the national standard that the IT system should be installed with insulation monitoring device, this paper addresses the detection and localization of insulation resistance under complex topology, comparing the detection principle and error analysis between DC injection and AC injection, On the basis of AC-DC mixed injection for IT system insulation monitoring analysis, using the DC injection to detect the fault resistance, to solve the problem of measuring the ground insulation resistance is greatly affected by the stray capacitance, using the AC injection method to solve the problem that the DC injection cannot be located. Finally, the fault detection and the fault selection method are simulated with MATLAB software.

Keywords: IT system · Fault detection · Fault line selection

1 Introduction

When faulty ground occurs of power system with IT grounding [1, 2], the fault current is very small, and the line voltage between the three phases is still symmetrical, there has no impact on the power supply of the load. Therefore, it is generally allowed to continue running for another 1–2 h. The grounding system is widely used in marine or port power supply networks, With the development of modern power grid towards large-scale and automation, the capacity of power system is increasing, especially the wide application of anti-radio interference capacitance, the ground-pair capacitance of the IT system has increased greatly, fire and serious short circuit caused by insulation reduction failure in time are common, Therefore, it is suggested that the IT system needs to be installed in the national standard for insulation monitoring devices.

When the neutral line of the IT power supply system has a grounding fault, there is no obvious electrical characteristic, so it is necessary to use the external power supply method to monitor the insulation state of the IT power supply system. At present, the insulation detection methods are mainly divided into DC injection method, AC

Z. Hu et al. (Eds.): AIMEE 2022, LNDECT 159, pp. 400–410, 2023.
https://doi.org/10.1007/978-3-031-24468-1_36

single-frequency injection method dual-frequency injection method [3]. Comparatively speaking, the DC injection method avoids the influence of the straycapacitance on the measurement, but the DC injection method is more difficult to achieve the fault positioning.AC single frequency injection structure is simple, the algorithm is intuitive, however, the calculation is obviously affected by the stray capacitance. And the current ship ac grid insulation monitoring device mainly uses dual frequency method principle. Same as dc injection method, the method with different frequency waveform injection can shield the influence of stray capacitor on fault detection, but dual frequency circuit need to inject two frequency monitoring signals, causing device structure becomes more complex, high frequency conversion control requirements, also reduce the reliability of device operation to a certain extent.

In this study, AC-DC mixed injection method is adopted. DC injection is used to detect the fault resistance. After detecting that the fault resistance is lower than the warning value, low-frequency AC injection is used to realize the fault line selection function. The method has high detection accuracy and flexible selection, and the algorithm is applicable to star and ring power supply topology.

2 AC-DC Mixed Injection

2.1 Insulation Resistance Detection

In order to realize the real-time monitoring of the IT system insulation, the power supply Ef can be injected into a certain phase of the power grid to generate a detection signal. The schematic diagram is shown in Fig. 1. Ea, Eb, Ec are the generator phase potential, $C_1,\ldots C_i$, Cn are the ground capacitances of the load branch, and R is the fault ground resistance.

2.1.1 AC Injection Signal

Assuming that the injection signal frequency is f, the schematic diagram is shown in Fig. 1. The current expressions of the normal line (formula 1) and the grounding fault line (Eq. 2) are respectively:

$$I_n = U_f / jX_{cn} = jU_f \omega C_n \quad (1)$$

$$I_n = U_f / R + jU_f \omega C_n \quad (2)$$

In the actual system, considering that the cable line-to-ground capacitance is generally in the μF level, the practical simplified calculation method is often adopted, and the grounding resistance R is directly calculated by the following formula:

$$\mathrm{R} \approx \frac{\dot{\mathrm{U}}}{\dot{\mathrm{I}}}\cos\Phi = \frac{\mathrm{R}}{1 + (\omega\mathrm{RC})^2} \quad (3)$$

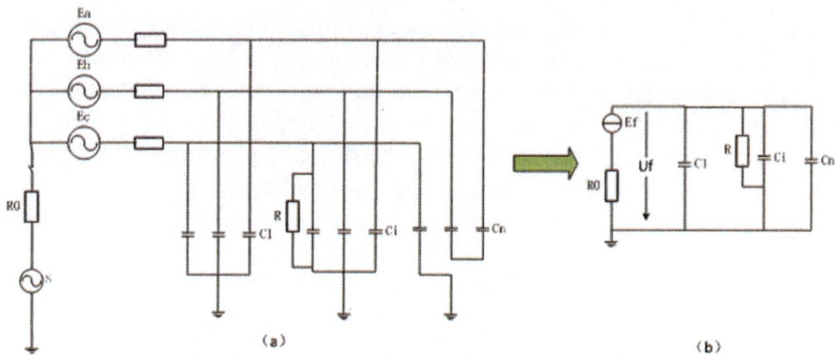

Fig. 1. Schematic of the insulation detection of the AC injection system

2.1.2 DC Injection Signal

When the injection signal frequency is a DC power supply, the schematic diagram is shown in Fig. 2, which can eliminate the influence of the capacitor of each line. The voltage method is used for the system insulation resistance detection. The main principle is that the partial voltage on the insulation resistance and the partial voltage on the measured resistance shall be equal to the voltage value of the injection source:

$$R_{ins} = R_a//R_b//R_c = \left(\frac{E}{U_0} - 1\right)R_0 \tag{4}$$

DC injection method can eliminate the influence of line capacitor on measurement, compared with single frequency method has better accuracy, but the DC injection method can't specific to a line of the whole grid, this paper using AC-DC mixed injection to realize the insulation monitoring function, namely using DC injection warning insulation resistance, insulation resistance after abnormal injection of low frequency ac current to determine the line fault section [5].

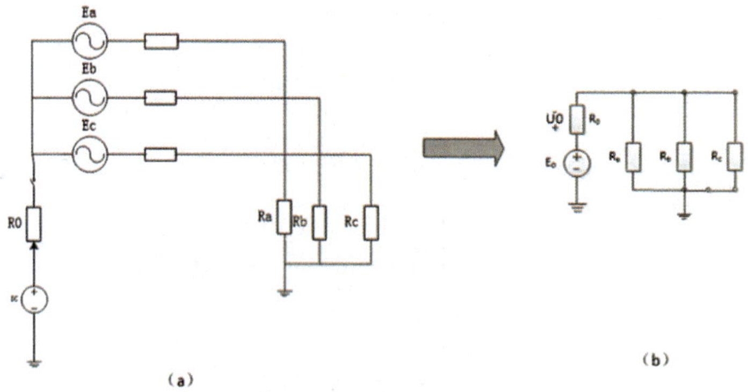

Fig. 2. Schematic of the insulation detection of the DC injection system

2.1.3 Error Analysis

The amplitude and phase errors may occur in the transmission and extraction process of the current and voltage in the actual system, which may adversely affect the calculation results. It is necessary to evaluate and analyze the impact of the measurement error on the performance of the detection method. For AC input signal, the relative error of ground resistance and voltage amplitude measurement meet the following relationship:

$$\frac{\Delta R}{R} = \left(\frac{U + \Delta U}{I}\cos\varphi - \frac{U}{I}\cos\varphi\right)/\frac{U}{I}\cos\varphi = \frac{\Delta U}{U} \tag{5}$$

The current amplitude measurement error meets the following relationship:

$$\frac{\Delta R}{R} = \left(\frac{U}{\Delta I}\cos\varphi - \frac{U}{I}\cos\varphi\right)/\frac{U}{I}\cos\varphi \approx -\frac{\Delta I}{I} \tag{6}$$

So, ground resistance measurement error is proportional to the voltage amplitude measurement error, when the voltage amplitude error increase or decrease, R detection value will increase or decrease accordingly, the influence of current amplitude, when the current amplitude error increase or decrease, R value detection value will be reduced or increase accordingly, ground resistance measurement accuracy is not only affected by current amplitude error also affected by the error of the voltage sensor.

When the system injects itself into the DC mode:

$$\frac{\Delta R}{R} = \left(\frac{E}{\Delta U + U} - \frac{E}{U}\right)/\frac{E}{U} \approx -\frac{\Delta U}{U} \tag{7}$$

The measurement accuracy of the ground resistance is only affected by the error of the voltage on the measured resistance. It should be noted that this analysis is based on the detection of the fault resistance, and in the fault positioning phase, the resistance current is only a few percent of the capacitive current in the most stringent circumstances. Its positioning requirements are not only related to the measurement accuracy, but also related to the signal processing and filtering algorithm [6].

2.2 Insulation Positioning

When the insulation resistance value of the system insulation resistance measurement is lower than the set alarm value, the fault line selection function is immediately activated. In order to solve the problem of fault line selection under the complex topology, the low-frequency AC injection method is adopted. After using AC injection, a low frequency resistance current is injected from the central point to the fault point. Due to the existence of the stray capacitor, the system insulation monitor also provides the capacitive current flowing through the stray capacitor, and the large diameter heart penetration sensor can collect the capacitive current after the superposition. When the fault point is located between the two transformers, the vector of the resistance current is not zero, this is the area fault; when the fault point is located outside the two transformers, the resistance current vector is basically zero, this is the area fault.

The positioning current generated by the injection current travels through the wire to the insulation fault position, flowing through the insulation fault point and returning to the injection point. After the injection source injects the fault location feature signal into the circuit, the zero-order current of the characteristic signal of the fault branch will increase, while the zero-order current of the non-fault branch is basically unchanged due to the large insulation resistance. After the fault loop is stable, the system insulation monitoring module issues instructions through the communication bus to enable each insulation fault positioning unit to collect the residual current signal of each branch at the same time [7, 8].

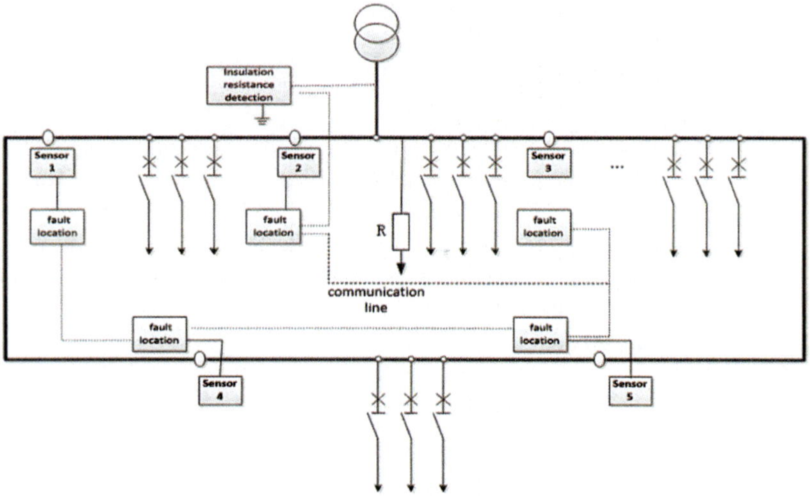

Fig. 3. Diagram of fault line selection system configuration

As shown in Fig. 3, the schematic diagram is configured for the fault line selection system. The positioning current feature signal is sampled by the measurement current sensor on the insulating fault path, and analyzed by the Fourier to extract the amplitude and phase of the characteristic current signal. The residual current information of each branch of the system is collected in the insulation fault module. After comparing the size and polarity of the resistance component in the current, it can be determined that the fault is located between two transformers. This principle is applicable for star and annular power supply topology.

After removing the power frequency current component of each transformer, the current phase of each transformer is different under different stray capacitance. The typical phase distribution is shown in Fig. 4. The capacitive component of each current is affected by the stray capacitance size, cable impedance and other factors, and the amplitude and polarity are different. The difference between the two sensor resistance components (I1-I2) of the fault interval is not zero, and the difference between the two sensor resistance components (I4-I5, I2-I3) or the sum (I3 + I5) outside the fault interval is zero (the positive direction of the sensor 3 and 5 is different). While the flow of the insulation detection is presented in Fig. 5.

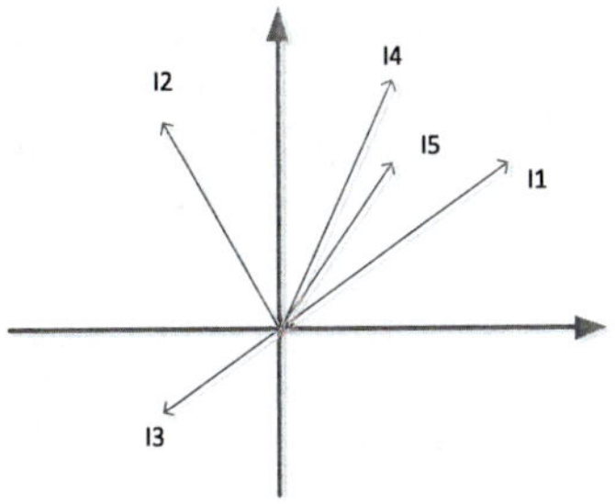

Fig. 4. Injection current vector diagram

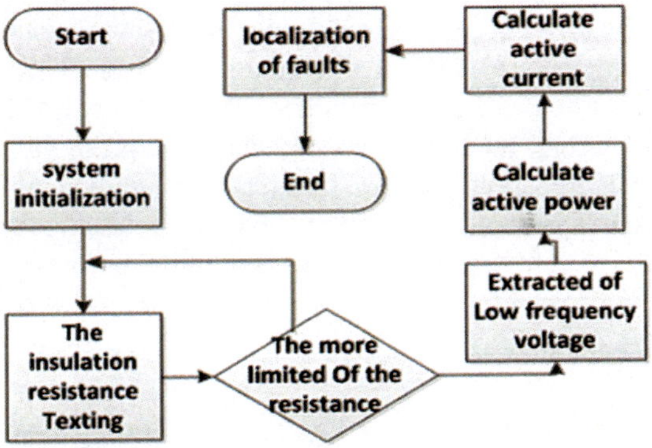

Fig. 5. Flow of the insulation detection

3 Simulation and Calculation

In order to test the performance of AC-DC injection method, Matlab/simulink is used as a simulation tool to build a typical power grid simulation model of single machine multi-circuit outlet with 0.4 kV voltage level with a certain power grid structure as an example, and the fault resistance detection and fault line selection function are simulated in the system. For small systems, considering the 300 m line, the branch capacitor unit parameter is 0.33 F/km, and the branch capacitor is set to 0.1 F.

3.1 Fault Resistance Detection

In the simulation analysis, different insulation reductions with a ground resistance of 10 Ω 10 kΩ were simulated. Resistance detection at the injection source frequency of 5 Hz and DC injection mode was analyzed and calculated, respectively [9].

Table 1. Simulation calculation results of the AC method

Actual value	Voltage	Current	Detection value	Error
10	1.089	0.109	9.901	0.99%
100	15.07	0.1525	98.8	1.2%
1000	16.39	0.016	1020.1	2.01%
10000	18.04	0.00183	9820.2	1.8%

The full week Fourier algorithm is used to extract the low frequency components of the bus to ground voltage and each branch current. The results are shown in Table 1, the AC injection method can accurately calculate the insulation resistance size of the line. By detecting the low frequency current or insulation resistance value flowing through each branch, the fault line can be determined by using the amplitude comparison and realize the fault positioning. It can be seen that when the grounding resistance is large, because the infuse voltage is low, the low-frequency current signal flowing through the line is weak, which has a certain impact on the accurate extraction of the current phasor, resulting in an increased calculation error. The amplitude of the input power supply can be increased accordingly, and the detection error can be reduced, which also shows that the theoretical analysis is correct.

Table 2. Test results of the DC method

Actual value	Detection value	Error
10	10.072	0.72%
100	98.99	1.01%
1000	1005.5	0.55%
10000	9890	1.1%

The detection results of the DC injection method are shown in Table 2. Compared with the AC injection method, the detection result accuracy is higher. When the fault resistance detection is lower than the alarm value, the fault line selection function is started.

3.2 Fault Selection Line Simulation

AC current at frequency 5 Hz was injected at the central point. The specific wiring is consistent with Fig. 3, but with an additional set of lines, there are 6 current sensors C1, C2, C3, C4, C5, C6 for partition detection. Set the line fault between current detection C4 and current detection C5, phase B has a ground fault at 0.3 s, and the insulation resistance is set at 1000 Ω. The voltage and current are detected by the sensor, the low frequency current is extracted by low pass filter, and then calculates the active frequency current of each node, then observes the current difference between the nodes to select the line fault. Figure 6 shows the fault current extraction and calculation model. Figure 7 and Fig. 8 are the voltage and current waveform, and Fig. 9 and Fig. 10 show the difference of node active current.

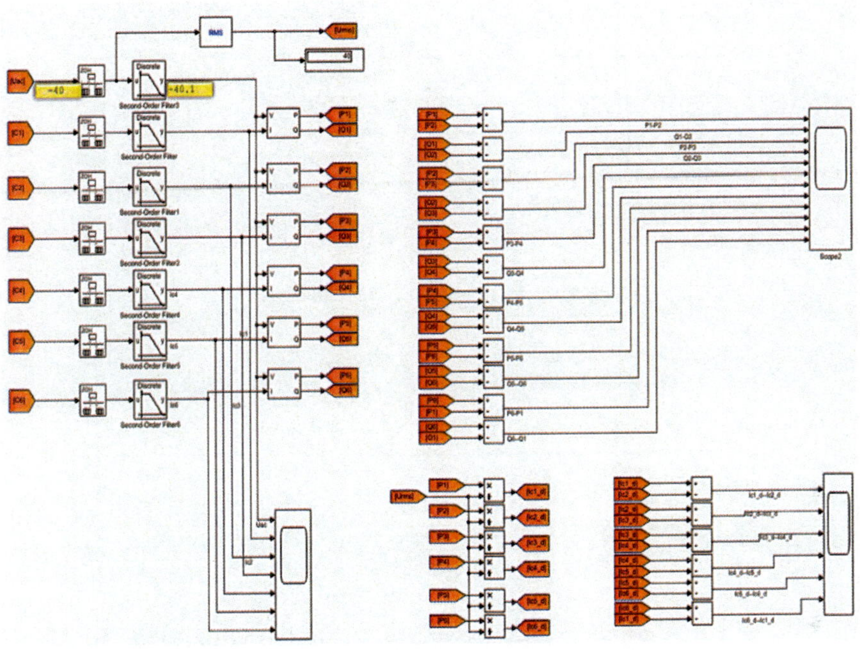

Fig. 6. Fault current extraction and calculation

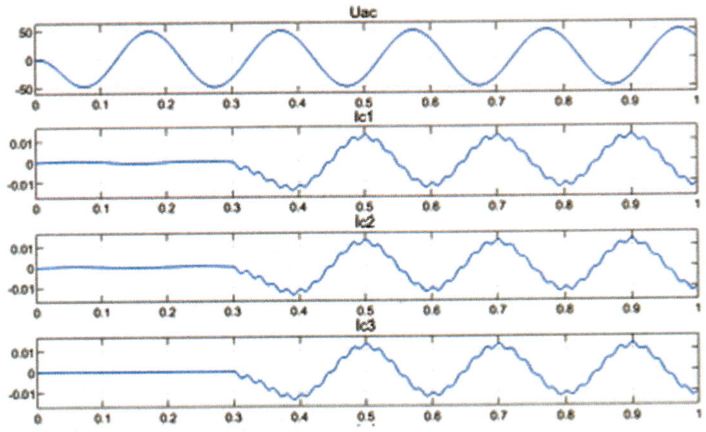

Fig. 7. Voltage and Current Waveform-1

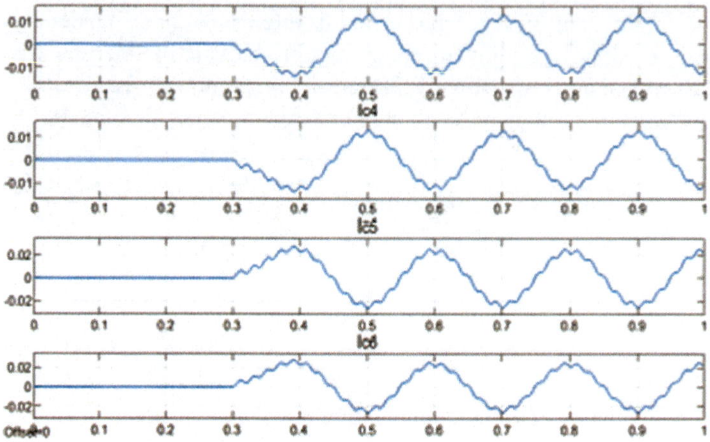

Fig. 8. Voltage and Current Waveform-2

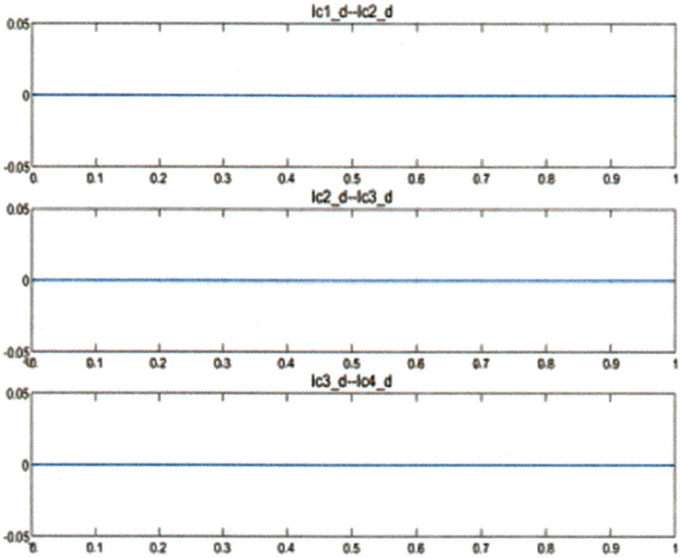

Fig. 9. Node active current difference-1

According to the simulation results above, the voltage and current waveform changes at 0.3 s. Judging by the current difference waveform, the grounding fault occurs between the current sensors C4 and C5.

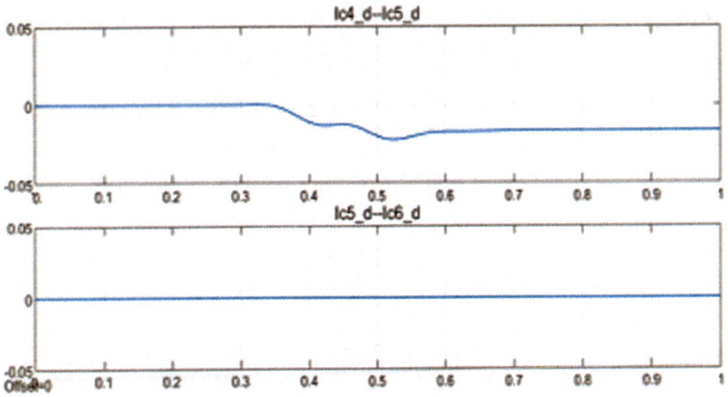

Fig. 10. Node active current difference-2

4 Conclusion

When the power supply system of IT has a grounding fault, there is no obvious electrical characteristic, so the external power supply method is needed to monitor the insulation state of the IT power supply system. This study adopts the AC-DC mixed injection method to realize the fault resistance detection and fault positioning of different topologies. The main conclusions are as follows:

1) The DC injection mode can eliminate the influence of stray capacitance on the fault resistance detection, with higher accuracy compared with the single-frequency method, simpler structure than the dual-frequency method, and good feasibility;
2) Using 5 Hz low frequency injection signal, the fault line selection positioning, by comparing the adjacent transformer between the resistance current vector and whether it is zero to determine the fault line, advanced detection method, good adaptability, can meet the different modes such as chain and ring power supply;
3) Simulation studies verify the correctness of the theoretical analysis and detection methods.

Acknowledgment. This project is supported by Hubei Provincial University Excellent Young and Middle-aged Science and Technology Innovation Team Program Project (T2021043).

References

1. Li, R., Gu, H., Hu, B., et al.: Multi-feature fusion and damage identification of large generator stator insulation based on lamb wave detection and SVM method. Sensors **19**(17), 3733 (2019)
2. Xu, J.: Study of Online Detection of Insulation Resistance. Xi'an University of Architecture and Technology, Xi'an (2008)
3. Ruihua, L., Hao, L., Bo, H.: Damage identification of large generator stator insulation based on PZT sensor systems and hybrid features of lamb waves. Sensors **18**(9), 2745 (2018)

4. Li, H., Li, R., Hu, B., et al.: Application of guided waves and probability imaging approach for insulation damage detection of large generator stator bar. IEEE Trans. Dielectr. Electr. Insul. **22**(6), 3216–3225 (2016)
5. Esteban, B., Sid-Ahmed, M., Kar, N.C.: A comparative study of power supply architectures in wireless EV charging systems. IEEE Trans. Power Electron. **30**(11), 6408–6222 (2015)
6. Ebadi, L., Shafri, H.: A stable and accurate wavelet-based method for noise reduction from hyperspectral vegetation spectrum. Earth Sci. Inf. **8**(2), 411–425 (2015)
7. Goldshtein, A., Vavilova, G., Rumkin, A., et al.: Detection of insulation defects in automated in-process testing of electric wire during its extrusion. Mater. Sci. Forum **94**(2), 141–150 (2019)
8. Rabeh, M., Petersen, J., O'Hearn, S.: Insulation optimization improves safety, reduces corrosion risks. Offshore **78**(2), 32–33 (2018)
9. Wang, P., Liu, H., Yuan, W., et al.: Carbon nanotubes regulated hollow skeleton carbon aerogel for effective thermal insulation and favorable mechanical behavior. J. Sol-Gel. Sci. Technol. **101**(1), 193–204 (2022)

Research on Talent Training and Platform Cooperation for Logistics Engineering Undergraduates Based on OBE Concept

Yanfang Pan[1] and Zhong Zheng[1,2](✉)

[1] College of Transportation, Nanning University, Nanning 530000, China
523201940@qq.com
[2] Thailand First Emperor University of Technology, Bangkok, Thailand

Abstract. Having a large number of high-quality logistics professionals is the basic guarantee for the development of the logistics industry. This paper focuses on the training of logistics engineering talents under the special background of intelligent upgrading and transformation, and analyzes the innovation of educational ideas and the reform of teaching methods. Guided by "multi platform combination + achievement orientation", this paper discusses the construction of an integrated platform of "production, teaching, learning, research, application, creation and competition", analyzes the integration of educational and teaching resources and practical teaching resources inside and outside the school, explores the construction of a curriculum system that is more conducive to the training of applied logistics engineering professionals, and uses AHP analysis method to make a quantitative analysis of the teaching index system supported by the platform, Strive to find the key factors of professional training in the cooperation mechanism.

Keywords: Talent training · Platform co-construction · Logistics engineering · OBE concept

1 Introduction

1.1 Subversive Transformation of the Logistics Industry, Multi-party Data Highlights the Trend of "Change"

The rapid development of logistics industry puts forward higher requirements for the training of logistics engineering professionals [1]. We should not only focus on the innovation of knowledge and technology, but also integrate curriculum ideological and political education and innovation and entrepreneurship education, so as to cultivate excellent talents who can adapt to intelligent upgrading and innovation development [2–4].

Under the background of the development of "Industry 4.0" and "Internet + ", my country's logistics industry has ushered in intelligent upgrading and transformation,

Z. Hu et al. (Eds.): AIMEE 2022, LNDECT 159, pp. 411–425, 2023.
https://doi.org/10.1007/978-3-031-24468-1_37

which plays a key role in effectively guiding and supporting the large-scale and intensive development of the logistics industry, and accelerating the transformation, upgrading and innovative development of logistics. Effect. The national government attaches great importance to it and has issued a number of policies to promote the rapid development of "smart logistics". In June 2020, the General Office of the State Council forwarded the "Implementation Opinions on Further Reducing Logistics Costs" by the National Development and Reform Commission and the Ministry of Transport, clearly proposing to promote the combination of reducing logistics costs and new infrastructure, and accelerating the promotion of a new generation of national smart roads and smart ports. The construction of integrated infrastructure such as smart logistics parks and the use of new technologies represented by big data, cloud computing, and artificial intelligence have become an important fulcrum to support the smart logistics ecosystem [5]. More and more enterprise logistics and logistics companies have transformed logistics into an enabling and value-added entry, and upgraded it to an ecological industry chain that can provide automated, intelligent, and intelligent full-process overall solutions [6].

In the Guangxi regional development strategy, the "dual-core drive" strategy of the Beibu Gulf Economic Zone and the Xijiang Economic Belt, the external thrust of the "Belt and Road" and the "Southbound Corridor", the "Development Plan for the Beibu Gulf City Cluster" and the "Western Land-Sea New Corridor" will be further promoted. The implementation of the Master Plan, the China (Guangxi) Pilot Free Trade Zone and the construction of the Nanning National Logistics Hub have brought new development opportunities and higher demand for logistics talents to the logistics industry [7, 8].

1.2 Industrial Development Puts Forward Higher Requirements for the Training of Logistics Talents

As the third profit source of multi-industry and multi-industry integration, the modern logistics industry is an export-oriented and value-added service industry with knowledge-intensive, technology-intensive, capital-intensive and labor-intensive characteristics [9, 10]. Talents should have professional technical ability and practical ability in logistics, trade, marketing, e-commerce, business management, computer science, finance, etc. The talent training of logistics engineering majors in colleges and universities should be reflected in the forward-looking formulation of talent training programs, the applicability of practical teaching, and the effective integration of logistics technical knowledge and multidisciplinary knowledge [11]. Under the new logistics industry environment, the undergraduate education of logistics engineering should focus on knowledge update, technological innovation and industrial integration, and focus on the comprehensive ability of students to speak, use, do and improve. On the one hand, it reflects the cross-industry and cross-field characteristics of the logistics industry [12].

Integrate ideological and political education and innovation and entrepreneurship education into the curriculum, cultivate and master the basic knowledge of engineering and management, systematically master the relevant professional theories and methods of logistics, master logistics system planning and design, supply chain solution design, warehousing and inventory control, transportation Basic skills such. as optimization design and control, logistics automation and intelligence, and strong engineering practice ability and innovative spirit in the fields of logistics technology application and logistics engineering management [13, 14].

2 Theories and Methods of the Research

2.1 Outcome Oriented Education

Outcomes-Based Education (OBE) is one of the core concepts of the Washington Accord for the accreditation of undergraduate engineering majors. The core of the OBE concept can not only be applied to the formulation and evaluation of the training target system, but also can guide the construction and optimization of the entire training and education process. The optimization of teaching methods and teaching models based on the OBE concept is a core of higher education research.

2.2 Analysis of Hierachy Process

Analysis of hierachy process (AHP), referred to as AHP, refers to a decision-making method that decomposes decision-related elements into goals, criteria, and plans, and then conducts qualitative and quantitative analysis on this basis [15, 16]. It refers to taking a complex multi-objective decision-making problem as a system, decomposing the objective into multiple objectives or criteria, and then decomposing it into several levels of multiple indicators (or criteria, constraints), and calculating the hierarchical single ranking (weights) and total ranking as a target (multi-indicator), multi-scheme optimization decision-making method [17].

3 Exploration on the Mode of Talent Training and Platform Cooperation for Logistics Engineering Undergraduates Based on the Concept of OBE

The core of the training of applied talents is ability training. In the professional setting, we should not only pay attention to the selection and teaching of professional theoretical knowledge, but also highlight the setting and implementation of experiments, and practice [18]. The selection of professional theoretical knowledge should rely on the actual production of enterprises. It has certain guiding significance for the control of the work process and the completion of work tasks; the setting of the experimental practice link should be gradually advanced around the deepening of professional theoretical knowledge, and the project teaching and case teaching modes are used to strengthen the understanding of individual work task scenarios. The "integration of theory and practice" teaching mode is adopted to form a virtuous circle of "professional theory guides operational practice, and operational practice expands theoretical knowledge". While solidifying the basic theoretical knowledge necessary for the job, students strengthen memory and knowledge digestion through practical links, so that students can master and apply the core knowledge of logistics. engineering [19] . As a result, a multi-platform co-construction-based logistics engineering talent training curriculum system is constructed, as shown in Fig. 1.

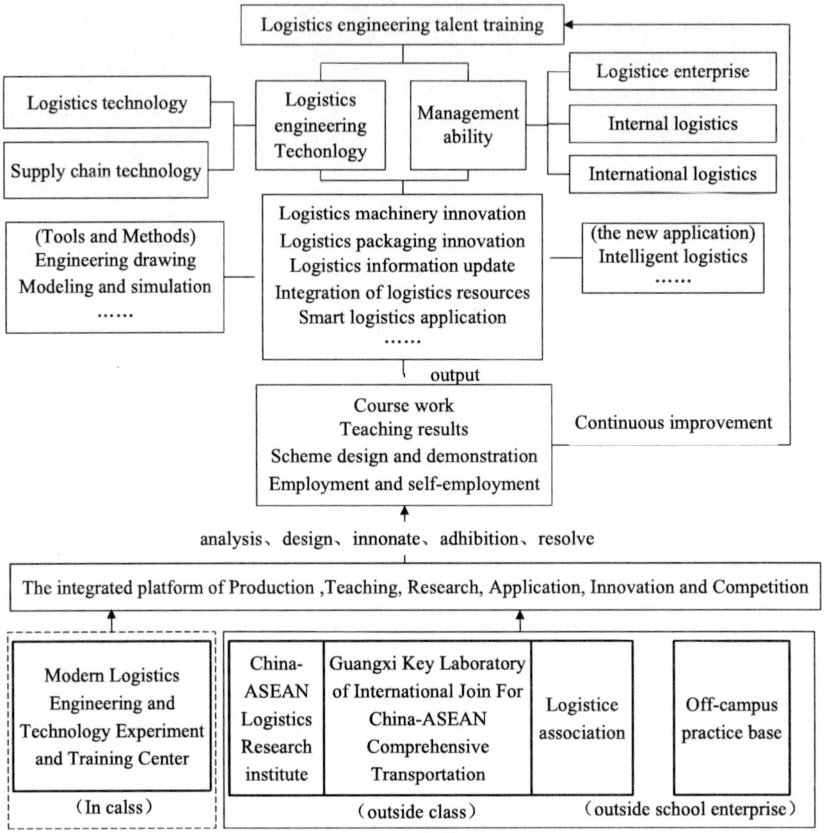

Fig. 1. The curriculum system of logistics engineering talents training based on "OBE concept + multi-platform co-construction"

(1) Schools and enterprises jointly build a practical teaching system for the application of multi-level logistics knowledge and technology

Aiming at the frontier fields and top level of the world's science and technology, realizing the professional chain connection with the industrial chain, deploying the innovation chain around the industrial chain, under the background of resources such as logistics big data mining and analysis, Internet + smart logistics, etc., in the course, practice, practice, In each teaching link of graduation design, through teamwork or teacher guidance, complete the multi-module learning and improvement of logistics machinery innovation design, logistics packaging innovation design, logistics system innovation, logistics information update, logistics data processing, logistics resource integration, and smart logistics application., learn to use scientific tools and methods to analyze, design, innovate, apply and solve

(2) "One hospital, one room, one meeting and one center" for collaborative communication to build an "integration of production, teaching, research, application, and competition" platform for professional teachers and students

The modern logistics engineering and technology experimental training center integrating virtual and real will give full play to the information integration and resource sharing functions of the experimental platform, and build a logistics engineering laboratory, a supply chain integration laboratory, and a smart logistics and technology innovation laboratory. Supporting course development and experimental textbook construction, the three-dimensional thinking focuses more on project application and case guidance, adopts the "OBE concept" and focuses on course achievement orientation, combines courses with papers, designs, patents, and competitions, and makes full use of the existing talents in the laboratory, Technology and equipment to carry out the integration of theory and practice teaching and simulation experiments. As an important platform for teacher training, innovation team training and scientific research project training, the China-ASEAN Logistics Research Institute and its affiliated logistics association student associations are an important way for the logistics teacher team and student team to communicate, communicate, and integrate research and innovation. The China-ASEAN Integrated Transportation International Joint Laboratory is a major project of scientific and technological innovation cooperation between the Guangxi Zhuang Autonomous Region Government and the Sichuan Provincial Government. Students participate in the application and research of "big transportation", "big logistics", "smart transportation", "big data" and other project directions. The three on-campus platforms of "one hospital, one room and one center" provide strong hardware and intellectual support for the curriculum construction and achievement cultivation of the logistics engineering undergraduate practice teaching system.

(3) Drive the construction of a training base for the integration of industry and education, and cultivate a "2 + 2" experimental course system to demonstrate new experiences

Combining the characteristics of the logistics industry, build an off-campus production-education integration base in multiple fields, industries, and regions, and focus on developing an experimental curriculum system of "cross-curricular + multi-stage + inside and outside school + inside and outside school", relying on professional teaching software, simulation, big data mining, enterprise research and post practice, etc., combined with the requirements of the development of new industries and new engineering disciplines to organize case extraction, to meet the needs of professional classroom teaching to extracurricular logistics association organization, from cognition visits to technical research, from course practice to post practice and other perspectives and multiple perspectives Level off-campus practice tasks, students participate in the production line of the enterprise, move the classroom to the production line of the enterprise, and the technical backbone of the enterprise and the backbone of the major classroom provide two-way technical guidance and practical teaching to the students.

(4) Build a "four-combination" double-qualified practice teacher team, and carry out a series of activities of "first research, second closeness, and three themes"

Optimize the organizational structure, and adopt the structural form of "combining professional and part-time jobs, combining school and enterprise, combining virtual and real, and combining main and auxiliary" to achieve the effect of complementing teachers' resources and clear responsibilities and rights. Through the three channels of "one hospital, one room and one center", integrate the resources of the industry, keep up with the times and the development of logistics, organize research activities of the teaching and research department, keep up with the seminar series activities of industry societies and the series activities of regional logistics development, and actively participate in and organize logistics disciplines Competition activities, new technology training activities and special academic lectures, and do a good job in the guarantee of teachers in the experimental teaching system.

(5) School-enterprise alliance + result-oriented, virtual simulation projects are changed from "virtual" to "real", and the operability and value are improved

Through the joint construction of courses with enterprises, oriented to solving practical problems, taking students as the center, real data collection, sorting, statistics and analysis, returning to the classroom for modeling, simulation and optimization, verification, real cases, simulation work Scenario setting, combining courses with graduation design, innovation and entrepreneurship, and subject competitions, helps to achieve the integration of curriculum creation, teaching and competition, and is closely in line with the school-running philosophy of the University of Applied Sciences and the talent training of logistics engineering.

4 Establishment of the Teaching Index System Under the Multi-platform Co-construction Model

4.1 AHP Analysis Hierarchy of First-Level Indicators

(1) Index selection and hierarchy establishment

The training of logistics engineering professionals is inseparable from the platform support of school-enterprise cooperation and school-school alliance.

According to the principle of comprehensive consideration and practicality, the key elements of teaching indicators are extracted: the first-level indicators are the realization of teaching objectives; the second-level indicators are 4 items, which are platform teaching team, platform teaching conditions, teaching organization, and achievement evaluation; the three-level indicators include teaching and research level, base construction, teaching project development, student evaluation of teaching, supervision and evaluation of teaching, etc. The model hierarchy is thus designed, as shown in Fig. 2.

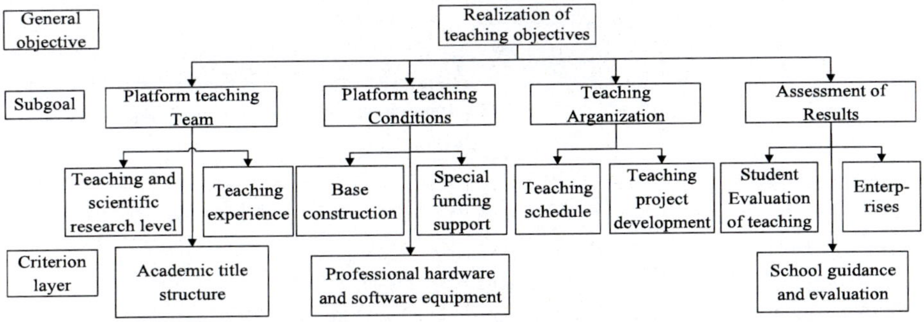

Fig. 2. Hierarchical structure of teaching index system

(2) Constructing a judgment matrix

According to the analysis of the relative importance of the four first-level indicators, the comparison matrix is obtained through the method of expert scoring. Table 1, 2, 3, 4 and 5 are the judgment matrix of indicators, teaching organization, achievement evaluation, platform teaching conditions and teaching team.

Table 1. Indicator judgment matrix

Index	Teaching organization	Achievement evaluation	Platform teaching conditions	Platform teaching team
Teaching organization	1	1	3	7
Outcome evaluation	1	1	4	3
Platform teaching conditions	1/3	1/4	1	3
Platform teaching team	1/7	1/3	1/3	1

Table 2. Judgment matrix of teaching organization

Teaching organization	Practice hours arrangement	Practice teaching project development
Practice hours schedule	1	1/9
Practice teaching project development	9	1

Table 3. Judgment matrix of achievement evaluation

Achievement evaluation	Student evaluation of teaching	School supervision and evaluation of teaching	Enterprise evaluation of teaching
Student evaluation of teaching	1	1/4	1/6
School supervision and evaluation of teaching	4	1	1/2
Enterprise evaluation of teaching	6	2	1

Note: The scale values of the two three-level indicators in the table are determined according to their relative importance: practical teaching project development > practical teaching hours arrangement, and the assignments are 9 and 1 respectively.

Table 4. Judgment matrix of platform teaching conditions

Platform teaching conditions	Practice base construction	Professional software and hardware equipment	Practical funding support
Practice base construction	1	6	6
Professional software and hardware equipment	1/6	1	1
Practical funding support	1/6	1	1

As shown in Table 1, according to the analysis of the relative importance of each index of the four first-level indicators, and through the method of expert scoring, a comparison matrix is obtained.

Note: The scale values of the four first-level indicators in the table are determined according to their relative importance: teaching organization > platform teaching conditions > achievement evaluation > platform teaching team, and the assignments are 7, 3, 1, and 1, respectively.

According to the relative importance analysis of each index factor of the secondary index, the comparison matrix is obtained by the method of expert scoring.

According to the relative importance analysis of each index factor of the secondary index, the comparison matrix is obtained by the method of expert scoring.

Note: The scale values of the three three-level indicators in the table are determined according to their relative importance: enterprise evaluation of teaching > school supervision and evaluation > student evaluation of teaching, and the assignments are 6, 4, and 1, respectively.

According to the relative importance analysis of each index factor of the secondary index, the comparison matrix is obtained by the method of expert scoring.

Note: The scale values of the three three-level indicators in the table are determined according to their relative importance: practice base construction = professional software and hardware equipment > practice funding support, and the assignments are 6, 6, and 1, respectively.

Table 5. Judgment matrix of platform teaching team

Platform teaching team	Academic title structure	Teaching and research level	Practical teaching experience
Academic title structure	1	1/3	1/8
Teaching and research level	3	1	1/6
Practical teaching experience	8	6	1

According to the relative importance analysis of each index factor of the secondary index, the comparison matrix is obtained by the method of expert scoring.

Note: The scale values of the three three-level indicators in the table are determined according to their relative importance: practical teaching experience > teaching and research level > academic title structure, and the assignments are 8, 3, and 1, respectively.

(3) Preliminarily determine the weight of the indicators

The weight value of each indicator is calculated by a simple weighted average. The specific calculation method is as follows: first, remove the sum of each column with the value of each item of the judgment matrix, and then add the new value obtained in each row of each row. A value is added to get the sum, and then the order is divided to calculate the value of the relative weight. According to the above formula, the consistency test CR is carried out. It can be seen that the second-level index has the largest weight in the achievement evaluation. Table 6, 7, 8, 9 and 10 is the weight table of indicators, teaching organization, achievement evaluation, platform teaching conditions and platform teaching team respectively.

The weight value of each index is calculated by a simple weighted average. The calculation method is the same as above, and the value of the relative weight is calculated. The consistency test CR is carried out according to the above formula. It can be seen that the teaching organization index has the largest weight in the development of practical teaching projects.

Table 6. Indicator weight table

Indicator	Practice teaching organization	Practice teaching Evaluation	Practice teaching conditions	Practice teaching Team	Weight	AW
Teaching organization	0.404	0.387	0.360	0.500	0.413	1.730
Outcome evaluation	0.404	0.387	0.480	0.214	0.371	1.573
Platform teaching conditions	0.135	0.097	0.120	0.214	0.141	0.595
Platform teaching team	0.058	0.129	0.040	0.071	0.075	0.304

Table 7. Weights of Teaching Organizations

Teaching organization	Practice hours arrangement	Practice teaching project development	Weight	AW
Practice hours arrangement	0.100	0.100	0.100	0.200
Practical teaching project development	0.900	0.900	0.900	1.800

Table 8. Results evaluation weight table

Achievement evaluation	Student evaluation of teaching	School supervision and evaluation of teaching	Enterprise evaluation of teaching	Weight	AW
Student evaluation of teaching	0.091	0.077	0.100	0.089	0.268
School supervision and evaluation of teaching	0.364	0.308	0.300	0.324	0.974
Enterprise evaluation of teaching	0.545	0.615	0.600	0.587	1.770

The weight value of each index is calculated by a simple weighted average. The calculation method is the same as above, and the value of the relative weight is calculated. The consistency test CR is carried out according to the above formula. It can be seen that the evaluation index of the achievement has the largest weight.

In Table 9, the weight value of each index is calculated by a simple weighted average. The calculation method is the same as above, and the value of the relative weight is calculated. The consistency test CR is carried out according to the above formula. It can be seen that the platform teaching condition index has the largest weight in practice base construction.

Table 9. Weights of platform teaching conditions

Platform teaching conditions	Practice base construction	Professional software and hardware equipment	Practice financial support	Weight	AW
Practice base construction	0.750	0.750	0.750	0.750	2.250
Professional software and hardware equipment	0.125	0.125	0.125	0.125	0.375
Practice financial support	0.125	0.125	0.125	0.125	0.375

Table 10. Platform teaching team weight table

Platform teaching team	Teaching and research level	Academic title structure	Practical teaching experience	Weight	AW
Teaching and research level	0.083	0.045	0.097	0.075	0.227
Academic title structure	0.250	0.136	0.129	0.172	0.523
Practical teaching experience	0.667	0.818	0.774	0.753	2.385

The weight value of each index is calculated by a simple weighted average. The calculation method is the same as above, and the value of the relative weight is calculated. According to the above formula, the consistency test CR is carried out. It can be seen that the index weight of the platform teaching team is the practice teaching experience.

(4) Consistency test

Check the consistency of the judgment matrix of the criterion layer and the target layer:

Maximum eigenvalue:

$$\lambda = \frac{1}{n}\sum_{i=1}^{1}\frac{(A\omega)_i}{\omega_i}(nisorder) \tag{1}$$

Consistency index:

$$CI = \frac{\lambda - 1}{n - 1} \tag{2}$$

The random consistency index RI can be obtained by looking up the table:

Order	1	2	3	4	5
RI	0	0	0.539	0.916	1.143

Consistency ratio:

$$CR = \frac{CI}{RI} \tag{3}$$

Through calculation, it is obtained that the index = *0.066* < *0.1*, the platform teaching team = *0.069* < *0.1*, the platform teaching conditions = *0.000* < *0.1*, the teaching organization = *0.000.* < *0.1*, the achievement evaluation = *0.009* < *0.1*, and the consistency test of the judgment matrix has passed.

4.2 Evaluation Results and Index Ranking of Comprehensive Indicators

Through the analysis and calculation of the four stages, the comprehensive indicators and ranking results shown in Table 11 are obtained.

It can be seen that in the case of platform co-construction and school-enterprise cooperation, more attention should be paid to the development of practical teaching projects, and attention should be paid to the evaluation and feedback of enterprises in the evaluation of results.

Table 11. Comprehensive indicators and ranking

Second-level indicator	Second-level indicator weight	Third-level indicator	Third-level indicator weight	Comprehensive weight	Sorting
Platform teaching team	0.075	Academic title structure	0.075	0.006	10
		Teaching and research level	0.172	0.013	9
		Practical teaching experience	0.753	0.056	5
Platform teaching conditions	0.141	Practice base construction	0.750	0.106	4
		Professional software and hardware equipment	0.125	0.018	8
		Practical financial support	0.125	0.018	8
Teaching organization	0.413	Practice hours arrangement	0.100	0.041	6
		Practical teaching project development	0.900	0.371	1
Achievement evaluation	0.371	Student evaluation of teaching	0.089	0.033	7
		School supervision and evaluation	0.324	0.120	3
		Enterprise evaluation and education	0.587	0.218	2

5 Conclusion

Education is a solid foundation for human progress. The orderly development of the logistics industry is a powerful driving force for the continuous advancement of the social economy. The establishment of an integrated platform of "production, teaching, learning, research, application, creation and competition" and the organic combination of educational and practical teaching resources inside and outside the school can achieve good results in the training of logistics professionals.

Market changes and corporate changes have a profound impact on the development of the logistics industry. As the front line of teaching, it is necessary to clearly grasp the pulse of the times, exert a keen sense of industry and workplace, and make full use of cooperation and communication with industry and enterprises to carry out diversified cooperation models. Carry out multi-party linkage platform construction to improve the quality of logistics engineering talents training in teaching organization and practice links.

Acknowledgment. This paper is supported by the "double qualified" teacher training base project of Nanning University - Guangxi Suning Tesco Sales Co., Ltd., the teaching reform project of Nanning University "reform and practice of virtual simulation practice curriculum of Logistics Engineering Major Based on OBE concept and school enterprise linkage" and the teaching reform project of Guangxi Department of Education (2022jgb440) "research and practice of logistics virtual teaching and research office construction from the perspective of" point, line, area and body "four-dimensional linkage".

References

1. Mtiaz Hussain Khan: A unified framework for systematic evaluation of ABET student outcomes and program educational objectives. Int. J. Mod. Educ. Comput. Sci. (IJMECS) **11**(11), 1–16 (2019)
2. Jian, L., Yuan, Z.: The exploration and practice in innovative personnel training of computer science and technology. Int. J. Educ. Manage. Eng. (IJEME) **2**(6), 47–51 (2012)
3. Xuejie, W., Sufen, H., Guoxiong, W.: Exploration and practice of the "423" collaborative training mode for innovative talents in automation majors from the perspective of new engineering. J. Higher Educ. **08**, 32–35 (2021)
4. Guobao, W., Jiajia, S., Xiaoxiao, Z.: Reform of applied talents training mode for undergraduate majors in logistics engineering based on social needs. Logistics Technol. **40**(06), 156–160 (2021)
5. Han, Z.: China's smart logistics into a new future. China's Nat. Conditions Nat. Strength **10**, 79 (2017). (in Chinese)
6. Liming, H.: Development trend of smart logistics in china. China Circ. Econ. **31**(6), 3–7 (2017). (in Chinese)
7. Zhongchang, W., Xiaojuan, L.: Construction of China ASEAN vocational education school enterprise community in the context of the "the Belt and Road." Vocat. Tech. Educ. **22**(41), 68–73 (2020). (in Chinese)
8. Zhida, J., Rui, W.: Progress and challenges of China ASEAN digital "the Belt and Road" cooperation. J. Pac. **28**(09), 80–91 (2020). (in Chinese)
9. Guirong, Y., Yamin, L.: Practice and thinking of multi-blended teaching of professional courses under the background of new engineering. J. Higher Educ. **09**, 128–131 (2021)
10. Yanhui, L., Xiaoguang, Z., Liang, F., Gang, L.: Research on the training mode of applied logistics engineering talents based on CDIO concept. China Market **15**, 37–38 (2016)
11. Qiang, W., Li, J., Wen, L., Peng, Z., Rong, L.: A new engineering talent training model for logistics engineering based on multidisciplinary integration. China Logistics Purchasing **15**, 28–29 (2021)

12. Shi, J., Dong, C., Hou, J.: Research and design of teaching evaluation system based on fuzzy model. Int. J. Educ. Manage. Eng. (IJEME) **2**(10), 45–51 (2012)
13. Liu'an, K., Xiaomei, W., Lin, Y.: The research of teaching quality appraisal model based on AHP. Int. J. Educ. Manage. Eng. (IJEME) **2**(9), 29–34 (2012)
14. Fangxiu, W.: The course design for algorithm analysis and research on improving the students' comprehensive ability. Int. J. Educ. Manage. Eng. (IJEME) **2**(2), 42–46 (2012)
15. Fei, L.: Analysis of the talent training program based on AHP. Mod. Vocat. Educ. **11**, 164–165 (2019)
16. Yanxi, Z., Peiling, H.: Evaluation of applied undergraduate teaching quality based on fuzzy comprehensive evaluation. J. Nanjing Inst. Technol. (Social Science Edition) **18**(04), 67–71 (2018)
17. Xinyi, Y., Keyi, J., Wei, S., Benwei, L., Yonghua, W.: Application of analytic hierarchy process to evaluate the teaching reform effect of "Engineering Thermodynamics" course. Univ. Educ. **12**, 111–113 (2015)
18. Rongxin, Z.: Construction of a practical teaching platform for the cultivation of innovative and outstanding talents in logistics engineering under the background of new engineering. Sci. Technol. Vis. **35**, 96–97 (2021)
19. Al-Hagery, M.A., Alzaid, M.A., Alharbi, T.S., Alhanaya, M.A.: Data mining methods for detecting the most significant factors affecting students' performance. Int. J. Inf. Technol. Comput. Sci. (IJITCS) **5,** 1–13 (2020)

"Student Centered" Undergraduate Course Teaching Reform Practice - Taking "Ship Maneuvering and Swing" as an Example

Lin Wang, Xinjian Zhang, Shan Fan, Juan Fan, Xichen Lin, and Yan Chen(✉)

Department of Mechanical and Electrical Engineering, Wenhua College, Wuhan 430074, China
525498463@qq.com

Abstract. Taking the core course of undergraduate major "ship manoeuvring and rocking" as an example, this paper discusses the teaching reform practice of "student-centered". This paper expounds the design and practice process of teaching reform in detail from several aspects, such as how to ensure students to learn effectively first, how to organize classroom teaching, how to carry out after-school tasks, and how to evaluate students' learning, analyzes the effectiveness of teaching reform, and expounds the problems encountered in the process of teaching reform. The practice of this paper shows that the "student-centered" teaching mode can improve students' learning effect, cultivate students' learning ability, and improve students' subject literacy. Through the in-depth research and practice of this teaching reform, we hope to provide some reference for the improvement of teaching methods of related professional courses and the improvement of talent training quality.

Keywords: Student centered · Teaching reform · Teaching mode · Learning ability

1 Introduction

Higher education plays a vital role in the development of the country [1, 2]. The real purpose of higher education is human development, which must be achieved through the pursuit of truth and the creation of knowledge [3, 4]. University teachers not only need to impart knowledge, but also need to lead students to explore unknown areas [5]. University teaching must be carried out on the basis of scientific research, not only to teach students to learn to learn, but also to teach students to learn to create [6].

However, at present, many classrooms in Colleges and universities in China still belong to the traditional teacher centered mode, that is, teachers focus on teaching objectives and content, teach in the form of lectures and PPT demonstrations, and evaluate students' mastery of classroom knowledge through examination papers and graduation thesis [7]. This teacher centered teaching model obviously cannot achieve the real purpose of higher education. Table 1 shows the evaluation of teachers' classrooms by the teaching supervision group of the university where the author works from 2021 to 2022. Seven teaching supervisors rated 364 classrooms in the whole school.

Z. Hu et al. (Eds.): AIMEE 2022, LNDECT 159, pp. 426–435, 2023.
https://doi.org/10.1007/978-3-031-24468-1_38

Table 1. Statistics of teaching supervision evaluation scores

	Teaching manners	Teaching contents	Teaching methods	Teaching Ability	Teaching effect	Total score
Full marks	20	30	20	15	15	100
Average score	18.16	26.64	17.32	13.58	11.82	87.77
Scoring rate	90.82%	88.81%	86.58%	90.51%	78.77%	87.77%

As can be seen from Table 1, among the first four scores, the scoring rate of teaching methods is the lowest. Most classes adopt traditional teaching methods, and students' learning enthusiasm and participation are not high enough. This teaching method directly leads to poor teaching effect. It can be seen that the scoring rate of teaching effect is less than 80%. In order to improve teaching effect, changing teaching methods is an important way.

In 1952, Carl Rogers first put forward the idea of "student-centered" teaching. In the mid-20th century, American scholars put forward the undergraduate education concept of "student-centered", which triggered a series of changes in the basic concepts, teaching methods and teaching management of undergraduate education, and brought great impact on higher education [8]. In 1998, UNESCO put forward at the first World Conference on higher education that higher education needs to turn to a new perspective and new model centered on students, requiring higher education decision makers to focus on students and their needs, and regard students as the main participants in education reform, including participating in the discussion, evaluation, curriculum and content reform of major education issues, policy formulation and university management [9]. In the 21st century, the educational concept of "student-centered" has attracted more and more attention of educational researchers all over the world [10–12].

The teaching concept of "student-centered" is centered on students' development and learning effect, and aims to promote students' development. Teaching is the means and learning is the purpose. Teachers should pay attention to the learning effect and help students adjust their learning in time [13].

The author came into contact with the teaching concept of "student-centered" in a lecture in 2017. After thinking and discussing in many ways, the author decided to implement the reform of "student-centered" teaching method in the course "ship manoeuvring and rocking" taught by the author. Ship manoeuvring and rocking is the core course of ship and ocean engineering major. The course is highly professional and has many formula derivation. Students are easy to feel boring when learning it. The author has been teaching this course for the eighth year since 2015, and has implemented the "student-centered" teaching method reform since 2018. Through comparison, it is found that this teaching method has many advantages over traditional teaching methods in improving students' classroom activity, stimulating students' learning interest and enthusiasm, and cultivating students' ability to discover and solve problems.

2 "Student Centered" Teaching Reform Practice

In order to facilitate the implementation of teaching reform, small classes are selected. In recent years, the number of students in this course is between 15–30. The mode of "learning before teaching" is initially adopted, and its general structure is to let students learn by themselves first, that is, "learning first" to cultivate students' self-learning ability; Then according to the students' learning results, teachers carry out targeted "post teaching" [14]. Classroom teaching uses the teaching mode of "271 efficient classroom", that is, 20% of the time in a class is taught by teachers, 70% of the time is students' autonomous learning, and 10% of the time is collation and review [15]. After class, I will assist some small tasks to help students consolidate.

In the classification of educational goals, American scholar bloom divides the goals of cognitive field into six levels: memorization, understanding, application, analysis, evaluation and creation. These six levels are progressive relations. Memorization, understanding and application are primary cognition, and analysis, evaluation and creation are advanced cognition, as shown in Fig. 1.

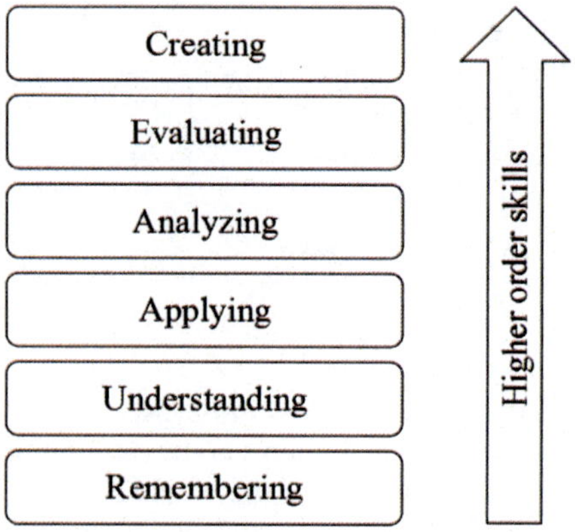

Fig. 1. Bloom classification of educational objectives

In the process of learning before class, students carry out this low-level cognitive behavior of memorizing and understanding; In the process of classroom teaching, we should understand, apply, analyze and evaluate this kind of cognitive behavior from low level to high level; In the process of completing the task after class, the high-level cognitive behavior of evaluation and creation is carried out, as shown in Fig. 2. The whole learning process is gradually deepened, and the cultivation of students' ability to explore and seek knowledge, analyze and synthesize, and innovate and create is emphasized.

In the process of teaching reform practice, we should focus on solving the following problems: (1) how to ensure that students effectively learn first; (2) How to organize classroom teaching; (3) How to carry out after-school tasks; (4) How to evaluate students' learning.

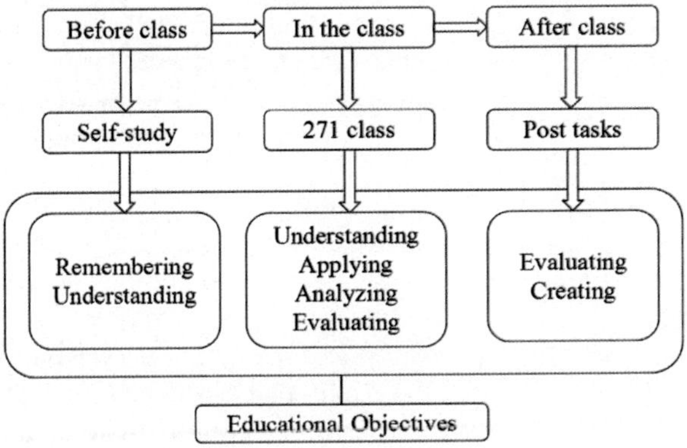

Fig. 2. Model of teaching reform

2.1 Ensure the Students Learn First Effectively

Students learn first plays an important role in the whole study. Only when students learn first, the subsequent teaching activities can be carried out smoothly.

In order to improve the efficiency of students' autonomous learning, teachers need to prepare guidance materials in advance to clarify the learning objectives, learning difficulties, learning ideas, inquiry questions, etc. of this class. The compilation of guidance materials should be concise and focused, otherwise students' reading of guidance materials has become a heavy task. The inquiry questions in the guidance materials should be in the order of the content framework of the knowledge points in this section, from easy to difficult, so that students can sort out the context of knowledge. Students learn independently in advance according to the guidance materials, find the answers to the problems they can solve, mark the problems that cannot be solved, and find out other questions in the process of self-study.

Combined with the actual situation of students in our school, students spend limited time on self-study before class. The purpose of self-study before class is to let students quickly understand what to learn, search for answers, and mark what they don't understand. It doesn't take much time, but you can know what you want to learn well.

2.2 Organize Classroom Teaching Effectively

2.2.1 Group Construction

In order to facilitate the development of cooperative learning and exploratory learning, we should first carry out group construction. According to the students' grades, there are three students in a group, including one with good grades, one with medium grades and one with poor grades. Each group selects a learning team leader, who is responsible for managing learning and maintaining discipline within the group.

Group construction is a very important part in the "student-centered" teaching reform. Group members cooperate with each other to win honors for the group. Learning is no longer a matter of themselves, but a matter with a sense of responsibility and a sense of honor of the group to stimulate students' initiative in learning. When completing group tasks, group members can assign roles according to their own characteristics, give full play to their strengths, and let everyone's value be reflected.

2.2.2 Cooperative Learning and Exploratory Learning

In the process of group cooperative learning, each group member exchanges their unsolved problems and newly discovered questions in self-study before class, and carries out cooperative learning among group members to jointly answer them. During the discussion of group members, the teacher should patrol the class and pay attention to the students' discussion. Give affirmation and praise to the groups that have a heated discussion in time, and actively guide the groups that lack discussion and communication.

After the group cooperative learning is completed, if there is still something you don't understand, each group will send representatives to ask questions. The teacher will write the collected questions on the blackboard, and the whole class will discuss and answer them together. If you can't solve the problems, the teacher will call and explain them. In the discussion of the whole class, students often have different opinions on different problems. The whole classroom atmosphere is relaxed, and everyone can express their views. In the continuous discussion, students will have a deeper understanding of knowledge. In this process, teachers should not only encourage discussion, but also provide timely guidance to prevent students from going further and further in the wrong direction.

2.2.3 Analysis and Summary of Knowledge Points

After all the questions are answered, give students time to review the knowledge learned in this class and summarize their own gains. Finally, the teacher selects 1–2 groups of representatives to the podium to summarize. Summing up is an important process. In this process, students can clarify the internal logical relationship between the whole knowledge through thinking, and understand why to learn this knowledge point, its relationship with the previous and subsequent chapters, and what application it has after learning.

The process of analysis and summary is a method to effectively exercise students' comprehensive evaluation ability. In this process, teachers should give timely evaluation to students' summary, and give timely guidance to students where the internal relationship between some knowledge points is relatively hidden, so as to help students find a breakthrough.

2.3 Carry Out After-School Tasks

Students' after-school tasks are divided into two parts, one is the test question to test the learning effect in class, and the other is the group project to broaden their horizons and cultivate their innovation ability, which requires investigation and research.

In order to facilitate the statistics of students' knowledge mastery, the after-school test questions are completed on the micro teaching assistant online. Students can complete the answer by mobile phone and see the answer analysis after submitting. The teacher can see the answer of each student and the answer statistics of each question in the background, as shown in Fig. 3. For many wrong questions, the teacher will explain them in the next class.

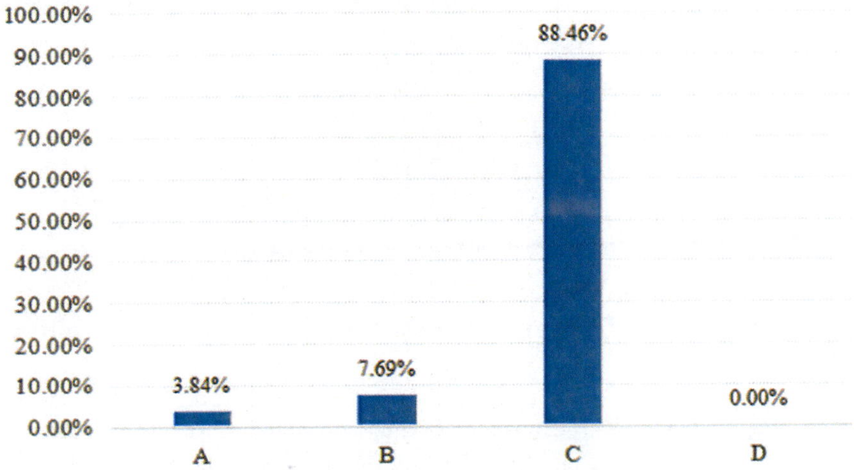

Fig. 3. Statistics of questions answered by micro teaching assistants

In the whole course teaching, group special tasks will be arranged regularly, such as special investigation and research related to the course, debate on hot issues in the industry, students' ppt report on the podium, etc. all group members will cooperate with each other to complete the tasks. The effect of completing the task will include group self-evaluation, group mutual evaluation and teacher evaluation, which will be included in the corresponding scores for each group to fully mobilize students' enthusiasm for learning.

2.4 Evaluate Students' Learning

Effective learning evaluation can mobilize students' enthusiasm. This curriculum reform increases the proportion of the usual grades of the curriculum.

The evaluation of students is no longer just the final exam, but a process oriented formative evaluation. The usual score and final exam score account for 50% respectively. The usual score consists of attendance, micro TA test questions and group scores, as shown in Fig. 4.

The most important part of the usual grades is the group score. Students can add points to their group by answering questions or asking questions in class. The scores of each group are recorded in each class. In the special task of the group, the group takes the group as the unit to conduct group self-evaluation, group mutual evaluation and teacher evaluation according to the performance of the students, and comprehensively score each group. Publish a score summary every four weeks to let each group know their ranking. At the end of the course, the score of each student in the first group is full, and the scores of other groups decrease according to the ranking.

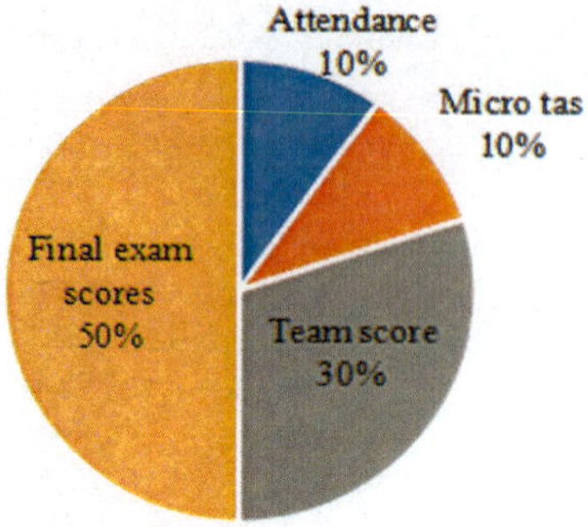

Fig. 4. Distribution of score composition

3 Analysis of the Effectiveness of "Student-Centered" Teaching Reform

After carrying out the "student-centered" teaching reform for a period of time, the results are remarkable. To sum up, it is reflected in the following aspects.

(1) Students' learning effect is significantly improved
Table 2 shows the results after the teaching reform. Class 1 is the class before the teaching reform, and class 2 is the class after the teaching reform. It can be seen from the table that after the reform, the proportion of ordinary scores and examination scores of more than 90 points has increased significantly, and the proportion of scores of less than 80 points has decreased significantly. From the classroom performance, teachers also clearly feel that the participation of students has increased, their interest in learning has been stimulated, and the scoring rate of questions with high ability in the test paper has increased more. Students' learning effect has been significantly improved.

Table 2. Comparison of results after teaching reform

	Usual score		Examination results	
	Class1 (26 people)	Class 2 (24 people)	Class 1 (26 people)	Class 2 (24 people)
≧90	7.69%	25.00%	19.23%	37.50%
80–89	26.92%	37.50%	34.62%	33.33%
70–79	57.69%	33.33%	42.31%	25.00%
60–69	7.69%	4.17%	3.85%	4.17%
≦59	0.00%	0.00%	0.00%	0.00%

(2) Students' learning ability has been effectively cultivated
The "student-centered" teaching mode requires students to learn independently after class to solve the exploratory problems in the guidance materials. In class, students learn in groups to ask, discuss and solve problems. In this process, students are always exploring knowledge by themselves. Through dialogue and interaction with students, teachers constantly guide students to explore ways to understand and discover knowledge, cultivate students' learning ability, and achieve the purpose of teaching people to fish rather than teach people to fish.
(3) Students' academic literacy has been improved

One of the important contents of undergraduate education is to cultivate students' discipline literacy. Discipline literacy is the professional thinking formed after long-term professional training. Through this professional thinking, we can promote the accumulation of basic knowledge, increase basic professional skills, and form basic professional experience. The "student-centered" teaching mode allows students to explore and master knowledge in multiple ways through autonomous learning, inquiry learning, cooperative learning, demonstration learning, etc., so as to deepen their understanding of discipline and professional knowledge. In order to solve special problems, students need to find a lot of data and broaden their subject horizons. In this process, students' professional thinking of cost science is imperceptibly shaped, and their discipline literacy is improved.

4 Problems and Reflections in the Process of Teaching Reform

In the process of promoting the teaching reform, we not only saw the achievements, but also found some problems as described below. Reflection on these issues will help us to clarify the direction of our future efforts.

(1) The content of some chapters is too difficult and self-study is difficult
Because the course "ship manoeuvring and rocking" is the core course of the major of ship and ocean engineering, it is highly professional. When students get the guidance materials and start self-study, they find it difficult to understand most of the contents. In view of this situation, teachers need to record micro class videos to explain these difficult knowledge points. Students watch micro class videos and complete the self-study process in combination with guidance materials.
(2) Students are too active and affect the classroom order
After the implementation of teaching reform, students' enthusiasm for class has doubled and the classroom atmosphere is active, but at the same time, it also brings some problems. For example, after a student raises a question, multiple students stand up at the same time to answer; When some students are answering questions, the students in a group are still immersed in the discussion of this question. These practices reflect students' enthusiasm for learning, but at the same time, they have a certain impact on classroom discipline, making the classroom "too active", which will have an impact on students' learning efficiency. By collecting students' opinions, the following measures were taken: when students answer questions, they need all members of the group to answer a round of questions before they are

qualified to answer questions again, unless no one answers the question; When a student answers a question, other groups should stop discussing the question. If there are different opinions, they can put forward a joint discussion after the student answers, and putting forward different opinions can add points to the group. Through these two measures, students' classroom discipline has been guaranteed, while maintaining everyone's enthusiasm for learning.

(3) The length of students' classroom discussion is uncontrollable, and the class hours are not enough
The "student-centered" teaching mode discussion learning is an important learning method, including group discussion and class discussion. Sometimes, different students will have different views on difficult problems, and the discussion is extremely intense. With the deepening of the discussion, the students will gradually approach the real answer step by step, but this process will take a long time, resulting in insufficient class hours. The author believes that this kind of discussion should be encouraged. For the problem of insufficient class hours, some easy to understand chapters can be completely self-taught by students, and there is no need to spend time in class.

(4) The time for teachers to prepare lessons has increased significantly

The "student-centered" teaching model seems to have less time for teachers to participate and students are the main body, but the actual time for teachers to prepare lessons has increased significantly. First of all, the compilation of guidance materials should be focused, concise and clear, so that students are willing to spend time reading. The design of inquiry questions should not only conform to students' cognition, but also stimulate students' thinking. Good guidance materials can play a very important guiding role in students' self-study, so teachers are required to devote themselves to writing guidance materials. The production of micro class videos and the preparation of online exercises for micro teaching assistants require teachers to spend a lot of time. Therefore, teachers spend more time in this teaching method. While, seeing that the students' learning effect is good, the author believes that the time spent is worth it. Since everything a teacher does should be for the students.

5 Conclusion

In the "student-centered" teaching mode, teachers have changed from "lecturers" to "guides", the focus of teaching activities has changed from "how teachers teach" to "how students learn", students have changed from "teaching objects" to "teaching subjects", and students' learning methods have changed from "passive learning" to "active learning". It can be said that this is an advanced educational thought, which reflects the return of education to the truth.

Using the "student-centered" teaching mode, students' learning effect has been significantly improved, their learning ability has been cultivated, and their subject literacy has been improved. Although the author encountered some problems in the process of implementation, due to frequent communication and discussion with students and experienced teachers, collecting everyone's opinions, changing and adjusting in time, such a teaching method was finally formed, hoping to provide some reference for the teaching of related professional courses.

Acknowledgment. This project is supported by the digital twin of intelligent logistics equipment of the outstanding young and middle-aged scientific and technological innovation team plan project (t2021043) of colleges and universities in Hubei Province.

References

1. Mtiaz Hussain Khan: A unified framework for systematic evaluation of ABET student outcomes and program educational objectives. Int. J. Mod. Educ. Comput. Sci. (IJMECS) **11**(11), 1–16 (2019)
2. Yufan, D.: Exploration on the precise education path of the integration of production and education in the new era of vocational education groups - Taking Guangdong Industry and Trade Vocational Education Group as an example. J. Shenzhen Vocat. Tech. Coll. **18**(4), 75–80 (2019). (in Chinese)
3. Jian, L., Yuan, Z.: The exploration and practice in innovative personnel training of computer science and technology. Int. J. Educa. Manage. Eng. (IJEME) **2**(6), 47–51 (2012)
4. Al-Hagery, M.A., Alzaid, M.A., Alharbi, T.S., Alhanaya, M.A.: Data mining methods for detecting the most significant factors affecting students' performance. Int. J. Inf. Technol. Comput. Sci. (IJITCS) **5**, 1–13 (2020)
5. Shi, L., Yao, Y., Xia, X., Xie, X., Wu, L.: Practical teaching staff construction under new situation. Int. J. Educ. Manage. Eng. (IJEME) **1**(2), 57–61 (2011)
6. Jianguo, W.: Ideas and Strategies of university classroom teaching reform. Higher Educ. Res. **39**(04), 66–70 (2018). (in Chinese)
7. fan, C.: How can student-centered teaching be possible-an empirical study based on the current situation of undergraduate classrooms in 51 universities. Higher Educ. Res. **38**(10), 75–82 (2017). (in Chinese)
8. Xianjun, L.: On "student-centered." Higher Educ. Res. **33**(08), 1–6 (2012). (in Chinese)
9. Kari, O., Kelly L.H., Michael, G., Alan R.B., Janet, B.: A tool for designing and studying student-centered undergraduate field experiences: the UFERN model. BioScience **72**(2) (2022)
10. Tara, L.: Strategies for student-centered learning. Educ. Leadersh. **77**(6) (2020)
11. Mohan, A.: A student-centered geography classroom. The Geogr. Teacher **15**(3) (2018)
12. juming, Z., Xiaohui, G.: Thoughts on the implementation of "student-centered" undergraduate teaching reform. China Higher Educ. Res. **08**, 36–40 (2017) (in Chinese)
13. Gubinxian.: Explanation of flipped classroom and its application in the course of "environmental management". New Curriculum Res. (mid-term issue). **06**, 9–11 (2018) (in Chinese)
14. Tengshaohong.: Investigation and research on the student-centered "271 efficient classroom" teaching model. Yantai: Ludong University (2015) (in Chinese)
15. Lin, W., Lisa, G.: Research on the teaching reform of "student-centered" undergraduate courses. New Curriculum Res. (mid ten day issue) **10**, 5–6 (2018). (in Chinese)

Team Construction and Cooperation Mechanism of Virtual Teaching and Research Office of Logistics Specialty in the Information Age

Zhong Zheng[1,2], Yanfang Pan[1(✉)], and Liwei Li[1]

[1] College of Transportation, Nanning University, Nanning, China
281340537@qq.com

[2] Thailand First Emperor University of Technology, Nakhon Ratchasima, Thailand

Abstract. Combined with the current situation and problems of the development of logistics professional teaching teams, this paper puts forward the urgency of developing virtual teaching and research room, defines the concept of virtual teaching and research room, probes into its development process, and analyzes its characteristics. Combined with the characteristics of logistics specialty, this paper studies the operation model of logistics virtual teaching and Research Office, excavates the main influencing factors affecting the team construction of logistics virtual teaching and research office through analytic hierarchy process, obtains the key factors affecting the team construction of logistics virtual teaching and research office through analysis, and studies the assessment and evaluation mechanism, incentive mechanism, management system and process of virtual teaching and research office, cooperation mechanism, etc., This paper puts forward the idea of establishing the cooperation mechanism of virtual teaching and Research Office, in order to provide a useful reference for the team construction of logistics virtual teaching and research office. Key words: virtual teaching and research room, cooperation mechanism, logistics, influencing factors, collaborative teaching and research.

Keywords: Virtual teaching and research room · Logistics major · Team building · Cooperation mechanism

1 Introduction

To find a more scientific teaching path and cultivate more excellent logistics talents is the direction of Logistics Majors and teachers [1–3]. The growth of the teaching staff is directly related to the quality of talent training, and then to the development of the entire logistics industry [4–6]. At present, the construction of grassroots teaching organizations has attracted more and more attention. With the advent of the Internet era, the mode and form of teaching organizations continue to develop and change. Virtual teaching and research office is an important exploration of the construction of new teaching organizations in the information age. It provides an important means for comprehensively improving teachers' educational ability and an efficient teaching organization to adapt to

Z. Hu et al. (Eds.): AIMEE 2022, LNDECT 159, pp. 436–447, 2023.
https://doi.org/10.1007/978-3-031-24468-1_39

the information age [7]. According to experts' research, the. virtual teaching and research office is a major trend and important support to promote the high-quality development of higher education [8]. At the same time, according to the new characteristics of talent training in the new era, from the analysis of talent classification characteristics of new engineering, cross integration and cross-border training is the inevitable choice of talent training mode of new engineering [9]. The establishment of virtual teaching and research room provides an important means for the training of new engineering talents. As a major with significant cross integration, it is imperative to establish virtual teaching and research room.

2 Problems Faced by Logistics Professional Educators and Requirements Put Forward by the New Situation

Currently, there are some problems in the teaching and team building of Logistics Majors in Colleges and universities. At the same time, the new situation also puts forward new requirements for logistics educators.

2.1 Problems Faced by Logistics Teaching Team

The problems existing in the teaching and team building of Logistics Majors in Colleges and universities mainly include the following aspects.

(1) There are few opportunities for cross-college exchanges and cooperation, and teaching resources are limited
Limited by the subject background and professional experience of teachers, teachers of a single subject have insufficient innovation in cross-integration. At the same time, some colleges and universities have insufficient teaching resources and limited conditions for teaching and scientific research. In addition, there is a disconnect between talent training and enterprises, and some teachers do not pay attention to the industry. It is related to the actual connection. It is only a closed-door study, lack of communication and exchange, and lack of opportunities to go deep into the front line of the enterprise. As a result, the teaching content cannot keep up with the development and changes of the industry, the teaching level is not high, and the ability of training students is difficult to match the requirements of the enterprise.
(2) The cooperative relationship between cross-institutions and cross-school enterprises is loose, and the effect is not obvious
At present, colleges and universities are fighting on their own, and the relationship is generally loose. Most of the short-term and superficial school enterprise cooperation. Some school teachers have made outstanding achievements, but due to the lack of cooperation and exchange mechanism, cross college cooperation is difficult to form normalization [10]. From the perspective of achievement protection, the exchange of teaching and scientific research between colleges and universities and between schools and enterprises only stayed on the surface, and there was no in-depth and effective cooperation. The joint application of topics was also terminated with the conclusion of the project, and the scientific research achievements were not further transformed into Teaching materials and knowledge base.

2.2 Virtual Teaching Team and the Opportunities It Brings to Logistics Teaching Team

The traditional teaching mode is difficult to face the new logistics industry, new technology and new mode.

New logistics, new formats, new models and new technologies emphasize the diversity and evolution of teaching objectives, the flexibility of the curriculum system, the iteration and agility of the teaching process, and the coordination of teaching resources. The curriculum system needs to be changed, and the teaching process and teaching management need to be changed accordingly, such as the implementation of the full credit system and the flexible school system [11]. This kind of change may be difficult for a single school/major to complete, and thus requires multi-school linkage and multi-discipline/professional linkage. Virtual teaching and research can provide better coping strategies.

Combined with the analysis of the bottleneck problems and restrictive factors of the current logistics teacher team, we should actively respond to the call of national policies according to the actual development needs of our own colleges and universities, and actively build a virtual teaching and research room for logistics. The cross-integration of colleges and teachers of different disciplines, the coordinated development of professional training program demonstration, the coordinated curriculum system design, the coordinated development of professional resources, and the coordinated development of teaching reform practice, etc., to achieve complementary advantages, and can better solve the problems existing in the above-mentioned teacher teams. At present, there are still few researches on the virtual teaching and research room. This paper focuses on the influencing factors of the team building of the virtual teaching and research room. Combined with the characteristics of colleges and universities with logistics-related majors in Guangxi, the team building research is carried out, focusing on analyzing the influencing factors of the virtual teaching and research room team building. Provide a reference for the establishment of an efficient logistics virtual teaching and research room.

3 Concept Definition and Development of Virtual Teaching and Research Room

3.1 Development History of Virtual Teaching and Research Office

Gestalt psychologist Kurt Lewin (1890–1947), through the study of group social behavior, found that there is mutual influence and mutual penetration between people in the group, and then produce various According to Lewin's basic point of view of group dynamic psychology, "team cooperation research teaching" has strong rationality [12]. By creating a virtual teaching and research room, members with different wisdom levels, different knowledge structures, and different ways of. thinking can inspire and complement each other. The collision of teaching and scientific research experience and wisdom can improve the level of thinking and even generate new ideas, which is extremely beneficial. Conducive to promoting the teaching team to complete the task creatively.

Summarizing the forms of collaborative teaching and research offices at home and abroad, it is found that there are currently two mainstream forms of collaborative teaching and research offices. The other is to use platforms that open teaching resources and resources such as courseware and teaching videos (such as MOOCs, Tencent Classroom and other online course platforms) to carry out collaborative education [13, 14].

The virtual teaching and research room is developed by further integrating resources on the basis of MOOCs (Fig. 1).

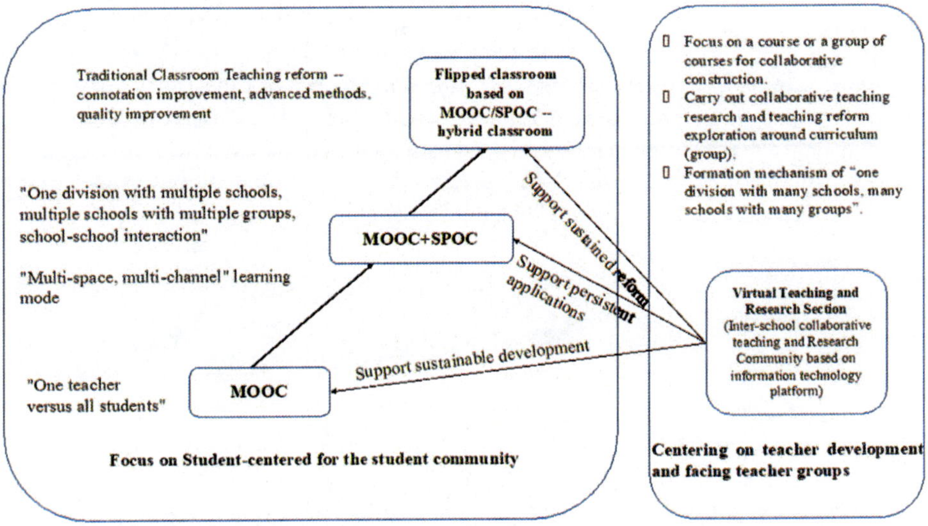

Fig. 1. Development history of virtual teaching and research

3.2 Development History of Virtual Teaching and Research Room

By comprehensively absorbing the research results of some experts [15, 16], this paper concludes that the definition of virtual teaching and research room is: the virtual teaching and research room is based on the modern information technology platform in the information society environment to improve teachers' education and teaching level and thus improve students' quality. The educational level is the goal, and it is a teacher community that is dynamically organized by teachers from different regions, schools, disciplines or professions to jointly carry out collaborative teaching research and reform practice.

The characteristics of the virtual teaching and research room mainly include:

(1) Virtual teaching and research section is a grass-roots teaching organization, a community of teachers who carry out collaborative teaching and research, and a new form of teaching and research section.
(2) The operation carrier of the virtual teaching and research room is the information platform, through which development, sharing and innovation are carried out.
(3) The virtual teaching and research room has the characteristics of openness, accumulation and co-construction and sharing.

3.3 Logistics Virtual Teaching and Research Office

Logistics is a comprehensive major, and the knowledge of the teaching team should not only stay in the field of logistics. On the basis of studying the characteristics of virtual teaching and research offices, combined with the characteristics of logistics disciplines, study and create the operation mode of logistics virtual teaching and research offices. The integration between school-enterprise and interdisciplinary, to create an efficient teacher team, through the virtual teaching and research room, further integrate "school, government, enterprise" and other resources to promote the deep integration of "industry-university-research-creation", and further explore the space for school-government-enterprise cooperation. Enterprise tutors enter the classroom and professional teachers reside in enterprises to carry out more cooperative scientific research projects, co-construction courses, joint student training and other activities to create a more three-dimensional professional teaching team (Fig. 2).

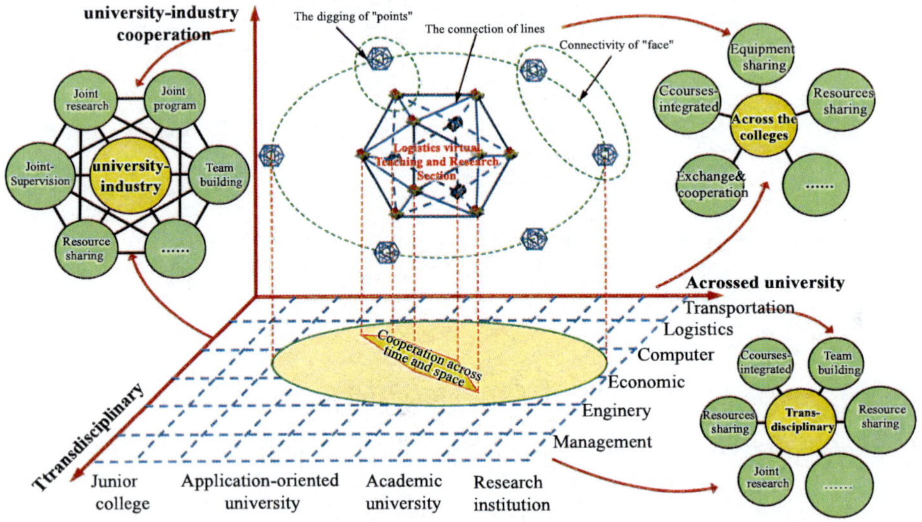

Fig. 2. Operation model of logistics virtual teaching and research office

Through virtual teaching and research, it can further promote the comprehensive and orderly development of students' "point, line, surface, and body". The logistics major. is a comprehensive discipline and needs to be integrated with different disciplines to break the isolated island of knowledge "points", and at the same time combine different knowledge "" "Point" and ability "point" are connected in series to become a knowledge "line", which is helpful for the cultivation of students. On the basis of the knowledge "line", comprehensive and sublimation contains multi-disciplinary knowledge such as engineering, economics and management." Therefore, it is necessary to further sort out the knowledge "face" of logistics engineering according to the positioning of application-oriented undergraduate universities, and integrate different disciplines according to different logistics scenarios., build the knowledge and ability of logistics engineering teachers from a more macro and systematic perspective, and build professional teachers' ability to meet the needs of different logistics scenarios according to the needs of the profession in serving the local economy and industry. Improving the teaching level can also serve the local economic development.

However, the cognitive level and ability of a single teacher at a certain time are limited. Only through the creation of virtual teaching and research can we further meet the needs of teaching and scientific research for various knowledge, and further help students cultivate students who are in line with the industry and enterprises. Ability, so as to improve the quality of teaching through the logistics virtual teaching and research room.

4 Analysis of the Factors Affecting the Team Building of the Virtual Teaching and Research Room

By synthesizing the existing literature research, and by studying the characteristics of the virtual teaching and research room, this paper concludes that the factors affecting the construction of the virtual teaching and research room mainly include three factors: equipment factors, institutional factors and team factors. The influence of factors on the virtual logistics teaching and research section is analyzed as follows.

4.1 Equipment Factors

(1) Hardware facilities
Teaching is mainly reflected in the equipment for recording courses, standardization issues, classroom basic adjustment conditions and standardized teaching hardware equipment; scientific research equipment is mainly used for research experiments and research teaching experiments. Whether the scientific research hardware is complete.

(2) Software factor
The virtual teaching and research room is an important feature of teaching informatization, and it is also a teaching grass-roots organization that has evolved to adapt to the development of informatization. It is a key factor in improving the quality of teaching by promoting the digitization of the teaching process. Team building has an important impact.

4.2 Institutional Factors

(1) Assessment and evaluation mechanism
The virtual teaching and research section can make teaching and research work more dynamic and open, free from time, space and geographical constraints, which makes up for the shortcomings of traditional teaching and research methods, but also increases the difficulty of assessment. Reasonable assessment of the teaching process and quality across schools and regions is the key to the virtual teaching and research section.

(2) Encouragement and incentive mechanism
On the basis of the assessment and evaluation mechanism, the virtual teaching and research office usually needs to formulate a perfect incentive mechanism for the team, provide the team with a spiritual or material mechanism, guide through goal incentives and performance incentives, and assist emotional incentives, Participation incentives, training incentives, growth incentives and other methods are used to encourage and guide team members by building a perfect incentive mechanism to make cooperation more sustainable. However, the virtual teaching and research room belongs to a cross-institutional and cross-departmental organization. How to concretize and operate the incentive mechanism is a key issue, and it is also a key factor affecting the deep cooperation of the virtual teaching and research team.

(3) Management system and process of virtual teaching and research room
The complexity of participating in the virtual teaching and research room also affects the construction of the team from some aspects. Does participating in the activities of the teaching and research room increase the burden of the usual teaching and research work? Does participating in virtual teaching and research often have meetings and discussions, and does it take up too much time? Is the process of uploading and maintaining data complicated? Such issues will affect the enthusiasm of the team, and these factors should be taken into account when building a virtual teaching and research room.

4.3 Team Factor

(1) Disciplinary differences among team members
Logistics belongs to a comprehensive discipline, which is a complex of science, engineering, and liberal arts. Some teachers come from engineering majors, and some teachers come from science or liberal arts majors. At the same time, the teaching styles of different schools are different, and the focus on logistics is also different., which enriches the knowledge of the virtual teaching and research room, but also brings certain challenges to cooperation.

(2) Teacher source structure
Some teachers in the logistics teacher team come from enterprises, such as "double-qualified" teachers, who are weak in theory and strong in practice and application; some teachers directly enter colleges and universities after graduation, lacking certain industry practice. The theory is strong, and the application and practice are weak. The diversity of teacher sources will provide a diverse knowledge structure for the logistics virtual teaching and research room, but it also increases the uncertainty of teamwork.

(3) Cooperation mechanism
The cooperation mechanism is mainly reflected in the sharing of teaching and scientific research equipment, the sharing of teaching resources, etc. Through the integration of equipment and resources, the effect of "$1 + 1 > 2$" is formed, which is conducive to improving the cohesion of the team. Regarding the sharing mechanism of facilities and equipment, different universities have different hardware and software resources. How to establish a good cooperation mechanism for resource integration and sharing is the key. Regarding the sharing of teaching resources, the

sharing of valuable teaching resources such as textbooks, courseware, examination papers, exercise banks, and teaching experience among different institutions is an important part of the virtual teaching and research section. The goal of building a virtual teaching and research section is to complement each other and integrate resources. Integrate the teaching resources between different colleges and universities to form a joint force to improve the teaching level. How to realize the common growth of team members through resource sharing is the key to the team cooperation of the virtual teaching and research room. Let every participant benefit from it, which can greatly enhance the enthusiasm of team members to participate in the activities of the teaching and research room, and by participating in the construction activities of the virtual teaching and research room It is an important factor for the long-term stable development of the team to further gain a sense of achievement and let the team building enter a virtuous circle (Table 1).

Table 1. Summary of factors influencing team building of logistics virtual teaching and research office

	Primary indicator	Secondary indicator
Influencing factors of team building in logistics virtual teaching and research room	Equipment factors (A)	Hardware equipment (A_1)
		Software equipment (A_2)
	Institutional Factors (B)	Assessment and Evaluation Mechanism (B_1)
		Encouragement and Incentive Mechanism (B_2)
		Virtual teaching and research room management system and process (B_3)
	Team Factors (C)	Team Member Disciplinary Differences (C_1)
		Teacher Source Structure (C_2)
		Cooperation mechanism (C_3)

Through the investigation and interview with a total of 20 people from 9 colleges and universities in Guangxi Logistics, the scores obtained are shown in Table 2 below:

Table 2. Score table of factors influencing team building of logistics virtual teaching and research room

Judgment matrix between criterion layer and target layer	A_1	A2	B_1	B_2	B_3	C_1	C_2	C_3
A_1	1	1/2	1/8	1/9	1/6	1/2	1/2	1/5
A2	2	1	1/7	1/8	1/5	1/2	1/2	1/4
B_1	8	7	1	1/2	3	5	6	3
B_2	9	8	22	1	5	7	7	5
B_3	6	5	1/3	1/5	1	4	3	1
C_1	2	2	1/5	1/7	1/4	1	1	1/4
C_2	2	2	1/6	1/7	1/3	1	1	1/5
C_3	5	4	1/3	1/5	1	4	5	1

The comparison matrix is obtained according to the scoring table:

$$A = \begin{Bmatrix} 1 & 1/2 & 1/8 & 1/9 & 1/6 & 1/2 & 1/2 & 1/5 \\ 2 & 1 & 1/7 & 1/8 & 1/5 & 1/2 & 1/2 & 1/4 \\ 8 & 7 & 1 & 1/2 & 3 & 5 & 6 & 3 \\ 9 & 8 & 2 & 1 & 5 & 7 & 7 & 5 \\ 6 & 5 & 1/3 & 1/5 & 1 & 4 & 3 & 1 \\ 2 & 2 & 1/5 & 1/7 & 1/4 & 1 & 1 & 1/4 \\ 2 & 2 & 1/6 & 1/7 & 1/3 & 1 & 1 & 1/5 \\ 5 & 4 & 1/3 & 1/5 & 1 & 4 & 5 & 1 \end{Bmatrix}$$

After normalizing by column, the Table 3 is obtained:

Table 3. Factor score after normalization

0.0286	0.0169	0.0291	0.0459	0.0152	0.0217	0.0208	0.0183
0.0571	0.0339	0.0332	0.0516	0.0183	0.0217	0.0208	0.0229
0.2286	0.2373	0.2325	0.2065	0.2740	0.2174	0.2500	0.2752
0.2571	0.2712	0.4650	0.4129	0.4566	0.3043	0.2917	0.4587
0.1714	0.1695	0.0775	0.0826	0.0913	0.1739	0.1250	0.0917
0.0571	0.0678	0.0465	0.0590	0.0228	0.0435	0.0417	0.0229
0.0571	0.0678	0.0387	0.0590	0.0304	0.0435	0.0417	0.0183
0.1429	0.1356	0.0775	0.0826	0.0913	0.1739	0.2083	0.0917

Calculated $\lambda_{max} = (A) = 8.3419$, the eigenvector corresponding to the largest eigenvalue of A is:

U = (0.0246, 0.0325, 0.2402, 0.3647, 0.1229, 0.0452, 0.0446, 0.1255)T,

Consistency test CR = 0.0346 < 0.1, passed the consistency test.

At this time, the weight vector can be regarded as the indicator weight. Therefore, the weights of the eight secondary capability indicators are: 2.46%, 3.25%, 24.02%, 36.47%, 12.29%, 4.52%, 4.46%, and 12.55%. Therefore, the assessment and evaluation mechanism (B1), the encouragement and incentive mechanism (B2), the management system and process of the virtual teaching and research office (B3), and the cooperation mechanism (C3) are the main influencing factors of the team building of the logistics virtual teaching and research office.

Because the virtual teaching and research room is a "temporary" organization that can enter and exit dynamically, the organizational form is relatively loose, so it is necessary to establish a good assessment and evaluation mechanism and encouragement mechanism in order to gather more teachers. The results of the analysis also show that the assessment and evaluation Mechanism is the primary factor affecting the construction of logistics virtual teaching and research room. Therefore, it is necessary to reasonably design the assessment and evaluation system and incentive mechanism, so that more teachers can actively participate in the construction of virtual teaching and research room. At the same time, due to the low pressure of teachers' scientific research and teaching work, the activities and participation process of the virtual teaching and research room should be more streamlined to reduce the burden on teachers. If participating in the activities of the logistics virtual teaching and research room takes up too much time and managers, many teachers will be discouraged or withdrawn. Therefore, it is necessary to design a reasonable management system and process. In addition, due to the different sources of teachers in the virtual teaching and research room, how to carry out cross-enterprise cooperation is the key. Therefore, it is necessary to take into account the characteristics of teachers in different colleges and universities to establish a cooperation mechanism, so that the virtual teaching and research room can achieve the effect of co-construction, sharing and common progress.

5 Conclusion

Combined with the existing research results of virtual teaching and research offices, combined with the characteristics of logistics major and disciplines, the paper analyzes the team building problems of logistics virtual teaching and research offices, and analyzes three factors that affect the team building of virtual teaching and research offices: equipment factors, institutional factors and team factors. The evaluation of the team building of the virtual teaching and research office by different efficient teachers was obtained through investigation, and the analysis was carried out through the analytic hierarchy process, and the key factors affecting the team building of the logistics virtual teaching and research office were obtained: assessment and evaluation mechanism, encouragement and incentive mechanism, virtual teaching and research office management system and process, Compared with technical equipment.

and platforms such as facility hardware and software, the cooperation mechanism is more important than the soft environment (management mechanism and team cooperation). Therefore, it is necessary to focus on the construction of the system and team cooperation mechanism when setting up a virtual teaching and research room.

The research of this paper provides a certain reference value for the team construction of the logistics virtual teaching and research department, and provides some ideas for promoting the reform and innovation of the logistics professional grassroots organization under the background of the information age. However, there are still shortcomings in the paper, the consideration of factors is not comprehensive enough, the survey sample is small, and the characteristics of colleges and universities outside Guangxi are not considered, and the characteristics of cross-regions are not fully considered. This is the focus of the next research.

Acknowledgments. This paper is supported by: (1) the teaching reform project of Guangxi Department of Education (2022jgb440), "research and Practice on the construction of logistics virtual teaching and research room from the perspective of 4D linkage of point, line, surface and body", (2) Guangxi zhe she project (17fjy014), (3) the teaching reform project of Nanning University (2020szsfk14 and 2020szsfk19) and "reform and practice of virtual simulation practice curriculum of Logistics Engineering Specialty Based on OBE concept and school enterprise linkage".

References

1. Mtiaz Hussain Khan: A unified framework for systematic evaluation of ABET student outcomes and program educational objectives. Int. J. Mod. Educ. Comput. Sci. (IJMECS) **11**(11), 1–16 (2019)
2. Liu'an, K., Xiaomei, W., Lin, Y.: The research of teaching quality appraisal model based on AHP. Int. J. Educ. Manage. Eng. (IJEME) **2**(9), 29–34 (2012)
3. Fangxiu, W.: The course design for algorithm analysis and research on improving the students' comprehensive ability. Int. J. Educ. Manage. Eng. (IJEME) **2**(2), 42–46 (2012)
4. Wang, X.D., Wu, Y.: Exploration and practice for evaluation of teaching methods. Int. J. Educ. Manage. Eng. (IJEME) **3**(29), 39–45 (2012)
5. Gilakjani, A.P., Leong, L.-M., Ismail, H.N.: Teachers' use of technology and constructivism. Int. J. Mod. Educ. Comput. Sci. (IJMECS) **5**(21), 49–63 (2013)
6. Kumar, S.: A fuzzy based comprehensive study of factors affecting teacher's performance in higher technical education. Int. J. Mod. Educ. Comput. Sci. (IJMECS) **5**(3), 26–32 (2013)
7. Zhiqin, L., Jianguo, F., Fang, W., Xin, D.: Study on higher education service quality based on student perception. Int. J. Educ. Manage. Eng. (IJEME) **2**(4), 22–27 (2012)
8. Zeng Jianchao, W., Shuqin, Z.C.: Virtual teaching and research office: innovation and exploration of primary teaching and research organizations in colleges and universities. China Univ. Teach. **11**, 64–69 (2020). (in Chinese)
9. Zhengliang, L., Ruijin, L., Lingyan, D.: Construction of new engineering majors: connotation, path and training mode. Higher Eng. Educ. Res. **02**, 20–24+51 (2018) (in Chinese)
10. Xiaofei, X., Dechen, Z., Yue, C., et al.: First Prize of National Teaching Achievement Award. Cross-regional and cross-school online open courses '1+M+N' collaborative teaching model innovation and practice (2018)
11. Yan, X., Wu, X.: Research on the teaching reform facilitated by digital technology in Chinese universities-based on the operation design of virtual teaching and research section. In: 2021 2nd International Conference on Information Science and Education (ICISE- IE). 1632–1636. IEEE (2021)
12. Zhizhong, H.: The evolution and reconstruction of grass-roots teaching and research offices in colleges and universities. Univ. Educ. Sci. **03**, 86–92 (2016). (in Chinese)

13. Lin, X., Hu, X., Hu, Q., et al.: A social network analysis of teaching and research collaboration in a teachers' virtual learning community. Br. J. Edu. Technol. **47**(2), 302–319 (2016)
14. Min, C., Ying, L.: Exploration and analysis on the incentive mechanism of virtual teaching team-incentive of knowledge-based employees based on psychological contract. Vocat. Educ. Forum **15**, 55–59 (2015). (in Chinese)
15. Dechen, Z., Lanshun, N., Dekai, T., Lijie, Z.: Virtual teaching and research office: a new form of collaborative teaching and research. Mod. Educ. Technol. **32**(03), 23–31 (2022). (in Chinese)
16. Korkmaz, Ö.: The effect of project-based cooperative studio studies on the basic electronics skills of students' cooperative learning and their attitudes. Int. J. Mod. Educ. Comput. Sci. (IJMECS) **5**, 1–8 (2018)

Consideration of IE-OBE-CDIO-Oriented Engineering Practice Innovation Curriculum Design

Xinjian Zhang, Jiajun Yang, Juan Fan, Min Tu, Yanqi Liu, and Shan Fan(✉)

Mechanical and Electrical Engineering Department, Wenhua College, Wuhan 430074, China
zxjchn@126.com

Abstract. In order to implement the transformation and development positioning of "building high-level application-based universities" that Ministry of Education sets for local regional universities, cultivate "high quality application-oriented talents who owns abilities of innovating and starting an undertaking in modern times as well as a high sense of social responsibilities, we have conducted extensive surveys on social needs and some preliminary practice. Based on those surveys and practice, the idea of "students-centered, learning outcome-oriented, continuous improvement", transformations of deeply carrying out project-based and flipped classroom teaching have made great achievements at some extent. This article only introduces engineering practice innovation curriculum design, expounds the principles of design and some specific ideas.

Keywords: CDIO · Engineering practice innovation · Curriculum design

1 Introduction

Education has always been a very important topic in the history of human development. In today's information age, education has become the focus of the whole society [1]. In order to implement the transformation and development orientation of the Ministry of education for local regional colleges and universities to "build a high-level applied university" and cultivate "high-quality applied talents with modern innovation and entrepreneurship ability and a high sense of social responsibility", we have conducted an extensive survey of social needs in recent years and carried out preliminary practice. On this basis, "student-centered, learning results oriented, continuous improvement", in-depth implementation of project-based teaching, flipped classroom teaching and other curriculum design reforms, and achieved some practical results. This paper only introduces the engineering practice innovation curriculum design, expounds the design principles and some specific ideas, and discusses the innovative road and development direction of education and talent training combined with the reform practice of our engineering practice curriculum design.

Z. Hu et al. (Eds.): AIMEE 2022, LNDECT 159, pp. 448–456, 2023.
https://doi.org/10.1007/978-3-031-24468-1_40

2 Historical Development Track of Personalized Education

The objects of education are different, and the individuals of the educated also differ in thousands of ways. Personality education has attracted great attention from almost all countries in the world, which has become the development trend of modern education [2, 3]. It can not only meet the needs of social development, but also is an effective educational method in line with the law of students' physical and mental development [4]. Personality education, or, more commonly call Individual Education, abbreviatedly, IE, is based on the development of each individual and the implementation of different programs of education, which deserves great attention [5–8].

More than two thousand years ago, Confucius, the great educator, put forward the educational concept of "teaching students according to their aptitude without discrimination". It refers to the large-scale education, which can educate all kinds of people, regardless of high or low, virtuous or foolish. At the same time, we should also pay attention to the individualization of education. The content and methods of education must vary from person to person, and be flexible and diverse. For thousands of years, the scale and personalization of education have been a pair of contradictions, and it is always difficult for people to achieve a balance.

In the industrial age, manual production was replaced by machines, and social productivity has been greatly developed. The mode of individualized or individual education in the era of agricultural society can no longer meet the requirements of social development. Comenius, Herbart, etc. advocate class teaching, and their corresponding teaching organization forms are: bell, class, standardized class, unified standard requirements, unified teaching materials, unified education methods and methods, and the assembly line arranged according to time "manufactures" a large number of standardized and available talents with unified specifications [9]. The central task of this kind of education is to impart scientific and cultural knowledge and combine it with social production practice. It is a stage that must pass through in the process of any technological development. Of course, it also has the function of subordinate superstructure and serves the politics and economy of its class [10, 11].

Entering today's intelligent society, various changes brought about by robots and artificial intelligence have an unprecedented impact on our production and lifestyle [12]. In this context, customer to business (C2B) has become the direction of change in various production fields such as automobiles, clothing, electronic products, etc. products are no longer mass production of one style, but multiple varieties and diversification. In the production and manufacturing process, in order to meet the personalized customized production needs of different models, batches and styles, it is required to carry out flexible manufacturing for products, and carry out compatible design for special machines, tooling and flexible body assembly of the production line, so as to meet the requirements of mixed flow production. The development trend of intelligent society has ended the history of standardization and made human life diversified. Education serving economic development also needs to cultivate all kinds of professionals in different fields, different types and different abilities required by xinshida reform. Education has also ended the uniformity of the industrial age and entered a new era of personalized education [13, 14].

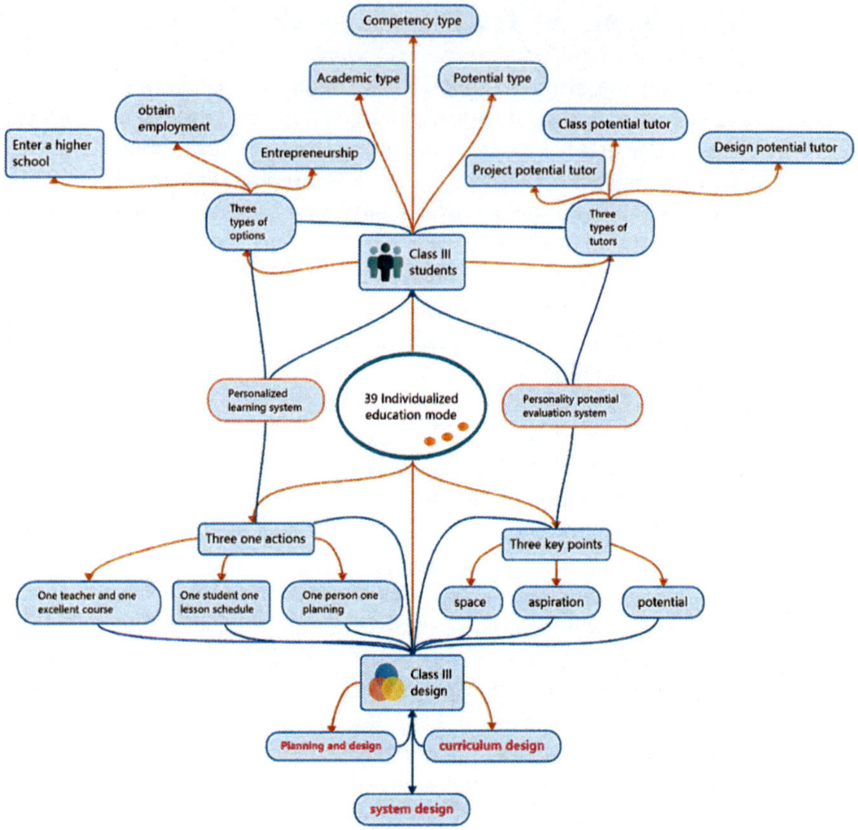

Fig. 1. "Three nine" individuality education mode

At present, educational reforms all over the world are advocating personalized teaching for students' individual differences. Personalized education has become a strong educational trend in the world. In 2006, the European Organization for Economic Cooperation (OECD) issued an opinion report, "school education for tomorrow: personalizing education", which took personalized education as an important education agenda in response to the era of change, and believed that "one size fits all" school knowledge and organizational arrangements were neither suitable for personal needs, nor incompatible with the development of the knowledge society. In the outline of the national medium and long term education reform and development plan (2010–2020) released by the Chinese government in 2010, it is proposed to "pay attention to students' different characteristics and personality. Differences, develop each student's advantage and potential", encourage personalized development, and provide each student with the concept of "suitable education" [15–17]. Therefore, the implementation of personalized education is an important starting point for promoting educational reform and innovation, an important measure for the implementation of quality education, a concrete embodiment of the implementation of the people-oriented scientific concept of development in education, and a guiding principle for educational and teaching reform [18].

Individuality education is the development direction of future education, Individuality education must be the fundamental instruction and milestone when designing courses. Under the leadership of Professor Xianjun Liu, Wenhua College took the lead in carrying out individuality education, after more than ten years of exploration and practice, a complete individuality education mode, which is named "three nine" was formed, as what is showed in Fig. 1, some great achievements have been made. Under the guidance of Individual Education concept, in the process of designing engineering practice curriculum, the author's department has drawn on the teaching principles of OBE and CDIO, combined the personalized process of talents' cultivation and the results of professional certificate of talent training, strived to explore how to provide suitable education for every student.

3 OBE Mode

As a teaching idea, "Outcome Based Education" was proposed by American scholar-Spady in 1981]. Up to now, a relatively complete theoretical system and complement mode has been formed after that, OBE has become the mainstream in education reform of America, England, Canada and so on. In 1989, America signed Washington Accord with England, Canada, Ireland, Australia and New Zealand. According to Washington Accord, engineering college graduates of countries and regions that have signed that accord should be regarded as they are qualified engage in preliminary engineering work in other contracting countries and regions. China signed Washington Accord and become preparatory member in 2013, finally, becoming a full member in 2016. Authorized by Ministry of Education and based on Washington Accord, China Engineering Education Accreditation Association has established an engineering education accreditation system that is international substantive equivalent, the fundamental target of doing so is to improving the quality of high-level engineering education in China.

According to OBE [19], when designing curriculum of engineering practice innovation education, some key points should be considered:

(1) Target: What are the learning outcomes we want our students to get?
(2) Demand: Why should we teach our students these things?
(3) Process: How can we effectively help students achieve these learning outcomes?
(4) Evaluation: How do we know that students have achieved these learning outcomes?
(5) Improvement: How do we ensure that students effectively achieve these learning outcomes?

Under the concept of OBE, there are distinctively differences between traditional classroom teaching design and OBE curriculum teaching mode, those differences are listed in Fig. 2. Obviously, OBE curriculum teaching design reflects the distinctiveness and pertinence of IE-"Individuality Education", which is students-centered and suitable for the laws of students' development.

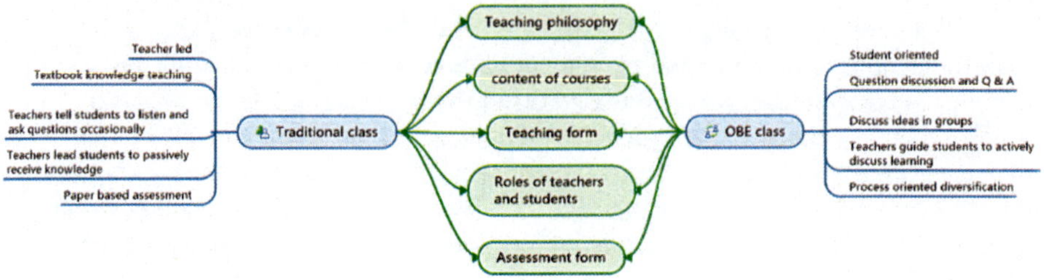

Fig. 2. Differences between traditional classroom teaching design and OBE curriculum teaching mode

4 CDIO Teaching Design Principles

CDIO stands for" Conceive, Design, Implement, Operate", which is based on actual project engineering. The best way to carried out OBE is systematically and organically combined curriculums, then integrating them with project design, so that students can be actively learn knowledge in the ways of learn engineering, improve the practice ability. CDIO is the concentrated generalization and abstract expression of "learning by doing" and "Project-based education and learning", it embodies the combination of systematic, scientific and advanced, it represents the developing trend of engineering.

OBE is the expected standard outcome orientation of Individuality Education, CDIO is the specific processing complement of OBE, integrating OBE with CDIO is helpful to cultivating students' abilities of engineering application and innovation. OBE is a students-centered mode of cultivating talents, moreover, it is the talent training target orientation of "providing suitable education for every student", which is under the Individuality Education concept. OBE indicates that there should be an expected, standardized and measurable result for learners' learning.

4.1 Ideas of Engineering Practice Innovation Curriculum Design under OBE

According to OBE, the overall goal of engineering practice is to cultivate students' comprehensive abilities, including the ability of mastering basic knowledge, the ability of using professional knowledge in application, personal quality, vocational ability, interpersonal ability and so on, which can be summarized in three aspects-knowledge, ability and quality [9]. When designing curriculum, these target abilities need to divided into and integrated with teaching goals of every practice curriculum project, in order to ensure that students can acquire learning achievements during cultivation. The target of curriculum design is to make students understand "what is learning?", reverse design and forward implement. The training objectives of the OBE practice course are showed in Fig. 3.

According to the idea of OBE courses design, based on professional training targets, requirements of graduate and courses correspondence matrix, curriculum teaching's knowledge goals, ability goals and quality goals of each course should be set from three aspects-knowledge, ability and quality. Taking an engineering actual training and practice course named Metal Heat Treatment as an example, its correlation matrix of

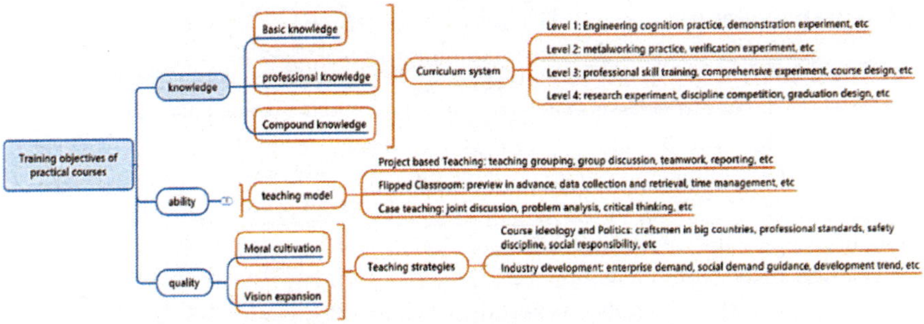

Fig. 3. The training objectives of the OBE practice course

course knowledge, ability and quality is showed in Table 1. Such students-centered design attaches greater importance on the application and innovation of knowledge, building teaching knowledge system under the instruction of cultivation goals, and the knowledge system supports the goal of training talents.

Table 1. The realized matrix of the knowledge, ability and quality of a practice course et al. Heat Treatment

No	Knowledge, ability, quality	Teaching method and measures
1	Basic knowledge of metal heat treatment helps students understand the relationship between material composition, microstructure and properties, material selection and technical solutions when analyzing material modeling	Analyzing and reconstructing relative knowledge about metal heat treatment "four fire". Organizing teaching according to the curriculum design projects, conducting teaching based on "knowledge-content module-teaching tasks-projects"
2	Software simulation and design capabilities	Explaining the simulation of Deform's metal heat treatment, including quench analysis, stress analysis, the temperature distribution uniformity of heating and cooling, etc
3	Autonomous learning capability	Panel discussion, team-work, previewing relative conferences and videos that sent by teachers in advanced or searching required topics, making sure your own work after interchange, submitting plan, pre-defensing, finishing your tasks, writing reports, defensing and evaluating projects, etc
4	Moral cultivation and vision expansion	Heat treatment technological programs in Sword of Goujian and Tian Gong Kai Wu, etc. Relations between modern technological programs and materials, occupational safety education, etc

4.2 The Implement of Teaching Design Based on CDIO Project

Based on the students-centered Individuality Education ideas, teachers are instructors of teaching, students are subjects of learning, if we want attract students' attention to teaching and make them be active in class, it's needed to tell students the clear learning goals, the relations between courses system and the correlation with our future jobs. In these processes, teaching design based on CDIO project is the most effective way. Implementing teaching based on CDIO project is usually divided into four parts-C, D, I, O and three phases-pre-class task assignment; in-class project teaching discussion, implement of project and defense report; after-class project implement and expansion. It will be best to combine teaching design of courses with online teaching platforms, such as Xuexitong and Yuketang, taking Xuexitong as an example, with the help of online teaching platform, the modular design of CDIO Individuality Education is easy to realize, as shown in Fig. 4. Case 1: Based on nonlinear state estimation.

When designing stage D of courses, completing the relevant discussion about the implement of course projects' teaching in class and finishing the defense and comment after projects are the main points. In teaching, teachers should mobilize students to join in projects in every class, they also need to conduct ideological and political education and make an overall mind maps of course project to show the relations between each project and corresponding curriculum knowledge. Targeting to each small project, the correlative mind map of small course project should be designed. The design of teaching projects' mind map is designed to make students understand why, what and how to learn before starting to learn. If students can realize the teaching ways and teaching effects before class, which means they have clear studying direction. In the concrete teaching process, Xuexitong can be used as an effective tool to realize varied teaching, according to what has been designed in advances, achieving the interaction of students-to-students or teachers-to-students. Teachers can lead students to discuss topics and divide them into groups. At the same time, with the help of Xuexitong, randomly ordering people, preemptive answer, vote and scoring can create a tense and competitive class atmosphere, which can attract students' attention, promote effective learning and complete the leaning about projects knowledge.

Stage I of courses, which is a stage that discussing the implement and practice of projects' plans. Deform can be used to conduct simulation operation of Metal Heat Treatment, such as the settings of furnace, the addition of mediators, the settings of technological program, etc. By conducting simulation operation, students can know the whole process of heat treatment, from heating, austenitization, quench, tempering to air cooling and so on; At the same time, students can inspect, understand and correct potential problems or defects before manufacturing. When conducting actual operation, students can directly heat the hacksow blade trough heating furnace, observe and analyze the experiment results under the instruction of knowledge "four fires", as well as complete the innovation design of project according to the innovative designing requirements of project. The implement of other theoretical curriculums is almost similar to this, discussing the problems in implement is the main part of class, implement will be conducted after class, such as the design switching power of electric power and electronics. Through these, the first class is naturally combining with the second class, the teaching expansion of curriculum is formed.

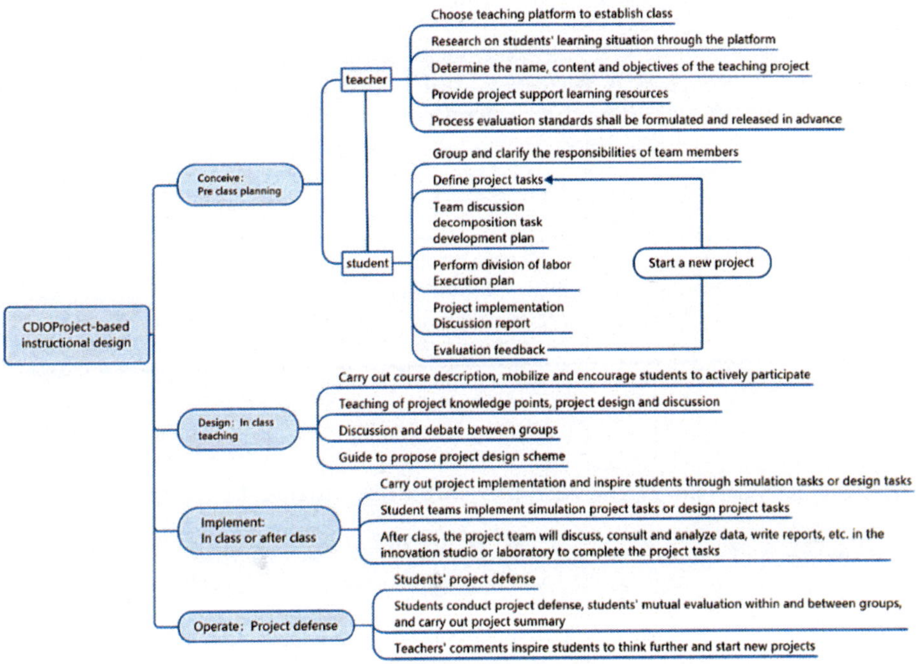

Fig. 4. The modular design of CDIO Individuality Education

5 Conclusion

The ability of engineering practice is the key point of cultivate applied talents, the teaching design and the complement of teaching organization are the important ways to realize the cultivation of applied talents, the recognition and update of teaching ideas are the core of high-level curriculum design and implements of teaching organization, the constructions of practice teaching platforms and systems are the base and support of cultivating applied talents.

All of these require us to consider the targets of practice innovation curriculums and the realized matrix of the knowledge, ability and quality of professional talents' cultivation goals in the beginning of constructing correlative practice platforms.

Constructing corresponding practice platforms, realizing the integration of talents training goals, curriculum knowledge system, course project teaching and practice teaching platforms, combing these with high-level curriculum design and finally, laying a solid foundation for realizing the goal of cultivating first-rate applied talents.

Acknowledgment. This project is supported by the outstanding young and middle-aged scientific and technological innovation team plan project (T2021043) of colleges and universities in Hubei Province.

References

1. Patrio, C.P.H., Tarloff, I.S.W., Hung, S.C.: Disciplinary variations in student perceptions of active learning classrooms. Int. J. Educ. Res Open **3**, 1–7 (2022)
2. Xianjun, L.: Personalized education model exploration. Res. High. Educ. **41**(01), 1–8 (2020)
3. Xianjun, L.: Ten concepts of personalized education. Res. High. Educ. **39**(09), 1–7 (2018)
4. Fangxiu, W.: The course design for algorithm analysis and research on improving the students' comprehensive ability. Int. J. Educ. Manage. Eng. (IJEME), **2**(2), 42–46 (2012)
5. Xianjun, L.: Exploration of personalized education system for undergraduate students. Res. High. Eng. Educ. **06**, 105–113 (2012). (in Chinese)
6. Zhili, C.: Vigorously promote personalized education. China Educ. J. **10**, 3 (2011). (in Chinese)
7. Mingyuan, G.: Innovation of personalized education and talent training mode. China Educ. J. **10**, 5–8 (2011). (in Chinese)
8. Li, K., Hua, W., Xiangming, F.: Thoughts on building a personalized university education system. China High. Educ. Res. **10**, 27–29 (2004). (in Chinese)
9. Xiting, S.: Principles of Education (Revised). Beijing Normal University Press, Beijing (2006). (in Chinese)
10. XinLiang, O., Wang, J., Liu, H., Wen, D., He, W., Pan, Y.: Research on the reform of practical teaching driven by practical training under constructivism. Int. J. Eng. Manufact. (IJEM) **1**(1), 77–83 (2011)
11. Ning, G.: Construction and implementation of innovation computer network practical teaching system. Int. J. Educ. Manage. Eng. (IJEME) **1**(2), 30–35 (2011)
12. Boyd, R., Holton, R.J.: Technology, Innovation, employment and power: does robotics and artificial intelligence really mean social transformation? J. Sociol. **54**(3), 331–345 (2018)
13. Howell, R.A.: Engaging students in education for sustainable development: the benefits of active learning, reflective practices and flipped classroom pedagogies. J. Clean. Prod. **325**, 1–11 (2021)
14. Zhiqin, L., Jianguo, F., Fang, W., Xin, D.: Study on higher education service quality based on student perception. Int. J. Educ. Manage. Eng. (IJEME) **2**(4), 22–27 (2012)
15. Jian, L., Yuan, Z.: The exploration and practice in innovative personnel training of computer science and technology. Int. J. Educ. Manage. Eng. (IJEME) **2**(6), 47–51 (2012)
16. Cen, G., Luo, J.: Design of a teaching process monitoring and management system. Int. J. Edu. Manage. Eng. (IJEME) **2**(4), 128–133 (2012)
17. Xianjun, L.: Exploration of personalized education mode. High. Educ. Res. **41**(01), 1–8 (2020). (in Chinese)
18. Nanxing, Z., Lian, Z., Xinfeng, W., et al.: Understanding OBE: origin, core and practice boundary- also on the paradigm shift of professional education. Res. High. Eng. Educ. **03**, 109–115 (2020). (in Chinese)
19. Sun, F.: Design of online virtual training project based on OBE concept. Exp. Technol. Manage. **38**(03), 210–213 (2021). (in Chinese)

Analysis of Learners' Satisfaction with Online Teaching in Applied Universities

Xiaofen Zhou[1], Yang Yan[1](✉), and Yi Zhang[2]

[1] College of Logistics, Wuhan Technology and Business University, Wuhan 430065, China
1142912067@qq.com
[2] School of Business, Sichuan University, Chengdu 610064, China

Abstract. This paper takes the learners of an applied university as the research object, regards the learners as the consumers of online teaching, and aims to explore the influencing factors of learners' satisfaction under the online teaching mode. This research is based on the core concepts of the American customer satisfaction model, combined with the influencing factors of online teaching design satisfaction to build a research theoretical model, collects data through questionnaires, and conducts validation analysis of the model. The research shows that the influencing factors of learner satisfaction in Application-oriented Universities under the online teaching mode platform learning experience, learner expectation, perceived quality and perceived value have a direct impact on learner satisfaction, learner expectation has an indirect impact on learner satisfaction, and learner satisfaction and platform learning experience have a significant positive impact on continuous learning intention. Finally, based on the conclusion of the model validation and the analysis results, this paper puts forward some suggestions and countermeasures to improve the learner satisfaction of application-oriented universities.

Keywords: Learner satisfaction · Online teaching · Structural equation model

1 Introduction

Since 2012, "Large-scale Online Open Courses" have sprung up worldwide, and online education has become another emerging hot topic in the field of higher education [1]. The strong support of the state for educational informatization has provided a strong driving force for educational reform and development, and promoted the organic integration of information technology and education and teaching. Online learning has become the focus of attention and has developed rapidly. Online learning has achieved extensive educational participation with its advantages of openness, free and autonomy [2, 3]. At present, domestic research on online courses focuses on its application in blended teaching, and pays less attention to learners' satisfaction with online courses. However, students in applied universities have relatively weak initiative, enthusiasm and concentration in learning [4, 5]. From the perspective of students' learning in application-oriented universities, the problem of satisfaction with online teaching is more worthy of exploration.

Z. Hu et al. (Eds.): AIMEE 2022, LNDECT 159, pp. 457–467, 2023.
https://doi.org/10.1007/978-3-031-24468-1_41

On the basis of literature analysis, this research constructs a model of students' satisfaction with online teaching in application-oriented colleges and universities, and uses the structural equation analysis tool LISREL and SPSS to verify the model, and puts forward suggestions to improve the satisfaction of online teaching in Applied Universities on the basis of this analysis.

1.1 Construction of Online Teaching Satisfaction Model

Satisfaction is an internal psychological activity, which is a pleasant feeling that arises after people's needs and wishes are satisfied. Satisfaction degree refers to the degree of satisfaction, which is a quantitative study of satisfaction [6]. This paper studies the experience and understanding of the online learning process of applied undergraduates, compares the effect of the learning experience formed with the traditional learning effect, and whether the service quality and effect of online learning can meet the needs of learners. Education is a social activity, and interaction is the most basic form [7]. The main subject of teaching interaction in this project is the undergraduates of applied universities. The purpose of the interaction is to improve the learning effect of application-oriented undergraduates and achieve teaching goals [8].

In this study, learners are regarded as consumers of online courses. Based on the core concepts of the American customer satisfaction model, combined with the influencing factors of online teaching design satisfaction, the theoretical model of this research is constructed as shown in Fig. 1.

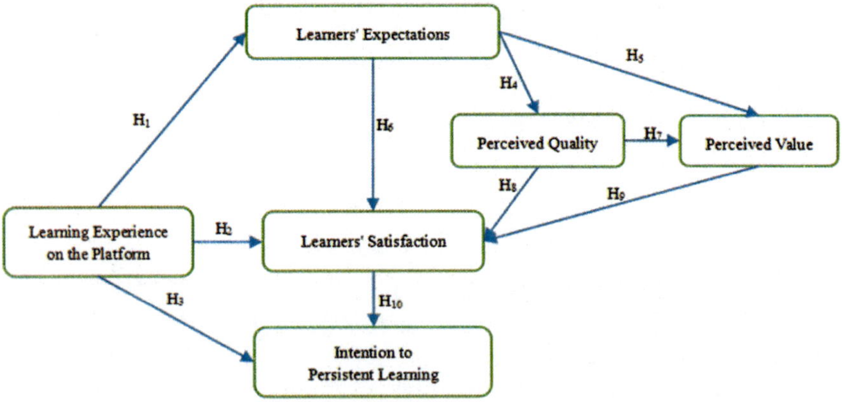

Fig. 1. Influencing factor model of online teaching satisfaction

The learning experience on the platform refers to the learning functions and experiences provided by various online platforms that may have an impact on the production of learners' satisfaction [9]. The variable of "Learning Experience on the Platform" designed in this project mainly refers to the overall impression of learners on the use of online learning platforms, the reputation of online learning, etc.

Learners' expectations refer to learners overall expectations of their online learning effects through online learning [10], mainly including learners' expectations of media interface interaction, teacher-student interaction, student-student interaction and learning resource interaction.

Perceived quality refers to learners' direct feelings about various interactive experiences designed in the teaching process through online teaching [11], including interaction design with platform interface, interaction design between teachers and students, interaction design between students and students, and interaction design between learners and learning resources.

Perceived value refers to the value that learners feel after comparing online learning and classroom learning [12], including the cost-effectiveness of the two teaching methods, learner input, etc.

Learners' satisfaction refers to the satisfaction of learners after comparing their expectations and perceptions of instructional design interaction before online learning [13], including four dimensions: interaction with platform interface, interaction between teachers and students, interaction between students and students, and interaction with learning resources.

Intention to persistent learning refers to learners' willingness to continue using online courses for learning after they are satisfied with the teaching design and learning experience in online learning.

1.2 Design of Online Teaching Satisfaction Questionnaire

The quantitative data in the paper model is collected by issuing questionnaires, so the formation of the scale is the key to the research. The design process of this research scale first draws lessons from the preparation of customer satisfaction or learner satisfaction questionnaires by domestic and foreign experts, explores the constituent elements of interaction under the teaching design in the online learning environment, combines the interview results of Undergraduates in application-oriented universities, and reviews by experts in related fields or learning participants with rich online experience, puts forward opinions, and modifies and improves the questionnaire again according to the feedback suggestions.

The questionnaire consists of two parts. The first part is to collect and analyze the basic personal information of learners, such as gender, grade, major, learning purpose, etc. The second part is the main part of the questionnaire, which investigates learners' satisfaction with each dimension of online teaching interaction. There are 35 questions, divided into 7 different variables. The specific measurement indexes and reference sources are shown in Table 1.

The questionnaire scale adopts Likert's 5° scale method. Learners score according to their online learning satisfaction [14]. The number 1 to 5 indicates that the degree from very disagree to very agree gradually deepens. Calculate the average value of the collected data as an indicator to determine the degree of learners' satisfaction. A score of more than 3 indicates that learners prefer to be satisfied, and a score of less than 3 indicates that learners prefer to be dissatisfied.

Table 1. Dimensions, number of items and reference sources of the questionnaire

Dimensions		Number of items	Reference sources
Learning Experience on the Platform		3	Zha Xianjin [15]
Learners' Expectations		4	Schmitz B &Hanke K [16]
Perceived quality	Interaction with Platform Interface	2	Sumi, Razia Sultana [17]
	Interaction between Teachers and Students	4	Sumi, Razia Sultana [17]
	Interaction between Students and Students	4	Zhang Jing [18]
	Interaction with Learning Resources	2	Zhang Jing [18]
Perceived value		3	Lin &Wang [19]
Learners' satisfaction		4	Parul Kumar [13]
Intention to persistent learning		3	Venkatesh, V.& F.D [20] Qin Hong-xia [21]

2 Data Collection and Model Verification

The survey data was collected through online channels, based on the Questionnaire Star platform, through mobile WeChat, mobile QQ and PC. 269 points of questionnaires were recovered. After review, 13 questionnaires did not meet the requirements, and 254 valid questionnaires were obtained, with an effective recovery rate of about 94.4%.

2.1 Reliability and Validity Analysis of the Questionnaire

2.1.1 Reliability Test of Questionnaire

SPSS 21.0 data analysis tool was used to test the reliability of the questionnaire data. The Cronbach'α results of each scale in this study is shown in Table 2.

For the convenience of data analysis, the coding only refers to the first letter of Chinese phonetic alphabet, which is written as a latent variable symbol. The reliability test standard is: when the Cronbach'α is greater than 0.7, it means that the reliability passes the test; when the Cronbach'α is greater than 0.8, it means that the reliability is good. From the test results, the potential variable Cronbach'α is greater than 0.8, indicating that the internal consistency of this questionnaire is good, and the scale has high reliability.

At the same time, in order to avoid the influence of the number of measurement items on the coefficient, this study supplemented the combined reliability measurement index to further test the inherent quality of the theoretical model. If the CR of the combination reliability is greater than 0.6, it means that it meets the reliability standard. Taking AVE value as the index of information inspection, the inspection results are shown in Table 3.

Table 2. Reliability measurement results of each dimension and the questionnaire

Variable name	Code	Number of items	Cronbach'α
Learning Experience on the Platform	ZXXX	3	0.829
Learners' Expectations	XXZQW	4	0.907
Interaction with Platform Interface	MTJM	2	0.800
Interaction between Teachers and Students	TSHD	4	0.890
Interaction between Students and Students	SSHD	4	0.908
Interaction with Learning Resources	ZYJH	2	0.823
Perceived value	GZJZ	3	0.858
Learners' satisfaction	XXZMYD	4	0.901
Intention to persistent learning	CXXXYY	3	0.858

Table 3. Values of factor loading, CR, AVE

Variable name		Code	Factor loading	CR	AVE	$\sqrt{AVE}$
Learning Experience on the Platform		ZXXX1	0.840	0.898	0.745	0.863
		ZXXX2	0.870			
		ZXXX3	0.879			
Learners' Expectations		XXZQW1	0.871	0.935	0.782	0.884
		XXZQW2	0.897			
		XXZQW3	0.877			
		XXZQW4	0.892			
Interaction with Platform Interface	Perceived Quality	MTJM1	0.913	0.909	0.834	0.913
		MTJM2	0.913			
Interaction between Teachers and Students		TSHD1	0.834	0.924	0.752	0.867
		TSHD2	0.859			
		TSHD3	0.9			
		TSHD4	0.875			

(continued)

Table 3. (*continued*)

Variable name		Code	Factor loading	CR	AVE	$\sqrt{AVE}$
Interaction between Students and Students		SSHD1	0.863	0.944	0.808	0.898
		SSHD2	0.984			
		SSHD3	0.89			
		SSHD4	0.892			
Interaction with Learning Resources		ZYJH1	0.871	0.895	0.739	0.859
		ZYJH2	0.861			
		ZYJH3	0.847			
Perceived value		GZJZ1	0.89	0.914	0.779	0.883
		GZJZ2	0.891			
		GZJZ3	0.867			
Learners' satisfaction		XXZMYD1	0.83	0.932	0.773	0.879
		XXZMYD2	0.897			
		XXZMYD3	0.9			
		XXZMYD4	0.888			

From Tables 2 and 3, The Cronbach'α is greater than 0.7; CR value is greater than 0.7; AVE is greater than 0.5. To sum up, the reliability of the scale has reached a very good level. Through these tests, the reliability of the scale is very good.

2.1.2 Validity Test of Questionnaire

Validity is an important indicator to test whether the results of survey data analysis can achieve the researchers' expected measurement goals. The validity test includes convergent validity and discriminant validity.

The convergence validity test usually requires the following conditions: a. the factor loading of the measurement index is > 0.7; b. AVE > 0.5; If the two are satisfied at the same time, it indicates that the convergence validity has passed the test. It can be seen from Table 3 that the standardized factor loading of all items is between 0.786–0.913, which is greater than 0.7; Meanwhile, AVE is between 0.691–0.834, greater than 0.5, which meets the index reference value. It shows that the convergence validity of the measurement model in this paper is good.

The discriminant validity test is judged by the size of the square root of AVE and the correlation coefficient. The square root of AVE is greater than the correlation coefficient, indicating that the discriminant validity is good. The correlation coefficient and the square root of AVE are shown in Table 4, and bold words on the diagonal represent the square root of AVE.

It can be seen from Table 4 that the minimum value of the square root of AVE (0.859) is greater than the maximum value of the correlation coefficient (0.839). So the discriminant validity passed the test.

Table 4. Correlation coefficient and square root of AVE

	ZXXX	XXZQW	MTJM	TSHD	SSHD	ZYJH	GZJZ	XXZMYD	CXXXYY
ZXXX	**0.863**								
XXZQW	0.773	**0.884**							
MTJM	0.675	0.737	**0.913**						
TSHD	0.678	0.769	0.669	**0.867**					
SSHD	0.638	0.747	0.676	0.809	**0.898**				
ZYJH	0.675	0.761	0.664	0.814	0.839	**0.859**			
GZJZ	0.647	0.73	0.692	0.71	0.767	0.803	**0.883**		
XXZMYD	0.629	0.743	0.623	0.786	0.754	0.776	0.76	**0.879**	
CXXXYY	0.689	0.737	0.644	0.76	0.76	0.768	0.787	0.784	**0.884**

Combined with the data results shown in Table 2, Table 3 and Table 4, there is sufficient evidence to show that the reliability and validity of the scale have passed the test.

2.1.3 Verification of Structural Model of Influencing Factors of Online Teaching Satisfaction

Through the verification of reliability and validity, the reliability and validity of the scale are very good. The paper uses LISREL 1.0 to carry out hierarchical linear regression on the structural model, sets gender, age, major, learning duration and learning platform as control variables, and verifies the path relationship of the structural direction. The standardized path coefficient of the research model is shown in Table 5.

Table 5. Standardized path coefficient of the model

Hypothesis	Path	Path coefficient	Significance	Hypothetical results
H1	Learning Experience on the Platform → Learners' Expectations	0.777	***	support
H2	Learning Experience on the Platform → Learners' satisfaction	0.627	***	support
H3	Learning Experience on the Platform → Intention to persistent learning	0.312	***	support

(continued)

Table 5. (*continued*)

Hypothesis	Path	Path coefficient	Significance	Hypothetical results
H4	Learners' Expectations → Perceived Quality	0.828	***	support
H5	Learners' Expectations → Perceived value	0.141	*	support
H6	Learners' Expectations → Learners' satisfaction	0.148	*	support
H7	Perceived Quality → Perceived value	0.682	***	support
H8	Perceived Quality → Learners' satisfaction	0.819	***	support
H9	Perceived value → Learners' satisfaction	0.670	***	support
H10	Learners' satisfaction → Intention to persistent learning	0.574	***	support

Notes: ***: $P < 0.001$; **: $P < 0.01$; *: $P < 0.05$

3 Research Conclusion

The research shows that Learning Experience on the Platform, learners' expectations, perceived quality and perceived value have a direct impact on learners' satisfaction, which confirms that learners' satisfaction is related to platform learning experience, learners' expectations, perceived quality and perceived value.

3.1 Conclusions on the Direct Influencing Factors of Learners' Satisfaction

According to the data in Table 4, the path coefficient of platform learning experience on learners' satisfaction is 0.627, the path coefficient of learner expectation and learners' satisfaction is 0.148, the path coefficient of perceived quality and learners' satisfaction is 0.819, and the path coefficient of perceived value and learners' satisfaction is0.670 respectively.

From the perspective of direct influencing factors, perceived quality has the greatest impact on learner satisfaction, and the path coefficient value is 0.819. Learners perceive that the better the teaching quality is through online learning, the higher the learning satisfaction is.

From the factor loading calculation of each dimension under the perceived quality in Table 2, it shows that students in applied universities pay the highest attention to student-student interaction, followed by teacher-student interaction and media interface interaction. Applied undergraduates pay more attention to cooperative learning and sharing learning ideas with peers (0.984) in the dimension of student-student interaction; In the dimension of teacher-student interaction, they are more concerned about whether teachers can participate in topic discussions and express opinions frequently (0.90); In the dimension of interacting with the media interface, they pay more attention to the smoothness (0.913) and simplicity (0.913) of the use of the media interface.

Learners' perceived value has a great direct impact on learners' satisfaction, and its path coefficient is 0.670. It shows that the greater the perceived value of learners, the higher the learning satisfaction. From the factor load values of each dimension in Table 3, we can see that in terms of perceived value, learners pay more attention to the efficiency, interest and effectiveness of online learning.

The learning experience of the platform also has a great direct impact on learners' satisfaction, and its path coefficient is 0.627. It shows that learners also have high requirements for platform learning experience. It can be seen from Table 3 that learners pay more attention to the service quality of the learning platform and the effectiveness of the courses of the learning platform.

3.2 Conclusions on the Indirect Influencing Factors of Learners' Satisfaction

From the analysis results in Table 5, it can be seen that the path coefficient of learner expectation on learner perceived quality is 0.828, which is greater than 0.8, indicating that learner expectation has a significant impact on learner perceived quality. The perceived quality has a significant impact on learning satisfaction, indicating that learner expectations also have a greater indirect impact on learner satisfaction. We also need to focus on learner expectations when we study learner satisfaction.

From the analysis results in Table 5, it can be seen that learner expectations are only affected by platform learning experience, and the path coefficient is 0.777, indicating that platform learning experience has a greater impact on learning expectations, and platform learning experience also directly affects learner satisfaction.

4 Countermeasures and Suggestions

Combined with the quantitative data analysis results in this study, the following aspects should be taken to improve adult learner satisfaction.

From the perspective of platform developers, it is necessary to improve the convenience and timeliness of online communication between students and students, and between students and teachers, from the perspective of the convenience and sharing of students and teachers. Hey need to increase platform visibility and usage, focus on service quality and the development and recommendation of high-quality courses, while improving the simplicity and comprehensiveness of the interface. They need to further sort out and subdivide the nature and types of courses, and make a detailed and comprehensive introduction to the course objectives, so that learners can have clear goals in the selection and learning process, and have clear gains through online learning.

From the perspective of teachers' teaching, teachers need to share and set more discussion topics or related assignments related to the course content during the course construction process, and enable the function of students' mutual evaluation at the same time. It is not only limited to the evaluation between students, but also pays attention to the exchange of opinions between students, and also actively participates in the discussion of students to increase the enthusiasm of students' discussion topics. Teachers should guide students to correctly deal with online learning methods, and classify the construction and use of courses of different nature and requirements. In terms of teaching methods, the course content of online teaching should be reasonably designed and adjusted to give full play to the advantages of online teaching and improve students' satisfaction with online learning.

Acknowledgment. This research was financially supported by the Distinguished Young and Middle-aged Program for Scientific and Technology Innovation in Higher Education of Hubei, T201938, "Smart Logistics Park Service Innovation and Support"; And supported by Hubei Education Science Planning Project (2018GB121).

References

1. Jun, R.: A study on promoting hybrid teaching mode reform in colleges. Mod. Educ. Technol. **27**(04), 74–78 (2017). (in Chinese)
2. Qi, D., Zhang, M., Zhang, Y.: Influence of participation and value co-creation on learner satisfaction of MOOCs learning: learner experience perspective. Asia Pac. Educ. Res. **31**(1), 61–70 (2022)
3. Nandi, D., Hamilton, M., Harland, J., Mahmood, S.: Investigation of participation and quality of online interaction. IJMECS **7**(08), 25–37 (2015)
4. Ying, L.: Study on evaluation of students' learning effect in application-oriented colleges and universities. Theo. Pract. Innov. Entrepreneurship **2**(05), 72–77 (2019). (in Chinese)
5. Leiye, Z.: Investigation on learning adaptability of local applied college students: taking Huangshan university as an example. The Sci. Educ. Art. Collects **4**, 38–40 (2019). (in Chinese)
6. Sachin, A., Puninder, K., Panda, S.N.: Identification of influencing factors for enhancing online learning usage model: evidence from an Indian university. Int. J. Educ. Manage Eng. (IJEME), **9**(02), 15–24 (2019)
7. Gomez, V., Patricia, E., Sandoval, B., Javier, O.: Structural analysis of interactions in a social network from an educational activity in higher education. Redes De Ingenieria-Rompiendo Las Barreras Del Conocimiento **7**(1), 6–15 (2016)
8. Tawafak, R.M., Sideir, A.A., Alfarsi, G.: E-learning vs. traditional learning for learners satisfaction. Int. J. Adv. Sci. Technol. **29**(3), 227–239 (2020)
9. Wang, L., Zhen, Z., Wo, T., Jiang, B., Sun, H., Long, X.: A scalable operating system experiment platform supporting learning behavior analysis. IEEE Trans. Educ. **63**(3), 232–239 (2020)
10. King-Spezzo, A.J., Hsiao, E.-L., Wiley, E.W., Wiley, L.P.: Comparing adult learners' expectations of ideal course environments focused on task orientation, teacher support, and student influence. J. Educ. Online **17**(01), 70–83 (2020)
11. Lanning, S.K, Pardue, K.T.: Early-learners' expectations of and experience with IPE: a multi-institutional qualitative study. Nurse Educ. Today **107**, 122–135 (2021)
12. Chikazhe, L., Makanyeza, C., Kakava, N.Z.: The effect of perceived service quality, satisfaction and loyalty on perceived job performance: perceptions of university graduates. J. Mark. High. Educ. **32**(1), 1–18 (2022)
13. Kumar, P., Kumar, N.: A study of learner's satisfaction from MOOCs through a mediation model. Procedia Comput. Sci. **173**, 354–363 (2020)

14. Aurangzeb, W.: Leadership processes and employee attitude in HEIs: a comparative study in the backdrop of Likert's systems theory. J. Humanit. Soc. Sci. Pak. **23**(3), 19–32 (2015)
15. He, Y., Xin, F.: Learning satisfaction of learners and curriculum design under different online teaching platforms. Int. J. Emerg. Technol. Learn. **17**(10), 237–249 (2022)
16. Schmitz, B., Hanke, K.: Engage me: learners' expectancies and teachers' efforts in designing effective online classes. J. Comput. Assist. Learn. (2021)
17. Sumi, R.S., Golamb, K.: Satisfaction of e-learners with electronic learning service quality using the servqual model. J. Open Innovation Technol. Mark. Complex. **7**(4), 19–33 (2021)
18. Jing, Z., Yingxiong, H.: Research on learning interaction status of MOOC in Chinese universities. Educ. Res. **27**(05), 73–80 (2021). (in Chinese)
19. Lin, W.S., Wang, C.H.: Antecedences to continued intentions of adopting e-learning system in blended learning instruction: a contingency framework based on models of information system success and task-technology fit. Comput. Educ. **58**(1), 88–99 (2012)
20. Venkatesh, V. F.D., Davis, A.: Theoretical extension of the technology acceptance model: four longitudinal field studies. Manage. Sci. (02), 442–452 (2000)
21. Hong-xia, Q., Jian-hua, Z., Zheng, L.: A study on the continuous use intentions differences of online teaching between teachers and students. J. Higher Educ. **42**(01), 83–93 (2021). (in Chinese)

Construction of Virtual Simulation Training Base and the Collaborative Development of Teaching Resources Between Schools and Enterprises

Kecheng Li, Quan Yang, Tong Xie, and HuaJian Xin(✉)

Guangxi Industrial Vocational Collage, NanNing 530001, China
493141592@qq.com

Abstract. Informatization and intellectualization are important national development strategies. For intelligent manufacturing and other majors, virtual simulation teaching is a teaching method with high practical value, which needs to be promoted under the support and guidance of information education strategy, as well as the cooperation and support of enterprises. This paper focuses on the talent training of intelligent manufacturing majors, faces the construction of advanced equipment manufacturing industry system and the construction of intelligent equipment demonstration virtual simulation training base, analyzes the current situation and progress of the base construction, defines the construction objectives and ideas, and discusses from the aspects of system design, open management, network operation, updating and improvement, specialty matching, school enterprise collaboration, etc., It aims to cultivate a high-level virtual simulation teaching team and project research and development team, and play a better role in the training of practical professionals.

Keywords: Intelligent manufacturing · Virtual simulation · School enterprise collaboration · Practical teaching

1 Introduction

In order to promote the construction and development of the country, a series of documents, such as the outline of the national innovation driven development strategy, the implementation plan for accelerating the modernization of education, and the implementation plan for the reform of national vocational education, have been issued successively [1, 2]. The Ministry of education has also launched the requirements of the action plan for educational informatization 2.0, the construction of a vocational education demonstration virtual simulation training base, and the improvement and excellence of vocational education, Guided by the needs of regional equipment manufacturing industry chain, and based on the principle of "integration of industry and education and cooperation between schools and enterprises", it is the future development direction to integrate

Z. Hu et al. (Eds.): AIMEE 2022, LNDECT 159, pp. 468–478, 2023.
https://doi.org/10.1007/978-3-031-24468-1_42

multiple resources to build a national intelligent equipment demonstration virtual simulation training base. Therefore, it puts forward new and higher requirements for Education, especially for higher education in intelligent manufacturing [3, 4].

It is the trend of internationalization and informatization to explore and establish a virtual simulation school enterprise collaborative development mode with college leadership, enterprise collaboration and different characteristics. Using virtual simulation, big data, cloud computing, vr/ar and other virtual simulation new technology innovation and construction modes, building a school enterprise cooperation bridge, and cultivating high-quality technical and skilled talents in the field of intelligent manufacturing for the regional industrial revitalization action plan is the foundation for local colleges and universities to build virtue and cultivate talents [5, 6]. Aiming at the teaching of intelligent manufacturing majors, this paper discusses the school enterprise collaborative development of virtual simulation training teaching resources under this background.

2 Background and Basic Ideas of Virtual Simulation Training Base Construction

2.1 Basic Conditions of Base Construction

Talent training is a responsible systematic project with many influencing factors and great responsibilities [7–9]. The teaching of the school is not related to many typical production environments, production links, technical links and post work links in the advanced equipment manufacturing industry [10]. The teaching scene is seriously disconnected from the production scene in the process of talent training, and there are also practical problems such as invisibility, inaccessibility, high cost and high risk in the process of training. Virtual simulation experiment teaching is an important practical content of the national information education strategy. We should develop virtual simulation training teaching projects that are the same as the real environment, explore new ideas, new models and new mechanisms of virtual simulation enabled vocational education teaching, and further promote the construction of national vocational education virtual simulation demonstration training base and vocational education demonstration virtual simulation training base, Boosting the digital upgrading of vocational education is an important direction for the development of higher education in China [11, 12].

The intelligent construction virtual simulation training center takes ar/vr/ai and other new technologies as means to build a realistic and visual operation environment and training objects, so that students can carry out efficient, safe and economic training in an open, autonomous, interactive and simulated virtual environment, so as to achieve teaching effects that real operation does not have or is difficult to achieve, innovate three-dimensional teaching, and realize cross hardware, cross specialty The guidance of cross course virtual training teaching [13–15]. The number of demonstrative virtual simulation experiment teaching projects carried out in colleges and universities from 2017 to 2021 is shown in Fig. 1. From 2017 to 2020, the construction of undergraduate schools is mainly focused, and from 2021 to 2023, the construction of higher vocational colleges is mainly focused. The classification of the first batch of undergraduate demonstration virtual simulation experiment teaching projects in 2017 is shown in Fig. 2.

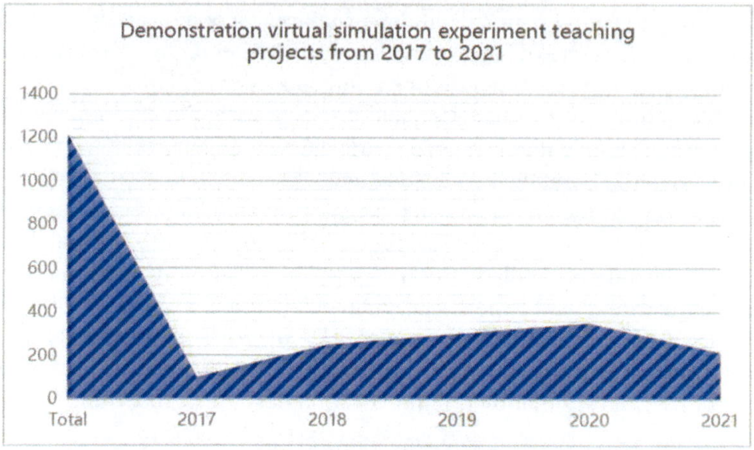

Fig. 1. Demonstration virtual simulation experiment teaching projects from 2017 to 2021

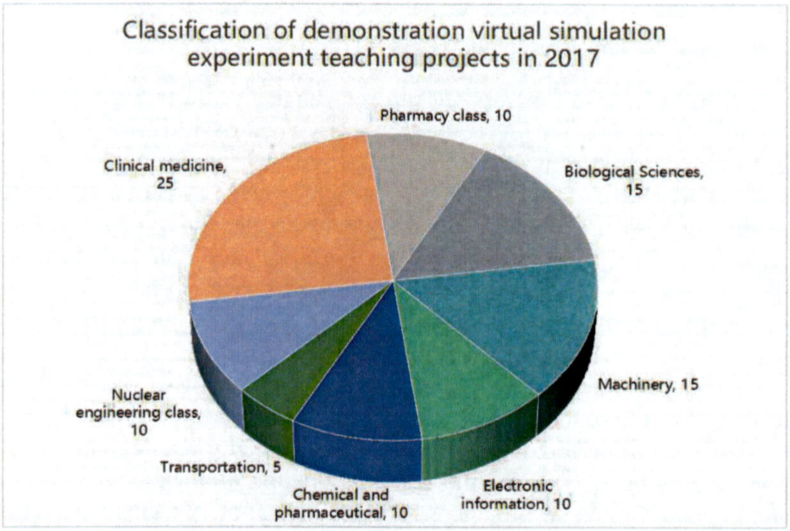

Fig. 2. Classification of demonstration virtual simulation experiment teaching projects in 2017

2.2 Basic Ideas of Base Construction Project

The overall goal of virtual simulation training base and resource construction is to serve the talent training of intelligent manufacturing majors. The basic idea is determined to achieve this goal.

(1) Docking enterprises and simulating production line scenarios.
   Talent training is to serve the society. Only by connecting with enterprises can our graduates meet the actual requirements of enterprises and society.

The construction of the virtual simulation training base will combine the actual business of relevant enterprises of the Institute of intelligent manufacturing industry, use the virtual simulation technology, "real with virtual, virtual with real", connect the industrial production process, professional standards, industrial standards and job specifications, take the real production line of the enterprise as the background, virtual simulate the intelligent manufacturing scene, and build courses and training projects such as robot operation, numerical control processing and electrical operation, Cultivate students' ability to operate, maintain, design and other aspects, empower the education science and technology ecosystem with a new generation of information technology, and reproduce the "real" factory of enterprises.

(2) Technology empowerment, simulation post operation.

The virtual simulation training base focuses on solving the dilemma that students are difficult to get to the real job scene. Through technical means, it captures industry and enterprise data and technologies for students to perform simulation operations and simulation results. The base pays attention to the construction of the overall application environment and the coordinated development of software, hardware and resources, so that it can really play a role in education and teaching and generate benefits. Taking teaching application as the guide, greatly improve the teaching effect and promote the leap-forward development of education.

(3) Use virtual to help reality, simulate the curriculum system.

The virtual simulation training base will update the content of the training courses and deepen the reform of the training courses. Centering on enterprises in the intelligent manufacturing industry, the simulation multi-post integration will integrate the production line business processes and cutting-edge content of new technologies of modern industry leading enterprises through "migration" interconnection, etc. With the integration of new formats, new industries and new technologies, the nodes, contents and characteristics are the starting points, and simulation technology is embedded in the curriculum system to establish a production line workflow that is "synchronized" with the enterprise. In the simulation, the students are guided scientifically in the craftsman spirit of devotion to work and excellence, and the combination of ideology and politics of the curriculum and science and technology is deeply explored.

(4) Transformation of achievements to serve local industries.

The virtual simulation base will take the industrial college as a platform to give full play to the talents of both schools and enterprises, promote the transformation of scientific research results, solve practical problems for the development of enterprises, promote the integration of industrial innovation and applied research, and open up professional teaching, experimental training, scientific research, etc. In the link, cultivate innovative and entrepreneurial technical and technical talents, promote the incubation of scientific research achievements, and achieve the effect of "integration of teaching, research and innovation". At the same time, according to the development plan of the enterprise, the professional teaching plan is adjusted reasonably, and the targeted output is highly suitable for high-quality talents that are highly matched with the job needs of the enterprise, so as to further improve the pertinence of the training base for personnel training.

3 Base Construction Objectives and Contents

By means of system design, advanced technology support, open management, network operation and continuous updating, matching the professional curriculum content, focusing on the post responsibilities generally implemented by enterprises, we will build professional virtual simulation training resources integrating teaching design, teaching materials, virtual training and teaching evaluation.

3.1 Construction of Intelligent Equipment Manufacturing Technology Exhibition Hall

Establish a science and technology exhibition hall dominated by many kinds of vr virtual reality simulation software, such as the red education exhibition hall, the monument to the founding of the people's Republic of China, the red research and interaction, the development history of intelligent equipment manufacturing technology, the history of industrial development, typical technical links, and typical production scenes, which will be displayed in the form of three-dimensional resources.

3.2 Build a Virtual Simulation Training Center That Meets the Needs of Talent Training

According to the development needs of enterprises and industries, schools and enterprises collaborate to build an artificial intelligence empowerment center. Every year, they jointly develop virtual simulation training programs required by enterprises and industries with personnel from enterprises and scientific research institutes, and update simulation content in a timely manner to keep up with technological updates of enterprises. Speed, promote the application of information technology in the teaching and training process, build a 020training system, introduce a domestic and foreign simulation skills appraisal and certification system, and undertake the study, training, experience and social-related positions of students from surrounding institutions and countries along the China-ASEAN route Skills training and certification needs.

3.3 Establish a Platform for the Transformation of Scientific and Technological Achievements

Relying on the intelligent equipment virtual simulation training center to form a platform for the transformation of scientific and technological achievements, focusing on the transfer of patented technology and the transformation of technological achievements to achieve the transfer and transformation of scientific and technological achievements, and promote the establishment of technology transfer and transformation innovation with enterprises as the main application body, market-oriented, and close integration of production, education, research and application. Mechanism. Through the joint application of scientific and technological projects with enterprises, the application of horizontal topics is promoted, and the transformation of teaching and scientific research achievements and the transformation of key technological scientific and technological achievements in the industry are promoted.

3.4 Build a Platform for Sharing Training Resources for Intelligent Manufacturing Professional Groups

Establish a new school-enterprise cooperation model with enterprises, and rely on the vocational education virtual simulation training base to develop a number of high-quality teaching and training resources. On the existing basis, develop virtual simulation training materials, develop high-quality virtual simulation training teaching projects, and apply teaching materials and teaching resources to teaching.

3.5 Activate a "Double-Trained" Teacher Empowerment Center for Intelligent Manufacturing

Create a virtual simulation teaching team with reasonable age, title, knowledge, and ability structure, with vocational education virtual simulation training course project development, virtual simulation training teaching, virtual simulation training application promotion, and training teaching management.

3.6 Set an Open Off-Campus Training and Social Service Platform

Intelligent equipment demonstration virtual simulation training base actively provides personnel training, training, and consulting services for enterprises in the intelligent manufacturing industry, and provides enterprises and schools with high-risk or extreme environments, unreachable or irreversible operations, high cost, high consumption, large-scale Or the visual, safe and economical training of comprehensive training and other projects, continuously enhance the ability to serve the economy and society, and help regional economic development.

4 School Enterprise Combination to Promote Base Construction

4.1 Significance and Role of School Enterprise Integration

With the help of the strength and resources of enterprises, the base construction can achieve the purpose of cultivating practical professionals for enterprises and society, which is the original intention of the base construction.

School enterprise cooperation is to interact the resources between colleges and universities and employers, combine the operation mechanism and job needs of enterprises with the talent training system and talent training objectives of colleges and universities, and achieve the purpose of seamless connection between the talent training achievements of colleges and universities and the talent needs of employers through long-term cooperation modes such as coordination, interaction and sharing. With the deepening of the teaching reform in Colleges and universities in China, school enterprise cooperation has become one of the important strategies for the development of higher education in China. School enterprise cooperation is a new direction of teaching reform, innovation and development in modern colleges and universities in China.

4.2 Implementation Strategy of School Enterprise Integration

Some effective measures should be taken to promote the combination of schools and enterprises in the construction of the base.

Strengthen collaboration with industry associations, enterprise institutions and other multiple subjects, form an organizational structure of joint construction and management, explore governance models such as councils and management committees, endow human rights, administrative and financial rights required for the integration of industry and education, and build a scientific, efficient and powerful institutional system.

Optimize the allocation mode of innovative resources, enhance the "self hematopoiesis" ability of school enterprise cooperation, and realize the deep integration of education chain, innovation chain and industrial chain. In order to keep the construction and operation of the base synchronized with the development of the latest technology, process, equipment and materials in the industry, and introduce advanced virtual simulation technology into the base in time. In the process of construction and operation management, the base will strengthen the coordination with industry associations, enterprise institutions and other multiple subjects, introduce regional leading units as participants, and jointly establish a base management committee to jointly carry out base construction, operation management and other work.

Formulate major matters such as the construction objectives, construction planning, operation mode, training mode, funding source and use of the base; The management committee holds regular meetings to jointly study or review some major issues in the process of base construction and management, and put forward guiding opinions.

Establish the practice teaching standard system of the base, schools and enterprises jointly formulate the curriculum standards, teaching standards and professional instruments and equipment specifications of the practice courses of the base, formulate the teaching guidance scheme and curriculum standards of the base training projects, and develop professional teaching standards and curriculum standards that are connected with international advanced standards. Combined with the educational function positioning of the base and the requirements of serving the regional industrial education objectives, draw lessons from and introduce the enterprise post norms, and formulate the talent training plan of the virtual simulation training base.

4.3 Achievements of Base Construction

After a period of efforts, we have achieved results in the construction of school enterprise coordination base. The distribution of the first batch of demonstrative virtual simulation experiment teaching projects in Higher Vocational Education in 2021 is shown in Fig. 3.

As can be seen from the figures, the virtual simulation training center and training projects of undergraduate schools have basically completed construction, but some projects have not completed construction and acceptance, and further research and construction paths are needed. The virtual simulation training in vocational schools is in its infancy. Only 215 projects will be established in 2021, and it needs to be vigorously developed in the future. Therefore, the research on virtual simulation training centers and projects in vocational schools is the development needs in the next few years. Based on

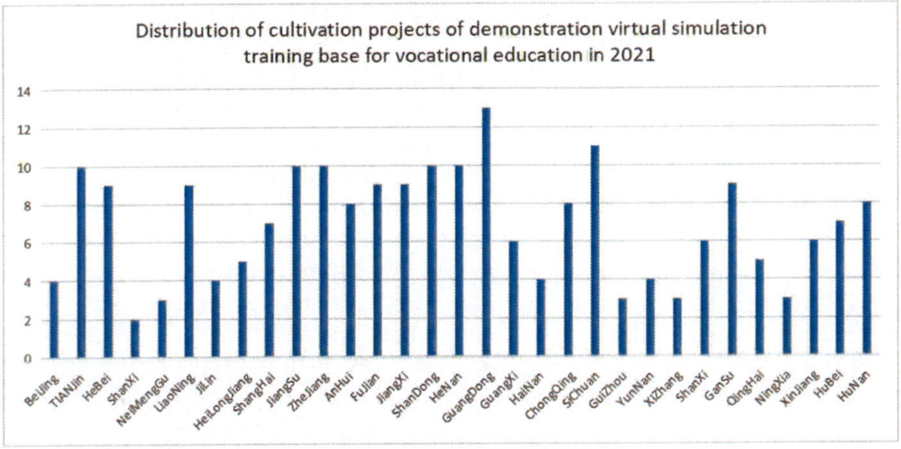

Fig. 3. Distribution of cultivation projects of demonstration virtual simulation training base for vocational education in 2021

school-enterprise the collaborative virtual simulation training mode is a guide to guide the development of the center.

While the number of base construction projects has made a breakthrough, the following aspects still need to be paid attention to.

(1) Integration of multiple training modes to meet diverse teaching needs.

The training base will be built around the five-layer structure of the existing intelligent equipment technology experiment and training system, combining virtual training resources with existing experimental training resources, and four levels of experimental training modes (resource cognitive learning mode, software The simulation training mode, the combination of virtual and real simulation training mode, and the hardware technology training mode) have been organically integrated, which fully meets the teaching links of resource learning, basic cognition, simulation operation, functional verification, practical operation, evaluation and assessment, etc., to build the virtual simulation training base of intelligent equipment into a comprehensive training base integrating teaching, practical training, training, scientific research, competition, popular science and other functions.

(2) Solve the difficulties of practical teaching with the combination of multi-dimensional virtual reality.

"Bringing the virtual with the real, using the virtual to help the real, and combining the virtual with the real", with the concept of combining the virtual and the real, and using the new technology and new means of digital twins, the virtual simulation and the real information are deeply interconnected, which not only meets the students' real job docking needs, but also It solves the experience and functional verification problems of various large-scale control scenarios and engineering scenarios that are difficult to implement, difficult to observe, and difficult to reproduce in the process of practical training and teaching.

(3) Integrate online and offline teaching data to create a hybrid teaching evaluation system.

The traditional paper-based assessment mode is replaced by information-based means, and an evaluation system for theory and simulation training is built through the online education service platform, so that teachers can be assigned teaching tasks, assessment tasks, and fault tasks. The statistics of key skill points, completion time, and completion progress in the simulation operation process, combined with the online subjective evaluation of teachers, finally generate comprehensive assessment results. Through the B/S and C/S fusion architecture, using big data and cloud technology, the offline assessment data and online assessment data are organically combined, the daily learning data, training data and assessment data are organically combined, and grades, classes and individuals are organically combined. The data is organically combined, and the data of students in the school and the data of social students are organically combined to form a comprehensive evaluation system that integrates teaching, learning, training, assessment and evaluation data.

(4) Combining education and training to carry out teaching reform and innovation.

Traditional teaching resources, such as animation, micro-lecture, simulation, etc., all require the support of a third-party development team for mass production. The three-dimensional classroom development tool allows teachers to freely organize materials to form micro-lectures or interactive materials, without relying on third-party development team. According to the actual course content, teachers can organically combine the "1+X" certificate system and the credit bank reform pilot content to prepare courseware and organize materials in a more intuitive, three-dimensional and relaxed way, and then record them as micro-lectures or interactive course packages, and then distribute them to designated students. Follow the concept of courseware video, and at the same time add more interactive effects on the basis of traditional video; students accept the courses assigned by teachers, and students can experience immersive learning, examination and training in a virtual three-dimensional environment The integrated learning process makes it easier to concentrate and improve learning efficiency, and innovates the teaching mode of practical training.

(5) Co-construction, research and sharing to create a green and sustainable operation and optimization platform.

As shown in Fig. 4, the shared virtual simulation integrated teaching cloud platform has a comprehensive architecture. The shared virtual simulation comprehensive teaching cloud platform integrates the on-site management platform, the simulation management platform, and the external open sharing platform to meet the offline on-site environment management, simulation system cluster management, VR/AR/MR form adaptation management, and external online simulation implementation, training and resource sharing. It integrates live broadcast and remote collaboration, and provides remote expert services. Relying on the management cloud platform, external experts, industry and enterprise experts can be invited to set up expert lectures, teacher quality courses and other content on the platform, so as to improve the radiation range of resources and improve the utilization of resources. Breadth and depth.

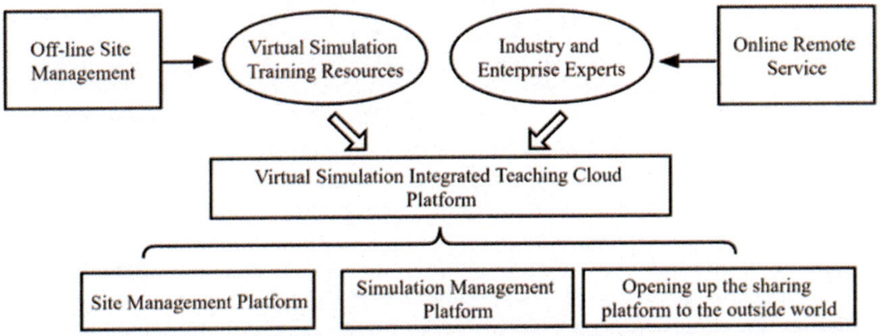

Fig. 4. Architecture of shared virtual simulation integrated teaching cloud platform

It integrates the two tools of "stereo classroom" and "mirror classroom" to realize the independent recording, production and development of course resources, and continuously expand the content and adaptability of the platform library; collect various virtual simulation resources to realize the remote online simulation function, and as the upper platform of software and tools, it integrates the functions of production, release, display, experience, learning, communication, collaboration, management, training, assessment, and development.

5 Conclusion

The virtual simulation experiment teaching base jointly developed by schools and enterprises not only has advanced technology and rich virtual simulation experiment teaching resources in the industry, but also has a high-level teaching team, which has formed a perfect teaching management operation mechanism and guarantee system. It can effectively cultivate the skills of students and teachers to explore new technologies, and effectively promote the reform of teaching content, teaching methods and teaching system.

The construction of the base takes the cultivation of new intelligent manufacturing talents as the main line, fully considers the characteristics of interdisciplinary cross training, and plans to build a highly integrated and shared intelligent manufacturing virtual simulation training center; Five regions of virtual simulation creation and R & D center that help virtual simulation resources, teaching, scientific research and development, and innovation and entrepreneurship incubation. The virtual simulation base builds a collaborative education mode between schools and enterprises, and forms a virtual intelligent management cloud platform with co construction, sharing and scientific operation. The research of the thesis and the construction of the base have been proved to be beneficial and effective for the cultivation of professional talents.

Acknowledgment. This project is supported by Guangxi Vocational Education Teaching Reform Research Project (GXGZJG2021A008).

References

1. Phadke, A.G.: Synchronized phasor measurements in power systems. IEEE Comput. Appl. Power **6**(2), 10–15 (2013)
2. Phadke, A.G., Pickett, B., Adamiak, M., et al.: Synchronized sampling and phasor measurements for relay and control. IEEE Trans. Power Delivery **9**(1), 442–452 (2014)
3. Qinjie, Y., Xiaoru, W., Jiaxun, Y., et al.: Equality constraints two-step state estimation model based on phasor measurements. Power Syst. Technol. **31**(10), 84–88 (2007). (in Chinese)
4. Xiaohui, Q., Tianshu, B., Qixun, Y.: A new method for hybrid nonlinear state estimation with PMU. Autom. Power Syst. **31**(4), 28–32 (2017). (in Chinese)
5. Al-Hagery, M.A., Alzaid, M.A., Alharbi, T.S., Alhanaya, M.A.: Data mining methods for detecting the most significant factors affecting students' performance. Int. J. Inf. Technol. Comput. Sci. (IJITCS) (5), 1–13 (2020)
6. Giret, A., Garcia, E., Botti, V.: An engineering framework for service-oriented intelligent manufacturing systems. Comput. Ind. **81**, 116–127 (2016)
7. Goncalves, R., Romero, D., Grilo, A.: Factories of the future: challenges and leading innovations in intelligent manufacturing. Int. J. Comput. Integr. Manuf. **30**(1), 4–14 (2017)
8. Cardin, O., Ounnar, F., Thomas, A., Trentesaux, D.: Future industrial systems: best practices of the intelligent manufacturing and services systems (IMS2) French Research Group. IEEE Trans. Industr. Inf. **13**(2), 704–713 (2017)
9. Zhao, S., Jia, X.: Research progress and trend of intelligent manufacturing and its core information equipment. Mech. Sci. Technol. **36**(1), 1–16 (2017). (in Chinese)
10. Khan, M.H.: A unified framework for systematic evaluation of ABET student outcomes and program educational objectives. Int. J. Mod. Educ. Comput. Sci. (IJMECS) **11**(11), 1–16 (2019)
11. Yufan, D.: Exploration on the precise education path of the integration of production and education in the new era of vocational education groups - taking Guangdong industry and trade vocational education group as an example. J. Shenzhen Voc. Tech. Coll. **18**(4), 75–80 (2019). (in Chinese)
12. Jian, L., Yuan, Z.: The exploration and practice in innovative personnel training of computer science and technology. Int. J. Educ. Manage. Eng. (IJEME) **2**(6), 47–51 (2012)
13. Catherine, A., Nathan, B., Erin, M.: The positive influence of inquiry-based learning teacher professional learning and industry partnerships on student engagement with STEM. Front. Educ. **6**, 319 (2021)
14. Patrio, C.P.H., Tarloff, I.S.W., Hung., S.C.: Disciplinary variations in student perceptions of active learning classrooms. Int. J. Educ. Res. Open **3**, 1–7 (2022)
15. Shi, L., Yao, Y., Xia, X., Xie, X., Wu, L.: Practical teaching staff construction under new situation. Int. J. Educ. Manage. Eng. (IJEME), **1**(2), 57–61 (2011)
16. Xiaoli, L., Jin, L., Li, Q.: The exploration and practice of the training system of automation professionals. Commun. Copyright **41**(10), 149–151 (2016). (in Chinese)
17. Guojin, C., Zhoushu, J.: Research on the experimental teaching system of intelligent manufacturing technical personnel training. Lab. Res. Explor. **35**(11), 189–194 (2016). (in Chinese)
18. Jian, L.: China's new engineering construction for the future. Educ. Res. Tsinghua Univ. **38**(2), 26–35 (2017). (in Chinese)
19. Ou, X., Wang, J., Liu, H., Wen, D., He, W., Pan, Y.: Research on the reform of practical teaching driven by practical training under constructivism. Int. J. Eng. Manuf. (IJEM) **1**(1), 77–83 (2011)

Mixed Teaching Mode of Micro Class and Project-Based Learning on the Course "ANSYS APPLication Practice"

Pan Zhou, Xinjian Zhang, Shan Fan, Yanmin Lu, Juan Fan, and Lili Cheng(✉)

Mechanical and Electrical Department of Wenhua College, Wuhan 430074, China
chenglilly@163.com

Abstract. With the continuous development of information technology in today's era, teaching methods and means are also constantly updated, helping to cultivate innovative talents. This paper analyzes the problems existing in the practical courses in the traditional teaching methods, and takes the "ANSYS application practice" course as an example, takes the micro course as the carrier of online teaching, and takes the project task as the driving force of offline teaching, and introduces the teaching mode, implementation process and evaluation method of the combination of micro course and project-based learning. A questionnaire with good reliability and validity is designed to measure the learning effectiveness of the teaching model from three dimensions: awareness, participation and satisfaction. The research of this paper shows that compared with the traditional teaching mode, the mixed teaching mode of micro class and project-based learning is conducive to enhancing students' self-confidence and creativity, and improving students' learning enthusiasm.

Keywords: Micro class · Project-based teaching · Teaching mode

1 Introduction

In the information age, with the rapid development of digital media technology, the new generation of young people has undergone qualitative changes in their ways of acquiring knowledge and thinking about problems. However, most college classrooms still cannot get rid of the old and backward educational concepts and are limited to the traditional teaching mode of "teachers teach and students passively accept". Teachers in the classroom rely on teaching plans to transmit knowledge, lacking effective teaching design and teaching activities, this traditional and outdated teaching method is difficult to cultivate students' critical thinking and ability to analyze and solve problems, and is no longer applicable. Vigorously promoting the reform of education has always been a problem worthy of attention [1, 2].

The sudden COVID-19 in 2020 changed the traditional offline teaching mode, and thousands of students received online education at home. Therefore, online education is one of the hottest topics in recent years. Online education is a kind of self-driven and

Z. Hu et al. (Eds.): AIMEE 2022, LNDECT 159, pp. 479–489, 2023.
https://doi.org/10.1007/978-3-031-24468-1_43

autonomous learning based on constructivism [3–5]. It can well realize personalized learning. Learners can choose learning content according to their own.

needs, knowledge background and personal preferences, and independently arrange learning progress, which helps to improve learning efficiency [6]. But it also has some disadvantages. Online education usually appears in the form of video courses, which cannot truly create a face-to-face learning atmosphere, and students are difficult to have emotional resonance Moreover, online courses are long and require very effective teaching management and self-discipline, otherwise it is easy to cause emotional boredom, inattention and other phenomena. When these emotions are mapped to the actual learning process, they will greatly affect learners' learning enthusiasm and cannot guarantee the effectiveness of learning.

Adopting the mixed teaching mode of micro class and project-based learning can effectively solve the above problems. Micro class originated from the 60 s course proposed by the University of Northern Iowa in the United States in 1993 and the one minute speech proposed by Napier University in the United Kingdom in 1995 [7]. Micro class is a teaching form of online course for the purpose of online learning, but it is essentially different from online learning video. Micro class mainly highlights the word "micro". It aims to explain a certain knowledge point, and explains it in 5–8 min [8]. Its advanced stage is "micro class", which should include corresponding teaching design and teaching activities, and has a complete teaching ecological environment. It is the sum of the teaching content and teaching activities of the knowledge points of a discipline [9–13].

There is an effective teaching method in offline Teaching - project-based learning. It is an action oriented teaching mode based on theory guiding practice. Teaching activities are based on inquiry learning and purpose driven [14]. The mixed teaching mode of micro class and project-based learning is to organically combine online micro class and offline project-based teaching. In this process, micro class is the carrier, project is the main line, and through the traction of project tasks, both give play to their respective advantages and effectively achieve teaching goals.

2 Research Background of Curriculum Reform

In recent years, graduates who master finite element analysis software are very popular with employers. Under this background, the ship and ocean engineering major of Wenhua University has opened a centralized practical course "ANSYS application practice". Such practical courses have some common problems: for example, the total class hours are small but the single learning time is long; Knowledge capacity is large, not easy to repeat, and the teaching content is easy to forget after class; Traditional teaching is one to many, and students' operation cannot keep up with the progress of teachers' explanation; It is difficult for teachers to solve all kinds of problems encountered by all students in learning in time, and it is difficult to achieve efficient and quality classroom learning. In order to further improve the teaching quality and teaching efficiency, this study takes the "ANSYS application practice" course as an example to explore the teaching strategy and effect of the mixed mode of micro class and project-based learning.

2.1 Analysis of Research Objects

Before teaching, we should analyze the characteristics of learners to understand their general characteristics, starting point learning ability, learning interest and style. The general characteristic analysis is mainly to understand the psychological status of learners and their adaptability to new learning; Starting point learning ability mainly analyzes learners' knowledge and skill reserves; Learning interests and styles mainly analyze learners' habits and preferences. These are helpful to understand the learners' learning preparation, so as to choose the appropriate teaching materials, adopt reasonable teaching strategies, and organize effective teaching activities.

The object of this study is the students who took the course of "ANSYS application practice" from 2019 to 2021. Through analysis, the students have a certain ability to read drawings, and can analyze the stress of simple engineering structures. In addition, they have had a short experience of pure online teaching and have a certain understanding of online teaching. They are generally tired of theoretical courses, prefer practical courses, and are more interested in online and offline hybrid teaching models, but their mastery of knowledge and professional skills remains at a relatively shallow level.

2.2 Preparation of Teaching Resources

At the beginning of the study, the micro teaching assistant platform was used to create the micro class resources of the course.

Table 1. Division of teaching units and projects

Teaching unit	Micro class teaching content	Project design
Basic software operation	Start, exit and save the software; Interactive operation interface environment and its composition; Common menu and dialog box operations; Screenshots and preservation; Coordinate system and workplane	Analysis of thin plates in tension
Solid modeling technology	Geometric modeling; Boolean operation; Dragging and rotation; Zoom, Move, Copy, Mirror; Merger; Delete;	Modeling of steel support; Modeling of bearing pedestal
Mesh generation technology	Meshing; Unit selection; Define material properties and real constants; The system of units in ANSYS; Define cell attributes; Meshing tool; Clear grid	Stress analysis of ship bottom plate beam composite structure

(continued)

Table 1. (*continued*)

Teaching unit	Micro class teaching content	Project design
Applied load and Solution Technology	Analysis type and solution process; Load classification and load step options; Apply load	Establishment of hull model with unit rib pitch
General post-processing technology	Introduction to post processor; Result documents and data analysis; Draw deformation map and ISOline map	Stress analysis of hull section structure

The micro class mainly serves students' autonomous learning and fragment learning. The duration is generally short. The knowledge module is divided into five teaching units and then subdivided. Offline classes are based on project-based learning, and the project design must fully consider the integration of theoretical knowledge and practical ability, as well as the hierarchical and progressive nature of the project. Around the target knowledge points, table 1 highlights one or several ability requirements according to the arrangement order of micro courses, and designs engineering practice projects that are from easy to difficult and closely combined with majors.

3 Implementation Method of Project-Based Teaching Based on Micro Class

The core of project-based teaching design based on micro class is how to reasonably integrate online micro class and offline project-based teaching [15–17]. Figure 1 gives the basic framework of project-based teaching based on micro class. Micro class is mainly used for pre class guidance and post class consolidation, which can simulate one-to-one teaching situations, paying attention to both teachers' teaching and learners' learning. Project-based teaching is used in the implementation of classroom teaching. A complete engineering project is designed for students to complete each class. Teaching activities are driven by project tasks [18].

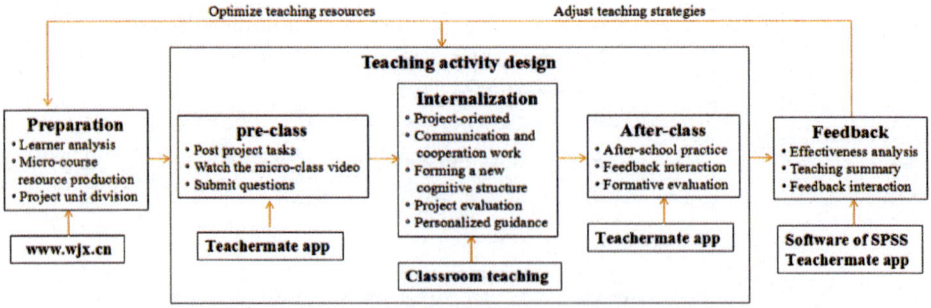

Fig. 1. Basic framework of project-based teaching method based on micro class

3.1 Micro Class Guidance

Pre class guidance can enable students to have a clearer understanding of teaching objectives, teaching contents and curriculum context, and can preview the knowledge needed to complete the project in class in advance. The teacher creates a micro class on the micro teaching assistant platform, and releases the learning tasks to the students before class. Each micro class task corresponds to an important knowledge point. After receiving the learning task, the students independently carry out learning activities in the form of resource viewing, textbook preview, extended reading, etc.,

and organize targeted pre class tests to master the learners' learning situation, so as to optimize offline teaching strategies.

3.2 Example of Project-Based Instructional Design

In the teaching link, take the project as the carrier to carry out effective teaching activities, combine the teaching unit knowledge points of the micro class, design the objectives to be completed in each class as teaching projects, divide the problems to be solved into several interrelated sub tasks, and clarify the duration of each sub task. In the teaching process, teaching methods such as demonstration learning, inquiry learning, autonomous learning, discussion and practice can be used to let students find and solve problems. This learning process has cultivated students' ability to solve practical technical problems and key abilities such as communication and cooperation. Taking the establishment project of the hull model of a chemical ship with unit rib distance as an example, Table 2 gives the list of sub tasks to be completed for the project.

Table 2. List of project subtasks

	Project subtask	Schedule	Teaching method
Part1	Read ship structural drawings	20 min	Publish task
Part2	Explain the basic structure of typical cargo ship sections with teaching aids	25 min	Demonstration learning
Part3	Count the types of plate frames and bone materials in the cargo hold	30 min	Group collaboration
Part4	Propose different modeling ideas	15 min	Guided learning
Part5	Build the model of ship fr79 cross section on the machine	25 min	Autonomous Learning
Part6	Use the editing operation to complete the establishment of fr81 cross section	20 min	Autonomous Learning
Part7	Edit and mesh, and complete the model with one rib pitch	30 min	Autonomous Learning
Part8	Students' self-evaluation and mutual evaluation, and teachers' project summary	15 min	Summary Comments

Development of Project-based Teaching:

1) Release task: release the ship structure drawings, students can read them, guide students to explore the laws of strong and weak cross sections in the ship cabin, and clarify the general task of the project based on this Law: complete the ship cabin model with a specified rib distance (Part1).
2) Demonstration learning: guide students to observe the ship section teaching aids and discuss its structural composition. The teacher will demonstrate the composition of the ship section structure to students in combination with the teaching aids: bottom plate, bilge, side, deck, longitudinal bone, beam and other structures. Guide students to preliminarily build a three-dimensional impression of the hull section structure by referring to the drawings (Part).
3) Group collaboration: let students divide into groups to count the number and size of the chemical ship's grillage and bone according to the drawings, and determine how to set the unit section. In this way, students' understanding of the ship's cabin model will be tested, and the problems found will be discussed and communicated between groups (Part3).
4) Question guidance: combined with the guided thinking questions set in the project, students will clarify the modeling ideas in the form of discussion, question and answer, and guide them to put forward at least two different modeling ideas (Part4).
5) Autonomous Learning: according to different modeling ideas, students can choose suitable methods independently, use solid editing technology and Meshing Technology to give plate frame and bone section parameters, and complete a ship segment model with one rib distance. The teacher gives personalized guidance to different students (Part5–7).
6) Summary and comments: students demonstrate their learning achievements, conduct self-evaluation and mutual evaluation, guide students to discuss, and propose how to extend the model of one rib distance to the whole cabin, leading to the learning objectives of the next class. Summarize and assign after-school thinking questions (Part8).

3.3 Course Assessment and Evaluation

After class, the micro teaching assistant platform publishes the corresponding exercise items and expanded learning materials to students. The platform can count students' learning situation, such as learning progress, learning time, answer time, answer accuracy, etc., and feedback to teachers with visual charts. Curriculum assessment should follow the principles of comprehensiveness, pluralism and objectivity to carry out formative evaluation on students. It is a kind of process evaluation, which evaluates students by stages in the teaching process and runs through the whole process of students' learning. Through various ways such as students' self-evaluation, group mutual evaluation and teachers' comments, students' learning input, knowledge and ability mastery are evaluated in a variety of ways, so as to encourage students to study independently, adjust and improve learning methods in time, and achieve the purpose of improving teaching effect.

4 Teaching Effect and Discussion

From 2019 to 2021, a total of 48 students participated in the course of "ANSYS application practice". The first group is the experimental group, in which 22 students adopt the mixed teaching mode of online micro class + offline project-based learning, and the second group is the control group, in which 26 students adopt the traditional offline classroom teaching mode. After the experiment, course questionnaires were distributed to the two groups of participants.

Table 3. Questionnaire design

Dimension	No	Issues	Average value
Awareness	Q2	I have been exposed to many online micro courses	3.509
	Q3	I have a deep understanding of project-based teaching	3.421
	Q4	I have a deep understanding of online and offline hybrid teaching mode	3.877
	Q5	I think online teaching should be combined with offline teaching	3.953
	Q6	I think the assessment method of this course is very reasonable	3.561
	Q7	The teaching resources of this course are rich and easy to obtain	3.526
Participation	Q8	I am very active in teaching activities	3.860
	Q9	I spent more time on this course before and after class	3.667
	Q10	I can complete all classroom projects within the specified time	3.368
	Q11	I can finish the corresponding exercises easily after class	3.561
	Q12	I am willing to spend time on extended knowledge reading	3.491
	Q13	I am willing to accept every pre class guidance and after class practice	3.877
	Q14	In the study of the course, I got more help and support from teachers	3.451
Satisfaction	Q15	Compared with group study, I like to study independently	3.351
	Q16	I prefer offline learning to online learning	3.544
	Q17	The teaching activities of the course have improved my self-confidence	3.772
	Q18	Course teaching activities have improved my creativity	3.965

(continued)

Table 3. (*continued*)

Dimension	No	Issues	Average value
	Q19	Learning this course is very rewarding and has mastered new skills	3.351
	Q20	The learning method of this course helps me learn other courses	3.491
	Q21	I like the learning method of this course	3.649

There are 21 questions in the questionnaire. Question 1 is the basic information. Question 2–21 investigates the three dimensions of participants' curriculum awareness, participation and satisfaction, as shown in Table 3. Students were asked to use the 5-level Likert scale to state, and use 5, 4, 3, 2 and 1 to represent the five responses of "very agree", "agree", "not necessarily", "disagree" and "very disagree".

Table 4. Reliability and validity test of the questionnaire

Dimension	Cronbach α	KMO	Cumulative % of variance (Rotated)
Awareness	0.923	0.908	72.275%
Participation	0.940	0.895	73.623%
Satisfaction	0.920	0.876	67.873%

From Table 4, we can see that all dimensions correspond to Cronbach α The values are higher than 0.8, indicating that the information of the research item is authentic. KMO values are greater than 0.6, indicating that the data can be effectively extracted information. The cumulative% of variance (rotated) interpretation rate after rotation is greater than 50%, which means that the amount of information of the research item can be effectively extracted. The questionnaire designed in this paper has good reliability and validity, and can be used for further data analysis of the sample.

From the average of the answers to the questionnaire in Table 3, the scores of 20 questions are higher than 3 points, and the average values of questions 5, 13 and 18 in the three dimensions are the highest, indicating that students have expressed positive views on the online micro class and offline project-based hybrid teaching mode; Willing to spend time and effort on the course; It is believed that the curriculum learning adopting this mode will help to improve creativity. The average value of questions 3, 10 and 15 is the lowest, indicating that students do not understand project-based learning enough; The course objectives and tasks are difficult to complete within the specified time; Compared with independent learning, students are more inclined to solve problems in groups. In order to study the impact of this model on students' curriculum awareness, participation and satisfaction, the data of the control group and the experimental group were analyzed by variance.

According to the research results in Table 5, there are significant differences between the experimental group and the control group in the cognition, participation and satisfaction of the course. Compared with the students in the control group, the students in the experimental group have a deeper understanding of the course, a higher willingness to participate in the course, can better complete the course tasks, master relevant skills, achieve the course objectives, and have higher satisfaction with the course. Through interviews with individual samples, we learned that the reason is that the mixed teaching mode of micro class and project-based learning gives learners more room for independent choice. Students of different levels can control the progress of micro class courses according to their learning foundation and acceptance, and micro class can study repeatedly, so that those students who usually respond slowly and are ashamed to ask questions can calmly and repeatedly watch, keep up with the learning progress, and improve their learning self-confidence. The project-based teaching makes students feel full of challenges. In the process of completing the project, students make efforts to overcome the difficulties they encounter. This process makes students full of a sense of achievement.

Table 5. Analysis of variance of control group and experimental group

Dimension	No	Mean ± S.D	
		control group ($n = 26$)	experience group ($n = 22$)
Awareness	Q2	2.81 ± 1.43	3.72 ± 0.78
	Q3	3.00 ± 1.51	3.71 ± 0.72
	Q4	3.09 ± 1.59	4.41 ± 0.56
	Q5	2.83 ± 1.44	3.71 ± 0.76
	Q6	2.91 ± 1.35	4.00 ± 0.95
	Q7	2.87 ± 1.25	3.97 ± 0.83
Participation	Q8	3.04 ± 1.66	4.41 ± 0.56
	Q9	3.00 ± 1.41	4.12 ± 0.84
	Q10	2.83 ± 1.44	3.74 ± 0.79
	Q11	2.87 ± 1.46	4.03 ± 0.87
	Q12	2.91 ± 1.38	3.88 ± 0.88
	Q13	3.13 ± 1.58	4.38 ± 0.60
	Q14	3.13 ± 1.58	4.53 ± 0.56
Satisfaction	Q15	2.70 ± 1.26	3.79 ± 0.95
	Q16	2.87 ± 1.49	4.00 ± 0.85

(*continued*)

Table 5. (*continued*)

Dimension	No	Mean ± S.D	
		control group ($n = 26$)	experience group ($n = 22$)
	Q17	3.00 ± 1.62	4.29 ± 0.58
	Q18	3.35 ± 1.67	4.38 ± 0.65
	Q19	2.61 ± 1.34	3.85 ± 0.82
	Q20	2.78 ± 1.48	3.97 ± 0.80
	Q21	2.96 ± 1.11	4.12 ± 0.81

5 Conclusion

Micro class plays a good role in teaching. It can provide students with an autonomous learning environment to meet the personalized learning of students' knowledge points of different disciplines. It can not only check the deficiencies and make up the omissions, but also strengthen and consolidate knowledge. As an important supplement and expansion of topic learning, it can also promote teachers to improve their professional level and obtain better development.

Based on students' interests and needs, project-based learning takes purposeful activities as the core of the educational process and the basis for effective learning, so that students can actively construct knowledge and formulate corresponding learning plans based on the knowledge and experience they have in their minds, and solve practical problems through practice.

Taking the course "ANSYS application practice" as an example, this paper expounds the implementation method of the mixed teaching mode of micro class and project-based learning, and discusses the organic combination of the two teaching.

modes, which provides a reference for how to effectively carry out online and offline course teaching in practical courses. A questionnaire with good reliability and validity is designed, and the course recognition, participation and satisfaction of learners are statistically analyzed. The results of variance analysis of the sample results of the experimental group and the control group show that the students in the experimental group using this teaching mode have a deeper understanding of the curriculum, a stronger willingness to participate, and a higher degree of satisfaction with the curriculum. This teaching mode is easier to mobilize learners' initiative, give play to their intelligence and subjective initiative in the learning process, and help cultivate learners' independent thinking ability and creativity.

Acknowledgment. This research is supported by the electrical engineering and automation construction project, mechanical engineering and automation professional construction project of the first-class undergraduate professional construction site in Hubei Province, as well as the teaching quality and teaching reform project of Wenhua College (j0900741220).

References

1. Khan, M.H.: A unified framework for systematic evaluation of ABET student outcomes and program educational objectives. Int. J. Mod. Educ. Comput. Sci. (IJMECS) **11**(11), 1–16 (2019)
2. Yufan, D.: Exploration on the precise education path of the integration of production and education in the new era of vocational education groups - taking Guangdong Industry and Trade Vocational Education Group as an example. J. Shenzhen Vocational Tech. College **18**(4), 75–80 (2019). (in Chinese)
3. Paudel, P.: Online education: Benefits, challenges and strategies during and after COVID-19 in higher education. Int. J. Stud. Educ. **3**(2), 70–85 (2021)
4. Nandi, D., Hamilton, M., Harland, J., Mahmood, S.: Investigation of participation and quality of online interaction. Int. J. Mod. Educ. Comput. Sci. (IJMECS) **7**(8), 25–37 (2015)
5. Nandi, D., Hamilton, M., Harland, J.: What factors impact student – content interaction in fully online courses. Int. J. Mod. Educ. Comput. Sci. (IJMECS) **7**(7), 28–35 (2015)
6. Bodyanskiy, Y.V., Tyshchenko, O.K., Deineko, A.O.: An evolving neuro-fuzzy system with online learning/self-learning. Int. J. Mod. Educ. Comput. Sci. (IJMECS) **7**(2), 1–7 (2015)
7. Parslow, G.R.: Commentary: The Khan academy and the day-night flipped classroom. Biochem. Mol. Biol. Educ. **40**(5), 337–338 (2012)
8. Zhongxing, Z.: Summary of research progress on micro-courses and micro-courses. China Med. Educ. Technol. **28**(06), 586–589 (2014)
9. Tiesheng, H., Mingyan, H., Min, L.: Three stages of micro-lecture development in my country and its enlightenment. Distance Educ. J. **31**(04), 36–42 (2013)
10. Wei, L.I., Ying-Zi, E.: Discussion on the application of micro course in the teaching reform of higher vocational education. J. Qiqihar Junior Teach. College **02**, 122–124 (2019)
11. Liu, X., Li, K.: Application research of the microlecture teaching model in the higher vocational education and teaching reform. In: International Conference on Information Technology in Medicine & Education, pp. 441–444. IEEE (2017)
12. Tang, W.F.: On the application of microlecture and flipped classroom in business English teaching in vocational colleges. J. Hubei Open Vocat. College **31**(23), 163–165 (2018)
13. Bilgini, I., Karakuyu, Y., Ay, Y.: The effects of project based learning on undergraduate students' achievement and self-efficacy beliefs towards science teaching. Eurasia J. Math. Sci. Technol. Educ. **11**(3), 469–477 (2015)
14. Jian, X.: Research on the application of micro class in the teaching reform of physical education in higher vocational colleges. Sci. Educ. Art. Collects **04**, 90–91 (2018)
15. Qiyan, X., Guotai, L., Yangming, G.: Research on project-driven teaching mode based on SPOC—Taking the course of "Building BIM Technology and Application" in higher vocational education as an example. Vocat. Tech. Educ. **41**(32), 52–57 (2020)
16. Dawes, J.: Do data characteristics change according to the number of scale points used? An experiment using 5-point, 7-point and 10-point scales. Int. J. Mark. Res. **50**(1), 61–104 (2008)
17. Ling, J., Jin, E.: Research on the effectiveness of online teaching and project-based learning integration practice-taking the course of "modern educational technology application" for normal students as an example. China Educ. Informatization (09), 81–87 (2021)
18. Ngereja, B., Hussein, B.: Does Project-Based Learning (PBL) promote student learning? a performance evaluation. Educ. Sci. **10**(11), 1–15 (2020)

The Empirical Study of Pandemic and War Influence on the University Students' Education

Olena Horina[1], Maryana Karkulovska[1], and Ivanna Dronyuk[2](✉)

[1] Department of General Physics, Lviv Polytechnic National University, Lviv 79013, Ukraine
[2] Artificial Intelligence Department, Lviv Polytechnic National University, Lviv 79000, Ukraine
Ivanna.m.droniuk@lpnu.ua

Abstract. The pandemic's impact on junior students' educational process was analyzed based on a survey conducted by the authors. The study is supplemented by an analysis of the effects of the war on learning based on open data. It is concluded that the conditions of natural disasters, which create both a pandemic and war, have caused and continue to cause significant damage to both the educational process and the psychological state of students. The ongoing online educational process is essential for continuing education and psychological support for young people throughout the educational process.

Keywords: Covid-19 · COVID-19 outbreak pandemic influence · Real war conditions · War Ukraine against Russian · Education at the University · Higher education · Online learning

1 Introduction

The entry of world civilization in a period of significant environmental challenges has led to the intensification of scientific research in this area. The announcement of the WHO pandemic and the significant restrictions caused by lockdowns has caused unprecedented challenges in the world education system; UNESCO created some recommendations supporting education in pandemic conditions [1]. Educators and scholars have been actively looking for ways to maintain an adequate level of education, ensuring equal access to educational opportunities for all population segments. Here an important role was given to the informatization of education. The transition from education to cyberspace during a pandemic has been somewhat easier in university settings, where the experience and motivation of learners are more significant than in high school. However, with the transition of the educational process to the online learning system, higher education institutions are forced to solve complex and diverse problems directly related to the technical aspects of implementing information technology and adapting online learning to the requirements and needs of modern students. All these problems have arisen now in higher education in Ukraine. The material for the study of the impact of the COVID-19 pandemic on the training of IT students was processed in early 2022. It is clear that with the beginning of the full-scale war of the Russian Federation on Ukraine on February 24, 2022, all Ukrainians, particularly teachers and students, faced

Z. Hu et al. (Eds.): AIMEE 2022, LNDECT 159, pp. 490–499, 2023.
https://doi.org/10.1007/978-3-031-24468-1_44

much more significant challenges than those caused by the pandemic, and the relevance of this material decreased sharply. During the month of this war, 869 educational institutions were damaged by bombing and shelling, 83 of which were utterly destroyed as of April 1, 2022. The worst situation is in Donetsk, Kharkiv, Mykolaiv, Sumy, Kyiv, Kherson, Zhytomyr, Chernihiv regions, and the city of Kyiv, according to the Office of the Prosecutor General of Ukraine [2]. However, studies were conducted as a reference to peacetime; when there was no immediate threat to human life, the authors decided to publish this material. The following studies will be devoted to studying the impact of war conditions on students' education. The method of investigation was a direct survey using Google Form. During the inquiry, there were about 180 students in the first year of schooling at the Computer Sciences and Information Technologies Institute at Lviv Polytechnic National University. The survey took place in January 2022. The authors prepared the questions for the study due to some previous publications [3, 4]. The system approach was presented in [5, 6]. The investigation of pandemic influence is improved by new empirical research and some facts about the impact of real war conditions on the high education process in Ukraine.

2 State of Art

The analysis of recent publications shows that pandemic conditions significantly influence education in high schools worldwide. The paper [7] presented a systematic literature review of studies covering online flipped classroom (FC) approaches in higher education during the pandemic. The FC is generally defined as a strategy that reverses the traditional education setting, i.e., the information transmission component of a traditional face-to-face lecture is moved out of class time. Therefore, the FC relies on technology and is suitable for online or blended learning, the predominant form of learning during the COVID-19 pandemic (March 2020–July 2021). The authors analyzed 205 publications in total and 18 in detail. The research has shown the successful implementation of online FC during online education.

As noted in [8], the most influence was on medical students' education because of their education process. This study examined the level of satisfaction of medical students with online lectures and identified factors that affect the assimilation of material remotely. The investigation method was a survey of 114 medical students and 17 lecturers. In [9], Saudi Arabia described the situation with medical specialties students. The 1984 respondents took placed a survey from May to June 2021. Students expressed concern about the inability to study medical specialties and the lack of clinical knowledge gained through online learning. However, some motivated students have coped with these challenges.

The subsequent analysis was realized at the University of Southampton, University of Edinburgh, and University College London [10]. The study convincingly proves that the closure of schools and universities and the rules of social distancing are tough for young people because they were applied during critical periods of social and emotional development in their lives. As a result, the authors argue that young people want to be heard and felt and to play a role in responding to the crisis caused by the COVID-19 epidemic. The paper [11] found that it is optimal to maintain a physical distance of one meter or more; face masks, respirators, and eye protection in public and medical institutions are crucial to prevent the spread of Covid-19. In the investigation [12], participants from 30 August to 30 September 2020 of 7 disciplines in all 16 higher technology colleges in the United Arab Emirates were involved. Research has shown that due to pandemic constraints, most students have a significant burden of psychological stress, which requires further preventive measures to maintain their psychological well-being. The influence of pandemic conditions was studied by Ukrainian scientists [13, 14] prognosis of the impact of pandemic conditions is presented in [18–21]. Multi-Criteria Decision Making based on an approach with the influence of pandemia is presented in [22, 23]. A brief review of the latest investigations shows that pandemic conditions significantly influence the education process in high school worldwide.

3 Investigation of Pandemic Influence

The study was conducted at Lviv Polytechnic National University (NULP). The experimental part of the work (online survey) was attended by 179 students majoring in 122 "Computer Science," mainly in the first and second year of study. The survey was conducted in Ukrainian at the end of January 2022, when students had an examination session according to NULP curricula. The survey was conducted online using Google Forms tools. It contained ten questions about the direct impact of full or partial lockdown conditions and other constraints caused by the pandemic on students' subjective perceptions of learning change.

The first survey question was: "How much has the pandemic increased the number of hours spent at the computer per day?" (Answer: Fig. 1 to 10). The most common changes are from 3 to 6 h (see Fig. 1). There are 86,6% that changes are equal or bigger than 3 h. Compared to the year before, in the authors' survey with 161 respondents, the same changes are comparable at 93,7% (see Fig. 2) [3].

We think that the result shows some adaptation of students to pandemic conditions. The correctness of this assumption is evidenced by the fact that the number of students with the maximum number of increased hours (10 h) decreased from 16,1% to 6,1%. The second question is:" Which area of your life has been most affected by the pandemic?" the answers were: study, work, family relationships, leisure, and group relationships (see Fig. 3).

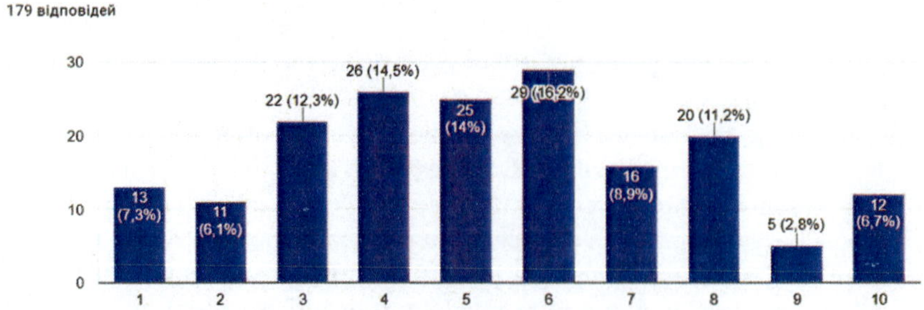

Fig. 1. Illustration of work hours changes with a computer for students during pandemia in January

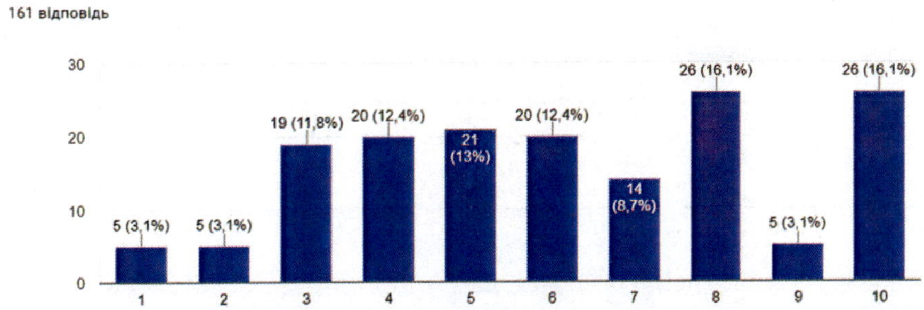

Fig. 2. Illustration of how long work hours are with computers for students after pandemia changes in January 2021 [3].

The results were: study –77,7%; leisure – 11,7%; family relationships –4,3%; group relationships –3,7%; work –2,6%. The results show the significant influence of pandemic conditions for study process for students.

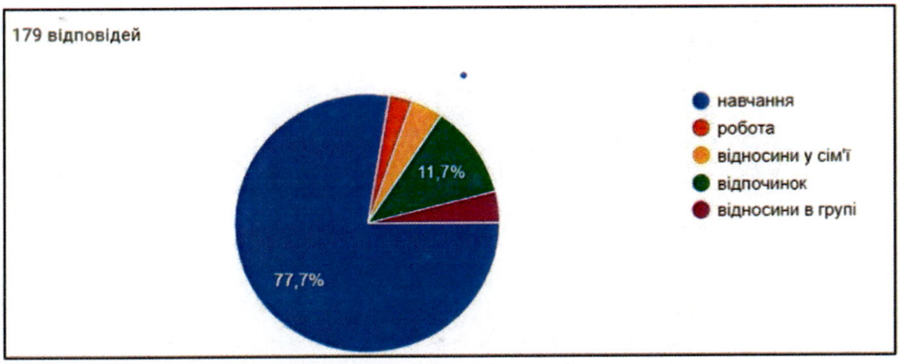

Fig. 3. Impact pandemic condition on students life area.

The third question was about vaccination: "What is your position on vaccination?" For the answer, there were three options: positive, negative, and "I don't know" (see Fig. 4). The positive responses were 57%, negative – 31,3%, and neutral – 11,7%. The answers show that there are many people in Ukrainian with antivaccination positions. So it is necessary to do a lot of explanatory work.

The fourth question was: "How many hours per day has increased your time for self-study or study material (number from 1 to 10)". The answers are presented in Fig. 5.

The maximum hours are three, and the result is better than presented in Fig. 1 because the most common changes are from 1 to 4 h. There are 60,8% that changes are equal to or bigger than 3 h.

The fifth question was about attitude to conducting an online exam. Results are presented in Fig. 6. According to the results, most students (95,6%) did not have an opposing position, and 76% had a favorable position on online controls.

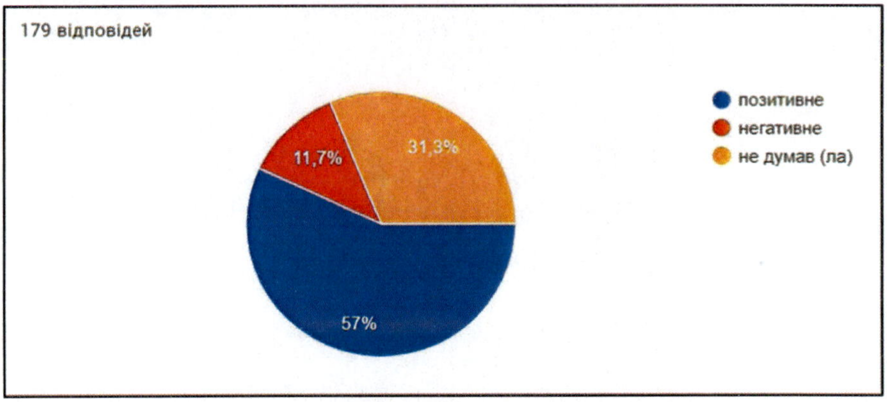

Fig. 4. Impact pandemic condition on students life area.

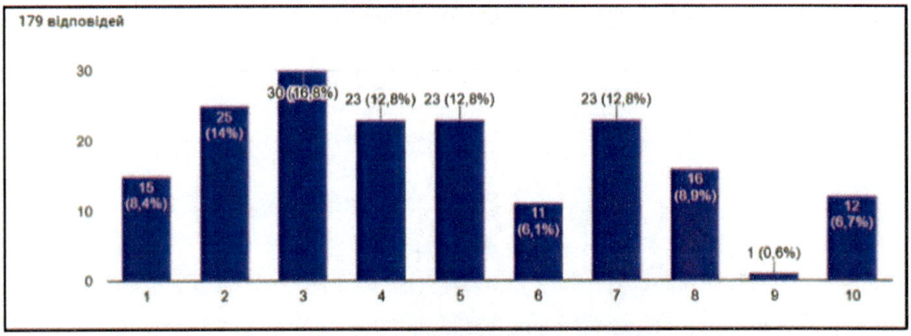

Fig. 5. Impact pandemic condition on students life area.

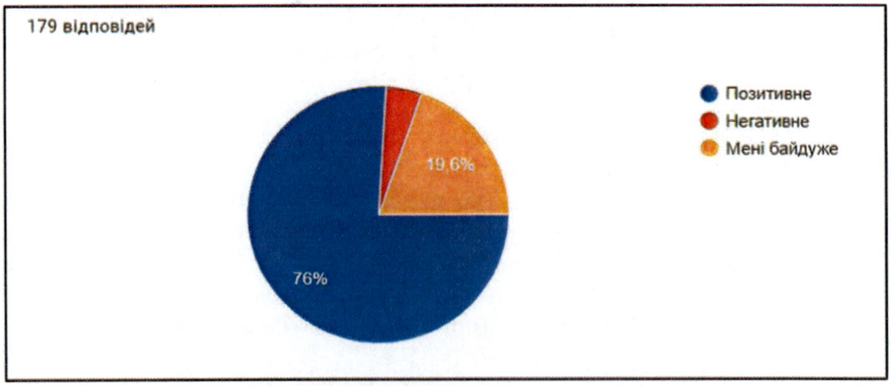

Fig. 6. Impact pandemic condition on students life area.

The sixth question was, "Has the pandemic affected streaming pages?" The goal of the question is to understand a new form of student resting during the pandemia. Results are presented in Fig. 7 and show that the time on streaming pages was increased by 34,4%.

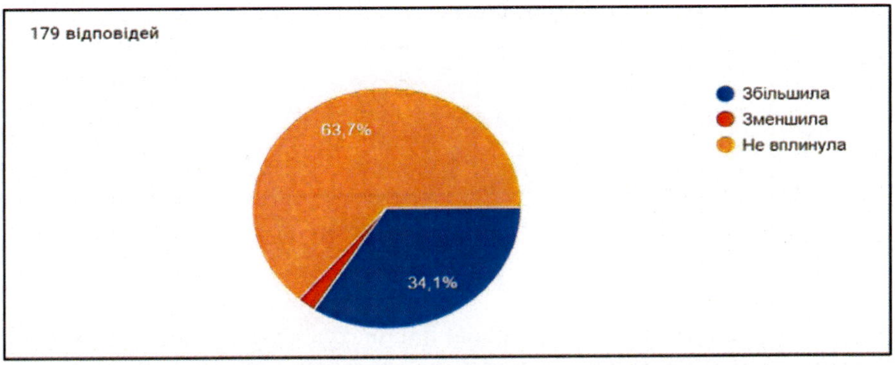

Fig. 7. Impact pandemic condition on students life area.

The seventh question: "Is there a sense of uncertainty about the future due to the pandemic?" 69,3% don't worry about the future, but 30,7% worry due to the pandemic. The eighth question: "What is most annoying in connection with quarantine?" Proposed answers with respondents answer are: Wearing masks (40,8%); Closed cafes, bars, restaurants (16,2%); Impossibility to travel (26,8%); Unable to attend concerts, music festivals (11,2%); Closed museums 5 (2.8%); Closed cinemas 4 (2.2%). Results are presented in Fig. 8.

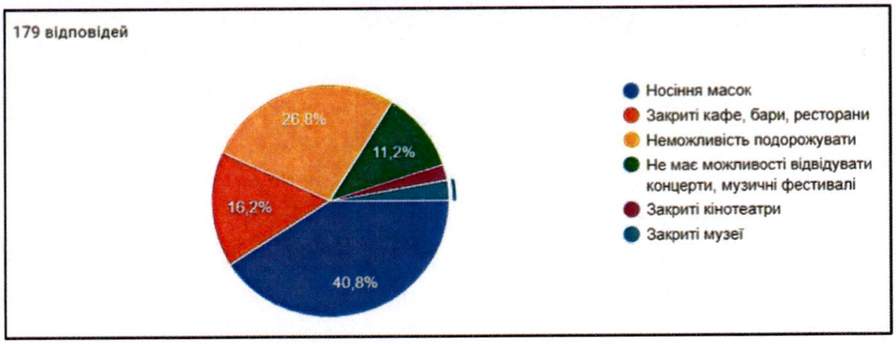

Fig. 8. Impact pandemic condition on students life area.

The mask-wearing is the most discomfort restriction in pandemic conditions. Also, things that connect with rest, vacations, and recreation are essential. For young people, especially students, the possibility of traveling is significant. The ninth question was: "Do you consider yourself sufficiently informed about methods to prevent the spread of coronavirus?" Answers visualization is presented in Fig. 9.

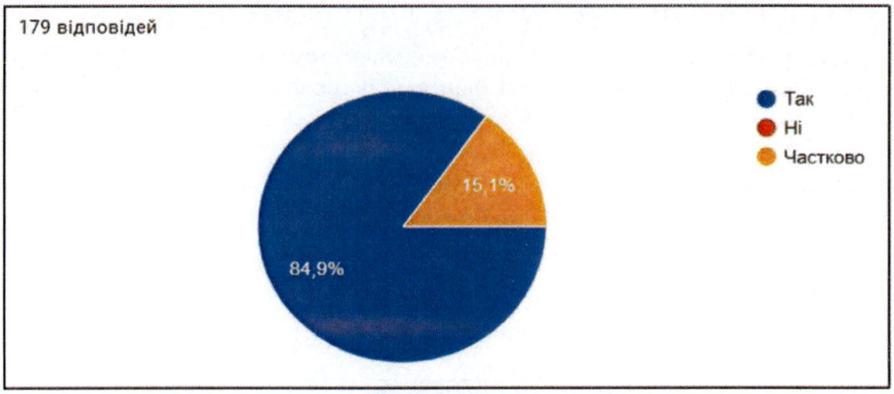

Fig. 9. Impact pandemic condition on students life area.

The answers show no students without information about the Covid-19 spread. In comparison with 2021 results the results are almost the same (yes-84,5%; no-0,6%; partially −14,9%) [3]. The tenth question was about an attitude to online learning; the results are in Fig. 10.

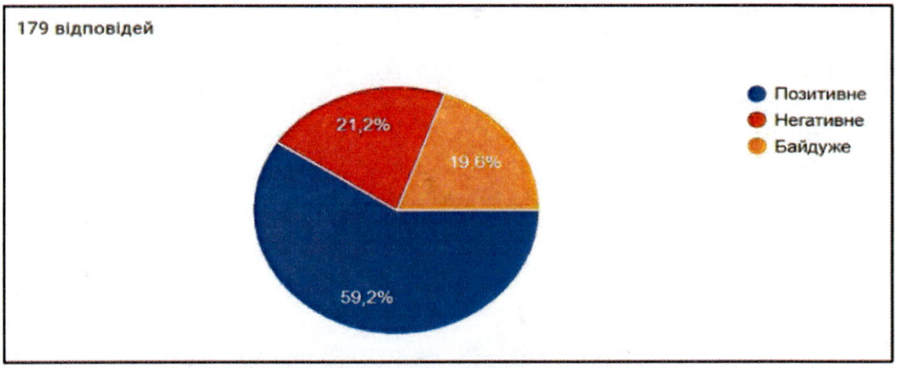

Fig. 10. Impact pandemic condition on students life area.

Most students are favorable to online education. The pandemic conditions have a significant influence on the system of high education. But students and lecturers can adopt the new requirements.

4 Real War Conditions Influence Education in Ukraine

For writing the chapter, the authors used the data from CEDAS monitoring from 24.02.2022 until 20.06.2022 actual during the natural war conditions, 1819 educational institutions were affected by bombings and shelling. One hundred and two of them were utterly destroyed (see Table 1). According to these data, the Kharkiv region suffered the most - 50 educational institutions were destroyed there. In the Donetsk region, 309 institutions were damaged; one was destroyed; in the Sumy region, 45 institutions were damaged; in the Kyiv region - 12 institutions were destroyed; and in the capital, 76 educational institutions were damaged. The data are connected with all educational institutions, not only high education. A severe loss for the higher education system in Ukraine is the partial destruction of almost all of the best Ukrainian Universities system.

Table 1. War impact on educational buildings [17] (date 20.06.2022)

Type of educational school	Damaged	Destroyed
Professional education institutions	801	105
Institutions of professional higher education	102	10
Higher education	28	2
Total all educational types, not only higher	1819	209

Here we presented the Top 5 Ukrainian higher education institutions included in the ranking [16]; four of them are damaged, and two of them in Kharkiv are seriously damaged:

35. Taras Shevchenko National University of Kyiv;-damaged;

63. National Technical University of Ukraine «Ihor Sikorsky Kyiv Polytechnic Institute»;- damaged;

69. Karazin Kharkiv National University; - destroyed;

105. National Technical University «Kharkiv Polytechnic Institute»; - damaged;

106. Lviv Polytechnic National University – not damaged.

The destruction is not only on building the main; the worst part is defeating and destroying people, especially people with high education. The Ministry of Education and Science of Ukraine recommended stopping the educational process from 25 February in educational institutions of all levels and sending students and educators on a two-week vacation. During this time, part of the territory of Ukraine was under temporary occupation, and several towns and villages (Mariupol, Chernihiv, Sumy, Kharkiv, and others) became the scene of active hostilities. More than 10 million people have been forced to flee their homes: 12 million within the country and 6 million abroad. UNICEF reports that more than half of Ukraine's children - 4.3 million out of 7.5 - have been forced to flee their homes due to the war [16, 17]. From March 14, the educational process in areas where the security situation is allowed began to recover—decided where and in what format to conduct classes and hosted regional administrations and educational institutions. For example, according to the Rector's order, the education at Lviv Polytechnic National University started on 14.03.2022 in online mode. Internally displaced persons (IDP) are placed in vocational, professional, and higher education institutions' dormitories. Also, IDPs can work and study at universities in the Western part of Ukraine.

5 Conclusion

The study convincingly proves that despite the challenging conditions of the coronavirus pandemic, the Ukrainian higher education system has overcome these difficulties and stabilized the situation as a whole. When the fight against the consequences of the pandemic had already entered the final stage, and students and teachers had adapted to online learning, Ukraine suffered the next natural disaster, the scale of which far exceeded the previous one. However, educators withstand the enemy's aggression and continue to work in harsh conditions. Huge material and moral damage were inflicted, especially in the East, South of Ukraine, and the city of Kyiv. It is still difficult to fully assess the effects of the war on Ukraine's higher education system. This will be the subject of further research.

Acknowledgment. The authors would like to thank the Armed Forces of Ukraine for providing security to perform this work. This work has become possible only because of the resilience and courage of the Ukrainian Army.

References

1. COVID-19 Educational Disruption and Response (n.d.). General format. https://en.unesco.org/covid19/educationresponse
2. Report of the Office of the Prosecutor General of Ukraine. https://www.facebook.com/pgo.gov.ua/posts/342264237936416
3. Shakhovska, N., Dronyuk, I., Shpak, Z., Klapchuk, M.: An empirical investigation of pandemic impact on IT students' educational schedule. In: Byrski, A., Czachórski, T., Gelenbe, E., Grochla, K., Murayama, Y. (eds.) ANTICOVID 2021. IAICT, vol. 616, pp. 35–40. Springer, Cham (2021). https://doi.org/10.1007/978-3-030-86582-5_4
4. Moiseenko, I., Shakhovska, N., Dronyuk, I., Datsko, O.: Social and economic aspects of the pandemic influence in Ukraine. Procedia Comput. Sci. **198**, 670–675 (2021). https://doi.org/10.1016/j.procs.2021.12.304
5. Kovtun, V., Izonin, I., Gregus, M.: Model of information system communication in aggressive cyberspace: reliability, functional safety, economics. IEEE Access **10**, 31494–31502 (2022). Institute of Electrical and Electronics Engineers (IEEE). https://doi.org/10.1109/access.2022.3160837
6. Trunov, A.: Theoretical predicting the probability of electron detachment for radical of cell photo acceptor. Paper presented at the 2017 IEEE 37th International Conference on Electronics and Nanotechnology, ELNANO 2017 - Proceedings, pp. 353–357 (2017). https://doi.org/10.1109/ELNANO.2017.7939776
7. Divjak, B., Rienties, B., Iniesto, F., Vondra, P., Žižak, M.: Flipped classrooms in higher education during the COVID-19 pandemic: findings and future research recommendations. Int. J. Educ. Technol. High. Educ. **19**(1) (2022). https://doi.org/10.1186/s41239-021-00316-4
8. Yagi, S., et al.: Clinical clerkship students' preferences and satisfaction regarding online lectures during the COVID-19 pandemic. BMC Med. Educ. **22**(1) (2022). https://doi.org/10.1186/s12909-021-03096-7
9. Almarri, F.K., Alshareef, R.I., Hajr, E.A., Alotabi, F.Z.: Impact of COVID-19 pandemic on Saudi medical students' career choices and perceptions of health specialties: findings from a national cross-sectional study. BMC Med. Educ. **22**(1) (2022). https://doi.org/10.1186/s12909-022-03224-x

10. Strömmer, S.T., et al.: Young people's experiences of COVID-19 messaging at the start of the UK lockdown: lessons for positive engagement and information sharing. BMC Public Health **22**(1) (2022). https://doi.org/10.1186/s12889-022-12755-3
11. Chu, D.K., et al.: Physical distancing, face masks, and eye protection to prevent person-to-person transmission of SARS-CoV-2 and COVID-19: a systematic review and meta-analysis. Lancet **395**(10242), 1973–1987. https://doi.org/10.1016/S0140-6736(20)31142-9
12. Mosleh, S.M., Shudifat, R.M., Dalky, H.F., Almalik, M.M., Alnajar, M.K.: Mental health, learning behavior and perceived fatigue among university students during the COVID-19 outbreak: a cross-sectional multicentric study in the UAE. BMC Psychol. **10**(1) (2022). https://doi.org/10.1186/s40359-022-00758-z
13. Chervinskaya, T., Petlenko, Y.: Problems and challenges of distance learning in higher education in Ukraine during the COVID-19 pandemic. Current Issues Philos. Sociol. **27**, 123–129 (2020). https://doi.org/10.32837/apfs.v0i27.932(InUkrainian)
14. Chernysh, N., Prykhodko, T.: What do students think about Ukrainian higher education in the context of the COVID-19 pandemic. Bulletin of VN Karazin KhNU. Series "Sociological research of modern society: methodology, theory, methods," vol. 45, pp.77–85 (2021). https://doi.org/10.26565/2227-6521-2020-45-08. (In Ukrainian)
15. The annual ranking of the best universities in Europe and Central Asia for 2022. https://mon.gov.ua/ua/news/opublikovano-rejting-krashih-universitetiv-krayin-yevropi-ta-serednoyi-aziyi-sho-rozviva-yutsya-na-2022-rik-sered-nih-41-universitet-ukrayini. (In Ukrainian)
16. Yu, N., Kohut, I., Zherobkina, T.: Education and the war in Ukraine (24.03–1.04.2022). https://cedos.org.ua/wp-content/uploads/osvita-i-vijna-v-ukra%D1%97ni.pdf
17. Education is in danger. https://saveschools.in.ua/. (In Ukrainian)
18. Peleshko, D., Rak, T., Izonin, I.: Image superresolution via divergence matrix and automatic detection of crossover. Int. J. Intell. Syst. Appl. (IJISA) **8**(12), 1–8 (2016). https://doi.org/10.5815/ijisa.2016.12.01
19. Izonin I, et al.: The combined use of the wiener polynomial and SVM for Material Classification Task In Medical Implants Production. Int. J. Intell. Syst. Appl. (IJISA) **10**(9). 40–47 (2018). https://doi.org/10.5815/ijisa.2018.09.05
20. Shakhovska, N., Montenegro, S., Kryvenchuk, Y., Zakharchuk, M.: The neuro controller for satellite rotation. IJISA **11**(3), 1 (2019). https://doi.org/10.5815/ijisa.2019.03.01
21. Santra, A., Dutta, A.: A comprehensive review of machine learning techniques for predicting the outbreak of Covid-19 cases. IJISA **14**(3), 40–53. https://doi.org/10.5815/ijisa.2022.03.04
22. Panda, D., Mukhopadhyay, S., Kumar, A., Roy, M.: Multi-criteria decision making based approach to assist marketers for targeting BoPs regarding packaging influenced purchase during Covid-19. IJISA **14**(3), 1–17. https://doi.org/10.5815/ijisa.2022.03.01
23. Pasieka, N., Kulynych, M., Chupakhina, S., Romanyshyn, Y., Pasieka, M.: Harmful effects of fake social media accounts and learning platforms. Paper presented at the CEUR Workshop Proceedings, vol. 2923, pp. 258–271 (2021)

Exploration on Teaching Reform of Railway Container Transport Under the Background of Intelligent Logistics

Qin Yang(✉)

Wuhan Railway Vocational College of Technology, Wuhan 43025, China
yangqin2097@163.com

Abstract. With the emergence of cloud computing, big data, intelligent hardware and other new information technologies and devices, the logistics enterprises need many professional talents. The vocational colleges should conform to the development of the times and cultivate skilled talents for the society. The traditional teaching mode is not conducive to the development of teachers and the progress of students. In order to improve the quality of classroom, the reform and exploration are carried out in the form, content and evaluation mechanism of curriculum. The practical and applied characteristics of the railway container transportation course determine the necessity of the teaching reform. Based on the perspective of intelligent logistics, this paper explores the reform measures of curriculum teaching, which has certain reference value for the teaching. The positive role of teachers, students and assistants in the curriculum has been brought into full play, and the higher order of the curriculum construction has been promoted. Practical results show that the curriculum reform has achieved good results, fostered students' practical ability, and effectively improved their vocational skills and innovative ability.

Keywords: Container transport of the railway · Intelligent logistics · Teaching reform · Implementation of the path

1 Introduction

Logistics industry, as the third profit source industry and the national economic support industry, has developed very rapidly in recent years. With the technological innovation, the change of business model, the development of information technology and the application and popularization of intelligent equipment, the intelligent logistics with "Internet + logistics" as the background arises at the historic moment [1]. There is a huge gap in modern logistics industry for new professional talents, but higher vocational colleges lag behind in the speed and quality of talent training, and cannot timely meet the social demand for logistics talents according to enterprise standards [2–5]. As logistics teachers in higher vocational colleges, we should pay attention to and think about how to train professional talents for social enterprises [6–8]. As the front-line executor of institutional policies and strategies, teachers' professional quality and ability of

Z. Hu et al. (Eds.): AIMEE 2022, LNDECT 159, pp. 500–509, 2023.
https://doi.org/10.1007/978-3-031-24468-1_45

students largely depend on teachers' teaching ability, and teaching effect depends on course teaching [9–11]. Therefore, teachers can use course teaching as the starting point and entry point, and strive to achieve better teaching effect through the optimization of course design, so as to achieve the purpose of cultivating good talents.

The quality of curriculum directly determines the quality of talent cultivation. The education and teaching system centered on student development has been established, and achievements have been made in building a community of teachers and students that focuses on learning and benefits students as well as students [12–14]. Based on the practice for the important role of cultivating the ability of vocational college of applied talents, the only real student centered on curriculum design, breaking the traditional teaching pattern, perfect curriculum system, the effective interaction between teachers and students, enhance students' autonomy in learning, to improve teaching quality, meet the target of talent training in the new period.

2 Background of Railway Container Course

The course of railway container transportation is one of the main core professional courses of railway logistics, and it is a necessary basic knowledge cognition course for railway logistics talents. The main contents of the course include basic knowledge of container, railway container transportation equipment, railway container transportation organization, railway container freight rate calculation, international railway freight combined transportation and container multimodal transportation, etc. This course enables students to deeply understand the basic knowledge of container and the railway container transportation organization theory, and at the same time to master the development law of railway logistics industry has a certain role. Around it with the "way" the implementation of the strategy of a link, "One Belt And One Road" contact between the countries increasingly close, international trade development in full swing, the international logistics development is swift and violent, especially the international railway freight volume growth spurt, Central European trains operation quantity increasing, number of trains operation in central Europe in 2021 reached 15000 columns. Intelligent logistics is more incisively and vividly reflected in international logistics, which plays a very important role in improving the utilization rate of bilateral trade funds. Again shows that the railway industry is very urgent to master the knowledge of container transportation industry talents.

3 Teaching Status of Railway Container Transportation Course

3.1 The Teaching Materials are Lack of the Development of Technology

Railway container transportation course is a new professional core course opened in recent years. There are only two kinds of textbooks on the market at present, and the contents are not systematic and incomplete. Both textbooks refer to the compilation of textbooks on container transportation to describe the definition, concept and classification of containers. The main deficiency of the textbook lies in the incomplete and systematic writing of the contents of the introduction of railway container types, railway container operation organization, international railway combined transportation organization and "hot metal combined transportation". A lot of knowledge points comb not in place, logical confusion. At the same time, with the development of intelligent logistics and technical innovation, the application of automation, intelligent railway logistics equipment, intelligent transportation car, automatic stacking devices such as robots and unmanned aerial vehicles, to popularize and apply in contrast teaching materials are not covered and included in the inside these advancing with The Times of knowledge introduction, clearly obsolete does not conform to The Times development. In summary, teachers are easy to be misled by teaching materials in the teaching process, and it is difficult to clarify the logical relationship between knowledge points. Moreover, due to individual differences of teachers, it is inevitable to ignore the infusion of new knowledge. Therefore, the teaching difficulty is relatively large, the teaching effect is poor.

3.2 Lack of Teaching and Training Resources

Belong to a whole course this course, theoretical part, such as the structure of the container, railway container transport equipment and unitized equipment combined with image understanding to absorb, railway logistics professional students with no mechanical course, cause most of the students to understand the important knowledge points, virtually increased the difficulty of teaching, students due to understand some principles for a long time, gradually lose interest in learning. The practical part of the course requires students to correctly understand container parts and structures, railway container handling machinery and railway container transport vehicles. On the one hand, due to the high price of these equipment, the school needs the equipment only for one major, resulting in low resource utilization rate and low input-output ratio. On the other hand, the relevant equipment of railway container is very large and covers a very large area, so the school cannot provide a large suitable site. Through analysis, it is impossible for colleges and universities to set up the training room base and conditions that match the course of Railway Container Transportation, not to mention the corresponding facilities and equipment of associated intelligent logistics. The nature of the course determines the difficulty of teaching.

3.3 Backward form of Assessment

Lots of courses adopt the method of "final exam score + usual score" to evaluate students' learning effect, and the assessment form is relatively simple. This kind of traditional assessment method has a lot of disadvantages. First of all, it cannot mobilize students' independent learning ability and enthusiasm, which is not conducive to cultivating students' ability to think, analyze and solve problems; Secondly, the lack of stage assessment, most of the students take the final exam before cramming knowledge points, and did not really understand the important knowledge points; In a word, traditional assessment methods cannot correctly and comprehensively evaluate students' mastery of professional knowledge and practical application.

3.4 Teaching Means are Backward and Practical Operation Ability is Lacking

Some teachers in colleges and universities is still in the traditional teaching mode, namely USES the non-computer majors in teachers as the main body of knowledge of classroom teaching mode, to the application of information technology also only stay in multimedia and PPT, one-way infusion teaching process, students mechanical passive to accept knowledge, in the long run, student interest is low, initiative and enthusiasm is not strong. The traditional teaching mode is not conducive to cultivating students' ability of independent thinking, analyzing and solving problems.

As the communicator of knowledge, teachers should not only enrich their own knowledge system, teaching methods and teaching means; It is also necessary to constantly strengthen practical operation ability. For example, the learning of railway container transportation operation process requires teachers to guide students to conduct process training in class. As most professional teachers are from school to school, lack of temporary training in railway related enterprises and practical operation experience, they can only follow the script in class, resulting in the teaching of knowledge "emphasis" on theoretical teaching, "light" on practical operation guidance, and are not able to guide students to practice the actual process according to the actual situation. In a word, the teaching effect is poor.

4 Train of Thought on Reform of Railway Container Transportation Curriculum

The reform idea of railway container transportation course is put forward based on its nature and type. This idea can be summarized as "one main line, four links and six measures", as shown in Fig. 1.

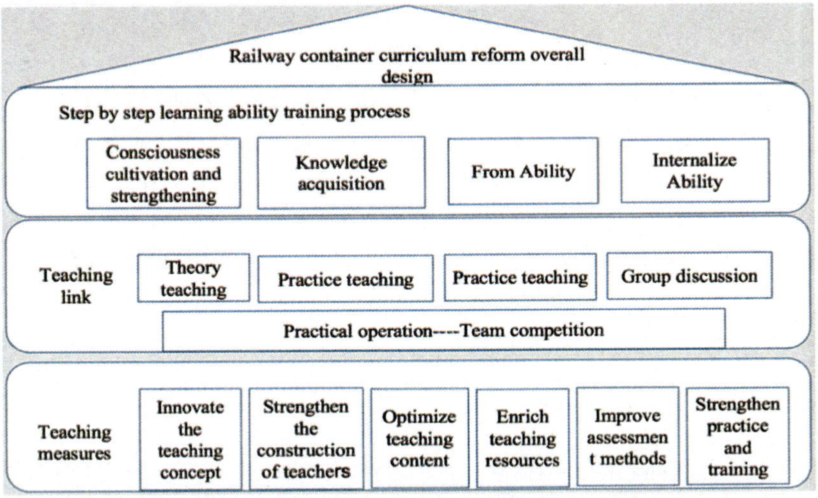

Fig.1. Railway container curriculum reform overall design

4.1 "Main Line"

Background of wisdom logistics the consciousness and strengthen, knowledge acquisition and accumulation, the ability to form and internalization and form a complete teaching activity chain, in view of the cultivation of the students learning ability and form a progressive regularity, the process is a gradual step the process of ascension, takes the process as the main line to the specific design of the course teaching reform.

4.2 "Four Link"

Essentially, curriculum reform is a system engineering problem, with systematic, relevance and other typical characteristics. In order to effectively guarantee and improve the effect of curriculum reform and avoid falling into the traditional mistake of "reform for reform's sake, the end of the course is the end of the reform", it is necessary to bring the closely related follow-up courses or the relevant teaching links of this course into the overall planning and systematic thinking. Specifically, it is necessary to follow the formation process and rules of students' learning and skills [15, 16].

4.3 "Six Measures"

In other words, the students' innovation and entrepreneurship ability can be improved step by step through 7 concrete teaching measures, such as reforming teaching concept, optimizing teaching content, improving the methods means of teaching, perfecting assessment methods, enriching teaching resources, strengthening the construction of teaching staff and strengthening curriculum practice.

5 Teaching Reform Measures of Railway Container Transportation Course

5.1 Smart Logistics Enrich and Optimize the Teaching Resources

There are problems such as the disconnection between theory and reality and the sublimation speed of theory cannot catch up with the development of practical technology which make the current teaching material content faulty. On the one hand, teachers should make full use of network resources to closely follow the development of the railway logistics industry, such as railway freight, logistics information and logistics salon by using wechat public account. The above public accounts will release the new industry trends, new technologies and the latest industry cases at home and abroad every day. Teachers can select the content that fits the course to expand relevant knowledge and teach students in class, so that students can timely understand the industry development trend and new technologies; On the other hand, teachers can collect pictures, equipment videos and enterprise practice cases and process, arrange and make these resources and embed them into classroom learning. Through the above two aspects of operation to improve the intuitiveness and experience of teaching, stimulate students' interest in learning, effectively enhance the teaching effect.

5.2 Diversified Information Teaching Improve the Assessment Methods

From the development of information network technology, teachers should make full use of online resource platforms such as "vocational Education Cloud" and "China MOOC". Teachers can divide the class into groups of 3–5 students to stimulate the students' sense of teamwork and enthusiasm for participation. The group unit is very convenient for the organization of teaching process, which is conducive to the efficient completion of classroom tasks. Before class, teachers can adopt new means and methods to prepare lessons, such as publishing the course guidance cases, key knowledge points and related videos and learning tasks on the platform, among which the guidance cases should adopt more current logistics industry news and practical intelligent logistics application cases of enterprises. In class, the teaching process adopts inquiry-based and open learning, with knowledge points, tasks or cases as the theme. Students report to the group representative according to the materials prepared before class, and explain their understanding of specific task knowledge points. Teachers explain and summarize the specific knowledge points according to the group report and the problems discussed. In addition, teachers should often throw out the latest problems in the industry in class, and guide students to explore the communication; After class, to consolidate and expand knowledge points, teachers should issue thinking questions closely related to the latest development of the industry, require students to complete the data collection and analysis of the problem, and submit the corresponding data conclusions, and consolidate and expand professional horizons through case exercises after class.

The railway container transportation course has a total of 60 credit hours and 3 credits. There are 28 h of theoretical teaching, 30 h of practical teaching and 2 h of review. The course assessment adopts the combination of process assessment and final assessment. The total score of the course is 100 points, and the final score ratio is: closed-book written examination (50%) + classroom performance (10%) + group discussion (10%) + operational ability (30%). as shown in Fig. 2.

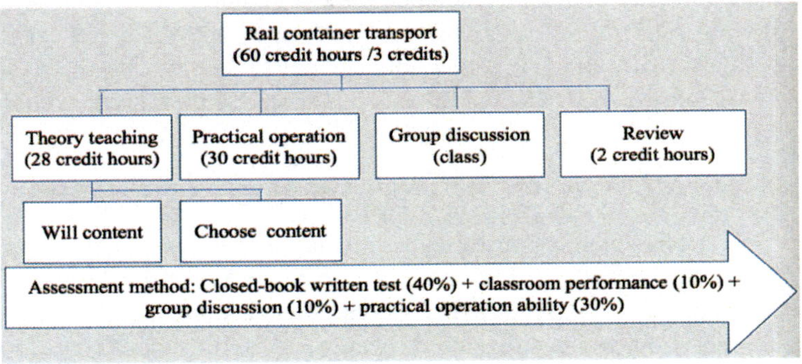

Fig. 2. Railway container transport course overall design drawing

5.3 Smart Training Base Need to Build With the Society

The railway container transportation involves a lot of equipment operation and application. For example, "Basic knowledge of container" requires a container model to explain the structural parts and structural parts of containers for students. Because the operation of "container lock device" is somewhat abstract, it needs "container lock" for real cognition and operation. The module of "railway container transport equipment" can establish a training base in the enterprise, and use the production equipment of the enterprise to carry out theoretical and practical teaching. Colleges and universities can cooperate with relevant enterprises, such as enterprises developing logistics simulation software and producing VR equipment, railway container center station, etc., to build teaching and training bases. Logistics simulation software enterprises can provide the latest software simulation system, students can learn equipment through the simulation system, or through wearable VR equipment, use virtual equipment to learn equipment operation. On the one hand, it can relieve the financial pressure of the school to a certain extent and get the maximum teaching effect with less cost. On the other hand, the co-built training base can also be open to the outside world and actively cooperate with social enterprises. The front-line skilled personnel of enterprises can also be trained in the training base. Teachers can frequently observe and study on site and improve their front-line operation and guidance ability efficiently.

5.4 Teachers Need to Keep Pace with the Times

Higher vocational colleges should cultivate the teachers. On the one hand, strengthen practical cooperation with enterprises, so that enterprises can provide more opportunities for teachers to participate in part-time training; On the other hand, relevant policies are issued to encourage teachers to attend high-level international academic conferences and master the latest information and development trend of the industry. Teachers constantly improve their professional quality through two aspects, and become real "teaching-oriented" and "scientific research" "double-qualified" teachers. Only with the improvement of teachers' ability, can they professionally impart theoretical knowledge to students and correctly guide students to practice and operate, so that colleges and universities can cultivate truly high-skilled talents in line with the needs of enterprises.

5.5 The Reform of Practice Course Should Center on Student Development

From current first-class course construction situation, to improve the quality of courses of intelligent professional practice courses, we need to set out from the practice of the course characteristics, set up the students' development centered guiding ideology, sum up experience and deficiencies, to ensure the quality of teaching to improve the quality of teaching, break the traditional teaching mode, out of the experience type teaching of the comfort zone, Constantly explore and improve teaching methods, improve course design, improve course evaluation system, truly achieve the unity of teaching and learning.

With the promotion of mixed teaching mode, rain classroom is more and more introduced into the theoretical class, and has been widely recognized. We introduced rain classroom teaching tools into the practical courses and designed the classroom in combination with the characteristics of intelligent professional practice, as shown in Fig. 3, effectively realizing the interaction between teachers and students in the whole course of "pre-class, in-class and after-class".

In this rain-based "pre-class and after-class" interaction, teachers and students have achieved effective communication and understanding, breaking the habit of teacher centered classroom, raising the classroom temperature, and making teachers and students make progress together.

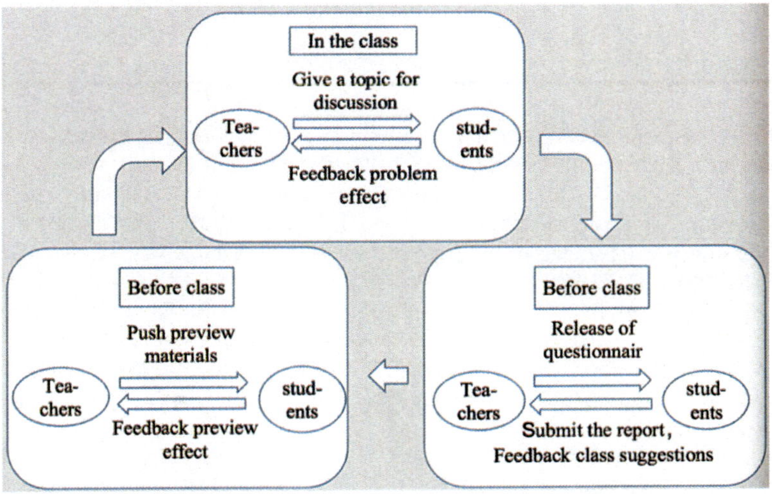

Fig. 3. Teacher-student interaction classroom design based on rain classroom

6 Conclusion

Based on the analysis of the teaching status and problems of "railway container transportation" course, the course positioning under the background of "intelligent logistics" is proposed, and the reform ideas of "1 main line, 5 links and 7 measures" are clarified. The curriculum reform plan of "Facility Planning and Logistics System Design" was designed from seven aspects: innovating teaching concept, optimizing teaching content, improving teaching methods/means, perfecting assessment method, enriching teaching resources, strengthening the construction of teaching staff, and strengthening the practice of innovation and entrepreneurship. And in our school 19, 20 grade railway logistics major for more than two years of teaching and education reform practice. The practice shows that the teaching reform has achieved a good effect, cultivating students' innovative and entrepreneurial consciousness and spirit, and effectively improving their vocational skills and innovative and entrepreneurial ability.

As a core course for integrating theory and practice, railway container transportation focuses on students' practical application and operation ability. In the era of rapid development of intelligent logistics, logistics enterprises have higher standards and requirements for talents in the industry. Higher vocational colleges should start from teaching materials, teachers and teaching conditions, and pay attention to cultivating students' subjective learning ability and practical operation and application ability. Guided by the demand of railway logistics enterprises, the teaching reform should strive to cultivate high-level skilled personnel for the society.

Acknowledgment. This paper is supported by the Projects "Research on Vocational Education Practice System from the Perspective of 'Double High School Plan' by China Vocational Education Society in Hubei Province. No. HBZJ2022050".

References

1. Yan, L.: Discussion on constructing "teaching integration" teaching mode in colleges and universities. Higher Educ. Res. Chengdu **3**, 40–44 (2012). (in Chinese)
2. Li, X., He, M., X, F.: Discussion on application-oriented talent training mode under the background of smart logistics. Logistics Eng. Manag. (03) (2018). (in Chinese)
3. Liu, J.: Research on strategies and patterns of research teaching. Int. J. Educ. Manag. Eng. (IJEME) **2**(12), 67–72 (2012)
4. Liu, S., Chen, P.: Research on cultivation of ethnic minorities IT Talents in Nationalities Universities. Int. J. Mod. Educ. Comput. Sci. (IJMECS) **6**(2), 33–40 (2014)
5. Li, W., Zhou, R., Deng, P., Fang, Q., Zhang, P.: Construction of case teaching model for management specialty supported by information technology. Int. J. Educ. Manag. Eng. (IJEME) **2**(9): 44–48 (2012)
6. Yifeng, N., Guozhong, Z.: On the construction of artificial intelligence curriculum system in colleges and universities. J. High. Educ. Res. **33**(2), 52–53 (2010). (in Chinese)
7. Hassan, M.M., Mirza, T., Hussain, M.W.: A critical review by teachers on the online teaching-learning during the COVID-19. Int. J. Educ. Manag. Eng. (IJEME) **10**(5), 17–27 (2020)
8. Khan, A.I., ul-Qayyum, N., Shaik, M.S., et al.: Study of blended learning process in education context. Int. J. Mod. Educ. Comput. Sci. (IJMECS) **4**(9), 23–29 (2012)
9. He, Y., Xi, R., Yang, J.: Exploration of "architecture structure" application undergraduate teaching reform. In: 8th International Symposium on Social Science, vol. 9, no. 2, pp. 58–64 (2022)
10. Su, Z.: Practice and discussion on teaching reform of situation and policy. In: 3rd International Symposium on Education Culture and Social Sciences, vol. 26, no. 1, pp. 108–117 (2021)
11. Shen, G.: A preliminary study on the teaching reform of analog electronic technology course in vocational college. In: 2nd International Symposium on Education and Social Sciences, vol. 35, no. 7, pp. 207–218 (2021)
12. Chen, P.: Reflections on the reform of electrical automation course in the transformation process of general undergraduate colleges and universities. In: International Conference on Machinery, Electronics and Control Simulation, no. 2, pp. 938–941 (2017)
13. Chen, P.: The research on the reform of the training mode of electrical automation in colleges and universities under the background of the new curriculum reform in Shandong. In: International Conference on Machinery, Electronics and Control Simulation, no. 2, pp. 935–937 (2017)
14. Kanglin, G., Mei, D., Shihu, A., Lifeng, J.: Research on the construction and reform of it practice teaching base in higher learning institutes of finance. Int. J. Educ. Manag. Eng. (IJEME) **1**(4), 53–58 (2011)

15. Zhang, Y.: Management information system. In: International Conference on Machinery, Electronics and Control Simulation, no. 2, pp. 290–293 (2017)
16. Chen, P.: Discussion on the curriculum reform of electrical automation major in newly built universities. In: International Conference on Machinery, Electronics and Control Simulation, no. 2, pp. 942–946 (2017)

Exploration and Practice of Flipped Classroom Based on Micro Teaching Assistant-Taking Ship Propulsion Course as an Example

Lisha Gao and Yanmin Lu(✉)

Mechanical and Electrical Department of Wenhua College, Wuhan 430000, China
gaolisha2022@126.com

Abstract. A classroom teaching interactive tool developed from the perspective of teachers, micro teaching assistants, which are characterized by simple operation, convenient operation and strong interest, have been applied in the course teaching of colleges and universities. In the context of Internet+, in order to cultivate application-oriented talents with strong independent ability, according to the course characteristics and course objectives of the ship propulsion course, the traditional "Teacher centered" teaching mode has been changed. The ship propulsion course has been reformed based on the micro teaching assistant platform, using the flipped classroom teaching mode, and using the micro teaching assistant to improve the interaction between teachers and students, Take "student-centered" and "personalized education" as the basic concepts of teaching. At the same time, introduce the latest developments in the industry and the deeds of experts and scholars in this major to guide students to establish a correct outlook on life and values.

Keywords: Teaching reform · Flipped classroom · Teaching model

1 Introduction

In the industrial age, in order to meet the economic needs, teachers are usually centered and standardized students are trained to meet the requirements of factories. With the development of the times and the rapid growth of information technology, the Internet has been integrated into our life and learning, and teaching activities are no exception. Online learning has become a teaching method that will be used in many cases [1–3]. In order to cultivate applied talents, it is necessary to change the traditional teaching methods and teaching means, implement personalized education [4], and cultivate students' autonomous learning ability, which has been widely valued by the educational community. All kinds of new methods and tools are constantly applied.

As an information platform, with its rich interactive teaching functions, micro teaching assistants have narrowed the distance between students and teachers [5–7]. It allows teachers to grasp the situation of students' preview before class, activate the classroom atmosphere in class, assign homework to students after class, and allow teachers to optimize the course content, coordinate homework assignment and evaluation. Students' are

Z. Hu et al. (Eds.): AIMEE 2022, LNDECT 159, pp. 510–521, 2023.
https://doi.org/10.1007/978-3-031-24468-1_46

learning before and after class can be recorded in the platform, which makes the evaluation of the teaching process more objective. Through micro teaching assistants, teachers can timely understand the students' mastery of the knowledge learned and adjust the teaching progress in time [8, 9]. Using micro teaching assistants can also effectively carry out flipped classroom teaching, which plays a positive role in improving the teaching effect [10].

2 Micro Teaching Assistant Platform and Flipped Classroom

2.1 The Meaning and Function of Flipped Classroom

Flipped classroom originated in the United States. At first, a chemistry teacher used screen recording software to explain ppt while recording videos in order to make up lessons for students who were absent from the classroom, and uploaded the videos to the network. Then they applied this teaching method to the traditional classroom. Before class, students were allowed to watch videos and listen to explanations at home. In class, the teacher tutored the students' questions, Later, this teaching mode was gradually promoted in American primary and secondary education, and attracted people's attention [11].

Flipped classroom has the following characteristics. For example, the videos recorded by teachers before class are relatively short and concise. They are recorded for specific knowledge points and specific problems, usually only for a few minutes, and there is less information about other distractions in the videos. At the same time, students can look back at the videos when they encounter problems they don't understand, and gradually figure out the problems. After watching the videos, You can also test whether you have clarified your knowledge points through questions, and cultivate the ability of self-study [12].

What cannot be ignored is that there are also some problems in flipped classroom. For example, teachers need to produce a large number of videos, need teachers to master such scientific and technological means, and consume a lot of time and energy; Students need to have a certain degree of autonomous learning ability, and different students have different abilities to internalize knowledge. At the same time, teachers cannot master students' learning situation; When students are studying, they cannot discuss with others and lack interactivity [13].

2.2 Flipped Classroom Based on Micro Teaching Assistant Platform

In the "Internet+ " environment, in order to solve the problems of less classroom interaction and poor learning atmosphere in the traditional classroom, the micro teaching assistant platform came into being. Micro teaching assistant is a classroom interaction tool jointly launched by the professional teams of central China Normal University and central China University of science and technology [14–16]. It does not require teachers and students to download software. As long as there is wechat, follow official account, and join the classroom, it can be used. The micro teaching assistant platform has a sign in function, which greatly facilitates teachers, saves the time for teachers to call the roll in the classroom, and can record and archive. The random roll call function of micro teaching assistants injects a dose of cardiotonic into the classroom, attracts students' attention, and brings students back to the classroom. The question answering function of micro teaching assistants helps teachers understand students' mastery of knowledge, while stimulating students' thirst for knowledge. The question ranking also stimulates students' desire to win. The function of micro teaching assistant discussion makes students' thinking more open, expands the teaching content of the classroom, and solves the problem that students in the traditional classroom cannot communicate in time.

For some professional courses with more content and less class hours, and the classroom teaching time is tight, the use of micro teaching assistants can also self-study part of the content of foreign students after class, only analyze and solve difficult problems in class, and effectively use the classroom time to achieve the best learning effect. By using the homework function of micro teaching assistants, we can timely understand the mastery of students' knowledge points.

Secondly, we can build a high-quality teaching team through micro teaching assistants. Using the micro teaching assistant platform, we can let all teachers work together to create a high-quality course, break the dilemma of teachers fighting alone, further optimize teachers' teaching content, so that high-quality resources can be shared, the function of question bank facilitates teachers, and also enriches teaching means.

Sing the micro teaching assistant platform, when designing flipped classroom teaching, you can upload the video of knowledge points to the system in advance before class, and the system automatically records students' learning time, so that teachers can understand students' learning situation. Similarly, other teaching activities of students can also be recorded in the platform, which helps teachers evaluate students in a process.

3 Characteristics and Teaching Status of Ship Propulsion Course

3.1 Characteristics of Ship Propulsion Course

The course "ship propulsion" is applicable to undergraduate students majoring in ship and ocean engineering. The semester of the course is the sixth semester. It is a compulsory professional core course for students majoring in ship and ocean engineering. Before learning this course, they need to learn basic courses of ship major such as ship hydrodynamics and ship resistance.

3.2 Current Teaching Situation of Ship Propulsion Course

The course "ship propulsion" is a comprehensive course combining theory and practice [17]. In the past, the traditional teaching method of "Teacher centered" was used for teaching. Teachers are the protagonists in the classroom and students are the audience. Often due to less class hours and more content, the interaction between teachers and students is poor, students' classroom participation is low, the classroom atmosphere is dull, and the learning effect is poor. There are also some problems in the traditional classroom course assessment method. The total score of the course is generally composed of the usual score and the final exam score according to a certain percentage, but this assessment method cannot comprehensively evaluate students' learning attitude, learning ability and learning effect.

4 Curriculum Reform and Practice of Ship Propulsion

4.1 Teaching Objectives and Implementation Objects o Ship Propulsion Course

The course objective of this course is to help students acquire the necessary basic knowledge of ship propulsion, understand the frontier of discipline development, and master the general methods of designing ship propellers; Guide students to accept the general training of engineering technology design and improve students' engineering technology literacy; At the same time, combined with the latest development of shipbuilding, the current situation of China's manufacturing industry and the examples of advanced figures, cultivate students' patriotism, establish a correct world outlook, outlook on life and values, and imperceptibly help students gradually clarify their learning goals and life goals.

The ship propulsion flipping class using the micro teaching assistant platform has been implemented in two semesters, 2018–2019-2 and 2020–2021-2. The number of students is shown in Table 1.

Table 1. Teaching objects of the course

Semester	Class	Number of people
2018–2019-2	Ship and ocean engineering 1601	28
2020–2021-2	Ship and ocean engineering 1801	15

4.2 General Idea of Teaching Reform of Ship Propulsion Course

The course "ship propulsion" adopts the micro teaching assistant platform to carry out the flipped classroom. In general, it takes "student-centered" and "personalized education" as the basic concept of teaching, and returns the learning subject to the students. The micro teaching assistant platform has built a communication platform for teachers and students, which can solve the problems of "Teacher centered" in the traditional classroom, students' low participation in the classroom and inactive performance, so as to improve students' subjective initiative and make students learn more actively, actively and effectively.

According to the characteristics and objectives of the course, firstly, modularize the content of this course [18], as shown in Table 2. At the same time, micro teaching assistants are used to group students to help solve the problem of weak social skills of some students. Students communicate in groups, discuss together, and jointly complete the tasks given by teachers. In the process of completion, students' interpersonal skills can be enhanced, and at the same time, they can lay a foundation for entering the society, and exercise students' organizational leadership ability and team cooperation ability.

This course adopts flipped classroom [19]. Teachers need to upload multimedia courseware, micro class, pre class exercises, etc. on the micro teaching assistant platform before class to help students preliminarily understand the learning content of the course.

In class, students can gradually master the content of the course by using the guidance questions released before class. After class, students can understand their mastery of the course through exercises and reports. Through the design of a series of teaching activities before, during and after class, the process learning evaluation of students is finally realized.

Table 2. Teaching module

Modular	Credit hours	Teaching objectives
1: overview of ship propeller development	2	Understand the development history and types of ship thrusters
2: geometric characteristics of propeller	2	Master the shape and name of the propeller, the spiral surface, spiral line, and the drawing method of the general drawing of the propeller
3: basic theory of propeller	4	Understand the theory of ideal propeller and ideal propeller, the force and torque acting on the blade, and the hydrodynamic performance of the propeller
4: open water test of propeller model and propulsion performance of real ship	8	Master the similarity law of propeller, critical Reynolds number and scale effect, propeller open water test method and data expression, and understand the self-propelled test of ship model and real ship performance prediction
5: propeller hull interaction	4	Master the various components of propeller wake, thrust derating, propulsion coefficient and propulsion efficiency, and the formula for estimating the interaction coefficient between propeller and hull
6: cavitation of propeller	4	Master the causes of cavitation, cavitation phenomenon of propeller and its impact on performance, and cavitation check
7: strength check of propeller	4	Master the verification method of the specification
8: propeller Atlas Design	12	Master the propeller design problems and design methods, design atlas and its application, hull propeller engine matching problems, and complete the propeller design of the selected ship type

4.3 Practice of Ship Propulsion Curriculum Reform

The reform of the course "ship propulsion" uses the micro teaching assistant platform to carry out inquiry learning guided by problems. Its general structure is to divide into groups first, let students learn by themselves in the form of groups, and cultivate students' self-study ability; Then, according to the students' learning results, the teacher will give targeted guidance, and finally the students will sort out and review the knowledge of this course by themselves. The following will be explained in detail from the aspects of group construction, learning guidance materials, evaluation mechanism, course question bank, course ideology and politics, and implementation steps.

4.3.1 Group Construction

Liu Xianjun proposed in an article that communication must be valued in education. Confucius said, "if three people walk by me, one of them must be my teacher." This fully illustrates the importance of learning knowledge in communication. "The book of rites-learning records" says: "learning alone without friends, lonely and unknown", which also explains the importance of communication in learning. Therefore, in the course construction, group construction is carried out first. According to the students' grades, there are three students in a group, including one with good grades, one with medium grades and one with poor grades [20]..

Each group shall have a clear division of labor according to the learning tasks provided by the teachers, and implement the team leader responsibility system. Under the leadership of the team leader, through the learning and mastering of basic knowledge, the team members take the task as the core, take the problem as the guide, combine the basic knowledge accumulated in the early stage, and accompany the communication and interaction within and between the groups, so as to finally realize the solution of problems and the completion of tasks, and form the internalization process of knowledge.

4.3.2 Compilation of Guidance Materials

In the concept of "personalized education". We should first help students clarify their learning objectives and mobilize their learning initiative. The premise of students' learning initiative is to know why, what and how to learn this course. This requires teachers to explain the purpose, main content and characteristics of this course, as well as its position in the curriculum system.

In order to improve the efficiency of students' autonomous learning, for each class, teachers need to prepare guidance materials in advance to clarify the learning objectives, learning difficulties, learning ideas, inquiry questions, etc. of the class. Students can study independently in advance according to the guidance materials, and if they encounter problems they don't understand, they can discuss in groups or search for materials online to improve their ability of autonomous learning. The compilation of guidance materials should be concise and focused, from easy to difficult, so that students can sort out the context of knowledge. Upload these materials to the micro.

teaching assistant. Through the micro teaching assistant, you can see the browsing situation of students and timely understand the learning dynamics of students.

4.3.3 Improvement of Evaluation Mechanism

This course not only enlivens the atmosphere of the course, improves students' learning initiative, but also changes the previous evaluation methods by using the functions of "micro teaching assistants", such as answering questions, discussing, and answering quickly every day. Reform the curriculum and increase the score proportion of the usual performance of the curriculum. The evaluation of students is no longer just the final exam, but a process oriented formative evaluation. If students answer questions or ask questions in class, they can be recorded in the micro teaching assistant platform in the "answer" mode; In class, after the end of each chapter, teachers will use the "daily quick answer" module to assess students' learning. The top five students will be rewarded; Some chapters require students to make ppt presentations on stage, and the team members cooperate with each other to complete the tasks. The effect of completing the task will include group self-evaluation, group mutual evaluation and teacher evaluation, which will be included in the corresponding scores for each group to fully mobilize students' enthusiasm for learning.

The revised scoring standard of this course is shown in Table 3 below. The overall peacetime score is increased from 30% to 40%, and the original attendance and homework are transformed into procedural evaluation standards. The distribution of final scores of students in the two sessions is shown in Fig. 1.

Table 3. Evaluation criteria

Scoring content	Score (%)		Total score (%)
Examination paper	60		100
Sign in	2	40	
Question answering	8		
Discuss	6		
Finish homework	8		
Student speech	8		
Complete daily quick answer	8		

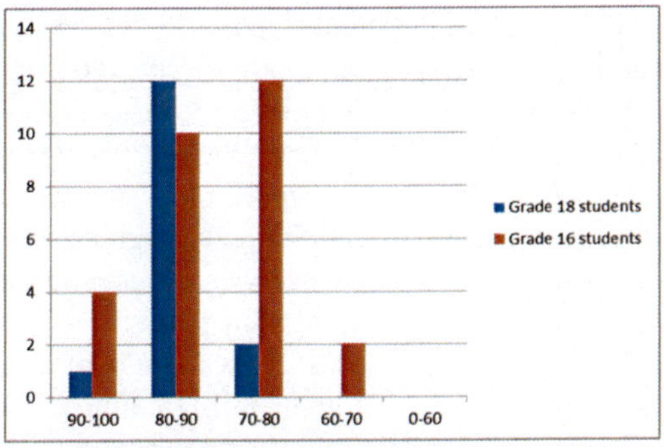

Fig. 1. Student performance distribution

4.3.4 Establishment of Course Question Bank

This course has established a rich question bank to test the effect of students' learning. The question types in the question bank include multiple-choice questions, blank filling questions, judgment questions, calculation questions and short answer questions. The specific proportion of question types is shown in Fig. 2. These questions are uploaded to the "answer" module and "homework" module of the micro teaching assistant, which greatly facilitates the teacher to count the students' usual class performance, and is also conducive to the students' final review. In class, students can see their answer sequence, and teachers can also project students' answers into the projection for students to discuss, which enlivens the classroom atmosphere.

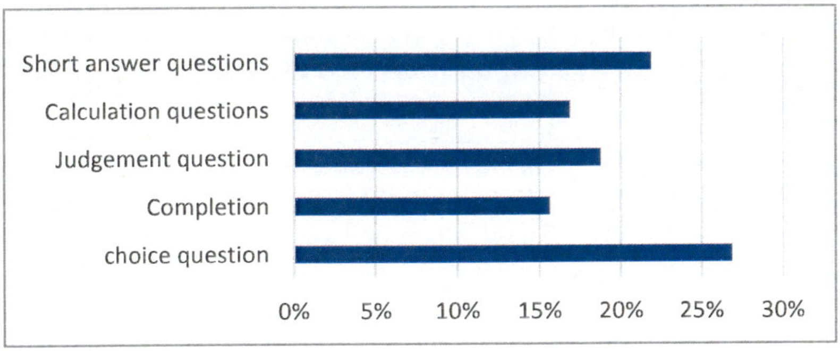

Fig. 2. Question type distribution

4.3.5 Curriculum Ideological and Political Introduction

In the article "personalized education in curriculum teaching", the author points out that the relationship between curriculum teaching and students' learning goals is reflected in two aspects: one is to help students clarify curriculum learning goals; Second, through course learning, help students gradually clarify their learning goals and life goals.

In the course of ideological and political construction, this course selects a large number of video materials, traces and expounds the development process and latest development trends of majors and disciplines from the perspective of scientific and technological progress and changes of the times, tells the career development trends and prospects, enables students to understand the great role of professional fields in national economic construction, scientific and technological progress and social development, and guides students to establish correct three outlooks and pursue their studies and careers with a positive attitude; Through introducing and praising the stories of Peng Shilu, Huang Xuhua, pan Jingfu and other predecessors who have made selfless contributions to the construction and development of China's military ships and submarines, as well as famous experts and scholars in their majors such as Academician Wu Yousheng and academician Ma Weiming, students are guided to cultivate a sense of destiny and historical responsibility of the times; Combined with the vivid and vivid engineering cases related to the contents of the major of ship and ocean engineering in the great power, students can better realize that majors are not only a means of making a living, but also reflect the value and pursuit of life, which will lead students to improve their spiritual realm, shape their feelings of family and country, strengthen social responsibility, and promote the spirit of responsibility.

4.3.6 Implementation of Reform Steps

The ship using the micro teaching assistant platform to promote the curriculum teaching reform is mainly divided into pre class, in class and after class:

Before class: the teacher uploads the guidance materials to the micro teaching assistant, and the students learn by themselves according to the guidance materials and textbooks and courseware, find out the answers to the guidance problems they can solve, mark the problems that cannot be solved, and find out other questions in the process of self-study, and then the team leader organizes them to discuss and solve the problems together.

In class: each group sends representatives to put forward difficult problems in the process of self-study, and the teacher writes the collected problems in the micro teaching assistant. Then, the group will answer the questions in the guidance materials and the difficult problems in the process of self-study. The whole class will discuss and solve the problems that cannot be solved, and then the teacher will call and explain them. At the same time, the students themselves check the knowledge learned in this class, summarize their own gains, and finally the teacher summarizes it, and release test questions with micro teaching assistants, such as multiple-choice questions and judgment questions, to test the students' knowledge mastery. At the end of the course, video is used to introduce the professional development process, the latest development trends or predecessors and industry experts who have made outstanding contributions, so as to arouse students' thinking and improve their spiritual realm.

After class: the homework is released through the micro teaching assistant. There are various forms of homework, such as short answer questions, calculation questions, and sometimes students need to find information online to supplement extracurricular knowledge.

4.4 Problems Encountered in the Implementation of Teaching Reform and Solutions

1) The classroom network is poor, which affects the teaching quality
In the process of using micro teaching assistants, when the classroom network is poor, students are prone to fail to open the operation page when submitting homework or answering questions, thus affecting the quality of teaching. In the later teaching process, you can use the wireless WiFi transmitter and use the network of the classroom desktop to send it to students to solve the above problems. The above problems can be taken into account when building smart classrooms.
2) The classroom atmosphere is too active, which affects the teaching progress
Through the random roll call, rush answer, class discussion, question answering and other functions of the micro teaching assistant, the students' enthusiasm and initiative are greatly stimulated, but it also brings some problems, such as the classroom atmosphere is too active and the discussion time is too long, which affects the teaching progress. For the above problems, I put some simple contents of the course after class, not in class. As the content that students need to learn by themselves after class, I use the micro teaching assistant to assign corresponding homework to understand the students' mastery, and several questions before the next class to test the students' learning.
3) Some students' attention is distracted, which affects the teaching effect
The micro teaching assistant platform is a small program on mobile phones. Some students are easily attracted by the media of "mobile phones" after using mobile phones to answer questions. The official account, circle of friends and QQ messages on mobile phones affect students' attention, thus affecting the teaching effect. To solve the above problems, teachers can draw students' attention back to the classroom and give students appropriate reminders and guidance through random roll calls or questions.

4.5 Teaching Effect Feedback

Using the micro teaching assistant platform to carry out the teaching reform of ship propulsion course has greatly activated the teaching atmosphere and stimulated students' enthusiasm for learning. At the end of the course, the students were investigated and their opinions and suggestions were collected. Most students feel that using the micro teaching assistant platform in the traditional classroom is conducive to their acceptance of new knowledge points, and the mode of group cooperation is adopted, which not only increases the communication between students, but also exercises their ability. Of course, there are also some problems in the implementation process. Some students believe that the classroom atmosphere is too active, affecting the normal progress of class; Some students are unwilling to participate in group communication and complete group tasks; Other students hope that the teacher can further refine the teaching tasks and incorporate the construction of three-dimensional model of propeller into the teaching content.

5 Conclusion

Micro teaching assistant platform is the product of education system informatization. It can help teachers set up a smart classroom in a very short time, carry out classroom teacher-student interaction, and improve teaching efficiency. At the same time of completing the teaching links, the evaluation of the teaching system is realized. It can also save the teaching content for convenient reuse.

Flipped classroom is a kind of teaching mode gradually popularized in the context of the popularity of online teaching. Students can use high-quality educational resources through the Internet, and no longer rely solely on teachers to teach knowledge. The traditional classroom where teachers teach alone has become a place for interaction between teachers and students and between students, so as to achieve better educational results.

Applying the flipped classroom teaching mode based on the micro teaching assistant platform to the actual classroom reflects the idea of "student-centered", making students become the masters of the classroom and teachers become the promoters of students' learning. Through the curriculum teaching reform, students' autonomous learning ability and teamwork ability can be improved, which has an important impact on students' future study and even their career. There is no end to learning. Only by improving their own teaching quality, enriching their own teaching means, constantly summarizing experience and constantly improving, can we achieve the teaching reform goal of "teacher led, student-centered".

Acknowledgment. The research of this paper is supported by the excellent young and middle-aged scientific and technological innovation team plan project "digital twin of intelligent logistics equipment" (t2021043) of Hubei Colleges and universities, as well as the guiding project of the scientific research plan of Hubei Provincial Department of education "Research on noise prediction and control of steel aluminum hybrid structure ships" (b2021337) and the first-class undergraduate professional construction site in Hubei Province "electrical engineering and automation professional construction project".

References

1. Nandi, D., Hamilton, M., Harland, J., Mahmood, S.: Investigation of participation and quality of online interaction. Int. J. Mod. Educ. Comput. Sci. (IJMECS) **7**(8), 25–37 (2015)
2. Nandi, D., Hamilton, M., Harland, J.: What factors impact student – content interaction in fully online courses. Int. J. Mod. Educ. Comput. Sci. (IJMECS) **7**(7), 28–35 (2015)
3. Bodyanskiy, Y.V., Tyshchenko, O.K., Deineko, A.O.: An evolving neuro-fuzzy system with online learning/self-learning. Int. J. Mod. Educ. Comput. Sci. (IJMECS) **7**(2), 1–7 (2015)
4. Brian Jones, G., Kratochvil, D.W., Nelson, D.E., Stilwell, W.E.: Student orientation to an individualized education system. J. Exp. Educ. **39**(3), 4–11 (2015)
5. Hong, L.: Reform and practical exploration of higher mathematics classroom teaching based on micro teaching assistants. Read. Writ. Early (2) (2018). (in Chinese)
6. Ning, X.: Teaching reform and practice of signal and system based on "micro teaching assistant." Electron. World **20**, 107–109 (2020). (in Chinese)
7. Youjun, L., Feng, P., Jiayin, W., Yanjun, L.: Teaching reform practice of data structure and algorithm based on micro teaching assistants. Teacher **19**, 96–97 (2021). (in Chinese)

8. Dong, W.: The application of micro-teaching assistant in large-class teaching. Educ. Teach. Forum **12**, 110–111 (2019). (in Chinese)
9. Wang, Y., Zheng, X., Zhang, T., et al.: Research on the application of "micro assistant" teaching method in imaging practice teaching under the wechat platform. China Continuing Med. Educ. **10**(23), 23–25 (2018). (in Chinese)
10. Hargie, O., Maidment, P.: Discrimination training and micro teaching: implications for teaching practice. Br. J. Edu. Technol. **9**(2), 87–93 (2010)
11. Norman, R.: Eliminating spurious error messages using exceptions, polymorphism, and higher-order functions. Comput. J. **5**, 360–372 (2018)
12. Khan, A.I., ul-Qayyum, N., Shaik, M.S., Maresh Ali, A., Bebi, C.V.: Study of blended learning process in education context. Int. J. Mod. Educ. Comput. Sci. (IJMECS) **4**(9), 23–29 (2012)
13. Yang, J.: Research on college PE teaching reform based on flipped classroom. Int. J. Educ. Econ. **4**(3) (2021)
14. Dong, W.: The application of micro-teaching assistant in large-class teaching. Educ. Teach. Forum (2019)
15. Chen, L., Xu, A., Chen, H., et al.: Research on BOPPPS teaching model of biochemistry integrated with micro-teaching assistant public platform. China Med. Educ. Technol. (2019)
16. Liu, Y.: Application of micro teaching assistants in college ideological and political theory course in Tibet Universities. The Guide of EnCE Educ. (2017)
17. Zhang, B., Zhou, H.: Ship resistance and propulsion" of the excellent engineer training program teaching reform and practice of the course. Mar. Eng. **45**(06), 158–160 (2016)
18. Liu, H.: Research on hybrid teaching reform of building construction and drawing course. Shanxi Constr. **44**(28), 233–234 (2018)
19. Xinglong, Z.: Knowledge internalization process and teaching model design in flipped classroom. Mod. Distance Educ. Res. **55**(02), 55–61 (2014)
20. Xianjun, L.: Personalized education model exploration. Res. High. Educ. **41**(01), 1–8 (2020)

Rational Regression of Student-Centered Classroom Teaching - A Comparative Study of Multimedia Teaching and Traditional Methods in Senior High School Chemistry

Yanyun Li[1(✉)] and Tangbin Liao[2]

[1] Wuhan Foreign Language School High School, Wuhan 430022, Hubei, China
164007607@qq.com

[2] School of Pharmacy, Hubei University of Chinese Medicine, Wuhan 430065, China

Abstract. With the support of computer information technology, classroom teaching is increasingly shifting from traditional blackboard presentation and on-site experiments to multimedia courseware teaching. In this paper, we study the teaching process of both of multimedia teaching and traditional methods in a lecture-based classroom. By studying the different psychological and behavioral characteristics of students in the two teaching methods, we analyze the characteristics and differences of the two teaching methods from qualitative to quantitative and discuss the advantages and shortcomings of each method. Using the principles of pedagogy, psychology, and educational psychology, we explain that multimedia courseware presentations can only be used as teaching aids in a student-centered classroom from multiple perspectives, including student psychology, student cognition, and classroom teaching art, while the rational return of traditional blackboard presentation should be an effective combination of scientific education and humanistic care.

Keywords: Student-centered · Classroom teaching · Multimedia teaching · Comparative research

1 Introduction

Conventional teaching and learning process has its own limitations and does not acquiring the requirements needed by the recent decade engineering education [1]. The rapid development of modern information technology and the widespread use of multimedia tools have opened up more and more possibilities for changes in teaching methods [2]. So active learning has been proposed as a solution for many challenges in the classroom including time management, engagement, motivation, and improved comprehension [3–5].

Z. Hu et al. (Eds.): AIMEE 2022, LNDECT 159, pp. 522–532, 2023.
https://doi.org/10.1007/978-3-031-24468-1_47

Due to epidemic prevention, the use of modern information tools, to some extent reduced the influence of all kinds of teaching through online courses, large scale online classes made the master class not limited to the top of the ivory tower [6, 7].

The most used IT tool in the lecture-based teaching is the slides with video and animated simulations. Having been spread rapidly and widely, the Internet has become an important platform for information dissemination and transmission. Every year, more universities are starting online programs [8–10]. What are the advantages and problems of using this kind of courseware or online programs in classroom teaching compared to the traditional teaching method which is based on the teacher's writing and explanation? In order to provide a basis for the reasonable application of appropriate teaching methods in actual teaching, the author takes high school chemistry lecture-based classroom teaching as the research object, obtains first-hand information through questionnaires, post-class feedback, peer exchanges, and centralized discussions, and conducts research through the aspects of teacher psychology, student psychology, teaching effectiveness, the development of students' learning habits and the cultivation of scientific attitudes, in order to provide a basis for the reasonable application of appropriate teaching methods in actual teaching [5, 11, 12].

2 Student-Based Comparison and Analysis

2.1 Comparison Based on the Student Perspective

Due to the particularity of teaching and research, it is impossible to achieve complete parallel comparative analysis [13–15]. Instead, it should be done in the ways shown in Table 1.

1) adopting the above different modes of teaching for two parallel classes when teaching the same content; and
2) adopting the above different modes of teaching for the same class at different stages of teaching. A comparative study was conducted.

According to the post-class survey and students' discussion, the main difference in students' listening habits between the two modes of teaching can be attributed to the difference in taking notes during the class [16, 17].

Table 1. Comparison of the two teaching models based on the students' perspective

		Conduct classroom teaching around multimedia courseware	Traditional blackboard presentation
Student	During The Classroom	1. Clear, but sometimes visually fatiguing 2. I am not worried that taking notes will interfere with my listening, I feel that I can do without notes and make a copy of them after class instead of taking notes 3. I wanted to take notes, but the teacher was talking too fast to keep up, so I had to give up taking notes and talk about it after class 4. I feel that the teacher has been staring at everyone, didn't dare to make small movements, but it feels easier to be distracted 5. Feels that the classes are following the same pattern, even the 'surprises' are pre-determined by the teacher	1. The teacher's handwriting sometimes is not very recognizable, but it is fine when you get used to reading it. Some teachers have very distinctive characters 2. It is too tiring to take notes, but I can't stop doing it 3. I can see the teacher's typo 4. The teacher writes and reads in bits and pieces, and I hear it all 5. The teacher turns his head and writes on the board, so I can make small movements
	After class	1. I felt that the teacher had talked a lot, but I had forgotten everything 2. I didn't have time to take notes, or I was too lazy to do so, and I had too many other things to do after class to go to the teacher to make a copy 3. I copied the teacher's multimedia classroom materials, but don't have the same feeling of teacher's teaching when review it	1. The teacher's handwriting and the knowledge to be remembered were blended 2. It was still convenient to make my own notes and see my own notes, and to recreate the class occasionally

2.2 Statistics and Comparison of Students' Behavior

Students are both the subject and object of education [18–20]. Teaching work should be carried out around students [21]. Statistical analysis of students' classroom note-taking was conducted in two types of classes in the same school under different teaching modes [22].

Students in Class A and Class B were students in high school classes classified according to their entrance scores. The enrollment scores of Class A classes were higher, with four classes totaling 202 students, and the enrollment scores of Class B classes were relatively lower, with four classes, totaling 186 students.

The summary statistics are shown in the Fig. 1 and Fig. 2.

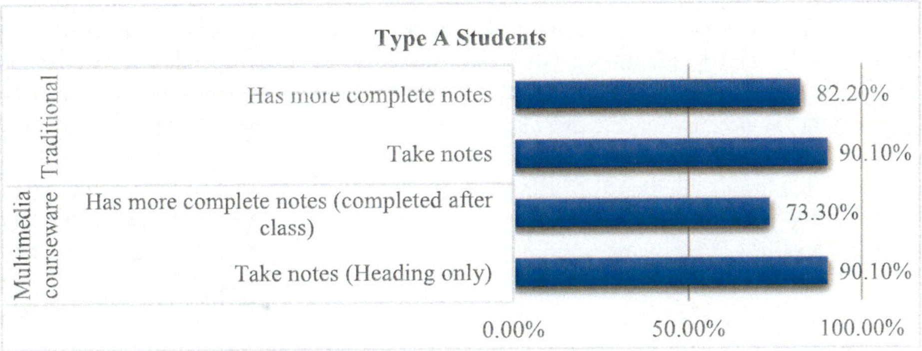

Fig. 1. Type A students' note-taking performances

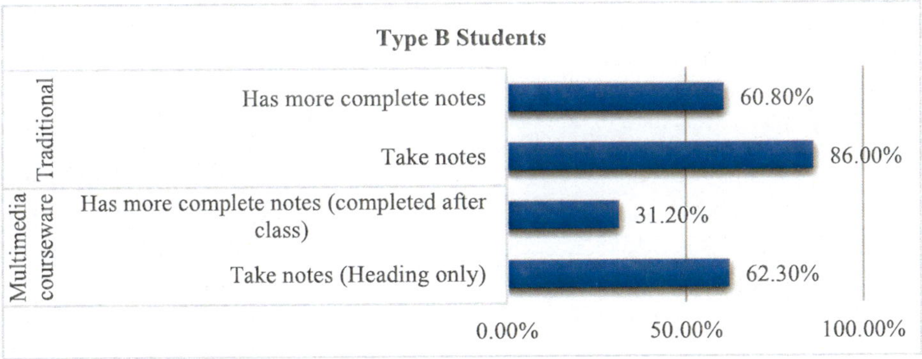

Fig. 2. Type B students' note-taking performances

The statistical comparison chart shows that in the traditional teaching mode, the proportion of students who can take good class notes is much higher than in the mode of using multimedia courseware to start teaching, especially for some students in Class B, who are relatively less self-motivated and capable of learning.

2.3 Analysis of Student Behavior and Learning Processes

According to Table 1 and chart 1, it can be seen that in the traditional teaching mode based on teachers' explanation and blackboard presentation, students' initiative to take notes is much greater than in the teaching mode based on the presentation of courseware. The main reasons are as follows: firstly, subjectively, because of the multimedia courseware can be preserved and repeated, which makes the students psychologically slack and dependent, believing that they can simply make a copy when necessary. The survey found that the greatest difference in classroom note-taking between the two modes of teaching was found among students with moderate levels of effort. Secondly, objectively, teachers' sense of time illusion, resulting in students' inability to complete the notes. The author recorded the writing speed of many teachers and students in classes: the teacher blackboard writing speed, on average, is about 30 to 35 words per minute, while the speed of students taking notes, is about 20 to 25 words per minute. If the slide directly expands on the key points in more than twenty words, the teacher needs to wait for the students to complete their notes for more than a minute. Even if the teacher intentionally allows time for notetaking, the atmosphere of the classroom and the teacher's state of mind at the lectern can affect the teacher's perception of time, making it difficult to achieve a minute of silence from time to time to allow sufficient time for students to complete the process. Making up notes after class, on the other hand, requires a high level of student self-motivation and is subject to teacher and student timetables. These contextual factors determining their choice of learning approach they choose to use when preparing and studying for the particular course based on their beliefs of how different types of knowledge might be more or less useful for achieving their learning goals.

Thus, the direct presentation of courseware reduces the need for students to take notes in class. In terms of learning strategies, listening while note-taking is a multi-sensory engagement process. From the perspective of information processing, note-taking is also a way of recoding knowledge and thus facilitating its recognition and internalization. In addition, teachers are used to briefly writing on the board or slowing down or repeating to emphasize important points in teaching, while students take notes to deepen their impressions and review their understanding. However, when teachers create multimedia presentations of difficult points, students often do not have enough time to take notes in class.

For some students, the habit of taking classroom notes is discarded, and often the concentration in class is greatly reduced, while the process of reproducing, understanding, and remembering knowledge in the note-taking process is also lost, and the effectiveness of copied classroom content in applying it to revision and consolidation is no better than that of the notes they have taken.

A state of learning is the unity of the hand and the brain. In the classroom, students have to hear, see, write and think, which takes time and process, and the teacher's own writing on the board gives a better visualization of the time it takes for students to complete this process, which in effect controls the pace of the class.

3 Teacher-Based Comparison and Analysis

3.1 Comparison Based on the Teacher's Perspective

Observations of teachers' regular classroom instruction revealed that teachers who had been in the profession for a shorter period relied more heavily on multimedia courseware. They would choose to use pre-made courseware to launch their teaching. In contrast, teachers who had taught for more than six years (two rounds of high school teaching) used significantly less multimedia courseware in their classroom teaching. Through interviews and questionnaires, teachers' general evaluations of the two teaching modes were collected here.

Table 2. Comparison of the two teaching models based on the teacher perspective

		Conduct classroom teaching with support of courseware	Traditional blackboard presentation
Teacher	Before class	1. Spend more time on courseware and look for online references to open up the mind 2. Need to check the courseware repeatedly to prevent writing and other mistakes 3. Need to be familiar with the order of courseware projection order and effect	1. Paper lesson plans should be slightly more detailed 2. It is necessary to do a more detailed blackboard writing design in advance to ensure that the classroom content can be fully, quickly and beautifully presented on the blackboard
	Classroom	1. Don't worry about careless typo 2. Easy to use, accurate charts and graphs to visualize microscopic models 3. Even the content of the parallel relationship must be carried out in the order of multimedia courseware performance 4. It was not easy to adjust when a better idea suddenly came to the classroom 5. Lack of flexibility for different classes 6.6.6.6.6.I always feel that the multimedia classroom is the main stage, and I am like the narrator of the multimedia classroom	1. Copying examples was a pain in the neck and took time and may have some typo. Some tables and diagrams were not well drawn 2. The focus and order of the lecture can be adjusted according to the students' situation and the reality of the class 3. Sometimes a better way of explanation and angle comes to mind and brings both you and the students into the best atmosphere 4. Can 'flip' the teaching at certain 'points' according to the actual situation of the students in the classroom
	After class	1. Adjust the multimedia courseware to address problems in the classroom is hard to save the track of the history wording to remind the teacher in the future teaching 2. Easy to summarize and organize	1. Because the paper teaching plan is more complete, according to the classroom teaching situation annotation, modification, there is a more perfect teaching practical material 2. The summaries are not organized systematically enough

3.2 Comparison Based on the Teacher's Perspective

Table 2 also reveals a few differences between the teacher's perspective of 'teaching chemistry around multimedia' and the 'traditional board-based, lecture-based chemistry'.

Macro-awareness and micro-analysis are one of the dimensions of core literacy in the chemistry curriculum. It is clear that the use of multimedia classroom materials is a good aid to secondary school students in their initial exploration of microstructures and can solve some of the problems of lengthy blackboard presentation; the diagrams presented on the PPTs are neat and tidy, and the relationships between substances in calculation problems can be sorted out more clearly, but they tend to rigidize the content of the class and restrict the form of teaching. Teaching with courseware more emphasis on tangible preparation for the lesson.

The use of traditional blackboard presentation as the main form of classroom teaching also has several advantages. It has the ability to demonstrate the thinking process well, the ease of showing the focus of the class in detail, and the flexibility to make various adjustments according to the classroom situation. In the case of exercises and more complex problems, the atmosphere in the classroom sometimes allows the teacher to come up with better ideas or better angles of analysis, so that the teacher can adjust the teaching methods in time.

Adequate preparation and the preset of the teaching process is a prerequisite for a good lesson. The modern view of teaching and learning emphasizes the creation of realistic scenarios, and the use of teaching courseware is a good way of showing this "thorough preparation" and "pre-determination".

'Teaching' and 'learning' are a process of mutual interactions. And class with the 'interactions' needs teachers to constantly adjust according to the actual situation of the teaching process and adopt the most appropriate teaching methods to achieve the best teaching effect. As each student is a different and thoughtful individual, a full preparation before class is necessary for the whole teaching process and the teacher should be able to make timely adjustments to the classroom situation. They can accomplish certain tasks and solve some problems together by communication, argument, and discussion, and then formatting a richer, more flexible understanding,. However, the use of teaching materials often makes the classroom teaching process bound by preparation and pre-determination, and the "dynamic" classroom is instead rigid because of adequate preparation and preset.

4 Overall Comparison

Here is a summary comparison of these two teaching models, combining the student perspective and the teacher perspective, is presented.

A questionnaire survey was conducted to find out students' feelings about the classroom and post-classroom effects of the two teaching modes. A total of 378 valid questionnaires were counted. Figures 3 and 4 show the preferences of different teaching methods in class and before exams.

Table 3. Strengths and weaknesses of the two teaching models

	Using multimedia courseware for classroom teaching	Traditional blackboard presentation
Pros	1. Clear, unambiguous, and fast 2. Standardized and accurate 3. Easy to reproduce content 4. Emphasis on tangible preparation for lessons 5. Can simulate the microstructure and microscopic particle movement and reaction process visually	1. Personalised writing is good for students' memory 2. More room for adaptation 3. Showing the process of thinking 4. Allowing space for students 5. Emphasize the teachers' intangible pre-class preparation and practice
Cons	1. Easy to make students inert 2. Easy to cause visual fatigue 3. It is difficult to adjust according to the actual class situation 4. Free time in class is difficult to control	1. Personal writing habits may cause confusion to the students 2. The charts are difficult to be standardized and neat 3. The presentation of examples takes a long time 4. Random typo rate is relatively high

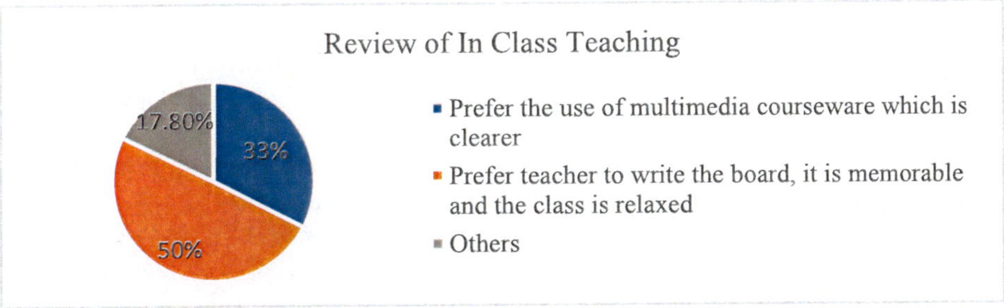

Fig. 3. Preference of different teach method in the class

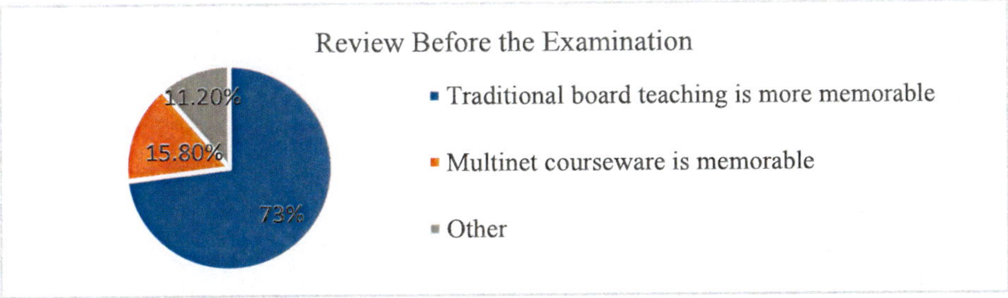

Fig. 4. Preference of different teach method before the exam

Appropriate "blank space" is the art of classroom teaching. The appropriate pause of teachers in class gives students the time and space to think about the knowledge point, and gives students the opportunity to adjust their thoughts, especially for the students who occasionally wander off and fail to keep up, which is the most practical care in class. When the teacher writes on the board, taking eyes off the students, also give the students time to adjust and release, so that the atmosphere of the classroom becomes relaxing, which is a wordless understanding and communication.

A right amount of beautiful, personalized blackboard writing will leave a lasting impression on students and is a silent communication with them. Every teacher has different writing habits and certain habits will leave a lasting impression on students, even to the point of becoming an organic part of their understanding and memorization of subject content. Teachers deepen the impression of students through the change of blackboard writing font, size and writing style, which often makes people feel vivid. Psychologically, it facilitates students' unconscious remembering, which is a long-lasting memory that is also easily achieved with the least burden on students.

Table 3 and Fig. 2 show that the two modes of teaching have their own strengths and weaknesses, but it is clear that in a lecture-based chemistry classroom, in terms of the overall effect of class notes and post-lecture revision, the traditional board book teaching is more advantageous, and courseware should only be used as an aid to teach.

5 Conclusion

The difference of opinion can be explained by the difference in course content. The traditional blackboard presentation is easier for the teacher to control the classroom and in most cases conforms to the cognitive characteristics of the students. In deep learning, course content is related to one's life context, new ideas are integrated with previous knowledge and experience and an underlying desire for intellectual mastery and understanding exists with less focus on assessment requirements.

The support of multimedia courseware can be used as an effective teaching aid, as a supplement and improvement to the blackboard script, and can play a better role in students' understanding of abstract microstructures and microscopic particle movements. At the present time of rapid development of computer science and information technology, while modern classroom education technology assists teaching, we should study students' learning psychology and cognitive characteristics, and look forward to the rational return of the traditional teaching model in a student-centered way.

Acknowledgment. This project is supported by Hubei Provincial Education Science Planning 2018 Key Project "Research on the Practice of Deep Teaching in High School Based on Core Literacy" (No. 2018JA095).

References

1. Fatima, S., Abdullah, S.: Improving teaching methodology in system analysis and design using problem based learning for ABET. Int. J. Mod. Educ. Comput. Sci. **5**(7), 60–68 (2013)
2. Shi, L.J., Yao, Y.J., Xia, X.Y., et al.: Practical teaching staff construction under new situation. Int. J. Educ. Manag. Eng. (IJEME) **1**(2), 57–61 (2011)

3. Kong, L.A., Wang, X.M., Yang, L.: The research of teaching quality appraisal model based on AHP. Int. J. Educ. Manag. Eng. **9**(29), 29–34 (2012)
4. Adam, L., Di Anthony, B.C., Roderick, B.J., Sebastian, Z.: Teaching: how to ensure quality teaching, and how to recognize teaching qualifications. Ind. Market. Manag. **100**(3), 1–5 (2021)
5. Heather, L., Kate, P.H.: Worlds collide: traditional classroom meets online learning. J. Hospitality Tourism Res. **45**(5), 924–926 (2021)
6. Zheng, X., Tingting, H., Hongyu, D., Xiaohui, Z.: Strategies for enhancing teachers' online teaching based on online learning acquisition for college students. In: Proceedings of the 17th International Conference on Innovation and Management **35**(8), 738–742 (2020)
7. Rusli, M.: The effects of various modes of online learning on learning results. Int. J. Adv. Comput. Sci. Appl. (IJACSA) **11**(4), 100–105 (2020)
8. Feng, Z., Zheng, F., Yang, D.: A new exploration into the educational function and educational mode of chemical experiment. Middle School Chem. Teach. Ref. **9**(1), 1–5 (2021)
9. Zhang, Q., Hu, X., Red Bayberry. Teaching strategies for developing "Scientific Attitude and Social Responsibility" literacy. Middle School Chem. Teach. Ref. **11**(9), 10–13 (2021)
10. Dada, E.G., Alkali, A.H., David, O.: An investigation into the effectiveness of asynchronous and synchronous e-learning mode on students' academic performance in National Open University, Maiduguri Centre. Int. J. Mod. Educ. Comput. Sci. (IJMECS) **11**(5), 54–64 (2019)
11. Rawiphon, C.: Development of an online learning management framework for vocational students in Thailand in the new normal under the COVID-19 situation. Int. J. High. Educ. **11**(1), 151–153 (2021)
12. Scully, J.R., Hutchison, M., Santucci, R.: EDITORIAL: the COVID-19 pandemic, Part 2: understanding the efficacy of oxidized copper compounds in suppressing infectious aerosol-based virus transmission. Corrosion -Houston Tx **77**(4) (2021)
13. Al-Hagery, M.A., Alzaid, M.A., Alharbi, T.S., Alhanaya, M.A.: Data mining methods for detecting the most significant factors affecting students' performance. Int. J. Inf. Technol. Comput. Sci. (IJITCS) (5), 1–13 (2020)
14. Newell, B.: Introduction: surveillance and the COVID-19 pandemic: views from around the world. Surveill. Soc. **19**(1), 81–84 (2021)
15. Voronov, V., Gerashchenko, L.: Cognitive barriers in training the students of higher education institutions, methodology for their elucidation and overcoming. Int. J. Mod. Edu. Comput. Sci. (IJMECS) **2**, 1–11 (2013)
16. Shani, Z., Tali, T.: "We know our students": teacher views of inquiry-based learning implementation in light of their working context. Proc. Singapore Nat. Acad. Sci. **15**(02), 91–104 (2021)
17. Coulter, R.W.S., Lindsay, O.: Using student-led discussions and snapshot lectures to stimulate active learning and accountability: a mixed methods study on teaching an implementation science course. Pedagogy Health Promot. **8**(1), 30–40 (2022)
18. Patrio, C.P.H., Tarloff, I.S.W., Hung, S.C.: Disciplinary variations in student perceptions of active learning classrooms. Int. J. Educ. Res. Open **11**(3), 1–7 (2022)
19. Howell, R.A.: Engaging students in education for sustainable development: the benefits of active learning, reflective practices and flipped classroom pedagogies. J. Clean. Prod. **325**(1), 1–11 (2021)
20. Dada, E.G., Alkali, A.H., David, O.: An investigation into the effectiveness of asynchronous and synchronous e-learning mode on students' academic performance in National Open University, Maiduguri Centre. Int. J. Mod. Educ. Comput. Sci. (IJMECS) **11**(5), 54–64 (2019)
21. Jessica, H., Kay, S.M.: Constructing goals for student learning through conversation. Math. Teacher **113**(11), 904–909 (2020)

22. Sal, C.: Practitioner research in a UK pre-sessional: the synergy between exploratory practice and student motivation. J. Engl. Acad. Purp. **57**(5), 1–6 (2022)
23. Kaviyarasi, R., Balasubramanian, T.: Exploring the high potential factors that affects students' academic performance. Int. J. Educ. Manag. Eng. **8**(6), 15–23 (2018)

Integrated Teaching Improvement Under the Background of Intelligent Manufacturing-Taking Engineering Graphics as an Example

Wenqi Wang, Yanan Ye, Sha Hu, Siyang Zhou, and Yan Bi(✉)

WenHua College, Wuhan 430074, China
18840846669@163.com

Abstract. The period from now to 2025 is a critical period for the vigorous development of China's new round of smart manufacturing industry. How to cultivate high-quality compound intelligent manufacturing engineering talents for the country is an important proposition that every higher education institution should consider in the current era. In the current situation of engineering education certification and manufacturing industry upgrading, the existing integrated teaching methods cannot fully meet the requirements for the training of compound applied talents. This paper takes the integrated teaching method of engineering graphics course as an example to improve the content of course construction and further enrich the theoretical connotation of engineering discipline, It also ponders and explores the optimization of the integrated teaching path of basic courses in the engineering talent training system of Application-oriented Colleges and universities, and puts forward a feasible scheme.

Keywords: Intelligent manufacturing · Applied undergraduate · Integrated teaching

1 Introduction

Intelligent manufacturing is the deep integration of the new generation of information technology and manufacturing industry. It is an advanced production process, system and management method. It has the functions of deep self-awareness, intelligent optimization and self-determination, fine management and self realization [1, 2]. With the emergence and interactive use of Internet of things, big data, cloud computing and other technologies, our manufacturing industry is undergoing great changes. At present, intelligent manufacturing industry has become the inevitable trend of the development of global advanced manufacturing industry [3–5].

It must be admitted that the existing talent training mode is not enough to meet the development needs of the intelligent manufacturing industry, nor can it meet its needs for the comprehensive analysis and practical ability and quality of technicians in the whole product life cycle. In order to achieve the goal of "made in China 2025", we must

Z. Hu et al. (Eds.): AIMEE 2022, LNDECT 159, pp. 533–543, 2023.
https://doi.org/10.1007/978-3-031-24468-1_48

formulate a strategic plan suitable for the training of engineering and technical talents in the future intelligent manufacturing industry [6, 7].

Under the background of intelligent manufacturing, based on engineering graphics, a general mechanical course, this paper will discuss how to reform and explore the course teaching for Application-oriented Undergraduate Colleges and universities, and find a practical scheme for the country to cultivate high-quality and compound intelligent manufacturing engineering talents [8].

2 Opportunities and Challenges Contained in the Background of Intelligent Manufacturing

Through data retrieval, it can be seen that the demand for talents in technical positions in the intelligent manufacturing machinery industry continues to increase with the development of industrial upgrading. The demand forecast of industrial talents mainly in key equipment, system integration and equipment application will reach 3.776 million person times by 2025 [9]. Judging from the current training scale of relevant professionals, the number of graduates is far from meeting the needs of enterprises.

At the same time, intelligent manufacturing enterprises put forward new requirements for the skills and knowledge of technicians. Among them, for technicians, they are required to have the knowledge background of cad/cam/cae integration. For production management positions, staff should master computer-aided tools. At the same time, higher lean needs have been put forward, and new technologies, new software and management methods that can realize cross-border collaborative work need to be organically integrated into the lean intelligent manufacturing production of enterprises [10, 11]. Therefore, for machine majors with application undergraduate courses, it is necessary to cultivate a group of professionals [12–14] who have certain professional knowledge and technical level and can engage in technology development, equipment design, production and operation management and application research in the machinery industry.

In addition, under the background of intelligent manufacturing, professional qualities such as responsibility, communication and expression ability, team cooperation ability have become important requirements for enterprises. In the actual teaching process, it is necessary to combine professional courses with humanistic quality, ensure the effective development of all-round education, and cultivate compound talents with a sense of social responsibility, good professional ethics, team spirit and practical skills.

3 Current Situation and Improvement Strategies of Integrated Teaching

Integrated teaching integrates offline and online teaching resources, and comprehensively cultivates students' professional ability and humanistic quality by means of results oriented and competition promoting teaching, which is suitable for the current developing talent training system of intelligent manufacturing industry.

Next, taking engineering graphics, one of the main basic courses of mechanical major, as an example, we will comprehensively explore the current situation and improvement strategies of integrated teaching in intelligent manufacturing system.

3.1 Multiple Resources Enrich Teaching

For integrated teaching, it is a common mode to combine online resources such as "learning link" and MOOC with the classroom, and realize the interaction of in class and extracurricular content through interactive software such as nailing. As colleges and universities explore and practice in many aspects: Zhou Rongan and others enhance real-time classroom interaction, adopt extracurricular "Online + offline" mode and take innovative design projects as teaching guidance [15]; Zhao Xinyue and others introduced virtual assembly technology in the classroom, and through the immersive virtual assembly teaching experimental platform, students can personally experience the process of assembling and disassembling ligands [16]; Deng Xiaozhen and others designed the teaching of "fragmented learning" for engineering graphics courses, making full use of spare time and improving learning efficiency [17]. However, some of the above methods are limited to the improvement of teaching effect, and some courses focus on traditional mechanical drawing, which is a certain distance from the current situation of industrial development.

In terms of intelligent manufacturing, the essence of the industry is to deeply combine information technology and manufacturing technology, and it is also the commanding height of "Internet + manufacturing". On the basis of fully integrating online and offline teaching resources, it is also necessary to reconstruct the course teaching content and resort out the existing knowledge system of traditional cartography courses. Based on IEET engineering education certification, it focuses on the cultivation of students' cross compound ability and informatization quality.

At the same time, in the process of integrating curriculum resources, according to the latest development of intelligent manufacturing technology, the teaching contents of intelligent CNC machine tools and industrial robots have been added. With the help of online related professional curriculum resources, we will integrate teaching implementation tasks, courseware, micro teaching video, off campus high-quality curriculum resources, etc. into the "one click on work map" teaching platform, build rich online and offline teaching content, and improve the integrity and comprehensiveness of the curriculum system. At the same time, this also makes the traditional graphics courses more suitable for the current situation of the development of intelligent industry, and realizes the integration with intelligent manufacturing technology.

3.2 Combination of Competition and Teaching to Promote Teaching

The traditional teaching content is based on the curriculum outline and the unilateral teaching mode dominated by teachers. This teaching method leads to students' learning enthusiasm and the overall teaching effect needs to be improved. In recent years, many colleges and universities have adopted organizing students to participate in provincial and national discipline competitions to promote teaching as a new means.

of teaching and learning. Through the competition, students' ability to integrate theory with practice and think independently is stimulated, and problems are found and solved through practice, forming a teaching mode of teaching through competition. The feedback obtained from the competition results will also enhance students' self-confidence in learning and work.

At present, the main professional competitions in the field of mechanical drawing include the CAD mechanical design project of the world skills competition, "Higher Education Cup" National College Students' advanced mapping technology and product information modeling innovation competition, and various competitions organized by provinces, cities and schools to investigate the professional basic knowledge and innovative application ability of the participants to a certain extent [18]. The evaluation overview of some skills competitions is shown in Table 1.

Table 1. Overview of evaluation of some mechanical drawing related skills competitions

Requirements	CAD mechanical design project of World Skills Competition	"Higher Education Cup" undergraduate advanced mapping technology and product information modeling innovation competition
Content	3D modeling, exploded view, demonstration animation, simulation animation, model projection part drawing and assembly drawing; Rendering; Data input and output; Sheet metal modeling; Part mapping	• Basic drawing knowledge and configuration • Part modeling, assembly modeling 3D drawing and 2D engineering drawing
Software	Inventor	Pro/E, SolidWorks, AutoCAD, ^
Submit	Drawing; Render; Animation	Drawing; Configuration test paper; 3D model; Exploded view;

It can be seen from Table 1 that the current skill competition not only comprehensively inspects the drawing specifications and model software, but also has a close relationship with the fields of mechanical design, production and manufacturing. This provides an effective window for college students to face the industry, the world and the future.

In the existing "teaching by competition" mode, the content related to discipline competition is integrated into daily teaching [19]. However, with the development of the current intelligent industry, the simple mapping or modeling ability is not enough to meet the industrial needs, because the intelligent manufacturing equipment itself is a complex equipment with high technical comprehensiveness, equipped with a variety of control software, and it is a relatively complex production system. Only by integrating the skills of multiple core courses can it support the core competence requirements of the post. Therefore, when learning the drawing specifications and modeling methods, we should also learn our own design, drawing, formulating processing technology and other aspects, and expand and extend the classroom teaching content with the help of the topics in the competition.

A very effective method has been proved to be the deep integration of competition, teaching and learning with the help of competitions, so that students can integrate the knowledge they have learned. In the training related to the competition, it is necessary to break the boundaries of the inherent disciplines. On the basis of mechanical drawing, combined with the professional basic contents such as mechanical principle, mechanical design and mechanical manufacturing, it is necessary to enhance the learning of knowledge and skills in the integration of cad/cam/cae, and supplement the corresponding background knowledge. The annual mapping competition is constantly approaching the reality of intelligent production, so it is also necessary to update teaching cases in time to keep pace with the times.

At the same time, in addition to drawing competitions such as mapping competitions, students can also be organized to observe and understand other related discipline competitions, such as college students' mechanical innovation design competition, "Challenge Cup" College Students' extracurricular academic science and technology competition, intelligent manufacturing challenge competition, etc., or parts design drawings, process procedures, and numerical control program lists involving the works of recent mechanical innovation design competitions are provided to students for after-school learning, Let students have full access to more contents related to engineering practice, and further enhance the practical ability of intelligent manufacturing industry.

3.3 Humanistic Quality Strengthening Teaching

Under the background of intelligent manufacturing, there are higher requirements and new changes in the professional quality of the machinery industry, and the professional quality such as sense of responsibility, expression ability and team cooperation consciousness has gradually become an important requirement. And in the talent training system, the teaching of professional knowledge needs ideological education as a support. In the actual teaching process, we need to organically combine the basic knowledge related to professional courses with humanistic quality education to form a synergistic effect between the two. In this way, we can truly realize the basic task of "Building Morality and cultivating people" and ensure all-round education.

In the current education and teaching, some teachers have introduced the elements of "Curriculum Ideology and politics" in order to improve students' Ideological and moral cultivation. However, the introduction of Ideological and political content of the curriculum alone will not fit the teaching content of cartography and not achieve the effect of moral education. In the process of teaching, humanistic quality education cannot be added rigidly, and students need to feel the beauty in the process of moral education, so as to accept rational moral norms more actively. We should really improve college students' professional awareness, improve their professional and technical quality, and cultivate excellent talents who are "knowledge-based, skilled and innovative".

The first is to integrate engineering graphics into daily teaching flexibly while organically combining the course content with the ideological and political content of the course. For example, taking the Song Dynasty's "construction method" as the starting point to tell the history of graphics, and leading students to explore the meaning and significance behind the compilation of Engineering monographs by ancient working people. Combined with the current era call of "made in China 2025", while enhancing students' cultural self-confidence and national pride, cultivate students' sense of responsibility and mission.

Secondly, it can also be integrated with traditional culture. For example, the drawing reading view part of the assembly is the key part of engineering graphics from geometric elements to stereoscopic projection, but many students are afraid of difficulties. The introduction of Tang poetry, which students are familiar with, that is, "look at the side of the mountain as a peak, and the distance is different". On the one hand, it shortens the distance between students and courses; On the other hand, it can also represent the three-dimensional structure of an object seen from one view, and combine multiple views to understand the high and low morphological changes of the plane it represents in the view. Combine literary knowledge with curriculum teaching to cultivate students' humanistic spirit and emotion.

In addition, when reading drawings, you can introduce the engineering aesthetics contained in different mechanical structures, such as configuration theory. On the basis of strengthening students' understanding of configuration, you can cultivate students' aesthetic awareness, help students avoid some wrong form choices and collocations in the design configuration, inspire students' design inspiration, make them have engineering aesthetic literacy, and bring a teaching environment in which aesthetic education promotes the coordinated development of moral education.

In the actual teaching process, it is also necessary to carry out targeted guidance and education according to the actual problems and needs of different students, encourage students to explore their own learning methods, and pay attention to the integrity and comprehensiveness of talent cultivation.

4 Case Design and Practice

Gear is the basic component of industrial equipment, and it is also a typical object to train students to use comprehensive knowledge to solve complex mechanical engineering problems. Next, we will take the preparation and development of the gear class as an example to carry out specific exploration and analysis.

4.1 Specific Case Design

First, according to the students' actual learning ability, the learning objectives are formulated in groups: 20% of the students complete the scale drawing of the basic cylindrical gear, 60% of the students complete the model establishment and engineering drawing generation, and finally 20% of the students carry out further innovation and expansion on this basis.

Pre class stage: 1–2 days in advance, push the micro video of gear structure and drawing method through the wechat applet of "one point connection of Engineering Drawings", so that students can have a preliminary preview. Get the preview feedback from students, understand in advance the problems that students will encounter in learning, such as the drawing of meshing area, the drawing of sectioning gear teeth and other difficulties, and make preparations to focus on them.

In class stage: first, describe the drawing method of gear ruler and gauge, and extend the drawing method of gear ruler and gauge to AutoCAD, SolidWorks and other two-dimensional and three-dimensional modeling software. At the same time, the auxiliary teaching can be carried out through UG NX and other software to introduce the gear processing method, so that students can further understand the main process of mechanical product design and processing according to the parameters, structure and other relevant information of the gear model.

At the same time, practical cases are introduced into the course. For example, the chip being "stuck" has become a common topic of discussion in the country, society and academia, and the accuracy problem in the manufacturing field is an important part of "sticking neck". As early as the 1960s, in order to complete the manufacturing task of high-precision and sophisticated national defense equipment and meet the challenge of ultra precision gear technology blockade, under the leadership of Wang Liding, an academician of the Chinese Academy of Sciences, China's ultra precision gears achieved "evolution" from level 4 to level 2 and level 2 to level 1. This shows the innovative spirit and national responsibility of the older generation of scientists who are brave to climb the peak and dare to be the first. At the same time, it can also encourage students to shape their professionalism and scientific literacy in pursuit of excellence.

After class stage: arrange the ruler and gauge exercise content of the specified drawing method of cylindrical gears, and customize the online exercise through the small program "one touch of Engineering Drawing".

Feedback stage: sort out the classroom feedback and evaluation contents according to the classroom observation and evaluation and the situation of answering exercises.

The specific process is shown in Fig. 1.

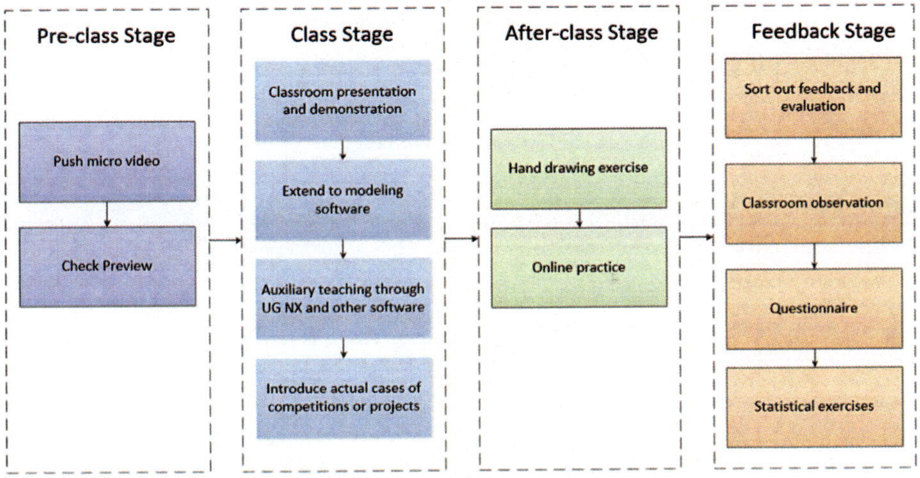

Fig. 1. Integrated instructional design process

4.2 Teaching Practice and Feedback

In the second semester of 2020–2021 academic year, a total of 63 people from two classes of machinery in our school were selected to carry out teaching reform and exploration. At the same time, 61 students of 2019 who use the traditional teaching method as the control group to analyze the differences in teaching effects.

The first is the stage of classroom observation and evaluation. From the aspects of students' learning, teachers' teaching, classroom nature and so on, the classroom situation of 63 people in two classes is counted, and the data are shown in Table 2.

Table 2. Classroom observation and evaluation

Evaluating indicator	Grade 2019 (%)	Grade 2020 (%)
Preview before class	17.54	52.63
Listen and see the drawing demonstration	100	100
Independent thinking	17.54	33.33
Take the initiative to answer questions	14.03	26.32
Associated with other disciplines	12.28	35.08

As can be seen from the data in the table, the improved integrated teaching effect has significantly improved in guiding students to think independently and connecting cartography with other disciplines in combination with professional background. In addition, due to the use of group teaching for students with different acceptance abilities, the application ability of students has been further improved on the basis of achieving the original cartographic teaching objectives.

Table 3. Evaluation of the experimental group on the teaching mode

Survey items	A	B	C	D	E	F
Satisfied with the teaching resources?	24	21	9	5	4	0.444
Difficult of the test appropriate?	21	27	8	5	2	0.477
Improved the understanding of knowledge?	18	33	6	4	2	0.484
Raised interest in learning?	25	18	17	2	2	0.492
Improved learning efficiency?	21	22	16	3	1	0.468
Increase professional skills?	21	27	12	2	1	0.516
Connected greater with other disciplines?	11	21	17	8	6	0.183
Enhanced communication between teachers and students?	13	19	21	8	2	0.262
Increased the burden?	4	19	19	16	5	0.007
No obvious help for learning?	4	8	40	7	4	0.007

Secondly, after class, a questionnaire is used to investigate the satisfaction of the teaching mode, including the quality of teaching resources, the difficulty of test questions, the initiative of autonomous learning, the effect of online learning, the effectiveness of teacher-student communication, etc. The questionnaire was evaluated with a five level scale (−2 means disagree, 2 means very agree), and the comprehensive attitude tendency coefficient f = $(2 \times N1 + 1 \times N2 + 0 \times N3 - 1 \times N4 - 2 \times N5)/(K \times N)$. N represents the number of samples, and *n1-n5* represents the attitude level. K is the absolute value of the highest grade score. The statistical results of the questionnaire are shown in Table 3. Among them, strongly agree, relatively agree, generally agree, disagree and disagree are recorded as A, B, C, D and E respectively.

The results of the survey show that the students who participate in the study recognize the effect of the course very much, especially for enhancing professional skills. Secondly, most students believe that they have improved their interest in learning and learning efficiency. Some students believe that the integrated teaching method increases the learning burden and takes up more time. Very few students think it is not helpful for learning.

At the same time, the examination results are sorted out and compared with the 2019 grade. The results are shown in Fig. 2.

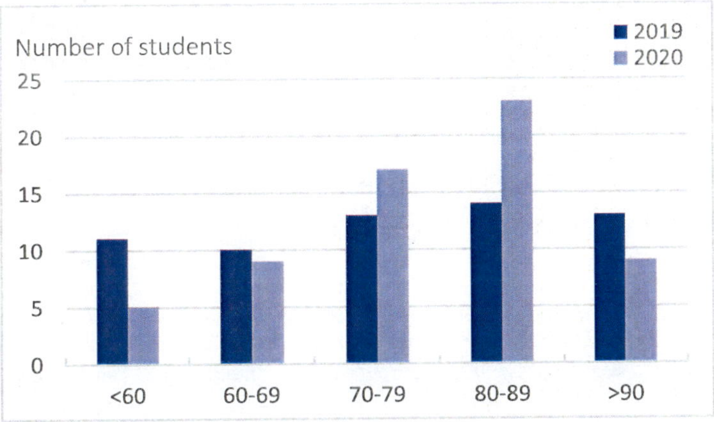

Fig. 2. Student achievement distribution chart

According to the data in the figure, the improved integrated teaching not only effectively promotes students' enthusiasm and initiative in learning, but also effectively improves the comprehensive evaluation results of the course. According to the comprehensive evaluation results of the class curriculum, the failure rate of the examination paper was reduced from 18.03% to 7.94%, the number of students with more than 80 points increased significantly, and the average score of the class also increased. In addition, students also achieved good results in the mapping competition. In 2021, the national competition won one second prize in the team and seven individual all-round awards. The winning rate of the 2022 provincial competition is 85.29%.

5 Conclusion

Intelligent manufacturing derived from artificial intelligence, including intelligent manufacturing technology and intelligent manufacturing system, needs to constantly enrich the knowledge base in practice. It also has the function of self-learning. It should be able to collect and understand environmental information and its own information, and have the ability to analyze, judge and plan its own behavior. Intelligent manufacturing has changed the traditional production mode, which not only improves efficiency, but also puts forward higher requirements for practitioners.

In the context of intelligent manufacturing, improving the existing integrated teaching method will not only enhance students' professional ability, but also improve students' quality. What is more important is to make the traditional basic courses more suitable for the development status of intelligent industry and realize the integration with intelligent manufacturing technology, so that students can meet the needs of society, regions and enterprises for high-end composite talents in the field of intelligent manufacturing.

As a basic course of industry and manufacturing, the teaching of engineering graphics needs to use modern educational ideas and strengthen the improvement of teaching methods, so that the trained talents can meet the requirements of the times and industry development.

Acknowledgment. This project is supported by the outstanding young and middle-aged scientific and technological innovation team plan project (t2021043) of colleges and universities in Hubei Province.

References

1. Tang Tang, T., Jie, W., Ming, C.: Fully realizing digitalization is the only way to intelligent manufacturing-interpretation of the road to intelligent manufacturing: digital factory. China Mech. Eng. **29**(03), 366–377 (2018). (in Chinese)
2. Orban, P.E., Gopalakrishnan, B., Gunasekaran, A.: Intelligent Manufacturing: 5263. SPIE Press (2004)
3. Giret, A., Garcia, E., Botti, V.: An engineering framework for service-oriented intelligent manufacturing systems. Comput. Ind. **81**, 116–127 (2016)
4. Goncalves, R., Romero, D., Grilo, A.: Factories of the future: challenges and leading innovations in intelligent manufacturing. Int. J. Comput. Integr. Manuf. **30**(1), 4–14 (2017)
5. Cardin, O., Ounnar, F., Thomas, A., Trentesaux, D.: Future industrial systems: best practices of the intelligent manufacturing and services systems (IMS2) French Research Group. IEEE Trans. Industr. Inf. **13**(2), 704–713 (2017)
6. Zhao, S., Jia, X.: Research progress and trend of intelligent manufacturing and its core information equipment. Mech. Sci. Technol. **36**(1), 1–16 (2017). (in Chinese)
7. Song, W., Cao, S., Yang, B., Song, K., Wu, C.: Development and application of an autonomous learning system for engineering graphics education. Int. J. Mod. Educ. Comput. Sci. (IJMECS) **3**(1), 31–37 (2011)
8. Zhang, Z., Wang, M., Wu, B., Liu, G.: Construction and teaching practice of "Online + offline integrated" engineering graphics course. J. Graph. **37**(5), 718–725 (2016). (in Chinese)
9. Ma, X., Chen, X., Xu, C.: Analysis on the matching between the talent demand of intelligent manufacturing machinery industry and the specialty setting of Vocational Colleges. China Vocat. Tech. Educ. **735**(11), 6–16 (2020). (in Chinese)
10. Rosen, D.W.: Thoughts on design for intelligent manufacturing. Engineering **5**(4), 609–614 (2019)

11. Farid, A.M.: Measures of reconfigurability and its key characteristics in intelligent manufacturing systems. J. Intell. Manuf. **28**(2), 353–369 (2017)
12. Jian, L., Yuan, Z.: The exploration and practice in innovative personnel training of computer science and technology. Int. J. Educ. Manageme. Eng. (IJEME) **2**(6), 47–51 (2012)
13. Catherine, A., Nathan, B., Erin, M.: The positive influence of inquiry-based learning teacher professional learning and industry partnerships on student engagement with STEM. Front. Educ. **6**, 319 (2021)
14. Chiu, P.H.P., Im, S.W.T., Shek, C.H.: Disciplinary variations in student perceptions of active learning classrooms. Int. J. Educ. Res. Open **3**, 100131 (2022)
15. Zhou, R., Fu, C.: Research on the application of integrated teaching mode in mechanical drawing course. J. Graph. **41**(6), 1039–1043 (2020). (in Chinese)
16. Zhao, X., He, Z., Liu, Z., et al.: Design of immersive teaching experimental platform for assembly guidance of Engineering Graphics Based on augmented reality. Exp. Technol. Manage. **39**(3), 195–199 (2022). (in Chinese)
17. Deng, X., Xiao, B., Wang, H., Tao, Z., Xi, J.: Research on ideological and political teaching of engineering graphics based on "fragmented learning." Equip. Manuf. Technol. **08**, 131–133 (2021). (in Chinese)
18. Xu, G., Zhang, M., Lan, C.: Rely on "skill competition and examination" to improve the innovative education of mechanical drawing. J. Graph. **36**(004), 631–637 (2015). (in Chinese)
19. Deng, X., Xiong, Z., Liu, L., et al.: CAD competition and analysis of the current situation of CAD education in China. J. Graph. **34**(001), 111–115 (2013). (in Chinese)

Construction of Practical Teaching System of College Broadcasting and Hosting Art Under the Concept of OBE

Bin Hao[1,2], Yahui Liu[1], and Ziye Wang[1](✉)

[1] Broadcasting and Hosting Art, Wuhan Sports University, Wuhan 430072, China
21257075@qq.com

[2] Business and Public Administration, Namseoul University, Cheonan 3102091, Republic of Korea

Abstract. In the 5G era of "everything is media", the specific manifestation of 5G technology to promote the change and development of media industry is the new technology and digital change, so it is urgent to explore the construction of a professional practice teaching system of broadcasting and hosting art that keeps pace with the times and realizes the close integration of social needs and universities. Under the concept of OBE education, the article proposes to construct the professional practice teaching system of broadcasting and hosting art with the orientation of results, and elaborates on six aspects: setting up professional practice courses, optimizing the design of practice teaching links, improving the assessment and evaluation of practice teaching, creating an innovative and entrepreneurial practice teaching platform, constructing teachers' team, and providing hardware guarantee for practice teaching, so as to promote the improvement of students' practice ability and the effectiveness of professional practice teaching.

Keywords: Practical teaching system · OBE concept · Broadcasting and hosting

1 Introduction

From December 30, 1940, the first broadcast of Yan'an Xinhua Radio, after more than eighty years of development and evolution, the major of broadcasting and hosting art has established a standardized discipline system and structure. Nowadays, science and technology are constantly developing and updating, bringing new directions for the major of broadcasting and hosting art, driving it to a more specialized, systematic and institutionalized direction. The major is highly practical, and its practical teaching part becomes an important part of students' development and teachers' teaching. At this stage, most of the applied majors in China's colleges and universities practice the OBE education concept, and the construction of a curriculum system corresponding to the professional talent training objectives is the key to realize the OBE concept, and an important part is the practical curriculum system [1, 2].

Z. Hu et al. (Eds.): AIMEE 2022, LNDECT 159, pp. 544–552, 2023.
https://doi.org/10.1007/978-3-031-24468-1_49

The OBE education concept proposed by Spady et al. in the 1990s (outcome-based education)is a new model of education based on learning outcomes, which requires student-centered, outcome-oriented teaching and learning in the education process,

focusing on the competencies and outcomes students acquire. In the OBE talent development concept, teachers need to teach the whole course in full accordance with the talent training objectives and students' specific learning processes and needs, and to assess students' overall course learning in a comprehensive, multidimensional and integrated manner after class to guide them to achieve the course goals. Students should be guided to achieve the teaching objectives of the course [3–6]. Traditional education philosophy is based on curriculum-based instructional design and implementation, focusing on how teachers "teach", while OBE education philosophy is based on output-based instructional design and implementation, focusing on how students "learn", emphasizing that everything in the education system should be focused on Everything in the education system is organized around ensuring that all students achieve the desired learning outcomes at the end of their learning [7].

2 Development Opportunities and Current Situation of Broadcasting and Hosting

Broadcasting and hosting majors must keep pace with the development of the times if they are to develop. In the face of the development of 5G technology, its communication tools and technical equipment conditions are constantly updated [8–10]. Therefore, in the context of the new era, the teaching system of the broadcasting and hosting art major will have to be updated. From the selection requirements of the 2019 China Central Radio and Television Corporation's host competition, we can see that the host competition has been comprehensively upgraded compared with previous competitions, breaking the previous traditional media assessment points for hosts' abilities, and making "the composite type with excellent orientation control ability, live news interaction ability, multimedia application ability and personalized communication ability" as new standard of the host of the integrated media era [11]. These requirements put forward new requirements for the professional teaching of colleges and universities. At present, according to the requirements of the professional standard of broadcasting and hosting art, all colleges and universities have offered practical courses in the curriculum system, including curriculum practice, such as radio broadcasting and hosting business, television broadcasting and hosting business, news program broadcasting and hosting, variety program hosting practice, etc.; internship, such as professional mid-term internship, professional graduation internship and graduation work creation, etc. [12–15]. However, according to the feedback from graduates and employers, it is found that students trained under the original practical teaching system show insufficient practical ability, lack of in-depth understanding and application of program adaptation and control, and relevant media technology.

Among the questionnaire statistics of graduates and undergraduates of this major, media practitioners, 120 questionnaires were issued and 101 valid questionnaires were issued. The frequency analysis is as follows (Table 1):

Table 1. Questionnaire statistics

Name	Options	Frequency	Percentage (%)
Identity	Undergraduate	32	31.68
	Graduate	36	35.64
	Media practitioner	33	32.67
Satisfaction with the practical class	Full Satisfaction	0	0
	Satisfaction	1	0.099
	Normal	100	99.01
	Dissatisfaction	0	0
	Very Dissatisfied	0	0
Professional practice course length arrangement	Long	27	26.73
	Short	27	26.73
	Some are longer, some are shorter, the length of the class is not appropriate	16	15.84
	The length of the class is appropriate	31	30.69
Repetition of professional practice courses	Some of them are repetitive	36	35.64
	Most of them are repetitive	37	36.63
	None of them are repetitive	28	27.72
For this major, there are practical courses that you think should be offered but are not	Yes, lots of them	99.01	99.01
	Yes, some of them	0.99	0.99
	No	0	0
	Don't know	0	0
Unreasonable aspects of practical courses	Too little/too much class time	0	0
	Skills are not learned or mastered	100	99.01
	The content difficulty of the course is too high or low than the current level of students	1	0.99

(continued)

Table 1. (*continued*)

Name	Options	Frequency	Percentage (%)
	No, I think the practice curriculum is reasonable	0	0
Whether the content of practice courses is out of touch with the media industry	Yes	55	54.46
	No	46	45.54
Whether the faculty of practice course is satisfied	Yes	52	51.49
	No	49	48.51
Have you gained anything from the practical courses	Yes	48	47.52
	No	53	52.48
Whether the practical courses will benefit your professional development	Yes, it plays a great role in promoting professional development	20	19.8
	Yes, it plays a great role in promoting professional development	31	30.69
	No, it has little effect on professional development	25	24.75
	Don't know	25	24.75

It can be seen from the above table that the respondents are evenly distributed among college students of this major, graduates of this major and media practitioners, the proportion is 31.68%, 35.64% and 32.67% respectively. In terms of satisfaction with practical courses, 99% of the respondents thought it was normal; In terms of the length arrangement of professional practice courses, the maximum proportion of appropriate length arrangement is 30.69%, followed by longer and shorter class hours, both of which are 26.73%; In the case of repetition of professional practice courses, some of them are repetitive, most of them are repetitive, none of them are repetitive, accounting for 35.64%, 36.63% and 27.72% respectively; For this major, there are practical courses that you think should be offered but are not, "Yes, lots of them" accounts for 99% at most; Unreasonable aspects of practical courses, "Skills are not learned or mastered" accounts for 99% at most; In terms of whether the teaching content of professional practice courses is disconnected from the media industry, there is little difference between the proportion of choosing yes and no, which are 54.46% and 45.54% respectively; 51.49% and 48.51% choose yes or no in terms of whether they are satisfied with the faculty of the professional practice courses; In the practical courses of this major, whether you have gained anything, the proportion of yes is 47.52, slightly lower than no 52.48%; In terms of the practical courses of this major, only 19.8% of the respondents believed that "Yes, it plays a great role in promoting professional development". Combining with the OBE education concept, we analyze the causes of this problem as follows:

2.1 Lack of Clarity in the Training Objectives of Practical Courses

At the present stage, the courses of broadcasting and hosting have set up practical courses and course hours that meet the requirements, but the teaching objectives are not clear in the real practical teaching link [16]. There is no clear setting of what kind of practical ability students can get, what kind of degree they can reach, and what kind of practical teaching methods they can use after the practical courses. Each practice course uses ambiguous words in determining the practical teaching objectives, and the practical ability cultivation objectives and specific differences between courses cannot be accurately expressed, which will lead the broadcasting and hosting art majors to take the practice courses as a kind of supplement to theoretical knowledge in the process of talent cultivation and neglect to think and reflect on how to realize the practical ability. The unclear positioning of practice cultivation goal and the lack of practice ability cultivation planning make the practice teaching not play a proper supporting role in the process of realizing talent cultivation goal and graduation requirements. It causes the phenomenon of low adaptability of students in the workplace and their inability to adapt to the realistic job requirements of the development of media industry.

2.2 The Content of Practical Courses is out of Touch with the Media Industry

Faced the rapidly changing media environment, presenters should both selectively adhere to and actively integrate, develop in inheritance and innovate in adaptation. At present, the broadcasting and hosting art majors carry out practical teaching generally with existing textbooks or limited by the school practice equipment and venues. The actual work scenes and contents and processes are far away from the students' practical content in school practice, and it is difficult for students to adapt quickly. Nowadays, the media industry is changing, new media is developing rapidly, and the era of full media brings infinite possibilities. To promote the development of media integration and build a full media has become an urgent issue for us. In order to adapt to the needs of the times and changes in media ecology, the teaching content of the original curriculum should be increased in the production of new media audiovisual programs, new media operations, new media user analysis and other related practical courses, in order to enhance the comprehensiveness and modernity of students' practical skills in the new media environment.

3 Construction of Practical Teaching System

The result-oriented teaching concept of OBE is designed and implemented closely around the "talent training objectives", so the practical teaching system of broadcasting and presenter art majors can be reconstructed based on the OBE concept. It is mainly constructed in six aspects: setting up professional practice courses, optimizing the design of practice teaching links, improving the assessment and evaluation of practice teaching, creating a platform for innovative and entrepreneurial practice teaching, building a team of teachers, and providing hardware guarantee for practice teaching.

3.1 Professional Practice Curriculum

The core of the practical curriculum is the practical teaching system, around the needs of media industry, the practical curriculum of broadcasting and hosting art majors can be set in the order of ability enhancement, from the first year to the fourth year of university focusing on cultivating students' humanities, innovative thinking ability, broadcasting and hosting ability, new media technology application ability, analysis and problem solving ability, teamwork and other six aspects of ability, and each ability cultivation goal can be achieved through multiple practical courses and practical ways (Table 2).

Table 2. Practical Skills and Practical Courses and Practical Methods

Practical skills	Practical courses	Practical methods
Human literacy	Appreciation of excellent Chinese and foreign broadcasting and hosting works, Appreciation of new media art	Centralized practice and decentralized practice during holidays
Innovative thinking skills	Improvised oral expression, art of recitation, film and television dubbing	Intensive course training, decentralized unit practice
Broadcasting and hosting ability	Broadcast announcing business, television announcing business, news program announcing, new media audio-visual program hosting, etc	Concentrated laboratory hours for courses, concentrated course training, decentralized unit practice
Ability to apply new media technology	Photography, audio production, television program creation, new media audiovisual program production	Intensive course design, decentralized unit practice
Analytical problem-solving skills	Thesis writing, graduation design, graduation internship	Intensive course training, decentralized unit practice
Teamwork skills	Various practical courses, mid-term internship, graduation design, graduation internship	Intensive course training, intensive practice

3.2 Optimize the Design of Practical Teaching Sessions

Under the concept of OBE education, it is necessary to continuously strengthen the construction of practical teaching system and clarify the clear path of application-oriented talents cultivation goal, which can not only improve students' practical innovation ability, but also continuously meet the development demands of young college students in the new era in the economic and social development, and improve the employment competitiveness and social adaptability. Therefore, the practical teaching resources should be actively expanded, the utilization rate of resources is constantly improved, and the comprehensive practical ability of students is further enhanced. In the practical teaching link setting of broadcasting and hosting art majors, we should strengthen the interaction with media, combine the development characteristics and new dynamics of media industry, adjust the practical teaching content and links in real time according to specific changes, and integrate the various resources that can be developed for important practical courses or popular industry demands, especially the changes in the industry development of new media, so that students can participate in the actual work in a complete way, so that students can move from school to the society without barriers, and cultivate to be the high-quality talents needed in the new era.

3.3 Improve the Assessment and Evaluation of Practice Teaching

An effective means to promote the quality improvement of practice teaching is to develop a scientific, comprehensive and reasonable assessment and evaluation system. Under the guidance of OBE education concept, the assessment and evaluation of practice teaching needs to be comprehensively sorted out and evaluated based on the level of practical ability acquired by students in broadcasting and hosting after completing practice learning. In the specific course practice teaching, the grade of practice learning needs to be given to the students' comprehensive situation of completing the practice tasks of this course, and at the same time, the total grade of this course is counted in a scientific and reasonable proportion; in the grade evaluation of professional practice, the whole process evaluation needs to be done, and as a practice course, the evaluation should focus on whether it achieves to promote the development of the evaluated person, by designing a comprehensive, scientific and reasonable evaluation It is especially important to evaluate the development of students' ability in professional practice scientifically through the design of comprehensive, scientific and reasonable evaluation criteria.

3.4 Faculty Building

Relying on the collaborative education platform, teachers of professional courses are organized to work part-time or on attachment in media and related media companies to improve their theoretical and practical abilities, and teachers with certain practical application abilities are hired or introduced to supplement and improve the teaching team. According to the development of the industry, we can also invite senior experts in the media industry to the school to carry out lectures or teacher training on industry dynamics, sharing of business skills, discussion of cutting-edge issues, development of teaching resources; to create a full-time and part-time faculty team of "dual-teacher and dual-competency" in application.

3.5 Create a Teaching Platform for Innovation and Entrepreneurship Practice

Integrate the existing practical teaching resources of the school such as studio, dubbing room, post-editing and packaging laboratory, off-campus internship practice base and talent training base, develop practical teaching syllabus of college students' innovation and entrepreneurship ability, project program development and project counseling according to the actual situation of students, and fully utilize various resources inside and outside the school to create a practical training platform for innovation and entrepreneurship.

3.6 Provide Practical Teaching Hardware Guarantee

In terms of hardware investment, colleges and universities should increase the construction of broadcasting and hosting art majors, build high-definition studios, recording rooms, editing rooms, filming equipment and other hardware facilities required for professional internship practice, provide students with scenes simulating media work, create virtual practice internship courses and platforms, so that students can adapt to the working environment in advance, learn to master the operation and use of various types of picking, editing and broadcasting equipment, improve their ability to adapt to work, and enhance the level of all-media business.

4 Conclusion

The goal of talent training of this major is to cultivate high-quality applied talents, and the practical teaching system is an indispensable part of its talent training process. With the continuous development of media technology, the requirements of the media industry for the talents of broadcasting and hosting art are getting higher and more strict. Therefore, this paper constructs a practical teaching system based on OBE education concept, actively improves the practical teaching concept and teaching methods from six aspects, builds a practical teaching system that adapts to the development of media industry, and cultivates applied talents that can meet the needs of media industry.

Acknowledgment. This project is supported by Wuhan Sports University 2019 Graduate Education Teaching Reform Project (YJSJG2019001).

References

1. Ou, X.L., Wang, J., Liu, H., Wen, D., He, W., Pan, Y.: Research on the reform of practical teaching driven by practical training under constructivism. Int. J. Eng. Manuf. (IJEM) **1**(1), 77–83 (2011)
2. Ning, G.: Construction and implementation of innovation computer network practical teaching system. Int. J. Educ. Manage. Eng. (IJEME) **1**(2), 30–35 (2011)
3. Liu, K., Sun, Y.: Research on the cultivation of university teachers based on OBE educational concept. Heilongjiang Res. High. Educ. **6**, 59–61 (2017). (in Chinese)
4. Gao, G., Xue, X.: The professional orientation of program hosts in the era of integrated media - inspiration from the China central radio and television corporation's 2019 host competition. News Writ. **02**, 91–95 (2020). (in Chinese)

5. Hao, G.: The strain and perseverance of hosts in the era of fusion media. China Newspaper Industr. **24**, 74–75 (2017). (in Chinese)
6. Zheng, X.: The Development trend of archives evaluation program in education and practice curriculum among teachers-taking the education major of J College of X University in the Southern of Henan Province as an example. J. Shaanxi Xueqian Normal Univ. **35**(05), 90–94+110 (2019). (in Chinese)
7. Ye, L.: Returning to the breakthrough-a pedagogical outline of "life and practice". East China Normal University Press (2015). (in Chinese)
8. Chang, J., Zhang, X.: Construction and practice of practical teaching system based on OBE concept – Electronic Information Engineering as an example. China Univ. Teach. **Z1**, 87–92+111 (2021). (in Chinese)
9. Li, S.: The teaching exploration and research on arts-professional integration of broadcasting and hosting specialty. J. Luoyang Normal Univ. **36**(09), 81–83 (2017). (in Chinese)
10. Liu, C., Lin, H.: Exploring the realization path of first-class undergraduate construction with OBE concept. Educ. Rev. **08**, 33–36 (2018). (in Chinese)
11. Yuan, H., Sun, Q.: Curriculum reform of construction technology of ship and ocean engineering based on OBE educational concept. J. Jimei Univ. (Educ. Sci. Edn.) **19**(4), 79–83 (2018). (in Chinese)
12. Wang, M., Liu, W., Zhang, Q.: Talking about the changes of the times and the development of the discipline of broadcasting and hosting art. Home Drama **09**, 180–181 (2014). (in Chinese)
13. Wang, Q.: The innovative development of broadcast hosting in the new media environment. J. Commun. **16**, 15–16 (2018). (in Chinese)
14. Ding, Y.: Analysis of the construction strategy of practical teaching system of radio and television choreography. J. Jilin TV Radio Univ. **11**, 66–67 (2017). (in Chinese)
15. Lu, J.: Research on the practice and reform of broadcasting and hosting under the perspective of integrated media. Media Forum **3**(02), 49+52 (2020)
16. Li, J., Zhou, Y.: The exploration and practice in innovative personnel training of computer science and technology. Int. J. Educ. Manage. Eng. (IJEME) **2**(6), 47–51 (2012)

Creditworthiness of Individual Borrowers Forecasting with Machine Learning Methods

Larysa Zomchak(✉) and Viktoria Melnychuk

Ivan Franko National University of Lviv, Lviv 79000, Ukraine
lzomchak@gmail.com

Abstract. Providing loans to individuals without a proper analysis of their creditworthiness is one of the reasons that can lead to risky activities of the lender and, consequently, to the weakness of the credit system as a whole. The investigation developed and implemented a model of gradient boosting regression as a method of assessing and forecasting the borrower's creditworthiness. A gradient boosting of the creditworthiness assessment of employees of IT company of Ukraine, which are provided with domestic credit. It was found that among the parameters that describe a person: age, department in which he/she works, position, qualification, city of residence in pre-war times, gender, marital status, availability of a guaranteed annual premium - the greatest influence on creditworthiness is the department and the qualifications of the employee. The model for assessing the creditworthiness of the borrower is adequate to reality, so its introduction into the credit system of the company will minimize the risks posed by lending to borrowers who are unable to fully meet their obligations to the lender. Further development of intensive implementation of the use of these models of gradient boosting, as well as their stacking with other methods of machine learning, can improve the functioning not only at the micro level, but also the optimal operation of the banking system of Ukraine as a whole.

Keywords: Forecasting · Creditworthiness · Risk · Bank · Machine learning · Gradient Boosting · Regression

1 Introduction

Assessing the reliability of the borrower is one of the important tasks of the credit system. Imperfect credit assessment mechanisms can lead to significantly negative consequences. In the worst case - to the bankruptcy of the creditor. In other, less critical cases, the consequence may be such a negative phenomenon as the liquidity of the institution providing the loan. The starting point in assessing the capabilities of the person wishing to obtain a loan - is to determine the ability of the borrower to repay the principal amount of the loan at a specified time. One of the main ways to avoid non-repayment of the loan is a thorough and qualified analysis of potential borrowers. The main means of such selection is the economic analysis and assessment of the borrower from the standpoint of his creditworthiness. Mathematical methods of machine learning are well suited to

Z. Hu et al. (Eds.): AIMEE 2022, LNDECT 159, pp. 553–561, 2023.
https://doi.org/10.1007/978-3-031-24468-1_50

solve this problem. The most popular among them are the method of random forests, boosting models, assembly training and logistic regression.

The artificial intelligence and machine learning methods application for solving lending problems has become widespread in recent years. D. Gupta and S. Goyal are using linear regression model and artificial neural network algorithms for creditworthiness prediction [1]. N. Mohammadi and M. Zangeneh also used artificial neural network for customer credit risk assessment [2]. A. Bazmara and S. Sardar Donighi used fuzzy expert system for bank customer credit scoring [3], A. Kaminskyi at all used machine learning methods for consumer banking problems, e.g. scoring models, neural network model and fuzzy clustering [4]. A. Kaminskyi and M. Nehrey used clustering methods for the credit risk analyzing of nonbank lenders [5]. Md.Mahbubur Rahman with coauthors used nearest neighbor classifier for decision making about loan by commercial bank [6]. Besides, machine learning methods application in banking and consumers creditworthiness is also discussed in [7–20].

2 Gradient Boosting as the Machine Learning Method

2.1 The Gradient Boosting Method Idea

Prevention or maximum reduction of credit risk remains an urgent problem today, so the methodology for assessing the creditworthiness of the borrower needs to be improved. Mathematical methods of machine learning are well suited to solve this problem. The most popular among them are the method of random forests, boosting models, assembly training and logistic regression.

Gradient boosting is a controlled machine learning algorithm which is widely used for both classification and regression problems. It is a complex technique that uses weak data positions to create a strong model for classification and regression. Unlike the decision tree method, gradient boosting begins with building one leaf (least) instead of a tree. When forecasting, the first leaf will have the average value of the numerical indicators of the initial series. Then the principle of gradient boosting builds a decision tree. Each subsequent tree is based on errors that were discovered earlier when constructing previous trees. It should be noted that gradient boosting regulates and limits the number of trees that can be created, as well as scales them in a certain way by the same size. In practice, the maximum number of leaves is from 8 to 32 pieces. The gradient boosting machine builds trees until a predetermined number of them is reached, or until the additional trees stop improving the result of the previous ones. Usually the tree is built on the basis of the difference between the observed and predicted values. When the target column contains continuous values, we use the Gradient Boosting Regressor, and when solving the classification problem, we use the Gradient Boosting Classifier. The only difference between them is the loss function. The goal is to minimize this loss function by adding weak components through a gradient descent. In this investigation, the problem was implemented using a gradient boosting regression algorithm.

2.2 The Gradient Boosting Regression Algorithm

The gradient boosting regression algorithm consists in sequentially performing the following steps, thus compensating for the errors in the previous step:

1. First of all, it is necessary to determine the input data (1) and the loss function, which is minimized (2):

$$\{(x_i, y_i\}_{i=1}^{n}, \tag{1}$$

x_i - parameters used for forecasting; y_i - the resulting values that relate to a particular person, or anyone in general; n is the sample size.

$$L(y_i, F(x)) \rightarrow min \tag{2}$$

y_i - is the observed value; x_i - parameters used for forecasting; F(x) is the predicted value.

The function (2) presented above is a differentiated loss function. In this case, it assesses how well we can predict our values. The loss function, which is most often used for regression with gradient boosting, is as follows (2):

$$\frac{1}{2}(y_{empirical} - y_{forecast})^2 \tag{3}$$

You can choose other loss functions, but the above (2) is the most popular for regression.

2. The next step is to initialize the model into specific constant values using the following function (4):

$$`F_0(x) = \underset{F(x)}{arg\ \min} \sum\nolimits_{i=1}^{n} L(y_i, F(x)) \tag{4}$$

x is the parameters used for forecasting; $F_0(x)$ - initial predicted value; y_i is the observed value; $F(x)$ - the predicted value; $L(y_i, F(x))$ - loss function; n is the sample size.

3. In the next step, find the initial values for each leaf of our decision tree. This means that there may be a case where 1 leaf receives more than one error, so you need to find the final output of all leafs. To find the output, we can take the average of all the numbers on the sheet, no matter if there is only 1 number or more. It is assumed that $h_m(x)$ is a decision tree made on these errors. Mathematically, this step can be represented as (5):

$$F_m(x) = \underset{F(x)}{arg\ \min} \sum\nolimits_{i=1}^{n} L(y_i, F_{m-1}(x_i) + F(x)*h_m(x_i)) \tag{5}$$

y_i is the observed value; x - parameters used for forecasting; F(x) - the predicted value; $F_{m-1}(x)$ is a preliminary prediction; $h_m(x)$ is a decision tree made on errors.

4. The last step is to update the forecasts of the previous model. This process can be displayed as follows (6):

$$F_m(x) = F_{m-1}(x) + \alpha * h_m(x) \tag{6}$$

x - parameters used for forecasting; $F_{m-1}(x)$ is a preliminary prediction; α-coefficient of learning speed; $h_m(x)$ is the last decision tree created taking into account the errors in the previous step.

That is, the final forecast will be equal to the average value calculated in step 1, plus all errors predicted by the trees that make up the forest multiplied by the learning rate α.

2.3 Advantages and Disadvantages of Gradient Boosting

Advantages of gradient boosting:

1) The predictive accuracy of the algorithm is in a high position;
2) Provides great flexibility and can optimize various loss functions, provides several parameters for adjusting hyperparameters, which make the function very flexible;
3) In most cases, pre-processing is not required, the gradient boosting algorithm works well with categorical and numerical data, processes missing data - imputation of the missing value is not required;
4) Own overhead costs of boosting are small, time of construction of model is almost completely defined by time of training of basic algorithms;
5) The ability to identify objects that are noise emissions;
6) Relative resilience to retraining;
7) Disadvantages of gradient boosting;
8) Gradient enhancement models will continue to be refined to minimize all errors. This can overemphasize the deviation and cause excessive tightness. Cross-checking should be used to neutralize.
9) The strategy of sequential addition leads to the construction of a suboptimal set of basic algorithms;
10) The process is very expensive in terms of calculations - it often takes a lot of trees (>1000), which can run out of time and memory;
11) High flexibility leads to many parameters that interact and strongly influence the behavior of the approach. This requires a large search grid during setup.

3 Assessing the Creditworthiness of the Borrower with Gradient Boosting Regression Method

The aim of the investigation is to build an economic and mathematical model for assessing the creditworthiness of the borrower using the method of machine learning, namely, gradient boosting regression method. To implement this task, statistical data on employees of the IT company of Ukraine were used. The main feature of the internal lending of this company is that each employee has the opportunity to obtain a loan. Building a regression model as part of the investigation should determine the amount of credit that the company can provide to internal users. Ukrainian economy has been hit hard by Russia's full-scale war in 2022, the lending system has also changed, with the deterioration of citizens' financial situation directly affecting credit relations, as borrowers are unable to meet their obligations properly. credit agreements. At the same time, Ukraine's IT cluster managed to consolidate and quickly reformat due to anti-crisis business plans. Most companies have retained customers and the volume of their contracts. Due to this, the industry remains financially stable. Thus, the IT company of Ukraine, whose activities are considered in this paper, has the opportunity to continue to provide loans to its employees during the martial law of Ukraine in 2022, and borrowers - to fulfill their obligations, as their salaries are not reduced. Data for the construction of a regression boosting model to assess the creditworthiness of the company's internal borrowers are current as of May 1, 2022 (Fig. 1) and contain information on: 1. age of the employee; 2. the department in which he works: • Administration • Engineering • Quality Assurance • Marketing • Finance. 3 position: • Trainee • Junior • Middle • Senior • Lead 4. the level of competence in the field in which the borrower works; 5. Gender; 6. city of residence (before the 2022); 7. number of years of previous experience in other companies; 8. the number of years of work in this company; 9. marital status: • married • divorced • single • civil marriage • widower 10. availability of the right to receive a permanent annual award; 11. level of employment in the company (rate).

```
import pandas as pd
data = pd.read_csv('/content/data.csv')
data.head()
```

	Date Of Birth	Job Family	Band	Grade	Gender Key	City	Previous Experience	GL Experience	Marital Status	MBO Eligible	Full time?	Credit
0	32	Quality Assurance	1	4	M	Kharkiv	4	1	Married	No	MONTHLY	16800.0
1	27	Engineering	1	4	M	Kharkiv	4	1	Single	No	MONTHLY	15600.0
2	29	Engineering	1	6	F	Kyiv	5	1	Single	Yes	MONTHLY	54000.0
3	23	Engineering	1	4	M	Kharkiv	0	1	Single	No	MONTHLY	15600.0
4	28	Quality Assurance	1	6	M	Lviv	3	1	Single	No	MONTHLY	25800.0

Fig. 1. Input data of the creditworthiness assessment model

To train the model of assessing the creditworthiness of the borrower also used data on the amount of credit granted and simultaneously paid by the employee in the previous period.

In order for the regression model of assessing the borrower's creditworthiness to be able to predict the amount of credit that can be provided to its employee, it must be trained with the data obtained from previous experience of the company. For this purpose, 75% of the total data sample was used (a total of 2,390 copies). The remaining 25% of the input data was used to test the implemented model. To assess the adequacy of the constructed model, a coefficient of determination is calculated, which shows what part of the variation of the dependent variable is explained by the model. In this case, this figure is 0.87 and this means that the built model with an accuracy of 87% can predict the amount of credit that the company should provide to its employee and he will be able to repay these loans in full without any delay.

The training of the model took place on the basis of data obtained as a result of successful lending only in the previous year. To build a more accurate and reliable model for forecasting the amount of credit that should be provided to the employee, it is necessary to conduct training using data on the experience of the company's credit system for at least the last 5 years. To increase the accuracy of the model, one can also change the number of trees, their depth, the number of samples to separate the internal node, which the algorithm must build. During the implementation of the model of gradient boosting using an experimental approach, it was found that the most accurate model can be built using the construction of 30 trees with a maximum depth of 10 leaves and a minimum separation of 5. It is worth remembering that no always building a larger number of more branched trees gives the best result, because overtraining of the model impairs its accuracy.

The implemented model of gradient boosting of the creditworthiness assessment of the company's employees gives the following forecasts (Fig. 2): for example, an employee number 199 of the company should be granted a loan of $51,254. US, but the employee number 2073 has a lower creditworthiness - 25 047 dollars. USA. The first has the following input parameters: is 29 years old, works in the engineering department, middle position at level 9 out of 22, married man who lived in Kharkov before the war, works full time only 1 year in this company and 4 years in another, and has no annual fixed premium. The second employee, who has a lower creditworthiness, aged 25, works in the process quality assurance department, Junior position with 6 levels of qualification, is a woman who lived in the pre-war period in Lviv, unmarried, works in this company full-time for 2 years and has 2 years of experience in another company, does not have a guaranteed annual bonus.

	Date Of Birth	Job Family	Band	Grade	Gender Key	City	Previous Experience	GL Experience	Marital Status	MBO Eligible	Full time?	predict
199	29	3	2	8	0	9	4	1	7	0	0	51253.956771
79	29	3	2	8	0	9	0	1	0	0	0	51768.740169
871	34	2	2	8	0	3	4	7	3	0	0	42027.670968
1652	34	3	1	6	0	3	12	3	7	0	0	41285.932478
1629	30	3	1	6	0	9	0	3	7	0	0	34630.057097
...	...	...	...	...	...	...	...	...	...	...	...	...
909	68	2	2	8	0	3	37	7	7	0	0	40488.446442
1661	38	3	3	14	0	3	13	3	3	1	0	74648.409114
2073	25	2	1	6	1	0	2	2	3	0	0	25046.779124
2098	34	3	2	8	0	3	9	2	7	0	0	53052.828232
114	44	3	1	6	0	3	3	1	7	0	0	33385.678535

Fig. 2. Assessment of the borrower's creditworthiness

Obviously, not all input parameters have the same effect on the assessment of the borrower's creditworthiness. Using the feature, importance function, the magnitude of the impact of each of the parameters describing the company's employee on the assessment of creditworthiness was calculated (Fig. 3).

```
Date Of Birth          0.041134
Job Family             0.061742
Band                   0.466329
Grade                  0.270804
Gender Key             0.011541
City                   0.024659
Previous Experience    0.035417
GL Experience          0.032233
Marital Status         0.007895
MBO Eligible           0.016636
Full time?             0.031612
dtype: float64
```

Fig. 3. The magnitude of the impact of input parameters on the assessment of creditworthiness

It turned out that the position (47%) has the greatest influence on the decision on the amount of credit a company can provide, and the level of skills within the position (27%). Family status has the least impact - 0.9% (Fig. 4).

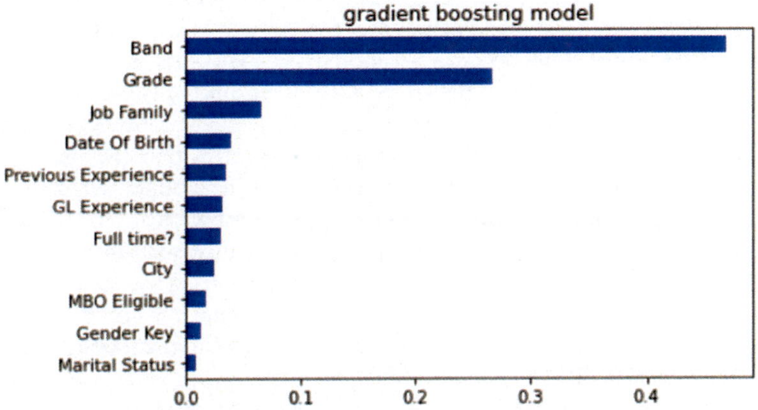

Fig. 4. Influence of input indicators on creditworthiness assessment

For more accurate results, it is necessary to conduct an experimental study with the combination of several models, the so-called stacking of models [21], which will combine the strengths of several algorithms, which can improve the results of the developed model of assessing the creditworthiness of employees.

4 Conclusion

The paper provides a theoretical generalization of the concept of credit and proposes a method of analytical support for assessing the creditworthiness of the borrower, which consists in building a mathematical model of gradient boosting. The results obtained during the study allow me to draw the following conclusions:

1) Based on the results of the analysis of scientific works, the theoretical foundations of the concept of creditworthiness are generalized and it is determined that creditworthiness is the ability of a borrower under specific lending conditions to pay its debt obligations exclusively in cash generated by the borrower. Normal activities in the current situation of the money market. In the context of the work, it is determined that the assessment of creditworthiness of borrowers is a system of assessment of certain factors, the influence of which or their acquisition of certain values determines the ability of the borrower to meet its debt obligations to the lender. It is emphasized that there is no single position in the economic literature on the relationship between the concepts of creditworthiness and credit risk. In this regard, it is proved that creditworthiness is a stimulant - high creditworthiness indicates the ability of the borrower to repay the loan on time and in full.

2) The general characteristic of mathematical methods of machine learning is given. The gradient boosting algorithm, which is often and successfully used to assess the borrower's creditworthiness, is described in detail.
3) A gradient boosting of the creditworthiness assessment of employees of IT companies of Ukraine, which are provided with domestic credit. It was found that among the parameters that describe a person: age, department in which he works, position, qualification, city of residence in pre-war times, gender, marital status, availability of a guaranteed annual premium - the greatest influence on creditworthiness is the factor position, the department in which he works and the qualifications of the employee. This means and reaffirms that these factors now directly affect the amount of wages and, consequently, the ability to meet obligations to the creditor, which in this study is the employer. The established economic and mathematical model for assessing the creditworthiness of the borrower is adequate to reality, so its introduction into the credit system of the company will minimize the risks posed by lending to borrowers who are unable to fully meet their obligations to the lender. Further development of intensive implementation of the use of these models of gradient boosting, as well as their stacking with other methods of machine learning, can improve the functioning not only at the micro level, but also the optimal operation of the banking system of Ukraine as a whole.

References

1. Gupta, D.K., Goyal, S.: Credit risk prediction using artificial neural network algorithm. Int. J. Mod. Educ. Comput. Sci. **10**(5), 9 (2018)
2. Mohammadi, N., Zangeneh, M.: Customer credit risk assessment using artificial neural networks. IJ Inf. Technol. Comput. Sci. **8**(3), 58–66 (2016)
3. Bazmara, A., Donighi, S.S.: Bank customer credit scoring by using fuzzy expert system. Int. J. Int. Syst. Appl. **6**(11), 29–35 (2014)
4. Kaminskyi, A., Nehrey, M., Zomchak, L.: Machine learning methods application for consumer banking. SHS Web Conf. **107**, 12001 (2021)
5. Kaminskyi, A., Nehrey, M.: Clustering approach to analysis of the credit risk and profitability for nonbank lenders. CEUR Workshop Proc. **2927**, 125–136 (2021)
6. Rahman, M.M., Ahmed, S., Shuvo, M.H.: Nearest neighbor classifier method for making loan decision in commercial bank. IJ Intell. Syst. Appl. **4**(8), 60–68 (2014)
7. Aniceto, M.C., Barboza, F., Kimura, H.: Machine learning predictivity applied to consumer creditworthiness. Fut. Buss. J. **6**(1), 1–14 (2020)
8. Munkhdalai, L., Munkhdalai, T., Namsrai, O.E., Lee, J.Y., Ryu, K.H.: An empirical comparison of machine-learning methods on bank client credit assessments. Sustainability **11**(3), 699 (2019)
9. Zomchak, L., Nehrey, M.: Economic growth and capital investment: the empirical evidence. In: Zhengbing, Hu., Zhang, Q., Petoukhov, S., He, M. (eds.) Advances in Artificial Systems for Logistics Engineering, pp. 645–652. Springer, Cham (2022). https://doi.org/10.1007/978-3-031-04809-8_59
10. Izonin, I., Trostianchyn, A., Duriagina, Z., Tkachenko, R., Tepla, T., Lotoshynska, N.: The combined use of the wiener polynomial and SVM for material classification task in medical implants production. Int. J. Intell. Syst. Appl. (IJISA) **10**(9), 40–47 (2018)

11. Shakhovska, N., Montenegro, S., Kryvenchuk, Y., Zakharchuk, M.: The neurocontroller for satellite rotation. Int. J. Intell. Syst. Appl. (IJISA) **11**(3), 1–10 (2019)
12. Derbentsev, V., Matviychuk, A., Datsenko, N., Bezkorovainyi, V., Azaryan, A.A.: Machine learning approaches for financial time series forecasting. In: Proceedings of the Selected Papers of the Special Edition of International Conference on Monitoring, Modeling & Management of Emergent Economy, M3E2-MLPEED 2020, Odessa, Ukraine, 13–18 July 2020, pp. 434–450 (2020)
13. Khrushch, N., Hryhoruk, P., Hovorushchenko, T., Lysenko, S., Prystupa, L., Vahanova, L.: Assessment of bank's financial security levels based on a comprehensive index using information technology. CEUR Workshop Proc. **2713**, 239–260 (2020)
14. Rubakha, M., Tkachyk, L., Hamkalo, O., Demkiv, K.: Integral assessment of banking activity effectiveness and rating of Ukrainian and Bulgarian banks. Econ. Stud. **28**(4), 70–96 (2019)
15. Hrytsiuk, P., Babych, T.: The cryptocurrencies risk measure based on the Laplace distribution. In: M3E2-MLPEED 2020, pp. 261–276 (2020)
16. Kichurchak, M.: Bank deposit activity in Ukraine: directions and factors of development activation. J. East. Eur. Central Asian Res. **6**(1), 145 (2019)
17. Kruppa, J., Schwarz, A., Arminger, G., Ziegler, A.: Consumer credit risk: individual probability estimates using machine learning. Exp. Syst. Appl. **40**(13), 5125–5131 (2013)
18. Kumar, M.R., Gunjan, V.K.: Review of machine learning models for credit scoring analysis. Ingeniería Solidaria **16**(1), 1–16 (2020)
19. Ampountolas, A., Nyarko, T., Date, P., Constantinescu, C.: A machine learning approach for micro-credit scoring. Risks **9**(3), 50 (2021)
20. Winahyu, R.K., Somantri, M., Nurhayati, O.D.: Predicting Creditworthiness of Smartphone Users in Indonesia during the COVID-19 pandemic using Machine Learning. In: 2021 International Seminar on Machine Learning, Optimization, and Data Science (ISMODE), pp. 223–227 (2022)
21. Izonin, I., et al. Stacking-based GRNN-SGTM ensemble model for prediction tasks. In: Proceedings of the 2020 International Conference DASA, pp. 326–330. IEEE, Bahrain (2020)

Research and Practice on OBE Teaching Reform of Automobile Theory Course-Taking the Automotive Service Engineering as an Example

Yi Wei(✉), Tianfeng Kuang, Lu Ban, and Daming Huang

School of Transportation, Nanning University, Guangxi 530200, China
1723872199@qq.com

Abstract. Cooperate with the professional certification work and the core curriculum construction, With the automotive theory course targeted, Research and practice of teaching reform based on OBE, Around the OBE's emphasis on addressing the core issues, The Student Center, Results-oriented, continuous improvement, To connect with the professional positioning, training goals and student reality, Determine the course learning objectives, Select the course teaching content, Building the course content framework, Form the five chapters, four cores, triple, two rings, one main line course content system, Integrate the course Ideological and Political OBE into the automotive theory course, Form the goal, element, implementation and effect as one of the curriculum ideological and political teaching design. It is pointed out that grasping the depth of teaching content and paying attention to the application of theoretical knowledge are the key to the OBE reform of application-oriented courses. Reform and innovating the teaching mode and implementing OBE+ are important ways to achieve the OBE goal.

Keywords: Automotive theory · OBE · Teaching reform · Automotive service engineering

1 Construction Background

Since 2015, the Ministry of Education and other three ministries and commissions jointly issued the *Opinions on guiding some local ordinary undergraduate universities to transform to application-oriented*, with the comprehensive promotion and deepening of the transformation, development, and reform of universities, application-oriented undergraduate education has become increasingly prosperous and vigorously. The application-oriented curriculum construction with the core characteristics of "student-centered, application-oriented and ability-oriented" has received great attention and attention, and has been listed as one of the main tasks to promote the application-oriented transformation and development of colleges and universities by the Ministry of Education. The document stresses that curriculum reform should be driven by social and economic development and industrial and technological progress, with a greater focus on cultivating learners' technological skills, innovation and entrepreneurship [1].

Z. Hu et al. (Eds.): AIMEE 2022, LNDECT 159, pp. 562–571, 2023.
https://doi.org/10.1007/978-3-031-24468-1_51

The *Automobile theory* is the theory that studies the performance of automobile. It is not only a theoretical application course, but also a basic course of professional technology [2]. Automobile theory based on automobile force analysis and experimental research, according to the external force on the car, analysis and automobile dynamics related to automobile performance, for automobile design, test, use and service provide theoretical basis, has important guidance to solve the problem of automobile complex engineering, is the basis of automobile professional theory level and innovation ability, in the highly qualified automobile talent training has extremely important position and significance, is the national standard of the important content of the automobile professional core knowledge.

Given this, the research group of *Automobile Theory* of Automobile Service Engineering major of Nanning University, in conjunction with the professional engineering certification work carried out by the school and the university-level core curriculum construction project of *Automobile Theory* carries out the research and practice of teaching reform based on OBE. Obviously do not tally with the applied course transformation and applied talents training requirements of defects and deficiencies, from setting out actually, geared to the needs of students, geared to the needs of engineering, geared to the needs of the application, by selecting the proper teaching content, using scientific teaching methods, the implementation of effective teaching organization, to achieve course teaching and talent training of organic coupling and seamless docking.

2 Curriculum Teaching Design Based on OBE

According to the requirements of THE OBE, the overall design of the course teaching revolves around the four problems that the OBE emphasizes solving four questions [3]:

(1) What learning outcomes are expected of the students;
(2) Why do students want to achieve such learning results;
(3) How to help students achieve these learning outcomes;
(4) How to know the students' learning outcomes.

Through extensive investigation and study, carefully summarize the teaching experience, fully absorb the reform achievement, follow the principle of "reverse design, is the implementation of", adhere to the industry characteristics and the reality of student characteristics, combining with the characteristics of applied talents training and applied course construction requirements, determine the course learning goals, choose teaching content, course content framework [4].

2.1 Determination of the Course Learning Objectives

According to "training can in automotive technology, marketing, insurance, transportation, and other fields, in the technical and commercial service of high.

quality applied talents" professional orientation, implement the school "core ability, employment orientation, position and use standard, all-round development" of the guidelines, combined with industry characteristics, working characteristics and professional characteristics, Docking the aim of training the talent training scheme and graduation requirements from focuses on helping students to be able to in the field of automobile service engineering, “professional performance evaluation”, “scientific analysis of the influence factors”, “the selection of reasonable management car”, in the related work, well done “car” “review car” “car” “tube” task needs, According to the concept and mode of OBE, the course objectives of automobile theory course are determined, as shown in Fig. 1.

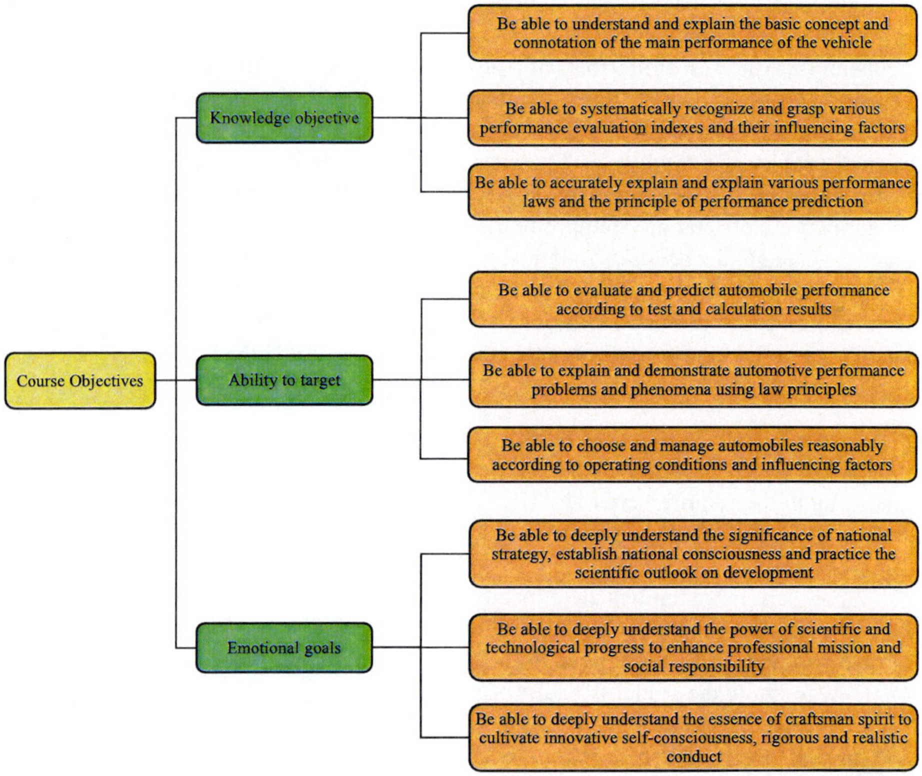

Fig. 1. Objective system of *Automobile Theory* course for automobile service engineering specialty

2.2 Screening of Course Teaching Content

OBE emphasizes student-centered, results-based, learning needs as the basis, learning results as the goal, planning students’ education and teaching process and content [5]. So the choice of course content, to learn the requirements analysis and student characteristic analysis as the foundation, from the industry demand characteristics and the actual conditions of students’ cognitive characteristics, combined with the requirement of applied talent training and applied course, based on the technical application ability and operation ability, based on technology application logic and action logic, reconstruction and optimization of curriculum content [6].

From learning demand, the main automobile service engineering is geared to the needs of the auto service field, develop the implementation of the auto serviceability, organization ability, management ability and the ability to communicate "service engineer", students' employment direction, mainly in automobile service enterprises engaged in technical and non-technical or supporting services. Therefore, the study of automobile theory should focus on cultivating students' ability to observe, analyze, evaluate and apply automobile performance and its rules, and the teaching content highlights the guidance of theory on the implementation, organization, and management of service work activities.

From the perspective of cognitive characteristics, Nanning College is a privately-run college. Influenced by students, it has low test scores, a weak knowledge base, poor symbolic thinking, poor abstract thinking ability, and strong utilitarian learning, which are the general situation and performance of privately-run college students. Automobile theory is to study the performance characteristics and laws of automobiles under the action of various external forces. The main characteristics of this course are many symbols, many concepts and many formulas, especially those involving dynamics equations that are more complicated and abstruse. Therefore, the course content should grasp the "depth" and pay attention to "application".

Based on the above analysis and understanding, based on the orientation of talent training, centering on the course teaching objectives, aiming at the cognitive characteristics of students, meeting the demand of employment posts, and responding to the ideological and political call of the course, the teaching content of automobile theory course is sorted out, selected and designed. There are mainly five directions and measures: One is to choose the content related to automobile test and application problems; the other is to avoid the derivation of abstract, complex, and esoteric dynamics problems; the third is to focus on theoretical application and qualitative analysis; the fourth is to set up independent courses for experiments; the fifth is to carry out curriculum ideology and politics.

2.3 Construction of the Course Content Framework

Implement the school's educational philosophy of "application-oriented, open and new experience" and the general goal of talent cultivation of "character, employment ability, entrepreneurial ability, the foundation for further study, and sustainable development" [7]. Based on the selection and design of teaching content, construct the curriculum content framework. At the macro level, it focuses on the transformation from knowledge to ability, from learning to application, and from profound to solid. In the middle view, focus on the basic concept, evaluation indicators, influencing factors, and solution learning; On the micro-level, emphasis is placed on the interpretation and application of performance laws reflected by the relationship between force on objects, variation of factors, and analysis of variables, forming the curriculum content system framework of "five chapters, four cores, three rings, and one mainline", as shown in Fig. 2.

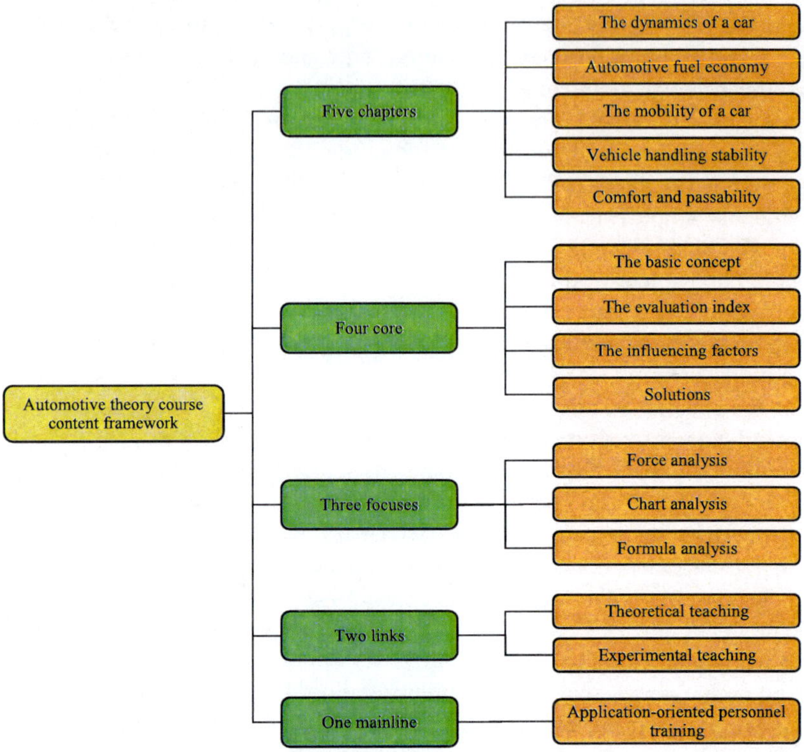

Fig. 2. Content and framework of automobile theory course of automobile Service Engineering major of Nanning College

3 OBE + Advanced Teaching Methods and Models

In accordance with the OBE concept, actively explore the "action learning" and "situational learning" of theoretical courses, carry out the integrated application of information technology and education and teaching, and flexibly and organically apply advanced teaching methods and effective organizational modes such as BOPPPS, PBL, and SPOC into the course teaching [8]. Change traditional curriculum pattern "teaching method of the single" and "knowledge transfer" one-way "teaching form is monotonous, the teacher-centered teaching into student-centered teaching, change passive learning into active learning, learning alone for collaborative learning, rote learning for inquiry learning, stimulate students learning motivation, promote the students' interest, improve teaching effect.

BOPPPSS is an interactive teaching mode, which can obviously mobilize students' learning enthusiasm, activate the classroom learning atmosphere, and improve the teaching effect when combined with OBE. Using the successful experience of BOPPPS for reference, the teaching method of BOPPPS is introduced in class teaching "according to the conditions of the class", and the teaching design is carried out selectively according to the requirements of the BOPPPS model, focusing on the three links of "introduction", "participation in learning" and "summary", and the ideological and political content of the course is integrated into it [9]. Through careful design, the "introduction" is wonderful, vivid, and fascinating, the "Participation in Learning" is rich in content and diversified forms, and the "summary" is concise, easy to understand, simple, and comprehensive.

Combined with the characteristics of the course, the PBL teaching mode is introduced into classroom teaching and experimental teaching respectively. The classroom teaching adopts the BOPPPS + PBL design. In the participative learning link, students are guided to carry out exploratory learning and cooperative learning by taking the student group as the unit, taking the topic discussion as the form, and taking the problem design as the guidance. In experimental teaching with pilot projects as the carrier, with "task drive + project design guide + situation, role-playing and group cooperation" as the organizational form, students, through the experiment project are expected to complete the LABS working condition, and the effect, the level of data obtained from the goals, rules and conclusions, PBL based on the concept and characteristics of project learning.

Actively explore the teaching design in the SPOC environment, and carry out the "SPOC+" mixed-mode integrating task-driven and problem-oriented teaching methods to achieve online and offline, in-class and extra-curricular complementarity, mutual benefit, and mutual promotion. Based on BOPPPS, SPOC online teaching is carried out through the combination of "Superstar Learning Pass + WeChat public account". Through video library, picture library, animation library, case library, question library, micro-lesson library, and so on, we can complete the teaching process of preview self-test, communication, and discussion, homework practice, answer and solve problems, examination, and examination, analysis, and evaluation on the learning platform. Combined with the construction of a virtual simulation experiment training teaching platform, carry out simulation experiment and virtual training.

4 Experimental Teaching and Curriculum for Ideological and Political Affairs OBE

4.1 The Experimental Teaching of the OBE

Automobile theory studies the various properties of automobile under the action of external forces [10]. It is a comprehensive course in which theory and practice complement and complement each other. Performance test is not only an important content but also an important application of automobile theory, so the experimental.

teaching with performance test as the main content is an important part of automobile theory teaching, is an important part of automobile theory teaching. In order to fully reflect the great importance of experimental teaching and give full play to the important role of experimental teaching in talent training, and highlight the characteristics of "open, situational, learning by doing" applied curriculum, automobile theory curriculum carries out the reform of independent experimental courses.

Automobile theory curriculum implementation independent experiment course, which is separated from the course of the experiment part, to the technology application ability and operation ability training as a starting point, around the implementation of the technical service and communication ability, organization, management, integrated courses experiment, carried out in accordance with the working technology application logic and action logic reconstruction, reconstruction and innovation, Form a relatively independent comprehensive experimental course of automobile performance based on comprehensive design experiment, achieve and strengthen students' complete, systematic and specialized training of technical methods and abilities or practice of working process and methods with a complete experimental course system and relatively independent content system.

4.2 Ideological and Political Theories Teaching in All Courses Base on OBE

Ideological and Political Theories teaching in All Courses is in comprehensive promote the moral education curriculum reform in colleges and universities in recent years, for the further implementation of xi Jinping, general secretary of the ideological and political work in colleges and universities across the country put forward at the meeting, "the ideological and political work through the education teaching process", "to use a good main channel of class teaching", "to make all kinds of courses with the direction of ideological and political theory, The formation of synergistic effect "and other important discussions, and the generation of important concepts and new ideas [11]. Combined with curriculum ideological and political construction, the research group of "Automobile Theory" earnestly implements the implementation opinions of the school's comprehensive promotion of curriculum ideological and political construction and professional ideological and political construction, and fully excavates ideological and political elements for curriculum ideological and political design through demonstration curriculum project establishment.

As shown in Fig. 3, with the national economy and people's livelihood, situation and policies, reform and development, achievement and progress, spiritual beliefs, thoughts, sense of responsibility, honor, and mission as the ideological and political elements, the government aims to establish national sentiment, enhance cultural confidence, cultivate awareness of rules, form a professional style, and promote the spirit of craftsman. The film and television pictures, news reports, topic discussion, and problem-solving, such as ways to carry out the curriculum education, help students to understand the national strategic significance, scientific and technological progress, and craftsman spiritual essence, guide students to establish national.

consciousness fulfill the scientific outlook on development, strengthen the professional sense of mission and social responsibility, and cultivate good moral excellence rigorous truth-seeking.

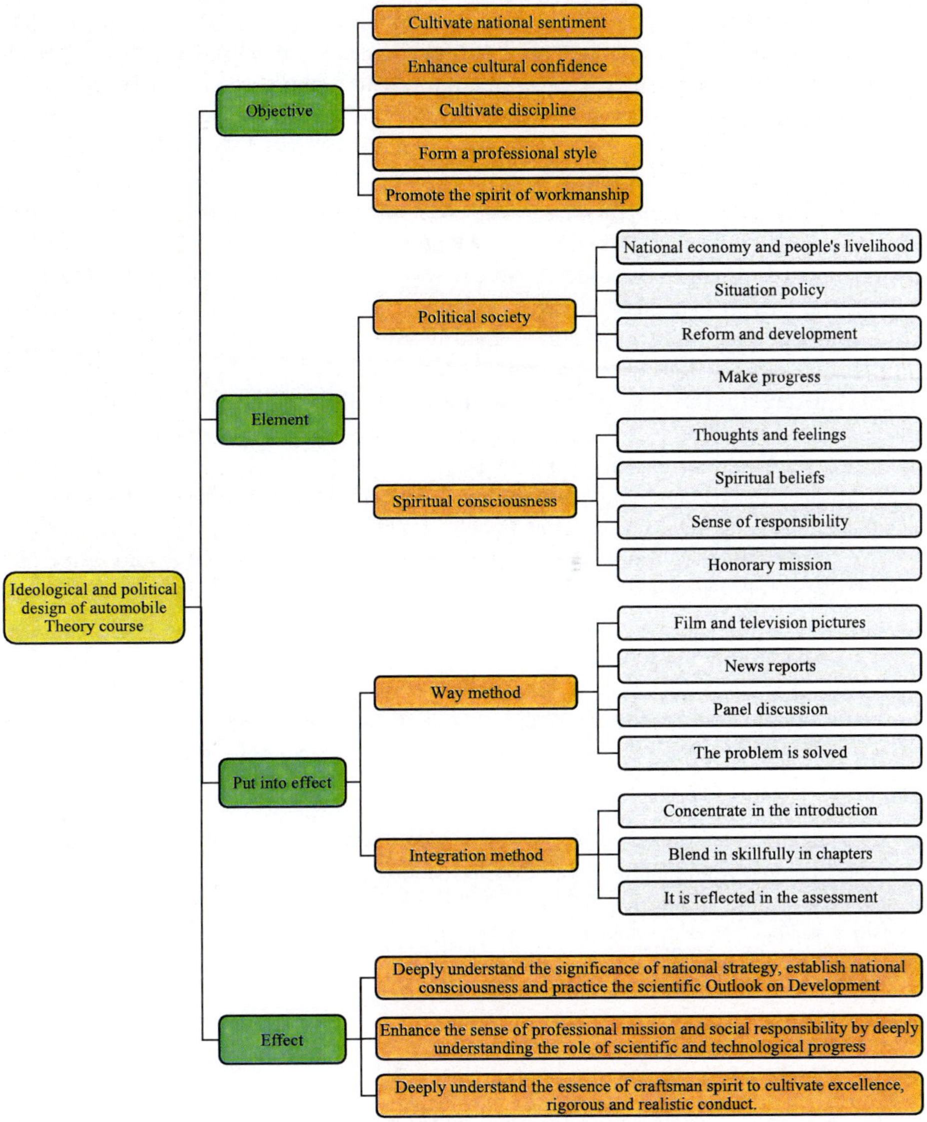

Fig. 3. Ideological and Political design of Automobile Theory course for Automobile Service Engineering major of Nanning University

5 Conclusion

Under the background of engineering certification, the results-oriented OBE concept has been widely concerned and applied in promoting the construction of applied undergraduates. OBE is a kind of advanced education idea, the "student center, results from orientation, continuous improvement," the core idea, not only for teaching reform to provide clear direction and goals, and to solve the problem of teaching is the effective path and method, both conform to the national education teaching reform train of thought, and fit application undergraduate course education development request, It is regarded as a good prescription to solve a series of contradictions and problems in the process of transformation construction and development of colleges and universities, and becomes an inevitable choice to cultivate high-quality applied talents.

Grasping the "depth" of teaching content and paying attention to the "application" of theoretical knowledge is the key to the OBE reform of the applied curriculum. The essence of OBE is "student-centered", so teaching must be student-centered and proceed from the actual situation of students, and strive to achieve the organic unity of course content and students' needs and abilities, so as to meet both the learning needs of the major and students' learning abilities. Learning requirements that are too low or too high in difficulty and out of touch with the actual needs of work are not conducive to stimulating and mobilizing students' learning motivation and enthusiasm, but will also make students bored or negative attitude, thus affecting the output of learning outcomes and the achievement of learning objectives.

To reform and innovate the teaching mode and implement "OBE+" is an important way to achieve the goal of OBE. "How to help students achieve learning results effectively" is an important core issue concerned by OBE "Student-centered, results-oriented". Therefore, the course teaching design based on OBE should not only pay attention to the selection of teaching content but also pay attention to the innovation of the teaching model. The implementation of "OBE+" is to promote the integration of information technology and course teaching, promote the combination of OBE with advanced teaching methods, positive teaching strategies, and effective teaching organizations on the basis of OBE ideological orientation and teaching design, so as to improve the teaching effect and help achieve OBE goals.

Acknowledgment. This project is supported by the Projects of Teaching Reform in Higher Education in Guangxi: Automobile Theory Course Ideological and Political Demonstration Course Construction Project (Grant no. 2021SZSFK04), the Projects of Teaching Reform in Higher Education in Guangxi: Automobile Theory Major Core Course Construction Project (Grant no. 2021BKHXK01).

References

1. The Development and Reform Commission of the Ministry of Education (2015) no. 7 Opinions on some local ordinary undergraduate universities to transform to Application oriented. Ministry of Education, Beijing (2015). (in Chinese)
2. Chen, J., Wang, X., Shen, T.: Research on the course content design and examination system of *Automobile Theory* based on OBE concept under the new engineering background. University **39**, 48–51 (2021). (in Chinese)
3. Chang, J., Zhang, X.: Construction and practice of practical teaching system based on OBE concept-taking electronic information engineering as an example. China. Univ. Teach. **Z1**, 87–92+111 (2021). (in Chinese)
4. Rajak, A., Shrivastava, A.K., Bhardwaj, S., Tripathi, A.K.: Assessment and attainment of program educational objectives for post graduate courses. Int. J. Mod. Educ. Comput. Sci. (IJMECS) **11**(2), 26–32 (2019)

5. Tan, W., Lu, B., Huang, D.: Research and practice of curriculum reform based on OBE+ BOPPPS-Take engineering drawing course as an example. Guide Sci. Educ. **04**, 76–78 (2022). (in Chinese)
6. Li, D., Zhu, W., Chen, Z., Wang, C.: Research of overall optimization based on a series of data structure courses teaching content. Int. J. Educ. Manage. Eng. (IJEME) **2**(11), 18–23 (2012)
7. Qin, G., Sun, S., Ye, H., Lu, D.: Strategies on teaching reformation for mechanical manufacturing technology. Int. J. Educ. Manag. Eng. (IJEME) **1**(6), 70–75 (2011)
8. Wang, D., Yuan, B., Zhu, R., Han, Q., Shun, J.: Teaching reform and exploration of "Advanced Manufacturing Technology" curriculum based on OBE concept. Teacher **07**, 87–89 (2022). (in Chinese)
9. Liang, P., Zhang, C., He, X., Zeng, Q.: On the teaching design of the space-time view of special relativity under the BOPPPS Mode. Guide Sci. Educ. **33**, 122–123 (2018). (in Chinese)
10. Yue, C., Chen, J.: Discussion on the teaching of ideological and political theory course based on OBE concept. Educ. Teach. Res. **36**(06), 58–67 (2022). (in Chinese)
11. Ma, F., Yang, P., Wang, X., Cui, Y.: Exploration and practice on reformation mode of undergraduate tutorial system in faculty of computer science. Int. J. Educ. Manage. Eng. (IJEME) **2**(4), 54–58 (2012)

Interactive Strategies of Chinese Cloud Classroom Based on the Perspective of Meta Universe

Haifeng Yang[1,2], Zhuoda Yang[3], and Ling Tao[1(✉)]

[1] Center for Teaching Chinese as a Foreign Language, Wuhan University of Technology, Wuhan 430070, China
taotaoxiruo@163.com

[2] Faculty of Letters, Arts and Sciences, Waseda University, Tokyo, Japan

[3] School of Chinese Studies and Cultural Exchange, Renmin University of China, Peking 100089, China

Abstract. With the American social software Facebook officially announced its name change to "Meta", the concept of "Meta Universe" has really entered the public field. The meta universe is an Internet social form based on the integrated application of various emerging information technologies. It has transformative value for the future development of all walks of life, including the education industry. Since the outbreak of covid-19, online Chinese learning of most foreign students has changed from a short-term accidental state to a normal state. In this way, we must pay attention to how to seize the opportunity of the development of meta universe technology, combine the meta universe cloud classroom with AI digital people, and create a new scene of educational application integrating multiple technologies under the new AI ecology. Only in this way can we build a new educational paradigm change, and solve some bottlenecks and difficulties in the current development of Chinese cloud classroom interaction for foreign students, so that classroom teaching can truly promote the development of students' knowledge, ability and emotion, and finally cultivate students' innovative spirit and practical ability.

Keywords: Meta Universe · AI digital people · Chinese cloud classroom · Strategies of interaction

1 Introduction

The concept of meta universe appeared firstly in the scientific novel named Snow Crash of Neal Stephenson in 1992. The meta universe refers in particular to a virtual world parallel to the real world, and everyone participates in relevant social activities by controlling their own virtual avatars in this world [1]. With the American social software Facebook officially announced its name change to "Meta", the concept of "Meta Universe" has really entered the public field [2, 3]. The meta universe is an Internet social form based on the integrated application of various emerging information technologies.

Z. Hu et al. (Eds.): AIMEE 2022, LNDECT 159, pp. 572–582, 2023.
https://doi.org/10.1007/978-3-031-24468-1_52

The meta universe is described as a new Internet ecology integrating a variety of new technologies by Professor Shen Yang of Tsinghua University, which can provide users with immersive experience. It can generate the economic system, social system and identity system of the virtual world through digital twin technology and blockchain technology, allowing users to produce and edit contents [4]. Although the current teaching methods of online classroom temporarily provide some assistance to the effective operation of the current Chinese classes, they also bring a series of new problems and challenges in class interaction. So with the help of meta universe, we can offer more immersive experience of Chinese classroom learning, and this is exactly what foreign students need [5]. Since 2020, facing the increasingly tense epidemic situation, most universities have adopted online teaching methods to carry out Chinese teaching. Although this teaching method has temporarily provided some help for the current international Chinese teaching, it has also brought a series of new problems and challenges in classroom interaction. What problems still exist in the integration of Chinese cloud classroom designed into the current online teaching interaction from the perspective of meta universe and how to promote the effective improvement of relevant interaction are the core issues of this paper.

2 Analysis of Interaction Methods of Chinese Cloud Classroom

2.1 Online Conference Platforms

With clear layout and simple operation, the online conference platform can be used on demand and meet the basic needs of online Chinese cloud classroom teaching. And TCSL teachers and students can also use some tools in the online conference platform to assist in teaching, making the interactive process more vivid and convenient [6]. The problems are sorted out in the Table 1.

Table 1. Summary of basic information about online conference platforms.

Names of platforms	Interactive tools	Limit to the number of users	Remarks
Tencent meeting	Chat area, Whiteboard, Electronic red Envelope	300	/
Dingding	Chat area, Whiteboard	30	/
Zoom	Chat area, Whiteboard, Questionnaire	100	Audience area can be set. The audience can only watch without interaction

However, the online conference platforms also have a series of drawbacks. On the one hand, these platforms can accommodate a limited number of users, and some large classes (more than 100 students) may not be able to operate due to the large number.

In addition, because of network and technical reasons, if more than 15 people turn on the camera simultaneously, it often causes network lag, delay of sound and picture. On the other hand, such platforms are designed to retain the content of the current class and if participants exit the meeting, these contents cannot be saved, which is not conducive to students' review of knowledge and teachers' reflection [7].

2.2 Course Live-Streaming Platforms

The advantage of the course live-streaming platform is that the interactive tools are more suitable for the actual needs of the classroom, and the choices are more diversified for teachers to use at will. It can imitate the real classroom teaching environment as much as possible, and let students devote themselves to the cloud classroom interaction. In addition, teachers can upload Powerpoints, reading materials, Word documents and other documents to interact with students orderly and hierarchically according to teaching chapters, thus these contents can be checked again in the history message record. These practices are convenient for students to preview before class and review after class, and also help teachers to collect the feedback of students' interaction information (such as signing in, signing out, group evaluation form, student speech record). The basic information of such platforms can be sorted out as follows:

(1) Tencent Classroom: It uses chat area, whiteboard, tool for raising hand, mock podium, mock examining room, questionnaire to interact with others. But it limits the number of users to 50.
(2) Classin: It uses chat area, whiteboard, laser pen, timer, tool for roll call, simulated teaching AIDS, tool for raising hand for the interaction. Also no more than 50 users can study with it together.
(3) Yu Class: It utilizes chat area, whiteboard, bullet comments, tool for roll call, classroom electronic red envelope, questionnaire within 50 students.
(4) Xuexitong: There is no limitation for using it. And everyone can make use of its chat area, whiteboard, tool for signing in, group discussion board.

The course live-streaming platform still has some shortcomings. Firstly, most of these platforms only have Chinese version, so the difficulty of language understanding leads to the low utilization rate of various teaching interactive tools for international students. Secondly, these platforms are usually limited to 50 students and are more suitable for small and middle class teaching, therefore, it is not conducive to organizing teaching activities such as classroom observation, large class teaching and lectures. Thirdly, these course live-streaming platforms were framed to serve traditional classes of Chinese students initially, so they lack some teaching tools and targeted methods in line with the characteristics of online Chinese courses, such as mock Chinese character cards and Pinyin chart.

2.3 Video Live-streaming Platforms

Video live-streaming platforms are represented by Bilibili, Tiktok and Youtube. The interaction of cloud classroom on these platforms has the following three advantages:

Firstly, it provides interesting and diverse ways of interaction, such as inputting bullet comments, sending small gifts, lighting up signboard, and voting, thus it effectively mobilizes the enthusiasm of students and lighten up the atmosphere. Secondly, there is no limit to the number of students involved in the platforms learning, which is convenient for Chinese teachers to organize large classes. Meanwhile, Chinese teachers can also share the live links of Chinese classes in other universities for international students to participate in interactive discussions, realizing the sharing of online teaching resources. Moreover, in the process of learning on such platforms, the behavior of searching from international students by entering keywords related to Chinese learning will guide the algorithm of the platforms to identify the content that meets their interests, and recommend a large number of learning resources such as relevant videos, high-quality bloggers to them, which can play an auxiliary role in classroom teaching, supplement extracurricular knowledge and expand learning channels. The basic information of such platforms can be sorted out as the Table 2 shows:

Table 2. Summary of basic information about video live-streaming platforms.

Names of platforms	Interactive tools	Limit to the number users	Remarks
Bilibili	Inputting bullet comments, voting connecting mic, making comments	No	/
Tiktok	Inputting bullet comments, sending gifts, lighting up signboards, connecting mic, making comments	No	Different kinds of gifts have different values, and users need to top up the platform account
Youtube	Inputting bullet comments, voting, making comments	No	/

However, there are still a series of problems with live-streaming video platforms in cloud classroom. On the one hand, the function of these platforms is entertaining, so they are not suitable for serious teaching occasions. On the other hand, the content of these platforms is usually characterized by fragmentation and simplicity, and they often use exciting sounds and colors to attract audiences. Therefore, the information generated on these platforms will have a certain interference effect on teaching. Students tend to be distracted by other information during the class and cannot concentrate on it, which is not conducive to participating in the interaction [8].

3 Problems of Interaction in Cloud Classroom

3.1 Questionnaire Survey

Aiming at the existing problems of interaction in cloud classroom, we selected 200 international students from School of Chinese Studies and Cultural Exchange of Renmin University of China, Institute of International Education of Wuhan University of Technology, Institute of International Education of Central China Normal University as samples for investigation, including 94 undergraduates, 60 master students and 46 doctoral students. According to their HSK levels, they are classified into 3 categories, namely 86 in primary stage, 64 in intermediate stage and 50 in advanced stage based on their HSK levels. Their online courses include listening and speaking, reading and writing, comprehensive courses and Chinese culture. Among these international students, the majority are from Asia (45%) and Europe (25%), the rest from Africa (15%), America (10%) and Oceania (5%).

In our survey, almost all students have used more than one platform for interaction, and their choices are as shown in Fig. 1.

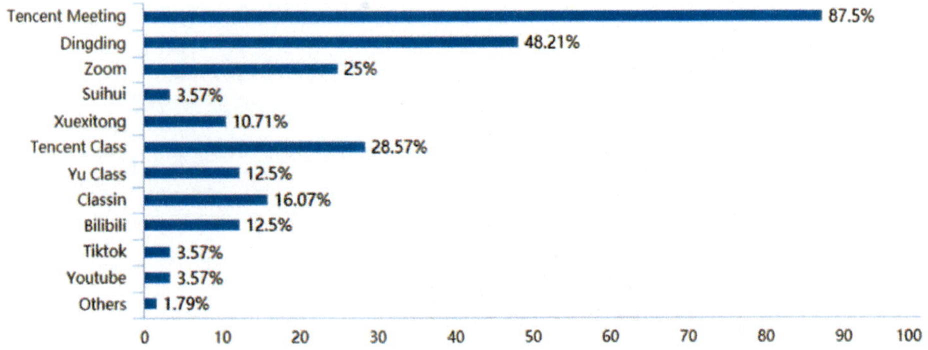

Fig. 1. Statistics of major online Chinese class interactive platforms for overseas students

Tencent Meeting and Dingding are the platforms used frequently by overseas students, the following are Tencent Classroom and Zoom. In the questionnaire, we paid special attention to the evaluation of the international students towards teacher-student interaction and student-student interaction in the online classroom. On the one hand, the author counted the overall frequency of the two kinds of interaction. On the other hand, the students were also asked to give the scores of the two kinds of interaction independently to get their perspective. Table 3 shows the score statistics of teacher-student interaction in online Chinese classes for international students.

According to the above statistical results, most international students are just content with or not very content with the teacher-student interaction and student-student interaction in cloud classroom, so their rating is basically around 3. Meanwhile, in terms of interaction frequency, the overall pace of cloud classroom is almost dominated by teachers. The percentage of "interacting regularly" is less than 20%. More than 60% of students think that there are few interactive links in cloud classroom, focusing mainly

Table 3. Statistics of international students' scores on teacher-student interaction in online Chinese classroom

Scores	0–1 score	1–2 scores	2–3 scores	3–4 scores	4–5 scores
Percentage	0	10.71%	39.29	33.93%	16.07%

on teachers' explanation, and other students are not willing to engage in interaction. It can be seen that the existing cloud classroom teaching still has major problems in terms of interaction frequency and rhythm, which need to be improved.

3.2 Existing Problems

3.2.1 Teacher-Student Interaction

The primary problem in teacher-student interaction of cloud classroom is that Chinese teachers are sometimes inconvenient or unwilling to turn on cameras and microphones, thus affecting the interaction effect. This accounted for 76.79% of the responses collected. In this way, it is difficult to transfer teaching-related information through facial expression, gesture simulation and other body language, and it is difficult to timely coordinate verbal information to give students teaching-related feedback, affecting the cloud classroom interaction effect further. In addition, 33.93% respondents indicated that some teachers designed too much preparation tasks for cloud classroom interaction for students, so students have to spend a lot of time and energy searching materials, watching online course videos and so on. Therefore, these students do not have time and energy to further engage in classroom interaction for in-depth learning. In the process of dealing with a large amount of preparation work, their interest in mastering new content through classroom interaction gradually faded, and ultimately failed to achieve satisfactory learning results [9].

3.2.2 Student-Student Interaction

In terms of student-student interaction of cloud classroom, the question with the most feedback is that they are inconvenient or unwilling to open cameras and microphones during interaction. The author believes that the factors affecting this problem mainly include external environment and learner psychology, which will be elaborated in the next section. Another problem is that 55.36% of overseas students believe that cloud classroom lacks a collective learning atmosphere, which makes them unable to concentrate on interactions. On the one hand, teachers are unable to understand and manage the interaction process of international students face to face because students do not study in a fixed offline classroom. On the other hand, international students are far away from each other in reality, and they cannot timely communicate with each other to solve their learning problems. Due to the avoidance of learning difficulties, their involvement in cloud classroom interaction will also be reduced [10]. In addition, the lack of interaction and the single form of interaction are still the crucial points that are difficult to be solved for a long time, and seriously affect the effect of interaction in teaching practice.

3.3 Analysis of Influence Factors

3.3.1 External Factors

The author thinks that the learning platforms and resource allocation are one of the factors that affect the interactive effect of cloud Chinese classroom. In online interaction, some platforms have complicated operation procedures and functions, and lack English version, which causes some inconvenience for international students. Meanwhile, different Chinese teachers prefer different learning platforms, or Chinese teachers of the same course require international students to use multiple learning platforms, which brings a lot of trouble to the international students, making the mistakes of locating learning resources and remembering interactive tasks.

3.3.2 Internal Factors

In terms of Chinese teachers, some of them are not adapted to the cloud classroom, which makes them unable to carry out innovative design of class interaction, sticking to the traditional lecturing method finally. In the questionnaire, 44.64% respondents said that the online Chinese classroom interaction was infrequent and monotonous, resulting in the loss of students' interest. The author thinks that the main reason is less time for Chinese teachers to prepare before class, ranging from teaching platform construction, course organization to class teaching and interaction arrangement. In addition, there is not much experience to refer to, and the teaching equipment is not enough, so teachers can only rely on the traditional teaching method to a large extent [11].

For international students, they must quit the original face-to-face offline classroom learning mode and adapt to the cloud classroom learning mode from a long distance across the screen, which is difficult for them. On the one hand, international students lack online learning experience, so it is difficult to cope with some new challenges in teaching interaction practice, such as lagged Internet and inconvenience of communicating with teachers and students. On the other hand, foreign students must overcome certain psychological factors in the interaction process, among which anxiety and avoidance are the most prominent.

Characterized by few forms of interaction and tight teaching pace, cloud classroom requires international students to absorb a large amount of Chinese knowledge in a short time, which causes strong psychological anxiety on students. Also, under the background of the increasingly serious global pandemic, international students' anxiety about their health, surrounding environment and future academic development is also easily projected into the classroom, making it difficult for them to focus on teaching and interaction. Furthermore, due to various inconveniences of cloud classroom and anxiety, many overseas students choose to sidestepping, such as avoiding answering questions and classroom interaction, making excuses for not turning on the cameras and microphones. Such avoiding psychology and behavior can weaken the effect of online Chinese classroom interaction, and it is difficult for both sides to get effective teaching feedback in time, ultimately not obtaining satisfactory teaching effect.

4 Meta Universe Assists Online Chinese Teaching

The above problems can be well solved in the meta universe. At present, the application of meta universe technologies in teaching Chinese as a foreign language is still in the exploratory stage, but it has great research prospects.

4.1 The Application of Meta Universe in Education

4.1.1 Create a 'Virtual' Learning Environment

Many governments are exploring ways to apply the meta universe to education. Through the use of more complex modeling, mapping, lighting and other technologies to achieve the true reproduction of the real environment, so that students can experience the real learning environment. Students enter the virtual space for classroom learning activities by wearing VR glasses and other devices. The meta universe provides immersive experience based on the extended reality technology, generates the mirror image of the real world based on the digital twin technology, and closely integrates the virtual world with the real world, providing a new model for educational reform. The meta universe makes the presentation of knowledge more audio-visual. For example, the operation of machinery can be directly observed by simulating the operation process of expensive mechanical equipment. It can let students experience historical and humanistic knowledge by simulating real historical and humanistic scenes, etc. Realize the real-time interaction of learning through virtual teachers. Through the creation of multi person 3D interactive space, the interactivity of learning can be realized. Meta universe can expand the depth and breadth of online education, achieve active interaction among teachers, students, resource developers and other aspects, and promote the improvement of teaching effectiveness, so as to create an "ideal classroom".

4.1.2 Constructing Rich Teaching Resources

The traditional classroom teaching resources are mainly the school or inter school cooperation resources. Auxiliary teaching resources are generally divided into physical teaching resources and image teaching resources. Physical teaching resources are subject to cost and other factors, and their richness is limited. However, the image resources are displayed in the form of pictures and videos, and there are some problems such as insufficient physical display. Meta universe can break through the cost limitation of physical resources and the two-dimensional display limitation of image resources. Virtualized teaching resources let students access to a large number of teaching resources that are closest to the real, and let students break through the limitations of space and time, and use these rich and high-quality teaching resources anytime and anywhere.

4.1.3 Promoting Autonomous Learning

The future talents are compound talents with the abilities of communication, cooperation, creation and innovation. On December 12, 2017, the organization for economic cooperation and Development released the pisa2018 global competency assessment standard at Harvard University in the United States. The assessment aims to test students' ability to conduct cross-cultural analysis of regional and global issues and the ability to respect cultural differences and interact with people from different backgrounds in an open, appropriate and effective manner. To test whether they have the ability to integrate into the society with a positive and open mind and interact with the external environment according to the actual needs of personal and social development, the meta universe plays an important role in the cultivation of these abilities. In the meta universe education model, more attention is paid to students' own independent inquiry ability, including the ability to collect, analyze and integrate resources and the ability to select personalized learning program and the ability to choose interdisciplinary learning mode, etc. Students can cooperate between virtual members who choose the same project in different fields and backgrounds. Meta universe provides a super platform for students' autonomous learning and self-realization.

4.2 The Application of Meta Universe in Chinese International Education

4.2.1 Creating Immersive Learning Scenarios

The most important condition for language learning is language environment. At present, affected by the epidemic, most overseas students are unable to study in China. They can't touch the real Chinese language environment. Online teaching is restricted by current technology, and it cannot provide sufficient Chinese social and cultural environment for foreign students. Meta universe can effectively solve this problem. It can make full use of AR/VR technology to create a 3D multi person interactive teaching space. By creating a virtual and real digital learning scene and a real language environment, students can learn language immersively. Creating a real environment of universities for students studying abroad in through virtual technology. So that students can clearly see the overall layout of the school, the surrounding area of the school and the face of the city where the school is located. In combination with the themes of "go shopping", "go to the bank", "go to the canteen" learned in the text, students can learn in the virtual real scene, have a dialogue with the virtual characters.

4.2.2 Creating Diversified Learning Resources

The current online classroom of Chinese international education is difficult to achieve the ideal teaching effect. One reason is that online learning resources are single and poor. Meta universe can gather high-quality resources for teaching Chinese as a foreign language in major universities around the world to create a rich learning platform. At the same time, meta universe can also formulate personalized learning programs and spaces according to students' individual differences to provide personalized learning resources that adapt to each student's learning characteristics, personality characteristics and thinking habits. For example, in addition to the common courses, students can choose individual courses according to their own professional characteristics, interests and practical needs, such as providing business Chinese learning resources for people engaged in commercial trade and tourism Chinese teaching courses for tourists.

4.2.3 Creating a Multi-channel Learning Approach

In traditional Chinese online courses, only in the course of class can teachers and students have full and immediate interaction. And after class, teachers and students can only communicate through chat software such as WeChat and QQ. There is no constant interaction between teachers and students. The teacher is unable to give timely feedback and answers to the students' problems at any time. This affects the students' learning effectiveness and enthusiasm [12]. However, in the meta universe, the birth of virtual teachers can provide an accompanying learning mode, which can always give learners emotional companionship which greatly improves learners' learning initiative and learning quality.

Multi person 3D learning area can be built in the meta universe to realize human-computer interactive learning. The highly interactive learning style strengthens the exchange and communication between students, and promotes the efficiency of joint learning, to improve students' ability of cooperation and communication.

Meta universe can also realize interdisciplinary learning and circular learning and provide a rich resource of experimental learning platform for students. In traditional Chinese international teaching, Chinese learning is separated from foreign students' professional learning. Meta universe can combine Chinese language knowledge with students' professional knowledge to achieve the purpose of interdisciplinary learning to assist foreign students to use Chinese for professional learning. It can also realize circular learning, test language level through human-computer interaction, so as to carry out circular learning and gradually consolidate learning content.

5 Conclusion

Through the questionnaire survey, this paper reveals a series of problems existing in the cloud classroom interaction in the current Chinese international teaching:

1) Online teaching platform cannot meet the actual needs of Chinese International Education.
2) Online teaching resources are very scarce, which is not enough to support students' in-depth learning and personalized learning.
3) The single online teaching method cannot meet the needs of students' learning and practice after class.

According to this, this paper believes that the best way to solve the outstanding problems of the current Chinese international online teaching and ensure the effectiveness of the Chinese international online teaching is to introduce the meta universe into the Chinese international teaching. The main functions are:

① Creating immersive learning scenes.
② Creating diversified learning resources.
③ Creating a multi-channel learning approach.

Acknowledgement. The program is supported by "The Fundamental Research Funds for the Central Universities" (WUT: 2022VI006-02). The Ministry of Education's Higher Education Department Industry-Academy-Education Project "Based on the Multi-terminal-based 'Introduction to Teaching Chinese as a Foreign Language' Premium Online Course Construction" (201801049043), and the research results of the National Social Science General Project "The Semantic Evolution of Chinese Willing Verbs" (21BYY138). The paper of this project is also supported by the National Study Abroad Fund.

References

1. Shen, Y.: Language changes in the Meta-cosmic era. Res. Lang. Strateg. **3**(7), 5–6 (2022). (in Chinese)
2. Iman, T., Nissrine, S.: Meta-analysis of systematic literature review methods. Int. J. Mod. Educ. Comput. Sci. (IJMECS) **11**(2), 17–25 (2019)
3. Brisebois, R., Abran, A., Nadembega, A.: A semantic metadata enrichment software ecosystem based on metadata and affinity models. Int. J. Inf. Technol. Comput. Sci. (IJITCS) **9**(8), 1–13 (2017)
4. Zhang, Z., Feng, J., Zheng, L.: AI Digital human cloud classroom based on the concept of meta universe. Chin. Mod. Educ. Equip. **7**(8), 7–11 (2022). (in Chinese)
5. Shi, J., Wang, L.: An investigation into online Chinese teaching of international students in universities under the pandemic Covid-19. Lang. Teach. Linguist. Stud. **1**(4), 23–33 (2021). (in Chinese)
6. Xiao, R., Zhao, J.: Research on effective interaction of online Chinese teaching in post epidemic era. J. Int. Chin. Teach. **3**(3), 90–96 (2021). (in Chinese)
7. Xu, L., Chen, Y., Shi, Y.: An analysis of international students' experiences of online courses: based on a survey of international students at a Chinese university. J. Int. Chin. Teach. **5**(1), 39–49 (2021). (in Chinese)
8. Wang, Y.: A study on interactional competence in Chinese as second language classrooms from a sociocultural perspective. J. Educ. Inst. Jilin Province **2**(2), 82–85 (2021). (in Chinese)
9. Li, Y., Li, B., Song, H., Bai, L., et al.: International Chinese language education under Covid-19: challenges and countermeasures. Lang. Teach. Linguist. Stud. **3**(4), 1–11 (2020). (in Chinese)
10. Sun, D., Li, M.: The construction principle of the interactive teaching models. Chin. Lang. Learn. **3**(3), 93–98 (2019). (in Chinese)
11. Chen, W., Liu, G., Li, J.: An empirical study of Chinese L2 learners' foreign language enjoyment and anxiety. J. Yunnan Normal Univ. (Teach. Study. Chin. Foreign Lang. Edn.) **2**(19), 44–53 (2021). (in Chinese)
12. Van Compemolle, R.A.: Interaction and Second Language Development, pp. 50–60. John Benjamins Press, Berlin (2017)

Current Situation and Trend Analysis of Domestic General Upgraded Education Research Based on CiteSpace

Yunyue Wu(✉), Liwei Li, and Xiaoli Teng

School of Transportation, Nanning University, Nanning 530200, Guangxi, China
W17736621947@163.com

Abstract. Upgraded education is an important form of higher education in China. In recent years, the country has expanded the enrollment scale of upgraded education. Using Citespace 6.1 software, taking CNKI database as the literature source and taking 2012–2021 as the time node, 464 journal papers related to domestic general upgraded education were retrieved. The volume of papers, journal sources, authors, institutional cooperation, research hotspots and research trends were analyzed and the corresponding maps were formed, so as to show the research hotspots and research trends of domestic general upgraded education. After analysis, the following results are obtained: the domestic general college to undergraduate education research shows a gentle trend, and the research content is mainly carried out from the aspects of teaching reform, talent training, student management and so on. For this purpose, the article summarizes and ponders according to the current research situation, in order to provide reference for the reform of domestic post-secondary education.

Keywords: Junior college students to undergraduates · CiteSpace · Talent training · Trends

1 Introduction

Education and talent training are matters that the world attaches great importance to. The progress of science and technology and the development of national economy are inseparable from the strength of the talent team [1, 2]. Education is crucial to the development of a country and society [3, 4].

In China's general higher education, graduates at the junior college level enter the undergraduate stage to continue their study through examination or selection, so as to achieve the improvement of their academic qualifications. This form of education is uniformly called General Higher Education (hereinafter referred to as general education). From the enrollment documents of various provinces, autonomous regions and municipalities directly under the central government, it can be seen that the enrollment object of ordinary college upgrading is fresh college graduates. In order to solve the employment problem of college graduates under the COVID-19, on February 25, 2020, the State Council deployed China's higher education to increase the number of students

Z. Hu et al. (Eds.): AIMEE 2022, LNDECT 159, pp. 583–595, 2023.
https://doi.org/10.1007/978-3-031-24468-1_53

enrolled from junior college to junior college. Subsequently, the Ministry of education made specific arrangements for the work of upgrading from junior college to undergraduate. In 2020, the enrollment scale of junior college to undergraduate will increase by 322000 people year-on-year. In 2021, the work of upgrading from junior college to basic college continued to advance, and the number of students enrolled reached 642000. Under the guidance of government policies, the number of fresh graduates at the junior college level who have passed the examination and selection to enter the undergraduate course for further study shows an obvious upward trend [5].

Some scholars have investigated the needs of college students to improve their academic qualifications, and 50% of college students hope to enter the undergraduate stage to continue their education [6]. The "junior college to undergraduate" education in ordinary colleges and universities meets the needs of high-level person times for the rapid development of China's economy and society. Some junior college students' continuing study can alleviate the pressure of employment, and also provide a promotion channel for junior college students to continue their further study and improve their academic qualifications. It is an epoch-making choice made by China's higher education to comply with the development of the times. With the expansion of enrollment scale, upgrading from junior college to undergraduate has also attracted the attention and high attention of all sectors of society. For example, Xiao Fuliu and Song Bei analyzed the dilemma of connecting junior college to undergraduate training in Applied Universities and put forward corresponding solutions [7]. Scholars such as Li Pingping discussed the learning situation of nursing students from junior college to undergraduate from the perspective of students' professional identity [8], Qi pan believes that there are many problems in the orientation of talent training and the connection of courses in the upgraded education. For this reason, he has carried out an integrated design of the professional courses of upgraded education [9], and Wang Jun started from the upgraded major of pharmacy, carried out teaching reform in specific courses, gave full play to the advantages of flipped classroom, and improved the learning effect of upgraded students [10]. After combing the relevant literature, it is found that at present, no scholars have discussed the research status and research hotspots of domestic full-time general post-secondary education.

In recent years, as an information visualization software, CiteSpace has become an important tool for academic circles to study the overall trend of a field. Its functions such as keyword co-occurrence, clustering, time zone view, cooperative network, etc. can intuitively show the research status of this field, which is more intuitive and scientific than the traditional literature method [11]. CiteSpace is a scientific literature analysis tool jointly developed by Professor Chen Chaomei of Drexel University and Dalian University of technology. Liu Zeyuan and Chen Yue took the lead in translating "mapping knowledge domains" into "mapping of scientific knowledge" [12]. Domestic scholars first used CiteSpace in the field of information science. With the deepening of research and the continuous improvement of software, various disciplines began to use this method to analyze the research status and research trends. In the field of pedagogy, scholars have also carried out a lot of attempts and practices. For example, pan Li made a quantitative analysis of eight CSSCI educational journals.

in China, including educational research, and the key words of the literature collected and published between 2000 and 2009, and obtained the hot line of educational research

[13], Wang Juan analyzed the research hotspots and trends in the field of educational big data based on CiteSpace in 2016 [14], and scholars used CiteSpace knowledge map software through the literature of web of science database, Explored the hot spots and trends of stem education research [15], and Zhou Meiyun took 1008 labor education research papers published on CNKI as the research object, forming a "knowledge map" of labor education research in China [16].

Through combing the literature, it is found that at present, no scholars have used CiteSpace software to study the current situation and hot trends of domestic general post-secondary education. Under this background, this paper analyzes the collected domestic relevant literature on general post-secondary education through literature metrology and knowledge map visualization methods. Through quantitative analysis of the number of publications, the source of periodicals, the cooperation between authors and institutions, research hotspots and research trends, the corresponding knowledge map is obtained, in order to grasp the research status and hot trends of general post-secondary education, and provide reference for the research of domestic post-secondary education.

2 Data Mining and the Distribution of Upgraded Studies

2.1 Research Methods and Data Collection

2.1.1 Research Methods

CiteSpace can find out the hidden knowledge of research in a certain field from the scientific literature, form a multi angle dynamic visual map through metrology and statistics, and present the structure, law and distribution of scientific knowledge through visual means. CiteSpace 6.1 is used to conduct quantitative analysis on the reference data in the field of domestic general upgraded education research, and explore the research status from the annual number of papers, source journals and colleges, literature citation, author cooperation, etc., and then through the analysis functions of keyword co-occurrence, clustering, etc., we can find the research hotspots and development trends of domestic general upgraded education.

2.1.2 Data Acquisition

This study takes the database of CNKI as the source of literature, and makes a quantitative analysis of the literature on the basis of literature retrieval. The specific operations are as follows: select the "advanced search" type in the CNKI database, and select the "subject" search. The search condition is "upgraded from junior college" as the subject, and the time range is "2012–2021". A total of 725 journal papers were searched, including 35 core Journals of Peking University and CSSCI. Due to the lack of core and CSSCI documents of Peking University related to college to undergraduate education, all journals were taken as the source of research documents in this quantitative analysis. After three experts screened the theme,

keywords and abstracts of the documents one by one, and excluded the documents not related to the research theme, as well as book reviews, news, enrollment brochures, etc., a total of 464 research documents were finally obtained.

2.2 Research Distribution of Upgrading from Junior College to Undergraduate

2.2.1 Spatial and Temporal Distribution of Research

This paper makes statistics on the relevant literatures published in China from 2012 to 2021. After searching, it is found that the total number of published documents is 464, as shown in Fig. 1. It can be seen from the figure that the domestic research on full-time general post-secondary education has shown a relatively flat trend in the past decade. The average annual number of papers issued from 2012 to 2021 was 44, and the highest number of papers issued in 2015 was 60.

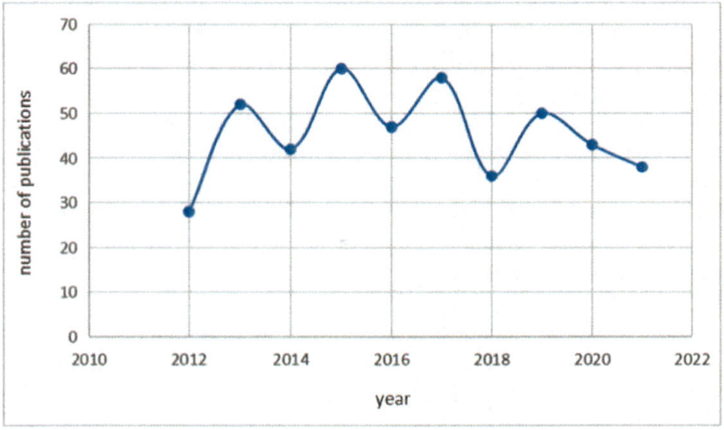

Fig. 1. Annual trend chart of domestic general upgraded education research literature from 2012 to 2021

2.2.2 Distribution of Source Journals

Journals reflect the quality of papers to a certain extent, and can also reflect the spatial distribution characteristics of the research field. By analyzing the number of journal publications, we can clarify the concerns and research groups in the research field [17]. Through the analysis of the papers on general upgraded Education published in China, it is found that health vocational education, education and teaching forum, education modernization, scientific and technological vision, intelligence, Journal of Hubei function University, education and occupation, western quality education, higher education forum, university education and other journals are the top ten journals that publish the most papers on upgraded education in China. From the perspective of the level of journals, the core journals do not pay much attention to upgraded education.

2.2.3 Author Statistics and Cooperative Network Analysis

By analyzing the amount of papers published by the authors and their cooperation, we can clearly understand the key figures in this field and determine the research authority in this field [18]. In this paper, the author uses the CiteSpace software to measure the authors and their cooperation, as shown in Table 1, and obtains the co-occurrence map of domestic general upgraded education research authors from 2012 to 2021.

Table 1. Top 10 journals in terms of the number of publications of domestic general upgraded education research literature from 2012 to 2021

Periodical title	Number of publications	Billing
Health vocational education	22	1
Education and teaching forum	17	2
Modernization of education	9	3
Science and Technology Vision	8	4
Wisdom	7	5
Journal of Hubei Correspondence University	6	6
Education and professional	6	7
Western quality education	6	8
Vocational and technical education	5	9
Tertiary education	5	10

The node represents the author, and the size of the node represents the number of papers published. If there is cooperation between authors, there will be a connection [19]. This paper selects the top ten authors (see Table 2).

From the co-occurrence chart of research authors in Fig. 2, it can be seen that Jin Ruihua, Wang Qinghua (all 6), Wang Hui (5), Zheng Jie, Ding Zhenzhu, Fang Xi (4), Wu Chen, Li Yonghong, Zhang Li, Zhang Yu have the largest number of papers. These are representative authors in the field of post-secondary education research.

It can be seen from Fig. 2, the author's cooperation network presents a state of "partial convergence, most of which are scattered". The most stable are the teams of Wang Qinghua and Jin Ruihua. Wang Qinghua's team mainly studies the learning ability and learning status of nursing students from junior college to undergraduate, while Jin Ruihua's team discusses the professional curriculum setting and reform from the perspective of nursing students from junior college to undergraduate, and most other scholars also explore from the two aspects of students and curriculum. However, most of the authors fight alone, and there is very little teamwork, especially Cross School and cross major cooperation.

2.2.4 Analysis of the Cooperation Network of the Issuing Agency

The co-occurrence map of the sending institution can show the academic status of the research institution in a certain field [20]. This paper makes quantitative statistics on the number of documents issued by the issuing institutions, and finally forms a co-occurrence map of research institutions (Fig. 3). It can be seen from the figure that

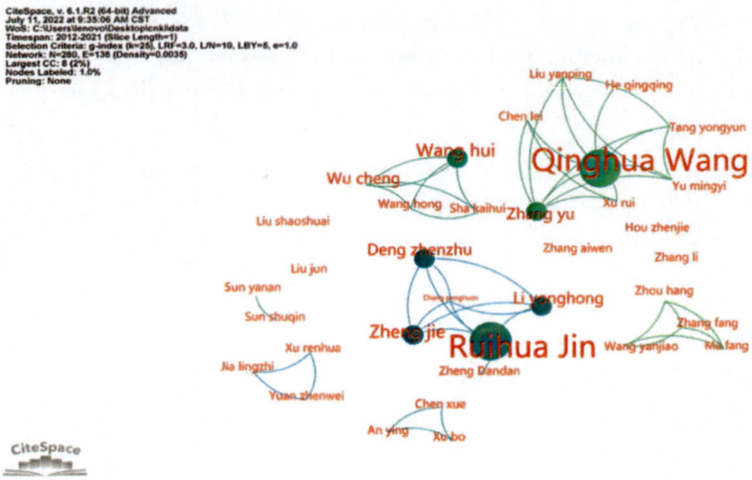

Fig. 2. Co-Occurrence of domestic general upgraded education research authors from 2012 to 2021

Table 2. Top 10 authors of domestic general upgraded education research literature from 2012 to 2021

Author	Number of papers	Rank
Ruihua Jin	7	1
Qinghua Wang	6	2
Wang hui	5	3
Zheng jie	4	4
Ding zhenzhu	4	5
Wu cheng	3	6
Li yonghong	3	7
Zhang yu	3	8
Chen lei	2	9
Wang hong	2	10

the sending institutions basically come from colleges and universities. The school of nursing of Binzhou Medical College ranks first in the number of documents sent, the school of nursing of Shanxi Medical University ranks second, followed by Binzhou Medical College and Beijing Union University, which have made great contributions to the research of post-secondary education.

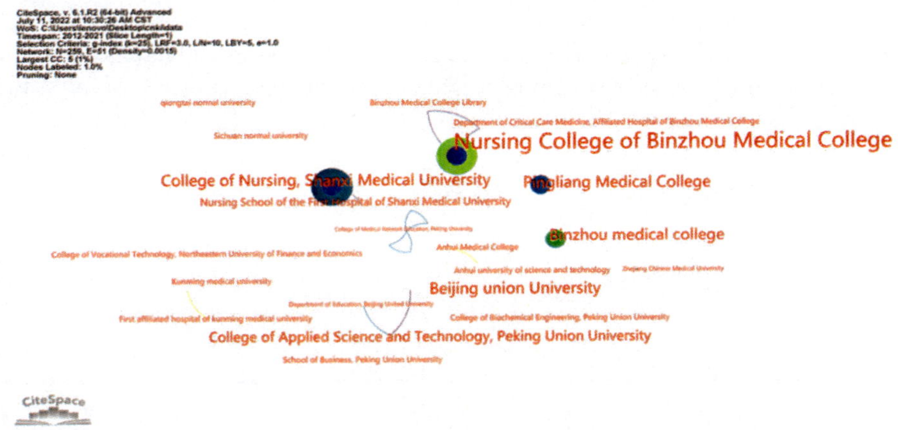

Fig. 3. Co-Current situation of domestic general post-secondary education research institutions from 2012 to 2021

3 Research Hotspot and Evolution Analysis of Upgrading from Junior College to Undergraduate

3.1 Hot Research Direction of Upgrading from Junior College to Undergraduate

Using the keywords in the bibliography, this paper uses CiteSpace software to analyze the keyword co-occurrence of the domestic general college upgrading paper, so as to determine its research hot areas.

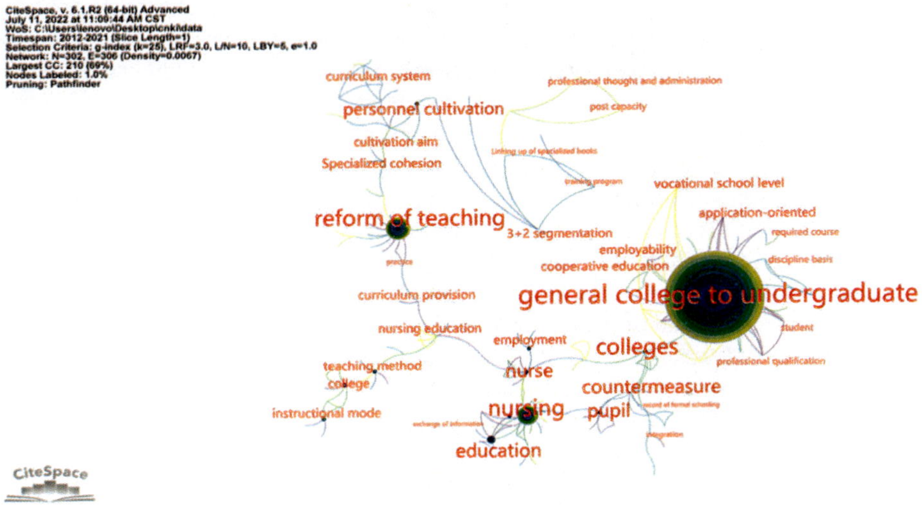

Fig. 4. Knowledge map of main research fields and research hotspots of domestic general post-secondary education from 2012 to 2021

Input a total of 464 collected domestic literature data related to general upgraded education into CiteSpace 6.1 software, run to get the knowledge map of research hotspots (Fig. 4), and sort out the information table of domestic general upgraded education research fields and research hotspots from the data table in CiteSpace 6.1 (Table 3).

Table 3. Statistics of domestic research fields and research hotspots of post-secondary education from 2012 to 2021 (top 10)

No	frequency	Centrality	Hot words
1	199	0.62	Upgraded from college to university
2	34	0.33	Teaching reform
3	21	0.17	Nursing
4	10	0.54	Higher Vocational Colleges
5	10	0.55	Advanced Nursing
6	10	0.11	Teaching Method
7	10	0.03	Education
8	8	0.05	Student
9	8	0.02	Take up an Occupation
10	8	0.20	Personnel Training

On the one hand, it can be seen from Fig. 4 and Table 3 that the largest node in the map is the "upgraded version", which has been running through the research in this field. In addition, "teaching reform", "nursing", "nursing students", "higher vocational colleges", "countermeasures" and other keywords are the most eye-catching, and their nodes are relatively large. From the perspective of disciplines and majors, nursing scholars pay the most attention to the post-secondary education of nursing students, which is closely related to the development of Nursing Higher Education in China. Post-secondary education is an important way for nursing students to improve their own level and knowledge and skills. In terms of research content, scholars focus on teaching reform and talent training, which is also the key content of higher education research in China. From the school level of research, higher vocational colleges are the main object of research.

On the other hand, as the key node of strategic role in the whole network, intermediary centrality can explain the size of information flow it controls. Table 3 shows that the highest degree of intermediary centrality is 0.62, which is at the core of this field. The following are "nursing students", "higher vocational colleges", "teaching reform", "talent training", "teaching methods" and "employment". From the analysis of knowledge map and centrality, we can draw the following basic conclusions. During 2012–2021, the research hotspots of post-secondary education in China are mainly concentrated in three aspects: "teaching reform", "talent training" and "student management".

(1) Teaching reform. Talent training should finally be implemented in actual teaching. As can be seen from Fig. 4, the teaching reform of upgrading from junior college to undergraduate course mainly focuses on two aspects. The first is the reform of the

teaching method of the course. For example, Li Yuhong and other scholars reformed the teaching of comprehensive nursing training by using the task driven case teaching method in the course teaching [21], and the crown prince used the "bisection classroom" to carry out the practice of financial management courses in the major from junior college to this major [22]. The second is the construction of the curriculum system. Wei yunnuan takes the marketing major as an example, takes the talent training goal as the setting of the positioning curriculum system, starts with how to effectively connect with the business management, innovation and entrepreneurship and other courses, focuses on improving students' core professional ability, and constructs a three-dimensional curriculum system [23]. Qi pan designed an integrated professional curriculum system of professional training by taking the electronic information engineering major upgraded from junior college to junior college in Guangdong Province as an example [24].

(2) Talent training. The talent training objectives of undergraduate courses are different from those of junior colleges. From the perspective of the mastery of theoretical knowledge, the requirements of undergraduate courses are broader and deeper than those of junior colleges. In terms of skills, junior colleges are more specific. College level talent training pays more attention to the basic skills that students must master in a certain position; The academic requirements of undergraduate level talent training are that facing the entire industry, we need to master deeper knowledge, more comprehensive skills, higher personal comprehensive quality requirements, and provide the basis for career development and promotion [25]. However, in the upgraded education, colleges and universities often regard the upgraded education as the same as the unified undergraduate education, and do not distinguish between the talent training, which leads to the unclear talent training objectives of the upgraded education and the poor connection of the courses, especially the knowledge structure fault, repeated courses and other problems exposed in the "upgraded education" [26], which greatly affects the training quality of the upgraded students. In this context, it is particularly important to establish the positioning of the training of talents from junior college to undergraduate. For this reason, scholars have done a lot of exploration and practice in the establishment of the talent training from junior college to undergraduate. For example, Jilin Institute of chemical technology, guided by the needs of local economic development, benchmarked professional standards and job needs, fully understood the requirements of employers for pharmaceutical talents, and established clear talent training objectives [26]. It can be seen from the practice in various places that the education from junior college to undergraduate is mostly consistent with the training objectives of undergraduate education. On this basis, it emphasizes the integration of production and education, and highlights the applicability and practicality.

(3) Student management. Junior college students have special attributes in knowledge base, learning ability, independent ability, ideological maturity and so on. In view of the characteristics of the students who have been upgraded to junior college, it is more necessary to pay attention to the analysis of the students' situation in the student management work. Some scholars investigated 124 nursing students who have been upgraded to junior college in Shandong Province, and found that their professional identity is not high, their learning burnout is at a medium level, and their sense of achievement is not strong [27]. Based on this, in the management work, we should grasp the psychological

needs of college students, suit the remedy to the case, and adopt flexible and refined management methods.

3.2 Research Trends

The keyword research Timeline view grasps the research hotspots and their evolution process in a field by analyzing the different time nodes, distribution and changes of keywords [28]. There are many literatures in a certain period of time, indicating that researchers at that time focused on this field. The node connection between periods indicates the inheritance relationship, and the tightness of the connection between the two periods is shown by the number of connections [29]. The hot time zone map of domestic general post-secondary education is generated through CiteSpace (Fig. 5). The abscissa is the year of publication of the paper, and the ordinate is the node and connection of keywords. It can be seen from the figure that there are fewer nodes in 2012 and more nodes in 2013–2017, which is consistent with the annual publication volume analyzed above. After 2018, there are relatively few new keywords, but post-secondary education is still attracting the attention and attention of researchers.

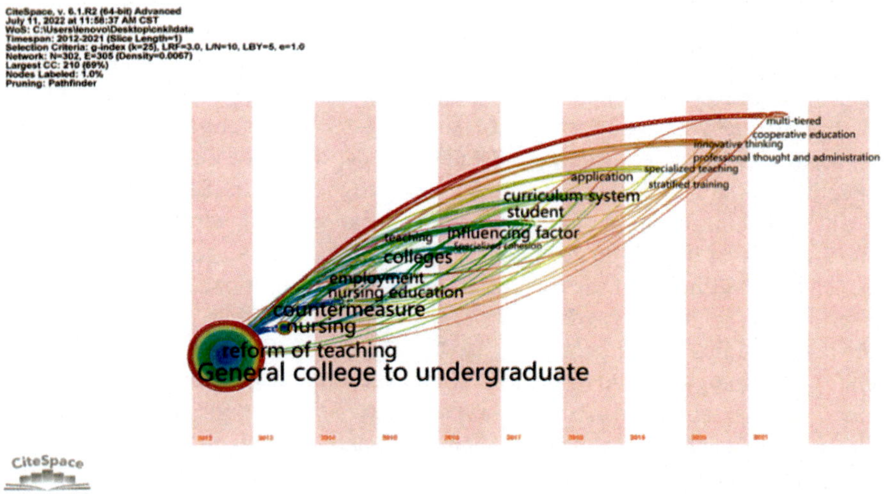

Fig. 5. Timeline view of domestic general post-secondary education research from 2012 to 2021

4 Development Trend Outlook

Using CiteSpace software, this paper analyzes the annual publication volume, discipline distribution, cooperation between authors and institutions, hot spots and development trends of the research literature on China's general post-secondary education collected in CNKI database from 2012 to 2021, and draws four conclusions:

First, from the perspective of the overall number of documents published, 2012–2021 is a period of steady development of domestic research on general post-secondary education, and the number of documents published remains stable.

Second, from the perspective of discipline distribution, the research on upgrading from junior college to college is mainly focused on nursing, followed by information engineering, chemical engineering and other disciplines.

Third, from the perspective of the cooperation between authors and institutions, the research institutions are basically colleges and universities, and the cooperation between various institutions is less, and there is no authoritative institution yet.

Fourth, from the distribution, clustering and time zone map of keyword co-occurrence, the research of general post-secondary education in China mainly focuses on teaching reform, talent training and student management, which is consistent with the concerns of higher education.

With the encouragement of national policies and the need to improve the academic qualifications of junior college students, in future research, expanding research fields and strengthening institutional cooperation can have a better development of junior college to undergraduate education. In addition, how to realize the curriculum connection of college to undergraduate, improve the ability of college to undergraduate students, and strengthen the effect of school enterprise cooperation and collaborative education will be the key content of college to undergraduate education research in the future.

5 Conclusion

In the eyes of most people, from college to undergraduate is a qualitative leap, in line with the values of today's society. Colleges and universities are quite rich in learning resources, human resources and so on. If you choose to study undergraduate, you can not only exercise yourself in the professional field, but also build and expand a higher-level communication circle, gain a good attitude, learning ability and learning habits, and accumulate valuable wealth in your life.

It has been 20 years since the beginning of upgrading from junior college to junior college. The results are obvious to all, which not only solves many employment problems, but also cultivates many professional and technical talents. It has become an important part of our national education system. With the rapid development of the country and the popularization of higher education, the social requirements for individuals, including knowledge, ability, quality, as well as academic qualifications and diplomas, are also gradually improving. Generally, with a higher degree, young people will have more opportunities and greater room for development. Upgrading from junior college to college has also triggered discussions among many scholars.

To explore the various problems in the process of upgrading from junior college to undergraduate and its development is to clarify the direction of future development, find the key points in the development process, and find the obstacles and even bottlenecks. Only in this way can we make scientific plans and Countermeasures for future work, arouse the great attention of the educational community, and cultivate more outstanding talents beneficial to the society.

Acknowledgment. This research is supported by the 2022 Guangxi Vocational Education Teaching Reform Research Project "Research and Practice on the Effective Connection and Curriculum System Construction of the Transportation Logistics Professional Group from Junior College to Undergraduate" (GXGZJG2022B181).

References

1. Rifat, M.R.I., Imran, A.A., Badrudduza, A.S.M.: Educational performance analytics of undergraduate business students. Int. J. Mod. Educ. Comput. Sci. (IJMECS) **11**(7), 44–53 (2019)
2. He, Y., Yang, M.: Study on quality evaluation system for university students. Int. J. Wirel. Microw. Technol. (IJWMT) **1**(3), 29–34 (2011)
3. Aliyu, I., Kana, A.F.D., Aliyu, S.: Development of knowledge graph for university courses management. Int. J. Educ. Manage. Eng. **10**(2), 1–10 (2020)
4. Wang, Q.-M., Zhao, X., Zhang, L.: A novel thought and method of university evaluation. Int. J. Educ. Manage. Eng. (IJEME) **3**(2), 59–65 (2013)
5. Zhang, Z.: Research on the working path of standardizing party construction by the Party branch of students from higher vocational colleges to undergraduate colleges. Sci. Consult. (Technol. Manage.) **07**, 26–27 (2021). (in Chinese)
6. Hui, Y.: Research on the willingness of higher vocational students to upgrade to college. Sci. Technol. **15**, 155–156 (2021). (in Chinese)
7. Xiao, F., Song, B.: Exploration on the training of talents connecting college and undergraduate courses in local applied universities under the background of "double high schools." Educ. Occup. **14**, 50–55 (2020). (in Chinese)
8. Li, P., Chen, L., Zhang, Y., Han, B., Wang, Q.: Analysis on the correlation between professional identity and learning burnout of nursing college students. Health Vocat. Educ. **38**(02), 105–107 (2020). (in Chinese)
9. Qi, P., Wu, Z., Xu, C., Xiao, M.: Research on the curriculum system of "three two sections" upgrading from junior college to undergraduate through training in electronic information engineering. Vocat. Tech. Educ. **40**(08), 33–35 (2019). (in Chinese)
10. Wang, J., Wang, S., Ding, L., Liu, J.: Flipped classroom teaching practice based on cloud class - Taking organic chemistry as an example. J. Jining Med. Coll. **44**(02), 141–144 (2021). (in Chinese)
11. Xia, L., Wang, K., Cheng, X.: Visual analysis of the evolution characteristics of knowledge map research in China. Inf. Sci. **37**(03), 9–16+74 (2019). (in Chinese)
12. Chen, Y., Liu, Z.: The quie rising Atlas of scientific knowledge. Sci. Res. 02, 149–154 (2005). (in Chinese)
13. Zhou, M.: Track, focus and trend: 70 years of labor education research. Contemp. Educ. Forum **03**, 106–113 (2020). (in Chinese)
14. Du, W.: Hot topics and characteristics of stem education research abroad. Audio Vis. Educ. Res. **39**(11), 120–128 (2018). (in Chinese)
15. Wang, J., Chen, S., Wang, L., Yang, X.: Research hotspot and trend analysis of education big data based on CiteSpace. Mod. Educ. Technol. **26**(02), 5–13 (2016). (in Chinese)
16. Zhou, M.: Track, focus and trend: 70 years of labor education research. Contemp. Educ. Forum **3**, 8 (2020). (in Chinese)
17. Pan, L., Wang, S.: Hot fields and cutting-edge topics of educational research in the past decade - econometric analysis based on the knowledge map of keyword co-occurrence of documents published in eight educational journals from 2000 to 2009. Educ. Res. **32**(02), 47–53 (2011). (in Chinese)

18. Du, L., Hao, Z.: Current situation, context and trend of household finance research – visual analysis based on WOS and CNKI journal papers. J. Southwest Univ. Nationalities (Humanit. Soc. Sci. Edn.) **40**(02), 114–124 (2019). (in Chinese)
19. Zhang, H., Hou, Y.: A summary of research on innovation behavior of domestic employees based on knowledge map. Sci. Technol. Prog. Countermeasures **34**(11), 153–160 (2017). (in Chinese)
20. Liu, S., Qu, B.: Analysis of influencing factors of Library Alliance in China based on keywords. Inf. Sci. **33**(05), 91–96 (2015). (in Chinese)
21. Li, Y., Liu, J., Zhong, Q., Zhang, L., Zhang, F.: Application of task driven case teaching method in the course of comprehensive nursing training. J. Chengde Med. College **36**(04), 348–351 (2019). (in Chinese)
22. Crown prince. The application and reflection of "bisection class" in the teaching of financial management from junior college to undergraduate. China Manage. Inf. **20**(22), 213–214 (2017). (in Chinese)
23. Wei, Y.: Construction of curriculum system of marketing major for Vocational Undergraduate Based on the mode of "specialized course through textbook." Time Honored Brand Mark. **08**, 163–164 (2021). (in Chinese)
24. Shi, Q.: Analysis of practical teaching reform under the background of transformation and development of local undergraduate colleges and universities. Jingchu J. **18**(06), 77–81 (2017). (in Chinese)
25. Hu, L.: Research on the "three part" talent training mode based on "junior college to undergraduate" education. J. Neijiang Normal Univ. **35**(01), 72–75 (2020). (in Chinese)
26. Qu, X., Zhang, Y., King, K., Zhong, F., Cheng, Z.: Research on the training mode of "3+2" upgraded talents in pharmaceutical majors. J. Jilin Inst. Technol. **37**(04), 22–25 (2020). (in Chinese)
27. Wu, C., Sha, K., Wang, H., Wang, H.: The relationship between clinical practice environment evaluation, professional maturity and professional identity of full-time nursing students from junior college to junior college. Chin. J. Med. Educ. Exp. **17**(07), 689–693 (2018). (in Chinese)
28. Feng, Y., Yan, C., Hu, C.: Structural characteristics and hot perspectives of domestic privacy research in the past 20 years. J. Inf. Resour. Manage. **10**(01), 65–74+101 (2020). (in Chinese)
29. Wu, Y., Li, Z.: Current situation and development trend of spatial metaphor research in China - knowledge map analysis based on CiteSpace. Psychol. Innov. **40**(04): 302–308+384 (2020). (in Chinese)

Method for Selecting Special Software for Course Projects in Economics

Nataliya Mutovkina(✉)

Tver State Technical University, Tver 170026, Russia
letter-boxNM@yandex.ru

Abstract. The article considers the problem of students choosing the optimal software for the development of course projects toward "Economics". It is established that currently there are many software environments that can solve problems of course projecting. Students are often limited in their choice of software tools. The software environment is offered by the teacher himself. But this state of affairs does not contribute to the development of students' creative approaches to design, the development of independent decision-making skills, etc. To solve the problem, the article proposes a method for choosing the optimal software environment for designing. The choice is made by the students of the study group. Because there are many specialized software products, and they all have certain differences, a clear point evaluation of particular preference indicators is difficult. Usually, this assessment is vague. The positive experience of using fuzzy set theory in solving selection and evaluation problems extends to the choice of software tools for course design. The method provides for the choice of one or more software tools, depending on the distribution of respondents' ratings. The experience of implementing the proposed method at the Department of Accounting and Finance of Tver State Technical University is presented.

Keywords: Design method · Course project · Fuzzy estimates · Fuzzy sets · Evaluation criteria · Software tools · Software product selection

1 Introduction

The educational process is a complex, multifaceted, resource-intensive phenomenon that requires a glorious return from both the teacher and the student. Currently, I can't imagine the educational process without the use of modern software, which is an indispensable tool for solving many educational, and subsequently, professional tasks. Thus, when teaching students of the economic direction, a wide range of various software tools is used, starting from a standard office software package and ending with special programs for automating accounting, forming financial statements, performing financial analysis, planning company activities, etc. Students use the software in classrooms when doing practical work and at home when doing tasks on their own. One type of educational work that is performed both in the classroom and outside it is course design.

Z. Hu et al. (Eds.): AIMEE 2022, LNDECT 159, pp. 596–606, 2023.
https://doi.org/10.1007/978-3-031-24468-1_54

A course project solves a specific task that has a high complexity and labor intensity. The analysis of the problem areas, as well as the description of the solution and the evaluation of the results obtained, is issued as an explanatory note. The explanatory note is a text-graphic document containing calculations and all the details (title page, contents, introduction, main part, conclusion, bibliographic list, appendices). An explanatory note is compiled under the methodological guidelines and recommendations issued by the teacher in Microsoft Word. Work on a course project is usually carried out in one or several software environments, depending on the discipline, considering the specifics of the goal and the tasks to be solved. And here there are several problems. First, the software environment chosen for course design should be accessible to all students. Every student should be able to install the program on their own computer. Second, the software environment must be modern, meet quality standards, and be compatible with the operating system and other programs. Third, the program should have a friendly interface, be easily customizable and be adaptive to user requests. And, of course, the program must ensure that the expected result is got.

As practice shows, the same task can be solved in different software environments. Therefore, both the student, especially in senior courses, and the teacher face choosing the optimal program for course design. The choice should be coordinated; i.e., the teacher and students should come to a common opinion regarding the design environment.

This article discusses possible ways to determine the optimal software environment for course design. Definition here means deciding which software environment students will study within the framework of an academic discipline and implementing their course projects in it in the future. The purpose of the study is to propose an adaptive method for choosing the software tool most suitable for course design. The application of this technique is considered within the framework of implementing the course project in the discipline "Project Management" by third-year students of the direction "Economics".

1.1 Background of the Study

In studying at a higher school, it is important for students to use special software to perform various works and projects. A significant place is occupied by course design, the purpose of which is to gain students the skills of independent work; the ability to analyze large amounts of information, identify problem areas, adequately assess problems, and offer effective solutions to eliminate problems.

Course projects in economics have special economic and financial specifics. This can be the development of a business plan, an investment project, a project for introducing an innovative product, etc. During course design, students learn to draw up a project plan, form an optimal set of resources for its implementation, select a project team, and monitor the timeliness of project tasks without going beyond the established budget.

Course design in the educational process is becoming increasingly important for the formation of professional competencies among students, and there are more and more software tools for implementing course projects [1, 2]. There is a problem in choosing the optimal software tool for the development of a particular project. This is important because it is necessary to consider that graduates will need skills to work in current software environments in their professional activities. It should be remembered that many educational tasks develop into professional tasks [3].

1.2 Literature Review

In the scientific literature, you can often find references to design problems that any organization may face. For example, in [4, 5], such problems include insufficient organization in the distribution of roles and responsibilities among the members of the project team and the lack of effective communication between them; violation of deadlines for project work; insufficient risk accounting. The project manager must organize the work on the project in such a way that the project is completed on time, with excellent quality and without attracting additional resources [6]. Effective project management is possible if the project manager has delved into the specifics of management at the stage of his training. Therefore, it is especially important to introduce theoretical and practical aspects of project management into the educational process [7]. Working on a course project, students learn to control the current project: plan the project structure, stages and deadlines for its implementation, budget; form a project team; select the most effective approach to project management to improve communication, and create a flexible working environment [8]. In addition, students develop a clear understanding of potential risks during implementing the project and the ability to take measures before undesirable incidents occur.

In the educational environment, there are methodological developments in course design. For example, a course design method for programming [9], an example of a course project for implementing a text editor of Chinese characters [10], a web application design method [11], and others. However, there are practically no methods, designed toward "Economics", in which the central place is occupied by applied programs and information technologies.

There is also no method for students to choose the software environment in which they would like to work. Usually students have no choice. Students work in the software environment in which the teacher tells them to. This does not contribute to the development of students' creative abilities, analytical thinking, and interest in obtaining a specific result. Therefore, the author of the article suggests the following solution to the problem of choosing a software tool for course projecting.

2 Method for Choosing a Software Environment

2.1 Formation of a List of Software Products for Course Projecting

Course projects on economic topics are carried out by students of all courses in this field of study. But, if in the junior courses (1–2) these works are of a general theoretical nature, in the senior courses students develop projects focused on practical implementation. Implementing such projects helps students merge and expand their professional knowledge; prepare for graduate projecting; and lay a solid foundation for further education, research and project development in their professional activities [12]. In course design, students actively apply basic economic theory and practical knowledge in financial mathematics, economic modeling, forecasting, statistics, econometrics, marketing, project management, etc.

The purpose of the course project is to merge theoretical knowledge and master the skills of applying the knowledge, skills, abilities, and competencies gained during the training process in the development and implementation of a specific socio-economic project. Such a project can be a business project; an investment project; a project to introduce some innovation, information system, software and hardware complex into the company's activities; a project to create a new department in the company, etc.

Course projecting toward "Economics" involves the use of several software products. Figure 1 shows the basic and specialized software used by students in course projecting.

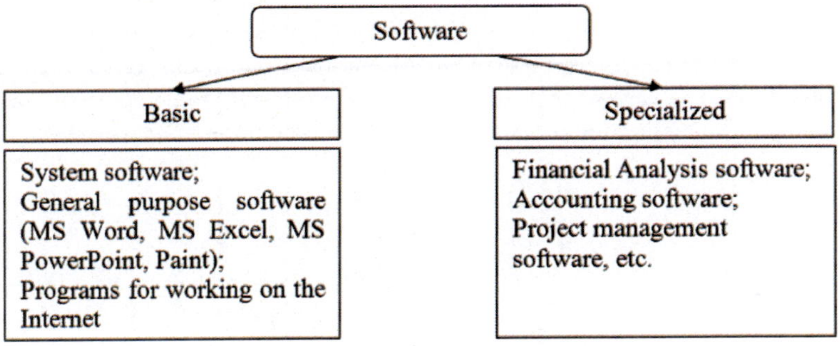

Fig. 1. Software classification

Currently, there are many computer programs designed to compile a project model and manage it. Such programs include MS Project, Primavera, Project8Libre, GanttProject and others.

From the moment the assignment is issued and within 3–5 academic days, students review the application software that can develop a specific course project. It is advisable to make a list of suitable software tools in a tabular editor, for example, in MS Excel. Each student makes such a list independently. Then the lists are compared and a set of alternatives N is formed from the most frequently repeated software tools ($i = \overrightarrow{1, N}$). There are M students ($j = \overrightarrow{1, M}$) in the student group.

2.2 Selection of Software Evaluation Criteria

Students together with the teacher form a representative set of evaluation criteria ($k = \overrightarrow{1, K}$) and student j assesses alternative α_{jk} for each of the criteria. Since the criteria for choosing a software tool are fuzzy numbers, it is proposed to use the fuzzy sets apparatus to assess students' preferences [13]. Among the criteria, there can be direct and inverse criteria. Direct criteria can be functionality, information security, and reliability (the higher, the better). The inverse criteria can be cost, time spent in the program, and complexity of the interface [14].

2.3 Selection of the Evaluation Scale

To compare estimates, it is necessary to switch from the values of different parameters to their fuzzy estimates measured on the same quantitative scale.

The evaluation scale is represented as a numerical segment [0, 1], then the $\alpha_{jk} \rightarrow \mu_k^j(i) \in [0, 1]$ score characterizes how much the software environment corresponds to the concept of "the best according to criterion k" [15, 17].

As a result, each alternative will be represented not by a set of parameter values α_{jk}, but by a set $\left\{\mu_1^j(i), \mu_2^j(i), \ldots, \mu_k^j(i), \ldots, \mu_K^j(i)\right\}$ of corresponding numerical estimates. Thus, for each criterion $k \in K$, there is a set $\left\{\mu_k^j(1), \mu_k^j(2), \ldots, \mu_k^j(i), \ldots, \mu_k^j(N)\right\}$, each element of which characterizes the correspondence of alternative i to the concept of "best" according to this criterion. This concept can be represented by a fuzzy set defined on a universal set of alternatives N:

$$\tilde{k} = \left\{\frac{\mu_k^j(1)}{1}, \frac{\mu_k^j(2)}{2}, \ldots, \frac{\mu_k^j(N)}{N}\right\} \tag{1}$$

with the membership function $\mu_k^j(i)$, which characterizes the compatibility of alternative i with the concept of "best".

2.4 Evaluation of Software by Criteria

Each of the students fills in an evaluation matrix (Table 1) estimates of N alternatives $(i = \overrightarrow{1, N})$ according to the $k = \overrightarrow{1, K}$ criteria. An important condition is that students do not confer, fill out tables independently of each other.

Table 1. Evaluation matrix

i	k					
	1	2	3	…	K-1	K
1	$\mu_1^j(1)$	$\mu_2^j(1)$	$\mu_3^j(1)$	…	$\mu_{K-1}^j(1)$	$\mu_K^j(1)$
2	$\mu_1^j(2)$	$\mu_2^j(2)$	$\mu_3^j(2)$	…	$\mu_{K-1}^j(2)$	$\mu_K^j(2)$
3	$\mu_1^j(3)$	$\mu_2^j(3)$	$\mu_3^j(3)$	…	$\mu_{K-1}^j(3)$	$\mu_K^j(3)$
4	…	…	…	…	…	…
N-1	$\mu_1^j(N-1)$	$\mu_2^j(N-1)$	$\mu_3^j(N-1)$	…	$\mu_{K-1}^j(N-1)$	$\mu_K^j(N-1)$
N	$\mu_1^j(N)$	$\mu_2^j(N)$	$\mu_3^j(N)$	…	$\mu_{K-1}^j(N)$	$\mu_K^j(N)$

Each of the criteria may have a specific importance (ν_k, $k = \overrightarrow{1, K}$), determined by a specially developed scale, for example, $\nu_k \in (0, 1)$, $\sum_{k=1}^{K} \nu_k = 1$. Considering the values of ν_k, adjusted estimates of alternatives i, $i = \overrightarrow{1, N}$ are calculated, for example, as shown in Table 2.

Table 2. Assessments based on the importance of criteria

i	k				
	1	2	3	...	K
1	$\nu_1\mu_1^j(1)$	$\nu_2\mu_2^j(1)$	$\nu_3\mu_3^j(1)$	...	$\nu_K\mu_K^j(1)$
2	$\nu_1\mu_1^j(2)$	$\nu_2\mu_2^j(2)$	$\nu_3\mu_3^j(2)$	...	$\nu_K\mu_K^j(2)$
3	$\nu_1\mu_1^j(3)$	$\nu_2\mu_2^j(3)$	$\nu_3\mu_3^j(3)$	...	$\nu_K\mu_K^j(3)$
4	...	...	...	...	...
N-1	$\nu_1\mu_1^j(N-1)$	$\nu_2\mu_2^j(N-1)$	$\nu_3\mu_3^j(N-1)$	...	$\nu_K\mu_K^j(N-1)$
N	$\nu_1\mu_1^j(N)$	$\nu_2\mu_2^j(N)$	$\nu_3\mu_3^j(N)$	...	$\nu_K\mu_K^j(N)$

2.5 Getting an Overall Score for Each Software Tool

To determine the overall score (α_i) for each matrix (Table 2) the "maximin" rule of composition of fuzzy sets is used:

$$\begin{aligned} \alpha_i &= \nu_k\mu_k^j(1) \circ \nu_k\mu_k^j(2) \circ \ldots \circ \nu_k\mu_k^j(N) \\ &= \max_i \left\{ \min_k \left[\nu_k\mu_k^j(1), \nu_k\mu_k^j(2), \ldots, \nu_k\mu_k^j(N) \right] \right\} \end{aligned} \quad (2)$$

The α_i grade reflects the preferences of the student j. The next step is to analyze the estimates of α_i and determine the final estimate of α^*.

2.6 Assessment Analysis and Decision-Making

To decide about the environment in which students will carry out course projects, it is necessary to analyze their preferences. To do this, a table is compiled, in the first column of which the IDs of students are shown; in the second column, the names of the software products selected by them are recorded, and their grades are entered in the third column.

The decision can be made by a simple majority of votes; i.e., the software tool most often found in the last table will be chosen as a course design environment.

If the preferences are divided into several approximately equal groups (two or more), then a decision can be made about designing in different software environments. When using several software tools, it is necessary to consider the possibilities of integrating products with each other and the ease of transferring data from one program to another [17].

3 Results and Discussion

3.1 Results of the Software Products Review

As an experiment, a course project was chosen for the academic discipline "Project Management", which was carried out by third-year students in Economics of the profiles "Accounting, analysis and Audit" (group No. 1), "Finance and Credit" (group. No. 2). Thus, two study groups were involved in the experiment.

Students received a design assignment in early February 2022. The task was that it was necessary to develop a project for a new division of the enterprise in a software environment with all the justifications: design deadlines, project participants, project costs, etc. According to paragraph 2.1 of the method, a week after receiving the assignment, students brought lists of software products that are suitable for design. The distribution of these programs by frequency of mentions is shown in Fig. 2.

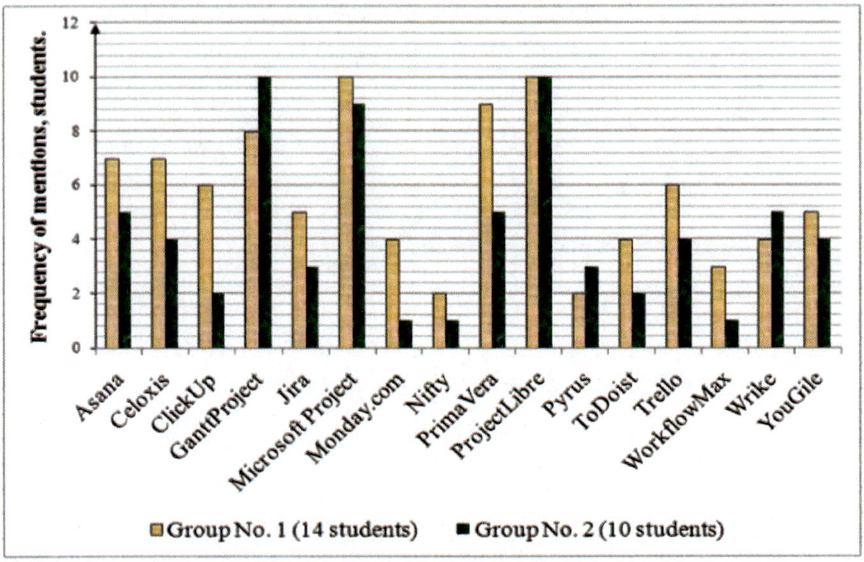

Fig. 2. Distribution of software tools by frequency of mentions

Obviously, three software tools are in the lead: ProjectLibre (20 mentions), Microsoft Project (19 mentions) and GanttProject (18 mentions). Next, students are asked to evaluate each of these software environments in order to choose one of them as a course project execution environment.

3.2 Choosing the Software Environment

The following evaluation criteria were considered: k_1 (0.3) is affordable, k_2 (0.3) is easy to install on the user's computer, k_3 (0.2) is functional and k_4 (0.2) is easy to use. The weight coefficients of the importance of the criteria are shown in parentheses. Further evaluation was carried out in MS Excel.

Considering clause 2.3 of the method, each student filled out a table of type 1, after which he received an automatic calculation of grades, considering the importance of criteria (Table 2) and the overall estimate according to the formula (2). Calculation examples are shown in Fig. 3.

According to rule (2), the first and third students from the first group chose ProjectLibre; the second student chose GanttProject.

Student 1.1

i \ k	1	2	3	4
1	0,9	0,9	0,85	0,9
2	0,1	0,5	0,9	0,75
3	0,9	0,8	0,8	0,85

i \ k	1	2	3	4	min
1	0,27	0,27	0,17	0,18	0,17
2	0,03	0,15	0,18	0,15	0,03
3	0,27	0,24	0,16	0,17	0,16
				max	0,17

Student 1.2

i \ k	1	2	3	4
1	0,8	0,6	0,5	0,75
2	0,35	0,6	0,8	0,6
3	0,95	0,6	0,6	0,75

i \ k	1	2	3	4	min
1	0,24	0,18	0,1	0,15	0,1
2	0,105	0,18	0,16	0,12	0,105
3	0,285	0,18	0,12	0,15	0,12
				max	0,12

Student 1.3

i \ k	1	2	3	4
1	0,98	0,7	0,55	0,8
2	0,3	0,3	0,85	0,7
3	0,95	0,7	0,45	0,7

i \ k	1	2	3	4	min
1	0,294	0,21	0,11	0,16	0,11
2	0,09	0,09	0,17	0,14	0,09
3	0,285	0,21	0,09	0,14	0,09
				max	0,11

Fig. 3. Example of implementing paragraphs 2.4 and 2.5 of the method

Figure 4 shows the estimates of α_i. Obviously, the choice is made in favor of ProjectLibre. This is evidenced by the frequency of mentions of this software product (7 out of 14 in the first group and 6 out of 10 in the second group).

The ultimate choice of the course project environment is made in favor of ProjectLibre.

Student ID	Selected software tool	Evaluation	Student ID	Selected software tool	Evaluation
Student 1.1	ProjectLibre	0,17	Student 2.1	GanttProject	0,16
Student 1.2	GanttProject	0,12	Student 2.2	ProjectLibre	0,15
Student 1.3	ProjectLibre	0,11	Student 2.3	ProjectLibre	0,17
Student 1.4	Microsoft Project	0,12	Student 2.4	ProjectLibre	0,17
Student 1.5	GanttProject	0,15	Student 2.5	GanttProject	0,15
Student 1.6	ProjectLibre	0,16	Student 2.6	Microsoft Project	0,17
Student 1.7	ProjectLibre	0,15	Student 2.7	ProjectLibre	0,13
Student 1.8	Microsoft Project	0,12	Student 2.8	ProjectLibre	0,17
Student 1.9	ProjectLibre	0,13	Student 2.9	Microsoft Project	0,16
Student 1.10	GanttProject	0,13	Student 2.10	ProjectLibre	0,13
Student 1.11	GanttProject	0,18			
Student 1.12	ProjectLibre	0,17			
Student 1.13	ProjectLibre	0,16			
Student 1.14	Microsoft Project	0,11			
ProjectLibre	7	1,05	ProjectLibre	6	0,92
GanttProject	4	0,58	GanttProject	2	0,31
Microsoft Project	3	0,35	Microsoft Project	2	0,33

Fig. 4. Analysis of estimates

4 Summary and Conclusion

Although the ProjectLibre software was chosen as the main design environment, the teacher considered the preferences of students and allowed them to carry out course projects in two other software environments.

At the end of the semester (May 2022), students completed coursework projects in the discipline "Project Management" and started defending projects. All works were successfully protected. At the defense, the following turned out:

1) Students who performed course design in the most popular environment (ProjectLibre) coped better with the task. This is because, during the design process, these students exchanged information and experience regarding the application of this software product, discussed its capabilities, and shared the results of their work.
2) Students who chosen other software products spent more time and effort to learn the basics of working in these software environments. This is because of a lack of communication regarding the use of GanttProject and Microsoft Project. They had to search for most of the information themselves.
3) At the defense of course projects, students made reports with presentations made in PowerPoint. The audience was most interested in reports on projects implemented in GanttProject and Microsoft Project. Therefore, the decision to design in different software environments is quite reasonable: students in working on a course project master one software environment, and in the last lesson, they get acquainted with the design possibilities in other programs. Thus, they can compare software products and draw their own conclusions about the possibilities of their application in the future.

4) Using specialized software allows students to gain experience for future graduation work, preparation of public speeches, business reports. This makes students competitive in the labor market.

In course, projecting, it is important that students themselves can make a choice in favor of a particular software, discuss their choice with the teacher; share the results of their work with classmates. This approach to learning stimulates the development of students' creative activity, interest in getting good work results, and also provides an opportunity to argue their opinion, justify their choice. The application of the developed method of software selection at the Department of Accounting and Finance in 2022 showed that the success of course design in both student groups increased by 17% compared to the last academic year. In the future, it is planned to use the developed method when choosing software products for performing coursework in other disciplines, practical classes, diploma projecting.

References

1. Fadel, M.A., Elrefaei, L.A.: Adjustments of methodology planning and assessment activities of senior projects in the computer science program. Int. J. Mod. Educ. Comput. Sci. **10**(2), 16–25 (2018)
2. Bilal, H.A., Amjad, S., Ilyas, M.: A comparative study of global software development tools supporting project management activities. Int. J. Educ. Manage. Eng. **7**(6), 32–39 (2017)
3. Qureshi, R.J., Alassafi, M.O., Shahzad, H.M.: Lean agile integration for the development of large size projects. Int. J. Mod. Educ. Comput. Sci. **11**(5), 24–33 (2019)
4. Shoukat, T., Majeed, M.N.: Impact of collaborative ALM on software project management. Int. J. Inf. Technol. Comput. Sci. **6**(6), 61–67 (2014)
5. Majeed, N., Shah, K.A., Qazi, K.A., Maqsood, M.: Performance evaluation of I.T project management in developing countries. Int. J. Inf. Technol. Comput. Sci. **5**(4), 68–75 (2013)
6. Zahra, S., Nazir, A., Khalid, A., Ayesha Raana, M., Majeed, N.: Performing inquisitive study of PM traits desirable for project progress. Int. J. Mod. Educ. Comput. Sci. **6**(2), 41–47 (2014)
7. Lavy, I., Rashkovits, R.: Motivations of information system students in final project and their implications to technology and innovation. Int. J. Mod. Educ. Comput. Sci. (IJMECS) **8**(2), 1–13 (2016)
8. Omri, M.N., Mribah, W.: Towards an intelligent machine learning-based business approach. Int. J. Intell. Syst. Appl. **14**(1), 1–23 (2022)
9. Yu, J., Zhao, Y., Li, Y., Kuang, H.: An example of course project of real time multitask programming. IJEME **2**(8), 65–71 (2012)
10. Ye, J., Yu, J.: A course project of Chinese character text editor for embedded system education. IJEME **2**(2), 47–53 (2012)
11. Bibi, R., Jannisar, M., Inayet, M.: Quality implication for prognoses success in Web applications. Int. J. Mod. Educ. Comput. Sci. (IJMECS) **8**(3), 37–44 (2016)
12. Wang, F.: The course design for algorithm analysis and research on improving the students' comprehensive ability. IJEME **2**(2), 42–46 (2012)
13. Parthasarathi, P., Rajnish, K.: A fuzzy based parametric approach for software effort estimation. Int. J. Mod. Educ. Comput. Sci. (IJMECS) **10**(3), 47–54 (2018)
14. Andreevsky, I.L., Sokolov, R.V.: Assessment of user preferences when choosing cloud software products. Bull. Samara State Univ. Econ. **5**(199), 26–32 (2021)

15. Zadeh, L.A.: Fuzzy sets. Inf. Control **8**, 338–353 (1965)
16. Bellman, R., Zadeh, L.: Decision-making in vague conditions. In: Questions of Analysis and Decision-Making Procedures, pp. 172–215 (1976)
17. Kuleshova, T.I., Tolstykh, T.O.: Automation of project activity and choice of software product. Bull. Voronezh Inst. High Technol. **4**(23), 200–203 (2017)

The Influence of Pictures, Animations and Videos in Teaching Courseware on Students' Listening Comprehension

Fengmei Chen[1,2], Wugang Li[1], Lanxiang Wei[1], Zhenpeng Qin[1], and Feng Ning[2(✉)]

[1] School of Intelligent Manufacturing, Nanning University, Nanning 530200, China
[2] School of Physics and Electronics, Nanning Normal University, Nanning 530001, China
wlxlwg2022@163.com

Abstract. The use of multimedia courseware is very common in modern classroom teaching. In response to the problem that many teachers are not aware of the role of the main elements of multimedia courseware, such as pictures, animations and videos, and their effect on students' listening effectiveness, a study on the effect of these elements of courseware on listening effectiveness needs to be conducted with Chinese science and engineering students. This was done by showing pictures, animations, and videos in the courseware to three groups of students of similar level. These students' perceptions, including memory, were then tested after the lesson, and the results were compared and analyzed. The paper concludes with comparative findings on the degree of difficulty of students' perception of physical motion, the degree of hindrance to the process of transforming and encoding information, and the magnitude of the impact of attention allocation, and from the results, several useful pedagogical insights on the production and use of multimedia courseware are presented. This provides ideas as well as suggestions for teachers in the process of creating courseware that will help them to use the above elements of courseware appropriately and effectively in their classroom teaching.

Keywords: Multimedia · Courseware · Listening effect · Comparison

1 Introduction

Information technology has been deeply integrated into classroom teaching, therefore, multimedia courseware has become a very important "document" for teachers' classroom lectures. However, there are many problems in the process of the widespread use of courseware. For example, Yan Zhao and Lin Yang found that there are not many coursewares that can both fully integrate course content and classroom teaching in reality and use multimedia features to enhance teaching effectiveness [1]. Yuguo Li found that most teachers use multimedia courseware inappropriately, leading to a decline in teaching quality [2]. Some teachers blindly take the use of courseware as the only criterion for judging the merit of a lesson, failing to consider the cognitive level of students

Z. Hu et al. (Eds.): AIMEE 2022, LNDECT 159, pp. 607–615, 2023.
https://doi.org/10.1007/978-3-031-24468-1_55

and the learning efficiency of students. They use courseware in order to complete the task of the lesson, which leads to a serious decline in teaching quality and failure to optimize the teaching classroom effect. To solve these problems and to enable teachers to make full use of classroom materials, teachers need to be clear about what and how the materials affect students' listening, the role of various elements in the materials, and so on. Lowe, R. K. found earlier that courseware has become increasingly popular in computer-mediated education in a variety of subjects [3], while he and Plotzner, R found that there is little research on how to design or present courseware in order for it to facilitate learning [4]. Richard E. Meyer [US] has proposed a multimedia learning theory more than a decade ago [5]. A review of learning theory research on multimedia courseware design and development has also been done by Cao Dianbo and Zhang Lijing [6]. There are other scholars who have done some studies on the impact and application of pictures, animations, and videos in courseware, respectively [7–13], but none of them have direct quantitative measurements and lack a scientific basis, so the research methods are not scientific enough. Therefore, it is necessary to conduct our own empirical research on Chinese students to address the above issues. In this paper, we will conduct a comparative study on the effects of pictures, animations, and videos in courseware on students' listening through classroom experiments, questionnaire tests, and individual interviews, and then correlate the findings and delve into the aspects of students' cognitive processes, memory, and comprehension involved.

2 Teaching Experiment

The teaching experiment was first conducted. The sample was selected to include science and engineering students, 39 students in the first class and 42 students in the second class and 42 students in the third class, for a total of 120 students.

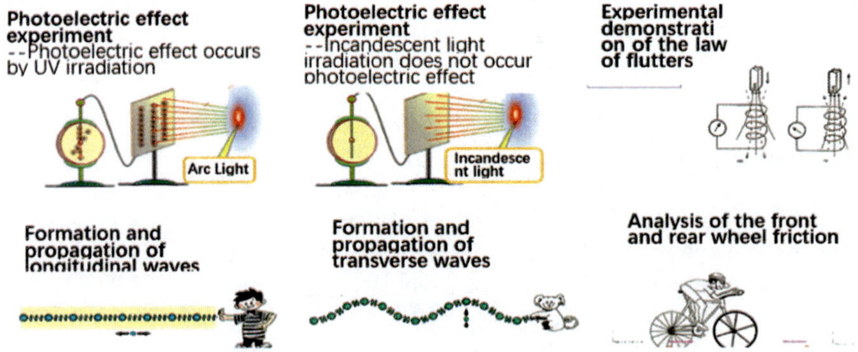

Fig. 1. Screenshot of the courseware in image form only

In the experiment, the teacher selected knowledge that was not learned and unfamiliar to the students, i.e., "photoelectric effect", "fluttering law", "formation and propagation of transverse waves", "formation and propagation of longitudinal waves", and "formation and propagation of longitudinal waves". The five knowledge points of "formation and propagation of longitudinal waves", and "analysis of front and rear wheel friction of bicycle" were prepared in advance. These materials are in the form of pictures, animations, or videos only (Fig. 1, Fig. 2, and Fig. 3). The first class was taught with pictures, while the second class was taught with animations converted from pictures in the first class, and the third class was taught with videos converted from pictures.

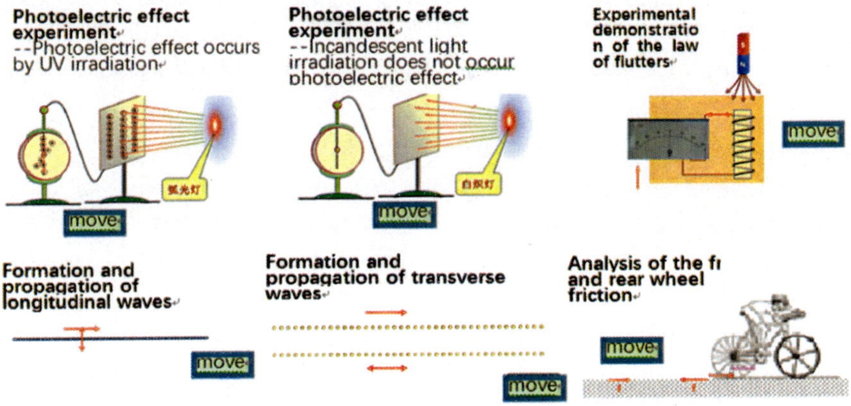

Fig. 2. Screenshot of the courseware in animation form only

Fig. 3. Screenshot of the courseware in video format only

3 Experimental Effect Testing and Analysis

3.1 Questionnaire and Analysis

After the lecture, I gave the students the relevant questionnaires prepared in advance and asked them to fill them in truthfully. The purpose was to test the students' knowledge transfer and application after the teacher's teaching process, and to understand their psychological process of learning. The specific findings are shown in Tables 1, 2, 3, 4, and 5.

Table 1. Responses to question 1 of the questionnaire

Question 1: How do you feel about the difficulty of the five knowledge you just learned?			
Form	A. Difficult	B. General	C. Easy
All the courseware is in the form of pictures	56.67%	33.33%	3.33%
All the courseware is in the form of animation	21.88%	53.18%	6.25%
All the courseware is in the form of video	17.29%	28.57%	21.43%

Table 2. Responses to questions 2 and 3 of the questionnaire

Question 2: What is the photoelectric effect phenomenon? Question 3: Why can the photoelectric effect occur with ultraviolet light, but not with incandescent lamps?	
Form	Answer to the main situation
All the courseware is in the form of pictures	Descriptions vary greatly, "UV light is required to conduct electricity", "UV light transfers positive charge to the detector", "It is the zinc plate that needs bright light for internal charge transfer to occur", "Only UV light can cause the photoelectric effect"
All the courseware is in the form of animation	Basically, it can be described that "only the zinc plate can be irradiated to be the photoelectric effect", "only ultraviolet light can be irradiated to be the photoelectric effect", and "only very strong light has to be irradiated to be the photoelectric effect"
All the courseware is in the form of video	Describing the difference is not too much, easy to catch the key point, "Zinc plates emit electrons when exposed to UV light"

Table 3. Responses to questions 4 and 5 of the questionnaire

Question 4: How to express the content of the law of flutters in a simple way? Question 5: How can the direction of the induced current be determined?	
Form	Answer to the main situation
All the courseware is in the form of pictures	Only magnets, magnetic fields and ammeters are seen, but the difference in magnetic lines of induction is known after attentive observation. The teacher can only conclude that the magnet lines of induction are changing after guidance
All the courseware is in the form of animation	Think of the words "same" or "opposite", but not "resist" or "obstruct", you cannot answer the question that the effect of induced current always resists the magnetic field that causes it
All the courseware is in the form of video	The answers are also relatively complete, with a few not getting the exact right answer

Table 4. Responses to questions 6, 7, 8, and 9 of the questionnaire

Question 6: How are transverse waves formed? Question 7: What about the direction of vibration and propagation of transverse waves? Question 8: What is the direction of vibration and propagation of longitudinal waves? Question 9: How are longitudinal waves formed?	
Form	Answer to the main situation
All the courseware is in the form of pictures	About half of the students answered more completely, and could only simply say that waves are moving, without knowing how to move specifically
All the courseware is in the form of animation	About 80% of the students answered more completely, formed the concept of "mass point vibration", and quickly understood that longitudinal waves are "sparse and dense vibration"
All the courseware is in the form of video	About 90% of the students answered the questions completely. Almost through the introduction of the video, they can quickly understand the concept of "mass vibration", and then understand that "the transverse wave starts from the wave source, and the mass in the medium vibrates in turn and drives the next mass to vibrate, thus forming an outward propagating wave"; they can also quickly understand that "The longitudinal wave is sparse and dense vibration", and then understand "the characteristics of longitudinal wave motion and propagation principle

Table 5. Responses to questions 10 and 11 of the questionnaire

Question 10: What is the front and rear wheel friction of a bicycle?
Question 11: Which way does the front wheel friction go? Which way does the rear wheel friction go?

Form	Answer to the main situation
All the courseware is in the form of pictures	Most students are able to know how to understand in the context of their life experiences and can easily come up with answers
All the courseware is in the form of animation	Almost more than half of the students were able to know how to use their life experiences to come up with answers easily
All the courseware is in the form of video	Almost all students were able to know how to use their life experience to quickly and accurately grasp the relevant knowledge and easily come up with answers

The above results show that when the courseware is presented in the form of pictures, students can receive the information of the pictures through the visual channel and can basically know and understand the five knowledge points shown in the courseware.

However, when students answered the corresponding questions, they could not see the whole process of the object's movement because it was only a flat static visual stimulus, which was rather abstract, and they could only rely on their previous familiar words to reason or express the relevant knowledge points, so the terms expressed were not accurate enough. This indicates that students use their previously acquired knowledge in their minds through the visual channel in the process of learning, and then form the final answer after certain transformation. The lack of completeness of the answer is a result of a certain degree of obstruction in the process of information transformation and encoding.

When the courseware is presented in the form of animation, students can receive the information of the animation through the visual channel and can basically understand and master the five knowledge points shown in the courseware. However, when students learn the more difficult knowledge of the photoelectric effect and the law of flutters, they are easily attracted to other external environments, which distracts their selective attention. When it comes to the simpler, easily understood and related to life experience knowledge, the rich visual stimulation of animation will stimulate and attract their attention and make their minds active. Although they cannot see the whole process of the movement of real objects, they can feel it through animation. They rely on their previous familiar words to reason or express that relevant knowledge point, which will be more effective than the completely static pictures. This shows that students can use their previous knowledge in their minds to transform and encode information more smoothly than pictures through visual channels in the learning process.

When the lesson was presented in the form of a video, they were able to receive the information from the video through the visual and auditory channels, and could grasp the five knowledge points shown in the lesson more accurately. However, when students answered the relatively difficult points of photoelectric effect and fluttering law, some of the wording was not accurate and complete. In contrast to the pictures and animations, more students mastered the video better. The rich visual and auditory stimuli of videos can stimulate and attract students' attention and make their minds active when they are learning simple, easy to understand and relevant to their life experiences. They can watch the entire motion of a real object through the video and thus perceive the entire motion process more easily. Therefore, they can reason or express the relevant knowledge point by relying on their familiar words, which is better than the effect of completely static pictures and animations. This also shows that students can use their previous knowledge in their minds to transform and encode information more smoothly than pictures and animations through visual channels in the learning process.

3.2 Interview Survey and Analysis

Through interviews with two students from each format of the experiment, we learned that learning with animated textbooks is less difficult and helps students to deepen their memory and recall what they have learned more quickly when they apply their knowledge. The students also preferred to listen to the teacher's lesson with the courseware during the class, but it would be better to combine it with the board. From the interviews with the four teachers, we know that teachers prefer to use a combination of multimedia materials, blackboard books, and students' hands-on experiments in class. However, only for the courseware, teachers would commonly use courseware in class, and would pay more attention to adding animations and videos to the courseware to arouse students' interest in learning and enhance their listening effect. The specific interviews are shown in Table 6.

Table 6. Main interviews of students and teachers

Student interviews	Faculty interviews
Compared with the pictures and animations on the courseware, the basic answer was more in favor of video, because video would be more memorable, more vivid and concrete, and could remember more and longer. Some students directly suggested "change the format of the courseware to video", while others said "add more videos and animations to it"	In the production of courseware, such as with some animations and games, it is almost believed that the pictures, animations and videos in the courseware will have an impact on the students' listening effect. Some excellent pictures, animations and videos will have a greater impact, with animations and videos having the greatest impact, especially during the epidemic internet lessons. These can make some experiments or difficult to understand points more visual and concrete

4 Results and Insights

Through practical investigations, it was found that, as previously studied, elements in the courseware have an impact on students' listening effectiveness pictures, animations and videos have an impact on students' listening effectiveness [14–16]. Specifically, they affect students' perception of the entire motion process when listening to the lesson, with the perceived ease of perception being video > animation > pictures. The courseware also affects students' transformation and encoding process of the information, with the degree of hindrance being pictures > animations > videos. The different formats of the courseware affect the problem of distributing students' attention, with the degree of impact being animation > video > pictures. The degree of visual images presented in different formats can also influence students' thinking patterns. However, the shortcoming is that students are not able to do deep processing of their thinking while listening to the lessons, which is not conducive to the development of thinking and logical abstract thinking. For knowledge that is in close contact with real life, students are able to grasp the key points quickly regardless of the form in which the lesson is presented, and can then master the knowledge. However, it was found that students who had learned through videos had the longest memory time, while those who learned through pictures only had the shortest memory time.

Based on the results of the above single factor in the lesson on the effectiveness of student listening, the following pedagogical insights were obtained.

1) In the process of making the courseware, the editing of the pictures, text, animations and colors, layout and so on should be as simple and clear as possible, not too fancy and artistic, otherwise it will affect the distribution of students' attention in the learning process and lead to the formation of wrong selective perception, thus making the listening efficiency low.
2) Under the premise of meeting the teaching objectives, animation and video should be inserted into the courseware as far as possible, which are easier for students to accept and love than static pictures, and are better for them to understand some abstract and difficult knowledge.
3) In the process of teaching, when the courseware is too concrete and formalized some knowledge, you can let students think and discuss and think and imagine first, and then provide the courseware to present them with pictures, animations or videos. This can help students develop their thinking and logical abstract thinking, thus improving their listening efficiency.
4) In the course of the lesson, when students encounter knowledge that is more familiar and related to their life experience, such as basketball-throwing motion, etc., the lesson material in any form for the lesson will have little impact on the application of knowledge after listening to the lesson, but it should be presented in the form of video or animation as much as possible so that students can maintain a relatively long memory duration.

5 Conclusion

In order to achieve good results in classroom teaching, it is very important to find different influencing factors on the basis of investigation and analysis, and then find effective methods and countermeasures. Through the classroom teaching experiment, this paper discusses the influence of pictures, animation and video on students' listening in physics class by means of questionnaire and interview; This paper analyzes the reasons for the influence of students' perception and memory, and obtains the relative difficulty of students' perception of physical movement, the degree of obstruction of information conversion and coding process, and the degree of influence on attention distribution. The following conclusions are drawn respectively, and some opinions beneficial to teaching are put forward. The research provides a scientific basis for teachers in the process of creating courseware that will help them to use the above elements of courseware appropriately and effectively in their classroom teaching.

References

1. Zhao, Y., Yang, L.: Experimental discussion on the influence of multimedia courseware design elements on students' cognition. Lanzhou J. Educ. **32**(2) (2016)
2. Li, Y.G.: How to Give Full Play to the Role of Courseware in Promoting High School Physics Teaching. Capital Normal University, Beijing (2005)
3. Lowe, R.K.: Interrogation of a dynamic visualization during learning. Learn. Instr. **14**, 257–274 (2004)
4. Plotzner, R., Lowe, R.K.: Dynamic visualisations and learning. Learn. Instr. **14**, 235–240 (2004)
5. Meyer, R.E.: Multimedia Learning, p. 61. The Commercial Press, Beijing (2006)
6. Cao, D., Zhang, L.: A review of learning theory research on multimedia courseware design and development. Shaanxi Educ. High. Educ. **3** (2014)
7. Wei, M., Tang, Q.: Experiment on the use of pictures in PPT courseware for elementary Chinese teaching. J. Yulin Norm. Coll. **39**(3), 98–103 (2018)
8. Li, R.: An Eye-Movement Study on the Effect of Picture Presentation in PPT Courseware Learning for Junior High School Students. Henan University (2016)
9. Wang, W.L.: Exploration on the use techniques of pictures and graphics in PowerPoint courseware. China Manag. Inf. Technol. **19**(10), 216–217 (2016)
10. Li, Q.: The design and application of animation in the multimedia courseware of secondary school physics. Sci. Technol. Innov. Herald **14**(6), 208–209 (2017)
11. Wei, M., Wang, F., Xu, J., Song, W.: The impact of custom animation on teaching effect in PPT courseware. Neijiang Sci. Technol. **33**(6), 192 (2012)
12. Wang, X.: The use of video clips in multimedia courseware. J. Hubei Correspond. Univ. **31**(7), 160–161 (2018)
13. Ren, C.: Introduction to the use of multimedia video in secondary school ideology and morality class. Educ. Technol. **12**, 247 (2015)
14. Zhao, C.: The Influence of Graphic Presentation and Type on Learning Effect. Nanjing Normal University (2017)
15. Shao, D.: Analysis of graphics, animation and video processing in multimedia courseware production and their application in teaching. J. Chaohu Coll. **9**, 161–162 (2007)
16. Zhao, G.X., Chen, L.F., Wang, D.N.: Research on the teaching effectiveness of video media application in multimedia courseware. China Educ. Technol. Equip. **6**, 11–12 (2017)

Application of Virtual Model Base of Engineering Drawing Based on Online Learning Platform in Teaching

Tianfeng Kuang(✉), Daming Huang, and Yi Wei

School of Transportation, Nanning University, Nanning 530299, China
fyql92@163.com

Abstract. With the rapid development of network information technology, the path of educational communication has changed significantly, and digital-intelligent online learning platforms have become an important channel for educational communication. Aiming at the development trend of online teaching, this study takes the OBE engineering education concept of as the guidance, combined with the construction of the core curriculum of engineering drawing, and in view of the shortcomings and problems existing in the teaching process of Engineering Drawing, studies the application reform of virtual simulation technology in the teaching of Engineering Drawing. In addition, the virtual model base of Engineering Drawing is combined with the "superstar learning system" to build a three-dimensional comprehensive online learning platform, and the combination of online teaching and offline testing is implemented, which promotes the reform of the teaching mode of participatory learning courses and provides a reference for the teaching reform of Engineering Drawing.

Keywords: Engineering Drawing · Virtual model · Online learning platform

1 Instruction

Engineering Drawing (ED) is known as the language of engineering. It is an engineer design language of expression and communication tools, especially for mechanical engineers, interpretation is a mechanical structure function principle and difficulty doubt important ability, drawing mechanical engineering drawing is a necessary basic knowledge and skills of engineers, the interpretation of ED is the basic ability that an engineer should have [1]. In the past, ED courses were mainly offline traditional teaching. In the traditional teaching process, the physical model is used to show the relevant content in the class. Physical model is widely used as an effective auxiliary tool for learning ED [2, 3]. However, due to the few kinds of teaching resources, they cannot meet the learning needs of structural diversity, and cannot adapt to the needs of progress changing with development, and the teaching lacks vitality and development. In addition, these models require a lot of money to manage, maintain and update. In this case, creating virtual model library can not only solve the above problems well, but also achieve the intended teaching purpose [4].

Z. Hu et al. (Eds.): AIMEE 2022, LNDECT 159, pp. 616–623, 2023.
https://doi.org/10.1007/978-3-031-24468-1_56

Outcome-Based Education (OBE) is a student-centered, result-oriented and continuous improvement engineering teaching concept. Based on the concept of OBE, many scholars have made many exploration in the teaching reform of ED [5, 6]. With the rapid development of online education technology and the wealth of digital teaching resources, online teaching has become the development trend of teaching mode [7, 8]. This study aimed at the development trend of online teaching, fully absorbing the existing education teaching reform research results. On the basis of thorough investigation and study, combined with the ED core curriculum construction, the concept of OBE is introduced into the teaching reform of ED course, students need to achieve the ability of reverse design teaching model. Computer-assisted software is used to build a virtual model base and optimize the models used in course teaching. In order to improve students' thinking ability and drawing ability, the virtual model base is combined with the online learning platform of "Super Star Learning System" to build a three-dimensional comprehensive online learning platform of "virtual mode + online learning". On this platform, students take the virtual model base as the learning focus, and complete the corresponding teaching processes such as preview, discussion, homework exercises, answering questions, examination and assessment, analysis and evaluation.

2 Design of Virtual Model Library

In this study, virtual simulation technology is introduced, and the virtual model library is established by computer drawing software according to the existing teaching materials and problem sets. The model library mainly includes typical structural models involved in ED courses, including basic body model library, assembly model library, component model library and typical structure model library. The model library is enough to meet the teaching and learning needs of teachers and students. In addition, virtual models can be modified and added in real time according to other requirements. The structure of the model library is shown in Fig. 1.

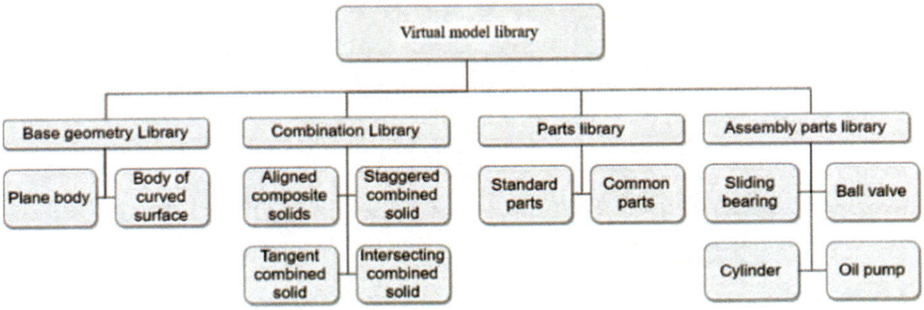

Fig. 1. Virtual model library

3 The Establishment of Virtual Model Library

In this study, the computer drawing software AutoCAD and the 3D solid software Creo are used to establish the virtual model library. According to the needs of ED teaching,

AutoCAD is used to draw two-dimensional graphics, and Creo's stretching, rotating, scanning, mirroring, chamfering and other functions are used to draw three-dimensional model libraries of basic body, assembly body, parts and components, and assembly body. The virtual model made is used to replace the traditional physical model used in teaching. These virtual models can be expressed in various ways through eDrawing output corresponding basic view, sectional view and sectional view, etc. The typical structure of the virtual simulation model library is shown in Figs. 2 and 3.

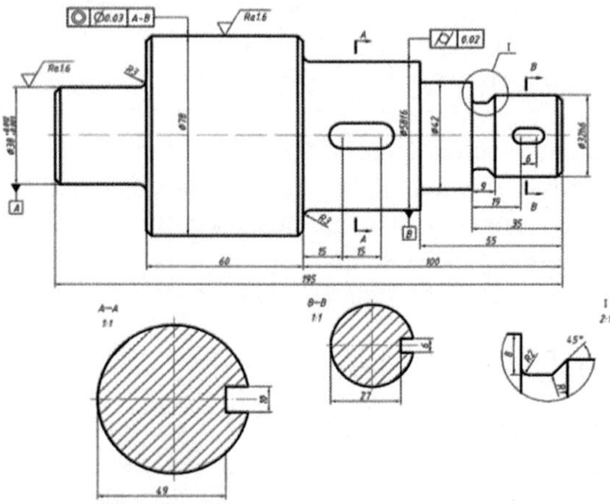

Fig. 2. 2D CAD models of the axis

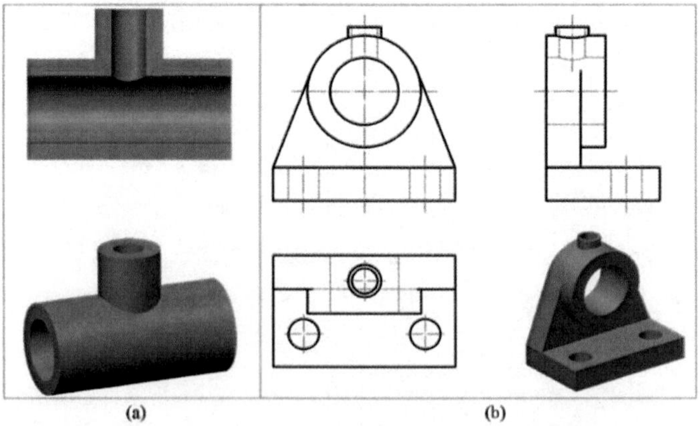

Fig. 3. 3D virtual model: (a) Sectional view of the combination; and (b) Three views of the combination

4 The Application of Online Learning Platform

4.1 Organizational Structure of Online Learning Platform

Based on the results-oriented OBE concept, this study explores the "SPOC+" online and offline hybrid teaching mode integrating task-driven and problem-oriented teaching methods in the SPOC environment [9]. On the basis of the virtual model library, an electronic resource library of ED virtual models including picture library, animation library, video library and problem library is established by using the "Superstar Learning System", forming a three-dimensional comprehensive online learning platform of "virtual model + online learning". On the learning platform, teachers and students can complete pre-test and post-test, communication and discussion, homework practice, answer and solve problems, assessment and examination, analysis and evaluation and other learning and teaching processes. The architecture of the learning platform is shown in Fig. 4.

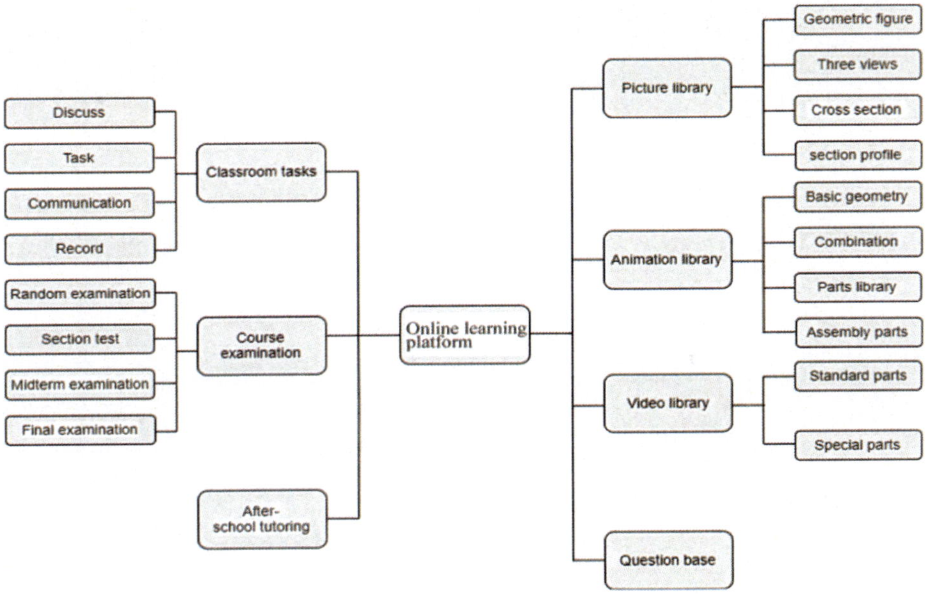

Fig. 4. Organizational structure of online learning platform

This study follows the principle of "student-centered, achievement oriented and continuous improvement" [5], provides diversified models through modification and supplement, and constantly enriches the model base, creating good conditions for students to realize autonomous learning, cooperative learning and inquiry learning. Using the learning platform, combined with task driven, problem oriented and other teaching methods, and taking "task driven + project guidance + Situational Design + Role Play + group cooperation" as the organizational form [10], build a learning interaction mechanism and knowledge transmission mechanism of "mutual help, mutual promotion and mutual progress", so as to promote students to learn more, practice more and think more.

As shown in Fig. 5, the virtual model of ED is mainly placed in the material module of the learning platform. This module sub-option includes virtual model e-learning resources such as question library, video library, picture library and animation library. The content of the material module is shared by teachers and students. When teachers use the online learning platform, they can add ED virtual models or other electronic resources to the corresponding model library according to their teaching needs. Virtual model electronic resources can also be updated in real time to meet the latest curriculum teaching needs.

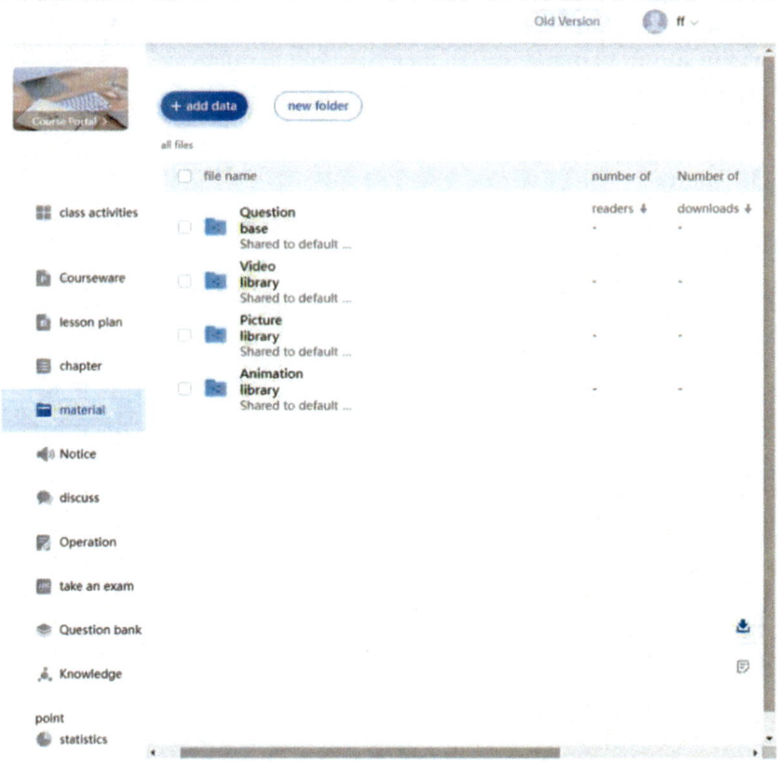

Fig. 5. Virtual model library of online learning platform

4.2 Online Teaching of Virtual Model

Reading engineering drawings and drawing engineering drawings are two important skills in ED learning, which are also the core objectives of the course. It is essential for engineers to be able to correctly describe the shape and structure of objects according to the view relationship in ED, and to identify and correct errors in defective drawings and views in ED. During the implementation of the course, it is found that there are not many excellent students in these two aspects, especially the deficiencies in projection relationship and object map transformation are very prominent, which is particularly obvious in the subsequent design courses.

Therefore, this study improves and optimizes the typical structural model according to the subsequent design courses. These structural models are helpful in training students to recognize mechanical drawings and drawing engineering drawings. The typical two-dimensional structural model of the virtual model base is shown in Fig. 6.

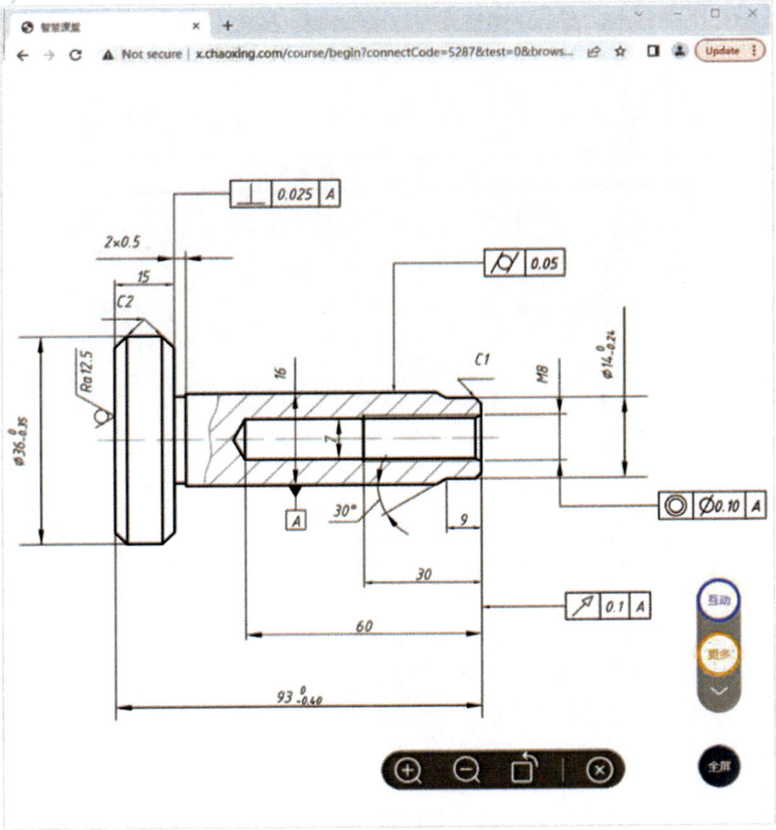

Fig. 6. 2D model demonstration

In the process of ED teaching, students are faced with simple abstract images, which is the weak link in students' cognitive ability. They need the help of objects to identify and construct spatial entity structure relations, which becomes the main obstacle of engineering graphics learning and the reason for students' low learning motivation. Therefore, in the virtual model library, an uncomplicated virtual model is established to enable students to gradually transition from simple model to complex model. As shown in Fig. 7, the 3D model library is demonstrated on the learning platform to help students develop spatial imagination. According to students' feedback, the virtual model has an obvious learning effect, and course learning becomes easier. The teaching evaluation of the course, the online learning platform has been well received by more than 90% of students.

Fig. 7. 3D model library demonstration

5 Conclusion

Integrated application of various software technologies. In the design of virtual model library and the construction of online learning system, a variety of software tools including computer aided design software and three-dimensional modeling software are integrated into the system. The model base is combined with a learning platform. In addition to the construction of virtual model base, the combination of online teaching and offline testing is implemented to promote the reform of course teaching mode of participating learning and build a multi-functional three-dimensional comprehensive online learning platform. The implementation of this study has produced the following results: Students give feedback that it is easier to master the course of ED by using virtual model library, and it is easier to complete the course learning, which is conducive to students' ability to achieve the course goals;

1) Help students to eliminate the psychological pressure of learning, improve their interest in learning, establish learning confidence, promote active learning, improve the degree of support for graduation requirements;

2) Help students develop good spatial imagination, creative thinking ability and the ability to solve problems independently, improve the teaching of education concept fit degree.

Acknowledgment. This project is supported by Guangxi Higher Education Teaching Reform Project (2021BKHXK01 & 2021XJJG05).

References

1. Azodo, A.P.: Attitude of engineering students towards engineering drawing: A case study. Int. J. Res. Stud. Educ. **6**(1), 61–74 (2017)
2. Faraje, F.H., Sadiq, K.M.: The concept of Engineering Drawing and its impact on engineering education. Eng. Technol. J. **37**(4), 454–459 (2019)
3. Zuo, Z., Feng, K., Chen, B.: The modern education mode for engineering drawing. JGG **7**(1), 121–128 (2003)
4. Violante, M.G., Vezzetti, E.: Design of web-based interactive 3D concept maps: A preliminary study for an engineering drawing course. Comput. Appl. Eng. Educ. **23**(3), 403–411 (2015)
5. Ross, V.: From transformative outcome based education to blended learning. Futures **44**(2), 148–157 (2012)
6. Zhang, W., Hu, Q.: Teaching reform of Engineering Drawing course against the background of emerging engineering education. J. Front. Educ. Res. **1**(5), 113–117 (2021)
7. Nian, L., Wei, J., Yin, C.: The promotion role of mobile online education platform in students' self-learning. Int. J. Cont. Eng. Educ. Life Long Learn. **29**(1–2), 56–71 (2019)
8. Sun, A., Chen, X.: Online education and its effective practice: A research review. J. Inf. Technol. Educ. **15** (2016)
9. Gurukkal, R.: Outcome-based education: an open framework. High. Educ. Future **7**(1), 1–4 (2020)
10. He, J.: Construction of "three-stage asynchronous" instructional mode of blended flipped classroom based on mobile learning platform. Educ. Inf. Technol. **25**(6), 4915–4936 (2020)

Innovation of International Logistics Talent Training Mode in Applied Undergraduate Colleges and Universities Based on the 'Belt and Road'

Xin Li[1](✉), Yong Wang[2], Qingying Zhang[3,4], Xiaofen Zhou[1], Xiaocui Wang[1], Hong Jiang[1], Yani Li[1], and Star Xie[1]

[1] School of Logistics, Wuhan Technology and Business University, Wuhan 430065, China
353930944@qq.com

[2] Evergrande School of Management, Wuhan University of Science and Technology, Wuhan 430070, China

[3] College of Transportation and Logistics Engineering, Wuhan University of Technology, Wuhan 430063, China

[4] Department of Production Engineering, KTH Royal Institute of Technology, Stockholm, Sweden

Abstract. Under the tide of world multi polarization and economic globalization, the internationalization of higher education has become the general trend. Implementing the 'going out' strategy and cultivating international logistics talents are the internal needs of Application-oriented Undergraduate Colleges and universities to adapt to the development trend of higher education, and also the internal needs of Application-oriented Undergraduate Colleges and universities to improve the level of running a school. At present, in terms of international logistics talent training, there are some problems in the undergraduate of application-oriented universities, such as the lack of international logistics talent training platform and funds, the lack of international teaching staff, and the imperfect teaching system of international logistics talent training. In view of these problems, this topic introduces the guiding ideology of the 'the Belt and Road' and builds an application-oriented undergraduate university, which can explore the cultivation of international logistics talents from the aspects of the construction of an international logistics talent training platform, giving full play to its own characteristics and advantages, improving the curriculum, teaching and evaluation system, and promoting the internationalization of professional construction, so as to continuously improve the cultivation level of international logistics talents.

Keywords: The Belt and Road · Applied universities · International logistics · Talent training · Innovation research

1 Introduction

The academic history and research trends of relevant research at home and abroad; Compared with the original academic value and application value of the existing research,

Z. Hu et al. (Eds.): AIMEE 2022, LNDECT 159, pp. 624–633, 2023.
https://doi.org/10.1007/978-3-031-24468-1_57

application-oriented undergraduate colleges and universities shoulder the important task of serving the regional economic development. The cultivation of qualified international logistics talents has become an important indicator to measure their education level. The 'the Belt and Road' strategy has put forward new requirements for the training of international logistics talents in applied universities.

Niu B (2019), believes that the 'the Belt and Road' initiative is a new path for China's transformation to growth [1]. Akbar, Bahari (2019) pointed out the vision, scope and objectives of the 'the Belt and Road' strategy in China, as well as its impact on the economic development of relevant countries [2]. Tunviruzzaman R (2020) proposed that China's 'the Belt and Road' initiative (BRI) is one of the most ambitious economic strategies in modern times. It has changed the economic, political and social relations between eastern and Western societies, and can improve infrastructure, required technology, management talents and greater contact with the world [3]. Bogojevi S (2021) proposed the 'the Belt and Road' initiative is a strategic measure aimed at expanding China's opening up in recent years. This strategy has played an important role in opening up a new pattern of comprehensive development. To cultivate a number of high-tech leading talents, international strategic scientific and technological talents, which is necessary for the young scientific and technological talents and high-quality talents [4].

With the rapid development of globalization and the in-depth implementation of the "the Belt and Road" initiative, international trade has become increasingly close, and a large number of intercontinental goods will be transported by sea. It can be seen that "maritime logistics" has become an important link connecting the trade of countries along the "the Belt and Road". Now, with the improvement of China's international status and the further implementation of the "the Belt and Road" initiative. The development of China's maritime logistics will have a great impact on the world's maritime territory. Therefore, it is particularly important to discuss the maritime logistics of the 21st century Maritime Silk Road. In order to better promote the development of the "the Belt and Road" maritime logistics and build an exchange platform in the field of international maritime logistics and supply chain management, it is very important for application-oriented universities to develop the "the Belt and Road" international logistics talents, aiming to contribute to the development of maritime logistics in China and the world [5].

1.1 Shortage of Logistics Talents

During the rapid development of logistics in China in recent years, a large number of logistics management talents have not been met. With the emergence of a series of new logistics operation and management modes such as logistics intelligence, new logistics retail and logistics platform, there will be a greater gap in international logistics talents. According to the statistics of relevant departments, 40.31% of the Logistics Majors' first jobs after graduation are primary logistics operation posts. The demand of logistics enterprises for such primary Posts accounts for 65% of the total demand for logistics talents, while 27% of the Logistics Majors choose to engage in other jobs other than logistics after graduation. This shows that not all logistics students can engage in logistics related work after graduation, and the logistics talents trained by colleges and universities at all levels cannot fully match the social needs. The shortage of international logistics talents of complex specialty has begun to restrict the iterative updating of China's logistics industry chain.

1.2 Good Employment Environment for International Logistics Talents

The first salary report of China's international shipping logistics industry was released, and the salary analysis of 15 posts in the international logistics industry is shown in Table 1.

Table 1. The salary analysis of 15 posts in the international logistics industry

Num.	Posts	Highest income	Highest income	Average income in the industry
1	Commissioner of Documents	48,733	108,039	76,674
2	Dispatcher	41,011	107,481	79,696
3	Pre-entry	38,341	75,353	55,847
4	Inspector	52,955	94,313	70,047
5	Handover officer	48,102	72,231	57,323
6	Bookers	57,721	147,398	90,126
7	Operations Specialist	72,908	139,279	101,496
8	Air Freight Manager	174,485	744,490	313,531
9	Route Specialist	72,822	12,839	90,702
10	Sales representative	36,406	210,086	113,995
11	Shipping Manager	149,152	795,143	303,742
12	Shipping Director	526,692	749,065	649,980
13	Transport Distribution Coordinator	54,954	128,816	94,003
14	Information Technology Specialist	73,589	220,365	118,945
15	Customs declaration	52,312	149,445	82,997

According to the above figure, as of April 2022, the salary of China's international logistics talents generally exceeds the industry average, the efficiency of China's logistics operation organization has been improved, and the comprehensive service capability has been greatly improved. It will inevitably lead to a large demand for international professional logistics talents. There are about 2,000 higher education institutions of all kinds in our country. Opened a major in material flow management, which can train about 150,000 talents for China's logistics industry every year. Various training institutions can train about 170,000 logistics talents every year. Various public training bases can provide about 140,000 logistics talents every year. The number of logistics professionals trained by various colleges and social institutions is far from meeting the needs of new logistics jobs every year, and vigorously cultivating international logistics professionals will help alleviate this social pressure [6].

2 Construction of Practical Teaching System for International Logistics Talent Military Training

According to the positioning of Application-oriented Undergraduate Universities and the characteristics of international logistics, the College of logistics has established the training goal of improving students' practical ability and innovative ability, and realizing the gradual cultivation of students' basic ability - professional application ability - research ability - comprehensive ability - practical work ability. After clarifying the training objectives, the research method of this topic adheres to the principle of combining qualitative and quantitative, comparative and comprehensive analysis. First, the research object is analyzed qualitatively by reading a lot of relevant literature, and then the data is collected, the relationship between the variables is empirically analyzed, and the characteristics of the research object are analyzed by comparison [7].

It mainly uses research methods such as literature research method, questionnaire survey method, SWOT-PEST matrix analysis method, SPSS data analysis method and comparative analysis method to collect information from students, experts and international logistics enterprises respectively by means of questionnaires, and uses statistical analysis method to determine the point of constructing practical teaching system, so as to ensure the effectiveness and practicality of the system. The composition of the system is shown in Fig. 1.

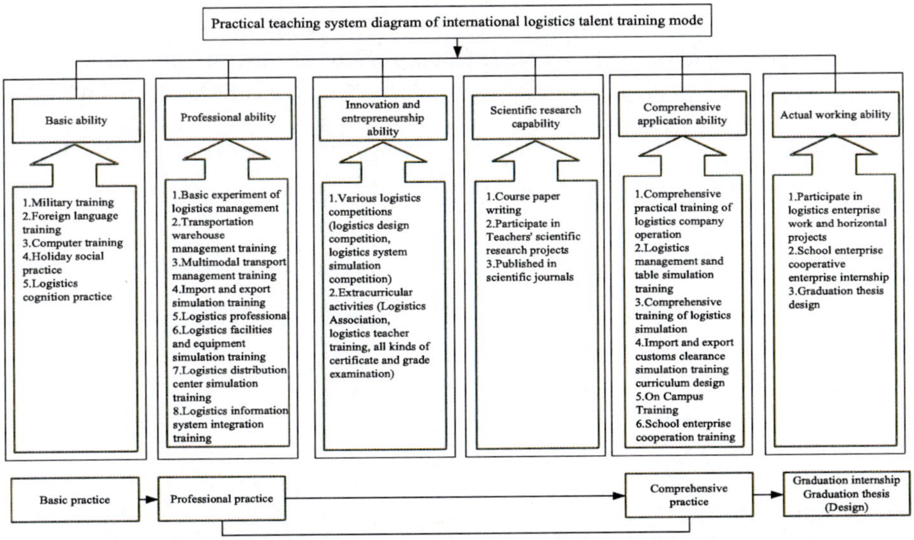

Fig. 1. Practical teaching system diagram of international logistics talent training mode

3 Implementation of Practical Teaching System for International Logistics Talent Training

The practical teaching system of international logistics talent training in Application-oriented Colleges and universities takes the on-campus laboratory, off campus practice base and industry university research cooperation project as the platform. This study takes the academic year and semester of undergraduate students' training as the sequence, and is divided into six stages: basic practice, professional practice, extracurricular practice, scientific research practice, comprehensive practice, graduation practice and graduation design. Among them, the basic practice stage is carried out in the first academic year and the first half of the second academic year; Professional practice and extracurricular practice are carried out in the second half of the second academic year and the third academic year after students carry out professional basic courses; In the stage of scientific research practice and comprehensive practice, it will be implemented in the first half of the Fourth academic year after students complete the study of major courses; The graduation internship and graduation design stage will be carried out in the second half of the students' Fourth academic year [8].

3.1 Implementation of Basic Practice Stage

Students majoring in international logistics mainly study basic courses in the first academic year, lacking the theoretical basis of practical teaching. Therefore, in the basic practice stage, we mainly carry out cognitive practice in the logistics industry, so that students can understand the current situation and development trend of logistics enterprises, and cultivate students' interest in professional learning. Arrange the teaching contents of this stage by visiting the school enterprise cooperative logistics enterprises (Jiuzhou Tong pharmaceutical logistics warehouse, JD Asia warehouse 1, DHL supply chain air freight warehouse and COSCO Shipping Container Port), expert training meetings and discussions, learning and MOOC teaching of smart tree app. Students visit the logistics infrastructure such as automated three-dimensional warehouses and drug distribution centers of enterprises, as well as the main activities of logistics systems such as air transportation and ocean container transportation [9].

3.2 Implementation of Professional Practice and Extracurricular Practice Stage

In the stage of professional practice and extracurricular practice, students have completed the study of basic courses and have the theoretical basis for studying basic professional courses. At this stage, students mainly study professional basic courses such as warehouse management, international cargo transportation, container and multimodal transport, logistics information management, modern logistics system, logistics professional English and logistics system simulation. The professional practice in this stage is the extension of theoretical teaching, and the purpose is to make students deepen their understanding of the professional basic theory of synchronous learning. The practice contents include: basic experiment of logistics management, transportation warehouse management training, multimodal transport management training, import and export simulation training, logistics professional, logistics facilities and equipment simulation training, logistics distribution center simulation training, logistics information system integration training. Combine professional basic course theory with practical training, so as to realize the training and cultivation of students' professional skills. The extracurricular practice content in this stage is implemented in the following three aspects:

1) Through setting up extracurricular scientific and technological practice activities, students' innovation awareness, entrepreneurship and practical ability can be improved. Such as organizing students to participate in college students' Entrepreneurship design competition, Challenge Cup competition, National College Students' logistics design competition and other college students' competition projects.
2) Through logistics associations and related institutions organized by students, activists, entrepreneurs and excellent college students will give a series of lectures, and part-time teachers with logistics practical experience and entrepreneurial experience will be hired to give special lectures. Such as "logistics enterprises and talent training", "international logistics and supply chain management frontier lectures", "College Students' international vision comprehensive quality training" and other special lectures.
3) Carry out various vocational qualification examination training, including academic education and social vocational qualification education, logistics engineers, customs brokers, purchasers, international freight forwarders, inspectors, internal auditors and external auditors, actively open up channels for students to obtain vocational qualification certificates, and improve students' ability to combine theory with practice.

3.3 Implementation of Scientific Research Practice Stage

After students complete their professional courses, they begin to implement the contents of the scientific research practice stage. Through the industry university research cooperation base established with COSCO or Sinotrans DHL international logistics enterprises, the off-campus tutors organize students to participate in the teacher's scientific research projects to complete this stage, and participate in the tutor's horizontal subject projects and declare relevant scientific research projects by themselves [10].

3.4 Implementation of Comprehensive Practice Stage

The comprehensive practice stage includes comprehensive training, on campus practice and off campus production practice. According to the international logistics talent training program, comprehensive training courses such as international logistics, logistics system simulation comprehensive experiment, import and export customs clearance comprehensive training, as well as on campus internships of "company creation and operation" are offered, and production practice is carried out in off campus internship bases. Through graduation internship in well-known logistics enterprises, students can master all operation links of enterprise logistics and collect first-hand data for graduation design. By combining students' graduation design with teachers' scientific research or production practice, students can complete the actual plan of enterprises and scientific research questions, and then cultivate students' independent thinking ability, practical ability and project coordination ability [11].

4 Optimization Measures for the Training of International Logistics Talents Under the Background of the 'Belt and Road'

In the implementation of the practical teaching system for international logistics talents under the background of the "Belt and Road", there are problems such as insufficient teachers' ability and quality, few practical teaching hours, incoherent practical teaching links, and lack of practical teaching materials. In order to better ensure the effective implementation of the practical teaching system, the following optimization measures are proposed.

4.1 Improving Teaching Environment and Innovative Practice

Table 2. Three types of curriculum models and development measures

Name	On-campus curriculum	Off-campus outreach programs	Vacation social practice
Development measures	1. Theoretical knowledge of the international curriculum 2. Systematic learning of professional knowledge	1. Foreign trainers come to the school to assist in teaching; 2. To expand students' professional knowledge, and to carry out oral listening and speaking training	1. At the end of the semester, the program team organizes students to go to COSCO for a one-month summer course; 2. Half a month of corporate course study at the School of Enterprise Logistics; 3. In Wuhan Hankou North, East Lake Comprehensive Insurance Zone and other major import and export trading markets to conduct half a month of market courses

The construction of the curriculum environment is an important condition for the implementation of the practical teaching system, and the applicable teaching materials, perfect practical syllabus and practical training guides are prepared according to the practical teaching system of the international logistics profession under the background of the

"Belt and Road". The practical syllabus or practical training guide includes the course experiment outline or practical training guide, the innovative experimental project outline or practical training guide, the production internship and graduation internship outline, the graduation design outline, etc. Sufficient school hours, multi-background and multi-national teachers, rich teaching [12].

The transformed classroom guarantees the cultivation effect of talents from more convenient, and students are employed after graduation, and they become industry professionals when they go out of school, which can be immediately transformed into productivity. The course is divided into three parts, as shown in Table 2.

4.2 Improving the Teaching Team of Teachers

Innovative practices in the faculty are an important guarantee for the implementation of the practical teaching system. The teaching team should be composed of college teachers and corporate trainers to form a high-quality teaching team, and the teachers who need to carry out practical teaching should not only improve their own teaching theory level, but also continuously improve their own practical operation level [13].

The school and the enterprise regularly hold international logistics talent training seminars, and feedback and solve problems that occur during talent development. Strengthen the teaching subject status of enterprises, students in the school learning stage, by the college teachers to carry out theoretical courses, by the enterprise trainer responsible for skill post learning. In the enterprise learning stage, the corporate trainer is fully responsible for the student's on-the-job learning and front-line practice. At the same time, enterprise teachers also participate in the process of logistics talent training, regularly go to the international logistics proofreading order class students for corporate culture training, and organize enterprise visits and exchanges. Only through personal experience can practically teaching teachers truly improve the quality of practical teaching and cultivate suitable talents for enterprises and society [14].

4.3 Constantly Refreshing the Teaching Concept

Improve the concept of education and strengthen exchanges and cooperation with foreign universities. Today's era is characterized by economic globalization, informatization and knowledge economy, for this reason, applied undergraduate studies need to broaden students' global horizons, so that students are good at thinking on the basis of multiculturalism, and good at cross-cultural communication, exchange and cooperation. Students need to learn to use information technology to select the information they need in the big data ocean when facing big data and artificial intelligence, and solve real problems through analysis; During their 4 years in college, students must not only learn knowledge, but also learn to learn and choose. The school of logistics should organically combine the domestic talent training program and the foreign talent training program, actively participate in and learn the teaching mode and teaching method of foreign countries through multiple channels and methods, and introduce and critically absorb the teaching concepts of foreign universities [15].

5 Conclusion

Under the "the Belt and Road", cultivating undergraduate international logistics talents in Application-oriented Colleges and universities is the development direction of colleges and universities in the future. The specific benefits will be gradually reflected in the implementation, and the purpose of this topic is also here. By constantly improving the practical operation ability, innovative spirit and problem-solving ability of international logistics students, they can truly meet the needs of society and enterprises. The relevant conclusions of this study are as follows:

1) In terms of the direction of use, it can provide reference for the "the Belt and Road" strategy and the development strategy, path and evaluation mechanism of undergraduate international logistics talent training of Applied Universities in Hubei Province.
2) In terms of expected social benefits, this study analyzes the international logistics talent training of Application-oriented Undergraduate Universities in the context of the "the Belt and Road" from the long-term and short-term perspectives, which is helpful to deepen the understanding of the international talent training of application-oriented undergraduate universities.
3) From the perspective of practice, it has a certain practical guiding significance for promoting the internationalization of applied universities, and provides a reference for colleges and universities in other regions to cultivate and develop international logistics talents.

Acknowledgment. This research was funded by of CSL (China Society of Logistics) Project (2022CSLKT3-132); Hainan Special PhD Scientific Research Foundation of Sanya Yazhou Bay Science and Technology City Project (HSPHDSRF-2022-03-032).

References

1. Niu, B., Xu, J., Lee, C., et al.: Order timing and tax planning when selling to a rival in a low-tax emerging market. Transp. Res. E: Logist. Transp. Rev. **123**, 165–179 (2019)
2. Akbar, B.: Practice model from theory to practice: CALL via focus on form approach and non-linear dynamic motivation to develop listening and speaking proficiency-Science Direct. Comput. Educ. **130**, 40–58 (2019)
3. Tunviruzzaman, R., Tahera, T., Zannat, T.: Economic & geopolitical opportunities and challenges for Bangladesh: One Belt-One Road (OBOR). Int. J. Res. Bus. Soc. Sci. **10**, 2147–4478 (2021)
4. Bogojevi, S., Zou, M.: Making infrastructure 'visible' in environmental law: The belt and road initiative and climate change friction. Transnat. Environ. Law **10**(1), 35–56 (2021)
5. Al-Hagery, M.A., Alzaid, M.A., Alharbi, T.S., Alhanaya, M.A.: Data mining methods for detecting the most significant factors affecting students' performance. Int. J. Inf. Technol. Comput. Sci. (IJITCS), **5**, 1–13 (2020)
6. Han, Z., Wei, B., Zheng, Y., et al.: Breast cancer multi-classification from histopathological images with structured deep learning model. Sci. Rep. **7**(1), 4172 (2017)

7. Jian, Q.: Effects of digital flipped classroom teaching method integrated cooperative learning model on learning motivation and outcome. Electron. Libr. **37**(5), 842–859 (2019)
8. Babincakova, M., Bernard, P.: Online experimentation during COVID-19 secondary school closures: Teaching methods and student perceptions. J. Chem. Educ. **97**(9), 3295–3300 (2020)
9. Yun, J., Zhang, G., Liu, R., et al.: The "Hand as Foot" teaching method in the anatomy and injury mechanism of ligaments around knee joint. Asian J. Surg. **44**(10), 1326–1327 (2021)
10. Brdulak, H., Brdulak, A.: Challenges and threats faced in 2020 by international logistics companies operating on the polish market. Sustainability **13**(1), 359 (2021)
11. Magnaye, R.B., Chaudhry, S.S., Sauser, B.J., et al.: Bridging the gap between practice and undergraduate teaching of operations management: the case of public liberal arts colleges. Int. J. Oper. Quant. Manag. **26**(1), 1–16 (2020)
12. Khan, M.H.: A unified framework for systematic evaluation of ABET student outcomes and program educational objectives. Int. J. Mod. Educ. Comput. Sci. **11**(11), 1–16 (2019)
13. Dada, E.G., Alkali, A.H., David, O.: An investigation into the effectiveness of asynchronous and synchronous e-learning mode on students' academic performance in national open university (NOUN), Maiduguri Centre. Int. J. Mod. Educ. Comput. Sci. (IJMECS), **11**(5), 54–64 (2019)
14. Li, J., Zhou, Y.: The exploration and practice in innovative personnel training of computer science and technology. Int. J. Educ. Manag. Eng. (IJEME), **2**(6), 47–51 (2012)
15. Zhiqin, L., Jianguo, F., Fang, W., Xin, D.: Study on higher education service quality based on student perception. Int. J. Educ. Manag. Eng. (IJEME), **2**(4), 22–27 (2012)

ISW Program Based on Teaching Ability Measurement Scale and Its Role in Improving Teachers' Ability

Jing Zuo(✉)

Center for Faculty Development, Nanning University, Guangxi 530001, China
zuojingnancy@qq.com

Abstract. The key to the quality of talents cultivation lies in teachers' teaching level and teaching ability. As an important way to develop teachers' teaching ability, training has been attached importance to by many colleges and universities, Instructional Skills Workshop, ISW, as a training project with rapid development in China in recent years, is also favored by many colleges and universities. In order to test whether ISW can effectively help teachers improve their teaching ability, teachers before and after participating in ISW were pre-tested and post-tested by compiling a teaching ability measurement scale. SPSS was used to conduct independent sample T test, and the results showed that participating in ISW could significantly improve teachers' ability of teaching design, teaching implementation, teaching evaluation and reflection. It is suggested that colleges and universities can refer to and imitate ISW program model when organizing other training programs, and integrate training, practice and reflection to better help teachers improve their teaching ability.

Keywords: Instructional skills workshop · Teaching ability · Independent sample T test

1 Introduction

The subject of talent cultivation in colleges and universities is teachers [1], and the key to improving the quality of talent cultivation lies in improving the teaching ability of teachers [2]. At present, more and more colleges and universities pay attention to and implement some measures to help teachers improve their teaching ability, such as training [3–5], establishing systems [6], organizing teaching competitions [7], improving informatization teaching ability through MOOCs teaching, etc. [8].

Among many ways to improve teaching ability, training is widely used [9]. As an effective way to promote the development of university teachers, Instructional Skills Workshop, ISW, an internationally recognized teacher development program [10], has also become a popular project in Teacher training in many Universities in China in recent years [11]. So, can attending ISW really improve teachers' teaching ability? Which dimensions are specifically improved?

Z. Hu et al. (Eds.): AIMEE 2022, LNDECT 159, pp. 634–643, 2023.
https://doi.org/10.1007/978-3-031-24468-1_58

Previous literatures were mainly based on observation and experience summary, and rarely discussed by empirical methods [12]. It is very important to use scientific methods to improve teachers' teaching ability to cultivate more excellent talents.

Therefore, this paper conducted a questionnaire survey of teachers before and after participating in ISW, to understand the influence and effect of ISW on teachers' teaching ability, and to provide empirical evidence for scientific improvement of teachers' teaching ability.

2 Instructional Skills Workshop and Teaching Ability

2.1 Instructional Skills Workshop Program

Instructional Skills Workshop originated in Canada in the 1970s, when the north American higher education development from the stage of popular education to popularize education stage, a rapidly increasing number of students, the number of college teachers and the teaching level needs to be improved, consequently Canada launched this program, which was designed to improve the effect of classroom teaching [13]. Since this program plays a great role in helping teachers improve their teaching skills, it soon became popular in North American countries and developed rapidly in China after entered China around 2004, many universities have taken it as an important item in teacher training [11].

ISW is a 24-h study program lasting for 3 days. During the training, 5–6 instructors (also called students) form a group led by 1–2 facilitators to form a relatively fixed study group. During the three-day study, there are abundant participatory learning activities, daily mini-lesson practice, and instant teaching feedback. Participants' reflection on complex teaching is based on the description of specific events and actions [11].

ISW is characterized by diversified learning activities, deeply involved practical training, and peer support learning mode. Diversified learning activities include ice-breaking activities, themed course training, question feedback circles, mini-lesson, formative questionnaires, and feedback, etc. These learning activities take on a variety of forms, presenting a structured but slightly different form, which enables instructors to have a lot of fresh experience during the three days of learning in the workshop. Deeply involved practical training mainly refers to the practice training of deep participation of instructors as active learners and facilitators of learning process, instead of the conventional passive listening and reporting. Peer assistance learning mode means that in the whole process of the project, instructors mainly rely on peer communication and feedback for self-improvement, because the assumption of the workshop is that "everyone is smart and intelligent, capable and willing to do things well" [14]. Therefore, they trust instructors' learning ability and regard each instructors as a resource for providing new information and demonstration cases in the workshop [15].

2.2 Teaching Ability

2.2.1 Composition of Teaching Ability

In terms of the composition of teaching ability, researchers divide teaching ability into several sub-abilities according to different classification standards. For example, according to the three elements of education (teaching objects, teaching influence and teachers), teaching ability can be divided into the ability to adjust, control, and transform the ability of teaching objects (students), the ability to adjust, control and transform the ability of teaching influence and the ability of teachers' self-regulation [16]. Some researchers have shown that, teaching ability can be divided into teaching selection ability, teaching design ability, teaching implementation ability, teaching evaluation ability and teaching innovation ability according to the three stages before, during and after teaching activities [17]. Some argue that, based on the consideration of multiple disciplines (pedagogy, psychology, and sociology), teaching ability can be divided into teaching design ability, teaching organization ability, teaching implementation ability, teaching cognitive ability, teaching operation ability, teaching monitoring ability, ability to deal with students' relationship and self-adaptation to social changes [18]. Some scholars also adopted the method of empirical research and proposed that the structure of teachers' teaching ability mainly includes: The ability to understand, transform and express teaching content, the ability to select teaching methods and media, the ability to integrate and design teaching elements, the ability to organize and communicate teaching, the ability to motivate students to learn actively, the ability to evaluate and reflect on teaching, the ability to research teaching and the ability to innovate teaching, etc. [19].

2.2.2 Ways to Improve Teaching Ability

According to the implementation of teaching ability improvement approach from the perspective of the main body, it can be divided into two parts: intervention from the organizational level and efforts from the individual level of teachers. The main measures to intervene in the improvement of teaching ability from the organizational and institutional level are as follows: building the centers for faculty development, and carrying out various training programs to improve teaching ability [3–5]; With system construction as the starting point, systematic establishment of overall coordination mechanism, mentoring mechanism, training and exchange mechanism, leading demonstration and competition mechanism, policy-oriented long-term mechanism [6]; Building integration and linkage in teaching concept, system design and practice mechanism [20]; Construct the concept of "teaching force" to realize the innovation of teacher system and the reconstruction of teaching culture [21]; Conduct student and staff training, academic and business seminars, business-oriented roundtable debates, and research [22]. The main measures for teachers to improve their teaching ability from the individual level are as follows: to temper their basic teaching skills by participating in "Teaching Competition" [7]; According to various teaching processes, the teaching ability should be improved in the aspects of pre-job preparation, learning from others, accumulation, careful lesson preparation, classroom teaching, summary and reflection, scientific research and teaching, practice and innovation, teaching research and continuing education [23].

2.3 Instructional Skills Workshop and Teaching

Some scholars conducted research on the teaching effectiveness of ISW, mainly based on the evaluation of participants' satisfaction level of ISW, and the evaluation results showed that ISW improved their teaching practice ability, enhanced their critical reflection on teaching, and improved their sense of self-efficacy [24].

Macpherson used a hybrid approach for her doctoral thesis, combining survey studies of ISW participants with follow-up interview data to determine whether components of ISW lead to transformative learning. Research results showed that ISW persuaded teachers to use more student-centered teaching methods in the classroom, and the program seems to lead to transformative learning [12].

Dawson et al. also conducted a comparative study on teachers who participated in ISW and those who did not through pre-and-post-test studies and interviews. The study assumed that after ISW participants participated in workshops, their teaching style would become more student-centered, and less teaching-centered. In the quantitative study, no significant changes were found before and after attending ISW, but in the qualitative analysis, participants thought more about teaching from the student-centered perspective [10].

In general, existing literatures mainly focus on the background of ISW program, organization mode and cultivation of facilitation skills; The research on ISW and teaching is mainly based on the satisfaction evaluation and the change of teachers' perspective from teaching-centered to student-centered. Many literatures point out that ISW has played a great role in improving teachers' teaching skills, but it is mainly based on observation and experience summary rather than empirical discussion. Therefore, this study conducted pre-test and post-test on teachers who participated in ISW to explore the role of ISW in improving teachers' teaching ability through empirical analysis.

3 Method

3.1 Participants

A total of 231 teachers (person-time) from 6 universities in Guangxi who have participated in ISW participated in this survey. Due to the small class size and quality of ISW, each session can accommodate only a small number of instructors, typically about 15 instructors per session. This study distributed questionnaires to 150 participants in 10 training sessions from 2020 to 2022 in Guangxi, divided into pre-training and post-training based on previous research methods [25]. Three days before the training, instructors received a questionnaire survey and answered the questions anonymously. A total of 145 people participated in the questions, with a questionnaire recovery rate of 96.7%. About half a year after ISW, the same questionnaire was sent out again. As there were only six sessions up to the time when the article was published, a total of 90 instructors received the post-training questionnaire, and 68 instructors participated in the survey, with a questionnaire recovery rate of 75.6%.

3.2 Development of the Instrument

The development of the questionnaire in this study is mainly divided into the following three processes:

First, based on literature research and discussion with experts in the field of pedagogy, we developed a teaching ability measurement scale (see Table 1), which includes four dimensions: teaching design, teaching implementation, teaching evaluation and reflection, and stimulating students' learning. All questions are based on an understanding of the concepts and definitions mentioned in the relevant literature. Secondly, some teachers in a university were selected to conduct a pilot test in a small range, and some items and expressions of the scale were adjusted according to the results of the pilot test. Thirdly, the reliability and validity of the scale are tested.

Table 1. Teaching ability measurement scale

Variate	Dimensions	No. of items	Sample items
Teaching ability	Teaching design	7	When doing teaching design, I will consider the following factors (the more I agree, the more I will consider when doing teaching design): the establishment of learning objectives
	Teaching implementation	4	I have a sense of time control in teaching, can control the teaching time well
	Teaching evaluation and reflection	5	I will design and carry out an assessment that is suitable for evaluating students' learning effect (outcomes of learning objectives)
	Stimulating students' learning	4	Inspire students' enthusiasm for the courses I teach through my teaching

4 Results

4.1 Data Analyses

Firstly, the reliability and validity of the scale were analyzed, and then the data before and after the ISW were divided into two groups. Independent sample T test was conducted to test whether there were significant differences in the four dimensions of teaching ability and teaching ability before and after the training.

4.2 Validity and Reliability Analysis

Factor analysis was first used to measure the structural validity of the scale. According to the measurement result of factor analysis, the factor load of one item in the teaching design dimension is lower than 0.5. After deleting this item, the KMO value is 0.868, and the Bartlett sphericity test p value is 0.000, so the factor analysis can be carried out. Through exploratory factor analysis, four initial characteristic roots were extracted showed in Table 2, and the cumulative variance contribution rate of common factors was 63.72%. The loading value of each item in its corresponding dimension was above 0.5, indicating good structural validity.

Table 2. Total variance of interpretation

Component	Initial eigenvalues			Rotate squares and load		
	Total	% of the variance	Accumulation %	Total	% of the variance	Accumulation %
1	6.666	35.086	35.086	3.811	20.055	20.055
2	3.162	16.644	51.730	2.937	15.460	35.515
3	1.200	6.316	58.046	2.692	14.169	49.684
4	1.078	5.673	63.720	2.667	14.036	63.720

Cronbach's alpha coefficient was used to analyze the reliability of each dimension (see Table 3). Teaching implementation was 0.789, with good reliability. The reliability of teaching ability and the other three dimensions is greater than 0.80, the reliability is very good.

Table 3. Reliability analysis

Variate and Dimension	N	Mean	SD	Cronbach's alpha
Teaching ability	213	4.1141	.36530	0.890
Teaching design	213	4.5180	.56433	0.88
Teaching implementation	213	3.9878	.45312	0.823
Teaching evaluation and reflection	213	3.8556	.48806	0.789
Stimulating students' learning	213	4.0951	.44516	0.81

4.3 Means, Standard Deviations and Independent Sample T-test

Table 4 shows the four dimensions of teaching ability of participants before and after ISW as well as the mean and standard deviation of teaching ability. To investigate whether there is a significant difference in teaching ability before and after training, an independent sample T-test was conducted. Correlation analysis shows that there are significant differences in the three dimensions of teaching design, teaching evaluation and reflection, teaching implementation and teaching ability at 0.1 level. After ISW, the teaching ability and its three dimensions (teaching design, teaching implementation, teaching evaluation and reflection) are significantly improved than before ISW. There was no significant difference in stimulating student's learning dimension before and after ISW.

Table 4. Independent sample T test was performed before and after training

Variate and Dimension	Group	N	Mean	SD	P
Teaching ability	Before ISW	145	4.0735	.37300	0.017
	After ISW	68	4.2008	.33468	
Teaching design	Before ISW	145	4.4667	.60368	0.052
	After ISW	68	4.6275	.45449	
Teaching implementation	Before ISW	145	3.9462	.44049	0.018
	After ISW	68	4.0765	.47003	
Teaching evaluation and reflection	Before ISW	145	3.8017	.50681	0.050
	After ISW	68	3.9706	.42657	
Stimulating students' learning	Before ISW	145	4.0793	.46277	0.452
	After ISW	68	4.1287	.40632	

5 Discussion

The independent sample T-test results of the Before ISW group support the view that ISW has an effect on improving teachers' teaching ability, which is mainly reflected in the improvement of teachers' teaching design ability, teaching implementation ability, teaching evaluation and reflection ability. The findings are closely related to the design and implementation of the workshop.

First, as the core theme course in ISW is the BOPPPS model, which provides a very useful teaching design framework. BOPPPS represents the six elements of teaching design, namely, bridge-in, objective, pre-assessment, participatory learning, post- assessment and summary [26]. Teachers practice and implement teaching design according to the model of BOPPPS. At the same time, the model also emphasizes the "Constructive Alignment" among these elements, that is, after setting an objective, the implementation of teaching (bridge-in, pre-assessment, participatory learning) should be carried out around this objective, teaching evaluation (post-assessment) should also ensure that the measurement of the objective is achieved. Through this workshop, instructors have mastered a highly structured and versatile teaching design model, which virtually helps instructors to quickly understand and master the essentials of teaching design and improve their teaching design ability.

Second, instructors need to carry out three mini-lessons during ISW, the mini-lesson only 10 min every time, but they are all full of "Experiencing-Reflecting-Thinking Acting" of the Experiential Learning Cycle [15], other instructors act as "students" participation in classroom learning process, after each mini-lesson, the teacher will receive how the "students" view the teacher's teaching from the perspective of students' learning, that is, the highlights and suggestions of this teaching, and the teacher will continue to carry out the next mini-lesson exercise after further improvement. It is through such repeated practice that teaching skills are integrated into practice drills, which improves teachers' confidence in teaching implementation, strengthens their teaching implementation skills, and thus improves their teaching implementation ability.

Third, after every mini-lesson, facilitator will talk to the instructor who act as teacher be one-on-one communication. Leading by questions, facilitator guide the teacher focus on the advantages and disadvantages of this mini-lesson, the degree of learning objectives achievement, and improving plan, etc. [26]. As mentioned earlier, "Constructive Alignment" is emphasized between the six elements, therefore in the students' feedback time, the degree of learning objectives achievement in this mini-lesson will be raised to guide students' thinking, etc. In this student feedback time, teachers also reflect on their entire teaching process again by listening to the feedback of their peers. Through this way to improve teachers' teaching evaluation and reflection ability.

Fourth, participatory learning is the main part of whole lesson, which emphasized the way of how teachers through multiple means to guide and encourage students to participate in the class, as a result, ability of stimulating student learning is also very important, but the research results show that for ISW was not effective promoted the ability, the possible explanation is that the instructors act as the students in mini-lesson, due to some characteristics of the mini-teaching: short course time (10 min), super small classes (1 teacher, 4–5 students), so the "students" participation rate is very high, even if the teacher did not take any measures to stimulate students, "students" is also actively involved in the mini-lesson so the ISW advocate the concept of" stimulates the student to participate", but lack of practical environment for this ability.

6 Conclusion

In general, ISW program is effectively improves teachers' teaching ability. The effectiveness of ISW is mainly reflected in the following four aspects:

First, ISW has training theme courses linked to each other. ISW contains three theme lessons. The first is the BOPPPS model, which is the main structured teaching design model and an important scaffold for teachers to carry out teaching design; The second lesson is participatory learning teaching methods, introducing some methods that can better guide students to carry out participatory learning, teachers are more focused on guiding students to participate in, rather than imparting knowledge; The third lesson is learning style, which enables teachers to realize that everyone has different learning styles and better understand why different students respond differently to the same teaching, so as to think more about how learning happens from the perspective of students. The training contents of the three theme courses are interlinked. The teachers who participate in the training first start from the basic teaching design model, then learn the methods to guide students to actively participate in the class, and finally understand different learning styles and reflect on how to improve their own teaching according to different learning styles.

Second, ISW attaches great importance to structured practice. BOPPPS model is adopted, and teachers are repeatedly trained in mini-lesson based on the model. This is a typical deliberate practice, which requires: (1) a long period of professional training; (2) with the guidance of experts; (3) Realize the change of learners' cognitive schema [27].

Through this deliberate practice mode in which students repeatedly practice a teaching mode with high intensity, students can fully apply it to their own teaching practice and make localization reconstruction according to the actual situation when they return to their own schools and classrooms [28].

Third, ISW attaches great importance to teaching reflection and peer support. In ISW, teachers and other teachers to attend training form a like-minded partners [28], discusses its teaching and the teaching improvement together, make the teachers in the peer cooperation experience teaching normative and mutual, and major changes on their own role, from a focus on their own teaching to focus on the students' learning, through peer feedback learning experience from the perspective of students, Teachers understand students' personality traits and learning difficulties, so as to reflect on how they should guide students to learn, rather than simply complete the task of teaching.

Fourth, the training courses, structured practice, and reflection above are nested and progressive. Such as after the first theme lesson, follow with the first mini-lesson, and then carry on the reflection, and then enter the second theme lesson, a second mini-lesson, reflection, such a nested loop, effective with the help of the teacher improving teaching design, teaching implementation, teaching evaluation and reflection ability.

To sum up, the successful experience of ISW program can help us think better about the design of training programs. When designing various training programs, colleges and universities can refer to the model of ISW program to break the simple form of training lectures and integrate training, practice and reflection, so that teachers can carry out learning through diversified learning activities, deeply involved practical training and peer mutual assistance.

Acknowledgment. This project is supported by 2020 Guangxi Higher Education Undergraduate Teaching Reform Project (2020JGZ163), Guangxi University young and middle-aged teachers scientific research basic ability improvement project (2019KY0948), The first batch of high-level teaching team cultivation project of Nanning College (BOPPPS mode teaching team).

References

1. Chen X.: Lane Change and Overtaking — Exploration on the Construction of New Applied Technology University, pp. 153–155. Guangxi Normal University Press, Guilin (2019) (in Chinese)
2. Wang, X., Li, H., Xu, Z., et al.: The key to improve the quality of talent training lies in improving the teaching level of teachers. China Univ. Teach. **12**, 18–21 (2013). (in Chinese)
3. She, Y.: Current situation and countermeasures of young teachers' teaching ability in local colleges and universities. High. Eng. Educ. Res. **5**, 150–155 (2010). (in Chinese)
4. Gustafson, K.: Instructional improvement centers in higher education. J. Instr. Dev. **7**, 2–7 (1984)
5. Liu, Y.: Discussion on teaching ability training of young teachers in Colleges and Universities. China After-School Educ. **6**(27), 48 (2012) (in Chinese)
6. Chen, M.: Strategies for improving the teaching ability of young teachers in universities. Jiangsu High. Educ. **3**(2), 110–112 (2016). (in Chinese)
7. Wu, X.: Practice and thinking of "youth education competition" to improve the teaching ability of young teachers in colleges and universities. Educ. Theory Pract. **38**(9), 41–42 (2018). (in Chinese)

8. Li, N., Zhang, Y.: Improvement and practice of secondary school geography teachers' informatization teaching ability based on the perspective of MOOCs. Mod. Educ. Comput. Sci. **8**(2), 11–18 (2009)
9. Breda, J., Clement, M., Waeytens, K.: An interactive training programme for beginning faculty: Issues of implementation. Int. J. Acad. Dev. **8**(12), 91–104 (2003)
10. Dawson, D., McIntyre, G., Meadows, K., et al.: The Impact of the Instructional Skills Workshop on Faculty Approaches to Teaching. The Higher Education Quality Council of Ontario, Toronto (2014)
11. Li, S.: Research on local practice of instructional skills workshop. Shanghai Teach. **7**(4), 53–60 (2021). (in Chinese)
12. Macpherson, A.: The Instructional Skills Workshop as a Transformative Learning Process. Kwantlen Polytechnic University (2011)
13. Foxe, J.P., Frake-Mistak, M., Popovic, C.: The instructional skills workshop: A missed opportunity in the UK. Innov. Educ. Teach. Int. **54**(02), 135–142 (2017)
14. Bens, I.: Facilitating with Ease. Electronic Industry Press, Beijing (2019). (in Chinese)
15. Zhou, L.: The path analysis of knowledge development of facilitators in instructional skills workshop. Shanghai Teach. **53**(4), 69–78 (in Chinese)
16. Chen, S.: A preliminary study on teaching ability. Curric. Teach. Mater. Teach. Method **8**(9), 53–58 (1988). (in Chinese)
17. Tang, Y.: Teacher Professional Development and Teacher Education. Anhui Education Press, Hefei (2008). (in Chinese)
18. Wang, S.: Multidimensional structure of teaching ability of college teachers. J. Shenyang Norm. Univ. (Soc. Sci. Edn.) **34**(1), 110–113 (2010). (in Chinese)
19. Zhang, Z.: On the teaching ability structure and development strategy of young college teachers: Empirical conclusion and logic analysis. China High. Educ. Res. **4**, 51–55 (2013). (in Chinese)
20. Su, Q., Wang, G., Tan, Z., et al.: The logical generation of young teachers' teaching ability improvement in colleges and universities. Educ. Res. Tsinghua Univ. **39**(4), 121–126 (2018). (in Chinese)
21. Fan, X., Wu, Q.: Improving the teaching ability of young teachers in research universities: Institutional innovation and cultural reconstruction. High. Educ. Res. **35**(9), 50–55 (2014). (in Chinese)
22. Kuzminykh, I., Yevdokymenko, M., Yeremenko, O., et al.: Increasing teacher competence in cybersecurity using the EU security frameworks. Int. J. Mod. Educ. Comput. Sci. **13**(6), 60–68 (2021)
23. Li, X.: Exploration and practice of improving teaching ability of young teachers in engineering colleges. Heilongjiang High. Educ. Res. **8**(12), 109–111 (2015). (in Chinese)
24. Day, R.: Instructional skills workshops: From grassroots initiative to international perspectives. In: Proceedings of the Poster Presented at the Meeting of STLHE in Ottawa, Canada (2004)
25. Yalcin, H., Demirekin, M.: Training program supporting language acquisition. Int. J. Mod. Educ. Comput. Sci. **13**(3), 1–12 (2021)
26. Fenrich, P., Johnson, R.: Instructional skills workshops: a model for training professors how to teach. Wu, W., Alan, S., Hebebci, M.T. (eds.). Research Highlights in Education and Science. ISRES Publishing (2016)
27. Rousmaniere, T.: Deliberate practice for psychotherapists: a guide to improving clinical effectiveness. Routledge (2016)
28. Chen, X.: Why is deliberate practice needed in learning to teach. Shanghai Teach. **22**(4), 79–82 (2022). (in Chinese)

Online Education: Obstacles and Improvements When Collaborative Learning Moves Online

Sishi Cao(✉)

University of Göttingen, 37073 Göttingen, Germany
sishicao@foxmail.com

Abstract. Collaborative learning has an important significance on students' study, teamwork and social interaction. However, in the specific context of the epidemic, it is a new start to move collaborative learning entirely online. The inconsistency of past offline classroom experience to online teaching and learning makes the survival and development of collaborative learning in the online environment uncertain. To solve this problem, this paper delicately examines the online implementation of collaborative learning in three areas based on the parties involved: online platforms, curriculum design and student feedback. Recognizing the obstacles to the implementation of online collaborative learning, two improvement strategies are proposed, one being the creation of an online community identity and the other being to design the online learning circuit as a complete system.

Keywords: Online · Collaborative learning · Online community · Online collaborative learning system

1 Introduction

The injection of technology into the education sector has opened up a new possibility for the teaching methods reform. During the time of covid-19, special contexts have accelerated this pace of comprehensive digital education, whether learners and educators are ready for it or not. Setting up new forms of online education to accommodate students' learning needs and to develop learning activities through their schools' online learning management systems (LMS) become an inevitable trend [1]. But what model can be used to adapt to the new learning conditions of learners and to improve the efficiency of learning is still in the process of exploring.

In response to this issue, collaborative learning, which treats learning as an active, constructive process, encompasses the development of students' participation, cooperation and social interaction, becomes a possible solution. There is a wealth of studies on collaborative learning already, with the significance of it being highlighted as early as 1992. Barbara Leigh Smith and Jean T. MacGregor advocated that collaborative learning "most center on students' exploration or application of the course material" and achieve pre-determined goals through students' proactivity [2]. On this basis, collaborative learning as a concept, its benefits [3], and the foundational.

Z. Hu et al. (Eds.): AIMEE 2022, LNDECT 159, pp. 644–655, 2023.
https://doi.org/10.1007/978-3-031-24468-1_59

elements [4] that should be considered in the learning process are gradually being refined. Since then, this concept has been used in the classroom and some scholars have developed specific practice models for this purpose [5]. In China, research on collaborative learning has become more specific, focusing on specific teaching designs and case studies, while some scholars have tried to incorporate new tools to enhance collaborative learning, such as mind mapping [6] and interaction analysis [7].

However, there is little attention paid to the special circumstances faced in transferring the full range of collaborative learning to the online environment. The implementation of online collaborative learning has been a new start, raises difficulties like the lack of socio-emotional interactions [8], but such difficulties require a multifaceted and comprehensive examination of the context. The purpose of this article is therefore to discuss the special circumstances that existed when collaborative learning was fully converted to online and the improved model for the new application, specifically divided into two questions: A). What parties are involved in the transfer of collaborative learning to online and what are the contexts in which it exist? B). What are the difficulties encountered in online collaborative learning and how can it be improved?

To address these issues, we interviews on students, who had online learning at different times during the epidemic about their feelings, opinions and experience as well as conducted statistics on the functions of online network. The implementation of online collaborative learning will be measured in three areas: learning platform, teacher class design and students' self-evaluation. Accordingly, we summarize the difficulties of online collaborative learning from the measured results and propose some models to help.

2 Analysis of the Present Situation of the Online Collaborative Learning

The study is designed to examine whether there is scope for collaborative learning approaches such as group discussion and material guidance online, and to check whether students are able to achieve the desired outcomes of collaborative learning. The platform for online learning has been built with a certain amount of foundation and time, but maintaining the effectiveness of learning and keeping students engaged at a distance across the camera is one of the key challenges that online education has [9]. It is hypothesized that collaborative learning approaches are developed in a variety of ways through the online platform, but the desired effect of collaborative learning, i.e. participation, cooperation and social contact are limited.

2.1 Analysis of the Online Platforms

A complete class process can be divided into six components: organization, presentation, interaction, questioning, feedback and evaluation [10]. The weighting of the six components varies according to the three models of teacher-centered, lecture-centered and student-centered classrooms [11]. Collaborative learning most places student exploration or application as its core, suggesting extensive context and dialogue interventions.

A prerequisite for the applications of online collaborative learning is whether these online platforms provide the ground for collaborative communication.

During the COVID-19 outbreak, course organization form in China completely required to be conducted online, many teachers choose a variety of network platforms to complement learning activities. In this regard, we have divided the network platforms into two types according to the learning process, one is for classroom sessions and the other is for post-course supplements. The former needs to enable educators to live-stream and organise classroom instruction, with students interacting in real time over 45–190 min to complete classroom learning. The latter is the post-course platform, where such applications are not used for live lectures but teachers upload teaching materials and information for students carring out discussions and submitting assignments. However, the distinction between these two types of software is not based solely on when they are used, but rather on the main purpose and method of implementation, which means

Table 1. List of classroom sessions platforms' function design

Category	Function design	Type of interaction		
		Individual	Double	Group
Tencent meeting	Vocalization			×
	Video			×
	Share Screen			×
	Member List + Chat	×	×	×
	Share Documents			×
Tencent classroom	Vocalization			×
	Share Screen			×
	Raise your hand	×	×	×
	Member List + Chat	×	×	×
DingTalk	Vocalization			×
	Video			×
	Share Screen			×
	Member List + Chat	×	×	×
Zoom	Vocalization			×
	Video			×
	Share Screen			×
	Response	×		×
	Member List + Chat	×	×	×
	Create whiteboards and edits		×	×
	Member List + Chat	×	×	×
	Group discussions		×	×

that some off-site software is used in the classroom several times, but it does not play a leading role. Based on this, we surveyed several of the most common online platforms currently available in market, grouping them together to see how these platforms work (See Table 1).

Four types of software were considered, although DingTalk and Tencent Meetings are not specifically designed for online classroom using, they all have the most basic functions: Vocalization, video and screen sharing. Of these, zoom was the most feature-rich, with the addition of collaborative editing for whiteboards and split screens for group discussions, both of them provide functional assistance for students work in teams, giving independent space for group awareness and discussion. The aggregation and discussion functions for the group format are the most considered in the web platform. Through sound, visuals and images, students are able to feel more connected to others in the online environment and therefore establish a social contact [12]. And this social contact is timely, with student-to-student, student-to-teacher discussions and exchanges, all done within class time.

As for post-course learning platforms, a few representatives were selected as examples, such as the online platform designed by the school itself, the integrated learning platform MOOC and the Super Star, which closely cooperate with universities. Similarly, we have documented the features of the software that can be implemented for collaborative learning, and have grouped the contacts provided by specific features into four categories based on the number of people. Taking into account the fact that the people students meet and interact with during lessons are mainly familiar teachers and classmates, but as shown in Table 2 that the post-course platforms are often open to the whole community, or at least other students in the school, strangers are added to the list of people with whom students can deal.

Table 2. List of post-course platforms' design

Category	Function design	Type of interaction			
		Individual	Group	Class	Stranger
THEOL	Question and Answer Discussion	×	×	×	×
	Course Questionnaire	×			
	Study Notebook	×	×	×	
Superstar	Statistical Rankings			×	
	Discussions	×	×	×	
	PBL		×	×	
MOOC	Discussion forum	×	×	×	×

All platforms are equipped with a discussion function. But we find that there are not many collaborative learning features supplemented by post-course online platforms, while they are focused on discussion forums. MOOC discussion forum is the most well built, it has the largest number of participants due to the open nature of the most course, which a University's internal courses have between 1,000 and 30000 participants, while public boutique courses often have more than 110,000 participants. And the content of the discussion has been categorized, including the teacher's Q&A section, posting queries that you would like to have answered by the teacher; the classroom communication section, presented in the courseware as a discussion of teaching content; and the general comments section, where you can post any experiences or ideas you would like to share with the class, on general topics such as study, work and life. The online discussion area motivates students to continue to engage in thinking and extending the classroom questions after the class. It is worth noting that the MOOC also opens up message boards about areas of work and life, where statements are more free and lively, promoting students to find community as individual identity, release of repressed emotions and making social interaction in a closed environment.

In general, collaborative learning on the platform is built to focus on classroom teaching and post-class discussions, however, most platforms are not set up with specific collaboration modes and all collaboration features are static by default. The design of these features does not allow teachers and students start their discussion fast. Instead, triggering by learning participants is necessary.

2.2 Analysis of the Online Classroom Designs

The most significant difference of collaborative learning is that the lecturer is no longer the primary knowledge deliverer but rather the learning designer [8]. The teacher's role in collaborative learning is crucial, and the effectiveness of his or her classroom design has a direct impact on the level of student engagement. Collaborative learning is a question-oriented process that leads students to think, discuss, correct and debate [13]. In this process, teachers need to use a variety of teaching techniques to keep students engaged in learning, most importantly how to conduct question-oriented learning and how to motivate students to participate in.

Clear and meaningful topics or questions can stimulate students' thinking and research. In the teacher's question design module, we found that the thematic framework established by the teachers was not closely focused on the students, but was based on the textbook content. This was particularly evident in the fact that the teacher's syllabus design and course chapters were developed and published well before the first lesson in which the students were engaged, and that there was no subsequent gathering of students' questions about the topic or revision of the course content. In addition, teachers rarely announce the list of topics or questions to students at the beginning of the class, leaving it up to students to choose and explore them according to their own interests. Although some online platforms provide boards with task lists, they are not used.

Besides, teachers often neglect the communication process in online classes and do not provide time for students to freely collaborate in completing their explorations. A proportion of teachers, focus on the outcomes of teaching, ask questions of their students despite using question orientation as a part of a lesson, however, the question has been self-defined already by the teacher and had a pre-determined answer. Therefore, the teacher consciously shifts to the 'right answer' to students' responses without asking them to think further and let them express their opinions. An opportunity for multiple attempts might have fostered more motivation to explore and practiced the students' ability to think and explain.

In addition, the approach teachers facilitate collaborative learning for students is single and direct. There are many approaches of collaborative learning already presented by scholars, for instance, subjects as law, finance or market, student engagement can be facilitated through the case study and discussion, and writing group where students collaborate on a particular task and provide feedback [14], are both effective ways to improve. Peking University has also proposed an online discussion scheme based on group work, with a 'big discussion + free discussion' and a set of incentives to quantify student engagement [15].

However, the students interviewed recalled that teachers made the incentives more straightforward and simple, engaging them in learning by incorporating discussion or presentation into marks in the most straightforward way, which is presented in Fig. 1. We surveyed 15 students about their classroom incentives, as well as looking at a number of online courses, and came up with the following encouragement models.

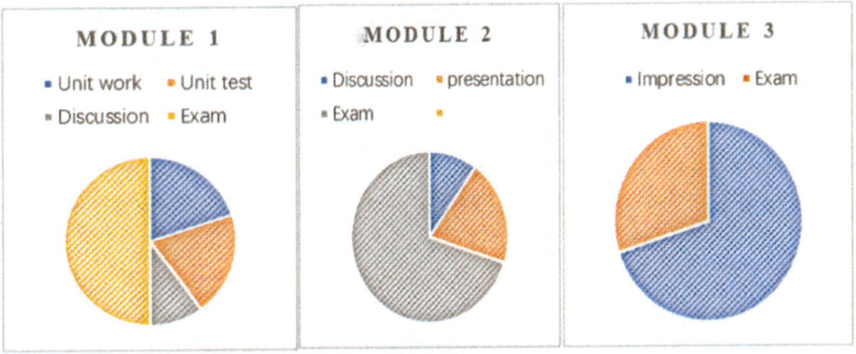

Fig. 1. Encouragement model from teachers

We found that teachers encouraged students to participate actively in the class, but the way they encourage is single, including class discussions and presentations in students' final grades. Some teachers adopted clear criteria for assigning marks, such as Model 1 and Model 2, with between 10%–20% of discussion marks. Some teachers also added ways for students to mark each other's work and evaluations to promote student-to-student communication, but this is generally anonymous and one-off. Only a small number of teachers, rather than informing students of a clear percentage of class discussion or presentation, encourage students to speak actively in each class and assign marks at the end on the basis of the impression they have made on the teacher. On the one hand, this format makes student participation in class less utilitarian, but on the other hand it is ambiguous, difficult for students to cope with and hurt their motivation.

2.3 Analysis of the Students Feedback

Online courses are more or less involved in collaborative learning, thanks to online platforms and teachers. However, it is unknown whether this collaborative learning actually helps students to be active or whether they are motivated and comfortable with collaborative learning online.

To get a true picture of the results of collaborative learning among students, 11 students were randomly selected. These 11 students were in higher education when they encountered the new epidemic and switched to online classes, 6 of whom were in the Masters program and 5 in the undergraduate program. They generally experienced long period of blocked status, studying online through a variety of channels, with the shortest time being 3 months and the longest student experiencing 2 years of online courses.

To quantify their learning outcomes, we designed three criteria for them to self-assess according to the goal of collaborative learning [13], where 5 is a perfect score and 0 is the lowest score. The three criteria are,

Involvement, do you bahave actively in the online learning process?

Teamwork, do you work together as a group to complete the project and take care of others doing during the online learning process?

Civic responsibility, are you giving your opinion and voice in the online learning process, showing your value in the community and listening to the views of others?

We then calculated the self-ratings of each student in the different sections, see Fig. 2.

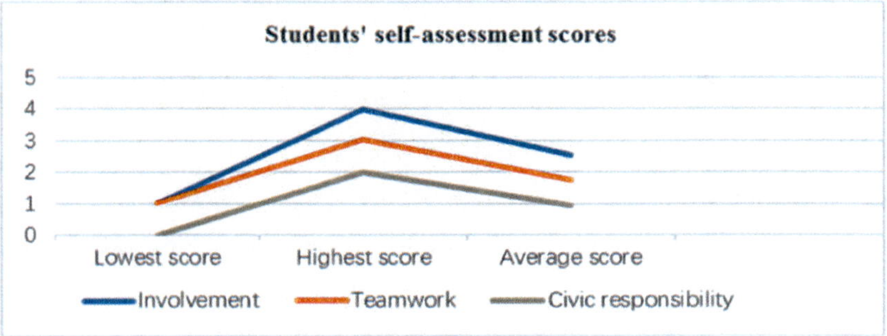

Fig. 2. Student self-assessment form

Interesting thing is, the average scores for these three criteria gradually decreased from top to bottom, with the majority of students giving themselves only 0–1 points in the measure of civic responsibility. This shows that in the current online education environment, neither schools nor teachers are paying attention to developing students' values and social responsibility. Collaborative learning, as a means of classroom engagement, only serves to increase student participation, and even in this component, the function performed is very limited.

In contrast, there was a strong self-motivation prevalent among the students interviewed that drove them to join in the collaborative learning process. When asked "What is the biggest difficulty you have encountered in collaborative learning online?", the psychological fear of speaking up and presenting themselves was a significant deterrent to collaborating with others and engaging in discussion. This fear of speaking was expressed by two students who said that it was mainly a language issue and they chose to be silent during English or foreign language class because they were worried about their accent and vocabulary. One student also said that it was not easy to concentrate in online classes, they could not catch the main points, and therefore feared that they would answer incorrectly.

However, despite their difficulties in engaging in online learning, students admitted inwardly that they wanted to be more engaged in class, and a small number of students made small cues and motivations to themselves. For example, students said, "I turn on the camera in my online class because firstly it is forced by the teacher, secondly it is necessary for the class presentation and lastly it is to make myself more focused" "I actually wish I could turn on the camera myself because it would give me a push and make me more involved in the class".

Overall, opportunities for online collaborative learning to take place remain scarce. For students, although collaborative learning is an important process for practising their speaking, thinking and social engagement, this part of the process is not well achieved due to teacher design limitations. Students have a strong desire and motivation for collaborative learning, helping them to engage spontaneously in classroom presentations and gradually improve their mental and technical shortcomings.

3 Improving Design of Online Collaborative Learning

In the section above, we have investigated the current state of online collaborative learning in terms of online platforms, teacher classroom design, and student self-feedback. The features and benefits of collaborative learning are now generally recognised though, and both platforms and teachers are consciously or unconsciously promoting collaborative sessions in the learning process. Unfortunately, educators and institutions still do not fully recognise the diverse approaches and social purpose of collaborative learning, but only use it as a pedagogical tool to promote student engagement in learning. As a result, there is still a lack of more proactive mechanisms for online collaborative learning, especially in the field of higher education, to help students not only acquire better knowledge but also develop the ability to collaborate, advocate and take on socially transformative roles [16].

3.1 Building an Online Community

Online communities have become a complement to people's social relationships in real life. Various social networking software, such as WeChat Friend Circle and Sina Weibo, share individual lives through text, sound and pictures, while encouraging individuals to help each other and exchange experiences with each other. On online education platforms, we have seen many software designs for social functions, such as chatting and emoji packs, showing a little hint of social interaction trend. However, for purposes of student management, more communication and exchange between students is not promoted. This, to a certain extent, curtails students' social skills and makes them psychologically fearful, depressed, etc.

The discussion forums in MOOC are rarely dedicated to students communicating and sharing space, yet such spaces are a necessary part of online learning. The creation of online zones can be organised by teachers and the use of online software to divide up class-based 'mini-communities', providing students with more opportunities to showcase themselves and meet others. Such small communities are safer and help students to establish their own community identity [17]. In order to build and maintain this identity, students will inevitably strengthen their image management and social skills, and in social encounters with others they will learn to express themselves more clearly, confidently, and to take more responsibility and affirm the worth of themselves and others.

3.2 Establishing Collaborative Learning System

A complete learning process corresponds to a collaborative learning process. The learning process consists of six parts: organisation, presentation, questioning, interaction, feedback and analysis [10]. Accordingly, by intervening in all six parts with the concept of collaborative learning, a system can be designed to develop the students' ability to learn collaboratively, see Fig. 3.

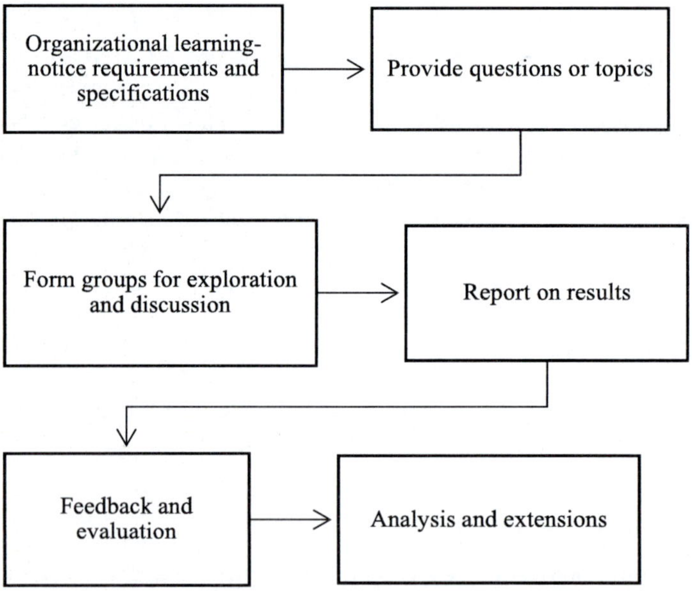

Fig. 3. Online collaborative learning system

The first is the organisation of the class. During the first session, the teacher needs to introduce the content of the course to the students, help students understand the content and purpose of the course. Meanwhile, teacher should also gather students' ideas and questions, and informing them of the rules of the online classroom. The requirements and criteria for attending classes must be, including student participation and cooperation are part of the final grades and how he or she will account for the results, so that students have a plan for what they will do and how they will behave before they formally enter the classroom. Secondly, the teacher, as a learning coach for the students rather than a

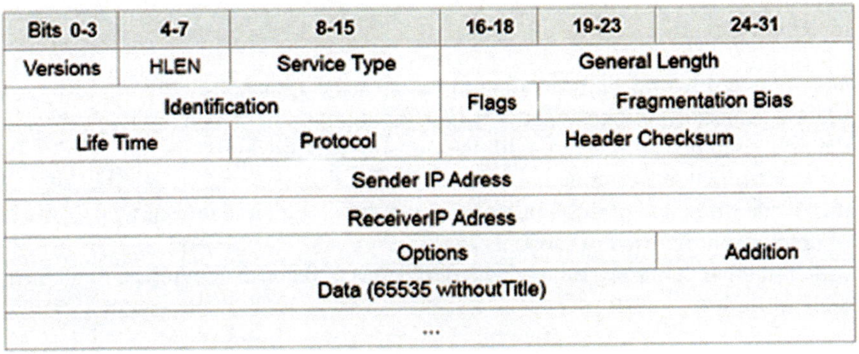

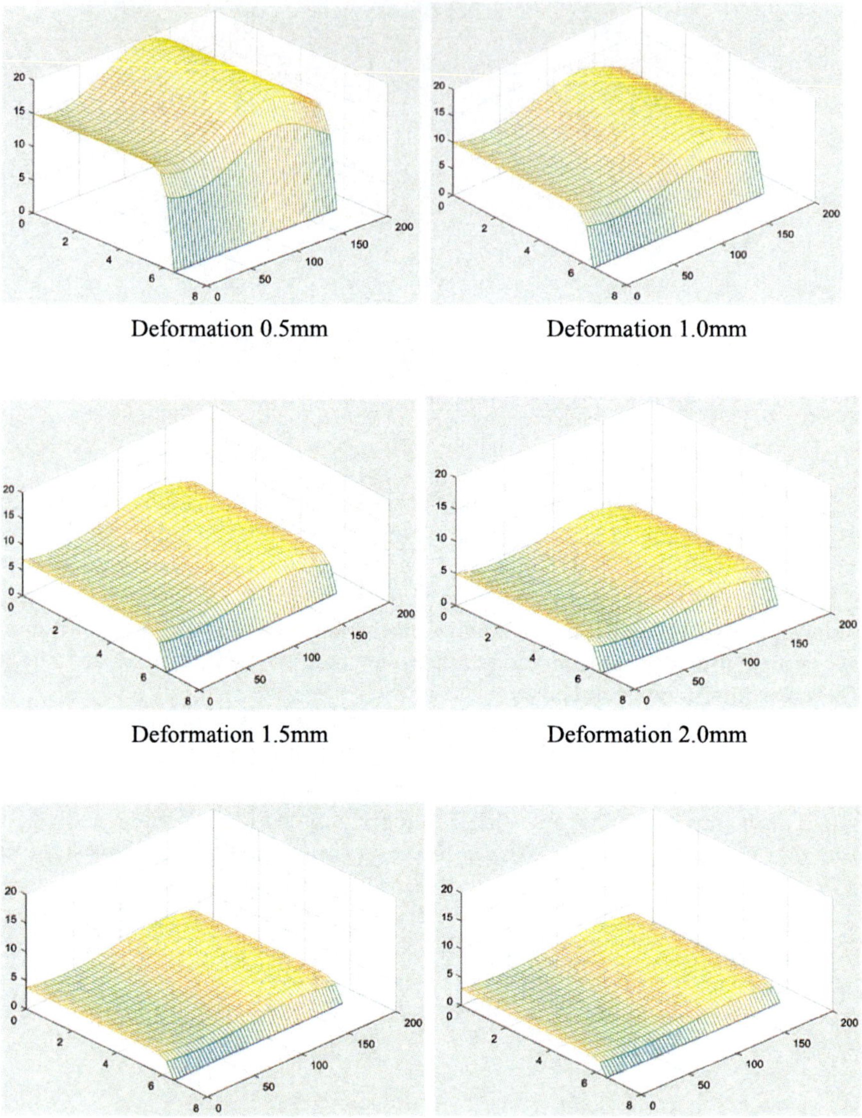

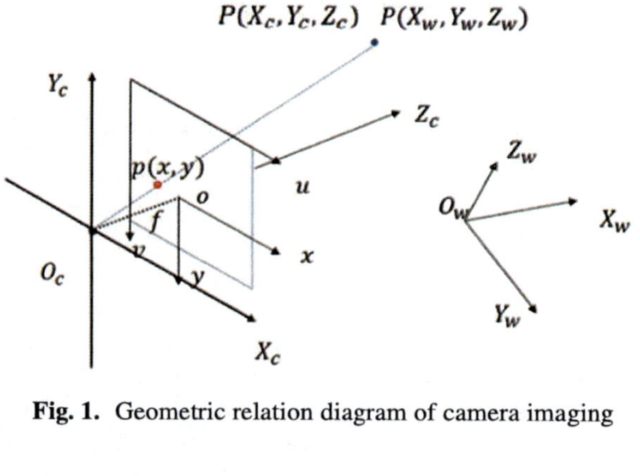

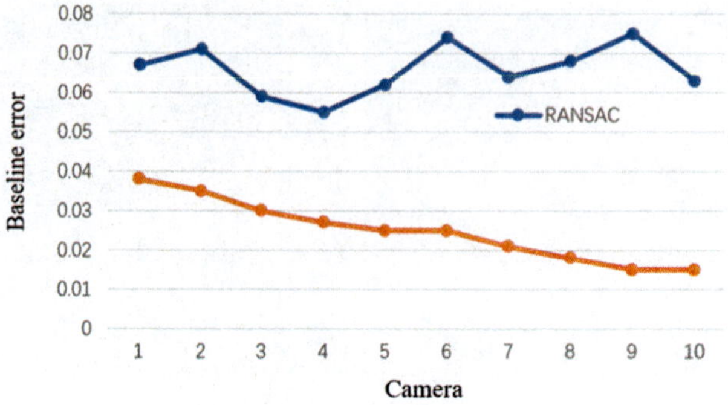

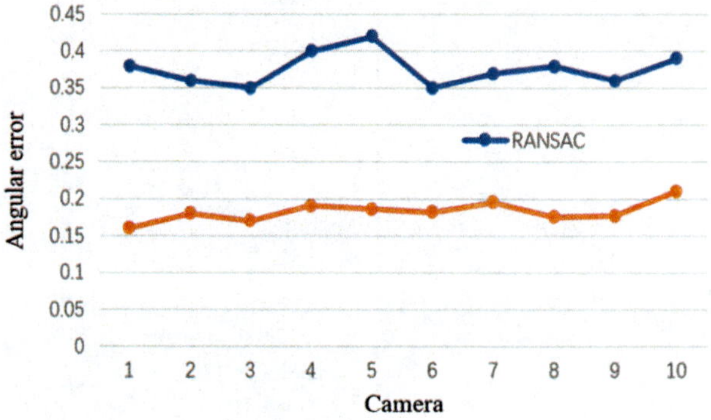

only numerically but also algebraically using special systems of the tensor (Kronecker) products using stochastic components. In these phenomenological rules, it was revealed that nature for some reason uses in genomic stochastics an algebraic operation, which is called by the author a tensor-unitary transformation and whose properties are described and discussed in this article. It should be added that the author didn't meet such systems of algebraic operations, which are based on tensor-unitary transformations, in other fields of mathematical natural sciences.

Why universal stochastic rules of genomic DNAs are connected with tensor-unitary transformations? What mathematical features are provided by these transformations? Could they be used for modeling the biological gestalt phenomena and the dualism "stochastics-and-determinism"? Why living nature prefers the usage of tensor-unitary transformations in genetic systems?

The main aim of this article is to study these questions and present the initial author's results and thoughts about appropriate algebraic tools for modeling the important biological phenomena, which were not been algebraically modeled early but now are been including in the field of algebraic biology.

2 Research Methods

In this study, methods of tensor-matrix analysis and modeling of stochastic organization of genomic DNAs are used jointly with formalisms of quantum informatics and multidimensional hypercomplex numbers. In particular, this article relies on the terminology of the classic book [7] and uses the generally accepted Dirac notations:

- $|\psi>$ Means a vector, which is known also as a ket-vector;
- $<\psi|$ means a vector, which is dual to $|\psi>$ And known as a bra-vector;
- $<\varphi|\psi>$ means scalar (or inner) product between the vectors $<\varphi|$ and $|\psi>$;
- $|\varphi>\otimes|\psi>$ means the tensor product of $|\varphi>$ and $|\psi>$

In quantum informatics, a system of "n" qubits is considered. The computational basis states of this system are written in the form $|x_1x_2....x_n>$; a quantum state of such a system is specified by 2^n amplitudes [7, p. 17]. In our model approach, we in general interpret each of the above-mentioned n-plet sequences of genomic DNAs as a multi-qubit quantum system.

Correspondingly, the representation of any genomic DNA sequences as a sequence of 4 nucleobases C, A, T, and G, whose probabilities in this DNA are denoted by %C, %A, %T, and %G, gives the following 2-qubit expression (1):

$$\Big|\psi_4>=(\%\ C)^{0.5}\Big|00>+(\%\ A)^{0.5}\Big|01>+(\%\ T)^{0.5}\Big|10>+(\%\ G)^{0.5}\Big|11> \qquad (1)$$

The representation of any genomic DNA sequences as a sequence of 16 duplets CC, CA, CT, CG, AC, …, whose probabilities in this DNA are denoted by %CC, %CA, %CT, %CG, %AC, …., gives the following 4-qubit expression (2):

$$\begin{aligned}\Big|\psi_{16}>=&(\%\ CC)^{0.5}\Big|0000>+(\%\ CA)^{0.5}\Big|0001>+(\%\ AC)^{0.5}\Big|0010>+(\%\ AA)^{0.5}\Big|0011>+\\&+(\%\ CT)^{0.5}\Big|0100>+(\%\ CG)^{0.5}\Big|0101>+(\%\ AT)^{0.5}\Big|0110>+(\%\ AG)^{0.5}\Big|0111>+\\&+(\%\ TC)^{0.5}\Big|1000>+(\%\ TA)^{0.5}\Big|1001>+(\%\ GC)^{0.5}\Big|1010>+(\%\ GA)^{0.5}\Big|1011>+\\&+(\%\ TT)^{0.5}\Big|1100>+(\%\ TG)^{0.5}\Big[1101>+(\%\ GT)^{0.5}\Big[1110>+(\%\ GG)^{0.5}\Big|1111>\end{aligned} \qquad (2)$$

learning leader, should design a series of questions or themes based on student feedback from the first lesson and their own lesson plans [18]. These question and topic sheets must be published and maintained in the online platform. The questions and topics must take into account the interest and level of the students; designing questions that are too difficult or too abstract can easily result in a loss of student motivation. And the questions should be designed as a range of alternative and preferably related to each other so that students can chose according to their interests.

As for problem exploration in a group patterm, it is appropriate for the teacher to help students to form teams. This does not mean that the teacher directly names or confirms the members, but providing them time to introduce themselves to each other, get to know each other and thus reach a consensus. In the process of getting to know each other, students begin the first steps of collaborative learning and develop their skills in dealing and communicating with strangers. However, according to the survey above, this process of group bonding should be designed according to the size of the class. For example, for a small class of 10–20 students, the teacher can use the classroom platform to allow students to introduce themselves briefly; however, for a large class of more than 20 students, given the time constraints of the class, the teacher can use the after-session platform to post a discussion forum for students to communicate.

And for students who are left alone, the teacher need help and guide them to join the team. The social attitudes and psychology of students are distinctly individual. If teachers do not help students to find their own groups in time, it is likely to undermine students' self-confidence and thus reduce the effectiveness of collaborative learning in the later stages. After several teams have been formed, the online platform can be used to allow them to register their questions or topics. In the following sessions, the teacher then allows them to discuss and explore on their own by opening different discussion spaces or boards. After a certain amount of preparation time, each team will report on their views, comments and solutions to the problem.

Last, evaluation and feedback is an important part of facilitating students' learning [19]. In the past, online feedback has often involved teachers simply acknowledging students' results with short words such as 'good' or 'okay', which does not really help students to aware of their own inadequacies, but dampens their enthusiasm. In the feedback part, different forms can be used, especially allowing students to evaluate and score each other. This is very easy to carry out through questionnaires on the online platform, and teachers can also give feedback with platform capabilities, but this must be timely and multi-dimensional. After class, the discussion section online still work actively, both students and teachers can continue to expand on the learning content, and students are also encouraged to ask the teacher about their questions in class, which helps students to further consolidate the learning content and expand new possibilities.

4 Conclusion

Collaborative learning is an effective way to facilitate students' learning and social interaction. In the current technological environment, online components can further implement collaborative learning, and the various approaches of collaborative learning are developed through platform features and courses design.

Through statistics on online learning functions and student feedback, we have looked at three aspects of the current status and difficulties of fully transferring collaborative learning to online development. The current network platform provides a section of collaborative learning to a certain extent, but these functions are limited to the discussion. And in the teacher's online classroom design, collaborative learning is fragmented into subtle components as a classroom aid to enhance student engagement, but this is achieved primarily through the inclusion of collaborative learning in student scores. And for the students, they felt that their collaborative learning was mainly reflected in their classroom participation and group activities, but they did not show great confidence in the effectiveness of their learning.

Although the research and practice of collaborative learning has been well established, but in the past it was mostly conducted offline, as an adjunct to classroom teaching. Now it is still a new beginning to move it completely online, as a pedagogical goal and as a whole. And this study strongly suggests strategies to improve some of the problems with this attempt, one being to build community identity on the online platform or media to help students develop their communication skills with others, and the other being to design the online learning circuit as a complete system to enable students to truly engage in collaborative learning in an online learning environment.

References

1. Kasim, N.N.M., Khalid, F.: Choosing the right learning management system (LMS) for the higher education institution context: A systematic review. Int. J. Emerg. Technol. Learn. **11**(6) (2016)
2. Yoo, M., Jin, S.-H.: Development and evaluation of learning analytics dashboards to support online discussion activities. Educ. Technol. Soc. **02**, 1–18 (2020)
3. Laal, M., Ghodsi, S.M.: Benefits of collaborative learning. Procedia Soc. Behav. Sci. **31**, 486–490 (2012)
4. Laal, M., Laal, M.: Collaborative learning: what is it? Procedia Soc. Behav. Sci. **31**, 491–495 (2012)
5. Suh, H.: Collaborative learning models and support technologies in the future classroom. Int. J. Educ. Media Technol. **5**(1), 50–61 (2011)
6. Cheng, L., Yang, T.: Research on the design and application of collaborative learning teaching mode for college students based on mind map. Comput. Knowl. Technol. **17**(35), 188–191 (2021). (in Chinese)
7. Zheng, L.: Interactive Analysis Methods for Collaborative Learning. People Post Press (2015) (in Chinese)
8. Kalmar, E., et al.: The COVID-19 Paradox of Online Collaborative Education: When You Cannot Physically Meet, You Need More Social Interactions. Heliyon (2022)
9. Cerro Martínez, J.P., Guitert Catasús, M., Romeu Fontanillas, T.: Impact of using learning analytics in asynchronous online discussions in higher education. Int. J. Educ. Technol. High. Educ. **17**(1), 1–18 (2020)
10. Zhang, W.: Classroom observation and enlightenment based on American NCTM classroom communication norms. J. Math. Educ. **22**(3), 49–52 (2013). (in Chinese)
11. Spooner, E.: Interactive student centered learning: A cooperative approach to learning. Rowman & Littlefield (2015)
12. Resta, P., Laferrière, T.: Technology in support of collaborative learning. Educ. Psychol. Rev. 65–83 (2007)

13. Smith, B.L., MacGregor, J.T.: What is Collaborative Learning, pp. 233–267 (1992)
14. Standl, B., Kühn T., Schlomske-Bodenstein, N.: Student-collaboration in online computer science courses–An explorative case study. Int. J. Eng. Pedag. **5** (2021)
15. Wu, R.: When "Discussion" moved to "Web Platforms". Diss. (2015) (in Chinese)
16. Dillenbourg, P., Schneider, D.: Collaborative Learning and the Internet (1995)
17. Sun, A., Chen, X.: Online education and its effective practice: A research review. J. Inf. Technol. Educ. **15** (2016)
18. Jiang, L., Sun, K.: Task-based approach in college English teaching of digital-based context. Int. J. Mod. Educ. Comput. Sci. **32** (2010)
19. Thurlings, M., Vermeulen, M., Bastiaens, T., et al.: Understanding feedback: A learning theory perspective. Educ. Res. Rev. **9**, 1–15 (2013)

Author Index

Z. Hu et al. (Eds.): AIMEE 2022, LNDECT 159, pp. 657–659, 2023.
https://doi.org/10.1007/978-3-031-24468-1